# Electron Microscopy and Analysis 1997

'an excellent source book for undergraduate projects in the final year... describes, clearly and concisely, the methods that must be available to, and understood by, everyone working in the world of condensed matter.'

(from a review by Mick Brown of Cambridge University in *Physics World*)

# Electron Microscopy and Analysis 1997

Proceedings of the Institute of Physics Electron Microscopy and Analysis Group Conference, Cavendish Laboratory, University of Cambridge, 2–5 September 1997

Edited by J M Rodenburg

Institute of Physics Conference Series Number 153
Institute of Physics Publishing, Bristol and Philadelphia

CODEN IPHSAC 153 1–688 (1997)

*British Library Cataloguing in Publication Data*

A catalogue record for this book is available from the British Library.

ISBN 0 7503 0441 3

*Library of Congress Cataloging-in-Publication Data are available*

Published by Institute of Physics Publishing, wholly owned by The Institute of Physics, London
Institute of Physics Publishing, Dirac House, Temple Back, Bristol BS1 6BE, UK
US Editorial Office: Institute of Physics Publishing, The Public Ledger Building, Suite 1035, 150 South Independence Mall West, Philadelphia, PA 19106, USA

Printed in the UK by Galliard (Printers) Ltd, Great Yarmouth, Norfolk

# Organizing Committee

The conference was organized by the Electron Microscopy and Analysis Group of the Institute of Physics and was co-sponsored by the Royal Microscopical Society, the STM Users Group and the Institute of Materials.

**Chairman of the Scientific Committee**
Dr C J Kiely (Department of Materials Science and Engineering, University of Liverpool)

**Scientific Programme Organizer**
Dr S McVitie (Department of Physics and Astronomy, University of Glasgow)

**Honorary Editor**
Dr J M Rodenburg (Cavendish Laboratory, University of Cambridge)

**Exhibition Organizer**
Dr C B Boothroyd (Department of Materials Science and Metallurgy, University of Cambridge)

**Local arrangements**
Dr K M Knowles (Department of Materials Science and Metallurgy, University of Cambridge)

**Advanced School Director**
Dr A Bleloch (Cavendish Laboratory, University of Cambridge)

**Poster sessions and poster prizes**
Dr W J Vine (DRA, Farnborough)

**Royal Microscopical Society representative**
Dr J L Hutchison (Department of Materials, University of Oxford)

**Electron Microscopy and Analysis Group Committee 1996–97**
Dr C J Kiely, Chairman (Department of Materials Science and Engineering, University of Liverpool)
Dr M Aindow, Secretary (School of Metallurgy and Materials, University of Birmingham)
Dr U Bangert (Department of Physics, UMIST)
Dr R Beanland (GEC Marconi Materials Technology)
Dr R M Brydson (School of Materials, University of Leeds)
Mr P Davies (JOEL UK Ltd)
Professor R E Palmer (School of Physics and Space Research, University of Birmingham)
Dr R Vincent (Department of Physics, University of Bristol)
Dr W J Vine (DRA, Farnborough)
Dr S McVitie (Department of Physics and Astronomy, University of Glasgow)

# Contents

## Section 3: High resolution electron microscopy

## Section 6: Advanced scanning probe techniques

## Section 7: Advanced scanning electron microscopy and surface science

## Section 8: Microanalysis and EELS

## Section 9: Catalysts

## Section 10: Semiconductors and superconductors

xiv

## Section 11: Ceramics and interfaces

## Section 12: Intermetallics

## Section 13: General materials analysis

# Preface

The 1997 biennial meeting of the Electron Microscopy and Analysis Group of the Institute of Physics (EMAG 97) was held in the Cavendish Laboratory of the University of Cambridge in the first week of September. The conference was attended by 205 delegates from 13 different countries. The subjects under discussion covered all areas of state-of-the-art electron imaging, electron energy loss and x-ray analysis, the scanning probe and electron beam microscopies and a wide range of applications of these advanced techniques in materials science, metallurgy and surface science. The meeting was preceded by a one-day advanced school on nano-spectroscopy attended by 35 delegates, and was run in conjunction with a major trade exhibition held in the same building as the conference. Delegates were accommodated in Churchill College, which also hosted the main conference dinner: other social events included an Exhibitors' buffet and a reception and dinner in Trinity College.

The four-day meeting was longer than previous EMAGs in order to accommodate a special set of plenary talks celebrating the centenary of the discovery of the electron by J J Thomson in Cambridge. It was also the first EMAG since the fiftieth anniversary of the founding of the Group in 1946. Unusually, therefore, this volume of Proceedings starts with two papers presented as plenary lectures which give a historical perspective of the subject: Peter Hawkes reviews the discovery of the electron and the very early development of electron microscopy; Ken Smith describes the pioneering contribution of Charles Oatley and his group in Cambridge to the development of the scanning electron microscope. Other plenary papers were given by Mick Brown (electron energy loss spectroscopy), John Spence (state-of-the-art imaging methods) and Colin Humphreys (the future of electron microscopy).

In addition to the plenary sessions, 12 invited papers and 67 contributed papers were presented orally in parallel sessions run in two lecture theatres, and a further one hundred papers were presented as posters. All papers submitted for publication in this volume were presented by their authors upon registration at the conference and were independently refereed during the course of the conference; I am grateful for the speedy but meticulous work of the 50 or so delegates who helped me with this task.

**J M Rodenburg**
September 1997

*Inst. Phys. Conf. Ser. No 153: Section 1*
*Paper presented at Electron Microscopy and Analysis Group Conf. EMAG97, Cambridge, 1997*

1

# Electron Microscopy and Analysis: the first 100 years

**P W Hawkes**

CEMES–LOE du CNRS, B.P. 4347, F–31055 TOULOUSE cedex 4

**ABSTRACT**: The electron emerged from obscurity in 1897 and it took just 30 years for it to acquire an "optics": in 1927, Hans Busch demonstrated that the motion of beams of electrons travelling in the vicinity of the axis of symmetry of an electromagnet could be described by the laws associated with light passing through a glass lens. In that year, then, electron ballistics gave way to electron optics and mentalities changed in a way that made the notion of an electron microscope conceivable. We follow the travels of the electron during its first 100 years, first through tubes and oscilloscopes, then through the electrostatic and magnetic lenses of the first primitive microscopes of the 1930s; the first scanning instruments were also built in that first decade of electron microscopy but only emerged from the shadows in the 1960s. In the 1940s, ways of correcting lens aberrations were proposed by Scherzer and Gabor and the first aberration correctors appeared in the 1950s but it is only now that they are being made to work and Gabor's holography too is still not in widespread use. With the centenary of the electron, we meet miniature electron microscope columns, constructed by electrons themselves and microscopes whose resolution enables us to see the structure of specimens, atom by atom. It is all the more disconcerting to recall that we still do not really understand how an electron beam can behave as a set of particles, which can be observed arriving one by one on a fluorescent screen, and as a wave field, producing interference fringes.

## 1. THE DISCOVERY

In 1858, Julius Plücker observed a green phosphorescence on the glass wall of a discharge tube in the vicinity of the negative electrode and "ascribed these phosphorescent patches to currents of electricity which went from the cathode to the walls of the tube and then, for some reason or another, retraced their steps" (Thomson 1897a). This observation is regarded as the discovery of cathode rays, as they soon came to be called, and was made possible through the glassblowing skills of Geissler, who made the evacuated tube.

During the following 40 years, repeated attempts were made to establish the nature of these rays. The German school of thought, notably Plücker's student Hittorf (Plücker & Hittorf 1865), Goldstein and Kaufmann, believed that the rays were vibrations of the æther of some kind, probably transverse, and that the laws governing their propagation remained to be discovered; this belief was reinforced by Hertz's erroneous finding (1883) that a transverse electric field did not deflect the rays. The English school thought it more likely that they consisted of charged bodies of some kind. Thus in 1871, Varley wrote that "This experiment, in the author's opinion, indicates that this arch is composed of attenuated particles of matter projected from the negative pole by electricity in all directions, but that the magnet controls their course...the author has been informed more than once, by captains of vessels, that when men have been struck by lightning a burn has been left upon the skin of the same shape as the object from which the discharge flew. In one instance, he was informed that some brass numbers attached to the rigging, from which the discharge passed to the sailor, were imprinted on his skin.

It is now seen that this is perfectly possible if the discharge be a negative one – that is if the man be + to the brass number." Towards the end of the century, it became urgent to find out what these rays really were, "owing to a remarkable property possessed by an offspring of theirs, for the cathode rays are the parents of the Röntgen rays" (discovered in 1896).

J.J. Thomson was well aware of the efforts that had been made to explain these rays and particularly admired the work of Crookes (1879). The results of his own experiments were presented at a Friday evening discourse of the Royal Institution in London delivered on 30 April 1897, during which he invited his listeners to "trace the consequences of supposing that the atoms of the elements are aggregations of very small particles, similar to each other; we shall call such particles corpuscles, so that the atoms of ordinary elements are made up of corpuscles and holes, the holes being predominant...I have endeavoured to get a measurement of the ratio of the mass of these corpuscles to the charge carried by them. [He finds] $m/e = 1.6 \times 10^{-7}$. This is very small" (Thomson 1897a). In the more formal report sent to *Phil. Mag.* in August 1897 (1897b), he wrote: "According to the almost unanimous opinion of German physicists, they [cathode rays] are due to some process in the æther to which – inasmuch as in a uniform magnetic field their course is circular and not rectilinear – no phenomenon hitherto observed is analogous; another view of these rays is that, so far from being wholly ætherial, they are in fact wholly material, and that they mark the paths of particles of matter charged with negative electricity....I can see no escape from the conclusion that they are charges of negative electricity carried by particles of matter. The question next arises, What are these particles? are they atoms, or molecules, or matter in a still finer state of subdivision? To throw some light on this point, I have made a series of tests of the ratio of the mass of these particles to the charge carried by it. [Here he gives $m/e = 0.4$ or $0.5 \times 10^{-7}$ ]

From these determinations, we see that the value of $m/e$ is...very small compared with the value $10^{-4}$, which is the smallest value of this quantity previously known, and which is the value for the hydrogen ion in electrolysis....The smallness of $m/e$ may be due to the smallness of $m$ or the largeness of $e$, or to a combination of these two."

The Germans remained sceptical up to the last minute. In an article dated 21 May 1897, Kaufmann, who appears to have been the most open-minded, concluded: "Ich glaube deshalb zu dem Schlusse berechtigt zu sein, dass die Hypothese, welche annimt, die Kathodenstrahlen seien abgeschleuderte Theilchen, zu einer befriedigenden Erklarung der von mir beobachteten Gesetzmässigkeiten allein nicht ausreichend ist." At that date, he had seen a report in *Nature* of a communication by Thomson to the Cambridge Philosophical Society but that was not really explicit. In a later paper in the same year (Kaufmann & Aschkinass, 1897) he realised that Thomson was right and he devoted the next few years to studying these new particles.

Nowhere in the published version of the Royal Institution lecture does Thomson use the word "electron" and he continued to eschew it in the 1897 article in *Phil. Mag.* and in a later one in 1899 to which we turn shortly (nor incidentally does he give the units of $m/e$, a practice that would today attract a very sharp editorial reprimand!). The word was, however, used repeatedly and with no explanation of its meaning by G.F. Fitzgerald in his introduction to Thomson's *Electrician* article and it was in common use soon after. Where then does the word come from? It first appears in the scientific literature in 1891, in both the abstract and the full version of a lecture delivered by G.J. Stoney to the Royal Dublin Society on 26 March and 22 May 1891: "...in electrolysis, a definite quantity of electricity , the same in all cases, passes for each chemical bond that is ruptured...the amount of this very remarkable quantity of electricity is about the twentiethet (that is, $1/10^{20}$) of the usual electromagnetic unit of electricity...This is the same as three-eleventhets ($3/10^{11}$) of the much smaller C.G.S electrostatic unit of quantity. A charge of this amount is associated in the chemical atom with each bond...These charges, which it will be convenient to call *electrons*, cannot be removed from the atom; but they become disguised when atoms chemically unite." It had, in all likelihood been in Stoney's mind before this for he was an early advocate of units based on natural constants rather than on such arbitrary quantities as the centimetre, gramme and second. His arguments in favour of units based on the velocity of light, the gravitational constant and an element of charge were presented in 1881 and published in full in *Phil. Mag.* and the *Proceedings of the Royal Dublin Society* (Stoney 1881, 1883) in irresistible language: "Hence we have very good reason to suppose that in $V_1$, $G_1$ and $E_1$ we have three of a series of systematic units that in an eminent sense are the units of Nature, and stand in an intimate relation with the work that goes on in her mighty laboratory". He had even used the word 'electron' in the title of an article in *Phil. Mag.* in

1894, which can hardly have escaped Thomson's notice. It therefore seems clear that Thomson deliberately chose not to use Stoney's word and the explanation offered by historians of science is that Stoney used the word to describe a unit, not the particle carrying it. Fitzgerald, who was Stoney's nephew, did not regard this as an objection. Another explanation is that Thomson preferred not to associate himself with a term introduced by a man in whose work originality sometimes came close to eccentricity; in the quotation above, we see Stoney's suggestion for a word meaning $10^{-n}$ and he also published a new musical notation, supposedly easier to learn than the standard notation, and a study of energy in bacteria. By 1913, the word was so familiar that the *Empire Review* could speak of "The imponderable electrons of sentiment and feeling which allow our far-away peoples and clans to cohere".

In 1897, Thomson was unable to decide whether his value of $m/e$ was so small because $m$ was small or $e$ larger than usual, though it is clear which way his sympathies lay. Two years later, however (Thomson, 1899), he was sure: "...we have clear proof that the ions have a much smaller mass than ordinary atoms; so that in the convection of negative electricity at low pressures, we have something smaller even than the atom, something which involves the splitting up of the atom, inasmuch as we have taken from it a part, though only a small one, of its mass....

From what we have seen, this negative ion must be a quantity of fundamental importance in any theory of electrical action; indeed, it seems not improbable that it is the fundamental quantity in terms of which all electrical processes can be expressed."

By a happy coincidence, 1897 saw not only the discovery of the electron but the invention of the cathode-ray tube by Braun (1897), who exploited the deflection of the rays by a magnetic field even before knowing what the rays were made of. It very soon came into use (e.g. Wehnelt and Donath, 1899).

## 2. THE FIRST FIFTY YEARS

The motion of charged particles in electrostatic and magnetic fields was studied in considerable detail for the next 30 years, notably by Störmer (1907a, b) and by Birkeland and Villard in connection with the Northern Lights and by a host of investigators attempting to improve the performance of CRTs. Störmer came close to discovering geometrical electron optics, as he pointed out ruefully in 1933: "...so habe ich in meiner Arbeit zur Theorie des Polarlichtes auch den Fall eines axialsymmetrischen elektromagnetischen Feldes behandelt". In fact, the geometrical optics of charged particles was not discovered until 1927, when Hans Busch observed that "Eine kurze Spule hat also die Eigenschaft, die Kathodenstrahlen nach der Achse zu um einen Winkel $\gamma$ abzulenken, der proportional der Achsenentfernung...des Strahles ist". The stimulus here was not the aurora borealis but the commercial pressure to understand how the cathode-ray tube worked and to develop better designs; see Rogowski (1920) for an appraisal of the situation at that time. Meanwhile, and somewhat ironically, Louis de Broglie had discovered the wave optics of electrons (de Broglie 1923, 1925) and electron diffraction experiments confirming his ideas were performed by G.P Thomson (J.J. Thomson's son) and by Davisson and Germer in 1927 and by Rupp (1928). Later still, when Walter Glaser began his theoretical work on electron optics in the 1930s, the "Hamiltonian analogy" was seen to be relevant: geometrical electron optics could have been "invented" by anyone familiar with Hamilton's work (published between 1828 and 1832) at any time since the discovery of the electron (or, indeed, at any time since the discovery of the laws of force on charged particles).

The seventy years between 1927 and today were punctuated by major advances, interspersed with less spectacular but no less indispensable developments. Interspersed too with a host of near-misses and dead ends. The great heroes of the 1920s are undoubtedly Busch and de Broglie but Gabor too very nearly achieved greatness. In his work on the cathode-ray tube, the nursery slopes of the electron microscope builders, he enclosed a magnetic coil in an iron yoke, thereby constructing the first modern magnetic lens – but alas, he could not understand how it worked and thus missed inventing the electron microscope a few years before Ruska and Knoll built their first model (1932).

The 1930s are the heroic age of the electron microscope and of electron optics. At the beginning of the decade, Ernst Ruska and Max Knoll, soon to be joined by Bodo von Borries (who became Ruska's brother-in-law), built the first transmission electron microscopes (Ruska, 1980) and by the end of the decade, Ruska, with the vigorous support of his brother Helmut, had persuaded Siemens to put a commercial model on the market (von Borries & Ruska, 1938; von Borries *et al.*, 1938). 38 instruments were produced in the space of about five years, an astonishingly high number (their fate is chronicled by Wolpers, 1991). Meanwhile, Ladislaus Marton had built a series of simple instruments in Brussels and obtained the first micrograph of an osmium impregnated biological specimen in 1934 (cf. Marton, 1968); soon after, Driest and Müller (1935) and Krause (1936) obtained images of unstained biological material with a Ruska instrument. Back in Berlin, Ruska was not the only microscope builder; in the AEG laboratories, electrostatic lenses and instruments using them were being explored and an emission microscope was the first to emerge. Moreover, it was there that some of the most important theoretical advances were being made, including Scherzer's unwelcome demonstration (1936) that the integrand in the formula for the spherical aberration constant of any electron lens (of traditional design) is positive-definite. Or rather, non-negative-definite for the other great genius of electron optics, Walter Glaser, found this result so unpalatable that he repeatedly attempted to overturn it (a note in his treatise published in 1952 shows that he had still not lost hope!). This coefficient can be written in the form $\int Ah^4 dz$ and in 1940 Glaser had the brilliant idea of setting the term $A$, which is a function only of the magnetic induction $B(z)$ on the axis and its derivatives (for a magnetic lens), equal to zero and solving the resulting differential equation for $B(z)$. Unfortunately, the field distribution that emerged is not capable of forming a real image of a real object and the corresponding potential for electrostatic lenses is no better. Before leaving the AEG laboratories, we must mention Hans Mahl, who introduced the replica technique that was very widely used until methods of producing thin objects were developed. A third group in Berlin was led by Manfred von Ardenne, an inventor of genius, who built the first true scanning microscope and even a primitive STEM, which he described at length in 1938 (von Ardenne, 1938a, b, 1985). He also appreciated the advantages of high-voltage operation and obtained images at 200 kV but it must be said that, at a time when it was impossible to make very thin specimens, this did not require very great acumen and several other such projects are to be found in the first decades of the subject: in Holland, in Russia and of course in France, where Gaston Dupouy was soon preaching the virtues of high voltage. Most of these projects, in which the role of the high voltage was to increase penetration, were abandoned with the development of the ultramicrotome in the 1950s (see Chapter IX of Sjöstrand 1967) – only Dupouy, whose dream was to observe living matter with electrons, persisted, for the electrons would then have to penetrate the atmosphere in which the specimen lived as well as the organic matter itself. His first images at more than 1 MV were published in 1960 (Dupouy *et al.*, 1960).

Still in the 1930s, L.C. Martin gave the first lecture in English on the foundations of electron optics, published in the newly launched *Journal of the Television Society* (Martin, 1934) and it was to his laboratory in Imperial College that Metropolitan Vickers delivered their first electron microscope in 1937 (Martin *et al.*, 1937). The first textbook in English on electron optics appeared soon after (Myers, 1939) and is remarkable not only for its masterly coverage of so young a subject but also for Myers' assertion in the Preface that "Without doubt the greatest proportion of experimental and theoretical work during the last few years has been carried out at the AEG Forschungsinstitut in Berlin, under the supervision of Dr Brüche...". Ruska's name is not mentioned and Knoll is merely thanked "for advice and encouragement". Since Myers was extremely well-informed about the German literature, we must assume that this is how it seemed to an objective student of electron optics in 1939; of course, the superiority of the magnetic lens was then not at all obvious. At the end of the decade, Ruthemann took the first steps towards EELS, which led Hillier to coin the word "microanalyser" in 1943.

# 3. MODERN TIMES

Soon after the war, Scherzer published two important papers: in the first (Scherzer 1947), he described several ways of correcting $C_s$ by abandoning one or other of the regular features of lens design (notably rotational symmetry and static fields) or by introducing a conductor on the axis or a potential discontinuity or a mirror configuration. Exploration of these and some other ideas continues today. In the other (Scherzer 1949), we meet the seed from which the transfer theory of electron imaging grew. In between Gabor (1948, 1949) invented holography, again in an attempt to circumvent the limitations imposed by spherical aberration. Or did he re-invent it? After reading Gabor's first announcement in *Nature*, Boersch wrote to him to draw his attention to his own "Zur Bilderzeugung im Mikroskop", published in 1938, a year earlier than a paper by Bragg (1939) describing a similar idea. Examination of these earlier papers shows that, like so many original suggestions, holography had a pre-history but that this in no way detracts from the novelty of Gabor's invention. Lastly, before leaving the 1940s, we recall that microscope performance was dramatically improved by the introduction of the stigmator, invented independently by Bertein (Bertein, 1947, 1948), Rang (1949) and Hillier and Ramberg (1947); priority has tended to be accorded on the basis of the writer's nationality! Nor should we forget that the first international congress on electron microscopy was held in Delft in 1949, albeit with no Germans present. It is noteworthy for the presence of an early paper by Raymond Castaing on what was to become the x-ray microanalyser (Castaing & Guinier, 1949) among many other distinguished contributions. And of course the club that we call EMAG was created in 1946 (Cosslett 1971)!

Arguably the most influential landmarks of the 1950s were the treatise on electron optics published by Walter Glaser in 1952 and its much improved abridgement of 1956. In between, Sturrock (1955) had published a masterly (if occasionally hermetic) account of the Hamiltonian approach to electron optics, building on and furthering Glaser's work. Tübingen became one of the main centres of electron optical expertise and in 1955, Möllenstedt and Düker described the electron counterpart of the Fresnel biprism (see Möllenstedt, 1991), which rapidly became the standard device for electron interferometry and, much later, for electron holography. In the 1950s and 1960s, numerous attempts to put Scherzer's suggestions for overcoming spherical aberration into practice were made (see Septier, 1966 or Hawkes, 1980 or Chapter 41 of Hawkes & Kasper 1989 for detailed accounts of these). Castaing's microanalyser (fully described in 1951) was improved by the addition of scanning coils by Peter Duncumb (Cosslett & Duncumb 1956, Duncumb 1958). From 1948 onwards, Charles Oatley oversaw the efforts of a series of students to develop the scanning electron microscope, a high point of which was the Stereoscan marketed by the Cambridge Instrument Co in 1965 (Oatley *et al.* 1965, 1985, Oatley, 1982). In 1956, Menter saw lattice fringes in copper and platinum phthalocyanine crystals and at the turn of the decade all the basic theory of crystal image formation was elucidated by Hirsch, Howie, Whelan and Hashimoto (see Hirsch *et al.* 1965). In 1965 and 1966a, b, Albert Crewe described his plans to build a scanning transmission electron microscope (STEM), the first results from which were presented in (Crewe *et al.*, 1968). The year 1965 also saw the first of a long series of publications by Karl-Joseph Hanszen in Braunschweig in which transfer theory of image formation was introduced into electron optics (Hanszen & Morgenstern 1965; Hanszen 1966) vividly illustrated by Thon soon after (1966).

The late 1960s also saw a change in thinking among electron microscopists. The development of computers was making new ways of using images possible: de Rosier and Klug (1968) performed the first three-dimensional reconstruction; the users of scanning microscopes quickly realised that the sequential nature of the signal from which the image was built up could be used to improve the image in various ways and to accentuate desirable features (MacDonald 1968, 1969; White *et al.* 1968); Gerchberg and Saxton devised an iterative procedure for solving what today we call the phase problem (1972, 1973); Schiske (1968, 1973) proposed a type of Wiener filter to reduce the undesirable effects of the electron microscope transfer function – it was now realised that since instrumental correction of $C_s$ was proving so difficult, it was worth tolerating poor images and correcting them subsequently by computer processing, an attitude which was at the

6

heart of Gabor's holography, though he of course pictured the second stage as being performed by analogue means (on an optical bench). Hoppe and Strube (Hoppe 1969; Hoppe & Strube 1969) suggested that the phase of the electron wavefunction could be obtained by suitably structuring the beam incident on the specimen; although "ptychography" in this form proved impractical for CTEM, it has been shown to be well adapted to the STEM by Rodenburg and colleagues (Rodenburg 1989), as indeed Hoppe himself foresaw (1982). 1968 is memorable for yet another reason: Möllenstedt and Wahl performed the first reconstruction of an off-axis electron hologram, followed soon after by Tonomura (1969) while in Japan, Tonomura and colleagues published the first example of Fraunhofer in-line holography (Tonomura & Watanabe 1968; Tonomura *et al.* 1968; Watanabe & Tonomura 1969), forerunner of a vast body of work on electron holography (see Tonomura 1993, Tonomura *et al.* 1995 and Tonomura 1997).

## 4. ENVOI

We have only space to mention one other memorable year: in 1986, the Nobel Prize was won by Ernst Ruska together with G Binnig and H Rohrer, the inventors of yet another kind of electron microscope.

## REFERENCES

Ardenne M von 1938a Z. *Physik* **109**, 553–572; 1938b Z. *Tech. Physik* **19**, 407–416
Ardenne M von 1985 On the history of scanning electron microscopy, of the electron microprobe, and of early contributions to transmission electron microscopy. *Adv. Electron. Electron Phys.* Supplement **16**, 1–21
Bertein F 1947 *Ann. Radioél.* **2**, 379–408; 1948 *Ann. Radioél.* **3**, 49–62
Boersch H 1938 Z. *Tech. Phys* **19**, 337–338
Borries B von and Ruska E 1938 Vorläufige Mitteilung über Fortschritte im Bau und in der Leistung des Übermikroskops. *Wiss. Veröffent. Siemens-Werken* **17**, 99–106
Borries B von, Ruska E and Ruska H 1938 *Wiss. Veröffent. Siemens-Werken* **17**, 107–111
Bragg W L 1939 A new type of 'x-ray microscope'. *Nature* **143**, 678
Braun F 1897 Ueber ein Verfahren zur Demonstration und zum Studium des zeitlichen Verlaufes variabler Ströme *Ann. Physik Chemie* **60**, 552–559
Broglie L de 1923 *C. R. Acad. Sci. Paris* **177**, 507–510, 548–550 and 630–632
Broglie L de 1925 *Ann. Physique* **3**, 22–128, republished in *Ann. Fond. Louis de Broglie* **17** (1992) 1–109
Busch H 1927 *Arch. Elektrotech.* **18**, 583–594
Castaing R 1951 *Thèse*, Paris
Castaing R and Guinier 1949 ICEM-1, Delft, 60–63
Cosslett V E 1971 *Phys. Bull.* **22**, 339–341 and EMAG 1971, 1–5
Cosslett V E and Duncumb P 1956 EUREM-1, Stockholm, 12–14
Crewe A V 1965 A scanning transmission microscope utilizing electron energy losses, Unpublished abstract, IoP Conference on Non-conventional Electron Microscopy, Cambridge
Crewe A V 1966a ICEM-6, Kyoyo, vol 1, 631–632; 1966b *Science* **154**, 729–738
Crewe A V, Wall J and Welter L M 1968 *J. Appl. Phys.* **39**, 5861–5868
Crookes W 1879 *Phil. Trans. Roy. Soc. London* **170**, 135–164 and *Phil. Mag.* **7**, 57–64
Davisson C D and Germer L H 1927 *Nature* **119**, 558–560
Driest E and Müller H O 1935 Z. *Wiss. Mikrosk.* **52**, 53–57
Duncumb P 1958 ICEM-4, Berlin, vol. 1, 267–269
Dupouy G, Perrier F and Durrieu L 1960 L'observation de la matière vivante au moyen d'un microscope électronique fonctionnant sous très haute tension. *C. R. Acad. Sci. Paris* **251**, 2836–2841.
Fitzgerald G F 1897 Dissociation of atoms. *The Electrician* **39**, 103–104 [Introduction to Thomson 1897a]
Gabor D 1948 *Nature* **161**, 777–778
Gabor D 1949 *Proc. Roy. Soc. London A* **197**, 454–487

Gerchberg R W and Saxton W O 1972 Optik 35, 237–246
Gerchberg R W and Saxton W O 1973 In *Image Processing and Computer-aided Design in Electron Optics* (P W Hawkes, ed) 66–81 (Academic Press, London)
Glaser W 1940 *Z. Physik* **116**, 19–33 and 734–735
Glaser W 1952 *Grundlagen der Elektronenoptik* (Springer, Vienna)
Glaser W 1956 Elektronen- und Ionenoptik. *Handbuch der Physik* **33**, 123–395
Hanszen K-J 1966 *Z. Angew. Phys.* **20**, 427–435
Hanszen K-J and Morgenstern 1965 *Z. Angew. Phys.* **19**, 215–227
Hawkes P W 1980 *Adv. Electron. Electron Phys.* Supplement **13A**, 45–157
Hawkes P W and Kasper E 1989 *Principles of Electron Optics* (Academic Press, London)
Hertz W 1883 Versuche über die Glimmentladung *Ann. Physik Chemie* **19**, 782–816
Hillier J 1943 On microanalysis by electrons. *Phys. Rev.* **64**, 318–319
Hillier J and Ramberg E G 1947 *J. Appl. Phys.* **18**, 48–71
Hirsch P B, Howie A, Nicholson R B, Pashley D W and Whelan M J 1965 *Electron Microscopy of Thin Crystals* (Butterworths, London)
Hoppe W 1969 *Acta Cryst.* A**25**, 495–501 and 508–514
Hoppe W and Strube G 1969 *Acta Cryst.* A**25**, 502–507
Hoppe W 1982 *Ultramicroscopy* **10**, 187–198
Kaufmann W 1897 *Ann. Physik Chemie* **61**, 544–552 and Nachtrag **62** (1897) 596–598
Kaufmann W and Aschkinass E 1897 *Ann. Physik Chemie* **62**, 588–595
Knoll M and Ruska E 1932 Beitrag zur geometrischen Elektronenoptik I, II *Ann. Physik* **12**, 6017–640 and 641–661
Krause F 1936 *Z. Physik* **102**, 417–422
MacDonald N C 1968 *EMSA Proc.* **26**, 362–363
MacDonald N C 1969 *Scanning Electron Microsc.* 431–437
Martin L C 1934 *J. Television Soc.* **1**, 377–383
Martin L C, Whelpton R V and Parnum D H 1937 *J. Sci. Instrum.* **14**, 14–24
Marton L 1934 Electron microscopy of biological objects. *Nature* **133**, 911.
Marton L 1968 *Early History of the Electron Microscope* (San Francisco Press, San Francisco); 2nd ed., 1994
Menter J W 1956 *Proc. Roy. Soc. London* A**236**, 119–135
Möllenstedt G 1991 *Adv. Opt. Electron Microsc.* **12**, 1–23
Möllenstedt G and Düker H 1955*Naturwissenschaften* **42**, 41
Möllenstedt G and Wahl H 1968 *Naturwissenschaften* **55**, 340–341
Myers L M 1939 *Electron Optics, Theoretical and Practical* (Chapman & Hall London)
Oatley C W 1982 The early history of the scanning electron microscope. *J. Appl. Phys.* **53**, R1–R13
Oatley C W, Nixon W C and Pease R F W 1965 *Adv. Electron. Electron Phys.* **21**, 181–247
Oatley C W, McMullan D and Smith K C A 1985 The development of the scanning electron microscope. *Adv. Electron. Electron Phys.* Supplement **16**, 443–482
Plücker J 1858 *Ann. Physik Chemie* **103**, 88–106 and 151–157, **104**, 113–128 and 622–630, **105**, 67–84 and **107** (1859) 77–113; 1861 Ueber die Einwirkung des Magnets auf die elektrische Entladung. *Ann. Physik Chemie* **113**, 249–280
Plücker J and Hittorf W 1865 On the spectra of ignited gases and vapours, with especial regard to the different spectra of the same elementary gaseous substances. *Phil. Trans. Roy. Soc. London* **155**, 1–29
Rang O. 1949 *Optik* **5**, 518–530
Rodenburg J R 1989 EMAG, vol. 1, 103–106
Rogowski W 1920 Neue Vorschläge zur Verbessung des Kathodenstrahl-Oszillographen. *Arch. Elektrotech.* **9**, 115–120
Rosier D J de and Klug A 1968 *Nature* **217**, 130–134
Rupp E 1928 Über Elektronenbeugung an einem geritzter Gitter *Z. Physik* **52**, 8–15
Ruska E 1980 *The Early Development of Electron Lenses and Electron Microscopy.* (Hirzel, Stuttgart)
Scherzer O 1936 *Z. Physik* **101**, 593–603
Scherzer O 1947 *Optik* **2**, 114–132
Scherzer O 1949 *J. Appl. Phys.* **20**, 20–29

8

Schiske P 1968 EUREM-4, Rome, vol 1, 145–146
Schiske P 1973 In *Image Processing and Computer-aided Design in Electron Optics* (P W Hawkes, ed) 82–90 (Academic Press, London)
Septier A 1966 *Adv. Opt. Electron Microsc.* **1**, 204–274
Sjöstrand F S 1967 *Electron Microscopy of Cells and Tissues* (Academic Press, New York & London)
Stoney G J 1881 On the physical units of Nature. *Phil. Mag.* **11**, 381–390
Stoney G J 1883 On the physical units of Nature. *Sci. Proc. Roy. Dublin Soc.* **3**, 51–60
Stoney G J 1891–2 On the cause of double lines and of equidistant satellites in the spectra of gases. *Sci. Proc. Roy. Dublin Soc.* **7**, 201–203 [Abstract of the following paper]
Stoney G J 1888–92 On the cause of double lines and of equidistant satellites in the spectra of gases. *Sci. Trans. Roy. Dublin Soc.* **4**, 563–608 [read 26 March and 22 May 1891]
Stoney G J 1894 Of the "electron" or atom of electricity. *Phil. Mag.* **38**, 418–420
Störmer C 1907a On the trajectories of electric corpuscles in space under the influence of terrestrial magnetism, applied to the aurora borealis and to magnetic disturbances. *Arch. Math. Naturvidensk. Christiania* **28**, No. 2, 47 pp and **32**, 1911, 117–123, 190–219, 277–314, 415–436 and 501–509; 1907b *Arch. Sci. Phys. Nat.* (Genève) **24**, 5–18, 113–158, 221–247 and 317–364
Störmer C 1933 Über die Bahnen von Elektronen im axialsymmetrischen elektrischen und magnetischen Felde. *Ann. Physik* **16**, 685–696
Sturrock P A 1955 *Static and Dynamic Electron Optics* (Cambridge University Press, Cambridge)
Thomson J J 1897a Cathode rays. *The Electrician* **39**, 104–109 [Text of a discourse delivered at the Royal Institution, 30 April 1997]
Thomson J J 1897b Cathode rays. *Phil. Mag.* **44**, 293–316
Thomson J J 1898 On the cathode rays. *Proc. Cambridge Philos. Soc.* **9**, 243–244; a brief report of this appeared in *Nature* **55** (1897) 453
Thomson J J 1899 On the masses of the ions in gases at low pressures. *Phil. Mag.* **48**, 547–567
Thomson G P 1927 *Nature* **120**, 802.
Thomson G P and Reid A 1927 *Nature* **119**, 890
Thon F 1966 *Z. Naturforsch.* **21a**, 476–478
Tonomura A 1969 *J. Electronmicrosc.* **18**, 77–78
Tonomura A 1993 *Electron Holography* (Springer, Berlin)
Tonomura A 1997 *The Quantum World Unveiled by Electron Waves* (World Scientific, Singapore)
Tonomura A and Watanabe H 1968 *Nihon Butsuri Gakkai-shi* [*Proc. Phys. Soc. Jpn*] **23**, 683–684
Tonomura A, Fukuhara A, Watanabe H and Komoda T 1968 *Jpn J. Appl. Phys.* **7**, 295
Tonomura A, Allard L F, Pozzi G, Joy D C and Ono Y A (eds) 1995 *Electron Holography* (Elsevier, Amsterdam)
Varley C F 1871 Some experiments on the discharge of electricity through rarefied media and the atmosphere. *Proc. Roy. Soc. London* **19**, 236–242
Watanabe H and Tonomura A 1969 *J. Cryst. Soc. Jpn* **11**, 23–25
Wehnelt A and Donath B 1899 Photographische Darstellung von Strom- und Spannungscurven mittels der Braun'schen Röhre. *Ann. Physik Chemie* **69**, 861–870
White E W, McKinstry H A and Johnson G G 1968 *Scanning Electron Microsc.* 95–103
Wolpers C 1991 *Adv. Electron. Electron Phys.* **81**, 211–229

*Inst. Phys. Conf. Ser. No 153: Section 1*
*Paper presented at Electron Microscopy and Analysis Group Conf. EMAG97, Cambridge, 1997*
© *1997 IOP Publishing Ltd*

# Charles Oatley: pioneer of scanning electron microscopy

**K C A Smith**

Engineering Department, University of Cambridge, Trumpington Street, Cambridge CB2 1PZ

**ABSTRACT:** Charles Oatley stands with Manfred von Ardenne as one of the two great pioneers of scanning electron microscopy. His involvement with the SEM began immediately after World War II when, fresh from his wartime experience in the development of radar, he perceived that new techniques could be brought to bear which would overcome some of the fundamental problems encountered by von Ardenne in his pre-war research on the instrument. Oatley's pioneering work led directly to the launch of the world's first series production instrument - the Stereoscan - in 1965.

## 1.    INTRODUCTION

The development of the SEM will forever be linked with the names of Manfred von Ardenne and Charles Oatley; but the manner of their contributions and the outcome of their work could not have been in more marked contrast. Although in 1935 M Knoll produced the first scanned image of a surface, it was von Ardenne who, in a relatively short burst of inspired activity just prior to World War II, laid the foundations of both transmission and surface scanning electron microscopy. His papers published in 1938 and chapters on the SEM in his book 'Elektronen-Übermikroskopie', published during the war (von Ardenne 1940), are the definitive works on the subject (McMullan 1988, 1995). His involvement with the electron microscope, however, came to an abrupt end when all his apparatus was destroyed in an air raid in 1944. After the war, circumstances dictated that he moved into other fields of scientific endeavour, and he played no further part in the development of the SEM. It was left to other research groups, as he wrote later, "to take up the baton once more" (von Ardenne 1985).

It was Charles Oatley who took up the baton in Cambridge in 1947, but in contrast to von Ardenne's relatively brief encounter with the subject, he spent the better part of two decades before reaching a successful conclusion with the world-wide acceptance of the SEM as one of the most powerful and productive methods of microscopy yet invented. In order to appreciate how Oatley arrived at a position which uniquely qualified him to take up von Ardenne's 'baton', we have to look at the events which took him to Cambridge immediately after the war.

## 2.    EDUCATION AND EARLY CAREER

Charles Oatley was born in Frome, Somerset in 1904, coincidentally, the year that mains electrical power was brought to the town. His father, William, was the enterprising owner of a flourishing bakery business, and he installed electrical power in his bakery immediately it became available. Although lacking any formal scientific education William Oatley was intensely interested in scientific matters and passed this enthusiasm on to his son. He gave Charles an electric motor on his sixth birthday - not a common toy in 1910 - and he possessed a fine Watson Royal microscope which he taught his son to use. Thus, Charles' natural scientific bent was nurtured from the earliest years. Up to the age of 12 Charles attended a local council school, and then proceeded to Bedford

Modern School as a boarder where he excelled in scientific subjects. From there he won exhibitions to St. John's College, Cambridge, to read Natural Sciences, entering the University in 1922.

At St. John's his supervisor was E V Appleton. A contemporary was J D Cockcroft whom he got to know quite well, although Cockcroft was 7 years older and reading for the Part II Mathematics Tripos. These two were to have a profound influence on Oatley's career. His first two years at Cambridge went well: a half-blue for swimming and a first in Part I of the Tripos. His third year began equally well; he was appointed captain of the University Swimming Team, and he had a successful interview with Rutherford who agreed to take him on to do research in nuclear physics, but things did not go according to plan. Appleton had left to take a chair at King's College, London, and his supervision and guidance was greatly missed by Oatley. For whatever reason, the expected First Class in Part II (Physics) failed to materialise, and with it went the chance of gaining a research grant to work with Rutherford.

With Appleton's advice and encouragement, Oatley obtained a job with Radio Accessories, a small company in Willesden manufacturing radio valves; here he gained valuable experience of manufacturing techniques and, as the only graduate in the company, was called upon to tackle a wide variety of physical problems. After two years, however, the company, no longer able to compete with the large valve manufacturers, went into liquidation. It was then, in 1927, that he was invited to return to academic life by Appleton, who offered him a Demonstratorship in the Physics Department at King's College, London. The next 12 years at King's were spent largely teaching and examining, with little time for research, but he produced some useful papers on a variety of subjects, some of which are listed in the references (Oatley 1931, 1936a, 1936b, 1939). He also became an acknowledged expert in the field of radio receivers; his Methuen monograph, 'Wireless Receivers', published in 1932, was read widely by enthusiasts at the time.

## 3.    WARTIME WORK IN RADAR

In the summer of 1939, by which time he held a Lectureship at King's, Oatley received a letter from Cockcroft, acting on behalf of the Air Ministry, inviting him to join a small party of university physicists to learn how they might be of help in the event of war. In the following weeks members of Cockcroft's party were initiated into the secret work on radar being carried out under the direction of R A Watson-Watt, and were briefed on the Chain Home stations then under construction around the east and south coasts of Britain. That winter, after war had been declared on 3rd September, found Oatley back at the Cavendish Laboratory in the old High Voltage Laboratory engaged in modifying Pye TV receivers for use in Coastal Defence (CD) radar sets, and constructing various pieces of test apparatus. A little later radar development was moved to the newly-formed establishment at Christchurch, later to become known as the Air Defence Research and Development Establishment (ADRDE), of which Cockcroft became Superintendent in 1941.

Oatley's first task at ADRDE was to investigate ways of improving the receivers used in the CD sets which operated at the then relatively high frequency of 200 MHz. The detection of weak radar echoes depended crucially on the noise performance of the first stage of amplification in the receivers. No commercial test equipment was available at this frequency, so he set about building a signal generator which incorporated for the first time a piston attenuator, a technique which he later extended to centimetric wavelengths. With this apparatus he was able to compare, quantitatively, the noise performance of the various types of valves available for the critical input stage of the receiver. The tests produced spectacular results, showing that one of the new valves produced by GEC specifically for this purpose was virtually useless. Thus, the importance of accurate quantitative testing of all components and equipment produced in the Establishment became generally recognised and led to the formation of Oatley's 'Basic Group' which served the whole of the Establishment throughout the war.

As the work of ADRDE grew, the need for improved liaison with other radar establishments and with the industrial groups designing and supplying equipment, became increasingly evident, and Oatley, with his extensive knowledge of the work being undertaken in all of the groups at ADRDE, became heavily involved in most of the committees that were formed to further co-operation. The importance of this work, and the recognition of his considerable administrative abilities, led to his appointment as Deputy to Cockcroft, and later when Cockcroft moved to take charge of the Canadian

Chalk River Nuclear Project, Oatley became Acting Superintendent. In effect, he was in sole charge of the organisation which by 1943 had moved to Malvern and numbered about 4000 personnel. By the end of the war in 1945 he possessed an unrivalled knowledge of the work being undertaken in radar and electronics, both in the government establishments and in industry, he was familiar with much of this work at a detailed technical level, and he knew all the key people involved; all of which would stand him in good stead for the next phase of his career. He was offered the post of Superintendent at the Establishment, but his heart lay in university life: a Cambridge Lectureship and Fellowship at Trinity College proved to be the far greater attraction.

Fig. 1. Sir Charles Oatley with former research students and colleagues at a symposium held in his honour on the occasion of his 90th birthday. Standing (left to right):   A N Broers;   W C Nixon;   R F W Pease;   T E Everhart;   D McMullan;   K C A Smith; O C Wells; C W B Grigson; A D G Stewart; P Chang; H Ahmed.

## 4.    RESEARCH AT CAMBRIDGE

Oatley was appointed to his Lectureship at the University Engineering Department with the specific remit from John Baker, the then Head of the Department, to introduce, for the first time, undergraduate teaching of modern electronics and to build up a strong research group in the subject. The SEM was one of a number of research projects that he set in train in the early years following the war. His interest in the instrument was first aroused when he learned of von Ardenne's pre-war work in Germany and that of Zworykin in the USA in the early 1940s. Although this work had produced inconclusive results, Oatley, fresh from his wartime experience in radar, appreciated that new techniques and methods were available which could be applied to the scanning concept. In particular, he was aware of work in the Cavendish Laboratory by A S Baxter on a new type of electron multiplier incorporating beryllium-copper dynodes, which could be repeatedly exposed to the atmosphere and thus used in a demountable vacuum system, and he saw that this could, at least in principle, solve the major technical problem encountered in the earlier instruments; namely, the noise-free detection of the pico ampere secondary electron currents obtaining in the SEM. (This was proposed also by von Ardenne in his book, Elektronen-Übermikroskopie, but this was unknown to Oatley at the time.)

Of equal importance to Oatley was that the topic appeared to offer the ideal avenue through which to exploit his strong background in electron physics and to stretch the minds of his future research students: "A project for a Ph.D. student must provide him with good training and, if he is doing experimental work, there is much to be said for choosing a problem which involves the construction or modification of some fairly complicated apparatus. I have always felt that university research in engineering should be adventurous and should not mind tackling speculative projects" (Oatley 1982). He decided, against the advice of many experts in the field who declared it to be a complete waste of time, that the SEM should be included in the topics undertaken by his research students.

His first research student, D McMullan, commenced in 1948, and a working instrument was produced by 1951 (Fig 2(a)). This incorporated three features which differentiated it from its predecessors: electrons scattered from the specimen surface were detected by means of an electron multiplier with beryllium-copper dynodes; the specimen was placed at a large angle to the electron beam; and a high primary beam energy was used to reduce the effects of surface contamination on secondary emission. In addition, images were viewed directly on a slow-scan radar-type CRT display, and recorded on film using a short persistence high resolution CRT. This instrument from the start produced the striking 'three-dimensional' images characteristic of the modern-day SEM (Fig. 2(b)).

(a)                                              (b)

Fig. 2. (a) McMullan's original microscope, SEM1. (b) Early micrograph (etched aluminium) obtained with SEM1. Horizontal field width: 37 μm.

The results obtained with SEM1 convinced Oatley, even at this early stage, that the SEM would turn out to be an important scientific laboratory tool; however, this was far from being the consensus among microscopists who, with notable exceptions, believed that this new instrument could never compete with the well-established electron microscopical techniques of the time; in particular, with the replica technique which offered much superior resolution. The new instrument was met with indifference and even, in some quarters, ridicule. Undeterred, Oatley decided to continue with the project. His main argument for doing so was that for a very large range of specimens, high resolution was not important and moderate magnifications sufficed to provide the all-important information. The SEM offered ease of specimen preparation, large depth of field, readily interpreted images, and great flexibility in the size and type of specimen that could be examined, including the observation of dynamically changing specimen properties. All of these advantages were appreciated by Oatley on the basis of the first results obtained with SEM1 - a full decade before microscopists in general were forced to the same conclusion. Two more research students were taken on: K C A Smith in 1952, who continued with the development of SEM1, and O C Wells a year later, who embarked on the construction of a new microscope, SEM2. L R Peters, a gifted technician, was assigned to the project.

Although the electron multiplier appeared to offer an elegant solution to the problem of detection in the SEM, its implementation proved to be technically extremely difficult. Detection of the low-energy secondary electron component required the input of the multiplier to be at a positive potential of several hundred volts, which in turn necessitated the output to stand at about 6 kV above ground. Because capacitors operating at high voltage generate significant noise, capacitive coupling of the signal to ground level at this stage was not possible. Consequently, a head amplifier floating at 6 kV, with all its attendant power-supply, screening and decoupling problems, had to be used to boost the signal to a level at which a capacitor noise became insignificant. Capacitive coupling also necessitated the use of beam blanking in the column and d.c. restoration circuits to maintain black level. All of these difficulties were eliminated at a stroke with the introduction in 1956 of the 'Everhart-Thornley' detector. With this detector the output of the multiplier is at ground potential, and d.c. coupling can be used throughout the whole signal chain.

How this detector evolved as the research in Oatley's Group progressed is illustrated in Fig. 3. It was while Smith was experimenting with an environmental cell that Oatley first suggested the use of a plastic organic scintillator coupled to a photomultiplier to detect the electrons transmitted through the cell. In Oatley's arrangement (Fig 3(a)) the rod-shaped scintillator forms one wall of the cell, and acts as a light-pipe to convey scintillations to the photocathode of the multiplier.

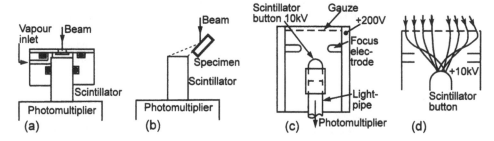

Fig. 3. Evolution of the Everhart-Thornley detector. (a) Oatley's arrangement for detecting electrons transmitted through environmental cell. (b) Detection of BSE component. (c) Everhart's SE detector. (d) Electron trajectories determined by Thornley.

This detector was used subsequently by both Wells and Smith to produce backscattered electron images of solid samples (Fig 3(b)). Everhart then took over SEM1 and extended the technique to the detection of the low-energy secondaries (Fig 3(c)). A little later Thornley conducted a detailed study of Everhart's detector using an electron trajectory tracer (Fig. 3(d)) - another research project initiated by Oatley (Sander 1951).

The wide-ranging nature of the programme of research and development that Oatley organised to demonstrate the power and flexibility of the SEM is perhaps best illustrated with reference to the work of his students who participated in the programme. A brief summary of their work is given below. The period spans from the late 1940s to the early 1960s, when the weight of evidence in favour of commercial production of the SEM finally became overwhelming. A full account of the work undertaken in this period, together with a complete list of references to published work, is given in Oatley et. al. 1985.

D McMullan: 1948. Constructed SEM1 and produced the first micrographs showing the 'three-dimensional' topographic image formation characteristic of modern-day instruments. Theoretical analysis of probe formation; measurements of distribution of backscattered electrons with angle; considered mechanisms of BSE contrast formation.
Dissertation: "Investigations relating to the Design of Electron Microscopes" 1952.

K C A Smith: 1952. Made improvements to SEM1, including efficient detection of the low-energy secondary electron (SE) component; considered SE contrast formation; extended theory of probe formation; investigated a wide range of applications; dynamic experiments, including chemical reactions at elevated temperatures.
Dissertation: "The Scanning Electron Microscope and its Fields of Application" 1956.

O C Wells: 1953. Constructed SEM2 and applied it to study of fibres. Explored new types of detector; used scintillator/photomultiplier detector in many novel configurations; established theory of

stereomicroscopy in the SEM; investigated ways of examining non-conducting specimens, including the use of positive ion bombardment; investigated atomic number contrast.
Dissertation: "The construction of a Scanning Electron Microscope and its application to the study of fibres" 1957.

T E Everhart: 1955. Continued improvements to SEM1; devised new detector ('Everhart-Thornley' detector). Studied contrast mechanisms in detail, including potential contrast; developed a new theory of reflection of electrons from solids.
Dissertation: "Contrast formation in the Scanning Electron Microscope" 1958.

P J Spreadbury: 1956. Constructed a simple SEM using a cathode-ray oscillograph as the display unit; made careful measurements of the performance of the electron gun. Constructed numerous pieces of electronic equipment which were used in other SEMs and projects in the laboratory.
Dissertation: "Investigations relating to the design of a Simple Scanning Electron Microscope" 1958.

R F M Thornley 1957. Made improvements to SEM2. Conducted detailed study of Everhart detector. Low-voltage operation for examination of non-conducting specimens; studies of frozen biological specimens.
Dissertation: "New applications of the Scanning Electron Microscope" 1960.

A D G Stewart 1958. Completed construction of a new microscope begun by Oatley (SEM4). Addition of ion gun allowed the direct observation of specimens while undergoing sputtering. Later moved to Cambridge Instrument Co. to work on 'Stereoscan' project (see section 5).
Dissertation: not submitted.

H Ahmed 1959 (supervisor A H W Beck). Used SEM2 to investigate activation processes of dispenser cathodes. Direct observation of emitting cathodes at temperatures exceeding 1300 K.
Dissertation: "Studies on high-current-density Thermionic Cathodes" 1962.

R F W Pease 1960. Designed a new microscope (SEM5) - first SEM to achieve a resolution of 10 nm. Several of these instruments were made in the Engineering Dept. and used in other Groups. Also supplied to F P Bowden's Group in the Cavendish Laboratory (see section 5).
Dissertation: "High resolution Scanning Electron Microscope" 1963.

A N Broers 1961. Made improvements to SEM4, including addition of magnetic lens. Added mass-filter to ion beam system to obtain pure ion beam species. Found that surface contaminants affected rates of sputtering and could thus be used to mask selected areas of the specimen surface. Used this effect to lay down patterns of gold wires and other structures - one of the earliest successful attempts at electron beam microfabrication and micromachining.
Dissertation: "Selective ion beam etching in the Scanning Electron Microscope" 1965.

Oatley was appointed to the Chair of Electrical Engineering and Head of the Electrical Division in the Department in 1960, and thereafter his direct involvement with the research programme declined; supervision of Pease and Broers was handed over to W C Nixon, who had joined the Department from V E Cosslett's Group in the Cavendish in 1959. On his retirement in 1972, however, Oatley again took up the research and continued to make contributions to the field well into his 80s (Oatley 1975, 1981, 1983, 1985). His book on the SEM was published in 1972.

## 5.    COMMERCIAL DEVELOPMENT: THE STEREOSCAN

Up to the mid-1950s there was little interest in the work of Oatley's Group, but a turning point in attitudes towards scanning electron microscopy came when D Atack, a member of the Pulp and Paper Research Institute of Canada (PPRIC), and J H L McAuslan, with Imperial Chemical Industries, learned of the work on the SEM. Both were on sabbatical leave at the time in

F P Bowden's Group in the Cavendish, and they decided to explore the potential of the instrument for their work. Using SEM1 Atack examined a range of pulp and paper specimens, while McAuslan studied the thermal decomposition of silver azide crystals. This work generated strong interest at the PPRIC, and it was subsequently arranged that the Institute would finance the construction of a new SEM in the Engineering Department. Smith undertook the task of designing this new instrument - designated SEM3 - as a post-doctoral research project in 1956.

Fig. 4. The prototype of the first Stereoscan supplied by the Cambridge Instrument Company to the du Pont Company, U.S.A. (Stewart and Snelling 1965)

At this stage Oatley persuaded Associated Electrical Industries (AEI) (formerly Metropolitan Vickers), a company then manufacturing both transmission instruments and electron probe microanalysers, to take an interest in the SEM. An understanding was reached that if the microscope appeared to be commercially viable, AEI would take it up. Later, in 1958, AEI received an order from Bowden, but for commercial reasons the instrument produced was based on their current production microanalyser, little use was made of the experience gained from the construction and operation of SEM3, and it was not successful. Bowden returned the microscope to AEI and later ordered a copy of the Pease-Nixon microscope from the Engineering Department. This ended AEI's attempt to enter the SEM market (Agar 1996; Brown et. al. 1996).

SEM3 was shipped to Canada in 1958 and used successfully by the PPRIC in their Montreal laboratories and by other companies which hired time on the instrument, including the du Pont chemical company of the USA. This, together with an accelerated flow of results from members of Oatley's Group - Nixon, Thornley, Stewart, Ahmed, Pease, and Broers - produced a change in attitudes towards the SEM as its advantages became better appreciated.

In 1961 Nixon and Smith made an informal approach to the Cambridge Instrument Company, suggesting that the Company should take up manufacture of the SEM as well as the microanalyser, which they were marketing at the time. Shortly afterwards Oatley reached a formal agreement with H C Pritchard, the managing director of the Company, and arrangements were made for the manufacture of two prototypes, one of which went to the du Pont Company (Fig. 4). In 1962 one of Oatley's former research students, A D G Stewart, joined the Company to take over development of the new SEM, and with Government backing, a batch of five microscopes was manufactured in 1965.

Oatley's 1982 paper, 'The early history of the scanning electron microscope', concludes with the following paragraph - a fitting testimony to his pioneering work:

> The first four production models, sold under the trade name "Stereoscan", were delivered respectively to P R Thornton of the University of North Wales, Bangor, to J Sikorski of Leeds University, to G E Pfefferkorn of the University of Münster, and to the Central Electricity Research Laboratories. By this time the Company had launched a publicity campaign and orders began to roll in. An additional batch of twelve microscopes was put in hand; and then a further forty........The scanning microscope had come of age.

## 6. ACKNOWLEDGEMENTS

The author is grateful to Alan Agar, Bernie Breton, Paul Brown, David Holburn, Dennis McMullan, Tom Mulvey, Laurence Smith and Oliver Wells, for many useful discussions and for help in the preparation of the manuscript.

## REFERENCES

Agar A W 1996 The Story of European Commercial Electron Microscopes, in The Growth of Electron Microscopy, ed T Mulvey. Advances in Imaging and Electron Physics, **96** (London: Academic Press) pp 415-584

Brown P D, McMullan D, Mulvey T, Smith K C A 1996 On the origins of the first commercial transmission and scanning electron microscopes in the UK, Proc. Roy. Microsc. Soc. **31**/2, 161

McMullan D 1988 Von Ardenne and the scanning electron microscope Proc. Roy. Microsc. Soc. **23**/5 283

McMullan D 1995 Scanning Electron Microscopy 1928-1965, Scanning, **17**, 175

Oatley C W 1931 The Theory of Band-pass Filters for Radio Receivers, Experimental Wireless and The Wireless Engineer, June 1931

Oatley C W 1932 Wireless Receivers (London: Methuen)

Oatley C W 1936a The Power-loss and Electromagnetic Shielding due to Flow of Eddy-currents in Thin Cylindrical Tubes", The Philosophical Magazine, **22**, Ser. 7, 445

Oatley C W 1936b The Design of Eddy-Current Heating Apparatus for Outgassing Electrodes in a Vacuum, The Philosophical Magazine, **22**, Ser. 7, 453

Oatley C W 1939 The Adsorption of Oxygen and Hydrogen on Platinum and the Removal of these Gases by Positive-ion Bombardment, Proc. Phys. Soc. **51**, 318

Oatley C W 1972 The Scanning Electron Microscope (Cambridge: CUP)

Oatley C W 1975 The tungsten filament gun in the scanning electron microscope, J. Phys. E **8**, 1037

Oatley C W 1981 Detectors for the scanning electron microscope, Jour. Phys. E.,**14**, 971.

Oatley C W 1982 The early history of the scanning electron microscope, J. App. Phys. **53**(2), R1

Oatley C W 1983 Electron currents in the specimen chamber of a scanning microscope, J. Phys. E **16**, 308

Oatley C W 1985 The detective quantum efficiency of the scintillator/photomultiplier in the scanning electron microscope, J. Microscopy **139**, Pt. 2, 153

Oatley C W, McMullan D and Smith K C A 1985 The Development of the Scanning Electron Microscope, in The Beginnings of Electron Microscopy, ed P W Hawkes. Advances in Electronics and Electron Physics Suppl. **16** (London: Academic Press) pp 443-482

Sander K F 1951 An automatic electron trajectory tracer and contributions to the design of an electrostatic electron microscope, Ph.D. Dissertation, Cambridge University

Stewart A D G and Snelling M A 1965 A new scanning electron microscope, Proc. 3rd European Conference Electron Microscopy, Prague 55-56

von Ardenne M 1940 Elektronen-Übermikroskopie (Berlin: Springer)

von Ardenne M 1985 On the History of Scanning Electron Microscopy, of the Electron Microprobe, and of early Contributions to Transmission Electron Microscopy, in The Beginnings of Electron Microscopy ed P W Hawkes. Advances in Electronics and Electron Physics Suppl. **16** (London: Academic Press) pp 1-21

*Inst. Phys. Conf. Ser. No 153: Section 1*
*Paper presented at Electron Microscopy and Analysis Group Conf. EMAG97, Cambridge, 1997*
© *1997 IOP Publishing Ltd*

# A Synchrotron in a Microscope

**L. M. Brown,**

Cavendish Laboratory, Madingley Rd., CAMBRIDGE CB3 0HE

**ABSTRACT:** Many laboratories are now equipped with electron spectrometers of high performance, giving about 0.3V energy resolution in parallel acquisition of spectra from a sub-nanometer probe. The electron energy-loss spectrum (EELS) can be regarded as containing information equivalent to that obtained from X-ray absorption spectra (XAS) using a dedicated synchrotron source. However, the EELS can be used to investigate absorption in the neighbourhood of the fundamental band-gap, as well as near-edge structure revealing the chemical state of particular ions. Furthermore, because of the high spatial resolution, the chemical state of ions in the neighbourhood of interfaces and other defects can be investigated. This opens up a whole new nanochemistry of the solid state. It goes almost without saying that either by direct imaging, or by microdiffraction techniques, the atomic structure can also be determined from the same area as the EELS. The aim of this paper is to discuss to what extent current capabilities in Electron Energy Loss Spectroscopy (EELS) supplant the facilities at major synchrotrons where the synchrotron radiation source powers several beam lines to provide XAS for a wide variety of applications.

## 1.    INTRODUCTION

I am indebted to Dr. J. Yuan for permission to publish Fig. 1, which shows his direct comparison between XAS and EELS arising from the study of thallium substituted copper-oxide superconductors, $Tl_{0.5}Pb_{0.5}Ca_{0.8}Y_{0.2}Sr_2Cu_2O_7$. The XAS was obtained at the Daresbury SRS, operating at 2GV, beam current 100mA, giving a photon count rate about $10^5$ $s^{-1}$ in a beam about 3mm in diameter. The acquisition time was about 20 min, using the total electron yield method, in which the secondary electrons resulting from the X-ray absorption are collected to produce a spectrum with an energy resolution of about 0.4eV. The EELS was obtained on the Vacuum Generators' HB501 operating at 100kV, using the McMullan acquisition system (McMullan et. al. 1992). The beam current was about 0.1nA, producing an electron count rate of $10^9$ $s^{-1}$, and a probe size about 1 nm. Acquisition procedure was to take 20 spectra which are then aligned by eye and superimposed to produce the spectrum shown, which therefore corresponds to a total acquisition time of about 20 s and an energy resolution about 0.3eV. For the EELS, the pre-edge background was estimated and removed in the usual way, leaving several thousand counts per channel in the peaks; the corresponding Poisson noise is about 2% per channel.

Specimen preparation was as follows: for the XAS, the sample was powdered and picked up on sellotape; thus the spectrum was obtained from many hundred randomly-oriented grains. For the STEM, the sample was crushed and the spectrum obtained by finding an electron transparent edge, perhaps 50nm thick. No attempt was made to orient the grain, although of course this can routinely be done by electron diffraction. In all, the time required to prepare the sample and acquire the spectra is about the same for the two techniques; one can conceive of acquiring spectra from ten samples per day.

The physical information in the two spectra is almost identical: for instance, the $L_3/L_2$ ratio is $3.2 \pm 0.2$, depending rather a lot on the numerical procedure. The ratio from the XAS seems slightly larger (perhaps 5%) than that for EELS; according to the argument in the next section, this

ay be a significant difference between electron and photon excitation, on the other hand, it may
flect multiple scattering which in EELS will increase the L2 peak relative to L1. The apparent
solution in the EELS is rather better than that in the XAS, but in both cases the peak widths are
ominated by the natural lifetime and will depend upon orientational effects, so that the XAS peaks
ay be broadened by sampling many crystallites. Certainly, there is a hint of structure in the L2
ak in the EELS which is absent in the XAS.

o summarize: the current EELS capability is comparable with the XAS capability, certainly in this
articular case.

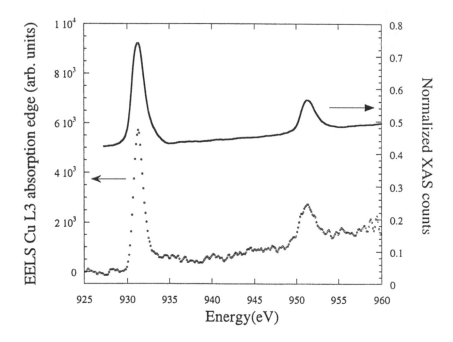

Fig. 1. XAS compared with EELS

## COMPARISON BETWEEN XAS AND EELS

Because photon excitation in XAS transfers so much less momentum to the struck electron
the sample than electron excitation in EELS, one must expect that the two spectra are in general
ther different, even when (as is usually the case) the dipole approximation holds in EELS, so that
e wavelength of the ejected core electron is much longer than the radius of the core orbit from
hich it comes. The differences between the two types of spectra are rather subtle.

### i)   Observation of Excitons

Batson and Bruley (1991) show that excitonic peaks in insulators at the threshold of edges
e more prominent under X-ray excitation than electron excitation. They argue that the steadily
ernating electric field of the X-ray permits the creation of a core hole which can be treated under
e 'sudden' approximation by time-dependent quantum mechanics, followed by a transition of the
cted electron to an excitonic state bound by the Coulomb field of the hole, although the strength
the Coulomb field is reduced by dielectric polarisation. By contrast, when excitation occurs by

the rapidly passing point electron, the exciton is distorted by the inhomogenous field with its wide range of excitation frequencies, and the simple 'sudden' approximation for the initial excitation is not such a good one. Furthermore, if the field of the fast electron is itself completely screened by the core electrons, no exciton can be formed because the ejected core electron is, as it were, carried away in the wake of the fast electron: Batson and Bruley here refer to the complicated picture of the partially screened fast electron developed by Echenique et. al. (1979). They present a simple argument which leads to an estimate of the magnitude of the correction, as follows: The fast electron traverses a specified region - say, the radius of the exciton orbit - in a time inversely proportional to its velocity, but the ejected electron traverses it much more slowly, in a time proportional to its orbital speed (which can be regarded as a consequence of the uncertainty principle). Thus the dielectric relaxation which can can be felt by the core electron is reduced by an amount estimated to be the ratio of its orbital speed to the fast electron speed, or $(E_{core}/E_0)^{1/2}$, which is always a small quantity. The potential felt by the ejected electron is thus approximately an unscreened Coulomb potential multiplied by $[\varepsilon^{-1} - (E_{core}/E_0)^{1/2}]$, and the fractional correction caused by relaxation is

$$\alpha = \varepsilon \left( \frac{E_{core}}{E_o} \right)^{1/2} \qquad\qquad 1.$$

The dimensionless number $\alpha$ gives an estimate of the reduction in oscillator strength at the excitonic energy loss which will be caused by electron excitation as opposed to X-ray excitation. This simple argument seems to agree roughly with observation and with careful additional theoretical modelling by Batson and Bruley. The exciton caused by a 100keV electron at the K edge in diamond, with $\varepsilon = 5.7$ and $E_{core} = 290eV$, is according to this reduced in intensity by 0.31, whereas the best fit suggests 0.30; in silicon at the L edge, the observed reduction is 0.60 whereas equation (1) suggests it should be 0.38. In general, the EELS results should resemble the one-electron DOS more closely than EXAFS, but excitonic effects cannot be neglected altogether if perfect agreement between theory and experiment is required. Of course, in metals, there are no excitons because the Coulomb potential of the hole becomes exponentially screened, and so can support no bound states. However, the general argument still holds, whereby the fast electron field is partially screened; the characteristic length is the metallic screening length within which there is no screening, so perhaps one should use eq. (1) with unity for the relative permittivity. This produces negligible correction for the Al L edge ($\varepsilon = 1$, $E_{core} = 72eV$, $\alpha = .03$), consistent with observation.

### 2(ii)   Observations of Extended Energy Loss Fine Structure

Ever since the pioneering work of Leapman and Cosslett (1976) it has been recognised that one can carry out the analysis of near-neighbour distances and population centered around a specified chemical species by EELS as well as by XAS, with, of course, the advantage of a nanometre probe in the latter case. A recent paper by Stern et. al (1994) presents a comparison of spectra from the two techniques for metallic Al; they obtain satisfactory first neighbour distances from EXELFS but poor information on the second shell of neighbours, in accordance with early work by Stephens et. al. (1981). Stern et. al. comment on a reduction in the amplitude of the oscillatory departures from background observed close to the threshold in EELS by comparison with XAS; the value of $\alpha$ for this case is 0.12, which might provide an explanation. The general conclusion is, that for edges below about 1.5kV, EXELFS spectra are fully competitive with XAS but electron excitation offers marked advantages, including spatial resolution, ancillary characterisation techniques such as high resolution imaging, and of course, it is an in-house technique.

### 2(iii)  Observation of Band Gaps

EELS can be used to measure the band gaps of insulators and semiconductors, but here one finds marked differences from spectra obtained by photon excitation because momentum transferred by the fast electron produces features in the spectrum associated with non-vertical transitions, particularly indirect gaps (Rafferty et. al., this conference).

### 3.   SPATIAL INFORMATION

The preferred method of deriving spatial information from EELS is to compare directly spectra from one probe position to the next, for example the 'spectrum-line' methods introduced by Colliex and co-workers (1993) and the changes in the EELS as the probe is moved from one atomic column to the next (Batson (1993), Browning et. al. (1993), Muller et. al. (1993). In spite of the complications in interpretation of spectra, subtraction of one spectrum from another taken from a neighbouring position is assumed to reveal the difference in the density of states accessible to the appropriate core electrons: thus changes in the d-band occupancy at grain boundaries in transition metals can be observed (Dray, 1995, Muller (1996), Botton and Humphreys (1994), Ozkaya et. al. 1995), and differences at the cores of dislocations studied. It is unlikely that complications such as excitons will prove difficult, since they cancel out in the difference spectrum. Such experiments seem poised to become an important new field of solid-state spectroscopy, unrivalled for spatial resolution and directness of approach. These experiments are scarcely possible in XAS with its unfocused beam.

The ultimate spatial resolution achievable by localised electron probe methods is limited by radiation damage. This is not surprising in view of the fact that typically in a specimen of thickness about half a plasmon mean free path the dose rate is approximately 5 TGy per second - $10^{12}$ times larger than the fatal human dose! Nevertheless, useful data can usually be achieved by sensible experimental methods. In aromatic polymers, C. A. Walsh (1995) has shown it possible by superimposition of individual spectra from adjacent areas of the specimen to produce high-quality results, rivalling or exceeding XAS in spectral quality, with micron (rather than nanometre) spatial resolution. However, even in robust metallic samples, radiation damage by the focused probe may be what limits observation of localised states at grain boundaries (Ozkaya et. al., 1995). However, in these cases, progress can be made by using the 'spatial difference method' where one acquires a spectrum from a large area containing the boundary, and then one from a neighbouring region but excluding the boundary. The difference between the two spectra reveals the extra empty states at the boundary, and of course any impurities which may be segregated there. Thus radiation damage is avoided but at the cost of reduced spatial resolution.

### 4.   COMMENTS ON XAS AND EELS

The concept of a national center for synchrotron radiation science and technology has found wide acceptance in many countries and has proved the effectiveness of concentrating limited resources for research. In England, the Central Laboratory of the Research Councils at Daresbury pursues a wide range of high quality applied and pure research. It is proposed to replace the aging synchrotron with a new dedicated machine, to be called DIAMOND. Users of the facility are being surveyed to help design the most useful array of beam lines, and to assess overall support. In round numbers, DIAMOND will cost about £100M, plus between £2M and £4M for each beam line, and there might be 20 such lines. If DIAMOND is staffed as the present facility, there might be 300 staff in all; the function of the staff ranges from technical development of novel beam lines, say, to routine support and secretarial services. Such a project is far more ambitious than any so far conceived for electron microscopy.

DIAMOND will produce somewhat different conditions for experiments at the different beam lines. Some are designed to achieve very high energy resolution (20meV) and small spot sizes (10$\mu$m or so). But typically the brightness at the specimen will be $10^{19}$ photons s$^{-1}$mm$^{-2}$ mrad$^{-2}$

for energy resolutions of 0.25eV. As Fig. 2 shows, *this performance is already matched by dedicated STEM.* The STEM is already a high-performance synchrotron in a microscope. Of course, there are many things STEM cannot do, such as X-ray topography of large specimens, so it cannot supplant a new synchrotron.

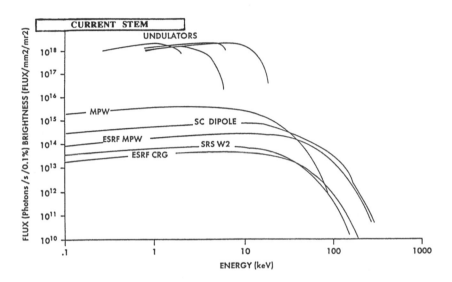

Fig. 2. The performance of current dedicated STEM (upper box) compared to DIAMOND with undulators; other machines and configurations shown.

Let us suppose that one tenth of the proposed cost of DIAMOND and its beam lines could be invested in a national center for STEM, perhaps sited at the Central Laboratory of the Research Councils to capitalise on the resources already deployed there. With £15M capital, and a staff of 30, one could imagine commissioning an array of specialised instruments including dedicated STEM with a performance far superior to any existing instrument: it would deploy multipole lenses according to Krivanek's design (Krivanek, et. al., this conference) which would remove spherical aberration and chromatic aberration; single-atom field-emitters with brightnesses still higher than any currently in use (James and Bleloch, this conference); and dedicated detectors to overcome the phase problem and produce atomic resolution images according to Rodenburg's concept of super-resolution (Coleman and Rodenburg, this conference). The use of co-incidence counting of X-ray photons and secondary electron emission from the focused probe will allow XPS to be carried out at high spatial resolution (P. Kruit, private communication). Now one does not want to get carried away, but it is very reasonable to expect half-Angstrom spatial resolution and 0.25 eV energy-loss resolution in a dedicated machine, operating at 200kV, achieved within say three years of commencement of work to build it. Experience shows that as the instrument develops, it would earn considerable income from industry, to look at particulates, catalysts, grain boundaries, semiconductor interfaces, etc. It would certainly be as good an investment as several beam lines on DIAMOND.

What is of great importance is that the community of electron microscopists should agree to such a programme. Perhaps EMAG could use its accrual account to fund a commission whose job it would be to put together a realistic, costed proposal; one which could be endorsed by the entire community and put to the Research Councils for support.

Is this proposal realistic? It needs to be systematically examined by the community of electron microscopists, working together, so that a unified project can be presented to the Research

22

Councils. One possible way forward is to use the accrual account of EMAG to support a working party which might develop realistically costed schemes, whose effectiveness can be carefully assessed by comparison with other major national investments in research.

## ACKNOWLEDGEMENTS

I am deeply indebted to all members of the Microstructural Physics Group, particularly Dr. J. Yuan, Dr. C. A. Walsh, and of course Prof. Howie who provides laboratory facilities as well as his invaluable contribution to the work.

Batson, P. E., 1993, Nature 366, pp 727-728.
Batson, P. E., and Bruley, J., 1991 Phys Rev Lett 67 , pp 350-353
Botton, G. A., Humphreys, C. J., Proc. of ICEM 13-Paris, B. Jouffrey, C. Colliex, Eds., Les Editions de Physique, Paris, p 631.
Browning, N. D., Chisholm, S. J., and Pennycook, S. J., 1993, Nature 366 pp 143-145.
Colliex, C., Lefevre, E., Tence, M., 1993 Inst Phys Conf Ser, 138 (EMAG Liverpool), Ed. A. J. Craven.
Dray, A. E., 1995, reported in Brown, L. M., Walsh, C. A., Dray, Ann, and Bleloch, A. L., Microsc., Microanal., Microstruct., 6, 1995, pp 121-125. A fuller account is available from the author.
Echenique, P. M., Ritchie, R. H., and Brandt, W.1979, Phys Rev B20 , pp 2567-
Leapman, R. D., and Cosslett, V. E., 1976, J. Phys. D. 9 L, pp 29-32.
McMullan, D., Fallon, P. J., Ito, Y., McGibbon, A. J., 1992, Proc. 10th Eur. Cong. on electron Microscopy. Granada, Eds. A. Rios, J. M. Arias, L. Megias-Megias & A. Lopez-Galindo, 1, p 103.
Muller, D. A., Near Atomic Scale Studies of Electronic Structure at Grain Boundaries in Ni3Al, Ph.D. thesis for Cornell University, 1996.
Muller, D. A., Tzou, Y., Ray, R., and Silcox, J., 1993, Nature 366, pp 725-727.
Ozkaya, D., Yuan, J., Brown, L. M., 1995, J. Micros. 180, pp 300-325.
Rafferty, B. E., and Brown, L. M., this conference        .
Stephens, A. P., and Brown, L. M., 1981, Proc. Conf. Manchester 1981 (The Metals Society) Book 277, pp 152-158.
Stern, E. A., Qian, M., and Sarikaya, M., 1994, Mat. Res. Symp. Proc. 332, pp3-14.
Walsh, C. A., 1995, reported in Brown, L.M., Walsh, C. A., Dray, Ann, and Bleloch, A. L., Micros, Miroanal, Midrostruct, 6, 1995, pp121-125. A fuller report is available on request.

*Inst. Phys. Conf. Ser. No 153: Section 1*
*Paper presented at Electron Microscopy and Analysis Group Conf. EMAG97, Cambridge, 1997*
© *1997 IOP Publishing Ltd*

# New electron microscopies - the state of the art

**J.C.H.Spence**

Department of Physics and Astronomy, Arizona State University, Tempe, Az. 85287, U.S.A.

**ABSTRACT:** A selection of the many exciting opportunities for the use of atomic resolution microscopy in materials physics is presented, with an emphasis on new ideas. Schemes for exceeding the classical resolution limit in atomic resolution microscopies are summarised, including the use of atomic columns as lenses, and use of a direct solution to the dynamical inversion problem for electron microdiffraction. A summary of work aimed at deducing atomic defect structure, band-structure, formation energy and migration energy from atomic resolution images is given, and work based on video recordings of kinks on dislocations in silicon is summarised. New results for Ge growth on Si(111) are reported using our Scanning Tunnelling Atom Probe, which allows atomic clusters seen in STM images to be transferred to a time-of-flight spectrometer for identification.

## 1. INTRODUCTION

In this paper I have tried to isolate some of the most exciting research opportunities in nanoscale microscopies, with emphasis on electron-beam methods and new ideas. Due to space limitations, it is not possible to review work on high resolution SEM or by field ion microscopy. Important techniques which operate at lower resolution, such as magnetic field imaging, vortex imaging, strain contrast imaging, acoustic imaging, X-ray holography and NMR imaging, amongst many other techniques, are also not discussed for similar reasons. The exciting recent developments in molecular imaging by AFM and related methods are not discussed (Chiang,1994) .

The rapid growth in new types of scanning probe microscopy over the last two decades has created and solved many problems. Most of these microscopies have two features in common : their "resolution" is not a property of the instruments alone (but also depends on the particular sample), and their resolution is often less than the wavelength of the radiation used. These factors greatly complicate image interpretation. The same period has also seen rapid growth in new modes for the operation of electron microscopes, better adapted to particular materials problems. We review some of the more promising developments in both areas below.

## 2. SUPER-RESOLUTION SCHEMES

The state-of-the-art in microscopy generally, for attainment of the highest interpretable resolution by any technique, (including all scanned probe microscopies and the atom probe) appears to be the images recorded on the JEOL ARM1250 instruments, operating at about 1 MeV. Using one of these, Ichinose, for example, has clearly resolved the individual columns of Silicon and Carbon in SiC with 0.109 nm separation (Ichinose and Ishida,1989) . Images recently recorded on the Vacuum Generators 300 kV STEM in the Z-contrast mode are highly competitive (Nellist and Pennycook,1997) .

To improve on this, a variety of super-resolution schemes have been developed. The situation in optics has recently been reviewed (Dekker and Van den Bos,1997) , and a review of related methods in electron microscopy can be found in (Smith,1997) . It is conventional to define a point resolution (Scherzer) limit, an information resolution limit set by electronic and mechanical instabilities, and resolution limits due to source size, intrinsic beam-energy spread, and wavelength (diffraction limit) (Spence,1988) . Which of these is limiting depends on the particular experimental arrangement, on methods of image processing, and on noise levels. Noise is the fundamental limit, not the Rayleigh criterion. In the absence of noise, the complete object spectrum may be reconstructed (beyond the lens cuttoff) by analytic continuation, if the object is small, at the expense of field-of-view. Incoherent imaging theory (to which the Rayleigh criterion applies) may, under certain approximations, be applied to Z-contrast STEM imaging, and has the desirable property that resolution is then independent of sample. It should be noted, however, that the resolution in phase contrast imaging is significantly better for points scattering in anti-phase than for incoherent imaging.

Near-field methods allow resolution to be obtained which is smaller than the wavelength of the radiation, as first pointed out by Synge (Synge,1928) . An instructive analysis, presumably with reference to the cavity magnetron, was given by Bethe (Bethe,1944) . The principle can be tested experimentally using microwaves. It will be found that 1 cm microwaves incident on a metal screen containing a 1 mm hole will generate a static bubble of intensity on the far side of the screen. By scanning this evanescent wavefield past an object and detecting scattered radiation, an image can be obtained whose resolution is limited by the hole size, not the wavelength of the radiation. The uncertainty principle is not violated, since the wavefield is not propagating. Similar principles explain the attainment of 0.2nm resolution using one volt electrons in STM. When combined with fluorescence imaging of living cells , near-field optical techniques hold great promise for biology, in view of their non-invasive, chemically specific and high resolution character (See Ultramic., Vol 61 for a review of the field). Images have been obtained with visible light showing variation of intensity on a 20 nm scale, however image interpretation is complicated and, as in STM and AFM, requires a detailed specification of the tip shape.

In electron microscopy many schemes have been developed for exceeding the point resolution of the instruments. In particular, recently we have seen the point resolution exceeded using off-axis electron holography (Ochowski, et al.,1995) (from 0.2nm to 0.12nm), using through-focus methods (Coene, et al.,1992) (from 0.24nm to 0.14 nm) and by synthetic-aperture beam tilting methods (Kirkland, et al.,1997) (from 0.2nm to 0.13nm). The first of these methods is limited by the number of pixels needed and the fineness of the carrier fringes (three fringes per point-resolution element, four pixels per fringe), the second and third by image registration problems and the correct manner of incorporating noise filters to control noise amplification during deconvolution (Schiske,1973) . All methods are limited by the errors in the experimentally measured parameters used to define the transfer function, and possibly by aberrations unaccounted for, such as three-fold astigmatism (Krivanek and Stadelman,1995) . The dark-field (Z-contrast) STEM mode is also capable of super-resolution, showing interpretable image detail out to 0.093 nm (Nellist and Pennycook, 1997) . The Ptychography method in STEM has recently been used to exceed the information limit (under the assumption of a periodic sample), from 0.33nm to 0.14nm (Nellist, et al.,1995) , a remarkable achievement. It is interesting to compare this method with the refinement of atomic scattering factors by CBED, which may be applied to similar data, and which allows for multiple scattering. Recently the CBED method has been used to refine the atomic coordinates in 4H SiC (Holmstad, et al.,1997) . In that case the method is perturbative, and a first guess for atom positions is required. Ptychography requires a knowledge of the crystal lattice, but not atom positions. It solves the phase problem using interference between overlapping orders, assuming single scattering. (Errors in Ptychography due to multiple scattering are estimated in (Spence,1978) ). CBED refinement uses the multiple scattering between different beams to provide phase information. Both are capable of super-resolution. A direct, non-perturbative solution to the dynamical inversion problem has been given which is based on Ptychography data (Spence, 1997). This method gives X-ray structure factors directly from dynamical microdiffraction intensities. This solves three problems, by determining atom positions directly, their atomic number (from the height of peaks in the charge density map) and providing super-resolution (since coupling reflections double the resolution).

There has been a resurgence of interest in the use of direct methods and maximum entropy techniques for phase extension - this topic is well reviewed in the recent text by Dorset (Dorset,1995) . Sayre's equation, the tangent formula and phase triplet sums have all been applied to combinations of images and diffraction data in order to improve resolution, with varying degrees of success. According to Baysian statistics, the prior used in the maximum entropy method is only one of many possible, and has no special status compared to other sharpening functions which might be used. The ad-hoc nature of many of the functions used, combined with errors due to multiple scattering, uncertainties (or variations) in thickness and the curvature of the Ewald sphere, and uniqueness problems, have all led to uneven (and occasionally spectacular) results (Dong, et al.,1992) from which it is difficult to generalise.

After decades of effort, (Rose,1994) aberration correction schemes using multipole corrector lenses appear to be on the verge of success, both for TEM (Heider and Zach,1995) and STEM (Krivanek and Spence,1997) . These will have a big impact on electron microscopy - reducing the influence of the spatial coherence damping envelope, improving point resolution and allowing larger working distances, X-ray detector take-off angles, and sample tilt in STEM. Higher order aberrations will be exposed, and the information limit (e.g. due to tip vibration in STEM) remains. The specific brightness of a STEM probe formed at optimum aperture and defocus is independent of Cs. In TEM, contrast mechanisms will have to be re-thought, since the method of balancing defocus and spherical aberration phase shifts against each other cannot be used if Cs = 0. The Projected Charge Density (PCD) approximation (reviewed in Spence (1988) may then be more useful than the weak phase object approximation.

The reconstruction of crystal structures in three dimensions from HREM images is an important milestone in electron microscopy (Downing, et al.,1990) . The development of field-emission guns, energy filters and new detectors is essential for accurate quantification of data, which might follow similar procedures to those adopted in the more mature and demanding areas of organic electron crystallography and CBED, where much experience in accurate quantification has already been gained (Spence and Zuo,1992) . A comparison of the new image plate digital imaging system with the slow-scan CCD camera can be found in Zuo et al. (1996) . In summary, the image plate is found to offer many more pixels and improved DQE at low dose, while avoiding the pixel cross-talk problems of YAG/CCD systems, but to lack the immediate image processing capabilities of the CCD. The dynamic range of image plate is limited by the plate reader, not by the image plate itself. We foresee that the accurate characterisation of electron detection systems will become essential in all fields of electron microscopy in future, as it has already become for CBED refinement.

Further progress in HREM requires an understanding of why the images consistently show less contrast than predicted by multiple scattering programs. This difference is revealed when quantitative comparisons between images and calculations are made (Mobus and Ruhle, 1994) . Boothroyd has made an exhaustive analysis of all the possible factors, including the contribution of energy-loss lattice images (Boothroyd,1997) .

A novel approach to super-resolution has recently been made involving the use of atomic columns as electron lenses (Cowley, et al.,1997) . Aberration coefficients scale with lens dimensions - a single heavy atom acts as a lens with a focal length of about 2 nm at 100 kV and has negative spherical aberration. A stronger lens can be made using a column of heavy atoms in a crystal. A STEM probe must be positioned over it, and a sample beneath it. Figure 1(a) shows the focusing effect of a thin gold crystal on a STEM probe about 0.2 nm in diameter, situated over one atomic column. Figure 1(b) shows how it emerges from the crystal, reduced in diameter to 0.04nm. In propagating beyond the crystal, the probe broadens to about 0.05 nm at 2nm distance beyond the focusing crystal. This very short working distance would require piezo-electric positioning devices for sample manipulation, and feedback control to stabilize the probe position.

A more practical scheme is suggested in figure 2, appropriate to a TEM instrument. Here the Fourier images formed beyond a thin focusing crystal are used as fine probes, which are scanned over the surface of the sample by tilting the incident beam. The focusing crystal consists of widely spaced heavy-atom columns. The Fourier lattice images form narrow waists in the electron stream, which repeat periodically with distance downstream. Calculations show that the array of Fourier "probes" incident on the second crystal (or on a general sample) may be much finer than 0.05 nm for favorable thicknesses of the focusing crystal. The wavefield exiting the sample is then imaged by a conventional TEM lens, whose resolution need only be sufficient to resolve the individual atomic columns in the focusing crystal. The resulting TEM images are recorded for each incident beam tilt, and this array of images then contains enough information to assemble an image of the sample with 0.05nm resolution. Multislice calculations show that these schemes work well, in addition to several others related by reciprocity. The Fourier images may be focused onto the sample by variation of electron wavelength, since their period along the axis is $2d^2/\lambda$, where d is the lateral period of the focusing crystal (Spence,1988) . Since the function of the TEM lens is simply to integrate over the response of the sample to the incident probe, the final image is independent of the aberrations and instabilities of this lens, provided it can resolve the atomic columns of the focusing crystal. We are about to start making sample structures to test this method, using Focused Ion Beam (FIB) and ion-implant stop-etch methods. In common with the "ALCHEMI" method, this method relies on the very fine structure of the dynamical wavefield inside a crystal to obtain improved spatial resolution. It is essential that the separation of the heavy atom columns in the focusing xtal be greater than the resolution of the TEM lens. Advantages of the method, in addition to improved resolution, include the small depth of field (about 1 nm at 100kV), which would allow three-dimensional imaging by optical sectioning, and microanalysis with higher spatial resolution.

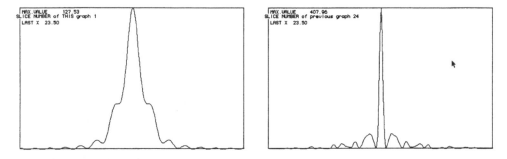

Figure 1(a) STEM probe (200 kV, Cs = 2mm) at entrance surface of thin gold crystal, located over column of atoms. Width of frame 2.35 nm, probe width 0.2 nm. (b) Intensity distribution on exit-face, thickness 6nm. Probe width has been demagnified to 0.04 nm.

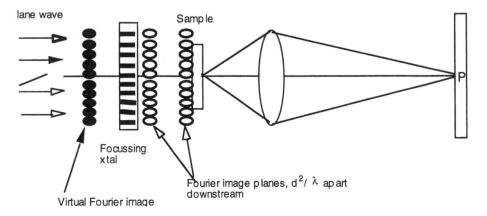

Figure 2. TEM super-resolution scheme. Fourier lattice images are formed as an array of fine probes beyond a focusing crystal . These are scanned over the surface of the sample by tilting the incident beam direction. TEM images are recorded for each tilt.

## 3. DEFECT ENERGIES AND DEFECT STRUCTURES FROM HREM.

Many properties of solids are controlled by the formation and motion of defects. Examples include interface reactions, dislocation motion, first-order phase transitions, crystal growth and diffusion.the ab-initio total energy method combined with lattice dynamics (quantum molecular dynamics) has proven powerful for understanding these processes. As one example, we have recently used this method to find the energy barrier to crack tip motion (and subsequent surface reconstruction) in silicon (Spence, et al.,1993) . It was thus possible to predict fracture toughness from first principles in good agreement with experiment. The first step is to determine the atomic structure of the defect. However, in most cases there are simply too many possible metastable defect structures for a unique atomic structure to be determined. HREM images can help here, by ruling out many defect models. This was perhaps first done by Northrup et al (Northrup, et al.,1981) , who treated the case of shuffle and glide partial dislocations in silicon. Here HREM images were used as the basis for pseudo-potential electronic structure calculations of the dislocation core structure. With modern computers and much higher resolution images, this general approach would seem to repay effort. The combination of these techniques with energy-loss spectroscopy is proving to be even more powerful (Batson,1993) , especially in the study of interfaces (Dehm, et al.,1995, Muller, et al.,1997) , where embedded atom methods (combined with HREM) are efficient (Medlin, et al.,1997) .

In addition to defect structures, it is also important to know the formation energy of defects, and the energy barriers to their motion. For thermally activated processes, these can be determined if atomic resolution movies can be made of the defect motion at known temperature, using an Arhenius analysis (Sinclair and Konno, 1995) . This will give an estimate of the migration energy. The ideal experiment would be one in which it was possible to quench-in the saddle-point structure of the defect, to use HREM images to solve this and then use this as the basis for computation. Again there are often too many possibilities for ab-initio methods to be useful - an extreme example is the difficult case of atomic diffusion. Entropy terms are also difficult to estimate accurately. The concentration of quenched-in defects will often give the formation energy. There are three main problems with the associated electron microscopy: 1. Beam induced effects on defect motion, 2. The effects of surface roughness, and 3. The projection problem.

For the case of kinks on the 30 degree partial dislocation in silicon, we have used ab-initio pseudo-potential methods in the local density approximation to estimate the energy barrier to kink motion (Huang, et al.,1995) , and to study the electronic band structure and charge density as the kink moves. This theoretical work was complemented by HREM lattice imaging of kink motion (Kolar, et al., 1996) , and appears to represent the first direct observation of dislocation kinks. The "forbidden" $(422)/3$ reflections generated by a stacking-fault (lying normal to the beam) between partial dislocations were used alone to form the lattice image. The kink structure could not be determined from these images (Spence, 1981) , but their location could be found to within about 0.3nm. Video recordings at $600^{\circ}$ C showed kink motion (and pinning) at the edges of the lattice image of the stacking fault. To address the problem of surface roughness on the atomic scale, difference images were formed, from which only details which change could be extracted. The effects of beam induced motion were minimized by turning the beam off during dislocation motion (except during the pinning studies), and using low-dose techniques in conjunction with image plates and CCD recording. From this work the kink unpinning energy could be found, together with the single kink formation energy (0.73 eV) and the migration energy (1.24 eV). The formation energy was obtained by applying the Hirth-Lothe nucleation-and-growth rate-equations to the observed concentration of saddle-point kink pairs under high stress conditions. Thus we find that, unlike metals, kink mobility rather than formation is the rate-limiting step for dislocation motion in the absence of obstacles. For segments longer than the kink mean free path, the sum of these values gives an activation energy for 30° partial dislocation motion of 1.97 eV, in excellent agreement with experiment. Because of the need to use very high beam intensities for the movies (which undoubtedly affects our unpinning energies), we cannot determine where the Hirth-Lothe theory or the Obstacle Theory of dislocation motion applies to silicon.

## 4. SPECIES IDENTIFICATION IN STM BY TIME-OF-FLIGHT - THE STAP.

Many processes in the solid state are controlled by foreign atom concentrations, from crystal growth to catalysis and Cotrell atmospheres around dislocations. To identify atomic species seen in STM images, we have constructed a new instrument, the Scanning Tunnelling Atom Probe (Spence et al.,1996) . Using this, atomic clusters of interest may be transferred to the STM tip by a small voltage pulse. The sample is then removed, leaving the tip aimed at a channel-plate, which thus forms a time-of-flight atom probe. Tips are cleaned by heating before use. A vertical-axis Inchworm STM is used of our own design, fitted with high voltage connections to the tip. Flight time data are collected using a digital oscilloscope under Labview control on a PC. In recent work we have applied this to the growth of Ge on Si(111) (Weierstall and Spence, 1997). Figure 3 shows STM images of 5X5 islands of Ge on Si(111) 7X7, with more than 2 monolayers coverage. Si (7X7) is seen in other regions. Figure 4 shows the effect on these islands of pulsing the tip, which transfers a few atoms to the tip. Figure 5 shows the time-of-flight spectrum of elements on this tip. They include a $Ge^{++}$ peak at a mass-to-charge ratio of 36.3. Spectra obtained after STM scanning without pulsing showed no Ge. The use of the instrument for microanalysis requires a full understanding of the factors affecting efficiency of atom transfer, which is not always 100%, and which depends on the materials system studied. In particular, for Ge on Si, we tend to see either large clusters transferred, or none at all, in contrast to Si on Si (111).

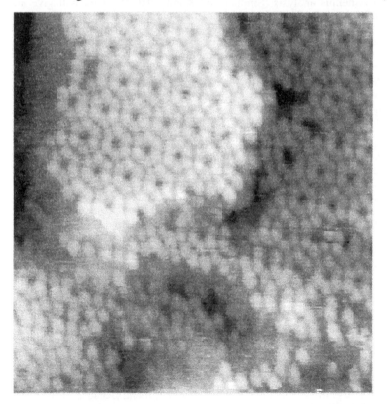

Figure 3. Three-dimensional island growth of Ge on Si(111). Ge (upper region) is 5X5.

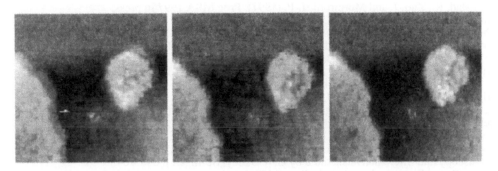

Figure 4. Ge STM images after pulsing tip at 5 V for 10 ms. Small modifications are visible.

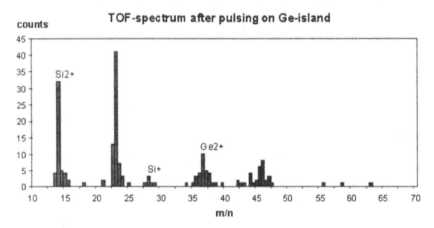

Figure 5. Distribution of flight times for the tip used in figure 4, plotted against mass to charge ratio. The main elemental peaks are indicated. The peak at 24 may be either $Ge^{3+}$ or $Ti^{++}$ from the sublimation pump.

## ACKNOWLEDGMENTS

This work was supported by NSF award DMR-9526100 (J.C.H.S. P.I.)

## REFERENCES.

Batson, P. (1993). Nature 366, 727.
Bethe, H. (1944). Phys. Rev 66, 163.
Boothroyd, C. (1997). J. Electr. Micros. In press,
Chiang, S. (1994). in "STM I", edited by Guntherodt, H. and Wiesendanger, R. (Springer, Berlin), p 258 - 267.
Coene, W., Janssen, G., Op de Beeck, M.,Van Dyck, D. (1992). Phys Rev. Letts 69, 3743.
Cowley, J. M., Spence, J. C. H. and Smirnov, V. V. (1997). Ultramic 68, 135 - 148.
Dehm, G., Ruhle, M., Ding, G. and Raj, R. (1995). Phil Mag B71, 1111-1124.
Dekker, A. and Van den Bos, A. (1997). J. Opt. Soc. Am. 14, 547.
Dong, W., Baird, T., Fryer, J., Gilmore, C., NacNicol, D., Bricogne, G., Smith, D., O'Keefe, M. and Hovmuller, S. (1992). Nature 355, 605.
Dorset, D. L. (1995). Structural Electron Crystallography. (Plenum, New York)
Downing, K., Meisheng, H., Wenk, H. and O'Keefe, M. (1990). Nature 348, 525.
Heider, M. and Zach, J. (1995) Proc. Micros and Microan. 1995, ed. G. Bailey, (Jones and Begall). New York, p 321.

Holmstad, R., Zuo, J. and Mornirolli, J.-P. (1997). Proc MSA 1997 in press,
Huang, Y., Spence, J. and Sankey, O. (1995). Phys. Rev. Letts. 74, 3392.
Ichinose, H. and Ishida, Y. (1989). Phil Mag A60, 555.
Kirkland, A. I., Saxton, W. O. and Chand, G. (1997). J. Electr. Mic. 1, 11.
Kolar, H., Spence, J. and Alexander, H. (1996). Phys. Rev. Letts. 77, 4031.
Krivanek, O., Delby, N., Spence, A. J., Camps, R.A., Brown, L.M. (1997) Proc MSA, 1997, ed. G. Bailey (Springer, New York) p.1171
Krivanek, O. and Stadelman, P. (1995). Ultramic 60, 103.
Medlin, D., McCarty, K., Hwang, R., Guthrie, S., Baskes, M. (1997).Thin Solid Films. In press.
Mobus, G. and Ruhle, M. (1994). Ultramic 56, 54.
Muller, D., Shashkov, D., Benedek, R., Yang, L., Seidman, D. and Silcox, J. (1997) Microscopy and Microanalysis.,edited by G.W.Bailey, Springer, p 647 - 648.
Nellist, P. and Pennycook, S. (1997). Proc. MSA in press,
Nellist, P. D., McCallum, B. and Rodenburg, J. (1995). nature 374, 630.
Northrup, J. , Cohen, M. , Chelikowsky, J., Spence, J., Olsen, A. (1981). Phys. Rev. B24, 4623-4628.
Ochowski, A., Rau, W. D. and Lichte, H. (1995). Phys. Rev. Letts. 74, 399.
Rose, H. (1994). Ultramic 56, 95.
Schiske, P. (1973). in "Computer aided design in electron optics", ed. Hawkes, P. W., (Academic press, London) p 82 - 90.
Sinclair, R. and Konno, T. (1995). Ultramic 56, 225.
Smith, D. J. (1997). Rep. Prog. Phys. in presss,
Spence, J., Weierstall, U. and Lo, W. (1996). J. Vac. Sci. Technol. B14, 1587.
Spence, J. C. H. (1978) Scanning Electron microscopy 1978, ed. O.Johari, I.I.T.R.I., p 61.
Spence, J. C. H. (1981). Proc. 39th Ann Meetng. Electron Micros. Soc. Am. (1981). G. Bailey, Ed. (Claitors, Baton Rouge). 120.
Spence, J. C. H. (1988). Experimental high resolution electron microscopy. (Oxford University Press, New York)
Spence, J.C.H. (1997). Acta Cryst. In press.
Spence, J. C. H., Huang, Y. M. and Sankey, O. (1993). Acta Met. 41, 2815-2824.
Spence, J. C. H. and Zuo, J. M. (1992). Electron Microdiffraction. (Plenum, New York)
Synge, E. (1928). Phil Mag 6, 356.
Weierstall, U. and Spence, J. C. H. (1997). Surface Science in press,
Zuo, J. M., McCartney, M. and Spence, J. (1996). Ultramic 66, 35-47.

*Inst. Phys. Conf. Ser. No 153: Section 1*
*Paper presented at Electron Microscopy and Analysis Group Conf. EMAG97, Cambridge, 1997*

# Electron microscopy and analysis: the future

Colin Humphreys

Department of Materials Science and Metallurgy, University of Cambridge, Pembroke Street, Cambridge CB2 3QZ, UK.

**ABSTRACT:** Electron microscopy has had a glorious past: in this paper we glimpse into the future. We predict that the resolution in TEM and STEM will be 0.5Å, achieved using aberration correctors. EELS will have a spatial resolution of 1Å (for high energy losses) and an energy resolution of better than 0.1 eV, achieved using monochromators or a new type of electron source. EDX will have a spatial resolution of better than 1nm (for high energy losses and thin specimens). STEM will be used for nanofabrication with atomic resolution. Future microscopes will have manual and automatic options, and be modular. Finally, the microscope of the future may sit on your desk.

## 1. INTRODUCTION

"The electron is not as simple as it looks".
W. L. Bragg, recounted by G. P. Thomson at the Electron Diffraction Conference, Imperial College, London, 1967.

"I have no doubt that in reality the future will be vastly more surprising than anything I can imagine."
J. B. S. Haldane, in Possible Worlds and Other Papers, 1927.

"It is bad enough to know the past; it would be intolerable to know the future".
W. Somerset Maugham, in Richard Hughes Foreign Devil, 1972.

Electron microscopy has had a glorious past: what about the future? As the above quotations show, predicting the future is notoriously difficult. In this paper we will try to predict some of the exciting developments which will occur based on present research needs and trends in microscope performance. Topics to be covered include resolution, aberration correctors, EELS, HAADF, monochromators, EDX, in-situ experiments, nanofabrication, automation and the integrated, multipurpose microscope.

## 2. RESOLUTION IN TEM

To resolve atoms in a wide range of perfect crystals in projections along important crystallographic directions requires a resolution of about 1Å (for example, atomic resolution in perfect silicon, viewed along <110>, requires a resolution of 1.36Å). Resolving individual atom positions in the cores of defects or at interfaces requires a resolution typically twice as good as that in perfect crystals, i.e. about 0.5Å. This would seem to be the ultimate resolution needed in electron microscopy. As is well known, the "point resolution" of a microscope, corresponding to the first zero of the contrast transfer function (CTF) at Scherzer defocus, is given by $d_0 = 0.66 C_s^{0.25} \lambda^{0.75}$. The simplest way to reduce $d_0$ is to decrease $\lambda$ while keeping $C_s$ as small as possible. For example, in theory a 1 MeV microscope with $C_s = 1.0$mm has a point resolution of 1.0Å (Humphreys 1981), and this has been achieved in practice (Phillipp et al 1994). However, using high voltages to reduce $\lambda$ can cause undesirable radiation damage by direct displacement of atoms. The optimum accelerating voltage to achieve maximum resolution and penetration yet minimum direct displacement damage is probably in the range 200-400 kV.

Increased resolution may be obtained by improving the "information resolution limit", $d_i$, of the CTF and using image processing. Humphreys and Spence (1981) showed theoretically that a 200 kV electron microscope with $C_s = C_c = 1$mm fitted with a field emission gun (FEG) would have a value of $d_0$ of 2.3Å, but a value of $d_i$ of 0.9Å ($d_i$ for a thermionic source being 1.7Å). Manufacturers have recently built 200 kV and 300 kV microscopes with field emission guns with $d_i$ of about 1Å. However, extracting information down to 1Å from such instruments using focal series and other methods is time consuming, particularly for images of defects, despite recent improvements in image processing (for example, Coene et al 1994; Rodenburg 1996; Saxton 1996). The future of high resolution probably lies in aberration correctors, see below.

## 3. ABERRATION CORRECTORS IN TEM AND STEM

Aberration correctors for TEM have been under development for many years, with very little success. However recently more substantial progress has been made (Uhlemann 1994 and papers at EUREM 1996) and it seems clear that a successful aberration corrector for a TEM should soon be available.

Aberration correctors for a STEM have been developed only recently, but it is now possible to correct fully for spherical aberration, so that $C_s = 0$ (Krivanek 1997). In order to obtain amplitude contrast from phase objects some spherical aberration is necessary and it is possible to de-tune the spherical aberration correctors to optimise the resolution and the contrast.

Hence a prediction for the future is that TEMs with a point resolution $d_0$ of 0.5Å, and STEMs with a probe diameter of 0.5Å will become available using aberration correctors. These instruments will enable the imaging of defects in a wide range of crystals with atomic resolution.

## 4. HIGH ANGLE ANNULAR DARK FIELD (HAADF)

HAADF imaging using a small probe in STEM mode has the advantage over HREM in a TEM that contrast reversals due to dynamical scattering effects are eliminated. The HAADF technique has been developed by Pennycook (e.g. Pennycook and Jesson 1992), and the method has a resolution of $0.44\, C_s^{0.25}\, \lambda^{0.75}$. A spherical aberration corrector producing $C_s \approx 0$ should enable a resolution of 0.5Å to be achieved in HAADF, thus enabling the atomic resolution imaging of defects in a wide variety of crystals provided that atomic columns around defects can be aligned parallel to the incident beam.

## 5. ELECTRON ENERGY LOSS SPECTROSCOPY (EELS)

The spatial resolution of EELS in STEM is determined by three main factors: the probe size, the slit size (which determines the effective magnitude of the beam spreading) and the impact parameter. For low energy losses the impact parameter is typically 10nm and hence dominates the spatial resolution if the probe size is small. However for high energy losses, the impact parameter is small and the probe size dominates. For high energy losses, if the probe size is only 1Å, then EELS from single atomic columns should be possible from a wide range of materials (EELS from a single atomic column has already been demonstrated in some favourable cases: Batson 1993; Browning et al 1993).

## 6. MONOCHROMATORS AND EELS

The energy resolution in EELS is typically 1 eV using a thermionic or $LaB_6$ emitter and 0.4 eV using a FEG. This is insufficient for detailed studies of the near-edge fine structure around characteristic losses. Studies of the near edge fine structure yields key information on the electronic structure and bonding in materials (Müller et al 1994; Botton et al 1996). An energy resolution of 0.1 eV would enable significant advances to be made in this work.

An energy resolution of 0.1 eV would also enable detailed studies to be made of the energies of defect states in the band gap of semiconductors. For example, such energy resolution would greatly facilitate work on defect states in GaN (Natusch et al 1997). Clearly a monochromator or a new type of source is required for 0.1 eV resolution (Batson 1988). Hence for the short-term future 0.1 eV resolution is required, but in the longer term meV resolution is required for phonon dispersion plots from localised regions of crystal (c.f. neutron diffraction which gives phonon dispersion plots from large single crystals).

## 7. X-RAY ANALYSIS

Quantitative X-ray analysis with high spatial resolution remains difficult due to beam spreading effects and stray scattering effects. However with optimised detectors, thin specimens and attention to secondary and tertiary scattering effects, better than 1 nm spatial resolution should become routine for higher energy X-rays for which the resolution is not limited by impact parameter effects.

## 8. IN-SITU EXPERIMENTS

Already a wide range of in-situ specimen stages are available (heating, cooling, straining, gas cell, etc). Dedicated in-situ microscopes will increasingly become available which will provide a "laboratory within an electron microscope" for studying a wide range of phenomena in-situ, for example growth by MBE, chemical reactions, etc.

## 9. ENERGY FILTERING

Energy filtering so that only the "zero-loss" electrons contribute improves all images and diffraction patterns. Hence the popularity of the Gatan Imaging Filter despite its significant cost. In addition the GIF contains a spectrometer for EELS work and it allows chemical mapping by forming images using loss electrons which have energies in a certain energy window. The GIF also facilitates quantitative imaging and diffraction work using elastically scattered electrons by removing all the inelastically scattered electrons except those with very low energy loss (e.g. phonons).

Manufacturers will increasingly develop in-column energy filters integrated into the microscope, for ease of operation. For specialist applications these will be combined with monochromators (see section 6 above).

## 10. NANOFABRICATION

An intense focused electron beam from a FEG can cut holes and lines in a range of materials, also externally machine certain materials, and form quantum wires and columns (Humphreys et al 1990; Chen et al 1993). Using an aberration corrector in a STEM a beam diameter of 1Å will be possible hence it may be able to remove a single column of atoms, or do external machining with single atom precision. This could have considerable applications in nanotechnology and the fabrication of nanostructures. Dedicated electron beam instruments devoted to nanofabrication may exist in the future.

## 11. AUTOMATION

Electron microscopes are becoming increasingly automated, with automatic astigmatism correction, automatic alignment, etc. Viewing screens are disappearing from TEMs and are being replaced by CCDs and other digital imaging systems. The TEM or STEM operator will increasingly become remote from the microscope and will operate the microscope from computer terminals, which may have voice control in the future. By computer networking, an operator in Australia, say, will be able to fully control a microscope in Cambridge provided someone loads the specimens.

34

Undoubtedly increased automation will happen, but there are attendant problems. Many experienced microscopists will prefer to sit at a microscope and drive it themselves and not via a computer, just as racing car drivers prefer to select gears themselves rather than have automatic transmission. For achieving ultimate microscope performance, the hands-on approach will continue to be favoured by some expert users, whereas many other users will favour more automation. Manufacturers will therefore be faced with difficult policy decisions. It is possible that two versions of microscopes will be developed: manual and automatic, rather like cars.

## 12   THE MULTIPURPOSE MICROSCOPE

Because of customer demand, the microscope of the future will be "modular", with customers ordering the modules they require. High resolution and analysis, once needing different microscopes, will increasingly be required in the same microscope. Customers will be able to choose from a huge range of accessories which the manufacturer will build into the microscope as a module, for example an holography attachment, imaging filter, monochromator, cathodoluminescence, EDX, etc.

## 13.   MAJOR BREAKTHROUGHS

This paper has concentrated on incremental improvements to microscopes. Step-function improvements are inherently unpredictable. However let us end this glimpse into the future by one speculation: what if room-temperature superconductors were discovered which could also carry a high current density? The microscope of the future would then have a resolution of 0.5Å, an in-column energy filter, ... and sit on your coffee table.

## REFERENCES

Batson P E 1988 Rev. Sci. Inst. **59**, 1132-1138
Batson P E 1993 Nature **366**, 727-729
Botton G A, Guo G Y, Temmerman W M and Humphreys CJ 1996 Phys. Rev. B **1996**, 567-572
Browning N D, Chisholm M F and Pennycook S J 1993 Nature **366**, 143-145
Chen G S, Boothroyd C B and Humphreys C J 1993 Appl. Phys. Lett, **45**, 1949-51
Coene W M J, Thust A, Op de Beeck and Van Dijck D 1994 Proc. 13th Int. Congress on Electron Microscopy, eds B Jouffrey and C Colliex (Les Ulis, France: Les Editions de Physique) **1**, 461-462
Humphreys C J 1981 J. Micr. et Spect. Elect. **6**, 18a
Humphreys C J and Spence, J C M 1981 Optik **58**, 125-144
Humphreys C J, Bullough T J, Devenish R W, Maher D M and Turner P S 1990 Scanning Microscopy Supplement **4**, 185-191
Krivanek O 1997, these proceedings
Muller D A, Tzuo Y, Raj R and Silcox J 1993 Nature **366**, 725-727
Natusch M K H, Botton G A and Humphreys C J 1997 Proc. Microscopy of Semiconductor Materials, Oxford, eds A Cullis and J L Hutchinson, Institute of Physics, in press
Pennycook S J and Jesson D E 1992 Acta. Metall. Mater. **40**, Suppl. S149-153
Phillipp F, Höschen R, Osaki M and Rühle M 1994 Proc.13th Int. Congress on Electron Microscopy, eds B Jouffrey and C Colliex (Les Ulis, France: Les Editions de Physique) **1**, 231-232
Rodenburg J M 1996 Proc. 11th Eur. Cong. on Microscopy, Dublin, eds D Cottell and M Steer, not yet published
Saxton W O 1996 Proc. 11th Eur. Cong. on Microscopy, Dublin, eds D Cottell and M Steer, not yet published
Uhlemann S, Haider M and Rose H 1994 Proc. 13th Int. Congress on Electron Microscopy, eds B Jouffrey and C Colliex (Les Ulis, France: Les Editions de Physique) **1**, 193-194

*Inst. Phys. Conf. Ser. No 153: Section 2*
*Paper presented at Electron Microscopy and Analysis Group Conf. EMAG97, Cambridge, 1997*
© *1997 IOP Publishing Ltd*

# Aberration correction in the STEM

**OL Krivanek\*, N Dellby\*, AJ Spence\*\*, RA Camps and LM Brown**

    MP group, Cavendish Laboratory, University of Cambridge, Cambridge CB3 0HE, UK
\*    now at: Dept. of Materials Science, Roberts Hall, U. of Washington, Seattle WA 98195, USA
\*\*  now at: Dept. of Applied Physics, Cornell University, Ithaca NY 14853, USA

**ABSTRACT:** A quadrupole-octupole spherical aberration corrector has been designed and built for a modified VG HB5 STEM. The corrector fits between the second condenser and the objective lens, and consists of 6 separate stages. Each stage contains one strong quadrupole, one strong octupole and 12 auxiliary trim coils. The corrector is provided with autotuning software that measures all the relevant aberrations on-line and tunes the corrector automatically. Preliminary tests of the corrector have verified the correction principle, and shown that it can readily compensate for parasitic aberrations arising from misalignment.

## 1. INTRODUCTION

Aberration correction in electron microscopy is a subject with a 60 year history dating back to the fundamental work of Scherzer (1936, 1947). There have been many partial successes, such as Deltrap's quadrupole-octupole corrector which nulled spherical aberration ($C_S$) over 30 years ago (Deltrap 1964a, 1964b). More recently, an electromagnetic sextupole-round lens-sextupole corrector built for a conventional transmission electron microscope (CTEM) by Haider et al. (1995), and a mostly-electrostatic quadrupole-octupole corrector of both spherical and chromatic ($C_S$ and $C_C$) aberration built by Zach and Haider (1995) for a scanning electron microscope (SEM) have both demonstrated an improvement of resolution in their respective microscopes. Nevertheless, the practical goal of attaining better resolution than had ever been reached by any other microscope operating at the same voltage appears to remain unfulfilled.

## 2. A QUADRUPOLE-OCTUPOLE CORRECTOR FOR A DEDICATED STEM

    A scanning transmission electron microscope (STEM) with corrected spherical aberration would produce a smaller probe size at a given beam current than an uncorrected STEM, and a larger beam current in a probe of a given size. $C_S$ correction in STEM is thus expected to provide more tangible benefits than in CTEM, where the optimum $C_S$ value for phase contrast imaging is non-zero and the "information limit" resolution is typically determined by chromatic aberration.

    Sextupole-round lens-sextupole correctors cannot be extended to $C_C$ correction, and electrostatic correctors are not suitable for beam energies above around 30 keV. Keeping in mind that $C_C$ will limit the resolution of a $C_S$-corrected microscope, the corrector we have designed and built for our VG HB5 dedicated STEM is an electromagnetic quadrupole-octupole $C_S$ corrector that is compatible with incorporating $C_C$ correction in the future. The VG HB5 has an objective lens with $C_S$ = 3.5 mm and $C_C$ = 3.5 mm. In the standard microscope operating mode, the lens only demagnifies the probe by 25x and gives a best dark field resolution of about 0.4 nm. The corrector has been built as a proof-of-principle instrument designed to impart negative spherical aberration to the electron beam and thereby compensate the positive spherical aberration of the microscope's other lenses. Improvements in the microscope performance beyond what can be achieved in the best current 100 kV STEMs are expected only when the objective lens is modified to provide smaller starting $C_S$ and $C_C$, and larger demagnification.

The corrector consists of 6 identical stages, each one of which comprises a strong quadrupole and a strong octupole. The quadrupole windings are distributed on the 12 poles in a way that approximates an ideal quadrupole up to 8th order, and the octupoles are ideal up to 6th order. There is also a separate weak auxiliary winding on each pole. Computer-controlled current supplies with 1 ppm stability are provided for each quadrupole, octupole and auxiliary winding. Software running under Microsoft NT operating system allows any number of the power supplies to be linked together with adjustable strengths and labeled as a new control, i.e. it provides the flexibility needed to compensate for imperfections of the principal quadrupoles and octupoles and also to create weak independent dipoles, quadrupoles, sextupoles and octupoles of arbitrary orientation. The electronics also sends currents to 4 sets of alignment coils and voltages to an 8-pole deflector that was used as an objective lens stigmator in the HB5 and now serves as an objective lens alignment coil, for a total of 96 computer-controlled power supplies. Because of the low power required by quadrupoles and octupoles and the high packing density of our design, the corrector electronics fits into an enclosure of 50x46x28 cm.

The quadrupole excitations are antisymmetric about the mid-plane of the corrector. The first-order trajectories are thus constrained by 3 independent parameters, which allows enough flexibility to couple the corrector to an entrance crossover situated within a range of distances D from the corrector (Fig. 1). The corrector produces a 1:1 magnification, pre-aberrated image of the crossover D away from the exit face of the corrector. The corrector is placed between the second condenser lens and the scan coils preceding the objective lens (Fig. 2). Because the scan coils are placed after the corrector, aberrations depending on the off-axis position in the image plane become unimportant. Varying the distance D of the exit crossover from the corrector allows the beam diameter inside the corrector and the post-corrector demagnification to be varied.

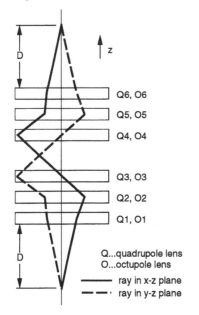

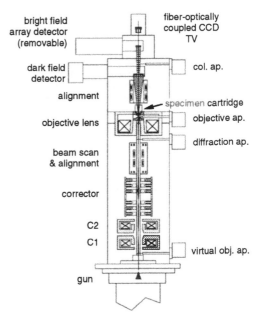

Fig. 1 Schematic diagram of the quadrupole-octupole corrector and of the electron trajectories through it.

Fig. 2 Placement of the corrector in the modified HB5 column.

Because quadrupoles and octupoles don't have the cylindrical symmetry of round lenses, three principal third order aberrations need to be considered: the normal spherical aberration ($C_{3,0}=C_s$), 2-fold astigmatism of spherical aberration ($C_{3,2}$) and 4-fold astigmatism ($C_{3,4}$). Octupole O3, which contains an x-focus line, acts on all three coefficients, in proportions 1, -4/3, 1/3. O4, which contains a y-focus line, also acts on all three, but in proportions 1, 4/3,

1/3. Exciting O3 and O4 equally thus makes it possible to adjust the spherical aberration coefficient to any desired value, but also results in 4-fold astigmatism. The 4-fold astigmatism can, however, be removed by any octupole in which the beam is round, without affecting the other two aberrations. This is exactly the case in octupoles O1 and O6, and approximately the case in octupoles O2 and O5. We excite these four octupoles approximately equally, as this gives about the same absolute excitation (and hence the same field strength and saturation characteristics) in all the octupoles of the corrector.

Further key components of our system are a fiber-optically coupled TV camera which captures the shadow image appearing in the detector plane, a retractable bright field detector using a YAP scintillator, and a computer-based scanned image acquisition system (Gatan DigiScan plus Power Macintosh) running under DigitalMicrograph image acquisition and processing software. The VG column has been modified by adding a second condenser lens, extra illumination alignment coils and of course the corrector. An overall system diagram is shown in Fig. 3.

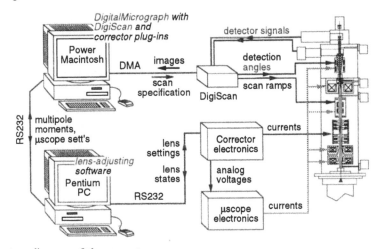

Fig. 3 System diagram of the corrector.

We have also modified the vacuum system of the upper column by replacing the VG diffusion pump, liquid $N_2$ trap and mechanical rotary pumps by an all-dry pumping system consisting of two ion pumps (270 and 25 l/s) plus a molecular drag pump backed by a diaphragm pump. Vacuum levels about 4x better than previously have been attained, especially in the critical gun valve region of the microscope, which is now pumped by the 25 l/s ion pump separately from the rest of the system.

## 3. ABERRATION DIAGNOSIS

Trying to correct spherical aberration without having a method for identifying and measuring all the aberrations up to third order can lead to a situation where one is faced with a fuzzy image, a large number (>10) of controls and no useful procedure for making the image sharper. In the STEM, two different methods for diagnosing the aberrations are available: visual examination of far-field shadow images (Ronchigrams), and acquiring images recorded with different detection angles and comparing them in the computer. The second method is related by reciprocity to TEM bright field autotuning using images acquired with different illumination angles (Krivanek and Fan, 1994).

Either method allows one to characterize the aberration function. Recording Ronchigrams requires simply an efficient two-dimensional detector such as our fiber-optically coupled TV. The second method requires software and hardware that can vary the detection angle using the microscope Grigson coils, record a tableau of the required images, analyze either the variation in their absolute position or in their apparent defocus and/or astigmatism, and

microscope, and found the manual Ronchigram method to be especially useful for initial set-up, and the automated image tableau method best suited for fine-tuning and precise measurement of the aberrations. Examples of results of both methods are shown in the next section.

When running the autotuning software on the microscope at probe currents 0.5 nA and greater prior to the incorporation of the corrector, we measured the $C_S$ of the entire microscope as 7 mm and larger (we obtained values as large as 120 mm). This is because in the large current mode the source is demagnified less, and the gun makes an appreciable contribution to the total $C_S$. A corrected microscope will therefore give dramatically smaller probes at 0.5 nA and larger currents, providing an extra reason for $C_S$ correction in the STEM.

## 4.    PRELIMINARY RESULTS FROM THE CORRECTOR

At the time of writing of this manuscript, the corrector had been operational for a total "beam time" of about 3 weeks. This has proved enough to verify the correction principle we use, but not enough to fully optimize the operation of the corrector.

First tests of the corrector verified that the magnetic fields produced by the principal quadrupoles of the 6 stages are concentric to each other within ±40 µm. This has been achieved without altering the number of turns on each pole to perfect the field shape. It is slightly better than what we expected from tolerance buildup and inhomogeneity of the polepiece material. It is very welcome as it allows us to reduce the strength of the trim coils from 10% to 3% of the excitation of the quadrupoles and the octupoles, and thus to decrease the influence of the trim coils on the stability of the total system.

When deciding on the best first order trajectories through our system, the single most important parameter is the ratio of the beam diameter in the corrector to the beam diameter in the objective lens (OL). Increasing the beam diameter in the corrector by 2x while keeping the diameter in the OL the same increases the corrector's effectiveness by 16x. The diameter ratio is determined by the position of the beam crossover between the corrector and the OL. In our system, locating the crossover at the diffraction aperture (see Fig. 2) would give a corrector capable of correcting $C_S$ of up to about 200 mm, but also result in increased instabilities. We therefore lower the crossover position roughly to the middle of the scan coils, and obtain a system capable of correcting $C_S$ up to about 8 mm with the present objective lens.

Once the crossover position is fixed, the next important choice is the overall demagnification of the "virtual" source (as seen by the first condenser). For high resolution operation we demagnify about 80x, for high current operation about 10x. Together with the requirements that the corrector quadrupoles be antisymmetric and that the entrance and exit crossover distances for the corrector be identical, this determines the required excitations of the two condensers and of the corrector quadrupoles.

Fig. 4 shows a shadow image (Ronchigram) obtained on the fiber-optically coupled TV when the corrector's quadrupoles were on and octupoles off. Positive spherical aberration of the microscope demonstrates itself by the circle of infinite azimuthal magnification (marked) and the circle of infinite radial magnification at 0.58 the radius of the azimuthal circle.

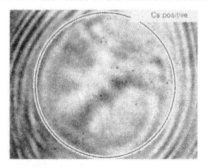

Fig. 4  Underfocus Ronchigram recorded with the corrector's quadrupoles on and octupoles off. Angular range ~ ±10 mrad.

Fig. 5 shows a Ronchigram obtained with both the quadrupoles and the octupoles on. The octupoles were stronger than required for $C_S$ compensation. As a result, the whole system had negative spherical aberration, demonstrated by an overfocused central part of the Ronchigram (revealed by the dark fringe inside the hole image near the center) surrounded by a circle of infinite radial magnification. (Had $C_S$ been positive, an overfocused central part could not have any infinite magnification circles around it.)

Fig. 6 shows a Ronchigram obtained with the octupole excitation scaled back to the value required for total system $C_S = 0$. Since a perfectly focused and aberration-corrected microscope gives an uninformatively uniform contrast in a Ronchigram, the objective lens was underfocused by about 400 nm. This has produced a Ronchigram that has uniform magnification, as expected for a $C_S$-corrected STEM free of parasitic aberrations. The microscope had been tuned manually simply by observing Ronchigrams and compensating for successively higher order aberrations using the appropriate controls, starting with first order aberrations (defocus and astigmatism) and finally fine-tuning the 3rd order coefficients ($C_S$, 2-fold astigmatism of $C_S$, and 4-fold astigmatism).

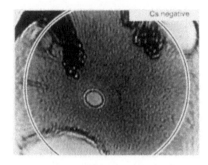

Fig. 5 Ronchigram recorded with the octupoles stronger than needed for $C_S$ correction. Angular range ~ ±10 mrad.

Fig. 6 Ronchigram recorded with the octupoles at the correct excitation. Angular range ~ ±10 mrad.

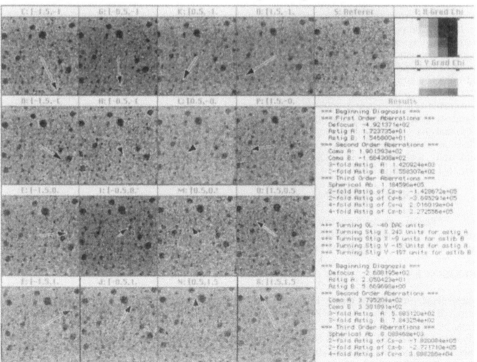

Fig. 7 Automated aberration diagnosis of the corrected microscope.

40

Fig. 7. shows the results of running the aberration diagnosis software on the microscope shortly after recording the Ronchigram shown in Fig. 6. The gold particles visible in the images are 3-5 nm in diameter. The injected detection tilt increment between neighboring images was 5 mrad. The arrows superimposed on the tableau images show the relative displacement of the images with respect to the reference image (after subtraction of a spurious shift diagnosed by detecting large angle dark field images under identical conditions). The X and Y components of the arrows are grouped together in the "X grad Chi" and "Y grad Chi" images. These are fitted with 2-dimensional third order polynomials, and the polynomial coefficients translated into aberration coefficients. The results for the present tableau are shown in the bottom half of the Results window, with all the aberrations in nm. The top half shows the results of the preceding diagnosis run with the same microscope set-up (to show the repeatability of the analysis). The results can be summarized as follows:

2nd order aberrations (coma a, b; 3-fold astigmatism a, b): all less than 1 μm
$C_S$: less than 0.12 mm
other third order aberrations: a-type: less than 0.1 mm; b-type: less than 0.4 mm

We made no attempt to correct the b-type third order aberrations, which should be zero in a system with perfect alignment, machining and material homogeneity. The fact that they are less than 10% of the initial $C_S$ again confirms the good mechanical precision that was achieved. With the b-type correction activated and the beam instability (see next paragraph) removed, we expect to be able to keep all 3rd order aberration coefficients easily below 0.1 mm.

Scanned images acquired so far have shown a resolution of about 3 Å in one direction and 5 Å in the perpendicular direction. The resolution loss is due to an instability that is also present with the corrector switched off and disconnected from its power supplies. We are now tracing the origin of this instability, and expect to have improved resolution images shortly.

## 5. CONCLUSION

The preliminary results described here show that thanks to the excellent flexibility and precision made possible by computer-controlled power supplies and to new methods of on-line aberration diagnosis, $C_S$ correction in our STEM has been achieved. Its benefits in terms of improved resolution and greater current into a given-size probe promise to be considerable.

Perhaps the most exciting aspect of this development is that it is open-ended, with $C_C$ and even $C_5$ correction using similar principles and techniques presenting themselves as the natural next (though harder) steps. The road should ultimately lead to an era of lab-sized STEM instruments able to routinely achieve 1 Å or even 0.5 Å resolution, and to deliver probe currents into atom-sized sample regions that are large enough for rapid and sensitive EELS and EDXS microanalysis. Such instruments will be able to image and determine the chemical types of individual atoms in many kinds of materials, and thereby bring benefits to the Materials Science community that are just as revolutionary as those brought to astronomy by the *aberration-corrected* Hubble space telescope.

## 6. ACKNOWLEDGMENT

We are grateful to the Paul Instrument Fund for financial support, the Cavendish Laboratory for provision of facilities, Gatan R&D for surplus equipment and Drs. S. von Harrach and A. Waye for advice on VG matters.

Deltrap JHM 1964a Ph.D. thesis, U. of Cambridge
Deltrap JHM 1964b Proc 3rd EUREM Congress (Prague) 45
Haider M, Braunshausen G and Schwan E 1995 Optik **99**, 167
Krivanek OL and Fan GY 1994 Scanning Microscopy Supplement **6**, 105
Scherzer O 1936 Zeitschrift Physik **101**, 593
Scherzer O 1947 Optik **2**, 114
Zach J and Haider M 1995 Optik **99**, 112

*Inst. Phys. Conf. Ser. No 153: Section 2*
*Paper presented at Electron Microscopy and Analysis Group Conf. EMAG97, Cambridge, 1997*
© *1997 IOP Publishing Ltd*

# Growth of AlN on (0001) α-Al₂O₃ using a novel ultrahigh vacuum transmission electron microscope with *in-situ* MBE

**M. Yeadon, M.T. Marshall and J.M. Gibson**

Materials Research Laboratory, University of Illinois, Urbana, Illinois 61801, USA

**ABSTRACT:** Wide bandgap III-nitride semiconductors have emerged as highly promising materials for the fabrication of optoelectronic devices operating in the blue and UV regions of the electromagnetic spectrum. Using a novel UHV TEM with *in-situ* MBE the early stages of the growth of AlN on (0001) α-Al₂O₃ substrates has been studied. Growth of device quality GaN on sapphire substrates typically involves nitridation of the substrate followed by the deposition of an AlN buffer layer. We present an overview of our instrument and some results from a preliminary investigation of the sapphire nitridation process.

## 1. INTRODUCTION

In the last few decades it has become clear that the development of growth techniques for electronic thin films, such as molecular beam epitaxy and e-beam evaporation, places stringent demands on vacuum and system cleanliness. Advances in vacuum technology now permit ultrahigh vacuum (UHV) conditions to be routinely achieved in the laboratory.

The transmission electron microscope (TEM) has emerged as a powerful tool for the study of microstructures and buried interfaces in electronic materials. Poppa (1965) achieved UHV in a TEM for *in-situ* growth studies by local cryopumping of the sample; more recently TEM has also become recognized as a powerful surface-sensitive tool. For example Yagi *et al.* (1979) built a UHV TEM and obtained the first reflection electron microscopy (REM) images of surface reconstructions. Since then a small number of other UHV TEMs of various designs have been constructed (e.g. Wilson and Petroff 1983, Swan *et al.* 1987, Marks *et al.* 1988 and McDonald *et al.* 1989). In this paper we describe a new UHV TEM in which the sample region has been replaced by an uncompromised UHV surface science chamber. We present some preliminary results obtained using this instrument from a study of the nitridation of (0001) α-Al₂O₃ substrates for III-V nitride thin film growth. Nitridation of the sapphire surface prior to the deposition of GaN improves significantly the properties of the GaN layer (e.g. Grandjean *et al.* 1996), however the nitridation process is not well understood.

## 2. EXPERIMENTAL METHOD

The instrument is a JEOL 2000EX TEM modified for UHV and *in-situ* MBE (Marshall *et al.* 1997). The original sample/objective lens region of the microscope has

Figure 1. (a) Photograph of the Surface High-energy Electron Beam Apparatus, (b) photograph of the sample manipulator, polepieces (upper and lower) and AES unit (left). (c) Photograph of a silicon wafer mounted in the sample cartridge.

been replaced by a 1000 liter UHV surface science chamber, isolated from the microscope gun and camera sections by differential pumping apertures. The base pressure routinely achieved in the chamber is $10^{-10}$ Torr and a photograph of the microscope is presented in figure 1. An internal view of the UHV chamber is shown in figure 1(b) together with a view of the sample cartridge in figure 1(c). The chamber is equipped with two effusion cells, an e-beam evaporator, low energy electron diffraction (LEED) system and Auger electron spectroscopy (AES) enabling deposition, analysis and imaging in the same vacuum chamber.

The samples, of $\sim 1 cm^2$, are mounted on a cartridge for transfer through an airlock to a combined sample manipulator/goniometer which permits continuous 360° primary rotation with a secondary tilt of ±90°. The configuration allows imaging and diffraction in both reflection and transmission modes. For the purpose of the present experiments an ammonia gas injector was fitted to the UHV chamber permitting direct gas injection to the chamber at the sample surface. The sample, a regular 3mm disc, cut from a sapphire wafer, was ion-milled to electron transparency and annealed at 1400°C in air. The sample was then bonded to a silicon support and mounted in the sample cartridge. After outgassing in the microscope the sample was heated to 950°C and exposed to a background ammonia pressure of $1.10^{-6}$ Torr. Periodic observation of the sample was made during the nitridation process. The pressure at the sample surface facing the injector nozzle was calculated to be $\sim 10^{-5}$ Torr.

## 3. RESULTS AND DISCUSSION

After annealing the sapphire sample for 12 hours in air at 1400°C, the presence of steps on the upper and lower surfaces were observed using the electron microscope. A dark-field image of a representative area of this sample is shown in figure 2(a). The terraces were typically ~1μm in width, and tens of microns in length.

The presence of epitaxial AlN islands was observed after 15 minutes nitridation of the sapphire surface. With further nitridation the islands increased in size until complete coverage of the surface was attained. A bright-field TEM image of the sample after 30 minutes nitridation is shown in figure 2(b); the corresponding selected area diffraction pattern is shown in figure 2(c). The AlN layer exhibited the epitaxial orientation relationship: $(0001)_{AlN} // (0001)_{sapphire}$, $[10\bar{1}0]_{AlN} // [11\bar{2}0]_{sapphire}$.

The atomic steps clearly visible in figure 2(a) were obscured after 15 minutes nitridation. Reduction of the background gas pressure to $1.10^{-7}$ Torr (and hence a pressure of $1.10^{-6}$ Torr at the sample surface facing the injector) resulted in a negligible rate of nitridation, and it was concluded that nitridation occurred primarily on the sample surface facing the gas injector, since the rear face of the sample was exposed only to the ambient background pressure.

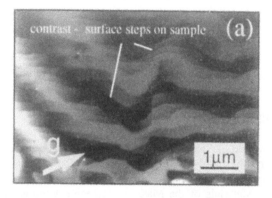

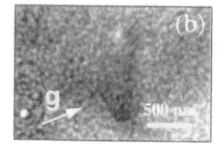

Figure 2　(a) Dark-field image of the annealed sapphire sample showing contrast due to thickness changes at atomic steps. (b) Bright field image of the sample after 30 mins. nitridation and (c) the corresponding SADP.

The reaction products resulting from the nitridation of the sapphire surface have been the subject of debate with speculation about the formation of (1) Al-O-N compounds and (2) epitaxial AlN. During the nitridation process we found no evidence of the presence of Al-O-N compounds; rather, we observed directly the nucleation and growth of epitaxial AlN. Our observations of AlN formation are in agreement with a recent *in-situ* RHEED study of the nitridation of (0001) sapphire by Grandjean *et al* (1996). Further nitridation of the sample (up to 2 hours) enabled the AlN growth rate to be determined; the rate was found to be well described by a simple diffusion model where grain-boundary diffusion is the dominant transport mechanism for reactants and products between the AlN free surface and the buried AlN/$\alpha$-Al$_2$O$_3$ interface (Yeadon *et al.* 1997). The success of the model is most likely a consequence of the 3-D island growth mode since a significant density of boundaries between coalescing islands are evident in figure 2(b).

## 4.    CONCLUSIONS

A UHV TEM with *in-situ* MBE and gas injection capability has been developed at the University of Illinois-Urbana and is now operational. Preliminary investigations of the nitridation of basal plane sapphire have been performed. These investigations have revealed directly the nucleation, growth and coalescence of 3D AlN islands. Electron diffraction patterns indicate the films to be fully epitaxial; the growth rate is well described by a simple diffusion model involving grain boundary diffusion as the dominant reactant/product transport mechanism between the film surface and the reaction interface.

## ACKNOWLEDGEMENTS

The authors would like to thank the Office of Naval Research for the support of this work under contract N00014-95-1-0324, monitored by Dr. R. Brandt. The authors thank Prof. H. Morkoc for useful discussions and acknowledge the use of facilities in the Center for Microanalysis of Materials of the Materials Research Laboratory at the University of Illinois, Urbana-Champaign.

## REFERENCES

Grandjean N., Massies J. and Leroux M., Appl. Phys. Lett. **69** (14) (1996), p. 2071
Marks L.D., Kubozoe M., Tomita M., Ukiana M., Furutsu T. and Matsui I., Proc. 46th Ann. EMSA Meeting (1987), p. 658, Ed. G.W. Bailey, San Francisco Press
Marshall M.T., McDonald M.L., Tong X., Yeadon M. and Gibson J.M., Rev. Sci. Instr. (1997), *in print*
McDonald M.L., Gibson J.M. and Unterwald F.C., Rev. Sci. Instrum. **60** (1989), p. 700
Poppa H., J. Vac. Sci. Tech. **2** (1065), p. 42
Swan P.R., Jones J.S., Krivanek O.L., Smith D.J., Venables J.A. and Cowley J.M., Proc. 45th Ann. EMSA Meeting (1988), p. 136, Ed. G.W. Bailey, San Francisco Press
Wilson R.J. and Petroff P.M., Rev. Sci. Instrum. **54** (1983), p. 534
Yagi K., Takayanagi K., Kobayashi K., Osakabe N., Tanishiro Y. and Honjo G., Surface Science **86** (1979), p. 174
Yeadon M., Marshall M.T., Hamdani F., Pekin S., Morkoc H. and Gibson J.M., *submitted to J. Appl. Phys.*

*Inst. Phys. Conf. Ser. No 153: Section 2*
*Paper presented at Electron Microscopy and Analysis Group Conf. EMAG97, Cambridge, 1997*
© 1997 IOP Publishing Ltd

# The teaching of TEM by telepresence microscopy over the internet

**E Voelkl, L F Allard, *J Bruley, *V J Keast and *D B Williams**

Oak Ridge National Laboratory, Bldg. 4515, 1 Bethel Valley Road, P.O. Box 2008, Oak Ridge, TN 37831-6064, USA
*Department of Materials Science and Engineering, 5 East Packer Avenue, Lehigh University, Bethlehem PA 18015-3195, USA

**ABSTRACT:** The teaching of transmission electron microscopy to students over the Internet has been successfully demonstrated. The advantages and disadvantages of such remote instruction are discussed. The need for complete computer control of the instrument to maximize the efficiency of the instructional process is described.

## 1. INTRODUCTION

The operation of a remote instrument via a computer network or the Internet (telepresence) combined with video and audio teleconferencing is a well-established concept in such fields as computer science and astronomy. Similarly, the technology exists to allow any electron microscope to be driven (at least in part) by a computer and thus the concept of telepresence microscopy (TPM) is also viable. Data compression techniques are sufficiently advanced that the images produced by the microscope can be viewed on a computer screen remote from the instrument, in real time. In fact, all the information from a microscope (images, diffraction patterns and spectra) can be digitized and transferred to local-area networks or to other locations via the Internet or phone lines.

A principal aim of TPM is to make dedicated instruments, such as HRTEMs and dedicated STEMs, available to users who do not have direct access to such instruments in their own laboratories. A further advantage of TPM is to optimize the use and the time spent on these expensive instruments, which spend much of their time idle, often because of time-zone differences in the USA. However, despite the technical capabilities and the obvious advantages of this approach to microscopy, TPM has only been demonstrated at a limited number of microscopy laboratories (e.g. Oak Ridge National Laboratory (ORNL) (Voelkl, 1995) Argonne National Laboratory (Zaluzec, 1995) and Lawrence Berkeley Laboratory (Parvin et al. 1995) and the University of California at San Diego (Ellisman et al. 1993, Fan et al., 1993).

While the research advantages of TPM have been demonstrated, there has been little emphasis on the instructional aspect, but the potential educational gains are as great as those for research. An undergraduate TPM class on HRTEM, has already been implemented between ORNL and Lehigh University (Voelkl et al. 1997). In this experiment, the Lehigh students operated a Hitachi HF-2000 field emission TEM at ORNL, TN from their classroom 800 miles away in Bethlehem, PA. Based on this experience, it is considered that TPM can revolutionize the teaching of TEM by linking a single instrument to a computer classroom with a central video display. Thus, many students can learn to operate a TEM simultaneously. This paper discusses the advantages and drawbacks of this new approach to the teaching of TEM.

## 2. EXPERIMENTAL DETAILS

The computer link between Lehigh and ORNL was based on standard communication software. Power Macintosh computers installed in the classroom at Lehigh were connected to the Macintosh computer running the HF-2000 at ORNL via TimbuktuPro, a program that allows the operation of one computer to be carried out on another computer. Video and audio communication between the students and the HF-2000 operator was established using Connectix digital cameras and CU-SeeMe freeware. Figure 1 shows the undergraduates running the HF-2000 over the Internet, with the image on the HF 2000 screen at ORNL projected on the classroom screen at Lehigh. Figure 2 shows the combined audio and video channels which are run on a separate Macintosh and the view of the Lehigh students as seen from ORNL. The Macintosh mouse and keyboard functions are fully functional at both Lehigh and ORNL. Details of the compete set-up have been reported elsewhere (Voelkl et al. 1997).

## 3. TECHNICAL LIMITATIONS TO TPM

There are two fundamental technical limitations to full implementation of TPM for teaching purposes. First, current commercial TEMs are not yet sufficiently automated for complete computer control and second, the Internet has bandwidth limitations. However, complete mouse-controlled operation of a TEM is technically feasible. Such a TEM would make teaching of the technique significantly easier, as discussed in the next section.

The limited bandwidth of the Internet restricts the quality of the live images, and the number of frames per second (fps) which can be transmitted live. As an example, two 256 by 256 images with 256 levels of gray (= 8 bit) require a bandwidth beyond 1 Mbps (much greater than a fast modem which runs at up to 56 kbps), if sent uncompressed. While image compression may help, it is not very effective for gray-scale images, in general. Furthermore, the resultant high-quality HRTEM images also need to be transferred "loss-less" to the remote site. It may take many minutes to transfer one full-size standard image (1k by 1k pixels with 14 bit data depth yields ~14 Mb per image). Such large images could be sent overnight through the Internet. Alternatively the images could be stored on a suitable device such as a CD, a magneto-optical disc or high-capacity magnetic storage disk (e.g. a Zip drive) and shipped overnight to the remote site.

The bandwidth problem will improve as Internet communications speed up by the availability of dedicated high-speed network. Plans are already underway in the USA for the development of Internet 2 which will be restricted to dedicated scientific users and will guarantee both access times and high data transmission rates (>100 Mbps).

### 3.1    Current TEM Teaching Limitations

The teaching of TEM has always been a challenge, in comparison to teaching SEM for example. There are two reasons for this. First the SEM is simpler in concept than the TEM, both in operation and in interpretation of the image information. Second the SEM image is usually observable in normal room lighting and can be displayed on one or more TV screens, easily viewed by many students. While not much can be done to ease the difficulties of understanding diffraction-contrast concepts, current TEM design and operation could be radically changed. The viewing and recording of TEM images is still largely analog and the operation of stages and apertures remains predominantly mechanical.

TEM operation is, in principle, no different to SEM operation. Once a specimen has been inserted (which is probably the most difficult step to automate) both instruments require only seven fundamental steps to operate: a) switch on the beam, b) traverse the specimen to a region of interest, c) select the appropriate magnification of the image (and/or diffraction pattern), d) select the required aperture(s), e) focus the image (or diffraction pattern), f) adjust the stigmation (illumination and/or image system) and g) record the image (or diffraction pattern. In many modern SEMs all the above steps are already under computer control. There is no fundamental reason why they cannot be similarly automated in a TEM. The technology is available to provide full TEM digital operation and computer control. Slow-scan CCD and high-definition TV cameras have already been incorporated in TEM viewing chambers, thus permitting the viewing of digital TEM images (and diffraction patterns) on one or more TV screens by multiple students. Similarly five-axis computer-controlled stages have been demonstrated by several manufacturers and computerized insertion and retraction of apertures

Fig. 1. An undergraduate at Lehigh describes the remote image from ORNL projected on the screen as as another undergraduate operates the microcope remotely via the Macintosh.

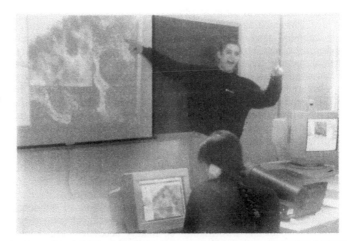

Fig. 2. The video channels show both the HF 2000 at ORNL, as seen from Lehigh and the student classroom at Lehigh as seen from ORNL. The audio control panel is also visible.

is already available in some SEMs

TEM teaching for has traditionally emphasized the arcane practical aspects of operation (e.g. centered dark-field imaging and tilting around a Kikuchi pattern) rather than the importance of the results. Significant manual dexterity and eye-hand coordination is still needed for such routine steps as aligning the objective lens rotation and voltage center, balancing the DF potentiometers, centering the apertures and correcting the objective-lens astigmatism. While the latter operation has been completely computer controlled in the rather specialized case of HRTEM, the routine operation of a TEM has not changed significantly in the last 30 years, since the introduction of electrical rather than mechanical alignments. If the technique is to step into the next century and, more importantly, remain a tool that attracts younger scientists, the instrument needs basic design changes. This fact is borne out by the reaction of students to the first TPM demonstration. While the teacher (representing the older TEM generation) was amazed by the ability to "see" atoms and control the images over a hundreds of miles, the students accepted this as being a routine capability of the Internet (i.e. the immediate presentation of information wherever it happens to reside).

Thus, in conclusion, the current generation of students in not enamored of the mechanical aspects of TEM operation and they expect mouse operation of instrumentation. They also expect that information be readily accessible and analyzable, whatever the source (e.g. direct comparison of on-screen Kikuchi patterns or convergent-beam patterns with simulations available on a remote web site). The next generation of TEMs should reflect those expectations and be completely computer controlled and linked directly to the Internet, making TPM feasible in any facility.

## 4. THE FUTURE

If TPM is to become a reality for teaching TEM, there is no need to stop at simple computer control of the basic operational steps. It is also possible to envisage the development of expert systems where the casual operators do not need (although they still may want) to master any aspects of TEM operation. On the contrary the user selects whatever operational mode is desired, (e.g. BF, CDF, $g$, $3g$ WBDF, two-beam or multi-beam phase contrast) and the computer simply sets up the necessary conditions. Only then will it be possible to bring TEM to uninitiated users, many of whom are daunted by the prospect of learning such a complex instrument. The teacher is then left with the task of teaching why you need to view a BF, CDF or WBDF image without having to teach how the operation is performed (differently on different commercial instruments) via multiple mechanical/electrical steps. Such skills should be obsolete. If, as listed above, the fundamental operations become simply traverse, choose magnification, focus, stigmate and record, a common interface can be envisaged for all electron microscopes. Having learned to operate the common interface, the students will, in principle, be able to operate *any* microscope.

Returning to the research aspect, TPM creates an alternative to the existing concept (in the USA) of national centers for electron very efficiently. A common user interface will make all instruments appear identical to the remote user. It is possible to contemplate a central booking facility where TPM will permit assignment of a user to *any* instrument on the network. The Department of Energy has funded these TPM efforts as a part of the DOE2000 initiative which supports, as a pilot project, the Materials MicroCharacterization Collaboratory (MMC). As part of its efforts, the MMC is supporting the educational aspect of a Collaboratory.

## REFERENCES

Ellisman M H 1995 Proc. 53rd Ann. Meet. MSA (New York: Jones and Begell Publishing) 66
Fan G Y, Mercurio P J, Young S J and Ellisman M H 1993 Ultramicroscopy **52**, 499
Parvin B, Agarwal D, Owen D, O'Keefe M A, Westmacott K H, Dahmen U. and Gronsky R 1995 Proc. 53rd Ann. Meet. MSA (New York: Jones and Begell Publishing) 82
Voelkl E 1996 Advanced Imaging (Oct. 1996) 31
Voelkl E, Allard L F, Bruley J. and Williams D B 1997 J. Microsc. in Press
Zaluzec N J 1995 Proc. 53rd Ann. Meet. MSA (New York: Jones and Begell Publishing) 14

*Inst. Phys. Conf. Ser. No 153: Section 2*
*Paper presented at Electron Microscopy and Analysis Group Conf. EMAG97, Cambridge, 1997*
© *1997 IOP Publishing Ltd*

# Demonstration of solid-state electron optical devices: Pixelated Fresnel Phase lenses

**Y Ito\*, A L Bleloch and L M Brown**

Cavendish Laboratory, Madingley Road, Cambridge CB3 0HE, UK.
\*Currently Department of Materials Science and Engineering, Lehigh University, 5 East Packer Ave. Bethlehem, PA, 18015 USA

**ABSTRACT** : A demonstration of solid-state Pixelated Fresnel Phase (PFP) lenses for electrons will be presented. The PFP lens consists of a set of concentric zones of radially varying depths so that the lens, as a whole, brings an incoming electron plane wave to a spot in its focal plane. Each zone consists of an array of holes. Using the sub-nanometre electron beam of the VG HB501 STEM, the PFP lenses, both convex and concave types, were patterned in a thin evaporated $AlF_3$ film supported on an amorphous carbon film.

## 1. INTRODUCTION

The ability to produce features of nanometer scale offers the possibility of phase manipulation of electron waves. The first conclusive results of phase manipulation by nanometer-scale diffraction gratings directly cut by the finely focussed electron beam in a scanning transmission electron microscope (STEM) has already been demonstrated (Ito et al., 1993). This was achieved by using a grating with a "wedge" profile. This produced an asymmetrical diffraction pattern. This violation of Friedel's law is expected only if the grating acts to a significant extent as a strong phase object. In this paper, a demonstration of solid state Pixelated Fresnel Phase (PFP) lenses for electrons will be presented.

## 2. IDEAL FRESNEL PHASE LENS and PIXELATED FRESNEL PHASE (PFP) LENS

For an ideal Fresnel phase lens, an incident plane wave (wavelength $\lambda$) travelling along the optical axis of a lens experiences the phase shift by the lens. The phase shift function of the emerged wave at the exit plane $\varphi_X(r)$ is expressed as:

$$\varphi_x(r) = k_0\left(f_x - \sqrt{f_x^2 + r^2}\right) < 0, \qquad\qquad 1$$

where $k_0 = 2\pi/\lambda$, $f_x$ is a focal length of the convex lens, and $r$ is the radius from the optical axis (Nishihara and Suhara, 1987). For a Fresnel phase lens, $\varphi_x(r)$ is modified to have a modulo $2\pi$ phase structure, and the phase distribution of a zone, $\varphi_{Fx}(r)$, becomes:

$$\varphi_{Fx}(r) = \varphi_x(r) + 2m\pi, \qquad\qquad r_m < r < r_{m+1}, \qquad\qquad 2$$

where $r_m$ is the inner radius of the $m$th zone.

$$r_m = \sqrt{2m\lambda f_x + (m\lambda)^2} \qquad\qquad 3$$

A concave lens can be formed by taking the negative of Eq. 1 and the exit wave from the concave lens becomes a diverging wave.

The idealised phase function of the PFP lens has Eq. 2 as an envelope function multiplied by a two-dimensional array or "comb" function which gives a phase shift value only at the position of holes in the grating (lens) and nil ($2\pi$) for the rest. In the real PFP lens, holes have finite diameters. The focal length and the number of zones achievable, hence the resolution are strongly influenced by the film thickness, minimum hole spacing and achievable maximum pattern size. For 200 keV electrons, an amorphous $AlF_3$ film 72 nm thick gives a phase shift close to $2\pi$ at the exit surface of the film. Any multiple of this thickness also satisfies the requirement for lens construction if the effects of inelastic scattering and large angle elastic scattering are ignored.

## 3. SIMULATION

In the numerical simulation, a wave is propagated for a given two-dimensional complex amplitude at the exit face of the PFP lens to a plane perpendicular to the PFP lens optic axis at some defocus. The illumination was a plane wave with unit amplitude. The effect of the magnetic field, in which the PFP lens was immersed, was ignored for simplicity. Some generalised imperfections were accounted for by convolution with a Gaussian function. This approximates to some blurring effects such as the effects of partially coherent illumination, inelastic scattering, multiple scattering, imaging characteristics of the microscope and instrumental instabilities.

Simulations of the on-axis intensity distribution of convex and concave PFP lenses are shown in Fig.1. Distinctive focal points exist for both of the lenses, i.e. real focal point for the convex lens and virtual one for concave lens. Also, simulations showed that the designs were robust to random phase errors (Fig.2). A deviation factor of 1.0 corresponds to a standard deviation of 100% for random phase error. In order to destroy the focusing effect, the deviation factor had to be 10 (equivalent to a standard deviation of 1000%). For the real lens fabricated by STEM, hole depth, i.e. phase shift at a hole, is likely to fluctuate around the designed value due to the proximity effect and the local inhomogeneity of the film. Therefore, this robustness to the random phase errors is the strongest point to support why the real lens could still exhibit a distinct focus.

## 4. RESULTS

The performance of the concave lenses was compared with the complementary convex lenses (Fig. 3) observed in CTEM (JEOL JEM 2000EX). Defocus values, $\Delta S$ (in arbitrary units related to the CTEM objective lens control), are proportional to the objective lens current and the in-focus position is taken as the origin ($\Delta S = 0$). At $\Delta S = 0$, the contrast of both (a) and (b) is minimal, indicating little amplitude contribution. With further overfocusing (+) of the objective lens, the intensity on the optic axis gradually increases with varying contrast patterns of the outer zones until it peaks at around $\Delta S = +880$ (~+1000 μm) for (a), then decreases for larger defoci. This behaviour supports the existence of the lens effect with characteristic focal lengths. The main feature of these through-focal series is the complementary behaviour of the convex and concave lenses to each other, particularly on the optic axis. At the focal point $\Delta S = +880$ on the overfocus side, the intensity of the convex lens (a) on the optic axis is more intense than that of the concave lens (b). However, at the virtual focal point $\Delta S = -1280$ (~-1000μm) on the underfocus side the intensity of the concave lens (b) on the optic axis becomes more intense than that of the convex lens (a). This indicates that the lens (a) acts predominantly as a converging lens and the lens (b) as a diverging lens. However, each of the lenses has still some components of the opposite characteristics i.e. diverging component for (a) and converging component for (b). This complementary behaviour could be explained by the simulation, which incorporated the proximity effect of the hole drilling process (effect of metal rich region around a hole) as an additional concave phase retardation of up to approximately $0.2\pi$ in the convex lenses and a variable hole diameter (Ito et al., 1997). The estimated FWHM of the focus of the PFP lens was approximately 8 nm.

51

## 5. CONCLUSIONS

The electron optical behaviour of convex and concave solid state PFP lenses has been demonstrated and their behaviour agrees very well with the simulation. Also, The simulation demonstrated that the design of the PFP lenses were robust for random phase lens errors.

ACKNOWLEDGEMENTS

The authors acknowledge the SERC for their funding of the project. YI acknowledges Telecom Australia Research Laboratories for financial support and leave to study at the Cavendish Laboratory and an ORS award.

REFERENCES

Ito, Y., Bleloch, A.L. and Brown, L.M. 1997 Submitted to Nature.
Ito, Y., Bleloch, A.L., Paterson, J.H. and Brown, L.M. 1993 Ultramicroscopy **52**, 347.
Nishihara, H. and Suhara, T. 1987 Progress in Optics **XXIV**, ed. Wolf, E., North-Holand, Amsterdam, 3.

(a)  (b)

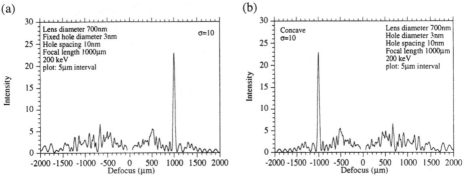

Fig.1 Simulated on-axis intensity distribution PFP lenses. (a) Convex PFP lens. (b) Concave PFP lens.

(a)  (b)

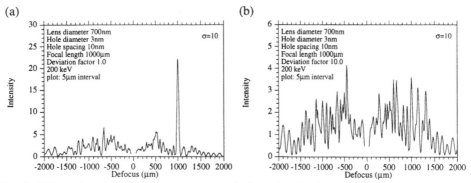

Fig. 2 Effect of random phase error on on-axis intensity distribution of convex lenses. (a) 100% standard deviation. (b) 1000% standard deviation.

52

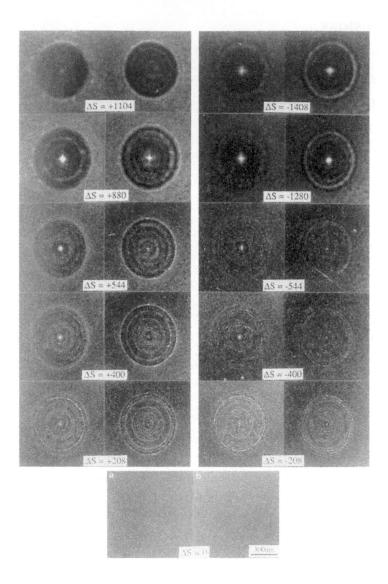

Fig. 3 Through-focal series of a pair of (a) convex and (b) concave PFP lenses. Designed focal length is |f|=1000 µm for both lenses. The original images at each defocus were recorded simultaneously on the same photographic plate.

*Inst. Phys. Conf. Ser. No 153: Section 2*
*Paper presented at Electron Microscopy and Analysis Group Conf. EMAG97, Cambridge, 1997*
© *1997 IOP Publishing Ltd*

# Towards the intelligent SEM

**N H M Caldwell, B C Breton and D M Holburn**

Department of Engineering, University of Cambridge, Trumpington Street, Cambridge CB2 1PZ.

**ABSTRACT:** The modern scanning electron microscope (SEM) is a fully software-controlled instrument with significant automation of individual instrumental parameters. Optimum performance still demands operator expertise. A knowledge-based approach to SEM operation is proposed to embed this expertise within the control software to provide global automation. An analysis of SEM investigations has resulted in a model of the control task with some of the associated knowledge being codified into a formal representation. A standalone prototype knowledge-based system has been designed and implemented. Ongoing and future directions of this research are presented.

## 1. INTRODUCTION

Recent advances in SEM instrumentation have integrated personal computers with microscopes to yield fully software-controlled instruments, with the control software predominantly executing in the Windows™ operating system environment. Significant emphasis has been placed by instrument manufacturers on ease of use and full software control has permitted innovations in terms of automating the setting of individual instrument parameters with routines to perform automatic focus, gun alignment, astigmatism correction and so forth now commonplace in the modern SEM. In addition to such "autofunctions", some instruments allow sequences of control instructions to be saved as "macros", which can be later invoked saving the microscopist time and restoring the instrument to a defined configuration, essential in many applications.

Despite the usefulness of such features, autofunctions only consider (at best) a few instrument parameters and are prone to fail when other related but unconsidered parameters have unusual values. Macros, being simply lists of instructions, cannot make sufficient use of feedback from the SEM to adapt to changes in the instrument, and so can set parameters inappropriately. The software is limited by a lack of knowledge in the areas of appropriate parameter initialisations, interactions amongst parameters, and what constitutes satisfactory and optimal images.

Hence optimum performance with modern SEMs still demands trained operators to supply the missing knowledge. Unfortunately the average skill level of SEM operators has declined. One novel means of redressing this expertise deficit is to embed human operational knowledge within the SEM control software by constructing a suitable knowledge-based system (KBS).

## 2. KNOWLEDGE-BASED SYSTEMS — AN OVERVIEW

A knowledge-based system (or expert system) is fundamentally a computer program which demonstrates expert performance on a well-defined task, by simulating human reasoning over formal representations of heuristic knowledge. The archetypal knowledge-based system consists of three components: a knowledge base, an inference engine, and a user interface. The knowledge base is constructed by modelling the task of interest, acquiring the associated "chunks" of knowledge from experts, textbooks, manuals, etc., and representing these "chunks" in a suitable formalism. The inference engine is designed to reason consistently over the knowledge and any data specific to a consultation in a logical fashion deriving and storing conclusions. The user interface is responsible for all interactions between the system and the user, including acquiring information, displaying results, and explaining results. More detailed introductions to this field can be found in Jackson (1990) and Gonzalez and Dankel (1993).

Knowledge-based systems have been successfully employed in many diverse applications including diagnosis, planning, analysis, configuration, process control, and identification in such fields as manufacturing, medicine, finance, agriculture, law and teaching. No previous use of expert systems in scanning electron microscopy is known, other than for instrument fault diagnosis by Caldwell et al (1996 and 1997).

## 3.    ANALYSIS AND MODELLING

From observation of instrument usage and the authors' personal experience, it was concluded that the typical SEM investigation could be decomposed into at most five phases:

- Specimen Preparation: The sample is prepared dependent on the nature of the application and placed on a stage within the specimen chamber.
- SEM Preparation: The instrument is readied for operation by initiating any necessary control software, enabling various instrument components and performing any required evacuation of the specimen chamber and/or the column.
- SEM Initialisation: The electron beam is switched on and a number of instrument parameters are initialised to preferred values dependent on sample type and other user constraints, obtaining an image of the sample as a result.
- Optimisation: The operator will adjust the settings of the principal instrument parameters (and others if necessary) in an iterative fashion improving the quality of the image is satisfactory and the desired information has been obtained.
- External Analysis: Dependent on the investigation, it may be necessary to use third-party software to analyse the image or to allow other hardware/software to control the instrument.

Not all SEM investigations will involve the full five phases: preliminary investigations of a sample will require only the first three phases; producing micrographs of journal quality will additionally require SEM optimisation; and X-ray microanalysis will demand all five phases. The scope of the research is SEM control and so concentrates on the task from SEM preparation to optimisation.

Each phase of interest and the instrument itself have been further analysed to identify the critical parameters which must be adjusted in each phase. The constraints on SEM imaging such as sample type, appropriate signal (secondary electron, backscattered electron, specimen current) and desired magnification affect the choice of initialisation value for the primary instrument parameters. Such knowledge has been codified in a rule-based representation. The formalism associates conditions (such as user constraints and existing parameter settings) with actions (changes to parameter settings) and the appropriate conclusions (new inferences to be drawn about the instrument state as a result of the actions). After experimental verification on the group's SEM, each new rule has been added to the knowledge base.

## 4.    DESIGN AND IMPLEMENTATION

A custom knowledge-based system was designed and implemented to utilise the operational knowledge base. To provide proof of concept, it is necessary for the system to control an SEM. The LEO 400 series instruments have the prerequisites of being fully software-driven and possessing accessible interfaces for other applications to control the SEM, and so the implementation is designed to operate LEO SEMs. The system could be adapted to control any modern software-driven SEM with minimal effort.

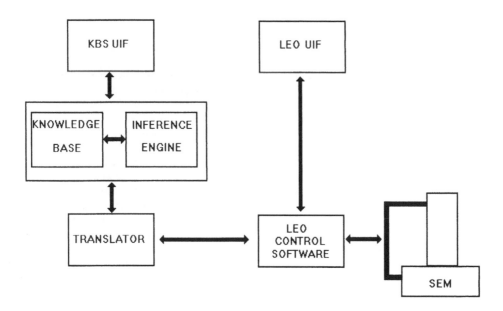

Fig. 1: An architectural schematic of the current system

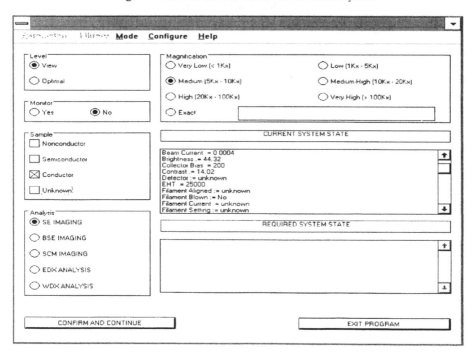

Fig. 2: Screenshot of the initial setup window

56

A schematic of the current system architecture is shown in Fig. 1. In addition to the knowledge base and inference engine, there are two other modules, namely the user interface (denoted KBS UIF) and the Translator module. The standalone nature of the current system means that the user interface is not intended as a replacement for the SEM user interface, rather it provides a means by which the operator can interact with the knowledge-based system. The operator still views images and other analyses via the SEM user interface. The Translator module is responsible for converting the actions proposed by knowledge base rules into equivalent instructions comprehensible by the LEO control software (and vice versa) and interpreting the results of such actions on behalf of the inference engine. As far as is possible, all instrument and manufacturer dependencies have been localised to the Translator module.

In operation, the system first displays a setup window (Fig. 2) to obtain the constraints on the consultation including any parameters which the operator requires to be set to a specific value. It then switches into the main consultation window. The inference engine searches the knowledge base for rules whose conditions match the current instrument state and the operator constraints. In some instances, the operator may be asked to provide the system with further information. Upon matching a rule, the associated actions will be converted by the Translator module and the success or failure of the actions will determine what conclusions, if any, should be derived. These conclusions in combination with previously known information may then invoke further rules. The cycle of rule processing continues until either the desired image quality is obtained or no more rules can be successfully matched. The knowledge base is partitioned such that each segment corresponds to a control phase. The inference engine remains within a segment until the criteria for completion of the given phase have been met. At any time during the consultation, the operator can alter the constraints on the system or change samples, whereupon the system will backtrack and reset itself to adjust to the new context.

## 5.    FUTURE WORK AND CONCLUSIONS

Although the system is currently a research prototype, it is sufficient to demonstrate "proof of concept" of the applicability of knowledge-based techniques to electron microscopy. It is not yet however a replacement for an expert microscopist. The knowledge base requires expansion to enhance the SEM Initialisation phase for uncommon and more challenging combinations of application constraints, and to codify the iterative sequences of parameter adjustments involved in image optimisation to improve images to the quality essential for publication and such applications as X-ray microanalysis. Research is ongoing into developing algorithms to objectively measure aspects of image quality for use in the optimisation phase.

The eventual objective is to encapsulate the knowledge base and the inference engine as an embedded subsystem within the conventional control software of an SEM, providing efficient expert assistance to the operator in a transparent fashion. The resultant system will truly deserve to be described as an intelligent instrument.

## ACKNOWLEDGEMENTS

The authors gratefully acknowledge the financial support of the Engineering and Physical Sciences Research Council, LEO Electron Microscopy Ltd., and the Isaac Newton Trust.

## REFERENCES

Caldwell N H M, Breton B C and Holburn D M 1996 Applications and Innovations in Expert Systems IV, eds A Macintosh and C Cooper (Cambridge: Cambridge University Press) pp177-88
Caldwell N H M, Breton B C and Holburn D M 1997 Scanning **19**, (3), 204
Gonzalez A J and Dankel D D 1993 The Engineering of Knowledge-Based Systems Theory and Practice (London: Prentice-Hall International)
Jackson P 1990 Introduction to Expert Systems (2nd Edition) (Reading MA : Addison-Wesley)

*Inst. Phys. Conf. Ser. No 153: Section 2*
*Paper presented at Electron Microscopy and Analysis Group Conf. EMAG97, Cambridge, 1997*
© 1997 IOP Publishing Ltd

# The development of an ultrahigh vacuum field emission electron microscope for the observation and analysis of crystal surfaces.

**M Takeguchi, T Honda, Y Ishida, A I Kirkland[*], M Tanaka[#] and K Furuya[#].**

Electron Optics Division, JEOL Ltd. 3-1-2 Musashino, Akishima, Tokyo 196, Japan.
[*] JEOL UK Ltd, JEOL House, Silvercourt, Watchmead, Welwyn Garden City, Herts, AL7 1LT, UK and University of Cambridge, Department of Chemistry, Lensfield Road, Cambridge, CB2 1EW.
[#] National Research Institute for Metals, 1-2-1 Sengen, Tsukuba, Ibaraki 305, Japan.

**ABSTRACT:** A 200kV UHV TEM equipped with a field emission gun (FEG) has been developed in order to obtain high resolution structural and compositional information from surfaces. The design of this instrument will be discussed and recent applications to the characterisation of surfaces will be presented.

## 1. INTRODUCTION

Conventional Transmission Electron Microscopes (TEM's) typically operate with vacuum levels of approximately $10^{-6}$ Pa at the sample which is insufficient for the direct observation of clean surfaces. In order to carry out direct imaging of clean surfaces in either plan view or profile modes a microscope capable of providing an Ultra-High Vacuum (UHV) environment at the sample together with the necessary specimen cleaning and pre-treatment facilities is required. This paper describes the design of a fully IHV compatible FEG-TEM and presents initial applications illustrating its use in surface imaging and diffraction.

## 2. CONSTRUCTION AND SPECIFICATIONS.

Figure 1. shows an overview of the JEM-2010FV together with the UHV specimen pre-treatment chamber. The new instrument is based on the conventional JEM-2010F (200kV FEG-TEM) but has been modified for UHV operation. The key design requirements for operation at UHV levels are the development of a fully UHV compatible double tilting goniometer, the modification of the microscope column to allow baking at 150° C and the incorporation of pre-evacuation and specimen pre-treatment chambers to facilitate specimen cleaning for observation under UHV conditions. The detailed design and operation of the UHV goniometer has been detailed elsewhere (Kondo et al 1994) and relies on the transfer of only the end of the specimen rod via magnetically coupled transfer arms between the pre-treatment chamber and the microscope column. In order to preserve UHV compatibility the shift mechanisms are located outside the vacuum and are coupled

58

via metal bellows thus eliminating all Viton "O" ring seals. In addition to the UHV goniometer the 2010FV features a UHV specimen pre-treatment chamber with 7 ICF UHV flanges for the attachment of standard UHV components such as evaporators, mass spectrometers and ion and electron guns for specimen cleaning. The microscope itself has been designed to accept both an EDX detector and PEELS spectrometer modified for UHV compatibility.

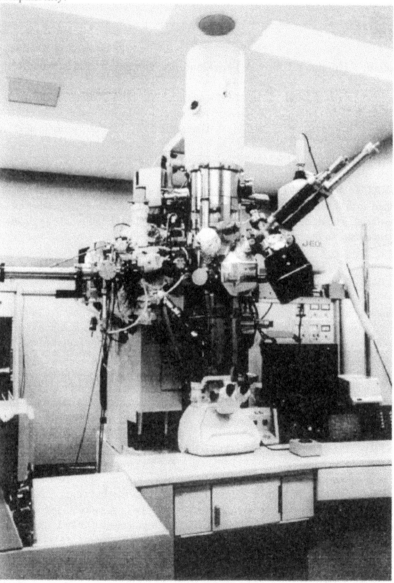

Figure 1. An overview of the JEM 2010FV 200kV UHV FEG-TEM together with the specimen pre-treatment and pre-evacuation chambers.

The principal specifications of the new instrument pertinent to high resolution imaging of crystal surfaces are listed below in table 1.

| Specimen Vacuum (Pa) | $<2 \times 10^{-8}$ |
|---|---|
| Spherical Aberration Coefficient (mm) | 0.7 |
| Chromatic Aberration Coefficient (mm) | 1.2 |
| Interpretable Point Resolution (nm) | 0.21 |
| 20% Information Limit (nm) | 0.15 |
| Electron Source | $ZrO_2$ (TFE) |
| Baking Temperature (°C) | 150 |

Table 1. The principal specification of the JEM-2010FV.

## 3. APPLICATIONS

Fig. 2 shows an electron diffraction pattern taken from a silicon single crystal in the [111] plan view orientation showing the (111) surface. In addition the principal reflections a series of superlattice reflections consistent with the 7x7 reconstruction reported by Takayanagi (1985) thus clearly demonstrating the UHV performance of the instrument.

Figure 2. Electron Diffraction pattern taken from a silicon single crystal in the [111] plan view orientation showing the 7x7 reconstruction of the (111) surface.

The growth and characterisation of "nano structures" is of fundamental importance to the next generation of electronic devices. Fig. 3 shows a high resolution image of a GaAs "nano wire" grown in situ under UHV conditions in the microscope. The characterisation of such structures requires the imaging of the true clean surface structure without contamination from carbonaceous deposits. In Fig. 3 the (100) and (110) surfaces are clearly resolved in an atomically clean state due to the enhanced vacuum conditions at the sample. A final application of the new instrument takes advantage of the availability of a small electron probe with high current density arising from the use of a thermal Field emission electron source coupled with the UHV specimen environment for direct electron beam lithography of metal and inorganic oxides. Fig. 4 shows a hole drilled directly in a <100> oriented gold foil using a focused electron probe. The hole is clearly imaged at high resolution and with minimal contamination due to the UHV specimen environment.

60

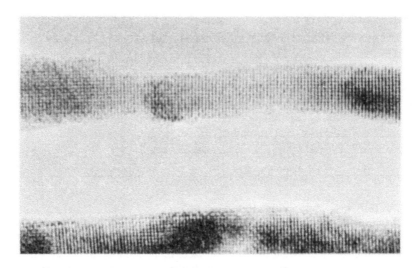

Figure 3. A GaAs "Nano Wire" grown in situ under UHV conditions exhibiting atomically clean (100) and (110) surfaces imaged in profile.

Figure 4. A high resolution image of a "Nano Hole" drilled directly in a <100> oriented gold foil using a focused electron probe.

## REFERENCES

Kondo Y, Kobayashi H, Kasai T, Numome H, Kirkland A I and Honda T 1994 ICEM-13, Paris, 269.
Takayanagi K 1985 J. Vac. Sci and Technol., **A3**, 1502.

*Inst. Phys. Conf. Ser. No 153: Section 2*
*Paper presented at Electron Microscopy and Analysis Group Conf. EMAG97, Cambridge, 1997*
© *1997 IOP Publishing Ltd*

# Utilisation of beam-blanking in electron beam-induced voltage contrast studies of high-Tc superconducting thin films

**S A L Foulds and J S Abell**

School of Metallurgy and Materials, Superconductivity Research Group, University of Birmingham, Birmingham B15 2TT, UK

**ABSTRACT:** Beam-blanking has been used to study the influence of electron beam pulse length and pulse current on voltage generation in a superconducting $YBa_2Cu_3O_{7-x}$ (YBCO) thin film microbridge. Pulse lengths from a few μS up to 400 μS have been studied. A peak in recorded voltage was found to occur for pulse lengths of 80 to 100 μS. This provides an upper limit for useful pulse lengths of approximately 100 μS. Further work is required to ascertain whether this is sample dependent and particularly whether it is dependent on sample thickness. Measured voltage across the microbridge was found to increase parabolically with beam current at temperatures well below the transition temperature (Tc). At temperatures close to Tc a less rapid rise in voltage as a function of beam current was observed. This is ascribed to the beam power, and subsequent heating, being sufficient to drive locally the superconductor normal and trace out the local resistivity curve.

## 1. INTRODUCTION

Electron beam-induced voltage contrast (EBIV), also referred to as low temperature SEM (LTSEM), is a useful tool for studying local critical current density (Jc) variations in thin films of high temperature superconductors (HTS). EBIV has been used to study a number of thin film structures and devices (e.g. Hollin (1994), Foulds (1995), Gerber (1996) and Foulds (1997)).

In EBIV the electron beam acts as a local source of heat. In this paper the use of beam blanking as an effective means of controlling the electron pulse length is discussed. The effects of electron pulse length and electron beam power on voltage generated, in an HTS test structure, are reported.

## 2. EXPERIMENTAL

### 2.1 EBIV technique

Superconducting samples are mounted on the cold finger of an Oxford Instruments constant flow helium cold stage which fits onto a JEOL 840A SEM. There are two thermometers on the cold finger: one for temperature control and the other for monitoring the sample temperature. Temperatures from ~30 K can be achieved at the sample. Four electrical contacts are made to samples. Two contacts are employed for applying a current to the sample. A dry cell and resistance box are used for applying current as they have been found to be less noisy than mains electronics. The other two contacts are used to monitor the voltage across the sample. The voltage is amplified in the column with a gain of 100 and then externally by an amplifier with a variable gain between 10 and 1000. The first stage of amplification is carried out close to the sample to reduce noise pick-up.

A framestore (Hall (1991)) controls the scanning of the electron beam while recording the sample voltage and secondary electron signal as a function of beam position. For a metallic sample the influence of the electron beam is to heat a volume with dimensions ~1 μm (Clem (1980)). This is the limit of the spatial resolution of the system. If an area of superconductor is at the foot of its superconducting transition then the action of the beam heating will be to cause a significant resistance change which will correspond to a significant voltage change. If the superconductor is well below or above its transition then no significant voltage change is recorded. In scanning the beam over the sample an image of voltage generation as a function of beam position is produced allowing the local critical current or transition temperature of the superconductor to be studied. In between electron pulses sufficient time must be allowed for the heat to dissipate. Prior to beam-blanking this was achieved by moving the electron beam to the edge of the field of view. Clearly at high magnifications this could result in heat being deposited into superconducting material in between measurement points. This problem has been overcome by the addition of a beam blanking unit. The beam blanking unit allows electron pulses as short as 1 μs in length.

## 2.2 Sample details

The sample studied is a $YBa_2Cu_3O_{7-x}$ (YBCO) thin films deposited by pulsed laser deposition (PLD) onto an MgO substrate. The growth time for the film of 20 minutes yielded a film thickness of approximately 1.5 μm. The film was patterned into a structure containing microbridges and feed tracks using standard photolithographic techniques and ion beam milling. Gold was deposited on feed tracks to allow electrical contact to be made to the microbridges. Fig. 1. shows a secondary electron image of one microbridge and a voltage contrast image. The EBIV image indicates that there is variation in contrast along the track.

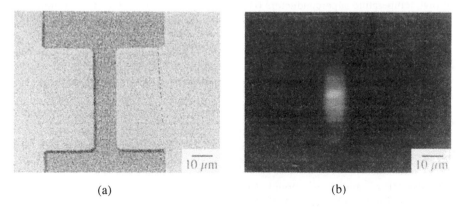

(a)                                    (b)

Fig. 1. (a) Secondary electron image of an YBCO microbridge on MgO substrate (b) EBIV image at 83 K and bias current of 20 mA (beam conditions 20 kV, 5 nA)

## 3. RESULTS AND DISCUSSION

In generating an electron pulse two times are specified, the pulse length and the time the beam is blanked for. The blanked time is important in allowing time for the heat from the last pulse to dissipate and for SEM scan coils and amplifiers to settle. The blanked time will not be discussed further but it has been kept sufficiently large in all measurements to ensure no artefacts arise.

The influence of pulse length on maximum voltage along the track is shown in Fig. 2. To determine the maximum voltage an average is taken over the track width. An initial rise in generated voltage with pulse length is observed corresponding to a greater increase in local heating and hence a larger local resistance change and generated voltage. However, a point is reached where further increases in pulse length cause a drop in measured maximum voltage. One would expect a situation to be reached where either, the thermal energy of the pulse is being dissipated and no further local temperature rise occurs with increased pulse length or the temperature rise induced by the pulse becomes sufficient to drive the local area normal. In both these cases the observed voltage would be expected to plateau as a certain pulse length is exceeded. The drop in observed voltage may arise from a spatial spreading in the thermal energy deposited by the electron pulse resulting in a lower temperature rise over a larger area. However, features in the voltage contrast along the length of the microbridge do not appear to be spatially smoothed as would be expected.

The form of maximum voltage versus pulse length shown in Fig. 2. was observed for a variety of sample temperatures and bias currents. The peak in maximum voltage was found to occur for pulse lengths in the range 80 to 100 μS. For this sample pulse lengths greater than this have a detrimental effect on the magnitude of the measured voltage. The influence of film thickness on this peak voltage merits further study.

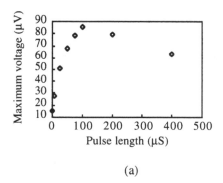 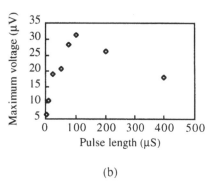

(a)  (b)

Fig. 2. Maximum voltage along the microbridge as a function of pulse length (beam conditions 20 kV, 5 nA) (a) 80.7 K, bias current 28 mA (b) 88 K, bias current 0.6 mA

Fig. 3. shows the maximum voltage along the microbridge as a function of beam current for a fixed heat pulse of 50 μS. On increasing the beam current a greater voltage is observed. This is consistent with the larger beam current resulting in a larger local thermal increase, a larger local resistance change and a corresponding increase in generated voltage. If the heating caused by the beam is proportional to the beam current then the voltage versus beam current is mapping out the local resistivity as a function of temperature. For sample temperatures below ~ 85 K the voltage was proportional to the square of the beam current and it is likely that the foot of the local resistivity versus temperature curve is being mapped. At temperatures closer to Tc (e.g. 87 K and 88 K) the heat from the higher beam energies is

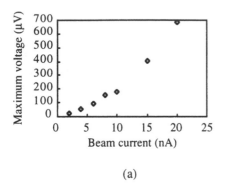 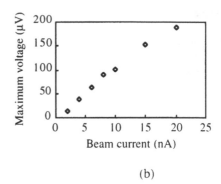

(a)                                                    (b)

Fig. 3. Maximum voltage along microbridge as a function of beam current (at 20 kV with pulse length of 50 μS) (a) 83.2 K, bias current 20 mA (b) 87 K, bias current 2 mA

sufficient to drive locally the superconductor normal. Here the initial rapid rise may correspond to the superconducting transition with the slower rise at larger beam currents resulting from the superconductor being driven normal (see Fig. 3b).

The influence of changing the beam current at a beam voltage of 10 kV was also investigated. No difference was detected in voltage generated for different accelerating voltages as long as the power delivered to the superconductor remained constant (i.e. a beam of 20 kV and 5 nA gave a similar voltage contrast image to one of 10 kV and 10 nA)

## 4. CONCLUSIONS

The influence of beam pulse length and beam current on the voltage generated in a superconducting microbridge biased close to Tc has been studied. Pulse lengths greater than approximately 100μS were found to produce a drop in the measured voltage and for this sample were of little use. Further work is required to determine if a pulse time of 100 μS is appropriate for other samples and particularly what influence changing the film thickness has. By changing the beam current the local resistivity as a function of temperature appears to be to map out. Beam conditions used in previous studies (Hollin (1994), Foulds (1995) and Foulds (1997)) of 5 nA and 20 kV seem appropriate.

## ACKNOWLEDGEMENTS

Thanks to Drs P Woodall and F Wellhöfer for film growth and Dr M Bari and Mr R Pflaumer for film patterning. Financial support form EPSRC is gratefully acknowledged.

## REFERENCES

Clem J R and Huebener R P 1980 J. Appl. Phys. **51** 2764
Foulds S A L, Sutton S D, Keene M N and Abell J S 1995 Inst. Phys. Conf. Ser. **148** 1581
Foulds S A L, Abell J S, Kang Dae-Joon and Tarte E J 1997 IEEE Trans. Appl. Phys. **7** 2142
Gerber R, Koelle D, Gross R, Huebener R P, Ludwig F, Dantsker E, Kleiner R and Clarke J
    1996 Appl. Phys. Lett. **68** 1555
Hall M G, Hulbert J K and Zarucki P N 1991 Scanning **13** 217
Hollin C A, Abell J S, Goodyear S W, Chew N G and Humphreys R G 1994 Appl. Phys.
    Lett. **64** 918

*Inst. Phys. Conf. Ser. No 153: Section 2*
*Paper presented at Electron Microscopy and Analysis Group Conf. EMAG97, Cambridge, 1997*
© *1997 IOP Publishing Ltd*

# A novel ultra-sharp field emission electron source demonstrated in a STEM

**E M James, A L Bleloch and J M Rodenburg**

Cavendish Laboratory, Madingley Road, Cambridge, CB3 0HE, United Kingdom.

**ABSTRACT:** Nanometre-sized protrusions have been grown at the apices of tungsten field emitters using a surface melting technique. Electrons were extracted from such a "nano-tip" installed as the source in a 100 kV STEM; the emission characteristics of this novel emitter were observed as a function of various gun lens settings and a high resolution image formed.

## 1.    INTRODUCTION

The properties of the electron source, and particularly its brightness, determine the extent to which interference effects are observable at the detection plane in the transmission electron microscope. These are the features which directly contribute to high resolution image contrast. Recently, novel new types of field emitter (FE) have been investigated in the low-voltage point-projection microscope (PPM) with emission confined to an area of atomic dimensions. The resulting reported beam properties include an extremely narrow angular beam spread (Vu Thien Binh *et al*, 1996), a brightness improvement of at least an order of magnitude over conventional sources (Qian *et al*, 1993, Fransen *et al*, 1996) and the possibility of a beam energy distribution of the order of 100 meV (Purcell *et al*, 1994 Yu *et al*, 1996). Experiments are described here which resulted in such an emitter being operated as the source in an unmodified 100 kV scanning transmission electron microscope (STEM).

In order to produce such a source the apex of a conventional tungsten emitter is manipulated so that it ends in an atomically sharp structure. Fink (1986) was the first to prepare a so-called nano-tip by sputtering an FE tip with Noble gas ions, followed by deposition of an atom at the apex. A different technique was pioneered by Vu Thien Binh and Garcia (1992) where nanometre-sized protrusions are grown on a tip by the action of heating and a strong electrostatic field. With a nano-tip installed in the low-voltage PPM, images have been acquired of molecular chain profiles; however, it is in the medium energy TEM that the brightness advantages of nano-tips would have the greatest impact.

## 2.    CHOICE OF MICROSCOPE AND TIP PREPARATION METHOD

A VG 100 kV STEM was chosen as the test instrument for installation of a nano-tip source. This microscope already uses a conventional cold field emission gun operating in UHV. Image information is available up to an upper spatial frequency limit set by the effective source coherence at the specimen plane. This is directly related to the finite virtual source area in the gun and the settings of the subsequent de-magnifying electron optics. Mechanical and stray field-induced instabilities also affect this limit. In order to detect interference effects above experimental noise levels, adequate current must reach the specimen. It is therefore source brightness which is the quantity to maximise to achieve optimum effective source coherence.

The main problem in incorporating a nano-tip in the STEM is the design of the gun electron optics. A converging lens is set up by the extraction and acceleration electrodes; to keep this weak, and minimise the resulting aberrations, the tip has to operate at an extraction voltage above around 2.5 kV. The process in which a nano-tip is formed involves reducing the emitting region to atomic dimensions. For most apex geometries this leads to a reduction in the extraction voltage to a value well below 1 kV. Incorporating such a tip into the VG STEM would produce a highly aberrated cross-over in the gun. For this reason, the method of nano-tip preparation due to Vu Thien Binh and Garcia (1992) was chosen. The growth of nanometre-sized protrusions can be achieved at the apex of a relatively blunt tip; the extraction voltage is then largely determined by the conventional base-tip radius.

Any reduction in virtual source size, and thence a reduction in the de-magnification factor needed to obtain a coherent probe, would have a favourable effect on the signal to noise ratio in high resolution experiments for both imaging and electron energy loss spectroscopy.

## 3.    GROWTH OF NANO-TIPS IN A TEST APPARATUS

A UHV field emission test apparatus was constructed. It was possible to mount a standard VG tip assembly on a bellows within the chamber and apply a variety of potentials to the <111> tungsten tip in conjunction with a heating current being passed through its support loop. Approximately 7 cm in front of the tip apex a two dimensional detector system was mounted. This consisted of a microchannel plate coupled to a YAG scintillator and fibre optic feedthough with a CCD camera externally attached. It allowed observation of emission patterns at high gain.

A review of the nano-tip preparation method has been given in Vu Thien Binh and Garcia (1992) and Vu Thien Binh *et al* (1996). Small pyramidal structures up to a few nanometres in height are formed and, under favourable conditions, a small current of atomic metal ion emission (AMIE) emanates form their apices. Figure 1a illustrates an AMIE emission spot, observed at the test apparatus detector for a tip heated red hot and with an applied electric field of the order of $10^{10}$ $Vm^{-1}$ (8 kV applied with opposite polarity to that for field emission). On reducing and reversing the applied potential, it was possible to observe electron emission from localised areas of the tip at room temperature; these closely corresponded to the positions of AMIE spots before cooling (fig. 1b). The beam FWHM was between 3° and 5°. For comparison, the emission pattern from the same tip prior to the sharpening process is shown in fig. 1c; the pattern is almost uniform over the small angular scale of the detector. The field emission properties shown in fig. 1b were reproducible and always showed the characteristics of high current stability up to a few tens of nano-Amps total extracted current; there was also a tendency to revert to the conventional emission pattern on application of a flash heating current. This behaviour correlates well to that described by Vu Thien Binh *et al* (1996) and strongly suggests the formation of protrusion nano-tips. Importantly, it was possible to repeatedly prepare nano-tips which emitted currents of around 1 nA at an extraction voltage of over 2 kV.

## 4.    A NANO-TIP SOURCE IN THE STEM

A nano-tip was successfully prepared *in situ* within the STEM gun chamber. Firstly, the approximate voltage and heating currents needed to yield a suitable nano-tip were established within the test apparatus. The tip was transferred to the STEM and the conditions necessary for AMIE reproduced using direct connections to the gun feedthough. The AMIE process was carried out blind. When a protrusion had by chance formed near the microscope optic axis, it was possible to observe an electron emission pattern at the diffraction screen (above the specimen plane) with the HT switched on. This was strikingly different to that obtained from a conventional emitter and a comparison between the two is shown in fig. 2. The microscope gun lens has been excited in order to focus a cross over such that the entire beam can pass through the differential pumping aperture between the gun and microscope chamber. Emission was confined to two cones near the edge of the first anode. By comparing the emission spot sizes to that of the first anode aperture shadow, the total beam opening angles were shown to be of the order of 5°. It was possible to align the microscope with only one cone emerging into the chamber, near the optic axis, and to form an annular dark field image (fig. 3). Large amounts of astigmatism correction were required to resolve image features at a resolution of around 1 nm. This was probably due to the close proximity of the emission cone to the first anode aperture edge.

The beam spectral width was estimated, using the STEM PEELS system, to vary between just over 0.4 eV to just under 0.5 eV as the extraction voltage was raised from 1.9 kV to 2.2 kV (corresponding to a twenty-fold increase in the emitted current). This variation may be due to emission from the band structure associated with a low protrusion (Vu Thien Binh *et al* 1996).

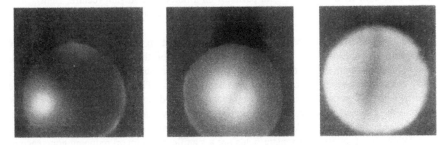

Figure 1: Emission patterns obtained in the tip testing apparatus. Left (a) AMIE emission spot from a negatively biased tip at approximately 1/3 of the melting temperature. Centre (b) - field emission pattern after growth of a protrusion. Right (c) - field emission pattern from a conventional tip. (the dark line across the images is due to a fault in the channel plate).

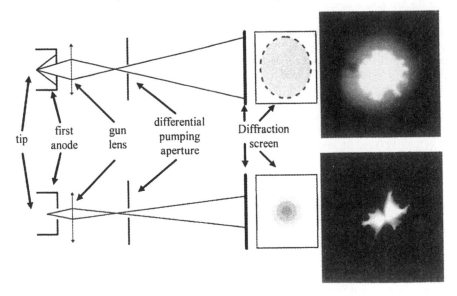

Figure 2: Ray diagrams (left) and diffraction screen emission patterns (right) for field emission from conventional (top) and protrusion (bottom) tips. The former shows a shadow image of a uniformly illuminated first anode; in the latter image, two narrow emission cones emerge from the gun.

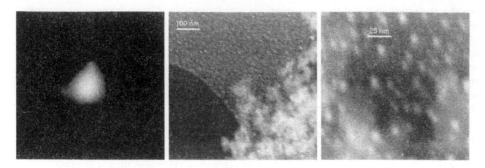

Figure 3: Annular dark field images of gold particles on a holey carbon film (centre, right). The probe was formed from an electron cone (shown left) emitted from a nano-protrusion.

Stability of the current emitted into the cones was excellent up to around the 1 nA level over an experimentation period of days, without a flash cleaning of the tip. At higher currents, instabilities became more prevalent and at 10 nA the protrusion structures were destroyed. A conventional coherent probe has a current of less than 50 pA. This figure could feasibly be increased using a protrusion nano-tip if the de-magnification needed is then less, because of a higher source brightness. The failure to observe this here was probably due to aberrations caused by a non-axial beam path in the gun. Also it is known from coherence measurements (James and Rodenburg, 1997) that stray electromagnetic fields are a significant factor in reducing coherence in the upper gun region of the microscope.

## 5.    CONCLUSIONS

For the first time a nano-tip electron source has been demonstrated in the medium-energy electron microscope and high resolution images obtained. By using the method of protrusion growth, it was possible to operate the microscope gun at a near standard extraction voltage and minimise aberrations in the gun system. This work has shown that correction of the gun aberrations, perhaps by a redesign of the electrostatics, and that more effective shielding from coherence-destroying instabilities could lead to a significant increase in probe brightness by using a nano-tip source.

The authors would like to acknowledge Dr C J Edgcombe, for many helpful discussions, and the financial assistance of the EPSRC, VG Microscopes, the European Community (grant ref. ERBCHRXCT930127) and St. John's College, Cambridge (EMJ).

## REFERENCES

Fink H-W 1996 *IBM J. Res. Dev.* **30** 460

Fransen M J, Damen E P N, Schiller C, van Rooy T L, Groen H B and Kruit P 1995 *Appl. Sur. Sci.* **94/95** 107-112

James E M and Rodenburg J M 1997 *Appl. Sur. Sci.* **111** 174-179

Purcell S T, Vu Thien Binh, Garcia N, Lin M E, Andres R P and Reifenberger R 1994 *Phys. Rev. B* **49** 17259-17263

Qian W, Scheinfein M R and Spence J C H 1993 *J. Appl. Phys.* **73** 7041-

Vu Thien Binh and Garcia N 1992 *Ultramicroscopy* **42-44** 80-90

Vu Thien Binh, Garcia N and Purcell 1996 S T *Adv. Im. & Electr. Phys.* **95** 63-155

Yu M L, Lang N D, Hussey B W, Chang T H P and McKie W A 1996 *Phys. Rev. Lett.* **77** 1636-1639

*Inst. Phys. Conf. Ser. No 153: Section 2*
*Paper presented at Electron Microscopy and Analysis Group Conf. EMAG97, Cambridge, 1997*
© *1997 IOP Publishing Ltd*

# Detection of BSE in the low voltage SEM with an electrostatic immersion lens and a field-free specimen

**J Hejna**

Institute of Material Science and Applied Mechanics, Wroclaw University of Technology, Smoluchowskiego 25, 50-370 Wroclaw, Poland

**ABSTRACT:** Detection of backscattered electrons in the scanning electron microscope equipped with an electrostatic retarding lens was studied theoretically and experimentally. The primary beam after retardation by the lens impinge on the specimen held at the potential close to the potential of the lower electrode of the lens. Low take-off backscattered electrons move initially in a field-free space and after acceleration are detected by a ring scintillation detector. Theoretical studies included analytical and numerical computations of electron-optical parameters of the retarding lens. Next, different configurations of low-voltage detectors were tested experimentally in the scanning electron microscope.

## 1.    INTRODUCTION

Advantages of the operation of the scanning electron microscope (SEM) at low beam energies are as follows:
-an improvement in the contrast of small features,
-a reduction in charging effects on non conducting specimens,
-a reduction in the depth of radiation damage,
-a reduction in the width of the edge effect in the secondary electron (SE) mode.
There are also some disadvantages when we reduce an energy of the beam:
-a decrease of the dependence of the SE signal on the surface tilt angle (a reduction in the magnitude of the tilt contrast),
-an increase of the sensitivity of the SE signal to the contamination of the specimen,
-an increase of the beam diameter and a decrease of the beam current, especially when thermionic electron guns are used.
Two first disadvantages are inherent in the use of the SE signal and can be eliminated in some extent by use of the backscattered electron (BSE) signal (Hejna 1995). The last disadvantage can be overcome by the use of brighter electron guns and in some extent by an operation of the gun at high voltages and deceleration of the beam in the column or below the objective lens (Yau et al. 1981, Zach and Rose 1988, Zach 1990, Beck et al. 1995).
The present paper deals with a study of BSE detection in the configuration with an retarding electrostatic lens below the objective lens and a field-free space over the specimen.

## 2. THEORETICAL ANALYSIS

Preliminary estimations of the influence of different parameters on performances of immersion lenses were done by analytical formulas derived by Hordon et al. (1995). Formulas are valid for two-electrode immersion lenses with flat, thin electrodes and small apertures. The analysis enabled to estimate dependence of an image position and a chromatic aberration coefficient on the distance between electrodes, on the immersion ratio and on the angle of convergence of the beam provided by the focusing magnetic lens. Detailed calculations were performed by numerical modelling. The ELD 3.60 (Electrostatic Lens Design) program from the Delft Particle Optics Foundation was used. It is the finite element method (FEM) program developed at the University of Delft by Lencova and Wisselink (1990). The program was used for an evaluation of distribution of the axial potential and the axial field, and for computations of the chromatic and spherical aberration coefficients of lenses.

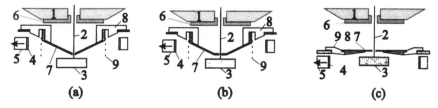

Fig. 1. Arrangements of detectors. In all configurations: 1- microscope objective lens, 2- primary beam, 3- specimen, 4- ring scintillator, 5- light-guide, 6- upper electrode. In (a) and (b): 7- bottom electrode, 8- isolator, 9- grid. In (c): 7- aperture, 8- aperture holder, 9- isolator.

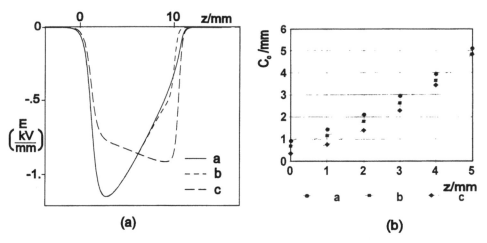

Fig. 2. (a) Distribution of an axial electric field in configurations a, b and c from Fig. 1, (z = 0 at the bottom surface of the microscope lens) (b) Dependence of an axial chromatic aberration coefficient on the distance below the bottom electrode of the retarding lens for three configurations of electrodes.

Fig. 1 shows configurations of detectors. Three shapes of bottom electrodes were studied: conical, tapered conical and flat. The diameter of the aperture in the bottom electrode is 1 mm in all configurations, the electrode thickness is 0.5 mm in configurations 1a and 1b, in the configuration 1c a microscope aperture of 0.25 mm thickness is used. BSE move in the field free space in their original directions and after passing the grid (configurations 1a and 1b) are attracted to the ring scintillator held at a ground potential, in the configuration 1c BSE strike the aperture and the aperture holder and create SE which are detected by the ring scintillator. Fig. 2 shows the distribution of an axial electric field and the dependence of an axial chromatic aberration coefficient on the distance below the bottom electrode. Computations were done for the immersion ratio $k = E_0 / E_1 = 10$ (beam energy $E_0$ = 10 keV and landing energy $E_1 = 1$ keV).

## 3.  EXPERIMENTAL RESULTS

Three configurations of BSE detectors were implemented in the Hitachi S-4000 field-emission SEM. Figure 3 shows images of the microscope grid #400, taken with detectors from Fig. 1. Distortions of the grid are not seen in Fig. 3a and 3b and the whole field of view is in focus. In Fig. 3c grid bars are distorted and only a part of the image is in focus, additionally there is a large change of magnification in this case. The electric field in the configuration 1c has a maximum at the bottom electrode and the aperture disturbs the distribution of the field more than in configurations 1a and 1b.

Fig. 3. Images of a microscope grid taken with three configurations of detectors from Fig. 1. The separation of grid bars is 63.5 μm. $E_0 = 10$ keV (a and b) and 11 keV (c), $E_1 = 2$ keV.

Fig. 4. Images of pyramids etched in a silicone wafer taken with three detector configurations from Fig. 1. $E_0 = 11$ keV, $E_1 = 2$ keV.

At low beam energies the interaction volume of the primary beam is small and a topographic tilt contrast predominates. The tilt contrast is higher in the case of detection of low take-off BSE than SE. Fig. 4 shows images of pyramids etched in a silicone wafer taken with three configurations of BSE detectors from Fig 1. Relatively good topographic tilt contrast is seen in all BSE images. Week shadowing in the image 4a indicates that also BSE emitted at

72

higher take-off angles can reach the detector and a rough surface of pyramids in the image 4c suggests that there is an influence of SE in this case. Results shown in Figs. 3 and 4 suggest that configuration 1b can be recommended for practical use.

Images of a magnetic tape taken at higher magnification with and without beam retardation are compared in Fig. 5. They were recorded with the configuration 1b. A worsening of the image with decreasing of the electron energy is very small.

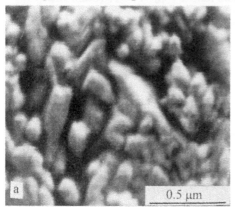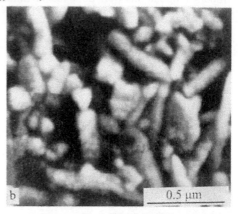

Fig. 5. Micrographs of a gold-coated magnetic tape. (a) $E_0 = 11$ keV, $E_1 = 2$ keV. (b) $E_0 = E_1 = 11$ keV.

## 4. CONCLUSIONS

The research shows that detection of low take-off BSE can be conducted in a configuration with retardation of the primary beam. The results are similar to those obtained with acceleration of BSE generated by a low-energy beam (Hejna 1995). The solution described in the paper can be useful in microscopes equipped with thermionic guns, where a worsening of electron-optical parameters at low energies is more pronounced than in the case of microscopes with the field-emission guns.

### ACKNOWLEDGEMENTS

This work was supported by the KBN (Committee of Scientific Research) research project No. 8T10C01909. Experiments were conducted at the University of Münster (Germany), the author is indebted to Prof. H. Kohl for providing the laboratory facilities.

### REFERENCES

Beck S, Plies E and Schiebel B 1995 Nucl. Instr. Meth. Phys. Res. **A363**, 31
Hejna J 1995 Scanning **17**, 387
Hordon L S, Boyer B B and Pease R F W 1995 J. Vac. Sci Technol. **B13**, 826
Lencova B and Wisselink 1990 Nucl. Instr. Meth. Phys. Res. **A298**, 56
Zach J 1989 Optik **83**, 30
Zach J and Rose H 1988 Proc. EUREM 88, Vol. 1 (Bristol: Inst. Phys.) pp 81-82
Yau YW, Iranmanesh A, Polasko K and Pease R F W 1981 J. Vac. Sci Technol. **19**, 1048

*Inst. Phys. Conf. Ser. No 153: Section 2*
*Paper presented at Electron Microscopy and Analysis Group Conf. EMAG97, Cambridge, 1997*
© 1997 IOP Publishing Ltd

# Design of an objective lens with a minimum chromatic aberration coefficient

**K Tsuno and D A Jefferson***

JEOL Ltd., 1-2, Musashino 3-chome, Akishima, Tokyo 196, Japan
*Department of Chemistry, The University of Cambridge, Lensfield Road, Cambridge, CB2 1EW

**ABSTRACT:** Limit of the axial chromatic aberration coefficient Cc at 200 kV is numerically calculated. The obtained minimum is Cc=0.63 mm and the corresponding spherical aberration coefficient Cs=0.39 mm for the objective lens with the gap length 1.2 mm and the bore diameter 0.4 mm. The estimated resolution of the information limit is 0.07 nm under the high voltage stability $1 \times 10^{-7}$ and the field emission gun. We consider a Wien filter monochrometer inside the gun to further improve the resolution.

## 1. INTRODUCTION

The highest Scherzer resolution achieved is 0.1 nm with a 1250kV high resolution, high voltage electron microscope(HVEM) [Philip et al.1994]. Computer simulation indicates that the limit of Scherzer resolution is 0.09 nm in HVEM [Tsuno 1993]. It is not possible to obtain 0.1 nm Scherzer resolution with microscopes less than 1000 kV. Aberration correction is the only method left for improving the Scherzer resolution.

On the other hand, the resolution of the information limit is mainly determined by the chromatic aberration coefficient Cc and incoherent effects such as the finite size of the source, beam divergence, energy spread, instabilities of the high voltage and lens current, and the vibration of the microscope. Due to the introduction of the field emission gun (FEG), the first three items of the incoherent effects have been greatly reduced. The information limit of 200~300 kV microscopes, whose Scherzer resolution is around 0.16~0.19 nm, approaches 0.1 nm. In this investigation, optimization of Cc of the objective lens has been done to improve the information limit. The resolution of the information limit is derived as a function of Cc under various incident beam conditions. Finally, a possibility of producing a monochrometer operating at 200 kV with a retarding Wien filter is discussed.

## 2. OPTIMIZATION OF Cc

A schematic drawing of the tip of the pole-piece is shown in Fig. 1 together with symbols of important dimensional parameters. We consider that the following conditions are essential in using the practical specimen holder and constructing lenses with practical machining tools: (1) The specimen position is in a middle of the gap. (2) The gap length (2S) must be equal or larger than 1 mm. (3) The minimum thickness of the pole-piece material at the top face of the pole-piece (D-b) is 0.1 mm. (4) The minimum bore diameter 2b is 0.4 mm.

Fig. 2 shows the result of the optimization of Cc for various S. The specimen position is fixed at the center of the gap. Cc has a minimum of 0.63 mm at S=0.6 mm, but Cs monotonically decreases with decreasing S. Under the minimum Cc condition, Cs is 0.39 mm. In Fig. 3, Cc and corresponding Cs dependence on the top face radius D is shown. Cc has not so strong

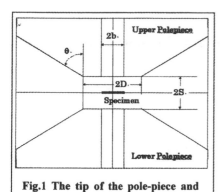

**Fig.1 The tip of the pole-piece and symbols of dimensions.**

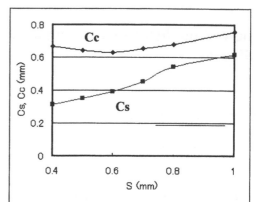

**Fig. 2. Half gap length S dependence of the minimum Cc and corresponding Cs.**

dependence on b, but largely depends on D. Because the smaller D give the smaller Cc, the smallest b and the smallest D within the above conditions ( b=0.2 and D=0.3 mm ) give the best result. There are no merit in using asymmetrical pole-piece, because the smallest b and D give the smallest Cc. In order to have asymmetrical dimension, one side pole-piece must have larger b and/or D than the other. Moreover, Cc does not so strongly dependent on the taper angle $\theta$ as shown in Fig. 4. Cc and Cs show nearly the constant values between $55< \theta <70°$

Fig. 5 shows Cc and Cs of the pole-piece giving the minimum Cc against the specimen position Zo. The coefficient Cc has a broad minimum at Zo=0~0.4 mm. On the other hand, Cs has a minimum at

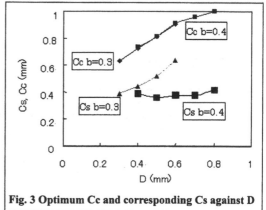

**Fig. 3 Optimum Cc and corresponding Cs against D**

Zo=-0.2 mm. It is concluded that the specimen at the center of the gap (Zo=0) is the best position in which both of Cs and Cc are small.

## 3   RESOLUTION OF THE INFORMATION LIMIT

The limit of the spatial resolution of an electron microscope is given by damping envelopes of the contrast transfer function. When a field emission gun (FEG) is used, the information limit is mainly determined by the chromatic envelope, because of its small source size. The chromatic damping factor Dc is expressed as [Mobus and Ruhle 1993]

$$Dc = (\pi\lambda\ Cc/2\ )[(\Delta V/V)^2 + (2\Delta I/I)^2 + (\Delta E/E)^2]^{1/2}.$$
(1)

Where, $\lambda$ is the wavelength of electrons. The former two contributions $\Delta V/V$ and $2\Delta I/I$ are the variation of the high voltage and lens currents and the third term $\Delta E/E$ is the energy spread of the electron source and the energy loss due to the

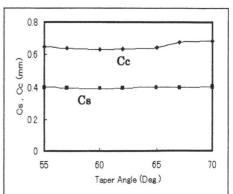

**Fig. 4   Taper angle $\theta$ dependence of Cc and corresponding Cs.**

specimen.

Figure 6 shows the relation between the resolution of the information limit d and Cc under various conditions. The resolution was estimated simply from the chromatic damping envelope.

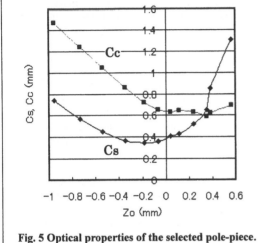

Fig. 5 Optical properties of the selected pole-piece.

The conditions of eq. (1) are listed in Table I. Here, the limit of the information, which is equal to the noise, is set at $E_\Delta = 1 / e^2 = 0.135$. If we use a $LaB_6$ cathode and assume $\Delta E/E = 5 \times 10^{-6}$, Cc=0.3 mm is necessary to get 0.1 nm resolution. On the other hand, if we use an FEG with the energy spread $\Delta E/E = 1.5 \times 10^{-6}$ (FEG1 in Fig.6), 0.1 nm resolution can be obtained for a wide range of Cc.

The curve FEG2 in Fig. 6 indicates d obtained by improving the stability of the high voltage and lens currents. A slight increase of the resolution is obtained. If we use a new objective lens pole-piece with Cc=0.6 mm, the resolution of 0.07 nm is achieved. The last curve indicates the resolution when the energy spread of 30 meV is achieved. The Wien filter monochrometer, which we have developed for the high resolution electron energy loss spectroscopy (EELS) instrument [Tanaka et al. 1992] reached 15 meV energy resolution at 60 kV. It will be more difficult to obtain the same level of energy resolution at 200 kV. If the energy spread of 30 meV is achieved, 0.03 nm resolution of the information limit is expected.

## 4. DESIGN OF A MONOCHROMETER

In order to improve the information limit, it is essentially necessary to reduce the energy spread of the source. One method is to use a new cathode material. Another method is to use a monochrometer. It has not been succeeded in constructing a monochrometer at 200 kV or higher accelerating voltages. Fig. 7 (a) shows a possible configuration of two monochrometers. The Wien filter monochrometer we developed previously was installed between condenser lenses similar to (a). The electrons are accelerated up to the

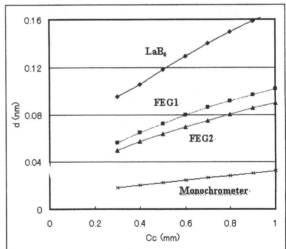

Fig. 6 Resolution d vs Cc for various incident beam conditions listed in Table I.

accelerating voltage Vo and forms a probe before the monochrometer. The beam is decelerated to 20~200 eV before the filter. After leaving the filter, the beam is accelerated again. It is possible to construct such the monochrometer if the accelerating voltage is lower than 100 kV. However, for microscopes above 100 kV, it is necessary to use accelerating tubes to avoid electric discharge. For generating 200 kV monochrometer, we need totally 3 accelerating tubes such as shown in Fig. 7(a). The total column length will be very high in such the system.

Fig. 7 (b) is a proposed model of the new monochrometer, in which the filter is installed between the first and the second anode. In this configuration, the monochrometer is a part of the electron gun. Because the extraction voltage of FEG is about 2~4 kV, it is necessary to decelerate electrons further to 20~200 eV. We need no extra accelerating tubes even if the final accelerating voltage is 200 kV. A problem is the position of the slit. Possible 2 slit positions A and B are shown in Fig. 7 (b). If the slit is set at the position A, the slit has to be floating on the high voltage and therefore we cannot touch of it. The slit must be controlled automatically and the power supply and control unit must also be floating on the high voltage. It is possible to construct such the high voltage electronics and mechanisms, but it is better to avoid it.

|  | A (LaB$_6$) | B (FEG1) | C (FEG2) | D (Mono) |
|---|---|---|---|---|
| $\Delta V/V$ | $1 \times 10^{-6}$ | $1 \times 10^{-6}$ | $1 \times 10^{-7}$ | $1 \times 10^{-7}$ |
| $\Delta I/I$ | $5 \times 10^{-7}$ | $5 \times 10^{-7}$ | $5 \times 10^{-8}$ | $5 \times 10^{-8}$ |
| $\Delta E/E$ | $5 \times 10^{-6}$ | $1.5 \times 10^{-6}$ | $1.5 \times 10^{-6}$ | $1.5 \times 10^{-7}$ |

**Table I High voltage stability $\Delta V/V$, lens current stability $\Delta I/I$ and the energy spread $\Delta E/E$ for various sources and instrumental conditions.**

In the slit position B, the slit is put on the earth potential and therefore can be controlled by hand. A problem which might be occurred is the asymmetry of the electron beam trajectory against the central plane of the filter. One of the important property of reducing second order aberrations of the filter is the symmetrical configuration of the electron trajectories. It is necessary to simulate how much aberrations occur and how much resolution can be attained in such the asymmetrical configuration.

## CONCLUSIONS

The pole-piece with Cc=0.63 mm and Cs=0.39 mm is now under construction in Cambridge together with a specimen holder and a goniometer. The expected resolution of the information limit is 0.07 nm ( FEG ) or 0.03 nm ( monochrometer with 30 meV energy spread) at 200 kV. After the monochrometer, the intensity of the beam is very weak. It is necessary to use a highly sensitive detector such as the imaging plate or a cooled CCD camera.

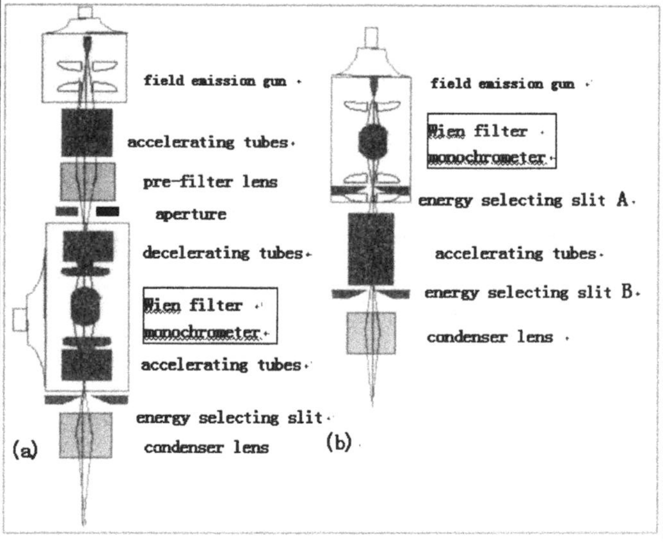

**Fig. 7 Comparison of two monochrometers. (a): A retarding Wien filter independent of gun. (b): A retarding Wien filter just after the first anode.**

## REFERENCES

Mobus G and Ruhle M 1993 Optik **93**, 108

Philip F, Hoschen R, Osaki M, Mobus G and Ruhle M 1994 Ultramicroscopy **56**, 1

TanakaM, Terauchi M, Kuzuo R, Tsuno K, Ohyama J and Harada Y, 1992, Proc. 50$^{th}$ Electron Microsc. Soc. America, ed. G W Bailey (San Francisco : San Francisco Press) pp940-41

Tsuno K 1993 Ultramicroscopy **50**, 245

*Inst. Phys. Conf. Ser. No 153: Section 2* ·
*Paper presented at Electron Microscopy and Analysis Group Conf. EMAG97, Cambridge, 1997*
© *1997 IOP Publishing Ltd*

# Design of a lens system to obtain standing electron wave illumination

**B M Mertens and P Kruit**

Delft University of Technology, Department of Applied Physics,
Lorentzweg 1, 2628CJ Delft, The Netherlands

**ABSTRACT:** In standing wave illumination, a specimen is illuminated by a fringe pattern created by the interference of two coherent plane waves. Objectives of this method are obtaining full elemental maps and full exit wave reconstruction. The measurement schemes can be realized by imaging a standing wave created by a beam splitter onto the specimen with a suitable optical system. Simulations of the first and third order optical properties show that the design of a three lens system for this purpose is possible.

## 1. INTRODUCTION

Standing wave illumination is the illumination of a specimen with an interference pattern (or fringe pattern) created by two coherent plane waves coming from different directions. This form of microscopy gives interesting opportunities in high resolution electron microscopy. It can be used to obtain full elemental maps beyond the fundamental limit in STEM at acceptable current levels (Mertens 1997); also, it is possible to do full exit wave reconstruction by phasing the diffracted beams (Buist and Kruit 1994), a method quite similar to ptychography (Hoppe 1969, Bates and Rodenburg 1989) but at much higher current levels. As an example, the method for obtaining full elemental maps is depicted in Fig. 1a.

The specimen is illuminated with a fringe pattern with a periodicity equal to the lattice spacing and the X-ray spectrum is collected as a function of the position of the fringe maxima. The result is a periodic function, Fig. 1b, also with a spatial frequency equal to the lattice spacing. From the offset and the phase of this periodic signal, the concentration and the relative position of both atomic lattice planes can be determined. The normalized amplitude gives information on the relative occupancy of a specific atom type on the respective lattice plane. The described experiment is limited, it is a typical ALCHEMI type of experiment (Spence and Tafto 1983); however, it can be extended for full elemental mapping by allowing sufficient flexibility in the fringe distance and orientation of the pattern on the sample. The reconstruction of the specimen function from the individual measurements involves only computer processing. In Fig. 1b, the method is compared to STEM in terms of current as a function of the resolution.

The above described measurement schemes, can be realized by imaging a standing wave created by a beam splitter onto the specimen with variable magnification and rotation. Therefore, a beam splitter as well as some optics between beam splitter and specimen needs to be incorporated in the illumination system of an electron microscope.

The design process starts with a set of optical demands for the system. The demands for the microscope are as follows:
- fringe distance d: between 0.1 nm and 2.0 nm;
- fringe orientation $\phi$: between $-\pi/2$ and $\pi/2$;
- fringe position x: a fraction of the fringe distance;
- field of view F: circular form, typical radius: 20 nm;

78

- Signal-to-Noise Ratio in the measurement: as high as possible;
- leave a possibility to operate the microscope in normal TEM mode for pilot experiments.

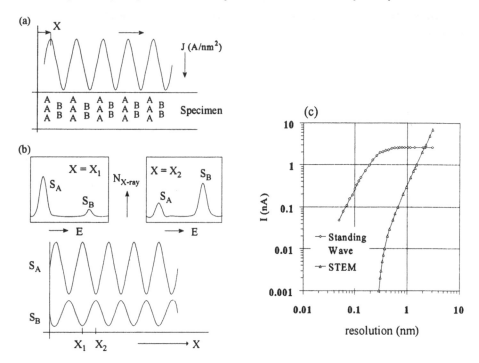

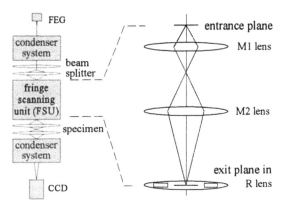

Fig. 1: The principle of analytical measurements with standing wave illumination, (a) and (b), and comparison between standing wave illumination and STEM © for a field of 100 fringes.

In the high resolution techniques mentioned above, the SNR is proportional to $v\sqrt{j}$ (Mertens 1997) where $v$ is the fringe contrast and $j$ the current density. There are two parameters which have influence on the SNR: the source brightness and the beam splitter. The source brightness influences the current density for a given illumination opening angle; the beam splitter determines the relation between the illumination opening angle and the fringe contrast. In order to optimize the SNR, a Field Emission Gun (FEG) will be used as an electron source combined with a crystal beam splitter (Marton 1952) instead of the more commonly used biprism (Möllenstedt and Düker 1956).

The consequence of the use of a crystal beam splitter is an extra objective lens, as fringe distances at the beam splitter level are small. Typical values for crystal lattice spacings are in the order of a few Angstroms. The extra objective lens with large magnification reduces the optical and mechanical demands on the optical system between the beam, splitter and the specimen, the Fringe Scanning Unit (FSU). To keep some flexibility in the beam splitter fringe distance to be used, the system will be designed for beam splitter fringe distances

Fig. 2: Schematic design of the microscope for standing wave illumination and design of the FSU.

between 0.2 nm and 0.4 nm. The results will be shown for a beam splitter fringe distance of 0.313 nm corresponding to the <111> reflections of Silicon. The schematic design of the complete microscope is depicted in Fig. 2. Also shown in Fig. 2, is the optical design of the FSU. It contains three lenses two of which are used to image the fringe pattern from the entrance plane of the FSU to the exit plane. The third lens, the rotation lens R, is located in the exit plane. Therefore, in first order it only rotates the image. Additionally, the use of such a lens simplifies the pilot experiment as for this purpose it can be used as the second condenser lens.

## 2.    FIRST ORDER PROPERTIES OF THE FRINGE SCANNING UNIT

The first order properties can be calculated by hand, but if thick lens effects are included, it is easier to use a software tool which has been developed especially for systems containing many lenses. This software tool is called POCAD, Particle Optics Computer Aided Design (Stam 1996). Using POCAD, suitable lens positions and geometries can be found. The excitations of the M lenses (for zero rotation lens excitation) as a function of the fringe distance at the specimen can be calculated, see Fig. 3a. As mentioned, the action of the rotation lens is to change $\phi$ without changing the magnification from entrance to exit plane. The action of this lens is only a rotation if the lens can be approximated by a thin lens. For rotations up to $\pi/2$, this approximation no longer holds. Therefore, in order to keep both the position of the exit plane and the magnification constant over the full range of orientations, the excitations of the M lenses need to be corrected. As an example, the influence of the rotation of the R lens on the excitations of the M lenses is depicted in Fig. 3b, calculated at a specimen fringe distance of 0.1 nm.

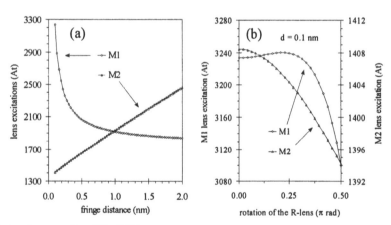

Fig. 3: First order properties of the FSU.

## 3.    THIRD ORDER PROPERTIES OF THE FRINGE SCANNING UNIT

The third order properties of the FSU are also calculated using POCAD. The relevant figures of merit in the evaluation are the global and local fringe contrast reduction and distortion. These effects would lead to systematic errors in the image reconstruction process as well as a reduction of the SNR. The allowable contrast variations and distortion values depend on the specimen, but typical values are estimated at a few percent contrast variation and a fringe shift of $\pi/4$. The calculation of the fringe contrast in the exit plane of the FSU is performed by putting a point source at the entrance plane of the FSU. This point source emits electrons off-axis, under an angle $\beta$ corresponding to the fringe distance. The illumination opening angle $\alpha$ is equal to the value which would lead to a contrast value of 0.8 if only the axial aberrations of the objectives would play a role. At optimum defocus, the image of the point source in the exit plane, a purely aberrated probe, has the typical form as depicted in Fig. 4a. The main aberrations are coma, proportional to $\beta\alpha^2$ and astigmatism, proportional to $\beta^2\alpha$. The fringe contrast can be calculated by convolving this aberration probe with the original fringe pattern. Fig. 4b shows the fringe contrast reduction as a function of the fringe distance at zero rotation lens excitation.

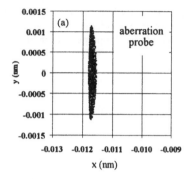 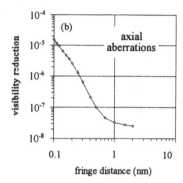

Fig. 4: Aberration probe at the specimen level (a) and global fringe contrast reduction (b)

The off-axial aberrations can be evaluated by shifting the position of the point source off-axis to a position h. Then, many aberrations play a role in the fringe contrast calculation. The distortion, proportional to $h^3$, can be evaluated directly with the correct transfer-coefficient from entrance to exit plane. The results are shown in Fig. 5 for a fringe distance of 0.1 nm.

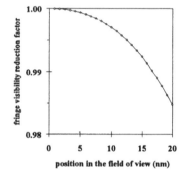 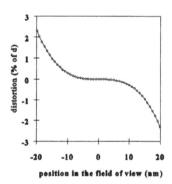

Fig. 5: Effect of off-axial aberrations of the FSU.

## 4.  CONCLUSIONS

Simulations have shown that a microscope for standing wave illumination with fringes as small as 0.1 nm is feasible. The effect of third order aberrations seems to be small enough.

## REFERENCES

Bates R H T and Rodenburg J M (1989) Ultramicroscopy **31** 303
Buist A H and Kruit P (1994), Proceedings ICEM 13, Paris, **1** 337
Hoppe W (1969), Acta Cryst. **A25** 495
Marton L (1952), Phys. Rev. **85** 1057
Mertens B M and Kruit P (1997), to be published in Scanning Microscopy International Supplement 11
Möllenstedt G and Düker H (1956), Z. Phys. **145** 377
Spence J C H and Taftø J (1983), J. Micr. **130** 147
Stam M A J van der , Barth J E and Kruit P (1996), Proceedings SPIE (editor: E Munro) **2858** 90

*Inst. Phys. Conf. Ser. No 153: Section 2*
*Paper presented at Electron Microscopy and Analysis Group Conf. EMAG97, Cambridge, 1997*
© *1997 IOP Publishing Ltd*

# Electrostatic in-line monochromator for Schottky Field Emission Gun

**H W Mook and P Kruit**

Delft university of Technology, Department of Applied Physics, Lorentzweg 1, 2628 CJ Delft, The Netherlands

**ABSTRACT:**   The spectral resolution in EELS is limited by the intrinsic energy spread of the electron source. The spatial resolution in scanning electron microscopy is limited by this spread via the chromatic aberration. The design of a Field Emission Gun monochromator with a resolution of 50 meV would be helpful. A new type of monochromator is introduced based on electrostatic deflectors. To minimize the brightness reduction of the electron source due to aberrations and Coulomb interactions, the beam potential, monochromator length and dispersion are balanced to get maximum current throughput.

## 1.    INTRODUCTION

The spectral resolution in electron energy loss spectroscopy  is limited by the intrinsic energy spread of electrons emitted from the source: for a cold field emission source at best 0.3 eV and for a Schottky emitter in practice 0.4 - 0.9 eV. For the analysis of chemical shifts it would be helpful to have a resolution of 0.05 - 0.1 eV. The spatial resolution in low voltage scanning electron microscopy is limited by chromatic aberration. By incorporating a monochromator the resolution of the SEM can be increased.

For a Schottky field emission gun, which is the most stabile high brightness electron source at the moment, a new monochromator design is proposed. Before this type of monochromator is discussed, more general considerations are given on the choice of the optical configuration. To reduce the negative influence of the monochromator on the source brightness, it is important to consider not only the dispersion and aberrations of the configuration, but also the Coulomb interactions, like trajectory displacement and Boersch effect. Whereas the dispersion and aberrations are reduced by lowering the beam potential in the monochromator, the Coulomb interactions are not and will for long monochromators and low beam potential have a negative effect on the brightness and the energy width of the filtered beam. An optimum can be found using a short monochromator operating on an optimized beam potential.

## 2.    OPTICAL CONFIGURATION

The composition of the electron optical elements, electron source, lenses, energy filter and selection slit, is called the optical configuration. The decision on the configuration must be taken baring two things in mind.

Firstly the required precision of the electrostatic and/or magnetic field of the monochromator is given by the ratio of the wanted energy resolution, dU, over the beam potential in the monochromator, $U_m$. This means the electrode fabrication precision and the voltage and/or current stabilities must be of the  order $dU/U_m$. To reduce these demands the beam potential must be kept low: below 5 kV for a reasonable ratio of $10^{-5}$. To cancel an additional retardation step in the TEM, the monochromator must be positioned right after the source where the beam is not yet accelerated to 200-400 kV.

Secondly the configuration must be designed and operated to have as little as possible negative influence on the most important beam characteristics, brightness and beam current. Three effects influence the brightness of the source and the maximum current throughput. The dispersion of the monochromator, the aberrations of the monochromator and lenses and the Coulomb interactions in the total configuration. Influence of the energy dispersion on the brightness is unavoidable for an energy filter. Because of the energy dispersion,

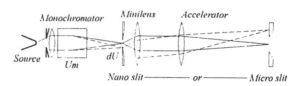

Fig. 1: Optical configuration

electrons with different energies are imaged over a large area. The dispersion is an increase of the effective source size and thus a reduction of the brightness. The minimum brightness reduction can be found when the required dispersion (for us 50 meV) is equal to the size of the source image. This is called the 'nice spot' condition. The aberrations limit acceptable aperture angle and thus the current throughput. The Coulomb interactions reduce the brightness directly via the trajectory displacement which also increases the source image. The Boersch effect (increase of energy width due to Coulomb interactions) also reduces the brightness when it occurs before the monochromator, but reduces the energy resolution when it occurs after the monochromator.

Using an decelerating gun lens, the monochromator potential, $U_m$, must be suited to make the dispersion fit the 'nice spot' condition (Fig. 1). For a Schottky field emitter the dispersion needed is very small. An aperture must be used to limit the beam current before the low potential section where the Coulomb interactions dominate. The energy selection can be done by a slit after the monochromator or after the accelerator. A Nano slit after the monochromator must be small since the source image is not yet magnified. One advantage of this configuration is that the low potential beam can be stopped by a very thin slit. By magnifying the source image (and dispersion) with an electrostatic minilens (diameter ~ 1 mm) through the accelerator to a Micro slit on ground potential, more reasonable slit sizes can be used.

## 3. RETARDING DEFLECTOR MONOCHROMATOR

The small dispersion needed to monochromatize a Schottky field emission source, makes it in principle possible to use an electrostatic deflector as energy dispersive element, but the dispersion of the first deflector will be compensated by a second when the beam is brought back on the axis. This means the slit or the source must be positioned off-axis, which makes it difficult to construct an in-line monochromator. By accelerating or decelerating the beam between the two deflections the large dispersion of the deflector on low beam potential will not be compensated by the deflection on high potential. Fig. 2 shows the suggested design. Somewhat similar is the idea to obtain a deflection away from the axis without having dispersion as suggested by Baranova and Yavor (1996).

The deflection angle, $\beta$, is fixed by the construction, because the lenses must be aligned to the beam. The potential, $U_m$, must be chosen to satisfy the 'nice spot' condition. A gun lens, $U_g$, is used to focus the beam on the selection slit. The aperture restricts the current to reduce the Coulomb interactions inside the low potential segment. By setting $U_m$ to the extraction voltage, $U_x$, the monochromator can be turned off.

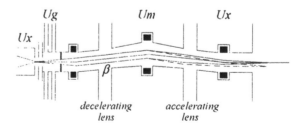

Fig. 2: Design of Retarding Deflector monochromator

## 4. OPTICS AND PERFORMANCE

### 4.1 First order optics

The source characteristics of a typical Schottky field emitter are chosen from Tuggle and Swanson (1985):

- source size: $s = 30$ nm
- extraction voltage: $U_x = 5$ kV
- angular current density: $I_\omega = 0.2$ mA/sr

Which indicates a brightness of $5.10^7$ A/sr/m²/V.

To limit the 'arm', L, needed, the retarding potential, $U_m$, is made as small as possible. For retardation larger than 1/5, the front focal point (at high potential of the decelerating lens) will, however, be too close to the lens field. Using the calculations of Harting and Read (1976) for a cylinder lens with D/d =1 (Fig. 3), the focal distance, f, and the coefficient for spherical aberration, $C_s$, can be expressed in the electrode dimension, d : $f = 5.5 \times d$ and $C_{so} = 91 \times d$. The dispersion as projected back to the source plane is given by:

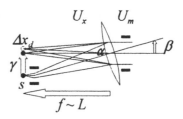

Fig. 3: Dispersion of virtual image

$$\Delta x_d = 2 \frac{\beta L}{U_x} \left( 1 - \sqrt{\frac{U_x}{U_m}} \right) dU$$

The deflection angle. $\beta$. is taken as large as possible, $\beta = 0.1$. Increasing the deflection angle beyond this point will change the initial simple deflector optics and thus the coefficient of second order aberration. To find the size of the configuration and the deflector arm, L, which is equal to the focal distance, f, the 'nice spot' condition must be satisfied: $\Delta x_d = s$. This shows the diameter of the lens, d, must be 2.2 mm and the distance to the source. L, 12 mm. To be able to position a deflector in the field free region between the deceleration and the acceleration lens, the low potential segment length, $L_m$, must be about 5.d, which is 11 mm.

## 4.2    Main aberrations

The spherical aberration of the deceleration (and acceleration) lens, $C_{so}$, was found to be 0.20 m. The diameter of the disk of least confusion, $\frac{1}{2} C_{so} \alpha^3$, must be kept smaller than the source size, giving an acceptance angle of 6.7 mrad. which allows a current of almost 30 nA. The contribution of the spherical aberration reduces slowly as the retarding potential is increased. This can be understood from the fact that the spherical aberration coefficient for higher retardation potentials decreases faster than the increase in needed object distance to satisfy the dispersion ( $I \times U_m^{1/3}$). For this configuration the second order aberration of the main deflector is found to be within limits.

## 4.3    Coulomb interactions

Two contributions of the Coulomb interactions can reduce the performance of the monochromator. The trajectory displacement. $\Delta x_t$. can reduce the brightness of the source by increasing the image size, and the Boersch effect, $\Delta U_B$, can increase the energy width above the wanted resolution (50 meV). From the current density and the beam radius and potential it can be deduced the current must at least be restricted to operate the monochromator in the Pencil beam regime using the equations from Kruit and Jansen (1997). From the formulae for the trajectory displacement and the Boersch effect the maximum allowable current can be deduced.

$$\Delta x_t = 2.6 \times 10^{30} \frac{L_m L^2 I^3 \alpha}{U^{\frac{5}{2}}}$$

$$\Delta U_B = 2.6 \times 10^{18} \frac{I^2 L_m}{U_m}$$

giving a restriction of 26 nA caused by the trajectory displacement and 42 nA for the Boersch effect. Although the Coulomb interaction limited current throughput can be increased by increasing the retarding potential or reducing the length of the low potential segment, the dependence is very small ( $I \propto (U_m/L_m)^{2/7}$ ).

## 5. DISCUSSION ON OPTIMIZATION

For the configuration discussed here the trajectory displacement restricts the current throughput. Increasing the retarding potential to reduce the influence of the trajectory displacement will only result in a slight increase of allow current throughput, because the influence of the second order aberration will increase faster with retarding potential. By changing the electrode design of the deceleration lens the increase of the spherical aberration can be reduced. Also the focal length of the lens will be increased by this improvement of the electrode shape keeping the focal point outside the lens field. This optimization will only go as far as Coulomb interactions allow. The optimum configuration limit by the trajectory displacement will have ,dependent on the source characteristics, about 25 nA maximum current throughput.

## 6. CONCLUSION

The fully electrostatic retarding deflector monochromator can be used a monochromator for a Schottky field emission source to yield energy resolutions down to 50 meV, with a current throughput of about 20 nA without unnecessary loss of brightness. This type of monochromator can be used to increase the EELS energy resolution in transmission microscopy and the spatial resolution in the SEM. The in-line design and the length of about 5 cm allows for installation of this monochromator right after the gun and before the accelerator of the microscope. The fact that the filter is purely electrostatic keeps the radius of the construction small (no coils needed) and makes fast operation possible. The reduction of brightness due to trajectory displacement must be limited by installing a fixed aperture in front of the monochromator, which reduces the current to about 20 nA.

## REFERENCES

Baranova and Yavor 1996 Achromatic deflection system - SPIE 1996
Harting and Read 1976 Electrostatic Lenses - Elsevier Scientific Publication Company
Kruit and Jansen 1997 Space Charge and Statistical Coulomb Effects - Particle Optics Handbook - Edt. Jon Orloff
Tuggle and Swanson 1985 Emission Characteristics of the ZrO/W Thermal Field Electron Source - J.Vac.Sci.Technol.B 3 (1) pp.220-223

*Inst. Phys. Conf. Ser. No 153: Section 2*
*Paper presented at Electron Microscopy and Analysis Group Conf. EMAG97, Cambridge, 1997*
© *1997 IOP Publishing Ltd*

# The need for care in providing magnetic material data in saturated FEM calculations

J P Davey

LEO Electron Microscopy Ltd, Clifton Road, Cambridge, CB1 3QH

ABSTRACT: A number of versions of Finite Element programs are in use for the calculation of electric and magnetic fields in electron lenses. In the course of time, various problem cases have been reported, usually associated with unusual features in the geometry under consideration. It has been discovered that an algorithm used by Munro for non-linear (saturated) magnetic lenses can break down for all geometries, if supplied with unsuitable B-H data.

## 1.    INTRODUCTION

Many of the programs in use for the calculation of magnetic fields in rotationally symmetric electron lenses derive from the work of Munro, whose 1975 publication contained three such programs, M11 for calculating scalar potentials in the pole-piece region, M12 for calculating vector potentials in the linear (unsaturated) case, and M13 for computing vector potentials in the non-linear (saturated) case. With increases in computer capability, the use of scalar potential calculation has diminished, but developments of the other programs have continued, both by the original author and others (Edgcombe and Roberts (1989), Hill (1979), Taylor (1990)).

There have been a number of cases reported in which the original algorithms have been shown to give significantly inaccurate results (Lencová and Lenc (1992), Tahir and Mulvey (1990), Tsuno and Honda (1983)), and various modifications to the algorithms have been proposed. The problems described so far, however, concern lens geometries which are in some respect unusual, and mostly affect both linear and non-linear computations. What is reported here relates solely to the non-linear case, but appears capable of affecting most lens geometries, and can make solution completely impossible, rather than inaccurate.

## 2.    ALGORITHM FOR NON-LINEAR FIELD SOLUTION

In the programs under discussion, magnetic materials are represented by the normal magnetisation curve (i.e. hysteresis is ignored). For each material, the data provided are a set of values of the flux density B as a function of the magnetic field H. Implicitly, the data include the point ( $B = 0$, $H = 0$ ), and are extended after the last data point supplied on the assumption of unity incremental relative permeability ( $dB/dH = \mu_0$ ). Within the data range, the curve is represented by a piecewise linear approximation.

86

On this basis, field solution involves a process very similar to the linear case, provided that, for each element within the magnetic materials, it is known which linear section of the B-H curve should be used. Initially, this information is unavailable. The algorithm as described by Munro (1973) performs an initial solution in which each element is set to the start of the B-H curve. The resultant solution is then used to assign the appropriate linear section to each element, and a set of residuals is calculated, which is used to calculate a new solution. The process is repeated until convergence is attained. Munro did not discuss the question of whether convergence would necessarily occur, though clearly it did for the examples which he presented.

## 3. DIFFICULTIES ENCOUNTERED WITH CONVERGENCE

In analysing a lens using a nickel-iron alloy for the magnetic circuit, this author found discrepancies between the measured and calculated fields at high excitations. In an attempt to increase the accuracy of calculation, the manufacturer of the alloy was asked for B-H data in the saturation region. When the extra data points were added, the program ceased to give any useful results, no longer converging to a solution. At first it was thought that the problem might be related to the specific lens geometry, or the particular version of the field solver (Taylor 1990), but both of these have been excluded.

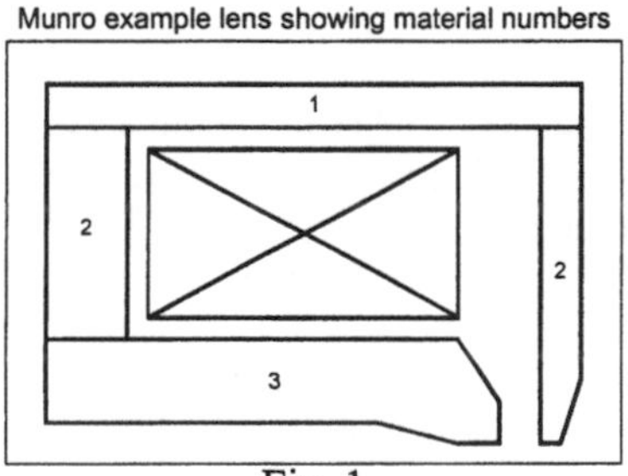

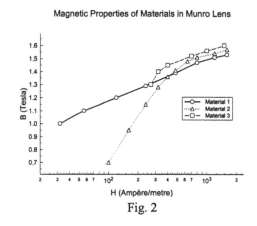

Fig. 1

Fig. 2

For the examples here, the lens geometry (Fig. 1) was taken from Fig. 18a of Munro (1975), and the field solver was a slightly modified version of program M13 listed therein, in which the iterative process has been forced to continue for a fixed number of steps. The normal convergence for the material properties (Fig. 2) used by Munro is shown in curve A of Fig. 3; curve B shows the effect of using material 3 for all components.

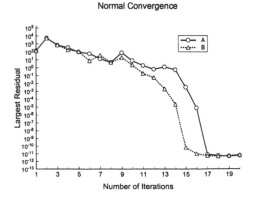

Fig. 3

The problem-causing B-H data are shown in Fig. 4, and their effect upon convergence, when used for all parts of the example lens, in curve A of Fig. 5. If the last data point is omitted, the amplitude of oscillation reduces, as shown in curve B, and if the last two data points are removed, convergence becomes normal, as in curve C.

This is, however, more of a demonstration of the weakness of the algorithm in use, than of the impossibility of finding a solution. If the program is modified to use under-relaxation, so that, at each iteration, the new solution changes by only half as much as at first appears necessary, the same data which were used for curve B now give curve D. This much slower convergence requires 53 iterations in order to reach the final residual level of curve C. The more extreme case of curve A still oscillates at this level of under-relaxation, but will converge if the change per iteration is reduced to an eighth of the calculated value.

Clearly, there will be large penalties in computational time for such changes to the algorithm, and preferably the data should be corrected.

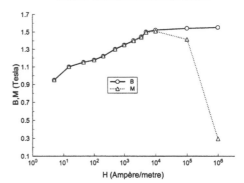

Fig. 4

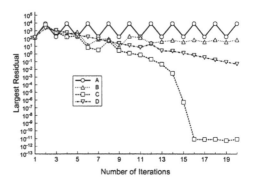

Fig. 5

## 4. DISCUSSION

The anomalous nature of the last two data points can be seen by looking at the curve for the magnetisation M ($M = B - \mu_0 H$) in Fig. 4. Clearly, since $dM/dH < 0$, they correspond to a relative incremental permeability of less than unity, something which is quite unlikely to occur in a ferromagnetic material. Unfortunately, data supplied for such materials frequently contain such erroneous points (Pissanetsky 1986), and this can lead to serious errors in computation.

Complete program packages usually contain some data files on ferromagnetic materials, and their authors will have verified that these cause no functional problems. However, when the user wishes to consider the use of some novel material, not already represented in their files, it is recommended that the data should be thoroughly checked, to verify that $dM/dH \geq 0$.

The author is grateful to Dr. J. Simkin of Vector Fields for drawing his attention to the work of Pissanetsky.

## REFERENCES

Edgcombe C J and Roberts D E 1989 Inst. Phys. Conf. Ser. **99** pp 181-4

Hill R 1979 Inst. Phys. Conf. Ser. **52** pp 49-52

Lencová B and Lenc M 1992 IEEE Trans. Mag. **28** pp 1107-10

Munro E 1973 Image Processing and Computer-aided Design in Electron Optics, ed P W Hawkes (London: Academic Press) pp 284-323

Munro E 1975 A Set of Computer Programs for Calculating the Properties of Electron Lenses (Cambridge: Cambridge University Engineering Department)

Pissanetsky S 1986 COMPEL **5** pp 41-56

Tahir K and Mulvey T 1990 T Nucl. Inst. Meth. **A298** pp 389-95

Taylor J H 1990 Ph D Thesis (Cambridge: Cambridge University)

Tsuno K and Honda T 1983 Optik **64** pp 367-78

*Inst. Phys. Conf. Ser. No 153: Section 2*
*Paper presented at Electron Microscopy and Analysis Group Conf. EMAG97, Cambridge, 1997*
© *1997 IOP Publishing Ltd*

# Use of Laplacian mesh for electron-optical calculations

**C J Edgcombe**

Department of Physics, University of Cambridge   and
Granta Electronics Ltd, 25, St. Peter's Road, Coton, Cambridge

**ABSTRACT:** Use of a Laplacian mesh permits electric field to be found by interpolation which is physically based and hence more accurate than standard polynomial interpolation. This type of mesh also permits smooth change of spacing between widely differing electrode sizes.

## 1. INTRODUCTION

In calculating the trajectories of electrons, values of the field components must be estimated at each step rapidly and accurately. We consider here only calculation of the gradient of a scalar potential. It is often convenient to define a geometric mesh between electrodes, then to find potential values at the mesh-points and field components by interpolation. The characteristic dimensions of electrodes in field-emission guns range from millimetres down to nanometres. For acceptable accuracy of the field estimates, it is desirable that the mesh spacing change smoothly between maximum and minimum feature sizes, but this can be tedious for the user to specify in detail. Also, the polynomial representations typically used in finite-element methods do not approximate well the behaviour of potential and field near a small spherical tip. It is desirable to use an interpolation method which corresponds to the physical behaviour, both to maintain accuracy of the trajectories over long paths and to ensure the correct overall change of kinetic energy.

## 2. THE LAPLACIAN MESH

When finite-element meshes are constructed for electrostatic systems, some mesh lines must coincide with the outlines of conductors, so as to define the electrode surfaces. This principle can be extended to require that every mesh-line (of one series) between electrode surfaces coincides with an equipotential. Then the obvious choice for a second series to intersect the first and so form mesh elements is a set of lines with the local direction of the gradient of potential (and electric flux). The choice of these two families of curves gives some valuable benefits:
- the required change of scale between electrodes is achieved easily, by choice of equipotentials;
- mesh elements (in longitudinal section) consist of curvilinear rectangles;
- the gradient of potential is everywhere normal to one set of mesh surfaces, and parallel to the other;
- the local direction of electric field is known from the quadrilateral co-ordinates;
- interpolation for field components can be carried out both consistently and accurately.

A mesh defined in this way is known as a 'Laplacian' mesh, as described by Binns et al (1992). An example for electrode shapes similar to those in a field-emitting gun is shown in Fig. 1. The mesh spacing changes smoothly, down to values similar to the tip radius.

## 3. INTERPOLATION

The Laplacian mesh can provide accurate interpolation for gradient. Where the field satisfies div $\mathbf{D} = 0$ and the local radius of curvature as seen in longitudinal section is $\rho$ and the local radius from the axis of symmetry is $r$, then by flux conservation $2\pi r\rho E_\rho$ is constant, so

$$dV = k\, d\rho / \rho\, r$$

from which

$$(V(\rho, r) - V_1)/(V_2 - V_1) = (\ln \rho/\rho_1 - \ln r/r_1)/(\ln \rho_2/\rho_1 - \ln r_2/r_1) \qquad (1)$$

For the special case $\rho_2/\rho_1 = r_2/r_1$, when the centre of curvature is on the axis, then

$$(V(\rho, r) - V_1)/(V_2 - V_1) = r_2(r - r_1)/(r_2 - r_1)r \qquad (2)$$

These relations provide accurate interpolation for $V$ and $E$. However, they require the local radius of curvature $\rho$ to be known. Where the mesh is Laplacian, $\rho$ can be found from the co-ordinates of corners of mesh elements, either accurately or by an approximation such as $\rho \approx ((1 - x)c + xd)/(d - c)$, where $x$ is fractional distance between equipotentials, and $c$ and $d$ are the lengths of sides of mesh elements on equipotentials 1 and 2.

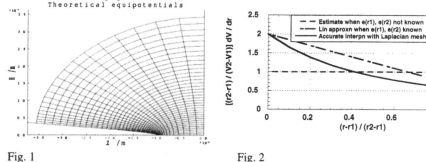

Fig. 1                                    Fig. 2

## 4. THE MESH IS THE POTENTIAL SOLUTION

Since the Laplacian mesh is a specification of the potential distribution, it cannot be defined before the potential solution is known. To make use of it, an independent method of finding the potential distribution is needed.

One way of generating a Laplacian mesh is by conformal mapping between a rectangle in co-ordinates $(u,v)$ and the given geometry in co-ordinates $(r,z)$. (Use of Laplace's equation in Cartesian co-ordinates produces closer spacing near the axis than the corresponding equation in cylindrical co-ordinates.) The equations

$$\partial^2 r / \partial u^2 + \partial^2 r / \partial v^2 = 0, \qquad \partial^2 z / \partial u^2 + \partial^2 z / \partial v^2 = 0$$

can be solved by minimising, subject to the given boundary conditions,

$$U = \int [(\partial r / \partial u)^2 + (\partial r / \partial v)^2 + (\partial z / \partial u)^2 + (\partial z / \partial v)^2]\, d(\text{volume})$$

and solution for $r$ and $z$ may provide suitable meshes for some geometries. Where the range of radii of curvature in the given geometry is large, as for field-emitting tips and anodes, this method can produce a mesh which folds over undesirably near the tip.

Another way of defining a mesh is to use the boundary element method to determine suitable equipotentials. This is currently being investigated and should enable a suitable mesh to be obtained by specification only of co-ordinates on the electrode surfaces.

## 5. CONCLUSIONS

For finite-element calculations with field emitters, use of a Laplacian mesh offers (1) greater accuracy than is possible with the usual polynomial interpolations, and (2) smooth, controllable change of scale between electrodes of different dimensions. A separate calculation of potential distribution is needed.

## REFERENCES

Binns K J, Lawrenson P L & Trowbridge C W 1992 The analytical and numerical solution of electric and magnetic fields (Wiley).

Inst. Phys. Conf. Ser. No 153: Section 2
Paper presented at Electron Microscopy and Analysis Group Conf. EMAG97, Cambridge, 1997
© 1997 IOP Publishing Ltd

# Comparison of Finite Difference, Finite Element and Boundary Element Methods for electrostatic charged particle optics

**D Cubric, B Lencova[†] and F H Read**

Department of Physics and Astronomy, Schuster Laboratory,
Manchester University, Manchester M13 9PL, UK
[†]Academy of Sciences of the Czech Republic, Institute of Scientific Instruments,
Brno, Czech Republic; Present address:Particle Optics Group, Department of Applied
Physics, University of Technology, Delft, The Netherlands.

**ABSTRACT:** The speed and accuracy of three computational methods used for solving problems in electrostatic charged particle optics have been compared. The methods considered are the Finite Difference, Finite Element and Boundary Element Methods, as applied to two dimensional systems. 'Benchmark' problems, all of which have accurately known solutions, are used for these comparisons

## 1. INTRODUCTION.

Many computer programs have been developed for solving problems in electrostatic charged particle optics, and almost all of them are based on either the Finite Difference Method (FDM), Finite Element Method (FEM) or Boundary Element Method (BEM). As far as we know there are no published comparisons of the relative speed and accuracy of these three methods. We have now carried out an initial and limited set of comparisons, using commmercially or publicly available programs. In each comparison the programs are applied to 'benchmark' tests that have accurately known solutions

## 2. THE PROGRAMS

We have used three programs each of which is familiar to one of the authors. As a representative FDM program, based on a square array of mesh points, we have used the SIMION program[†]. To represent the FEM method we have used the ELENS program[††], which uses an array of mesh points that has a general quadrilateral spacing and graded mesh (Lencova 1995). Finally for the BEM method we have used the CPO-2D program[†††].

## 3. THE BENCHMARK TESTS

The first benchmark test is the ideal spherical deflection analyzer, consisting of two concentric spheres that in the present instance have radii of 0.75 and $1.25cm$ and potentials of 5/3 and $3/5V$ respectively. The median trajectory for an electron of energy $1eV$ lies in the middle of the gap, at a radius of $1cm$, where the field is $2V/cm$. The first part of the test consists of computing the magnitude of the electric field strength $E(\theta)$ in the middle of the gap, at $n$ values $\theta_i$ of the relevant angular coordinate, and then deducing the relative errors in the mean and root-mean-square values:

$$\bar{\varepsilon} = \frac{1}{2}\bar{E} - 1, \qquad \varepsilon_{rms} = \frac{1}{2}E_{rms} \qquad (1)$$

In the second part of the test 5 trajectories are integrated from an initial position at $r = 1cm$, $\theta = 0$ to an exit plane at $\theta = 180°$. The first of these is the median trajectory, the next two measure the angular aberration and have launch angles $\alpha$ of +/- $0.1rad$ in the plane of motion, while the last two measure the energy dispersion and have initial energies $E_i$ that differ by +/- 5% from that of the median trajectory. For the ideal analyser the change in $r_f$ for the second and third trajectories should be 0.019737 while the difference in $r_f$ for the last two should be 0.200501. The relative errors in the values obtained with the programs are then used as a measure of the accuracy.

The second benchmark test concerns the field in the vicinity of a circular hole in an infinite flat sheet that divides two regions of different field strength. For simpicity a single parameter, the potential at the centre of the hole, is used to represent the accuracy. For a hole of radius 1 and with field strengths of 0 and 1 on the two sides of the plate this potential should be $1/\pi$ (Hawkes and Kasper 1989). The relative error in this potential is presented.

The third and final benchmark test used in the present study is that of the double cylinder lens with zero wall thickness, a gap to diameter ratio of 0.1 and with applied voltages of 1 and 10 (so that an electron increases in energy by a factor of 10 in passing through the lens). In the first part of the test the accuracy of the field is represented by the field at the centre of the lens, the relative error of which is presented. The exact value of this field is not known analytically, but it can be obtained with high accuracy by an extrapolation technique (Read 1997). Using the CPO-2D program for this, we find that the value is 11.8250(1). In the second part of the test 4 trajectories that start on the axis at $z = -4$ (where all distances are measures in units of the lens diameter) are integrated to an exit plane at $z = 4$, where the extrapolated positions $z_f$ at which the axis is crossed are evaluated. The first two measure the spherical aberration of the lens and have initial angles of 0.025 and 0.05 $rad$, while the last two measure the chromatic aberration and have an initial angle of 0.05 $rad$ and initial energies of 0.95 and 1.05 $eV$. The accurate values of $z_f$ for the 4 trajectories are found to be 1.99188(2), 1.88511(2), 1.89824(1) and 1.86988(1). The mean errors of the first and second pair are presented.

## 4. THE COMPARISONS

Figures 1, 2 and 3 show the results of the comparisons for the three benchmark tests. In each comparison the relationship is shown between the achieved error $\varepsilon$, always expressed as a percentage, and the computing time needed to achieve this level of error. The times are normalised to the computing speed of a 100 MHz Pentium PC. In the FDM and FEM programs the level of error and the computing time are determined by the choice of the number and spacing of the mesh points and the inaccuracy parameter for the iterative solution of the mesh potential equation. In the BEM programs these levels depend on the number and distribution of the boundary segments, as well as on the inaccuracy parameters for the evaluation of the potentials and fields and the integration of the trajectories.

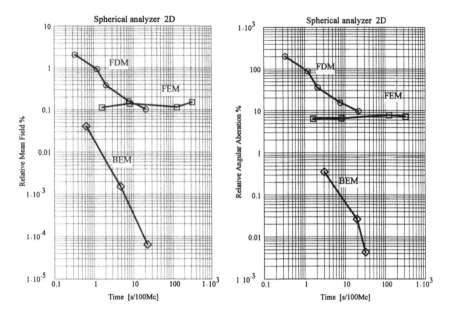

**Figure 1** Percentage error in the computed mean field and angular aberration for an ideal spherical deflection analyser.

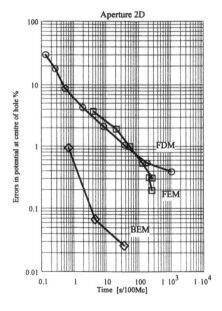

**Figure 2** Percentage error in the computed potential at the centre of a circular hole in an infinite flat sheet.

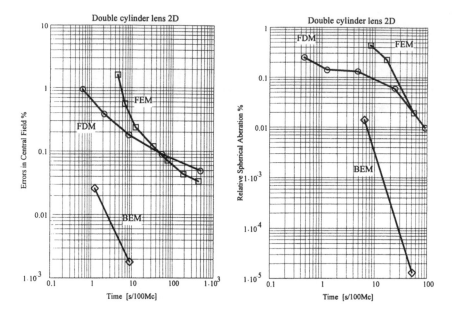

**Figure 3** Percentage error in the computed field at the centre and the relative spherical aberation of a double-cylinder lens.

To summarize, it can be seen from the figures that for these problems the FDM and FEM programs used in the present study have the same order of accuracy for the same computing time, while the BEM program is usually one or two order of magnitude more accurate (the geometric mean for the present results being 50 times more accurate).

[†]SIMION Program, information available at http://www.srv.net/~klack/simion.html
[††]ELENS Program, information available at http://cpo5.tn.tudelft.nl/dpof
[†††]CPO Programs, programs and information available at http://cpo.ph.man.ac.uk/

**REFERENCES**

Hawkes P H and Kasper E 1989 *Principles of Electron Optics* (Academic Press, New York)
Lencova B 1995 *Nucl.Inst Meth.* 363, 190
Read F H 1997 *J Computational Physics* 133, 1

Inst. Phys. Conf. Ser. No 153: Section 3
Paper presented at Electron Microscopy and Analysis Group Conf. EMAG97, Cambridge, 1997
© 1997 IOP Publishing Ltd

# Resolving atoms: what do we have?, what do we want?

D Van Dyck[1], E Bettens[1], J Sijbers[1], A J Den Dekker[1,2], A Van den Bos[2], M Op de Beeck[1], J. Jansen[1,3], H. Zandbergen[3]

[1] Department of Physics, University of Antwerp (RUCA), Groenenborgerlaan 171, B-2020 Antwerpen, Belgium
[2] Department of Applied Physics, Delft University of Technology, Lorentzweg 1, NL-2628 CJ Delft, The Netherlands
[3] National Centre for HREM, Delft University of Technology, Rotterdamseweg 137, NL-2628 AL Delft, The Netherlands

ABSTRACT: The state of the art and the future in quantitative high resolution electron microscopy are discussed in the framework of parameter estimation. Reconstruction methods are then to be considered as direct methods to yield a starting structure for further refinement. With the increasing flexibility of the instruments, computer aided experimental strategy will become important.

## 1. INTRODUCTION

During the last decennium, a variety of new electron microscopic techniques has emerged which push the resolution down to atomic or subatomic dimensions. To name a few: High voltage HREM, off-axis holography, focus variation holography, ptychography, STEM (Z contrast), aberration correction, etc., ... Whereas thus far these techniques have been developed on different instruments, we believe that the electron microscope of the future will be a versatile TEM - STEM instrument in which most of these options (apart from the high voltage) can be chosen under computer control, without compromising.

An ideal electron microscope should be an instrument with a maximal number of degrees of freedom (controllable settings).

As shown in Figure 1, information about the object can be deduced by knowing the electron wave at the entrance plane of the object, and by measuring the electron distribution at the exit plane.

A twin condensor-objective type of instrument with a field-emission source, with flexibility in the illumination conditions and with a configurable detector would allow to choose the form of the incident wave freely in real or reciprocal space (STEM, TEM, hollow cone, standing wave, ...) as well as the plane and area of detection (image, diffraction pattern, HAADF, ptychography ...).

An ideal detector should combine high quantum efficiency (i.e., ability to detect single electrons), high dynamic range, high resolution and high speed. Thus far these requirements are not yet met in the same device but developments are promising. If the instrument is furthermore equipped with an energy filter before and after the object, one could in principle acquire all the information that can be carried by the electrons. At present the energy resolution is still limited to the order of 1 eV so that information from phonon scattering or from molecular bonds cannot yet be separated.

Secondary particles (X-ray photons, Auger electrons, etc.) can yield complementary information and if combined with coincidence measurements, complete inelastic events in the object can be reconstructed.

The most important feature of the future electron microscope is the large versatility in experimental settings under computer control, such as the selection of entrance wave, the detection configuration and

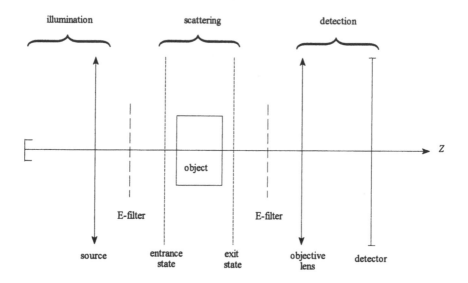

Figure 1. Ideal experimental set-up.

many other tunable parameters as focus, voltage, spherical aberration constant, specimen position, orientation etc. The only limiting factor in the experiment will be the total number of electrons that interact with the object during the experiment or that can be sustained by the object.

The major problem that has still to be overcome is the reliable interpretation of the experimental data. Isn't it an anachronism that the interpretation of electron microscopic data is mostly done qualitatively (i.e., visually) whereas for most instruments of comparable cost, the data are handled quantitatively? As compared to X-ray crystallography, HREM is still in its infancy.

The potentialities of the improved experimental techniques and the flexible use of the various experimental settings will never be exploited fully if not combined with a quantitative analysis of the experimental data. Furthermore, there is need for a guided experimental strategy (design) which optimizes the experimental settings so as to maximize the information content per incident electron in function of the problems under study.

Questions such as: what is the best object thickness to visualize light atoms?, what are the best focus values?, does it pay off to develop an instrument with 0.01 nm resolution?, etc., will have to be answered by experimental strategy.

In this work we will discuss the aspects of quantitative interpretation and experimental strategy and resolution in a more general framework.

## 2. QUANTITATIVE STRUCTURE DETERMINATION

The ultimate goal of high resolution electron microscopy is to determine quantitatively and accurately the atomic structure of an object. In this respect the electron microscope can be considered as an information channel that carries this information from the object to the observer. High resolution images and diffraction patterns are then to be considered as data planes from which the structural information has to be extracted.

However, this structural information is usually hidden in the images and cannot easily be assessed. Therefore, a quantitative approach is required in which all steps in the imaging process are taken into account.

Ideally quantitative extraction of information should be done as follows: one has a model for the object, for the electron object interaction, for the microscope transfer and for the detection, i.e., the ingredients needed to perform a computer simulation of the experiment. The object model that describes the interaction with the electrons consists of the assembly of the electrostatic potentials of the constituting atoms. Since for each atom type the electrostatic potential is known, the model parameters then reduce to atom numbers and coordinates, Debye Waller factors, object thickness and orientation (if inelastic scattering is neglected). Also the imaging process is characterised by a number of parameters such as defocus, spherical aberration, voltage, etc. These parameters can either be known a priori with sufficient accuracy or not in which case they have to be determined from the experiment. The model parameters can be estimated from the fit between the theoretical images and the experimental images. What one really wants is not only the best estimate for the model parameters but also their standard deviation (error bars), a criterium for the goodness of fit and a suggestion for the best experimental setting. This requires a correct statistical analysis of the experimental data. The goodness of the fit between model and experiment has to be evaluated using a criterium such as likelihood, mean square difference or R-factor (cfr. X-ray crystallography). For each set of parameters of the model, one can calculate this goodness of fit, so as to yield a fitness function in parameter space. The parameters for which the fitness is optimal then yield the best estimates that can be derived from the experiment. In a sense one is searching for a maximum (or minimum depending on the criterium) of the fitness function in the parameter space, the dimension of which is equal to the number of parameters.

The probability that the model parameters are $\{a_n\}$ given that the experimental outcomes are $\{n_i\}$ can be calculated from Bayesian statistics as

$$p\big(\{a_n\}/\{n_i\}\big) = \frac{p\big(\{a_n\}\big)\,p\big(\{n_i\}/\{a_n\}\big)}{\sum_{\{a_n\}} p\big(\{a_n\}\big)\,p\big(\{n_i\}/\{a_n\}\big)} \tag{1}$$

$p\big(\{n_i\}/\{a_n\}\big)$ is the probability that the measurement yields the values $\{n_i\}$ given that the model parameters are $\{a_n\}$. This probability is given by the model. For instance, in case of HREM, $p\big(\{n_i\}/\{a_n\}\big)$ represents the probability that $n_i$ electrons hit the pixel $i$ in the image given all the parameters of the model (object structure and imaging parameters) i.e., $n_i$ then represents the measured intensity, in number of electrons, of the pixel $i$. $p\big(\{a_n\}\big)$ is the prior probability that the set of parameters $\{a_n\}$ occurs. If no prior information is available all $p\big(\{a_n\}\big)$ are assumed to be equal. In this case maximising $p\big(\{a_n\}/\{n_i\}\big)$ is equivalent to maximising $p\big(\{n_i\}/\{a_n\}\big)$ as a function of the $\{a_n\}$. The latter is called the Maximum Likelihood method (ML). It is known (e.g. Van den Bos, 1981) that if there exists an estimator that obtains the Minimum Variance Bound (or Cramer-Rao Bound), it is given by the ML. (The least squares estimator is only optimal under specific assumptions.)

In practice it is more convenient to use the logarithm of (1) called the loglikelihood $L$.

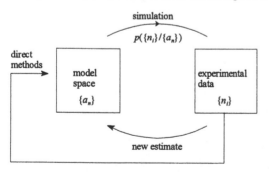

Figure 2. Refinement procedure.

$L$ can then be considered as a fitness function. In principle the search for the best parameter set is then reduced to the search for optimal fitness in parameter space. This search can only be done in an iterative way as schematised in Figure 2. First one has a starting model, i.e. starting value for the object and imaging parameters $\{a_n\}$. From this one can calculate the experimental outcome $p\big(\{n_i\}/\{a_n\}\big)$. This is a classical image simulation. (Note that the experimental data can also be a series of images and/or diffraction patterns) From the mismatch between experimental and simulated images one can obtain a new estimate for the model parameters (for instance using a gradient method) which can then be used for the next iteration. This procedure is repeated until the optimal fitness (i.e., optimal match) is reached.

## 3. DIMENSIONALITY PROBLEM

A major problem is now that the effect of the structural parameters is completely scrambled in the experimental dataset. Due to this coupling one has to refine all parameters simultaneously which poses a combinatorial problem. Indeed, the dimension of the parameter space becomes so high that even with advanced optimisation techniques such as genetic algorithms, simulated annealing, tabu search, etc. one cannot avoid ending in local optima. The problem is only manageable if the number of parameters is small, as is the case for small unit cell crystals. In some very favourable cases, the number of possible models, thanks to prior knowledge, is discrete and very small so that visual comparison is sufficient. These cases were the only cases where image simulation could be meaningfully used in the past. The dimensionality problem can be solved by using direct methods. These are methods that use prior knowledge which is generally valid irrespective of the (unknown) structure of the object and that can provide a pathway to the global optimum of the parameter space.

Examples of direct methods in electron microscopy and diffraction are:

### i. electron diffraction

In electron diffraction the amplitudes of the diffracted beams can be measured directly but the phases are unknown. Prior knowledge based upon the fact that the electrostatic potential of the atoms is real, positive and point-like, allows to derive relations between the phases of the diffraction beams, so as to reduce strongly the number of independent parameters. These phasing methods have been very successful in X-ray crystallography and can be applied to some extent to electron diffraction (in the kinematical regime). Ptychography provides another way for measuring the phase information directly.

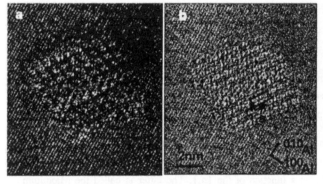

Figure 3. HREM image (left) and phase of the experimentally reconstructed exit wave (right) of an Mg/Si precipitate in an Al matrix.

### ii. High resolution electron microscopy

In high resolution electron microscopy, the dimensionality problem can be solved by deblurring the information, so as to unscramble the influence of the different object parameters in the image. In this way the structural parameters can be uncoupled and the dimension of the parameter space reduced. This can be achieved in different ways: high voltage microscopy, correction of the microscopic aberrations or direct holographic methods for exit wave and structure reconstruction. An example of an exit wave, retrieved with the focus variation method, is shown in Figure 3. However these methods will not yield the final quantitative structural model but an approximate model. This model can be used as a starting point for a final refinement by fitting with the original images and that is sufficiently close to the global maximum so as to guarantee convergence. Recently an hybrid method has been proposed, the

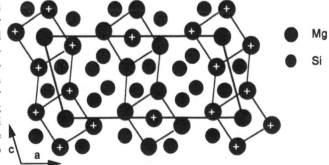

Figure 4. Structure model obtained with MSLS from the fitting procedure described in the text.

Multislice least squares refinement (MSLS), to determine the structure of unknown microcrystals by first obtaining an approximate model using focus variation exit wave reconstruction which afterwards could be refined using several selected area diffraction patterns.

An application of MSLS refinement is shown in Figures 3 and 4. Figure 3 (left) shows an HREM image of a Mg/Si precipitate in an Al matrix. Figure 3 (right) shows the phase of the exit wave which is reconstructed experimentally using the focus variation method. From this an approximate structure model can be deduced. From different precipitates and different zones, electron diffraction patterns could be obtained which were used simultaneously for a final fitting with MSLS.

For each diffraction pattern the crystal thickness as well as the local orientation were also treated as fittable parameters. An overview of the results is shown in Table 1. The obtained R factors are of the order of 5%, which is well below the R factor using kinematical refinement, that do not account for the dynamical electron scattering. Figure 4 shows the structure obtained after refinement. Details of this study have been published by Zandbergen et al.

Table 1

| zone | number observed reflections | thickness (nm) | crystal misorientation | | | R-value (%) | |
|------|------|------|------|------|------|------|------|
| | | | h | k | l | MSLS | kinematic |
| [010] | 50 | 6.7(5) | 8.3 | 0 | -2.3 | 3.0 | 3.7 |
| [010] | 56 | 15.9(6) | 2.6 | 0 | -1.8 | 4.1 | 8.3 |
| [010] | 43 | 16.1(8) | -1.7 | 0 | 0.3 | 0.7 | 12.4 |
| [010] | 50 | 17.2(6) | -5.0 | 0 | -1.0 | 1.4 | 21.6 |
| [010] | 54 | 22.2(7) | -5.9 | 0 | 2.5 | 5.3 | 37.3 |
| [001] | 72 | 3.7(3) | -3.9 | 4.5 | 0 | 4.1 | 4.5 |
| [001] | 52 | 4.9(6) | 3.6 | -1.9 | 0 | 6.8 | 9.3 |

## 4. ACCURACY

At present, the accuracy of structure models obtained from fitting with HREM data alone is not yet comparable to that of X-ray diffraction work. Especially the "contrast" mismatch between experimental and theoretical exit waves of known objects rises to 20%-30%. We believe that this is due to the inaccuracy of the image simulations. However, as shown above the refinement based on electron diffraction reaches R factors below 5%. Hence, the accuracy should only occur in images and not in diffraction patterns. It can also not be in a constant background in the image since this is taken care of in the focus variation method. Hence the effect should be periodical in real space and diffuse in Fourier space. Since in the diffraction refinement (Jansen et al., 1995) also multislice calculations were used we do not believe that the inaccuracy stems from the calculation of the elastic scattering such as the accuracy of the scattering factors or the multislice algorithm itself. Furthermore the mismatch is still present in energy filtered images so that it is not due to inelastic scattering with energy loss above 1 eV (plasmon, core excitation). Moreover, it has been observed (Zandbergen, private communication) that the electron intensity in 10 nm thin crystals is about 80% of that at the crystal edge. Hence, 20% of the electrons are scattered outside of the objective aperture. Based on these reasons we are inclined to believe that the inaccuracy has to be sought in the contribution of phonon scattering which, due to the focussing effect of the channelling of the electrons in the atomic columns parallel to the incident beam may be much larger than intuitively expected for these crystals. These effects are eliminated by background subtraction in electron diffraction refinements (yielding low R factors) but not in HREM.

It should be worthwhile to adapt the holographic reconstruction method so as to correct for the effect of phonon scattering.

## 5. RESOLUTION

If the building blocks of matter, the atoms, can be seen, the useful prior knowledge about the object is large, i.e., it consists of atoms, the form of which is known. Hence, the only unknown parameters of the model are the atom positions. Now, the concept of resolution has to be reconsidered as the precision with which an atom position can be determined, or the distance at which neighbouring atoms can still be resolved.

Let us simplify the problem as follows. Let us consider the "image" of an atom, spread by the imaging system, as a one-dimensional Gaussian peak with width $\sigma_0$, located at position $a$, i.e.

$$p(x_i/a) = f(x_i, a)$$

The ML estimate of $a$ is located where the loglikelihood function is maximal. It could be shown (for instance Van Dyck et al., 1997) that the standard deviation is

$$\sigma = \frac{\sigma_0}{\sqrt{N}}$$

where $N$ is the number of counts. As expected, the standard deviation is both function of $\sigma_0$, the "resolution" of the instrument, including the width of the atom itself, and the electron dose.

Resolving two atoms that are close together becomes a yes or no problem. If the dose is infinite, one can in principle resolve atoms even when they are infinitely close together. However, for a finite dose, for each interatomic distance, one can define a probability of resolution. There is a finite probability that the atoms will not be resolved, i.e., where the fitting procedure only finds one atom, so that the atoms in a sense "collapse". This probability increases with decreasing distance between the atoms. The resolving power of an instrument is then to be considered as the number of parameters per unit area that can be determined with a certain probability by fitting. If the real number of unknown parameters is higher than the resolving capacity of the instrument, the parameters will collapse.

## 6. EXPERIMENTAL STRATEGY

Based on a well established theory for experimental design (Fedorov, 1972; Van den Bos, 1982) it is possible to get expressions for the variances (error bars) of the parameters that are determined by the fitting procedure. From these expressions one can optimise the settings of the microscope so as to minimise the variance of particular parameters of interest. For instance, if one is interested in the position of one particular atom one can choose the tunable parameters of the experiment (focus, etc.) so as to minimise the variance of that particular position.

Buist (1995) has used in this way the theory of experimental design to optimise all the focus values in a focus variation experiment. We believe that, in view of the flexibility of future electron microscopes, on-line decisions based on computer aided experimental design will become important to obtain the highest accuracy in structure determination.

## REFERENCES

Buist A H 1995 Ph.D. Thesis, Delft University of Technology
Fedorov V V 1972 Theory of Optimal Experiments (New York: Academic Press)
Jansen J, Tang D, Zandbergen H W and Schenk H 1995, in Proceedings ECM16, Lund p. 45
Op de Beeck M, Van Dyck D, Coene W 1996 Ultramicroscopy **64**, 167-183
Van den Bos A 1981 in Handbook of Measurement Science, Volume 1, ed. P.H. Sydenham (New York: John Wiley & Sons)
Van Dyck D and Op de Beeck M 1996 Ultramicroscopy **64**, 99-107
Van Dyck D, Bettens E, Sijbers J, Op de Beeck M, Van den Bos A and Den Dekker A J 1997 Scanning Microscopy, in press
Zandbergen H W, Anderson S and Jansen J Science, August 1997

*Inst. Phys. Conf. Ser. No 153: Section 3*
*Paper presented at Electron Microscopy and Analysis Group Conf. EMAG97, Cambridge, 1997*
© *1997 IOP Publishing Ltd*

# Aberrations due to localized potential defects on apertures

**F H Read, L A Baranova**[†]**, N J Bowring, J Lambourne and T C Whitwell**

Department of Physics and Astronomy, Schuster Laboratory, University of Manchester, Manchester M13 9PL, UK
[†]A F Joffe Physico-Technical Institute, Russian Academy of Science, St. Petersburg 194021, Russia

**ABSTRACT:** The near-axis aberrations are considered for electrostatic systems that are axially symmetric except for the presence of a localized potential defect near the edge of an aperture. When a beam fills the aperture the resulting aberrations cannot be characterized in terms of coefficients, and so a new characterization of the aberrations is defined. Numerical evaluations are carried out for a representative type of potential defect. Relationships are presented that enable the increase in spot size of a focussed beam to be deduced from the measured or estimated mean deflection of the spot.

## 1. INTRODUCTION.

Apertures in thin flat plates are frequently used to limit the size or range of angles of beams of charged particles. When such an aperture is situated in a field-free region it is usually assumed that it removes the outer parts of a beam a without disturbing the inner part that passes through the aperture. It is known however that insulating layers can sometimes accumulate on the areas of the plate that are struck by the beam. This type of potential defect causes what are often referred to as **'patch fields'** or **'parasitic fields'**, the origins and time characteristics of which have been discussed in detail by Langner (1990). Potential defects can also be caused by changes in the structure of the surface.

An example of a localized potential defect is illustrated in Figure 1 by the shaded area, situated at the edge of a circular aperture in a thin flat plate. The area of this defect is 3% of the area of the aperture.

## 2. THE WEAK-FIELD IMPULSE APPROXIMATION.

A localized potential defect would usually have only a small effect on a beam and would usually act over only a short distance, and so in the terminology used for lenses (Hawkes and Kasper 1989) it could be described as a 'weak-lens, short-lens' system. In the present study we assume that the defect is of this type, so that the transverse force experienced by each electron in the beam is weak and that it exists over a distance that is small compared with the object or image distances, producing a change in its transverse momentum that is small compared to its total momentum. Using a terminology more appropriate to the case of a localized potential defect we shall refer to this as the *'weak-field impulse approximation'*.

In the weak-field impulse approximation it becomes possible, and more convenient for practical purposes, to work in terms of the change in direction of a general trajectory as it passes through the deflector. The deflections $\delta\theta_x$ and $\delta\theta_y$ depend only on the coordinates $x_d$ and $y_d$ of the trajectory at the central plane of the aperture, and can be represented by:

$$\delta\theta_x = \frac{V_d}{\phi}\left[(d + ax_d + c\frac{1}{2}(x_d^2 - y_d^2) + ...\right], \qquad \delta\theta_y = \frac{V_d}{\phi}(-ay_d - cx_dy_d + ...) \qquad (1)$$

102

where we define the new aberration coefficients

$$d = \frac{1}{2}\int_{z_o}^{z_i} f_1 dz \qquad a = \int_{z_o}^{z_i} f_2 dz \qquad c = 3\int_{z_o}^{z_i} f_3 dz \qquad (2)$$

Higher order terms in equation (2) would contain the slopes $x_i'$ and $y_i'$, but they are not considered in the present study.

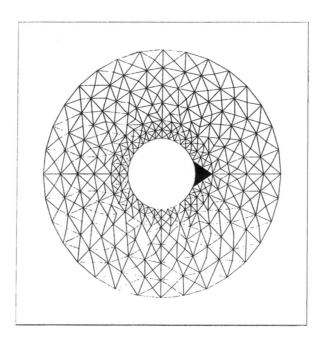

Figure 1. A localized potential defect at the edge of a circular defining aperture. The defect occupies the shaded area. The plate has been subdivided into triangular segments by the CPO-3D program. In this example the area of the defect is 3% of the area of the aperture.

If the particles of the beam have relativistic energies the potential $\phi$ must be replaced by the corrected potential

$$\phi_{rel} = \phi(1 + \frac{e\phi}{2m_0c^2})(1 + \frac{e\phi}{m_0c^2})^{-1} \qquad (3)$$

To validate and also exemplify the analytical treatment we present computational results for representative defects, using the CPO-3D program, which is based on the Surface Charge Method (Harting and Read 1976). The computational part of the present study has been made possible by the ability of this program to accurately simulate electrostatic fields near the edges of thin electrodes in 3-dimensional systems. These fields are strongly non-uniform in such regions and so they usually cannot be well simulated by programs based on the Finite Difference and Finite Element Methods. The CPO-3D program has been used to evaluated the coefficients $d$, $a$ and $c$, and also to carry out a direct integration of a set of trajectories through the aperture illustrated in figure 1, for an aperture radius $r_a = 0.5mm$, a defect potential difference $V_d = 0.1V$ and an electron energy $\phi = 1keV$. The resulting deflections $\delta\theta_x$ and $\delta\theta_y$ were then multiplied by the factor $\phi/V_d$, to produce a generalized representation of the aberrations, which is shown in figure 2.

In practice the potential, size and position of the defect are usually not known and cannot easily be estimated. The mean deflection $\theta_{mean}$ caused by the defect can however often be estimated from the corrections that have to be applied by deflectors elsewhere in the system. It is therefore of interest to compare the size of an aberrated image with its mean deflection. We find that the root-mean-square deviation from the mean deflection is related to the mean deflection by

$$\theta_{rms} / \theta_{mean} \approx \frac{1}{2} r_b \tag{4}$$

where $r_b$ is the radius of the beam at the plane of the aperture.

## 4. ABERRATIONS OF DEFINING APERTURES.

When a beam fills an aperture of radius $r_a$ the deflections can be expressed as

$$\delta\theta_x = \frac{V_d}{\phi_{rel}} f_x(x_d / r_a, y_d / r_a) \qquad\qquad \delta\theta_y = \frac{V_d}{\phi_{rel}} f_y(x_d / r_a, y_d / r_a) \tag{5}$$

where

$$f_w = -\frac{1}{2} \int E_w(x_d, y_d, z) dz \tag{6}$$

and where $E_w$ is the field in the $w$ direction for a unit potential difference on the localized defect. Because of the usual scaling rule (Hawkes and Kasper 1989) the functions $f_x$ and $f_y$ depend on the shape and size of the defect relative to the area of the aperture but are independent of the area of the aperture itself. Unusually however, in the weak-field impulse approximation the deflections depend on the coordinates of the trajectory at the plane of the aperture but not on the slopes of the trajectory. Figure 2 is a two-dimensional representation of $f$ for the particular defect shown in figure 1.

As with the near-axis aberrations discussed above, the potential, size and position of the defect are usually not known and it is then helpful to know the ratio of the size of an aberrated image to its mean deflection, which we find to be

$$\theta_{rms} \approx \theta_{mean} \tag{7}$$

As a specific example, consider a $1MeV$ beam of electrons that is converging to a point a distance $30mm$ from an aperture, filling an aperture that has a defect of the form shown in figure 1, with a potential difference of $10V$. The scaling factor $V_d/\phi_{rel}$ (where the relativistically corrected potential is now used) is $1.49 \ 10^{-5}$, and the root-mean-square spreads are approximately $60nm$ and $30nm$ in the $x$ and $y$ directions respectively and the mean deflection is $35nm$ in the $x$ direction. If only the middle 10% (by area) of the aperture is filled then it can be seen in figure 2 that the diameter of the innermost circle is approximately $0.061rad$ and so the rms spreads are approximately $10nm$ in both directions.

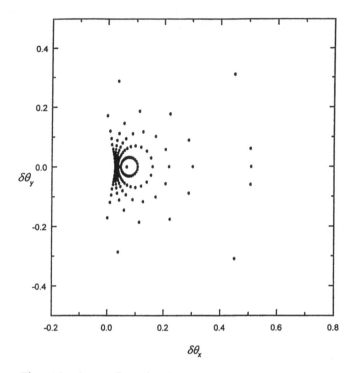

Figure 2. A two-dimensional representation of the deflections $\delta\theta_x$ and $\delta\theta_y$ experienced by the trajectories in a beam which fills an aperture that has a single localized defect of the type shown in figure 1, at the $+x$ edge of the aperture. The data have been generated (see text) by integrating the trajectories of a uniform converging beam, and the deflections have been renormalized to $V_d/\phi = 1$. There are five sets, each consisting of 40 trajectories that start from a circle in object space. The deflection is also shown for the single trajectory that is initially directed along the axis. The circles represent the envelopes of the inner 10%, 30%, 50%, 70% and 90% of the beam that fills the aperture.

**REFERENCES**

CPO Programs, available on the Web site http://cpo.ph.man.ac.uk/
Harting E and Read F H, 1976 *Electrostatic lenses*, (Elsevier Publishing Company, Amsterdam)
Hawkes P H and Kasper E 1989 *Principles of Electron Optics* (Academic Press, New York)
Langner G, 1990 *J. Vac. Sci. Technol.* **B8**, 1711

*Inst. Phys. Conf. Ser. No 153: Section 3*
*Paper presented at Electron Microscopy and Analysis Group Conf. EMAG97, Cambridge, 1997*
© *1997 IOP Publishing Ltd*

# Super resolved microscopy and aberration determination in the TEM.

**A I Kirkland, W O Saxton\* and R Meyer\***

University of Cambridge, Department of Chemistry, Lensfield Road, Cambridge, CB2 1EW
and JEOL UK Ltd., Silvercourt, Watchmead, Welwyn Garden City, Herts, AL7 1LT, UK.
\* University of Cambridge, Department of Materials Science and Metallurgy, Pembroke
Street, Cambridge CB2 3QZ, UK.

**ABSTRACT:** This paper presents the results of a super resolved reconstruction of an
[010] oriented crystalline fragment of $H-Nb_2O_5$ using a tilt azimuth data set. Particular
attention is given to the interpretation of both the phase and modulus of the reconstruction
by means of multislice simulations of trial structural models.

## 1. INTRODUCTION

The continued improvement in the performance of modern high resolution Transmission
Electron Microscopes (TEM's) and in particular the advent of instruments equipped with
field emission guns has lead to the development of "indirect microscopy" with the aim of
recovering the specimen exit surface wavefunction at high resolution from the processing of
either focal (Coene et al. 1992) or tilt azimuth image series (Kirkland et al. 1995). However
although the potential benefits of such methods are clearly established the interpretation of
the data obtained still requires detailed investigation in terms of the structural information it
provides.

In this paper we consider a tilt azimuth reconstruction of $H-Nb_2O_5$ paying particular
attention to the information provided in both the phase and the modulus of the reconstructed
wavefunction. By means of multislice image simulations using the effective Transfer
Function determined during the reconstruction process it is shown that information about
both the cation lattice positions and the details of the anion sublattice can be recovered.

## 2. DATA COLLECTION

A specimen consisting of crushed fragments dispersed on a holey carbon support film of
the high temperature form of $Nb_2O_5$ consisting of metal-anion octahedra ($a$=2.12 nm,
$b$=1.94nm, $c$=0.38nm, $\beta$=119.8°) was used to investigate the capabilities of this approach.

A set of six images was recorded using a JEOL JEM-2010F[1] ($C_s$=0.5mm, 200kV), four
with two mutually orthogonal pairs of opposite tilts together with initial and final axial
images. For this data set the images were recorded using photographic film although we note
that the data acquisition and subsequent processing would be considerably simplified by the
use of a CCD camera. The microscope was controlled via an external PC connected to the

[1] We are grateful to JEOL Ltd for making time available on this instrument, and in particular to Mr Hosokawa
and Dr Oikawa for valuable assistance.

RS232 serial interface; in the context of a programme of comprehensive automation of high resolution microscopy, we have recently developed dedicated software running under Windows NT 4.0 to automate the acquisition of the required tilt azimuth data. In this case the initial focus level of the microscope and the coma free axis were set manually; this process will also automated shortly and incorporated in a single experimental procedure. An injected tilt with magnitude approximately $\theta = 0.8(\lambda/C_s)^{\frac{1}{4}}$ was chosen, and the focus was adjusted so as roughly to meet the achromatic condition $D = C_s\theta^2$, which provides optimal transfer under tilted illumination. Subsequently 4096×4096 pixel regions including the thin edge of a crystal and an adjacent narrow region of disordered support film were digitised to 8 bits, using an Eikonix 1412 densitometer, at a sampling interval of approximately 0.03nm. The microscope aberrations were determined by fitting the defocus and astigmatism for each diffractogram by simple visual comparison of calculated with fitted patterns, and are listed in table 1.

| Sampling interval: | 0.0308±.0001nm |
|---|---|
| RMS focus spread: | 5±1nm |
| RMS sectional beam divergence: | 0.2±0.1mrad |
| RMS sectional specimen vibration: | 0.03±0.01nm |
| Defocus: | −22.4±1.0nm |
| Astigmatism components: | −0.9,−0.8±0.3nm (magnitude 1.2nm) |
| Beam misalignment components: | 0.41, −0.08±.05mrad (magnitude 0.42mrad) |
| Three-fold astigmatism components: | 290,60±30nm (magnitude 300nm) |

Table 1. Imaging conditions inferred from the fitted diffractograms, with estimated standard errors. Aberrations and beam tilts are measured w.r.t. the coma-free axis.

The images were registered using the crystalline area itself using a Transfer Compensated Cross Correlation which was additionally processed to remove the strong crystal contrast as discussed in detail elsewhere (Kirkland et al. 1997).

## 3. IMAGE RECONTRUCTION AND SIMULATION

Following the determination of the imaging conditions and the registration of the images, which we reiterate could be simplified given fully automated accurate microscope control and digital imaging the reconstruction itself is relatively straightforward and has been described in detail elsewhere (Saxton 1987). Figure 1 shows both the modulus and phase of the reconstructed wavefunction together with the original axial image for comparison. It is apparent from Fig. 1 that the reconstructed modulus exhibits dark spots at a higher resolution than in the axial image corresponding to the cation sites. Furthermore the reconstructed phase shows bright spots at the cation sites with fainter spots located between them. In order to interpret these features, we have carried out a series of multislice simulations both for H-$Nb_2O_5$ using the published x-ray structure and for a hypothetical structure with the oxygen atoms removed but with the cation lattice undisturbed (Fig. 2). Simulations were carried out for both the axial imaging conditions determined from the experimental axial image and for the super resolved imaging conditions achieved on reconstruction. In the latter case the complex exit surface wavefunction was calculated using the multislice formalism and this was then multiplied with the Transfer Function effective in the super resolved reconstruction.

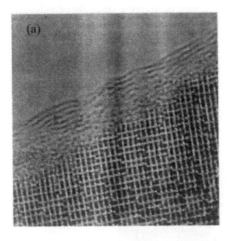

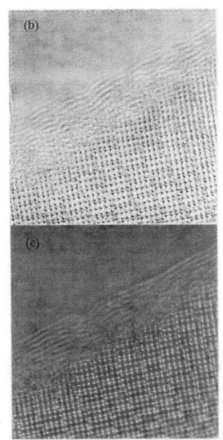

Figure 1. Corresponding views of
(a) The initial axial image.
(b) Modulus of the reconstructed wavefunction.
(c) Phase of the reconstructed wavefunction.

Figure 2. Projected potentials of (a) an idealised structure based on H-Nb$_2$O$_5$ with the oxygen atoms completely removed and (b) fully oxygenated H- Nb$_2$O$_5$.

Simulated axial images calculated for a specimen thickness of 30.5nm (which gives the best visual match to the experimental image) and for the determined defocus of -22.4nm show no difference between the models, so that details of the oxygen anion sublattice are not detectable in such images. For this particular specimen thickness certain large underfocus values do show differences between the two models; however in practice since these values are themselves thickness dependent, recording images to visualise the oxygen anion sublattice presupposes a knowledge of either the structure or of the specimen thickness. In contrast Fig. 3 shows the phase and modulus of the super resolved reconstruction simulated using the same structural models. In this figure, it is apparent that for both models the modulus reveals details of only the cation lattice as expected but that significant differences between the models exist in the phase. In particular the fully oxygenated model shows faint white spots between the cation sites corresponding to those observed experimentally, confirming that the super resolved reconstruction does indeed reveal details of the anion lattice not available via the conventional axial image.

Figure 3. Modulus (a) and (c) and phase (b) and (d) of the simulated super resolved reconstruction, H-Nb2O5 with oxygen atoms removed (a) and (b) and fully oxygenated (c) and (d). Both (a) and (c) reveal the cation lattice whereas (b) and (d) show significant differences due to the oxygen anion sublattice. Note all images are scaled to the same black and white levels.

## REFERENCES

Coene W, Janssen G, Op de Beeck M and van Dyck D 1992 Phys. Rev. Lett. **69**, 3743.
Kirkland A I, Saxton W O, Chau K-L, Tsuno K and Kawasaki M, 1995 Ultramicroscopy **57**, 355.
Kirkland A I, Saxton W O and Chand G 1997 J. Electron Microscopy **1**, 11.
Saxton, 1987 Scanning Microscopy Supplement 2: Proc. 6th Pfefferkorn Conf., Niagara 213.

*Inst. Phys. Conf. Ser. No 153: Section 3*
*Paper presented at Electron Microscopy and Analysis Group Conf. EMAG97, Cambridge, 1997*
© 1997 IOP Publishing Ltd

# Direct structure determination by atomic-resolution incoherent STEM imaging

**P D Nellist, Y Xin\* and S J Pennycook\***

Cavendish Laboratory, University of Cambridge, Madingley Road., Cambridge CB3 0HE, UK.
\*Oak Ridge National Laboratory, Solid State Division, PO Box 2008, Oak Ridge, TN 37831-6031, USA.

**ABSTRACT:** Use of a large, annular dark-field (ADF) detector in a scanning transmission electron microscope is shown to give images that can allow direct structure determination, being a convolution between the illuminating probe intensity and an object function localised at the atomic column positions. The ADF image is also shown to resolve crystal spacings more than twice smaller than the phase contrast point resolution limit of the microscope used, with sub-angstrom structural information being retrieved.

## 1. INTRODUCTION

In the determination of atomic structures by conventional high-resolution transmission electron microscopy (HRTEM), there are two important difficulties. First, because the electron interacts strongly with the electric potential within the crystal, strong multiple scattering occurs that can complicate the image intensity and in general renders it impossible to interpret the image intuitively in terms of the projected structure. Second, spherical aberration limits the angle of scattered electrons that can be focussed by the objective lens, giving a resolution limit that can limit the accuracy to which the positions of atomic can be determined, and can also give rise to artefacts in the image.

It is now more than 100 years since Lord Rayleigh (1896) considered the incoherent mode of imaging for light optics. He noted that illuminating the specimen with a large incoherent source renders the specimen effectively self-luminous, destroying interference effects between spatially separated parts of the specimen. In addition, he noted that the incoherent mode could resolve features with half the spacing of that for axial plane wave illumination. By the principle of reciprocity a large detector in a scanning transmission electron microscope (STEM) is equivalent to a large incoherent source in a conventional microscope. Because few electrons are absorbed in a thin specimen, an infinite detector that collected all transmitted electrons would give no contrast, so we add contrast by introducing a hole, giving an annular dark-field (ADF) detector (Fig. 1). An ADF image from a $\{1\bar{2}10\}$ domain boundary in wurtzite GaN viewed along <0001> is shown in Fig. 2. The positions of the atomic columns are clear in this image and agree with the structure proposed by Xin et al. (1997). Here we describe briefly how such structure images arise, and how quantitative information may be retrieved from them.

## 2. INCOHERENT IMAGING FROM COHERENT SCATTERING

We start be considering the situation for a stationary lattice, so that the scattered electron wavefield can be assumed to be completely coherent, with purely Bragg diffraction for a crystalline specimen giving Bragg diffracted discs as depicted in Fig. 1. The Fourier transform of the ADF image intensity can be written in terms of the Fourier components of the

110

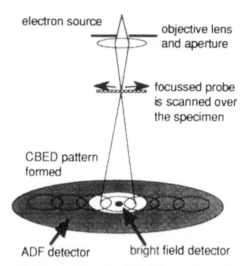

Fig. 1. A schematic of ADF image formation in a STEM

Fig. 2. A {1 2̄ 10} domain boundary in wurtzite GaN viewed along <0001>. The atom columns are 0.19 nm apart.

Bloch wave solutions of the Schrödinger equation (Nellist and Pennycook, 1997a). The advantage of such a formulation is that it is possible to perform the summation over the Bragg beams incident on the ADF detector *before* summing over the Bloch waves. It is then possible to determine which Bloch states contribute to the ADF image. A calculation for GaAs viewed along <110> at 300 kV shows that for a 30 mrad inner detector radius, the image contrast is dominated by the contribution from the tightly bound 1s-type states (Fig. 3). It is only the 1s states that have the sufficient transverse momentum to scatter the electrons to the detector. Even if we now include the effect of lattice vibrations, phonon scattering is not able to provide the high-transverse momentum required, and only helps to destroy interference effects between the high-angle elastic scattering from different heights within a single column. We can therefore think of the ADF detector as a Bloch state filter.

Since the excitation and form of the 1s-type states, which dominate the ADF image contrast, vary only slowly as a function of the incident beam angle, all the coherent partial plane waves in the probe-forming incident cone of illumination can contribute with similar weight. In this

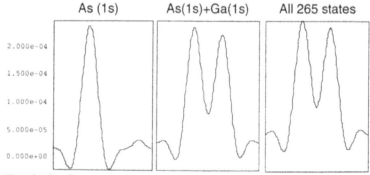

Fig. 3. Profiles through calculated object functions of GaAs<110> at 300 kV including various numbers of Bloch states. Fourier components up to and including {004} were included to form the object function.

situation it is straightforward to show (Nellist and Pennycook, 1997b) that the image can be treated as a convolution between an object function formed from the highly localised 1s type states and the probe intensity,

$$I(\mathbf{R}) = \int |P(\mathbf{R}' - \mathbf{R})|^2 O(\mathbf{R}') \, d\mathbf{R}' \tag{1}$$

where $P(\mathbf{R})$ is the complex amplitude of the illuminating probe, and $O(\mathbf{R})$ is the object function. Since the intensity of the probe is a real and positive quantity, it can be immediately deconvolved from the image intensity, in contrast to the coherent imaging process where in general it is necessary to solve for the phase of the image-plane before a deconvolution can be performed.

## 3. DECONVOLUTION AND RESOLUTION

Equation (1) can be used in two ways: If the probe intensity function is known, then it can be deconvolved to give information on the object function that allows greater accuracy in structure determination. This is useful because in any resolution limited imaging system, the peak intensity positions will not necessarily be located at the atomic column positions. Alternatively, if the object function is known, which often happens in practice because a typical image will contain a defect surrounded by regions of perfect crystalline matrix, then it is possible to find an estimate for the probe intensity (Nellist and Pennycook, 1997b). Having determined the probe intensity, we need to consider how it may be deconvolved from the experimental image intensity. Fig. 4 shows raw data from an image of GaAs<110> and the result when the probe intensity function is deconvolved from it using a Wiener filter. Although the deconvolution has reduced the noise level in the image, the resulting object function is not localised, and there are stronger artefacts in the tunnels between the dumbbells. The reason that an object function with localised peaks at the atoms columns has not been reconstructed is because the deconvolution has not been able to reconstruct information beyond the resolution limit of the microscope. There is not one unique object function that fits the data, and to find the atomic locations accurately a further constraint is required (Nellist and Pennycook, 1997).

For the accurate determination of structures, therefore, the achievement of higher resolution in TEM is an important goal, even when the projected interatomic spacings are within the resolution limit of the microscope. For the same imaging conditions, the ADF image is able to resolve spacings half as small as conventional bright-field imaging, as indicated in Fig. 1 where the bright-field image would not show any contrast. For ADF imaging, using a large objective aperture with a highly underfocussed lens can pass information well beyond the conventional Scherzer incoherent resolution limit, albeit with

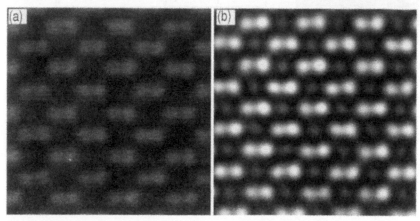

Fig. 4. (a) Raw image data of GaAs<110> and (b) the image when the probe intensity function has been deconvolved using a Wiener filter.

112

some loss in the intuitive nature of the image. Fig. 5 shows the magnitude of the Fourier transform of an image of Si<110> taken using an objective aperture 180% larger than the optimum Scherzer size, and with the objective lens approximately 130 nm underfocussed. A (444) spatial frequency may be seen, which indicates information transfer to 0.078 nm. Data taken from a sample of CdTe<11$\bar{2}$> under similar conditions is shown in Fig. 6. A (444) spatial frequency can be seen in both the image intensity profile and the magnitude of the Fourier transform, thus resolving the 0.093 nm spacing between the Cd and Te columns in this projection. The phase relationship between the (11$\bar{1}$) and (444) image spatial frequencies should allow the positions and polarity of the Cd and Te atomic columns to be determined.

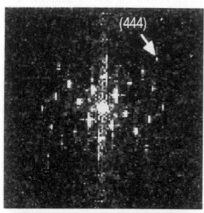

Fig. 5. The magnitude of the Fourier transform of a highly undefocussed ADF image of Si<110> using an oversized objective aperture.

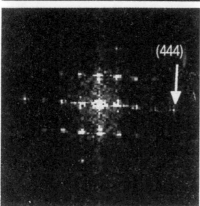

0.093 nm spacing resolved, and showing polarity information

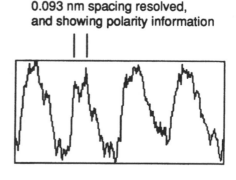

Fig. 6. The magnitude of the Fourier transform of an image of CdTe<11$\bar{2}$> taken using similar conditions to Fig. 5 along with an image profile along the [111] direction..

Lord Rayleigh 1896 Phil. Mag. (5) **42**, 167
Xin Y, Brown P D, Humphreys C J, Cheng T S and Foxon C T (1997) Appl. Phys. Lett. **70**, 1308
Nellist P D and Pennycook S J (1997a) to be published
Nellist P D and Pennycook S J (1997b) J. Microsc. in press
This research was sponsored by the US DOE under contract no. DE-AC05-84OR21400 with LMER corp. and by a Post-doctoral appointment to ORISE.

*Inst. Phys. Conf. Ser. No 153: Section 3*
*Paper presented at Electron Microscopy and Analysis Group Conf. EMAG97, Cambridge, 1997*
© *1997 IOP Publishing Ltd*

# The determination of the ionicity of sapphire using energy-filtered high resolution electron microscopy

S J Lloyd, R E Dunin-Borkowski[†] and C B Boothroyd

Department of Materials Science and Metallurgy, Pembroke Street, Cambridge CB2 3QZ, UK
† Center for Solid State Science, Arizona State University, Tempe, AZ 85287-1704, USA

**ABSTRACT:** The sensitivity of high resolution images to the degree of ionicity of sapphire is assessed through a comparison of experimental [$2\bar{1}\bar{1}0$] images with simulations. It is shown that, while the degree of ionicity does indeed affect image contrast, its quantification requires the determination of microscope imaging parameters to an unfortunately high degree of accuracy. Significantly, in contrast to previous results, the contrast of the experimental high resolution images is found to be *comparable* to that of the simulations. The reasons for this unusual lack of discrepancy are discussed, as are the potential advantages of electron holography rather than high resolution electron microscopy for determining the ionicities of such materials.

## 1. INTRODUCTION

The success of convergent beam electron diffraction (CBED) for measuring bonding charge densities in pure materials has resulted not only from the sensitivity of low order structure factors to the distribution of valence electrons (e.g. Midgley and Saunders, 1996) but also from the insensitivity of the fitting procedure to the magnitudes of certain microscope and specimen parameters whose determination plagues similar quantitative matching for high resolution images. However, alternative approaches to CBED that are better suited to the characterisation of charge transfer and bonding at interfaces, and which do not rely on the use of a focused probe that may damage beam-sensitive samples, are required. Accordingly, here we continue the investigations of Stobbs and Stobbs (1995) and Gemming et al. (1996) in an assessment of the sensitivity of high resolution images of [$2\bar{1}\bar{1}0$] sapphire ($Al_2O_3$) to ionicity and the importance of including ionic scattering factors in simulations. The suitability of sapphire for such an investigation results in part from its large unit cell, while in the [$2\bar{1}\bar{1}0$] orientation the amplitudes of certain beams that contribute to the image are sensitive to differences between the scattering factors of Al and O as opposed to their sums. The high resolution image simulations presented below were carried out using the tabulated X-ray scattering factors of Rez et al. (1994) and incorporated representative parameters for a JEOL 4000FX operating at 397kV ($C_s$=2mm, $\Delta f$=15nm, beam semi-angle 0.5mrad, objective aperture radius 20.6mrad). Debye-Waller factors for $Al_2O_3$ were taken from Kirfel and Eichhorn (1990).

## 2. RESULTS AND DISCUSSION

Fig.1 shows the sum of the electron scattering factors for Al and O for one formula unit of $Al_2O_3$, as calculated for neutral and fully ionic species from the tabulated X-ray scattering factors of Rez et al. (1994), while the amplitudes of the Fourier coefficients of the projected potential are plotted as a function of spatial frequency (with s=sin$\theta$/$\lambda$) for both neutral and ionic scattering factors in fig.2. Similarities are apparent between figs. 1 and 2 in that the differences between the neutral and ionic calculations extend to ~5nm$^{-1}$. In fig.2, the difference is greatest for the spatial frequencies that depend on the difference between the scattering factors of Al and O (e.g. 01$\bar{1}$2 and 0006) rather than on their sum (e.g. 03$\bar{3}$0), the greatest fractional difference occurring for the 0006 beam (for which the sense of the change as a function of ionicity is opposite to that for 01$\bar{1}$4 and 01$\bar{1}$2).

114

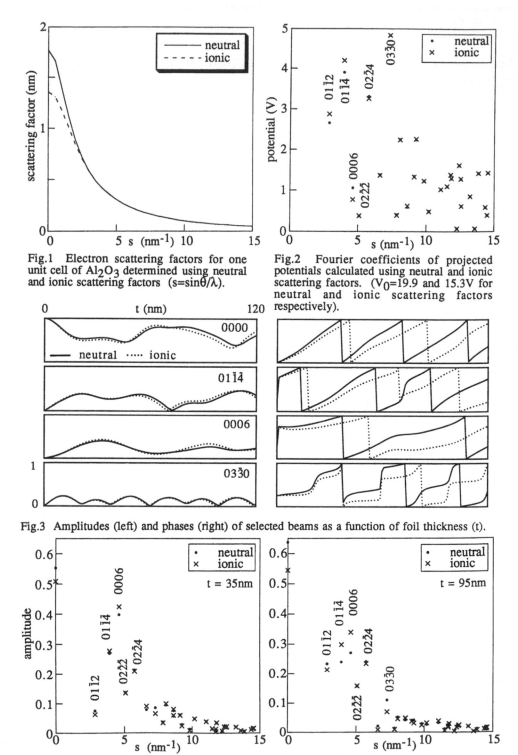

Fig.1 Electron scattering factors for one unit cell of $Al_2O_3$ determined using neutral and ionic scattering factors ($s=\sin\theta/\lambda$).

Fig.2 Fourier coefficients of projected potentials calculated using neutral and ionic scattering factors. ($V_0=19.9$ and 15.3V for neutral and ionic scattering factors respectively).

Fig.3 Amplitudes (left) and phases (right) of selected beams as a function of foil thickness (t).

Fig.4 Fourier coefficients of image contrast as a function of s at foil thicknesses of 35 and 95nm.

Fig.3 shows the amplitudes and phases of selected beams plotted as a function of foil thickness. The fact that the effect of ionicity on the beam amplitudes increases with foil thickness as a result of dynamical diffraction can be seen clearly in fig.4, which shows beam amplitudes plotted as a function of s for two foil thicknesses. At the lower thickness of 35nm, the amplitudes follow the trends exhibited by the Fourier coefficients of potential in fig.2, with the exception of the 0006 and 0330 beams whose intensity is largely governed by double diffraction. (In fact, for 0006 the sense of the change in amplitude as a function of ionicity is reversed relative to fig.2). At the greater foil thickness of 95nm, dynamical effects are now dominant and the beam amplitudes are more sensitive to ionicity than at 35nm. (Stobbs and Stobbs (1995) found even greater differences in beam amplitudes as a function of ionicity for an accelerating voltage of 1200kV, although ionic scattering factors were included differently in their simulations.)

Fig.5 shows a comparison of energy-filtered experimental images (obtained with an energy window of 10eV centred on the zero-loss peak) and simulations for foil thicknesses of 35 and 95nm. The $Al_2O_3$ specimen was made by dimpling a disc of single crystalline sapphire that had a $[2\bar{1}\bar{1}0]$ surface normal, ion-milling to perforation, removing organic and metallic contaminants by washing in aqua regia, and then annealing the sample for 8 hours at 1450°C to allow reconstruction of the surface (Tietz et al. 1992). The simulated images show that at 35nm the difference in the 0006 beam amplitude as a function of ionicity is not sufficient to result in significant image pattern changes. Although the image pattern is distinguishable between the simulations at a thickness of 95nm, particularly at a defocus of -96nm, the discrepancy between the experimental pattern and both sets of simulations is greater than that between the two simulations, as a result of the extreme sensitivity of the

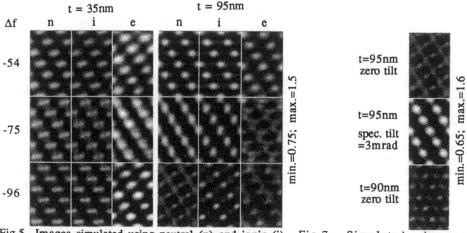

Fig.5 Images simulated using neutral (n) and ionic (i) scattering factors, as compared to experimental (e) data as a function of defocus (Δf) for two foil thicknesses.

Fig.7 Simulated images compared for different imaging conditions (Δf = -96nm)

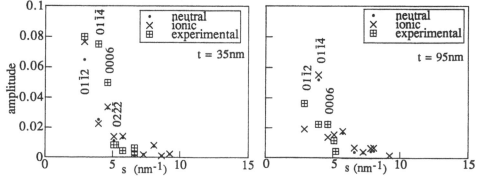

Fig.6 Fourier coefficients of experimental and simulated image contrast for foil thickness of 35nm and 95nm and a defocus of -96nm.

116

image pattern to the experimental imaging conditions. Fourier coefficients of the image contrast are plotted as a function of s in fig.6 for both the experimental and the simulated images for a foil thickness of 95nm and a defocus of -96nm. While the details of the variation in contrast with s are not reproduced by either set of simulated images, it is significant and unusual that the magnitude of the contrast in the experimental and simulated images is comparable (see Boothroyd, 1997). The sensitivity of the image pattern to the imaging conditions is a consequence of the strongly dynamical nature of the scattering. This is illustrated in fig.7, which shows the effect of a small change in specimen tilt and foil thickness on one of the simulated images. Paradoxically, it appears that the close match between the experimental and simulated image contrast results from the use of a clean specimen that has no amorphous surface layer, whereas this attribute makes the accurate determination of defocus and specimen tilt difficult. The solution may lie in the use of an experimental tilt series to determine the imaging parameters accurately (Saxton, 1995).

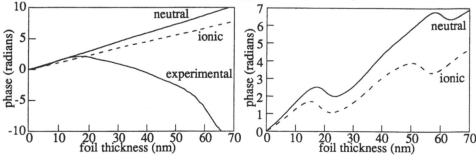

Fig.8 Experimental and simulated phase of holographic interference fringes as a function of thickness (see text for details).

Fig.9 Simulated phase of holographic interference fringes at the [0001] zone axis,

It was clear from fig.3 that the phases of the beams show a strong dependence on ionicity - and there is a correspondingly large difference in the mean inner potential $V_0$ for $Al_2O_3$ of 23% between values calculated using neutral and ionic scattering factors. While neither high resolution imaging nor CBED can be used to determine $V_0$ for a pure material, we now discuss the use of electron holography to determine $V_0$ and hence ionicity. A preliminary experimental hologram of a sapphire sample with an [0001] foil normal was obtained using a Philips CM200 FEG operated at 200kV with a voltage of 100V applied to the biprism. The foil was tilted from the [0001] zone axis to a weakly diffracting orientation, and the resulting phase of the holographic fringes is shown in fig.8 as a function foil thickness, alongside full multislice simulations of the phase for the experimental conditions using both neutral and ionic scattering factors. The experimental phase profile is only consistent with the simulations in the thinnest regions of the foil, probably as a result of charging of the specimen due to its lack of carbon contamination. Both dynamical effects and absorption were included in the simulations and so cannot be responsible for this discrepancy. (Fig.9 shows that the simulated phase does not reverse in the manner observed experimentally even for a foil oriented exactly at the [0001] zone normal, at which dynamical effects should be strongest). However, the sensitivity of electron holographically-measured phase to ionicity at both weakly diffracting and more dynamical orientations, as well as the simplicity of such a measurement as compared with the quantitative analysis of high resolution image contrast, is beyond doubt.

We are grateful to Trinity Hall, Cambridge for financial support and to D J Smith and M R McCartney for the use of laboratory facilities at Arizona State University. We also thank A R Preston for help with the calculations and P A Midgley for useful discussions.

**REFERENCES**

Boothroyd CB 1997 J. Microscopy, in press
Gemming T, Exner M, Wilson M, Möbus G and Rühle M 1996 Presented at EUREM Dublin
Kirfel A and Eichhorn K 1990 Acta. Cryst. A46, 271
Midgley PA and Saunders M 1996 Contemp. Phys. 37, 441
Rez D, Rez P and Grant I 1994 Acta Cryst. A50, 481
Saxton WO 1995 J. Microscopy 179, 201
Stobbs S H and Stobbs W M 1995 Inst. Phys Conf Ser. 147, 83
Tietz L A, Summerfelt S R and Carter C B 1992 Phil. Mag. A 65, 439

*Inst. Phys. Conf. Ser. No 153: Section 3*
*Paper presented at Electron Microscopy and Analysis Group Conf. EMAG97, Cambridge, 1997*
© *1997 IOP Publishing Ltd*

# Super-resolution STEM imaging of crystals with large unit cells

**C P Colman and J M Rodenburg**

Cavendish Laboratory, Madingley Road, Cambridge CB3 0HE

**ABSTRACT:** This paper models and compares two methods of reconstructing the complex specimen function in a scanning transmission electron microscope at super-resolution. Using dynamical calculations, the accuracy of both the 'ptychography' and 'quadrant' detector methods in reconstructing the projected potential is compared for increasing specimen unit cell sizes and their suitability is discussed.

## 1. INTRODUCTION

The method of super-resolution imaging (Rodenburg and Bates, 1992) is a means of reconstructing the complex specimen function to a much higher resolution than the conventional bright-field image. It involves collecting a set of coherent convergent beam electron diffraction (microdiffraction) patterns for various probe positions in a STEM. The time taken to grab all the data causes problems due to specimen drift to manifest themselves (Colman and Rodenburg, 1995).

To eliminate such problems, simplifications are made to the algorithm for particular types of specimen. In order to find out the next step in the experimental development, two such simplified methods of specimen reconstruction are compared using dynamical calculations. These two methods are known as the 'ptychography' and 'quadrant' reconstruction methods.

The ptychography method assumes the specimen is crystalline, producing a microdiffraction pattern with a discrete number of discs, and has been investigated by Plamann and Rodenburg (1995) for different thicknesses of InP. Nellist et al (1995) used this method to resolve silicon along the <110> direction to atomic resolution in a STEM with a point resolution of 4.2Å – a resolution improvement by a factor of three. This method has since been used to reconstruct gallium phosphide along the <110> direction to atomic resolution in the same microscope.

The quadrant detector method processes only the central disc of the microdiffraction pattern using a 4-sector detector and can theoretically reconstruct a general, weak scattering specimen to double the point resolution of the microscope (Landauer et al, 1995).

Since the ptychography and quadrant reconstruction methods are applicable to small and large unit cell specimens respectively, the knowledge of the accuracy of both reconstruction methods with increasing unit cell size would allow a practical reconstruction algorithm to be chosen for a particular specimen. In this paper we compare the performance of both methods in reconstructing the complex specimen function for the case of increasing unit cell size in the presence of dynamical scattering.

## 2. THE RECONSTRUCTION SCHEMES

### 2.1 Ptychography reconstruction

Referring to Nellist and Rodenburg (1994), the ptychography method relies on the specimen being crystalline and hence producing a discrete number of beams in the microdiffraction pattern.

This assumption allows for a one-dimensional set of probe positions to be used – thus reducing the quantity of data and collection time. If the beams are singly overlapping then unaberrated information beyond the resolution limit of the microscope can be obtained. If one considers a set of microdiffraction patterns collected for each probe position along a line-scan, then the interference between two overlapping discs $a$ and $b$ can be written as:

$$\left|M(r',\rho)\right|^2 = \left|\Psi_a\right|^2 + \left|\Psi_b\right|^2 + 2\left|\Psi_a\Psi_b\right|\cos\{\angle\Psi_b - \angle\Psi_a + \chi(r'-b) - \chi(r'-a) + 2\pi\rho\cdot(a-b)\}$$

where $r'$ is a pixel in the microdiffraction plane, $\rho$ is the position of the probe, $\angle\Psi_b$ - $\angle\Psi_a$ is the phase angle between the two diffracted beams and $\chi$ is the lens aberration containing defocus and spherical aberration terms.

By Fourier transforming the above formula with respect to the probe position and assuming the aberration term is centro-symmetric, the relative phase of the beams can be obtained from the centre of disc overlaps (the $\chi$ terms cancel). In this way beams can be reconstructed beyond the point resolution of the microscope.

As the unit cell size/complexity increases, the discs in the microdiffraction pattern start to overlap more and so the extraction of unaberrated information from disc centres would be expected to be more difficult.

## 2.2    Quadrant reconstruction

Landauer et al (1995) describe a method of reconstructing a general weak specimen using a quadrant detector in a STEM. Since only 4 pixels are collected from each microdiffraction pattern using a customised detector, the quantity of data and collection time are much lower than the full super-resolution method. Reconstruction is via an optimal (Wiener) filtering process, which is based on the assumption that the estimated specimen function is a linear combination of the Fourier transforms of the four different images. This technique can reproduce the complex specimen function at up to twice the conventional resolution of the microscope and has so far been applied to amorphous specimens (Landauer and Rodenburg, 1995).

Since this method reconstructs at double the point resolution of the microscope, it is limited to 2.1Å reconstructions in the HB501 STEM with an analytical pole piece ($C_s$=3.1mm).

## 3.    SIMULATIONS

To test the reconstruction methods, simulations were performed on a InP structure which had a lattice parameter of 5.87Å, 7.64Å and 9.4Å corresponding to 'dumbbell' spacings of 1.47Å, 1.91Å and 2.35Å respectively. Initially, a multislice calculation was performed for a thickness of ~25Å to produce a projected potential along the <110> direction. This potential was then used to produce simulated microdiffraction patterns for different probe positions. The microscope was assumed to be a VG microscopes HB501 STEM operating at 100KV with $C_s$ = 3.1mm and using a 50µm objective aperture. In the case of the quadrant detector, a two-dimensional raster scan of probe positions was performed and the central disc processed. In the case of ptychography, a line-scan was chosen to separate the spatial frequencies and information from the centre of discs and disc overlaps was processed. Fig. 1 shows the magnitude of the relevant coherent CBED patterns along with the phases of both reconstructions as the unit cell size is increased. The effective real-space sampling of the reconstructions varies so that the dumbbell features appear on the same scale.

CBED pattern      'Ptychography' reconstruction    'Quadrant' reconstruction

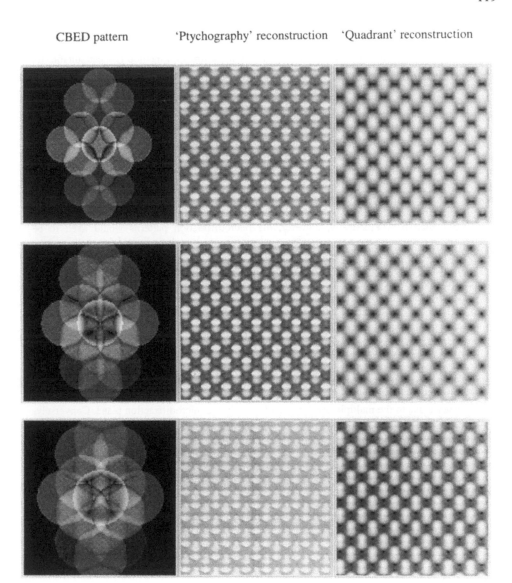

Fig. 1. Successive reconstructions of an InP type structure along the <110> direction. The left column shows the relevant coherent CBED pattern expected and the middle and right columns show the phase of the 'ptychography' and 'quadrant' reconstructions respectively. From top to bottom, the lattice parameter is 5.87Å, 7.64Å and 9.4Å corresponding to 'dumbbell' spacings of 1.47Å, 1.91Å and 2.35Å respectively. The reconstructions have been re-scaled for comparison.

## 4. DISCUSSION

It can be seen from Fig. 1 that the ptychography reconstruction method is most suitable for crystals with small unit cells - in this case reconstructing the specimen function at up to three times the point resolution of the microscope. However, as the unit cell size increases so the multiply overlapping discs cause difficulty in extracting unaberrated information and a degraded reconstruction results.

In addition, as the unit cell becomes more complicated, so different types of disc overlaps are visible in the microdiffraction pattern. This has further implications for the ptychography method since it relies on choosing a line-scan direction which will separate the different spatial frequencies once the dataset is Fourier transformed. As the number of spatial frequencies increases then so it becomes more difficult to choose a line-scan direction to separate them. The use of two separate line-scans in different directions should alleviate this situation.

In theory, using different size objective apertures would allow the discs to be phased out with reduced aberration. Practically, however, the microdiffraction patterns would need to be stigmated and focussed and the post-specimen spectrometer quadrupoles adjusted for each aperture used and the specimen tends to move in the meantime.

The quadrant reconstruction method is more suitable as the unit cell size increases and as the complexity of the unit cell increases. Since it is restricted to a double resolution reconstruction, atomic resolution imaging is not practicable in a 100kV STEM – though atomic imaging of a material such as InP <100> should be possible. One would expect the method to be very useful at imaging weak specimens in a 300kV microscope.

## 5. CONCLUSIONS

It has been shown through dynamical simulations that the ptychography method of reconstructing the projected specimen potential breaks down as the complexity and size of the unit cell size increase due to the multiple overlapping of discs in the microdiffraction plane. Conversely, the quality of the quadrant detector reconstructions increases to a point where this method is preferable over ptychography. It is suggested that the quadrant reconstruction method is a practical solution to high-resolution imaging of weakly scattering specimens with large and/or complex unit cells.

## ACKNOWLEDGEMENTS

C.P.C. acknowledges the EPSRC for financial support. J.M.R. is grateful to the Royal Society and the EPSRC for support.

## REFERENCES

Colman C P and Rodenburg J M 1995 Proc. EMAG95, ed D Cherns 107-110
Landauer M N and Rodenburg J M 1995 Proc. EMAG95, ed D Cherns 281-284
Landauer M N, McCallum B C and Rodenburg J M 1995 Optik **100** 1 37-46
Nellist P D and Rodenburg J M 1994 Ultramicroscopy **54** 61-74
Nellist P D McCallum B C and Rodenburg J M 1995 Nature **374**
Plamann T and Rodenburg J M 1995 Proc. EMAG95, ed D Cherns 117-120
Rodenburg J M and Bates R H T 1992 Phil. Trans. Roy. Soc. **339** 521-553

*Inst. Phys. Conf. Ser. No 153: Section 3*
*Paper presented at Electron Microscopy and Analysis Group Conf. EMAG97, Cambridge, 1997*
© *1997 IOP Publishing Ltd*

# Construction of a Database of HREM Images on the World Wide Web

**D Shindo, A Taniyama, Y Murakami, K Hiraga[*], T Oikawa[#] and A I Kirkland[+]**

Institute for Advanced Materials Processing, Tohoku University, Sendai 980-77, Japan.
[*] Institute for Materials Research, Tohoku University, Sendai 980-77, Japan.
[#] Electron Optics Division, JEOL Ltd., 1-2 Musashino 3-Chome, Akishima, Tokyo 196, Japan.
[+] JEOL UK Ltd., Silvercourt, Watchmead, Welwyn Garden City, Herts, AL7 1LT, UK and University of Cambridge, Department of Chemistry, Lensfield Road, Cambridge, CB2 1EW.

**ABSTRACT:** The rapid development of the World Wide Web has opened up many opportunities for the electronic distribution of data from various Electron Microscopes to a global audience. The implementation of a data base of High Resolution Images distributed across the Web will be discussed together with potential future Web based developments of interest to Electron Microscopists.

## 1. INTRODUCTION

The World Wide Web (www) has expanded sufficiently rapidly in recent years such that it is now a genuinely global repository of information that can be readily accessed and searched with relatively simple and freely available hardware and software. The potential to utilise the Web as a platform for collating microscopy related information is therefore of considerable interest and a number of microscopy web sites have already been established. An additional use of the Internet lies in the provision of "telemicroscopy" where access via the Web allows users to control microscopes in remote locations ( Voelkl et al 1997); this latter use being of particular relevance to the education of the microscopy community.

In this paper we will discuss the implementation and contents of a microscopy based Web site with the pseudonym EMILIA (Electron Microscope Image Library and Archive) which was established in 1986 with the aim of providing an electronic database of High Resolution Images from a wide variety of samples. At the current time a visual collection of some 47 images from a range of materials can be browsed. In the future catalogues of frequently asked questions (FAQ's) generated via electronic postings to the Web site and common microscopical techniques will also be available together with a comprehensive database of images accessible by subscription.

## 2. IMPLEMENTATION

EMILIA is implemented on a dedicated server attached to the intelligent computer network at Tohoku University (Super TAINS) which is linked by high speed (100Mbps) connections to the EMILIA Server, the High Voltage EM Centre and to the ACOS-S3900 super computer. Fig. 1 shows a schematic diagram of the interconnections between EMILIA, Super TAINS and other local network components.

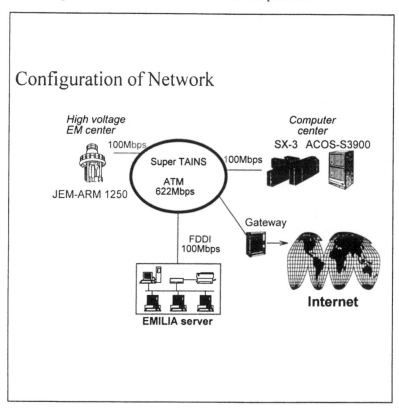

Figure 1. A Schematic diagram of the network connectivity of the EMILIA Web server.

The detailed configuration of the EMILIA Web server is given below in Table 1.

| URL | http://asma7.iamp.tohoku.ac.jp/EMILIA |
|---|---|
| EMILIA Web Server | Sun SS5 compatible, CPU: Turbo SPARC 170 Mhz. |
| Server Program (httpd) | Apache v 1.20. |
| Disk Space available for HTML documents. | 1.3GB |

Table 1. The detailed configuration of the EMILIA Web server.

Statistics for access to EMILIA are given in table 2.

| Requests received | 5397 |
|---|---|
| Bytes transmitted | 62464403 |
| Average requests received daily | 90 |
| Average bytes transmitted daily | 1041073 |

Table 2. Access statistics for the EMILIA Web server for the summary period January 8th 1997 to march 14th 1997.

## 3. CONTENTS

The EMILIA Web server is divided into a number of clearly defined areas (some of which are still under development and not yet available for public access). The principal areas are as follows:

a) A welcome page providing basic information on the services provided by EMILIA.

b) The EM Gallery, an example collection of high resolution images grouped according to material type. Each image is accompanied by a full description of the imaging conditions under which it was recorded together with details of the specimen preparation.

c) The File Library (under construction). A full subscription service database containing a more comprehensive collection of images than is available via the EM Gallery together with a dedicated search engine.

d) EMILIA Q and A (under construction). A collection of frequently asked questions about EMILIA together with a list of EMILIA visitors, contributors and subscribers.

An illustration of the index page for the EM Gallery together with an example gallery from the superconductors section are illustrated in Fig. 2. To illustrate the quality of the stored images Fig. 3 shows an extracted image from the superconductors section of the EM gallery

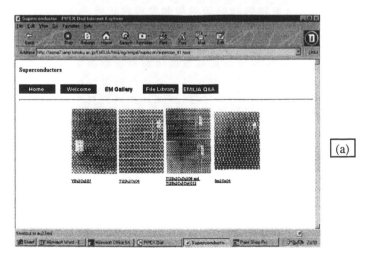

124

Figure 2. Example pages from the EM gallery. (a) (previous page) The base index page listing available materials by category, (b) the superconductors section showing thumbnails images of four superconductors.

Figure 3. An example image of $Tl_2Ba_2CuO_6$ viewed along the [010] direction extracted as a bitmap file from the superconductors section of the EM Gallery.

## REFERENCES

Voelkl E, Allard L F, Bruley J, Williams D B, This conference, 1997.

Inst. Phys. Conf. Ser. No 153: Section 3
Paper presented at Electron Microscopy and Analysis Group Conf. EMAG97, Cambridge, 1997
© 1997 IOP Publishing Ltd

# Chi-squared differences in Fourier space: comparing experimental with simulated images

W O Saxton

Department of Materials Science & Metallurgy, Pembroke St, Cambridge, CB2 3QZ

**ABSTRACT:** The chi-squared difference has advantages over the most other measures that have been used to quantify the difference between experimental and simulated images. In particular, if reliable estimates of the noise statistics are available it can directly quantify the plausibility of an experimental image given a particular specimen and imaging model. Applied in *Fourier* space, it is the *only* measure that remains interpretable when the image noise is not normally distributed or when – as is normally the case – the noise in neighbouring image pixels is not independent.

## 1. INTRODUCTION

Numerical measures of the difference between pairs of images are important for image interpretation at high resolution – for example, in comparing experimental images with those predicted by particular specimen models – since more sensitive difference measures allow finer structural inferences to be drawn about the specimen. Some such comparators are set out here, and their inter-relationships pointed out.

Quantitative conclusions as to the plausibility of an experimental image given a particular predicted (simulated) image may be drawn only from particular comparators in carefully defined circumstances; in particular, it will normally be necessary to apply these in Fourier rather than real space.

A fuller account of this work, with mathematical details, is given by Saxton (1997b).

## 2. REAL SPACE COMPARATORS

There are two rather different kinds of question that we may want to pose concerning the comparison of images. We may ask *"Which of a set of simulated images is most like one particular experimental image?"*, in which case a large number of comparators are perfectly adequate. Those in most frequent use are the RMS difference (RMSD) $D_{fg}$ between two images $f$ and $g$, defined via its square $D_{fg}^2 = \overline{(f-g)^2}$, and its close relative the cross mean $C_{fg} = \overline{fg}$ which measures similarity rather than difference. Where, as is often the case in practice, image data have an unknown gain and intensity offset, the useful measures are the RMSD $D'_{fg}$ and cross mean $C'_{fg}$ between *normalised* forms of the images – versions rescaled to have zero mean and unit standard deviation. $C'_{fg}$ is more familiar as the 'correlation coefficient'. $D'$ is zero, and $C'$ is one, only when the images are linearly related (i.e. when $f=ag+b$). The hyper-angle $\theta$ between the image vectors in a Hilbert space spanned by their pixels has become popular recently (Ourmazd et al., 1990), but although very differently conceived it is in fact virtually identical to the RMSD $D'_{fg}$. The following relationships exist between the various measures mentioned so far:

126

$$D^2 = \overline{f^2} + \overline{g^2} - 2C; \quad D'^2 = 2 - 2C; \quad \theta = \cos^{-1}(C') = D'.$$

The MSD between two unnormalised images can be conveniently partitioned (e.g., Hÿtch & Stobbs, 1994) into three terms that separate clearly the contribution made by the difference between the image means (background levels) and the difference between their standard deviations (contrast levels) from the residual difference between the normalised images in their patterns:

$$D^2 = (\bar{f} - \bar{g})^2 + (\sigma_f - \sigma_g)^2 + D'^2 \sigma_f \sigma_g.$$

For a comparator is to be open to verification by independent researchers, its definition must be supplemented with a statement of the exact field of view compared and the sampling lattice used. Coarse sampling obviously misses fine detail and so reduces high frequency contributions to the RMSD; the precise effect of the sampling is however neither simple nor isotropic. In addition, any filter applied to images before comparison (e.g. to eliminate high frequency noise) must be specified exactly.

## 3. FOURIER SPACE COMPARATORS

The difference measures defined above for real space need slight adaptation to accommodate complex Fourier space data. Where small unit cell images are concerned, there is a useful reduction in the number of data to be compared; other reasons for preferring this space emerge later.

Transforms of finite images are necessarily of infinite extent, so all Fourier space comparators must be restricted to a specified finite region of the space. The RMSD between two image transforms $D_{FG}$ is defined via its square $D_{FG}^2 = \overline{|F - G|^2}$, and the cross mean requires slight generalisation, being defined as $\mathrm{Re}\{C_{FG}\} = \mathrm{Re}\{\overline{FG^*}\}$. $D'$ and $\mathrm{Re}\{C'\}$ are defined between normalised transforms – rescaled to have unit RMS, and with the region compared redefined to exclude the zero order component (the image mean); $\mathrm{Re}\{C'\}$ is the correlation coefficient (though $C'$ itself may also be called that). $D'$ is zero, and the cross mean $\mathrm{Re}\{C'\}$ is 1, only when the transforms differ by no more than a (real) multiplicative constant. The hyper-space angle $\theta$ between the two transforms is again virtually identical to the RMSD $D'$. The following relationships exist between the comparators:

$$D^2 = \overline{|F|^2} + \overline{|G|^2} - 2\mathrm{Re}\{C\}; \quad D'^2 = 2 - 2\mathrm{Re}\{C'\}; \quad \theta = \cos^{-1}(C') = D'.$$

The MSD between two unnormalised transforms can be partitioned into two terms separating the contribution made by the difference between the RMS values from the residual difference between the patterns:

$$D^2 = (F_{RMS} - G_{RMS})^2 + D'^2 F_{RMS} G_{RMS}.$$

## 4. SIGNIFICANCE TESTS IN REAL SPACE

A different question, also important, is *"Is a particular experimental image f plausible given a particular predicted image g?"*. To answer this, we need to be able to say what range of values of $D$ (or the other comparators) is plausible at random.

We define the image noise as the difference between the observed image $f$ and the predicted image $g$, i.e. $f_i = g_i + n_i$, (so that all noise may be said to be additive). Initially, we make two assumptions about the noise in a given pixel $i$: that it is normally distributed with zero expectation and variance $\sigma_i^2$, and that the noise in different pixels is independent (or at least uncorrelated). In these circumstances, the MSD is then also normally distributed with a mean and variance found by simply adding the summing those of individual squared noise values. For example, if the noise in all pixels has the same variance $\sigma^2$, then the MSD is distributed with a standard deviation $\sigma^2 \sqrt{2/M}$ around a mean value $\sigma^2 / M$ ; values more than twice this SD from the mean are unlikely to occur at random and imply an implausible specimen or imaging model.

$D^2 = \Sigma_i n_i^2 / M$ is already almost a $\chi^2$ difference; in the latter, the noise in each pixel is measured in units of its own standard deviation $\sigma_i$, giving more weight to more accurately observable pixels. $\chi^2$ has acquired a probably unfair reputation for being difficult to use because of its complicated dependence on the number of degrees of freedom when these are few; applied to images with large numbers of pixels, it is actually very simple to interpret without any use of tables. The mean chi-squared difference (MCSD), $D_\chi^2 = \Sigma_i (n_i^2 / \sigma_i^2) / M$ , is normally distributed with mean 1 and standard deviation $\sqrt{2 / M}$ ; values further than about $3 / \sqrt{M}$ from 1 are unlikely at random (fig. 1).

The first assumption made above does not require signal-independent or position-independent noise, nor low contrast images; only low-dose recording conditions (under which the noise is not normally distributed) require a more elaborate treatment.

However, the second assumption can rarely be made of image data: even though pixels some distance apart are unrelated, the point response of the recording system and re-interpolation of the image to achieve a geometrical match with the simulated image normally introduce significant correlation between neighbouring pixels. Such correlation increases the variance of the MCSD, making the $3 / \sqrt{M}$ threshold given above misleadingly small. As a simple example, if a given noise image is re-sampled at half the original spacing using nearest neighbour interpolation, the MCSD is unchanged, but the threshold level is halved, making the experimental image appear less plausible than is actually the case. To solve this serious problem, we turn to Fourier rather than in real space.

## 5.    SIGNIFICANCE TESTS IN FOURIER SPACE

We can define noise in Fourier space in the same way as in real space, as the difference between the transforms $F$ and $G$ of the experimental and predicted images: $F_i = G_i + N_i$. Crucially, however, there is usually no correlation between the noise in different transform pixels even when the image pixels are inter-dependent. In particular, there is no correlation when the real-space noise correlation properties are spatially invariant, which is usually the case, and when the noise variance is the same at all pixels, which is the case when the image contrast is low and windowing (e.g. De Ruijter, 1994) is avoided. In addition, the real and imaginary parts of $N_i$ (which are always independently distributed) are normally distributed regardless of the distribution of $n_i$, accommodating low dose images too (Saxton, 1997a).

These two properties together mean that a MCSD, once adapted to Fourier space, does indeed involve a sum of squares of standard normal variates for any image, and virtually *independent* variates for most images; the Fourier space MCSD is therefore directly interpretable in terms of the probability of so large a value arising at random, in marked contrast to the position in real space.

In Fourier space, the noise variance is defined via the (ensemble) mean square modulus, $\mathrm{var}(N_i) = \overline{|N_i|^2} = \sigma_{Fi}^2$, and the MCSD is defined in terms of this by

$$D_\chi^2 = \frac{2}{M} \sum_{i \in HP} \frac{|N_i|^2}{\sigma_{Fi}^2} ;$$

the sum is extended over one the $M/2$ pixels in one half-plane only, as the transform data appear in complex conjugate pairs. This is distributed exactly like the real-space expression of the previous section, and we may again treat values more than about $3 / \sqrt{M}$ from unity as suspicious (fig. 1 again) – but with much less restrictive requirements of the image noise statistics.

The normalisation of individual pixels with respect to their own standard deviations, helpful without being essential in real space, is indispensable in Fourier space because of the wide variation in orders of magnitude between different Fourier components, all of them still important for characterising the structure.

It should not be necessary to stipulate a maximum frequency for the region used for MCSD calculation, as changes in this will be reflected in appropriate changes to the range to be considered plausible, provided the transform noise estimates are reliable.

## 6.    DISCUSSION

In principle, it is now clear that the Fourier space MCSD is the best comparator of those considered, and the best candidate for quantitative interpretation in the long term. At present, it is however probably still premature to expect such interpretation, for two reasons.

Firstly, rather more accurate estimates of experimental noise levels are required than we are accustomed to making – and possible more accurate estimates than we can expect to make easily; this process deserves further attention.

Secondly, and more seriously, predicted images at present frequently do *not* agree closely with experimental images.   Experimental contrast levels in particular are commonly two or three times lower than those predicted (e.g., Hÿtch & Stobbs 1994), and MCSD values reported so far in practice (e.g. Zhang et al., 1995) have been well outside the range shown here to be plausible at random.   Until these problems are resolved, much of the exploration reported here must be regarded primarily as preparing for the future.

I am grateful to New Hall, Cambridge, for financial support, and to my department for laboratory facilities.

## REFERENCES

De Ruijter W J  1994  J. Comp. Assisted Microsc. **6**, 195
Hÿtch M J & Stobbs W M 1994 Ultramicroscopy **53**, 191
Ourmazd A, Baumann F H, Bode M & Kim Y 1990 Ultramicroscopy **34**, 237
Saxton W O 1997a Transform noise statistics and Fourier component estimation.   In *Signal and Image Processing in Microscopy and Microanalysis* (ed Hawkes P W) Scanning Microscopy Suppl.  Chicago: Scanning Microscopy International, in press
Saxton W O 1997b  J. Microscopy, in press
Zhang H, Marks L D, Wang Y Y, Zhang H, Dravid V P, Han P & Payne D A 1995  Ultramicroscopy **57**, 103

Fig. 1 Is there any significant difference between the simulated and noisy 'experimental' images left? MCSD values say there is none.   Left: simulated image of h-$Nb_2O_5$ 100 pixels square, and 'experimental' image in fact created by addition of (uncorrelated) noise to the former.   Right: transforms of these, shown as log modulus. In both cases, the range of MCSD expected at random is 0.97–1.03; the calculated values are 1.021 between the images and 1.022 between the transforms.

*Inst. Phys. Conf. Ser. No 153: Section 4*
*Paper presented at Electron Microscopy and Analysis Group Conf. EMAG97, Cambridge, 1997*
© 1997 IOP Publishing Ltd

# Quantitative Convergent Beam Electron Diffraction (CBED) measurements of low-order structure factors in metals

**M Saunders and A G Fox**

Center for Materials Science and Engineering, Department of Mechanical Engineering, Naval Postgraduate School, Monterey, CA 93943, USA

**ABSTRACT:** The quantitative zone-axis CBED pattern matching technique developed by Bird and Saunders is applied to low-order structure factor refinements from the metals nickel and copper. A new method is proposed for determining which structure factors can be refined from a specific data-set. The validity of the widely used correction potential for thermal diffuse scattering (TDS) is considered by comparing CBED refinement values of the imaginary structure factors to those given by the Einstein model. In addition, the importance of second-order TDS corrections in bonding studies is investigated.

## 1. INTRODUCTION

In recent years quantitative CBED techniques have been developed which permit bonding charge density studies through the accurate refinement of low-order structure factors (Holmestad, et al, 1995; Saunders, et al, 1995). The zone-axis pattern matching technique of Bird and Saunders (ZAPMATCH) minimizes the sum-of-squares difference between an elastic filtered CBED pattern and a many-beam simulation by adjusting the low-order structure factors, sample thickness and various constant terms (see Saunders, et al, 1995 for details). Despite its early success, some fundamental questions still exist about its application.

For example, the choice of the number of structure factors that can be refined from a single zone-axis pattern remains something of an arbitrary choice. While it has been realized that the sensitivity of the data to the structure factors must change as a function of the sample thickness, rules which govern the number of structure factors we can refine from a given data-set have yet to be developed. Here we consider this sensitivity question for structure factor refinements from <110> zone-axis patterns of nickel and copper.

The effects of thermal diffuse scattering (TDS) are generally introduced via an imaginary potential given by the Einstein model of independent vibrating atoms (Bird and King, 1990). These correction terms are refined as part of the ZAPMATCH procedure but no attempt has previously been made to extract information from the optimized values. Here we test the validity of the Einstein model by comparing our refined values for the imaginary structure factor corrections for nickel and copper with those given by Bird and King (1990).

The Einstein model also produces a second-order correction term in phase with the bonding charge contribution to the low-order structure factors (Anstis, 1996) which may interfere with our bonding measurements in metals. We must, therefore, consider this additional correction if we are to investigate bonding in these materials.

## 2. SENSITIVITY

The data-set used for the pattern matching calculations consists of intensities extracted from the (000), {111} and {200} reflections at the <110> zone-axis. The only single-scattering paths that exist for scattering from the direct beam into another reflection in the data-set are for the 111 and 200 structure factors. The data should therefore be very sensitive to these parameters. All higher-order structure factors either scatter out of the data-set or produce scattering into the data-set via multiple-scattering paths. This restriction on the higher-order structure factors will dramatically reduce the sensitivity of the data-set to them.

The sensitivity to the various low-order structure factors must also be governed by the sample thickness which determines the relative scattering probabilities. As the sensitivity to the 111 and 200 structure factors is dominated by single-scattering it should be possible to refine these accurately for data acquired from thin samples, say a few hundred Ångstrom or more. Conversely, the dependence of the higher-order structure factors on multiple-scattering means that much thicker samples are required before we can refine these with any accuracy.

The sensitivity as a function of the sample thickness has been investigated by carrying out repeated pattern matching calculations with increasing numbers of variable structure factors to two room temperature nickel <110> zone-axis patterns acquired from 800Å and 1400Å samples. Bonding effects are small and reduce as a function of scattering angle. Thus, the additional structure factor introduced in each fit is expected to deviate little from the neutral atom value at which it was fixed in the previous fit. By monitoring the variation in the results as we introduce the additional variable we should be able to detect the point at which sensitivity is lost, i.e. introducing structure factors to which the data is insensitive will give the fit too much freedom producing significant changes in the best-fit structure factors.

The real and imaginary best-fit structure factors for successive fits are shown in Tables 1 and 2. The 800Å data shows a clear loss of sensitivity on the introduction of the 222 structure factor as a variable. This is reflected both in the real and imaginary structure factors. The data from the thicker sample, however, maintains reasonable sensitivity out to 400 (Fit 4).

| g | Fit 1 (220 varies) | | Fit 2 (113 varies) | | Fit 3 (222 varies) | | Fit 4 (400 varies) | |
|---|---|---|---|---|---|---|---|---|
| | 800Å | 1400Å | 800Å | 1400Å | 800Å | 1400Å | 800Å | 1400Å |
| 111 | 20.47 | 20.46 | 20.48 | 20.46 | 20.49 | 20.46 | 20.48 | 20.46 |
| 200 | 19.13 | 19.12 | 19.14 | 19.12 | 19.13 | 19.13 | 19.13 | 19.13 |
| 220 | 15.47 | 15.47 | 15.45 | 15.47 | 15.66 | 15.54 | 15.68 | 15.52 |
| 113 | | | 13.65 | 13.66 | 13.82 | 13.62 | 13.37 | 13.62 |
| 222 | | | | | 13.03 | 12.90 | 13.27 | 12.87 |

Table 1. Best-fit real structure factors for <110> room temperature nickel.

| g | Fit 1 | | Fit 2 | | Fit 3 | | Fit 4 | |
|---|---|---|---|---|---|---|---|---|
| | 800Å | 1400Å | 800Å | 1400Å | 800Å | 1400Å | 800Å | 1400Å |
| 111 | -0.22 | -0.21 | -0.22 | -0.21 | -0.22 | -0.21 | -0.22 | -0.21 |
| 200 | -0.17 | -0.19 | -0.19 | -0.18 | -0.18 | -0.18 | -0.19 | -0.18 |
| 220 | -0.14 | -0.16 | -0.11 | -0.18 | -0.18 | -0.18 | -0.18 | -0.19 |
| 113 | | | -0.09 | -0.15 | -0.08 | -0.14 | -0.07 | -0.13 |
| 222 | | | | | +0.02 | -0.11 | 0.00 | -0.12 |

Table 2. Best-fit imaginary structure factors for <110> room temperature nickel.

## 3. THERMAL DIFFUSE SCATTERING (TDS)

The effects of TDS are included in our calculations using the Einstein model of independent atomic vibrations. The ATOM program of Bird and King (1990) is used to calculate an imaginary correction to each structure factor. The imaginary parts of the variable low-order structure factors are then refined alongside their real counterparts while the higher-order terms remain fixed at the Einstein model values. The validity of the Einstein model in quantitative CBED calculations has never been questioned and the refined imaginary structure factors are usually ignored.

A comparison of the refined low-order imaginary structure factors and the Einstein model values is given in Table 3 for nickel (at both room and liquid nitrogen temperatures) and copper (at liquid nitrogen temperatures only). In all cases the best-fit results are higher than the Einstein values but the magnitude and the trend as a function of scattering angle are similar to the model. Thus, the Einstein model, while not being perfect, is certainly a reasonable approximation.

| g | RT Nickel | | $LN_2$ Nickel | | $LN_2$ Copper | |
|---|---|---|---|---|---|---|
| | Einstein | Fit | Einstein | Fit | Einstein | Fit |
| 111 | -0.16 | -0.21(1) | -0.10 | -0.14(1) | -0.11 | -0.16(1) |
| 200 | -0.15 | -0.19(1) | -0.10 | -0.11(1) | -0.11 | -0.15(2) |
| 220 | -0.14 | -0.16(2) | -0.09 | -0.09(2) | -0.10 | -0.10(4) |
| 113 | -0.13 | -0.13(3) | -0.09 | -0.07(3) | -0.10 | -0.08(4) |

Table 3. Comparison of the refined imaginary TDS structure factor corrections and those given by the Einstein model.

The imaginary TDS structure factor considered above is intended to correct for the effects of elastically scattered electrons whose final scattering interaction is with a phonon. This term is only the first in an expansion series where other terms describe electrons which have undergone multiple phonon scattering. These additional terms are small and are usually ignored in simulations. However, in a recent paper, Anstis (1996) suggested that the accuracy achieved with quantitative CBED techniques such as ZAPMATCH may be sufficient to detect the effects of the second-order TDS correction. This is significant as the second-order correction is real and therefore indistinguishable from the bonding contributions in the low-order structure factors. Thus, if we are to make accurate bonding measurements we must investigate the importance of these second-order TDS corrections.

The results of ZAPMATCH calculations for room temperature nickel both with and without second-order TDS corrections are given in Table 4. The results are averaged from four data-sets at a range of thicknesses from 600Å to 2500Å. For comparison, neutral atom structure factors from Doyle and Turner (1968), solid-state theory calculations using the WIEN95 FLAPW code (Blaha, et al, 1995) and critical voltage measurements (Fox and Fisher, 1988) are also included. Both the experimental techniques and the theoretical calculations are in excellent agreement. Repeating the ZAPMATCH calculations with the second-order TDS corrections lowers the measured temperature factor (Saunders and Fox, in these proceedings) but does not alter the refined low-order structure factors within errors.

Our results for liquid nitrogen cooled copper are shown in Table 5 compared to neutral atom, solid-state theory and critical voltage values. The two experimental results are again in excellent agreement but the theoretical calculations now lie outside the experimental errors.

| g | Neutral Atom | Theory FLAPW | Critical Voltage | ZAPMATCH | ZAPMATCH 2nd-order TDS |
|---|---|---|---|---|---|
| 111 | 20.54 | 20.45 | 20.45(1) | 20.47(2) | 20.46(2) |
| 200 | 19.25 | 19.12 | 19.12(2) | 19.13(3) | 19.12(3) |
| 220 | 15.53 | 15.48 | | 15.44(5) | 15.46(4) |
| 113 | 13.66 | 13.66 | | 13.67 | 13.67 |

Table 4. Comparison of best-fit elastic structure factors for room temperature nickel with neutral atom values (Doyle and Turner, 1968), solid-state theory calculations (WIEN95, Blaha, et al, 1995) and CV measurements (Fox and Fisher, 1988).

| g | Neutral Atom | Theory FLAPW | Critical Voltage | ZAPMATCH |
|---|---|---|---|---|
| 111 | 22.05 | 21.71 | 21.76(1) | 21.77(2) |
| 200 | 20.70 | 20.37 | 20.44(2) | 20.44(3) |
| 220 | 16.75 | 16.75 | 16.71(14) | 16.71(5) |
| 113 | 14.75 | 14.75 | | 14.77 |

Table 5. Comparison of best-fit elastic structure factors for liquid nitrogen cooled copper with other techniques (as in Table 4).

## 4. CONCLUSIONS

Quantitative zone-axis CBED has been applied successfully to low-order structure factor measurements in metals. The sensitivity of the data can be investigated by repeating the calculations with increasing numbers of structure factor variables and monitoring the behavior of the best-fit structure factors. Using the Einstein model to include the effects of TDS is satisfactory but the refinements always lead to increased imaginary correction terms. The use of second-order TDS corrections does not significantly improve or degrade the results.

## ACKNOWLEDGEMENTS

This research was carried out while one of the authors (MS) held a National Research Council-Naval Postgraduate School Research Associateship. The authors would also like to thank Geoff Anstis for his help in calculating the second-order TDS corrections.

## REFERENCES

Anstis G R 1996 Acta Cryst A52, 450.
Bird D M and King Q A 1990 Acta Cryst. A46, 202.
Blaha P, Schwarz K, Dufek P and Augustyn R 1995 WIEN95, Tech. University of Vienna.
Doyle P A and Turner P S 1968 Acta Cryst. A24, 390.
Fox A G and Fisher R M 1988 Aust. J. Phys. 41, 461.
Holmestad R, Zuo J M, Spence, J C H, Høier R and Horita Z 1995 Phil. Mag. A72:3, 579.
Saunders M, Bird D M, Zaluzec N J, Burgess W G, Preston A R and Humphreys C J 1995 Ultramicroscopy 60, 311.

*Inst. Phys. Conf. Ser. No 153: Section 4*
*Paper presented at Electron Microscopy and Analysis Group Conf. EMAG97, Cambridge, 1997*
© 1997 IOP Publishing Ltd

# A convergent beam method for determination of structure factor phases and amplitudes from center disk intensity

C R Birkeland, R Høier, R Holmestad and K Marthinsen*

Department of Physics, Norwegian University of Science and Technology (NTNU), N-7034 Trondheim Norway and *SINTEF Materials Technology, N-7034 Trondheim Norway.

**ABSTRACT:** Quantitative Convergent Beam Electron Diffraction (QCBED) is used to determine selected structure factor amplitudes as well as three-phase structure invariants in non-centrosymmetric InP. The method used is based on center disk intensities only. Particular parameter sensitive pixels in the pattern are identified theoretically. With this selected area QCBED method, structure invariants can be determined with an accuracy of a few degrees and the structure factor amplitudes with an accuracy better than 0.5%.

## 1. INTRODUCTION

Quantitative Convergent Beam Electron Diffraction (QCBED) is the quantitative analysis of CBED patterns recorded in transmission electron microscopes (Spence and Zuo 1992). In earlier work, emphasis has been put on the analysis of Bragg diffracted disks (Holmestad et al 1995, Holmestad and Birkeland 1997), because intensities in the center disk are subject to strong dynamical many beam effects and are generally more complex to analyse. However, developments in theory, computer technology and improved experiments have made it possible to reproduce intensity contrast variations in the center disk with great accuracy. In this study, the direct beam of a CBED pattern taken from the non-centrosymmetric crystal InP is compared quantitatively with theoretical simulations. Through a least square fitting procedure, the incident beam direction and structure factor phases and amplitudes are extracted from intensity variations in the direct beam of the CBED pattern studied. Furthermore, a new method for increasing the sensitivity of the QCBED procedure is introduced.

## 2. EXPERIMENTAL PROCEDURE

A Convergent Beam Electron Diffraction (CBED) pattern of the non-centrosymmetric crystal InP has been obtained in a Zeiss 912 microscope equipped with an Omega energy filter. The experimental pattern, see Fig.2a, is recorded at liquid $N_2$ temperature with a slow scan CCD camera, and the specimen is tilted close to the $\bar{1}50$ zone axis. A schematic illustration of the diffraction conditions and the indexing of the direct beam are shown in Fig.1.

134

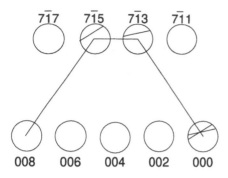

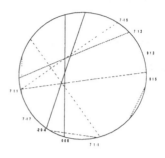

$\overline{7}1\overline{7}$ $\quad$ $\overline{7}1\overline{5}$ $\quad$ $\overline{7}1\overline{3}$ $\quad$ $\overline{7}11$

008 $\quad$ 006 $\quad$ 004 $\quad$ 002 $\quad$ 000

**Fig.1:** Diagram showing the diffraction condition corresponding to the incoming beam direction for the test case studied, i.e. 008, $7\overline{1}3$ and $7\overline{1}5$ are at or near the Bragg condition.

## 3. REFINEMENT OF THE INCOMING BEAM DIRECTION

From an indexed experimental CBED pattern, see Fig.1, one can determine qualitatively the incoming beam direction through visual comparison of the experimental image and geometric line patterns. The incoming beam direction is further refined using traditional QCBED techniques. Through a least square fitting procedure, the incoming beam direction is varied until the best match between theoretical calculations and the experiment is found. Because the pattern consists of many sharp HOLZ lines, the best match is well defined and the QCBED procedure is robust. Furthermore, the incoming beam direction is determined with great accuracy. Fig.2a and Fig.2b show the experiment and the simulated image respectively. Notice that all HOLZ lines are properly reproduced and that grey-scales match perfectly.

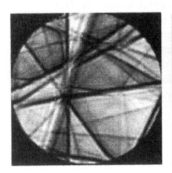
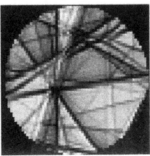

**Fig.2a:** Center disk of the experimental CBED pattern of InP near the $\overline{1}50$ zone axis.

**Fig.2b:** Simulated theoretical pattern. Notice the excellent match in grey-scales.

**Fig.2c:** Window function used in selected area QCBED. Details given in the text.

In the simulations, the theoretical intensities $I_i^{theory}$ are calculated from dynamical theory, diagonalising on average 20 beams and including 164 additional beams using Bethe perturbations. The beam selection criteria are taken from Birkeland et al (1996). Structure factor values are determined for neutral atom using the algorithm of Bird and King (1990). Debye Waller factors given by Reid (1983) are used, i.e. $B_{In} = 0.3226$ Å$^2$ and $B_P = 0.2652$ Å$^2$. The thickness of the specimen is approximately 1400 Å.

## 4.    THE SELECTED AREA QCBED TECHNIQUE

The selected area QCBED technique refers to a method of increasing the sensitivity of the QCBED algorithm to particular parameters refined. The principle is to identify limited areas of the CBED pattern that are sensitive to the refined parameters. The fit factor $\chi^2$ used in QCBED is given by

$$\chi^2 = \frac{1}{n}\sum_{i=1}^{n} W_i \frac{\left(I_i^{exp} - cI_i^{theory}(a_1, a_2, \ldots, a_m)\right)^2}{\sigma_i^2} \qquad (1)$$

where $I_i^{exp}$ and $I_i^{theory}$ are experimental and calculated intensities respectively, $\sigma_i$ are the corresponding standard deviations, $n$ is the number of points in the pattern, $(a_1, a_2, \ldots, a_m)$ are the unknown refined theoretical parameters and $c$ is a normalisation constant. $W_i$, the window function is defined

$$W_i = \left\{ \begin{array}{ll} 0 & \text{if point } i \text{ is outside a sensitive area} \\ 1 & \text{if point } i \text{ is inside} \end{array} \right. \qquad (2)$$

This definition of the fit factor $\chi^2$ gives a QCBED algorithm that is much more sensitive to changes in the refined parameters. A step by step description of the method is: **1.** Refine the incoming beam direction until it is determined with sufficient accuracy. **2.** With fixed incoming beam direction, calculate a theoretical image (reference image) using the initial values $(a_1, \ldots, a_m)$. **3.** Calculate new theoretical images after perturbating the parameters $(a_1, \ldots, a_m)$. The suggested approach is to change each parameter in turn to allow the analysis of single parameter perturbation effects. This method will also help identifying correlations between refined parameters. **4.** Calculate difference maps between images calculated in point 2 and 3. For each parameter, these maps are plotted to illustrate how each parameter affects the total CBED pattern. **5.** A threshold value is set for each point $i$ in the CBED pattern. If the difference map intensity is higher than the threshold, set $W_i = 1$, else disregard point $i$ ($W_i = 0$). This step is repeated for all difference maps, giving a set of different window functions $W_i$. The union of these window functions $W_i$ is used in the $\chi^2$ refinements.

## 5.    RESULTS AND DISCUSSION

The selected area QCBED method is used to determine structure factors in InP from the direct beam of the CBED pattern described earlier and shown in Fig.2a. Structure factors in non-centrosymmetric crystals are generally complex quantities, with both an amplitude $|U_g|$ and a phase $\theta_g$, i.e. $U_g = |U_g|\, e^{i\theta_g}$. The phase $\theta_g$ depends on the choice of origin of the unit cell, but the three-phase invariant $\Psi = \theta_g - \theta_h + \theta_{h-g}$ is independent of the origin of the unit cell. Here $g$, $-h$ and $h - g$ are reciprocal lattice vectors that form a closed loop (Høier et al 1993). The determination of these three-phase invariants is in general of greatest interest, as they may serve as important input in direct methods for structure determinations.

The aim of this study is to determine the structure factor amplitudes $U_{002}$, $U_{713}$ and $U_{715}$, and corresponding phases $\theta_{713}$ and $\theta_{715}$. The selected QCBED algorithm described previously is used to generate the window function that identifies areas in the pattern that are sensitive to the parameters refined. This window is shown in Fig.2c. Note that most points are taken close to the crossing of the 713 and 715 HOLZ lines, where the effects of the structure factors refined are most dominant.

The QCBED refinement procedure based on the selected point intensities is repeated for various initial starting structure factor amplitudes and phases. The results are summarised in Table 1. InP has a known crystal structure, and the neutral atom structure phases are $\theta_{002} = 0.0$,

$\theta_{713} = 19.58$ and $\theta_{715} = -19.88$. The corresponding three-phase structure invariant is $\Psi = 39.5$ deg. In Run 1, only the three structure factor amplitudes are refined, and the result is used as a reference in subsequent runs. The refinements were performed for a wide range of initial starting values, three of the runs are listed in Table 1. The initial high $\chi^2$ values demonstrates the sensitivity of the method. Furthermore, all refinements converged to the same absolute minimum, showing that the algorithm is very stable. A thorough analysis of the table shows that the three-phase structure invariant can be determined with an accuracy of approximately 1 degree. Also, assuming that the structure factor amplitudes found in Run 1 are accurate, all refinements determine the structure factor amplitudes with an accuracy better than 0.5%.

| Run | Description | $U_{002}$ | $U_{713}$ | $U_{715}$ | $\theta_{713}$ | $\theta_{715}$ | $\Psi$ | $\chi^2$ |
|---|---|---|---|---|---|---|---|---|
| 1. | Neutral atom values | 0.023288 | 0.011116 | 0.008923 | 19.58 | -19.88 | 39.46 | 48.4 |
|  | Refined values | 0.023334 | 0.010920 | 0.008322 | – | – | – | 47.5 |
| 2. | Start values | 0.023288 | 0.011116 | 0.008923 | 19.58 | 45.88 | -26.3 | 77.0 |
|  | Refined values | 0.023331 | 0.010873 | 0.008298 | – | -20.73 | 40.31 | 47.4 |
| 3. | Start values | 0.023288 | 0.011116 | 0.008923 | 65.00 | 65.00 | 0.00 | 117.4 |
|  | Refined values | 0.023393 | 0.010845 | 0.008352 | 23.10 | -16.58 | 39.68 | 47.5 |
| 4. | Start values | 0.025 | 0.013 | 0.011 | 0.00 | -19.88 | 19.88 | 79.4 |
|  | Refined values | 0.023411 | 0.010951 | 0.008297 | 20.44 | – | 40.32 | 47.4 |

**Table 1:** Results of the refinement procedure; Structure factor amplitudes and phases, the three-phase structure invariant $\Psi = \theta_{002} - \theta_{715} + \theta_{713}$ and $\chi^2$ values for selected refinements.

## 6.    CONCLUSION

This QCBED study demonstrates how much crystal information can be extracted from the center disk beam of a CBED pattern. Furthermore, the important three-phase invariants have been extracted from the pattern using a novel selected area QCBED algorithm, allowing refined parameters to be determined with high accuracy. More work is in progress, refining three-beam structure invariants with inaccurate input data (Birkeland, Høier and Marthinsen 1997).

## REFERENCES

Høier R, Bakken L N, Marthinsen K and Holmestad R 1993 Ultramicr. **49**, 159-170.
Bird D M and King Q A 1990 Acta Cryst. **A46**, 202.
Birkeland C R, Holmestad R, Marthinsen K and Høier R 1997 Ultramicr. **66** 89.
Birkeland C R, Høier R and Marthinsen K 1997 Proc. International Centennial Symposium on the Electron 1997 Cambridge UK.
Holmestad R and Birkeland C R 1997 submitted to Ultramicr. A.
Holmestad R, Zuo J M, Spence J C H, Høier R and Horita Z 1995 Phil.Mag. **A 72** 579.
Høier R, Bakken L N, Marthinsen K and Holmestad R 1993 Ultramicr. **49**, 159-170.
Reid J S 1983 Acta Cryst. **A 39** 1.
Spence J C H and Zuo J M 1992 Electron microdiffraction, Plenum.

*Inst. Phys. Conf. Ser. No 153: Section 4*
*Paper presented at Electron Microscopy and Analysis Group Conf. EMAG97, Cambridge, 1997*
© *1997 IOP Publishing Ltd*

# Quantitative CBED studies of SiC 4H

**R Holmestad[*#], JP Morniroli[+], JM Zuo[#], JCH Spence[#] and A Avilov[§]**

[*]Dept. of Physics, University of Trondheim, NTNU, N-7034 Trondheim, Norway
[#]Dept. of Physics and Astronomy, ASU, Tempe, AZ 85287-1504, USA
[+]Physical Metallurgy, URA CNRS 234 , University of Lille I, 59655 Villeneuve d'Ascq., France
[§]Institute of Crystallography of RAS, Leninsky prospect 59, 117333 Moscow, Russia

**ABSTRACT:** We report a preliminary study of the SiC 4H structure by quantitative convergent beam electron diffraction. A full dynamical treatment of HOLZ line positions is used to measure lattice parameters at -165°C, which give a = 0.3085 nm and c = 1.0089 nm. Electron microdiffraction show that SiC 4H has the space group P6₃mc. The aim is to refine higher order structure factors for accurate measurement of the distance between Si and C double layers.

## 1. INTRODUCTION

Silicon carbide (SiC) is a widely used ceramic material, with many structural and electronic applications (see i.e. Spencer et al 1993). It exists in many polytypes, which differ only by the stacking sequence of close packed double layers of Si and C atoms. All SiC polytypes can be considered as an assembly of tetrahedra. The c/a ratios of the different polytypes do not exactly correspond to the value of ideal tetrahedras (GomesDe Mesquita 1967) and the exact positions of the Si and the C layers in the c-direction are adjustable parameters.

The aim of this work is to see whether we can determine the atomic positions in the c-direction in SiC 4H by quantitative convergent beam electron diffraction (QCBED) (Spence and Zuo 1992). QCBED is based on pixel by pixel comparisons between experimental and simulated dynamically diffracted intensities using a pre-defined goodness of fit parameter, such as least squares. Until now it has mainly been used for accurate low order structure factor determination (Zuo and Spence 1991, Bird and Saunders 1992, Holmestad et al 1994, Zuo et al 1997). We will examine its potential for measurements of higher order structure factors and atomic positions. The goal is to develop a general refinement approach for structure and bonding charge determination using electron diffraction. The small probe used in CBED makes it ideal for studying materials such as SiC, which is often full of stacking faults (mixed with other polytypes) as shown by Pirouz and Yang (1993). This is also very useful for many newly synthesised materials, which are available only in small quantities and/or mixed with other phases, making X-ray diffraction methods difficult. In this paper we show the first steps in a general refinement approach; how to find accurate values of lattice parameters and to confirm the space group. These are important tasks which have to be done before structure factors can be refined.

## 2. EXPERIMENTAL METHODS

For CBED studies of non-cubic materials, it is essential to have TEM specimens with crystals in many orientations. To obtain this, we use a novel specimen preparation method. The SiC single crystal is crushed and mixed with Al powder. This mixture is rolled between two metal plates to make a very thin film, which is directly thinned by ion milling. Also a single crystal is mechanically thinned and ion milled. This preparation method gives less defects, but it is more difficult to find all orientations. For the CBED studies, we use a Leo 912 electron microscope with Omega energy filter, equipped with a Gatan CCD camera model 679 and Fuji imaging plates (IP). High voltage of the microscope is found to be 119.52 kV. The advantage of IPs is the large number of pixels compared to the CCD. The point spread function (PSF) of the CCD/IP was experimentally measured and removed from the patterns using image restoration techniques (Zuo 1996, Zuo et al 1996, Zuo 1997). Energy filtering and removal of PSF are necessary for quantitative intensity comparisons. The diffraction patterns for space group determination are taken on a Philips CM30 electron microscope operated at 300 kV.

## 3.  MATERIAL AND SPACE GROUP

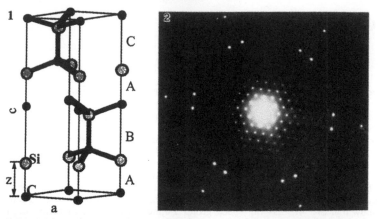

Fig. 1.Crystal structure of 4H SiC; the distance z is the adjustable parameter.
Fig. 2.[0001] zone axis showing hexagonal 'net' and 'ideal' symmetry.

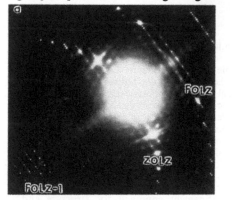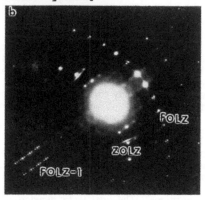

Fig. 3.a) [11$\bar{2}$0] zone axis tilted to FOLZ-1 orientation. There is no periodicity difference
     between ZOLZ and FOLZ. b) [10$\bar{1}$0] zone axis tilted to FOLZ-1 orientation. There is a
     periodicity difference between ZOLZ and FOLZ.

The 4H polytype has a hexagonal structure with a stacking sequence of ABACABAC... of
the double layers as shown in Fig 1. Fig. 2 and 3 show microdiffraction patterns required for
crystal structure identification. The procedures described by Morniroli and Steeds (1992) and
Redjaimia and Morniroli (1994) are followed. Fig. 2 shows the [0001] zone axis with 'net'
symmetry 6mm, thus the system is hexagonal. The next point is identification of Bravais lattice
and glide planes. In a FOLZ-1 orientation, the incident beam is tilted so the first negative layer
becomes tangential to the Ewald sphere. A glide plane will make a periodicity difference between
ZOLZ and FOLZ nets in a microdiffraction pattern. Fig. 3a) shows no periodicity difference
between ZOLZ and FOLZ for the [11$\bar{2}$0] zone axis, giving a partial extinction symbol P.-. Fig.
3b) shows a periodicity difference between ZOLZ and FOLZ for the [10$\bar{1}$0] zone axis, which
give a partial extinction symbol P. . c. Combining these two extinction symbols give P.-c, for
which there are 3 possible space groups:     P6$_3$mc     P$\bar{6}$2c        P6$_3$/mmc
with corresponding point groups:     6mm        $\bar{6}$2m        6/mmm
From Fig. 2 we can find that 6mm is the 'ideal' symmetry of [0001], this eliminates $\bar{6}$2m. In
Fig. 4 there is a difference in intensity between (0004) and (000$\bar{4}$). Thus the structure is non-
centrosymmetric and the space group is P6$_3$mc.
We have now confirmed that the space group of SiC 4H is P6$_3$mc. This space group, with
the given Wyckoff positions, requires systematic absence of reflections (hk.l) with h-k=3n;
l=2n+1. In addition we find extinctions of h-k=3n; l=4n+2. Without violating these  extinction

conditions, the only way to adjust the z-coordinates of Si and C is to change the distance z between the Si and C layers shown in Fig. 1. Thus SiC 4H is a one-parameter structure.

Fig. 4. [10$\bar{1}$0] zone axis showing evidence of non-centrosymmetry. Intensity in reflections (0004) and (000$\bar{4}$) is different.

## 4.    LATTICE PARAMETERS

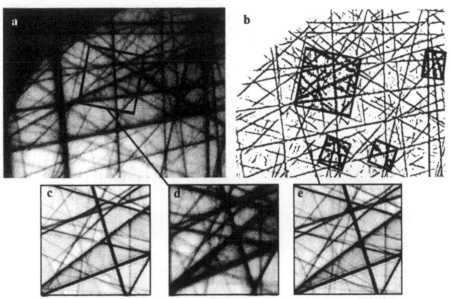

Fig. 5.a) Part of experimental CBED centre disk [$\bar{3}$210] orientation. Thickness ~350 nm.
b) Binary image with HOLZ line positions from experiment. Boxes show overlay from best fit line positions, giving a = 0.3085 nm c = 1.0089 nm. c) Part of calculated image with given LPs, a = 0.3053 and c = 1.0073 nm. from Wyckoff (1982) d) Corresponding part of experiment. e) Corresponding part calculated with refined lattice parameters.

Lattice parameters (LPs) for SiC 4H at room temperature are given to be (Wyckoff 1982) 0.3073 nm and 1.0053 nm for a and c respectively. Positions of higher order Laue zone (HOLZ) lines are sensitive to lattice parameters. Our experiments at -165°C do not correspond to calculated patterns with the measured X-ray lattice parameters, as shown in Fig. 5c) and d). Kinematical refinements of lattice parameters (Zuo 1992) were tried, but in this case, the results did not converge below the 1% accuracy level, as they strongly depend on the zone axis and which lines were included in the fit. This indicates a strong dynamic influence on the HOLZ line positions. The inclusion of dynamic effects is needed to refine lattice parameters with better accuracy. The intensity in the centre disk is affected by many other parameters, such as thickness, absorption, Debye-Waller factors (DWFs) and low order structure factors which are still not well known. Therefore, it is difficult to refine LP alone based on the comparison between intensities. Considering this, we have employed image processing techniques for the extraction of the line positions, and converted a grey-scale image into a binary image of 0 and 1's, 1 where a line exists (three connecting points define a line). The same procedure was applied to the calculated pattern based on the Bloch wave method. The resulted experimental and theoretical patterns are correlated and the best fit is chosen with maximum correlation. To save computer time, only regions of high sensitivity to experimental parameters are calculated. Fig. 5a) shows the recorded experimental pattern. In Fig. 5b), the best fit which gives a=0.3085 nm and c=1.0089 nm is superimposed with the processed experimental pattern. Fig. 5e) shows the theoretical pattern with the refined lattice parameters.

## 5. STRUCTURE FACTORS AND ATOM POSITIONS

Tab. 1.    Calculated $U_{000l}$ for neutral atoms (procrystal) and % change from ideal value for different values of z. High voltage 119.52 kV. Debye-Waller factors $B_{si}=0.20$ and $B_C=0.25$ $Å^2$.

| z-value | $U_{0004}$ [$Å^{-2}$] / % change | $U_{0008}$ [$Å^{-2}$] / % change | $U_{00012}$ [$Å^{-2}$] / % change | $U_{00016}$ [$Å^{-2}$] / % change | $U_{00020}$ [$Å^{-2}$] / % change |
|---|---|---|---|---|---|
| ideal value $0.1875$ ($\frac{3}{16}$) | 0.05620 | 0.007685 | 0.01154 | 0.009342 | 0.004608 |
| 0.1850 | -2.79 | +2.70 | +6.79 | -0.676 | -13.1 |
| 0.1800 | -9.11 | +18.4 | +16.4 | -6.32 | -52.5 |
| 0.1900 | +2.58 | +2.70 | -8.53 | -0.676 | +9.38 |

Table 1 shows the sensitivity of the structure factors along the (000l) row. Generally, higher order structure factors are more sensitive to atomic positions than lower order ones; they are also less affected by bonding and can be used to determine the atomic positions. The DWFs give an overall damping effect. Fig. 6 shows an example of a CBED pattern with extracted line scans. From Fig. 6 it is possible to extract three structure factors. Additional measurements are needed to determine the unknown structure parameter z and DWFs for Si and C. The greatest challenge is to develop a robust procedure for measuring high order structure factors with many unknown structure parameters. Efforts are under way to study the stability of least-square criteria against a certain percentage of intensity points with significant systematic errors, and the choice of diffraction patterns minimising effects from other beams.

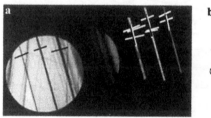 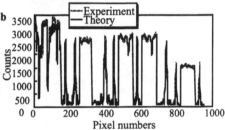

Fig. 6.a) Energy filtered CBED pattern close to [11 $\overline{7}4$0], with the systematic (0004) row.
b) Best fit of extracted lines after refinement of thickness, position and 3 structure factors.

## 6. CONCLUSIONS

It is possible to measure lattice parameters from strongly dynamic electron diffraction patterns by comparing geometric line features using image processing and correlation. Lattice parameters of 4H SiC at -165°C were measured to a=0.3085 nm and c=1.0089 nm compared to 0.3073 and 1.0053 previous reported X-ray values. Space group of SiC 4H is found to be P6₃mc.

REFERENCES
Bird DM and Saunders M 1992 Acta Cryst. **A48**, 555
Deiniger C, Necker G and Mayer J 1994 Ultramicr. **54** 15
GomesDe Mesquita AH 1967 Acta Cryst. **23**, 610
Holmestad R, Zuo JM, Spence JCH, Høier R and Horita Z 1994 Phil Mag. **A72**, 579
Morniroli JP and Steeds JW 1992 Ultramicr. **45**, 219
Pirouz P and Yang JW 1993 Ultramicr. **51**, 189
Redjaimia A and Morniroli JP 1994 Ultramicr. **53** 305
Spencer MG et al 1993 Silicon Carbide and related Materials IOP conf. ser. 137
Spence JCH and Zuo JM 1992 Electron microdiffraction (Plenum Press)
Wyckoff WG 1982 Crystal Structures (Krieger Pub. Comp.)
Zuo JM 1992 Ultramicr. **41**, 211
Zuo JM 1996 Ultramicr. **66**, 21
Zuo JM 1997 Proc.EMSA-97, 55. Ann.meeting, Ed. Bailey et al 1091
Zuo JM and Spence JCH 1991 Ultramicr. **35**, 185
Zuo JM, McCartney M R and Spence JCH 1996 Ultramicr. **66**, 35
Zuo JM, O'Keeffe M, Rez P and Spence JCH 1997 Accepted, Phys. Rev. Lett.

RH is supported by the Norwegian Research Council; JMZ and JCHS are supported by NSF grant DMR 9412146; AA and JCHS thank the CRDF award number RP1 -208. We are grateful to M. O'Keeffe (ASU) for fruitful discussions and P. Pirouz for providing the SiC single crystal.

Inst. Phys. Conf. Ser. No 153: Section 4
Paper presented at Electron Microscopy and Analysis Group Conf. EMAG97, Cambridge, 1997

# Quantitative measurement of symmetry in CBED patterns

**T D Walsh and R Vincent**

H H Wills Physics Laboratory, University of Bristol, Tyndall Avenue, Bristol BS8 1TL

**ABSTRACT**: Group theory and linear algebra are used to develop a quantitative measure of the deviation from ideal symmetry in a CBED pattern. The pattern is represented as a function vector, where projection operators are used to construct maps of the orthogonal irreducible representations, and the centre is defined as the position which maximises the norm of the identity representation. In typical patterns, errors of 5% are observed, attributed to a combination of specimen strain and distortions in the electron optics.

## 1. INTRODUCTION

Although the symmetries within a CBED pattern are listed by Buxton et al (1976) in the diffraction group tables, an experimental pattern always exhibits deviations from these ideal symmetries, attributable to a combination of strain, inclined surfaces, incomplete unit cells and imperfections in the electron optics. Given the trend towards quantitative interpretation of the intensity distributions in CBED patterns (see, for example, Holmestad et al (1995) and Saunders et al (1995)), it is useful to develop the basis for a single quantitative measure of the deviation from ideal symmetry, combined with a procedure for mapping the local distribution of asymmetries within a pattern referred to any point group.

As a simple example, we consider first the intensities $f_1$ and $f_2$ at two points related by a mirror or a diad axis. For ideal symmetry, the two intensities are equal, and it is natural to define the overall deviation from symmetry for experimental data by taking the squared sum over both points of the deviations from the mean intensity $f$. This squared measure of the asymmetry is written as $D^2$, where normalisation is imposed by setting the squared sum of the experimental intensities equal to unity. For a continuous normalised image, an equivalent measure for $D^2$ is found by taking the sum over all pairs of symmetry-related points. However, this approach is somewhat arbitrary, takes no account of the group structure and provides no recipe for systematic mapping of the symmetry errors or locating the optimum centre for the group operators. Below, we outline an approach based upon group theory and linear algebra which remedies these defects, and is applicable to any point group in any number of dimensions.

## 2. QUANTITATIVE MEASURE OF SYMMETRY

For simplicity, the pattern is described by a continuous intensity distribution, $f(x,y)$ or $f(\mathbf{r})$, referred to an arbitrary origin and display axes. By converting integrals between $\pm\infty$ to summations over pixels, the results are immediately applicable to sampled distributions. The intensities may be represented as a function vector $\mathbf{f}$ in a configuration space of infinite dimension for continuous variables x and y, or of dimension equal to the number of pixels for a sampled function. Following standard notation in linear algebra, the 2-norm of $\mathbf{f}$ is written as $\|\mathbf{f}\|$, and is defined by

$$\|\mathbf{f}\|^2 = \iint f^2(x,y)\,dxdy, \tag{1}$$

and the inner product between two functions f and g is defined as

$$(\mathbf{f},\mathbf{g}) = \iint f(x,y)g(x,y)\,dxdy, \tag{2}$$

being equivalent to a vector dot product. Below, it is assumed that $f(x,y)$ is square-integrable, and that its norm $\|\mathbf{f}\|$ is set equal to unity.

An element of the abstract group $G$ of order g is written as $G_a$, where we consider an isomorphic representation of the group as a set of unitary matrices acting on the orthonormal display vectors for f(r) in the image plane. The new basis vectors induce a transformation in f, where

$$T(G_a)f(r) = f(G_a^{-1}r) \qquad (3)$$

is the standard definition for an induced representation of f(r) referred to the original axes. Equivalently, the group operators acting on **f** generate a set of non-orthogonal function vectors $T(G_a)\mathbf{f}$ within an invariant subspace, L. The maximum dimension of L is equal to g, but may be less if some or all of the induced vectors are identical. For a pattern with ideal symmetry, **f** is invariant under $T(G_a)$, and the induced vectors are identical.

The next step is to use standard group theory to project **f** onto the 1- or 2-dimensional subspaces which correspond to the orthogonal, irreducible representations of **f** for one of the 10 crystallographic, 2-dimensional point groups. The details of the projection operator and the associated group algebra are not described here, but we discuss the relevant equations for two simple point groups. A complete account and an annotated version of the character tables is given elsewhere (Vincent and Walsh 1997), based upon section 4.19 in the text by Elliott and Dawber (1979).

The projected components of **f** are labelled $\phi_A$, $\phi_B$ etc. in standard group notation, where $\phi_A$ corresponds to the identity representation with ideal symmetry. For $\|\mathbf{f}\| = 1$, we have

$$\sum_\alpha \phi_\alpha = \mathbf{f}, \qquad \sum_\alpha \|\phi_\alpha\|^2 = 1 \quad \text{and} \quad (\phi_\alpha, \phi_\beta) = 0, \qquad (4)$$

where the summations are taken over all representations and $\alpha \neq \beta$. A measure of the symmetry S relative to a designated centre and point group is defined by $S = \|\phi_A\|$, and the deviation from symmetry is given by $D^2 = 1 - S^2$. After some manipulation, we obtain a simplified expression for S, where

$$S^2 = g^{-1} \sum_\alpha (\mathbf{f}, T(G_\alpha)\mathbf{f}), \qquad (5)$$

being the mean value of g inner products between **f** and its replicas induced by the group operators. It can be shown that this expression is equivalent to the intuitive definition discussed in the Introduction.

To illustrate these results, consider the point group m, isomorphic to the abstract group with 2 elements. The symmetric and anti-symmetric representations are given by

$$\phi_A = \tfrac{1}{2}(f_1 + f_2) \qquad \text{and} \qquad \phi_B = \tfrac{1}{2}(f_1 - f_2), \qquad (6)$$

where $f_1 = T(E)\mathbf{f} = \mathbf{f}$, $f_2 = T(m)\mathbf{f}$, E is the identity operator and m is the mirror operator. Then $\phi_A$ and $\phi_B$ satisfy the conditions $\mathbf{f} = \phi_A + \phi_B$ and $(\phi_A, \phi_B) = 0$. The equivalent vector geometry is sketched in Fig. 1. The corresponding expressions for point group 2mm are

$$\phi_{A_1} = \tfrac{1}{4}(f_1 + f_2 + f_3 + f_4) \qquad \text{2mm}$$

$$\phi_{B_1} = \tfrac{1}{4}(f_1 - f_2 - f_3 + f_4) \qquad \text{2'mm'} \qquad (7)$$

$$\phi_{A_2} = \tfrac{1}{4}(f_1 + f_2 - f_3 - f_4) \qquad \text{2m'm'}$$

and $\quad \phi_{B_2} = \tfrac{1}{4}(f_1 - f_2 + f_3 - f_4) \qquad \text{2'm'm,}$

where the primed symbols in the symmetries listed on the right-hand side of eq. (7) imply sign inversion, and the functions $f_1$ to $f_4$ are generated from f(x,y) by the identity, diad and mirror operators in group 2mm. It can be shown by direct substitution that these expressions satisfy eqns. (4).

The optimum centre and alignment for a point group is defined as the global maximum of $S(x,y,\theta)$, where x and y are the offsets between an arbitrary origin for the image and the point group centre, and $\theta$ is the rotation angle applied to the image. For a CBED pattern, a software operator is applied just inside the boundaries of the discs to define the relevant pixels. Only the areas of mutual overlap between the images generated by the group operators contribute to S, and therefore f(x,y) must be renormalised for every position of the point group centre (Fig. 2). In the limit of an image with uniform intensity, $S = 1$ for apertures of any shape or size and for all point groups in any position or alignment.

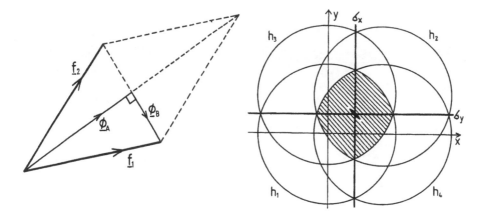

**Fig. 1** Projection of $\mathbf{f}_1$ onto the orthogonal representations for a diad or mirror operator.

**Fig. 2** Shaded area represents the mutual overlap of 4 circular masks $h_1$ to $h_4$ (see text).

## 3. RESULTS AND DISCUSSION

The steps required to process a typical image are illustrated by analysis of the zero layer reflections in a Si<110> zone axis CBED pattern acquired with a 2nm focused probe at 200kV and 100K. Inelastic electrons were removed with an imaging filter, followed by deconvolution of the detector point spread function. In the unprocessed pattern Fig. 3(a)), slight deviations from the ideal 2mm projection contrast were visible in the dark field (DF) discs, combined with a small offset between the apparent pattern centre and the centre of the bright field (BF) disc. A masking function $h_i(x,y)$ was set equal to unity for pixels within circular areas just inside the physical boundaries of the discs, with $h_i=0$ for all other pixels. The nominal pattern centre was bracketed by a cluster of pixels, and the functions $f_1$ to $f_4$ were calculated for each pixel acting as the point group centre. The corresponding Boolean product of the masking functions $h_1$ to $h_4$ was used to label pixels for inclusion in the mutual overlap area (Fig. 2). The symmetry error D was calculated for each pixel, followed by pattern rotation to find the global minimum of D as a function of x, y and θ.

The corresponding projections are shown in Fig. 3, where (b) is the identity representation $\phi_{A_1}(x,y)$ with perfect 2 mm symmetry and a norm of S=99.4%. A more useful measure of the deviation from ideal symmetry is given by the functions $\phi_{A_2}$, $\phi_{B_1}$ and $\phi_{B_2}$ with respective norms of 6.3%, 7.1% and 5.0% (Figs. 3(c) to (e)), and symmetries 2m′m′, 2′mm′ and 2′m′m (cf. eqs. (7)). These patterns show lines of zero intensity along the anti-mirrors. The sum of (c), (d) and (e) gives the total error function (Fig. 3(f)) with a norm of D=10.8%, equivalent to the root mean square error.

For the pattern shown in Fig. 3, the errors were evenly distributed between the 3 anti-symmetric representations, and also were delocalised across the BF and DF discs. When the calculations were repeated for the BF disc alone, the total error was reduced from 10.8% to 6.1%, although the distribution of errors within the BF disc remained similar to Fig. 3(f). None of the error maps showed much evidence of residual noise, implying that the loss of symmetry follows from a combination of residual strain and pattern distortions associated with the electron optics in the projector lenses and spectrometer.

The analysis of similar zone axis patterns from Si, Ni and Cu confirmed that absence of ideal symmetry even within the BF disc must be attributed to inhomogeneous strains. Symmetry errors of ~5% were found even in patterns which appeared to be perfect and had been used for calculation of the bonding charge distribution. It follows that a reduced $\chi^2$ statistic close to unity is a reliable guide to the accuracy of pattern matching calculations only in the limit where the symmetry error map shows mostly Poisson noise.

144

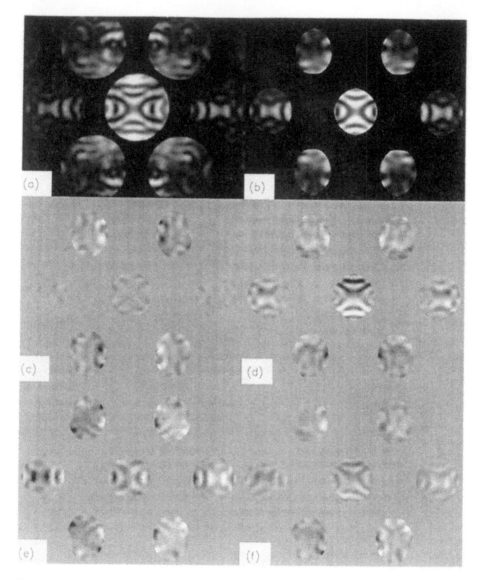

Fig. 3(a) Original unmasked Si<110> pattern (cf. Fig. 2). The identity representation $A_1$ is shown in (b), followed by representations $A_2$, $B_1$ and $B_2$ in (c), (d) and (e), and the overall error function in (f). In (c) to (f), the grey background between discs represents zero intensity.

## REFERENCES

Buxton B F, Eades J A, Steeds J W and Rackham G M 1976 Phil. Trans. R. Soc. **281**, 181
Elliott J P and Dawber P G 1979 Symmetry in Physics, Vol 1 (London:MacMillan) pp 74-77
Holmestad R, Zuo J M, Spence J C H, Høier R and Horita Z 1995 Phil. Mag. A 72, 579
Saunders M, Bird D M, Zaluzec N J, Burgess W, Preston A R and Humphreys C J 1995
   Ultramicroscopy **60**, 311
Vincent R and Walsh T D 1997 Ultramicroscopy (in the press)

*Inst. Phys. Conf. Ser. No 153: Section 4*
*Paper presented at Electron Microscopy and Analysis Group Conf. EMAG97, Cambridge, 1997*

# Quantitative Convergent Beam Electron Diffraction (CBED) measurements of temperature factors in metals

**M Saunders and A G Fox**

Center for Materials Science and Engineering, Department of Mechanical Engineering, Naval Postgraduate School, Monterey, CA 93943, USA

**ABSTRACT:** The quantitative zone-axis CBED pattern matching technique developed by Bird and Saunders for the refinement of low-order structure factors is now applied to the determination of temperature factors in metals. Two alternative strategies are considered. The first involves the optimization of a goodness-of-fit parameter as a function of the fixed temperature factor used for the structure factor refinement. The second compares the variation of the refined structure factors with their neutral atom equivalents as a function of the temperature factor used for the fit.

## 1. INTRODUCTION

At previous EMAG meetings we have shown how the quantitative zone-axis CBED pattern matching technique (ZAPMATCH) of Bird and Saunders (1992) can be used to refine the low-order structure factors in semiconductors with sufficient accuracy to study the bonding charge redistribution (Saunders, et al, 1995). The principle of ZAPMATCH is that the sum-of-squares difference between an elastic-filtered CBED pattern and a many-beam simulation can be minimized by adjusting the scattering potential, sample thickness and various constant terms while the temperature factor is kept fixed (see Saunders, et al, 1995 for details). To convert the refined scattering potential into an interpretable charge density requires knowledge of the temperature factors (often referred to in electron diffraction as Debye-Waller factors) which describe the temperature dependent vibrations of the atoms.

Accurate experimental measurements of these temperature factors are difficult to find at the liquid nitrogen temperatures often used in quantitative diffraction work. Those working with quantitative CBED techniques have therefore sought to develop new methods of temperature factor determination to obtain measurements under electron diffraction conditions. These new approaches have included the use of HOLZ excess line intensities (Holmestad, et al, 1995), electron precession patterns (Sleight, et al, 1996) and high-index systematic row refinement (Nüchter, et al, 1997).

Here we propose two alternative methods involving zone-axis CBED patterns. Both require low-order structure factor refinements to be repeated at a range of fixed temperature factor values. In the first method we consider the behavior of the best-fit sum-of-squares as a function of the fixed temperature factor. In the second we monitor the behavior of the fitted structure factor values. In both cases the ultimate aim is to refine both temperature factors and low-order structure factors from the same data-set.

146

## 2. $\chi^2$ MINIMIZATION METHOD

The variation in the optimized sum-of-squares parameter for fits to a 1300Å room temperature nickel <110> zone-axis pattern is shown in Fig. 1. The fits include the four low-order structure factors out to {113} as variables. The solid curve is a quadratic function fitted through the experimental data. It is evident that there is an optimum value of the temperature factor at which the sum-of-squares finds a minimum and that within the search range local minima are not important.

We expect the low-order structure factors contributing to the diffraction pattern to be modified by the effects of the bonding charge redistribution. The ZAPMATCH technique varies a small number of low-order structure factors while fixing many more at their neutral atom values. Consequently, the optimized temperature factor is likely to be offset by bonding effects when the number of variables is such that some of the fixed structure factors are meant to exhibit bonding modifications. In order to investigate this effect we repeat the temperature factor minimization with a range of variable structure factor numbers from zero, i.e. only sample thickness and constant terms vary, up to six where all structure factors out to {400} are included as variables. A plot of the optimum temperature factor as a function of the number of structure factor variables is shown in Fig. 2. The plot is seen to converge as bonding effects are added by making the low-order structure factors variables in the fit. However, as more structure factors, to which the data has little sensitivity, are added to the fit the result wanders away from the optimized value.

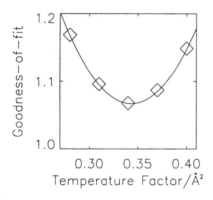

Fig. 1. Plot of optimized sum-of-squares as a function of the fixed temperature factor used in the fit for a 1300Å room temperature nickel <110> zone-axis

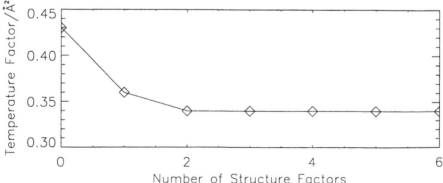

Fig. 2. Plot of the optimized temperature factor as a function of the number of variable structure factors in the fit for a 1300Å room temperature nickel <110> zone-axis pattern.

## 3. LINE CROSSING METHOD

Another alternative is to analyze directly the refined low-order structure factor values obtained from each of the fits. The optimized low-order structure factors are observed to vary as a function of the temperature factor used for the calculation. This occurs because the low-order terms compensate for changes in the fixed higher-order structure factors which are dependent on the temperature factor. It has previously been proposed that monitoring the behavior of the low-order structure factors may permit measurement of the temperature factor (Burgess, 1994).

The variation of the refined {111}, {200}, {220} and {113} structure factors as a function of the temperature factor used for the fit is shown in Fig. 3. The dashed line in each plot is the neutral atom structure factor as given by Doyle and Turner (1968). Bonding effects are most pronounced in the lowest-order structure factors so that temperature factor and bonding effects cannot be separated. Comparison of the fitted and neutral atom values provides no information about the temperature factor in these cases. For higher-orders where bonding effects are small, e.g. {113}, the intercept of the fitted and neutral atom lines should occur when the correct temperature factor is used. It must be realized that in other material systems, where bonding effects are more dramatic, the comparison must be made for higher-order structure factors to avoid introducing systematic errors due to bonding. Not all zone-axis patterns will show sufficient sensitivity to these higher-order terms. Thus, temperature factors obtained from thin samples using this method could be misleading.

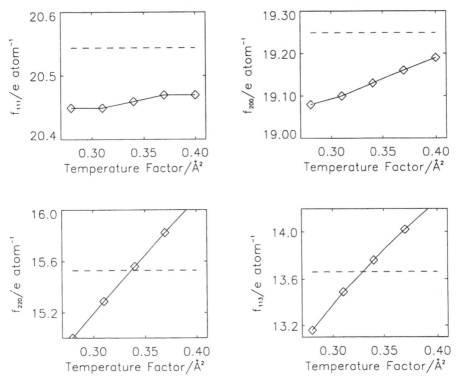

Fig. 3. Comparison of the best-fit low-order structure factors with their neutral atom equivalents (dashed lines) for a range of temperature factors; (a) {111}, (b) {200}, (c) {220} and (d) {113}. The experimental diffraction pattern is the same as in Figs. 1 and 2.

Values for the temperature factors for nickel (at both room and liquid nitrogen temperatures) and copper (at liquid nitrogen temperatures) have been determined using both the CBED sum-of-squares optimization and {113} line intersection methods. In each case the results from four different CBED patterns acquired at thicknesses in the range 500Å to 2500Å are considered. The average values (with error estimates in parenthesis) are given in Table 1. The predictions of Sears and Shelley (1991) who investigated the phonon density of states by inelastic neutron scattering are included for comparison. Clearly there is excellent agreement between the line crossing results and those of Sears and Shelley.

| Material | CBED optimization | CBED line crossing | Sears and Shelley |
|----------|-------------------|--------------------|--------------------|
| Ni RT | 0.35(2) | 0.34(2) | 0.34 |
| Ni LN$_2$ | 0.15(2) | 0.13(2) | 0.13 |
| Cu LN$_2$ | 0.24(2) | 0.21(2) | 0.21 |

Table 1. Comparison of the zone-axis CBED temperature factor measurements (for both optimization and line crossing methods) with the predictions of Sears and Shelley (1991).

## 4. CONCLUSIONS

Two alternative methods of using zone-axis CBED patterns for temperature factor refinement have been considered. Of these, the most successful appears to be the use of line crossings between fitted structure factors and their neutral atom equivalents. The results obtained from the {113} structure factor in nickel and copper are in excellent agreement with those predicted from the phonon density of states work of Sears and Shelley (1991). The technique works well for metals where bonding only significantly alters the first two or three low-order structure factors such that higher-order structure factors are adequately described by neutral atom values. The universal application of this method may be limited when bonding effects extend to higher-order structure factors. In addition, the limited sensitivity of data acquired from thin samples to the important mid-order structure factors may place a lower limit on the sample thickness for temperature factor measurements.

The refinement of temperature factors from the optimization of the sum-of-squares fitting parameter appears more prone to error and consistently produces higher values than the line crossing method. It is possible that this problem stems from our use of a fixed temperature factor in each fit. Making the temperature factor a variable in the full pattern matching calculation may improve the situation, however the difficulty in separating the effects of bonding from sensitivity to the temperature factor may be prohibitive of this approach.

## REFERENCES

Bird D M and Saunders M 1992 Ultramicroscopy **45**, 241.
Burgess W G 1994 PhD Thesis, University of Cambridge.
Doyle P A and Turner P S 1968 Acta Cryst. A24, 390.
Holmestad R, Zuo J M, Spence, J C H, Høier R and Horita Z 1995 Phil. Mag. A72:3, 579.
Nüchter W, Weickenmeier A L and Mayer J 1997 Acta Cryst. A, in press.
Saunders M, Bird D M, Zaluzec N J, Burgess W G, Preston A R and Humphreys C J 1995 Ultramicroscopy **60**, 311.
Sears V F and Shelley S A 1991 Acta Cryst. A47, 441.
Sleight M E, Midgley P A, Saunders M and Vincent R 1996 in proceedings EUREM96.

*Inst. Phys. Conf. Ser. No 153: Section 5*
*Paper presented at Electron Microscopy and Analysis Group Conf. EMAG97, Cambridge, 1997*

# The effect of objective-lens aberrations on EFTEM, STEM and SEM images

R.F. Egerton

Department of Physics, University of Alberta, Edmonton, Canada T6G 2J1

ABSTRACT: The point-spread function (PSF) representing chromatic blurring in an energy-filtered TEM image is an arctan×hyperbolic or a hyperbolic function, depending on the specimen thickness. The radius containing 50% of the electrons is typically 10% - 30% of the total radius, and can be even less in the case of spherical aberration. The same PSF describes the current density distribution in an SEM probe, particularly at low incident energy where chromatic aberration predominates.

## 1. INTRODUCTION

Lens aberrations are discussed in many publications on electron microscopy, but mainly in terms of the overall diameter of the aberration disk. A more quantitative measure of their effect on image contrast and resolution is in terms of the point-spread function. The PSF is the image-intensity distribution produced by each scattering point in the specimen, referred to the specimen plane by dividing the radial coordinate by the image magnification. Here we discuss analytical expressions for the chromatic and spherical-aberration PSF for two important practical cases: core-loss TEM images (as employed in elemental mapping) and SEM images (particularly at low electron energy, where chromatic aberration is important).

## 2. CHROMATIC ABERRATION OF AN EFTEM IMAGE

An energy filter selects electrons within a narrow range of energy loss (E) to form the TEM image. If objective-lens focussing has been done using the zero-loss image (E=0) and the filter excitation is then changed to image electrons of energy loss E=ε, the chromatic-aberration PSF has the same functional form as the angular distribution of scattering at the energy loss ε , a consequence of the fact that the radial displacement $r = C_c \, \theta \, \varepsilon/E_0$ (referred to the specimen plane) is proportional to scattering angle θ . Because the angular divergence of the electrons decreases down the TEM column, chromatic aberration arises mainly in the objective lens, whose aberration coefficient is $C_c$ ; the incident-electron energy is denoted $E_0$.

For a very thin specimen, which is necessary for core-loss imaging, the inelastic-scattering angular distribution (dI/dΩ) is Lorentzian with a halfwidth $\theta_E$ and the PSF is:

$$dI/da = (E_0^2 \, C_c^{-2} \, \varepsilon^{-2}) \, (dI/d\Omega) \propto (C_c E/E_0)^{-2} \, (\theta^2 + \theta_E^2)^{-1} \propto (r^2 + r_E^2)^{-1} \qquad (1)$$

where $r_E = C_c \theta_E (E/E_0) \approx 0.5 \, C_c (E/E_0)^2$. In this case, the chromatic PSF is a Lorentzian function with a halfwidth at half maximum (HWHM) equal to $r_E$ . If we take the image resolution $r_i$ as the separation of two points whose intensity distribution falls by 23 % between each peak (approximately the Rayleigh criterion), this point resolution is given

(Egerton and Wong, 1995) by $r_i = 2.2\, r_E \approx 1.1 C_c (E/E_0)^2$ , assuming the semiangle $\beta$ of the objective aperture to be greater than $\theta_E$ , a reasonable assumption for most core-loss images.

In practice, inelastic images are usually obtained by changing the microscope high voltage rather than the spectrometer excitation. If the objective lens is adjusted to give a Gaussian focus with the zero-loss beam at the *centre* of the energy-selecting slit, the core-loss image will be in focus for electrons which pass through the centre of the selected energy window. For these electrons, the chromatic PSF is a delta function. However, for electrons whose energy is $\pm\varepsilon$ relative to this central value, the PSF is a Lorentzian whose halfwidth is $r_E = C_c\theta_E(\varepsilon/E_0) \approx (E/2E_0)(C_c\varepsilon/E_0)$. In order to obtain the overall point-spread function, for a transmitted-electron current I , Eq.(1) must be integrated over $\varepsilon$ to give (Egerton and Crozier, 1997):

$$\frac{dI}{da} = \frac{I}{\pi\,\Delta\,\ln(1+\beta^2/\theta_E^{\,2})}\frac{4E_0^2}{C_cE}\left[\tan^{-1}\left(\frac{E}{2\beta E_0}\frac{r_c}{r}\right) - \tan^{-1}\left(\frac{E}{2\beta E_0}\right)\right]\left(\frac{1}{r}\right) \qquad (2)$$

for $|r| < r_c$ and zero otherwise. This point-spread function is plotted in Fig. 1a ; it consists of a central maximum surrounded by long tails extending to $r = r_c = (C_c/E_0)(\Delta/2)\beta$. Note that the overall radius $r_c$ is proportional to the width $\Delta$ of the energy-selecting slit whereas the *shape* of the PSF depends on the angular distribution of the inelastic electrons and is therefore a function of the selected energy loss E.

Because the $(1/r)$ function in Eq.(2) tends to infinity at $r = 0$ , it is not possible to define a halfwidth for this point-spread function. However, one can define a radius $r_x$ containing a given percentage x of the electrons. The solid data points in Fig. 1a show x (vertical axis) plotted against $r_x/r_c$ (horizontal axis) for different values of the energy loss E .

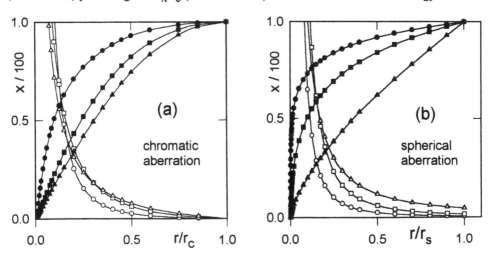

Fig. 1. Point-spread function dI/da (open data points) and percentage x of the total current contained within a radius r (solid points), for the case of (a) chromatic aberration and (b) spherical aberration. Circles are for $E/E_0 = 10^{-3}$ and squares for $E/E_0 = 10^{-2}$, calculated for a thin specimen according to Eq.(2); triangles are for a thick specimen, calculated from Eq. (3). Each PSF is shown only for positive values of r , since it is symmetric about the optic axis.

Values of $r_{50}$ are typically a factor of 4 to 8 smaller than the total radius $r_c$, reflecting the fact that only electrons which pass close to the edges of the energy-selecting slit (large $\varepsilon$) contribute to the tails of the PSF.

If the specimen is *not* very thin, the angular distribution of scattering $dI/d\Omega$ contains an appreciable contribution from plural scattering, principally mixed (elastic + core-loss) scattering. In the case of a thick amorphous specimen, $dI/d\Omega$ approximates to a rectangular function and the chromatic-aberration PSF becomes:

$$dI/da = (I/\pi\beta\Delta)(2E_0/C_c)[1/r - 1/r_c] \tag{3}$$

for $|r| < r_c$ and $dI/da = 0$ outside this range. In this case, the radius containing a given percentage x of the electrons is: $r_x = r_c[1-(1-x/100)^{1/2}]$, giving $r_x \approx 0.3\ r_c$ for x = 50.

## 3.    SPHERICAL ABERRATION OF AN EFTEM IMAGE

A point-spread function for spherical aberration can be defined in similar way to that for chromatic aberration. In this case, however, the scattering angle $\theta$ and equivalent radius r in the object plane are *not* linearly related. Instead, $r = C_s\,\theta^3$ , where $C_s$ is the coefficient of spherical aberration. For the core-loss image of a very thin specimen, where single inelastic scattering provides the major contribution to the image, the point-spread function is:

$$\frac{dI}{da} = \frac{I}{3\pi\ \ln(1+\beta^2/\theta_E^2)}\ \frac{1}{r^2+\theta_E^2C_s^{2/3}r^{4/3}} \tag{4}$$

for $r < r_s$ (where $r_s = C_s\beta^3$ is the total radius) and zero outside this range. This function tends to infinity for $r\to0$ (see Fig. 1a). Cutoff of the angular distribution of inelastic scattering (by the objective aperture) causes the PSF to fall *abruptly* to zero at $r = r_s$ , unlike the case of chromatic aberration. As a result of the $\theta^3$ dependence of r and the extended tails of the Lorentzian angular distribution, the radius containing 50% of the total intensity is typically between 2 % and 10 % of the total radius, as shown by the solid data points in Fig. 1b.

For a thicker specimen, in which plural (elastic+inelastic) scattering fills the objective aperture, the point-spread function simplifies to:

$$dI/da = (I/3\pi\beta^2)C_s^{2/3}r^{4/3} \tag{5}$$

for $r < r_s$. In this case, the radius which contains x% of the total intensity I is given by: $r_x/r_s = (x/100)^{3/2}$, showing that half the electron intensity is contained within a radius $\approx 0.35\ r_s$ .

## 4.    THE INFLUENCE OF DIFFRACTION AND DELOCALIZATION

The previous discussion was based on geometrical optics, treating the electron as a particle. Wavelike aspects can be introduced in terms of diffraction at the lens aperture, described (for a uniformly filled aperture) by the Airy point-spread function:

$$dI(r)/da = dI(0)/da\ [J_1(\rho)/\rho]^2 \tag{6}$$

where $\rho = (2\pi/\lambda)\beta r$, $\lambda$ is the electron wavelength and $J_1$ is a first-order Bessel function. In the absence of lens aberrations, half of the electrons would appear to come from an object diameter given by $d_{50} = 0.52\ \lambda/\beta$. In incoherent imaging theory, the effect of both diffraction and lens aberration is represented by a two-dimensional convolution of the appropriate point-spread functions. In Fig. 2a , we show the result of convolving the Airy function with the chromatic-aberration PSF given by Eq.(2). The effect of diffraction is to broaden the PSF and give it a finite value at $r = 0$.

Figure 2a also shows the intensity distribution of scattering from a single atom, calculated using wave-optical dipole theory (Berger and Kohl, 1993) for the same instrumental parameters. This distribution (triangular data points) differs in shape and is somewhat broader than the PSF (square points), reflecting the fact that the atom is not a point scatterer. In other words, under conditions where the chromatic broadening is comparable to or smaller than 1 nm, the classical PSF is broadened as a result of delocalization of the inelastic scattering (Schenner et al., 1995).

A simple expression which provides an intuitive and fairly accurate description of the object diameter $d_{50}$ due to delocalization and diffraction is (Egerton, 1996):

$$(d_{50})^2 \approx (0.6\ \lambda/\theta_{50})^2 + (0.6\ \lambda/\beta)^2 \approx (0.5\ \lambda/\theta_E^{3/4})^2 + (0.6\ \lambda/\beta)^2 \qquad (7)$$

Figure 2b shows the separate contributions arising from chromatic aberration, spherical aberration, diffraction and delocalization (calculated for a 500eV image and a 200keV TEM

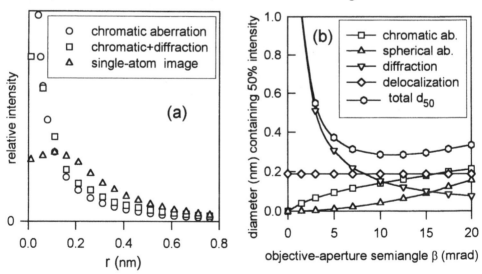

Fig. 2. (a) Point-spread functions for chromatic aberration (circles) and for chromatic aberration plus diffraction (squares), calculated for a Zeiss EM912 EFTEM ($E_0 = 120$ keV, $C_c = 2.7$ mm, $\beta = 7.8$ mrad, $E = 540$ eV). Triangular data points represent the image intensity derived from single-atom calculations (Berger and Kohl, 1993).
(b) Object diameters containing 50% of the electron intensity, for four different sources of image broadening, calculated for a thin specimen using Eqs. (2), (4), (6) and (7) with the parameters: $C_c = 1$ mm, $C_s = 0.47$ mm, $\Delta = 20$ eV, $E = 500$ eV and $E_0 = 200$ keV.

with high-resolution polepieces) together with the total image broadening obtained by adding all components in quadrature. Assuming a thin specimen, the predicted resolution is $\approx 0.3$ nm for $\beta$ in the range 10 - 20 mrad (objective-aperture diameter 20-40 $\mu$m), being limited mainly by delocalization and chromatic aberration. Decreasing the width of the energy-selecting slit from 20 eV to 10 eV would provide 0.2nm resolution, with a twofold reduction in intensity. Increasing the incident-beam energy would also reduce the chromatic term, but would barely affect the delocalization component which, according to Eq.(7), is approximately $17$nm/$E^{3/4}$ (with E in eV) for $E_0$ in the range 100 keV to 1 MeV. This E-dependence suggests that atomic resolution should be possible for ionization edges above 500 eV, although the weak core-loss intensity and associated statistical noise (Berger and Kohl, 1993) would require long recording times, making specimen drift an important consideration (Krivanek et al., 1995).

## 5.    RESOLUTION OF A STEM IMAGE

Because the lenses of a scanning transmission electron microscope operate only on the incident electrons, inelastic and zero-loss images have essentially the same spatial resolution, determined by the intensity distribution in the incident probe. Colliex and Mory (1984) have used wave-optical theory to calculate incident-probe current-density profiles, taking into account spherical aberration and diffraction of the probe-forming (objective) lens, and with consideration given to chromatic aberration (dependent on the energy spread of the incident electrons) and the electron-source size. Ignoring the latter, the Gaussian-focus profile is the PSF of the probe-forming system and could be calculated from Eqs. (3) and (5) for a rectangular distribution of the angles of incidence (the incident intensity being constant up to an angle $\beta$ relative to the optic axis) or from Eqs. (2) and (4) for a Lorentzian angular distribution (of halfwidth equivalent to $\theta_E$ but falling abruptly to zero at an angle $\beta$).

## 6.    RESOLUTION OF AN SEM IMAGE

Similar considerations apply to the current-density profile in an SEM probe which, apart from some broadening associated with the interaction volume in which secondary and backscattered electrons are produced, determines the spatial resolution of the SEM image. In the case of a low-voltage SEM, this interaction volume is small and the spherical-aberration and (for a field-emission source) source-size terms are relatively low; therefore the current-density distribution in an optimally-focussed probe is given approximately by Eq. (3), convolved with the Airy diffraction function.

It should be emphasized that the PSF provides more complete information about SEM image contrast than a single number representing 'resolution' or 'spot size'. However, this PSF is not easy to measure; scanning electron microscopes are not equipped with postspecimen lenses which could magnify the probe (for measurement of its current-density distribution) and scanning the probe itself across a 'sharp edge' (in front of an electron detector) fails in the case of very small probes. Therefore, probe size remains a useful concept, particularly if defined as the radius or diameter containing a given percentage of the electrons.

Different contributions to the probe size (each calculated as a diameter containing 50% of the incident current) are compared in Table 1 for three different types of SEM operated under typical conditions, at high and low accelerating voltage. For a traditional

154

(tungsten-filament) instrument operated at 20 kV, spatial resolution is limited by the electron-source size (even for large demagnification 1/M) and by spherical aberration. Operated at lower voltage (e.g. 5 kV), the resolution becomes worse because of chromatic aberration.

FEGSEM instruments have the advantage of very small source size ($\approx$ 5 nm) and can achieve sub-nanometer probe size at $E_0$ = 20 keV. In addition, the energy spread of a field-emission source is smaller, allowing probes of only a few nm diameter at accelerating voltages as low as 2 kV. Future improvements in lens design may reduce this probe size even further. Also in favour of increased resolution at low accelerating voltage is the fact that the interaction volume (including backscattering) falls to nanometer dimensions for $E_0$ < 2 keV (Joy and Joy, 1996).

In practice, electronic noise is also important in SEM imaging and depends on the recording time (limited by stage drift or by radiation damage, including contamination) and on the incident-probe current. Because of its higher electron-optical brightness, a field-emission source provides increased current in very small probes.

| type of SEM ( + example) | $C_s$ (mm) | $C_c$ (mm) | $\beta$ (mrad) | 1/M | $E_0$ (keV) | ss (nm) | diff. (nm) | sph. (nm) | chr. (nm) | tot. (nm) |
|---|---|---|---|---|---|---|---|---|---|---|
| W-filament (Philips 505) | 40 | 15 | 9.1 | $10^3$ | 20 | 20 | 0.5 | 28 | 9.0 | 36 |
|  |  |  |  |  | 5 | 20 | 1.0 | 28 | 36 | 50 |
| snorkel-lens FEG (Hitachi S4500) | 4.2 | 3.2 | 7 | 10 | 20 | 0.5 | 0.6 | 1.0 | 0.3 | 1.3 |
|  |  |  |  |  | 2 | 0.5 | 2.0 | 1.0 | 2.7 | 3.5 |
| in-lens FEGSEM (Hitachi S900) | 1.9 | 2.5 | 7 | 10 | 20 | 0.5 | 0.6 | 0.5 | 0.2 | 0.9 |
|  |  |  |  |  | 2 | 0.5 | 2.0 | 0.5 | 2.1 | 3.0 |

Table 1: Contributions from source size (ss), diffraction (diff), spherical aberration (sph) and chromatic aberration (chr) to the total diameter (tot) containing 50% of the incident electrons, for a tungsten source (energy width 1.5 eV) and field-emission source (energy width 0.4 eV).

## REFERENCES

Berger A and Kohl H 1993 Optik **92**, 175
Colliex C and Mory C 1984 Quantiative Electron Microscopy (SUSSP: Edinburgh), Ch. 5
Egerton R F 1996 Electron Energy-Loss Spectroscopy in the Electron Microscope (London and New York: Plenum) p 348
Egerton R F and Wong K 1995 Ultramicroscopy **59**, 169
Egerton R F and Crozier P A 1997 Micron **28**, No.2
Joy D C and Joy C S 1996 Micron **27**, 247
Kohl H and Rose H 1985 Adv. Electron. Electron Phys. **65**, 175-200.
Krivanek O L Kundmann M K and Kimoto K 1995 J. Microscopy **180**, 277
Schenner M, Schattschneider, P and Egerton R F 1995 Micron **26**, 391

*Inst. Phys. Conf. Ser. No 153: Section 5*
*Paper presented at Electron Microscopy and Analysis Group Conf. EMAG97, Cambridge, 1997*
© *1997 IOP Publishing Ltd*

# Atom images and diffraction patterns of high Tc superconductors photographed by Ca-L2,3 shell loss electrons

H. Hashimoto, Z.P. Luo, M. Kawasaki*, F. Hosokawa* and E. Sukedai

Okayama University of Science, Okayama 700, Japan.
* JEOL, 3-1-2, Musashino, Akishima, Tokyo 196, Japan.

**ABSTRACT:** Using Ca L2,3 (350 eV, 346 eV) shell loss electrons, images and diffraction patterns of high Tc superconductors containing single, double and triple Ca atomic columns were recorded with 6eV energy window using 200 kV field emission electrons and drift free goniometer. The images which were recorded with 60 seconds exposures had heavy quantum noise. Using image processing in real space and Fourier space, images of Ca-atoms were revealed. Characteristics of the nature of electron diffraction patterns are discussed.

## 1. INTRODUCTION

The characteristic energy loss electrons can be used for the identification of atoms in thin crystals (Hashimoto et al 1992, Krivanek et al 1992, Makita, Hashimoto et al 1993). Endoh, Hashimoto et al (1994) delivered the localization width of the inelastic scattering potential to be about 0.08 nm from the inelastic scattering factors for SiL3 and AlL1 ionization. They discussed wave mechanically the image contrast and corresponding diffraction patterns formed by the core loss electron of Si and Al atoms at various depths of crystals with consideration of multiple elastic scattering before and after inelastic scattering and also successive processes of image formation by the conventional lens (Endoh, Hashimoto et al 1994). Endoh et al (1996) recently checked the value of the cross section of inelastic scattering of Si-L loss electron and derived the localization to be about 0.3 nm which is larger than the values derived in 1994. Even if the estimated values of localization width change with such amount, the quality of the calculated electron microscope images shown in 1994 will not be big changed. The resolution limit of the images formed by the core loss electrons may be simply estimated for very thin crystals by the root mean square of localization width, chromatic error and spherical aberration (Shuman et al 1986, Kurata et al 1996).

The above work suggests that it is possible to get images of characterized single atoms by core loss electrons. However if the atoms are arranged regularly, the images of the atoms in crystals formed by the core loss electrons have the contrast similar to the structure image formed by the interference of elastic scattering, which has taken place before or/and after the inelastic scattering. It was shown by Makita, Hashimoto et al (1993) that the production of the diffraction pattern can be the measure of the order of the disturbance to the inelastic image contrast due to the elastic scattering before or/and after the inelastic scattering. It was emphasised that it was important to take the diffraction patterns formed by the core loss electrons as well as the electron micrograph to confirm the image nature. This was also shown by the calculations of the image contrast and the intensity of diffraction patterns (Endoh, Hashimoto et al 1994).

With increasing the energy of core loss electrons, the localization width becomes small and thus the resolution of the images becomes high but the intensity of electrons forming the images and diffraction patterns becomes small which inevitably makes difficult to record the image in atomic level and corresponding diffraction pattern. The present paper concerns to take images of Ca atoms using rather high energy loss electrons of Ca-L2,3 (350 eV, 346 eV)

156

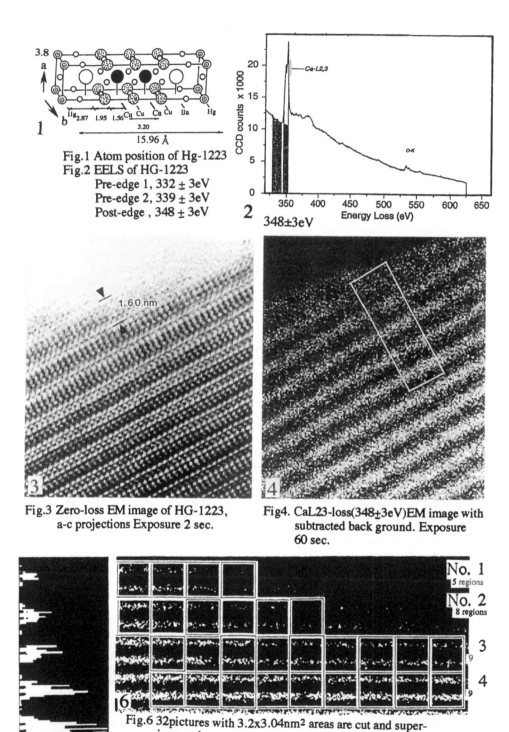

Fig.1 Atom position of Hg-1223
Fig.2 EELS of HG-1223
Pre-edge 1, $332 \pm 3eV$
Pre-edge 2, $339 \pm 3eV$
Post-edge , $348 \pm 3eV$

2   $348 \pm 3eV$

Fig.3 Zero-loss EM image of HG-1223,
a-c projections Exposure 2 sec.

Fig4. CaL23-loss(348±3eV)EM image with
subtracted back ground. Exposure
60 sec.

Fig.6 32pictures with 3.2x3.04nm² areas are cut and superimposed.

Fig.5 Intensity profiles of Fig.4 along c-axis made to the summation of 60 pixels
along a-axis a pair of dots indicate the position of Ca atoms.

## 2. EXPERIMENTALS

### 2.1 Observations of Images

The specimens used for taking the images of Ca atomic column were high Tc superconductors Hg-1212, Hg-1223 and Hg-1234 , which contain single, double and triple Ca atom columns. Their unit cells are a=b=0.385 nm and c=1.276 nm (Hg-1212), c=1.596 nm (Hg-1223) c=1.895 nm (Hg-1234) as shown in Fig. 1 on Hg 1223 (Hg $Ba_2Ca_2Cu_3O_{8+\delta}$) in which separation of Ca atom in c direction is 0.32 nm. Energy loss spectrum near Ca $L_{2,3}$ is shown in Fig. 2. Using JEOL 2010 F microscope with a field emission gun, a drift free goniometer and a Gatan image filter with CCD camera, the images of Ca atoms were photographed with energy window of ±3eV. The calculated resolution of the image for very thin crystals is about 0.2 nm for Cs=0.5 mm, Cc=1.1 mm, objective aperture =10mrad. Localization width R=0.17 nm which is based on the equation by Pennycook (1982).

In this imaging condition, the intensity of electron contributing to the images of Ca atoms is 1000 ~ 10,000 times smaller than the one of illuminating electrons. Thus it is necessary to give the exposure of about 60 minutes for taking the image of Ca atoms, which are equivalent to the quality of the bright field images taken with two seconds exposure. The core-loss electron images of each specimens were recorded by subtracting the images formed by the background of the spectrum which was extrapolated by two pre-edge intensities, which are shown in Fig.2. Figs.3 and 4 are zero loss and Ca $L_{2,3}$ core loss electron images with exposures of two seconds and 60 seconds respectively. In Fig.3, the extinction contrast due to the dynamic scattering is seen with increasing the thickness. In Fig.4, the bright line images of lined Ca column atoms become wide and strong in intensity with increasing the specimen thickness. Fig.4 has heavy quantum noise due to the very limited exposures.

### 2.2 Image Processing

In order to estimate the improvement of image quality of Fig.4 by image processing, the intensity profiles of the images along c axis were made by summing the intensity in 60 pixels along a axes in the area marked by rectangle in Fig.4 and are shown in Fig.5. The profiles in Fig.5 are equivalent to those of the images obtainable with the exposure of one hour. By referring the Ca atom position indicated by marks at the bottom of profiles, it is seen clearly that two peaks of the profiles appear at Ca atom position. Figs.6 and 7 show the two dimensional image processing. 30 pictures with 3.2 x 3.04 nm² areas were cut from the images of thin regions and were superimposed 8 pictures in the four times large area ie. 6.4 x 6.08 nm². The elimination of the noise of the images` was carried out in Fourier space as shown in Fig.7(b) by using slit windows with enough width not to introduce the artifact. Fourie transform of Fig.7(b) is shown in Fig.7(c), in which Ca atoms with the separation of about 0.32~0.38 nm in c and a directions- are seen. Distortion of recorded image disturbes the quality of the processed images.

### 2.3 Observation of Diffraction Pattern

Figs.8 and 9 are the diffraction patterns from thin and very thin ( a few unit cells thickness ) crystals respectively. Fig.8 shows the diffraction patterns by (a)40eV, (b)Pre-edge 2 (339±3)eV, (c)Post-edge (348±3)eV, and (d)Ca$L_{2,3}$-loss electrons with subtracted background. In Fig.8(a), the discreate and diffuse diffraction spots due to plasmon scattering are seen. Since the thickness of the crystal is not very thin, multiple scatterings of plasma loss electrons produce the pattern shown in (b). The very weak diffuse bands in (d) are seen to be the diffraction pattern of core loss electrons which is formed by the elastic interaction with the nearest neighbouring atoms.

For the case of very thin crystal shown in Fig.9, no diffraction spots can be seen. The intensity distribution in Fig.9(d) is comparable with theoretical inelastic scattering intensity

158

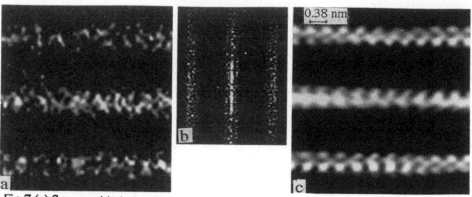

Fig.7 (a) Superposition of pictures selected from Fig.6    (c) Fourier transform of (b)
      (b) Fourier transform of (a) with selecting mark

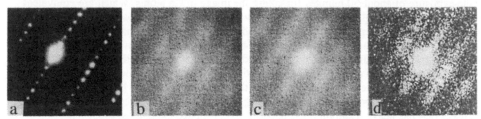

Fig.8 Diffraction pattern by (a) 40eV loss (b) pre-edge 2 loss. (c) post edge loss. (d)
    CaL23 loss electrons.

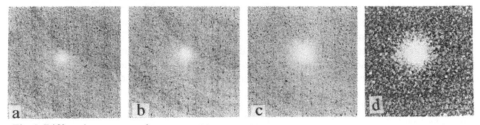

Fig.9 Diffraction patterns from very thin crystal by (a) pre-edge 1 loss. (b) pre-edge 2
    loss. (c) post edge loss. (d) CaL23 loss electrons.

from the potential region which produced the CaL23 loss electrons. Center of the pattern might
have the effect of after image on CCD camera.

## REFERENCES

Endoh H., Hashimoto H. and Makita Y.,1994 Ultramicroscopy 54 pp351-356,56 pp108-120.
Endoh H. and Kumao A, 1996 Abstract of meeting of Phys. Soc. Japan Oct1, M2.
Hashimoto H., Makita Y. and Nagaoka N., 1992 Optik 91 pp119-126
Krivanel O.L., Gubbens A.J., Delby N., Mayer C.E., 1992 Micross. Microanal. Microstruct.
3.p187
Kurata S., Moriguchi S., Isoda S. and Kobayashi T., 1996 J. Electron Micros. 45 79-84.
Makita Y., Hashimoto H., Nagaoka N. and Kumao A., 1993 Micross. Microanal.
Microstruct. 4 pp143-152.
Pennycook S.J.,Contempt. Phys. 23, pp371-400
Shuman H., Chang C.F. and Somlyo A.P.., 1986 Ultramicroscopy 19, pp121-134.

*Inst. Phys. Conf. Ser. No 153: Section 5*
*Paper presented at Electron Microscopy and Analysis Group Conf. EMAG97, Cambridge, 1997*
© *1997 IOP Publishing Ltd*

# Interpretation of Electron Energy Loss Spectra from Hard Elastic Carbon and Related Materials

J Yuan, N Menon, G A J Amaratunga*, M Chhowalla*, C J Keily**

Cavendish Laboratory, Madingley Road, Cambridge CB3 0HE, U.K.
*Department of Electrical Engineering and Electronics and **Department of Material Science and Metallurgy, University of Liverpool, Liverpool L69 3BX, U.K.

**ABSTRACT:** By incorporating a helium jet in the filtered arc ion deposition system previously used to produce hard diamond-like amorphous carbon film, carbon composite materials consisting of both amorphous carbon and nanoparticles can be deposited, some of which are both hard and elastic. These remarkable mechanical properties have been attributed to the possible $sp^3$ linking of $sp^2$ dominated fullerene-like nanotube. We investigate issues involved in the determination of atomic bonding character of carbon atoms in such structured materials, using electron energy loss spectroscopy.

## 1. Introduction

Carbon can exist in at least three elemental forms: graphite, diamond and fullerene exhibiting soft, hard and elastic mechanical properties. One of the challenges in material science is to produce composite materials with the desired combination of these properties. Recently, hard *and* elastic films have been reported in thin films of carbon and its alloys (Sjöström *et al.* 1995; Amaratunga *et al.* 1996). The origin for such remarkable mechanical properties remains a subject of great interest. Preliminary electron energy loss spectroscopy (EELS) study suggests that the fullerene-like nanoparticles in these carbon samples are covalently linked by tetrahedral $sp^3$ bonding, supporting the idea that $sp^3$ linked graphite fragments could be responsible for the remarkable mechanical properties. In this paper, we discuss some of the issues involved in such a determination and describe the search for evidence for interlinking in carbon based nanostructured films. High resolution electron microscopy (HREM) study of these films is reported in the paper by Alexandrous et. al. (1997).

## 2. Sample preparation

The carbon film is prepared inside a vacuum arc carbon plasma in the presence of a local high pressure gas background. Except for the gas jet which creates a high pressure in the vicinity of arc spot whilst keeping the pressure in the rest of the deposition chamber at 6 mTorr, the apparatus is very similar to that used for deposition of highly tetrahedral amorphous carbon (ta-C) films with diamond-like mechanical properties (Veerasamy *et al.* 1993). The presence of background carrier gas in the carbon plasma is known to encourage the formation of fullerene-like nanoparticles. The film deposited on Si substrates at room temperature (RT) appears largely amorphous, but detailed examination reveals numerous carbon nanoparticles such as that shown in Fig. 1a which is taken in a 100 keV VG HB510 scanning transmission electron microscope (STEM). This shows that fullerene-like structures are generated in the plasma and arrive at the substrate with a density high enough for possible cross-linking. The film deposited on the substrate at high temperature (HT) of 350°C shows an apparently amorphous phase with little evidence for extended fullerene-like objects (Fig. 1b). This is taken to be the evidence that the carbon nanoparticles are broken upon impact and fragments are incorporated into the film. The nanoindentation test suggests that the HT sample has a hardness comparable to diamond--like ta-C films, but with nearly 80% elastic recovery.

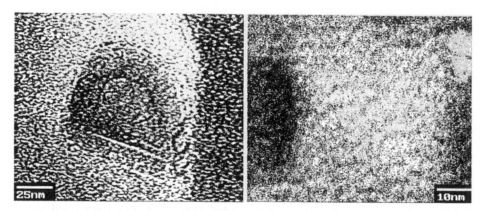

Fig. 1a A STEM bright field image of the carbon film grown with the substrate held at room temperature, showing a fullerene-like nanoparticle.

Fig. 1b A STEM bright field image of the carbon film grown with the substrate held at 350°C, without visible fullerene-like structure.

## 3. Electron energy loss spectroscopy

### 3.1 EELS analysis of carbon nanostructure

Fig. 2 shows the spatially resolved EELS spectra from different parts of the carbon nanoparticle shown in Fig. 1a. We found that the spectra are very graphite-like, showing a pre-edge peak at 285.5 eV which is characteristic feature of the $sp^2$ bonded materials. But closer examination shows that the pre-edge intensity is smaller at the centre of the particle than at the edge. Another characteristic of the graphite EELS spectrum is the doublet structure at 292 eV which marks the onset of 1s-2p($\sigma^*$) transition. This doublet is also observed here with its intensity anti correlate with that of the pre-edge intensity. The presence of graphite signal at the centre of the particle shows that the particle is not a hollow nanotube with its axis parallel to the beam. Rather both the tilting experiment and line scan spectra from EELS and STEM annular dark-field signals suggest that the particle is round 'onion' or polyhedral with hollow core.

To detect the rehybridisation of the p($\pi$) and $sp^2$($\sigma$) bonds to form $sp^3$($\sigma$) bonded region in the possible cross-linked region, we need to understand the intensity variation of the 1s-2p($\pi^*$) transition at different positions in this structured material. We note that a similar intensity variation of the characteristic 1s-2p($\pi^*$) transition at 285.5 eV has also been observed in a nanotube by Cullen et al. (1993) and Stephan et al. (1996), and the authors have correctly related the variation to the orientation of the local graphene plane in the nanotube, by comparison with the EELS taken from the graphite standard. However, Cullen et al. (1993) did not comment why the 1s-2p($\pi^*$) excitation is enhanced when the electron probe axis is perpendicular to the local c-axis of the graphene plane. Stephan et. al. (1996) mentioned, also without elaboration, that the mean momentum transfer in their case is perpendicular to the beam axis. Using TEM, Dravid et. al. (1993) have shown EELS spectra of C K-edge absorption which has the opposite orientation dependence. We attribute this to the role played by the experimental conditions such as to the convergence and collection angles deployed in the experiment. This is confirmed by experimental result from single crystal graphite (Menon and Yuan, 1997). The effect of collector size of the EELS spectrometer on EELS of uniaxial materials has been investigated by Browning, Yuan and Brown (1993) for the case of plane wave illumination which is appropriate for TEM. Recently, Menon and Yuan (1997) have refined the analysis to take into account the use of a convergent probe which is more appropriate here for the STEM case.

The critical parameter in such an analysis is the characteristic scattering angle $\theta_E$ (=$\Delta E/2E_0$) which has a value of 1.43 mrad. for C K-edge excitation at 285 eV. Even if the electron beam is parallel to the c-axis, the mean orientation of the momentum transfer in the EELS would be tilted towards the perpendicular direction, making the excitation of $\varepsilon^{\parallel}$ less likely if either the convergence or the collection angle is much bigger than $\theta_E$. The latter is almost always true in the imaging mode of STEM, this explains the fact that the excitation of the 1s-2p($\pi^*$) transition is always subdued when the beam is parallel to the local c-axis of the graphic plane in STEM.

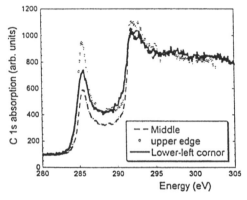

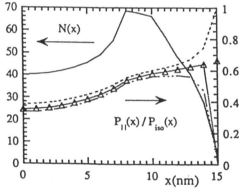

Fig. 2 Spatially resolved EELS taken from the middle, the top side and the left lower corner of the particle shown in Fig. 1a.

Fig. 3 The simulated linescan profile across a carbon nanotube or onion for isotropic (N(x)) and anisotropic 1s-2p($\pi^*$) transition signals ($P_{\parallel}/P_{iso}$). x is measured from the particle centre.

To account for the wrapping of the graphite plane in the carbon nanoparticle, we have carried out a simulation of the intensity variation expected for the carbon 1s-2p($\pi^*$) transition in different positions (x, measured from the centre) of a nanotube or spherical onion using the formulation of Menon and Yuan (1997) for an anisotropic material (Fig. 3). The carbon particle is modelled by a series of concentric spheres of appropriate radius. The simulated line scan of normalized excitation probability ($P_{\parallel}/P_{iso}$) of carbon 1s-2p($\pi^*$) transition shows $A+Bx^2$ dependence characteristic of the single shell fullerene particle when moving away from its centre. It changes roughly to a linear increase as the probe passes through outer rims of successive graphene shells. The number of C atoms probed is also shown as N(x) and is the linescan profile expected from the isotropic part of the EELS spectra ($P_{iso}$) or the signal of the annular dark field. The characteristic shapes of both line profiles have been observed in nanoparticles shown in Fig. 1a and in nanotubes. Also shown in the Fig. 3 is the modified linescan profile expected if the outermost layer is amorphous carbon of mostly sp2 bonding character (profile rises sharply) or diamond-like character (the profile drops sharply). The former situation is observed in some of nanotubes investigated and is related to the presence of surface carbon contamination. The latter is very close to our expected signature for the cross-linked region between nanoparticles. This shows that the detection of sp3 cross-linking is possible if the electron beam direction is very close to perpendicular to the line linking the centre of the two particles. An alternative way to eliminate the orientation effect of wrapped graphite is by manipulation of convergence and collection angles which has been shown to be

162

possible (Menon and Yuan, 1997). This is particularly suitable to a microscope with post-specimen lenses such that this experimental condition can be fine-turned precisely.

3.2 EELS analysis of hard and elastic carbon film.

The Carbon 1s absorption spectrum from hard and elastic film produced at elevated substrate temperature is plotted in Fig. 4 and compared with spectra taken from the matrix phase of the room temperature sample and that from ta-C film. It is clear that the pre-edge 1s-2p($\pi^*$) transition is intermediate in the HT sample, suggesting some of the carbon may be $sp^3$ bonded as speculated. The sharpness of the pre-edge peak also suggests that the atoms with $sp^2$ bonding character may retain local graphite atomic structure. However, the lack of sharply defined doublet at 292 eV shows that the graphene planes are highly distorted and furthermore, any such local graphitic structures can not be perfect fullerene molecules such as $C_{60}$, $C_{70}$ or $C_{84}$ as the quantum confinement splits the pre-edge peak into multiplet which is not observed.

Fig. 4 C 1s absorption spectra taken from highly tetrahedral carbon film (ta-C), and from nanostructured film (RT and HT) investigated here.

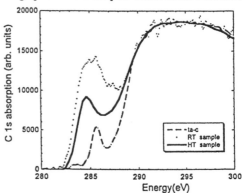

## 4. Summary

We have presented an EELS analysis of the carbon film produced by ion beam technique similar to that for the diamond-like ta-C film, but in the presence of carrier gas background. We detected the presence of the graphitic particles in 'onion or polyhedron forms in the room temperature sample and the hint of its fragments in the high temperature sample. We have shown that the interpretation of the $sp^2/sp^3$ ratio of the carbon atoms in structured film needs to take into account the experimental condition deployed and the position of the probe. With suitable modelling, the effect of cross-linking can be detected.

## Reference

Alexandrous I, Pringle S D, Munindradasa A I, Chhowalla M, Lim G, Kiely R and Amaratunga G A J 1997 at this conference
Amaratunga G A J, Chhowalla M, Kiely C J, Alexandrou I, Aharonov R, Devenish R M, Nature, Vol 383, p321-323, 1996
Browning N D, Yuan J and Brown L M 1993 Phil. Mag. A67, 261-271
Cullen S L, Botton G, Kirkland A I, Brown P D, Humphery C J 1993 *Inst. Phys. Conf. Ser.* No. 138, p79
Dravid V P, Lin X, Wang Y, Wang X K, Yee A, Ketterson J B and Chang R P H 1993 Science, 259 1601-1607
Menon N and Yuan J 1997 to be submitted
Sjöström, H, Stafström S, Boman M, and Sundgren J -E, Phys. Rev. Lett. 75(7) p1336-1339, 1995
Stephan O, Ajayan P M, Colliex C, Cyrot-Lackmann F and Sandre E 1996 Phys. Rev. B 53(20) 13824-13829
Veerasamy V S, Yuan J, Amaratunga G A J, Milne W I, Gilke K W R, Weiler M, Brown L M 1993 Phys. Rev. B 48, 24, 17954-17959

*Inst. Phys. Conf. Ser. No 153: Section 5*
*Paper presented at Electron Microscopy and Analysis Group Conf. EMAG97, Cambridge, 1997*
© 1997 IOP Publishing Ltd

# A study of electron energy loss spectroscopy with the electron beam parallel to an interface

**Jørgen Houe Pedersen**

Aalborg University Esbjerg, Niels Bohrs Vej 8, DK-6700 Esbjerg, Denmark

**ABSTRACT:** The model to be used consists of a semi-infinite medium and an electron moving parallel to the surface outside the medium. Surface excitations are of special interest. In the present non-local approach a number of surface eigenmodes are excited at the medium surface. In two important limits, the *local* and *electrostatic* limits, electron beam excitations of surface modes are briefly discussed.

## 1. INTRODUCTION

Dielectric response theory has been applied for many years to describe different aspects of electron beam specimen interaction. This response theory can be divided into four categories: 1) Electrostatic and local, 2) Electrostatic and non-local, 3) Electrodynamic and local, and 4) Electrodynamic and non-local. Normally the electrostatic versions are sufficient to describe the experiments properly, especially at low voltages. At higher electron beam energies an electrodynamic approach seems to be adequate. When more details about surface excitations are needed a non-local model should be used.

## 2. THEORY

The model to be used consists of a semi-infinite medium and an electron moving in a distance $z = -z_0$ from the medium surface which coinside with the x-y-plane. The z-axis is perpendicular to the medium surface. The medium occupies the half-space $z > 0$.

The interaction between the electric field caused by the moving electron and the matter is to be based on dielectric response theory, i.e. the Maxwell equations. A constitutive relation between the induced current density and electric field is needed. In this case a non-local approach is to be employed.

The basic idea is to find the electric fields inside and outside the matter and then use the electromagnetic boundary conditions in order to determine the remaining constants. The electric fields must obey the wave equation, which can be derived from the Maxwell equations. Outside the medium the external current density is given by $\vec{J}_{ext}(\vec{r},t) = Qv\vec{e}_x\delta(x - vt)\delta(y)\delta(z + z_0)$ where $Q$ is the particle (electron) charge, $v$ is the electron speed, $\vec{e}_x$ is a dimension less unit vector along the x-axis, and $\delta$ is the Dirac delta function.

The concept of non-locality is to be implemented via the following constitutive relation between the electric field $\vec{E}$ and the displacement field $\vec{D}$ :

$$\vec{D}(\vec{r},t) = \varepsilon_0 \int\int \int\int_{-\infty}^{\infty} \ddot{\varepsilon}(\vec{r}_{\parallel} - \vec{r}'_{\parallel}, z, z', t - t') \cdot \vec{E}(\vec{r},t) d^3 r' dt' \tag{1}$$

where $\varepsilon_0$ is the vacuum permittivity and $\ddot{\varepsilon}$ is the tensorial dielectric function. Note that translational invariance in space coordinates parallel to the surface $\vec{r}_{\parallel}$ and in time has been assumed in eq. (1). These are the relevant equations to solve inside and outside the matter. In the so-called Semi Classical Infinite Barrier (SCIB) model the complicated z-dependency can be managed by utilizing the so-called "mirror condition", which states that the component of the electric field parallel to the surface is an even function in z, whereas the perpendicular field component is an odd function in z, viz. $E_{\parallel}(z) = E_{\parallel}(-z)$ and $E_{\perp}(z) = -E_{\perp}(-z)$ (Kliewer and Fuchs 1968).

Using the SCIB-model the Fourier analysis of the above-mentioned equations becomes possible and it yields for the Fourier components $\vec{E}(q_{\parallel}, \omega, z)$ of the electric field inside the matter

$$\vec{E}(q_{\parallel}, \omega, z) = \frac{1}{2\pi} \int_{-\infty}^{\infty} \ddot{\Xi}(\vec{q}, \omega) \cdot \vec{g} e^{iq_{\perp} z} dq_{\perp} \tag{2}$$

and for the electric field outside the matter

$$\vec{E}_0(q_{\parallel}, \omega, z) = \frac{1}{2\pi} \int_{-\infty}^{\infty} \ddot{\Xi}_0(\vec{q}_0, \omega) \cdot \left\{ \vec{g}_0 - G e^{iq_{\perp 0} z_0} \vec{e}_x \right\} e^{iq_{\perp 0} z} dq_{\perp 0} \tag{3}$$

where $G = i\omega\mu_0 Q 2\pi \delta(q_{\parallel} - \omega/v)$ and the tensorial quantities are given by

$$\ddot{\Xi}(\vec{q}, \omega) = \frac{1}{N_T} \left( \ddot{1} - \frac{\vec{q}\vec{q}}{q^2} \right) + \frac{1}{N_L} \frac{\vec{q}\vec{q}}{q^2} \tag{4}$$

and

$$\ddot{\Xi}_0(\vec{q}_0, \omega) = \frac{1}{N_{T0}} \left( \ddot{1} - \frac{\vec{q}_0 \vec{q}_0}{(\omega/c)^2} \right) \tag{5}$$

in which $N_T = N_T(\vec{q}, \omega) = (\omega/c)^2 \varepsilon_T(\vec{q}, \omega) - q^2$, $N_L = N_L(\vec{q}, \omega) = (\omega/c)^2 \varepsilon_L(\vec{q}, \omega)$, and $N_{T0} = N_{T0}(\vec{q}, \omega) = (\omega/c)^2 - q_0^2$, where $\varepsilon_{T,L}$ are dielectric functions. Note that $N_T = 0$, $N_L = 0$ and $N_{T0} = 0$, gives the bulk wave dispersion relations for transverse (T) and longitudinal (L) wave propagation inside and outside (0) the matter. The quantities $\vec{g}$ and $\vec{g}_0$ contain the two integration constants $B_y(0^+)$ and $B_y(0^-)$, viz. $\vec{g} = (2i\omega B_y(0^+), 0, 0)$ and $\vec{g}_0 = (-2i\omega B_y(0^-), 0, 0)$. Only the denominators from the explicit expressions for $B_y(0^+)$ and $B_y(0^-)$ are needed here, the denominator is namely the same for both. If this denominator is to be set equal to zero the non-local surface wave dispersion relation appears, i.e.

$$\int_{-\infty}^{\infty} \frac{1}{q^2} \left\{ \frac{q_\parallel^2}{N_L} + \frac{q_\perp^2}{N_T} \right\} e^{iq_\perp 0^+} dq_\perp = \frac{q_{\perp 0} c^2}{\omega^2} \int_{-\infty}^{\infty} \frac{q_\perp}{N_T} e^{iq_\perp 0^+} dq_\perp \qquad (6)$$

From the equations above it is realised that excitations of surface waves will have a significant effect on the electric field at the site of the moving electron. Hence, the electron energy loss spectrum will be altered as well.

The surface wave dispersion relation can be simplified by assuming that only collective medium excitations takes place. It can be further reduced by using the hydrodynamic approach in describing these excitations. In the hydrodynamic approach the dielectric functions take the form $\varepsilon_T = 1 - \omega_p^2 / [\omega(\omega + i/\tau)]$ and $\varepsilon_L = 1 - \omega_p^2 / [\omega(\omega + i/\tau) - Dq^2]$ where $1/\tau$ is the collision frequency, and the diffusion constant $D = (3/5)v_F^2$, where $v_F$ is the Fermi velocity. Doing this, it can be shown that in the frequency range from $\approx \omega_p/\sqrt{2}$ to $\approx \omega_p$ ($\omega_p$ being the plasma frequency) three distinct eigenmodes can exist to the surface wave dispersion relation. Outside this frequency range the three eigenmodes collapse into two. These can be recognised as the surface wave dispersion relation in the *local* and *electrostatic* limits, respectively. In Fig. 1a the three non-local surface eigenmodes are obtained on the basis of the hydrodynamic approach with a finite collision frequency. For comparison the *decoupled* modes, i.e. the *local* ($D \to 0$) and the *electrostatic* ($c \to \infty$) modes, are obtained in Fig. 1b, also on the basis of the hydrodynamic approach with a finite collision frequency.

## 3. DISCUSSION

The local ($D \to 0$) dispersion relation is given by

$$q_\parallel = \frac{\omega}{c} \left\{ \frac{\varepsilon_T(\omega)}{1 + \varepsilon_T(\omega)} \right\}^{1/2} \qquad (7)$$

where the transverse dielectric function can be taken as $\varepsilon_T(\omega) = 1 - (\omega_p/\omega)^2$, in the limit $\tau \to \infty$. The electrostatic ($c \to \infty$) dispersion relation is given by

$$q_\parallel = \frac{2\omega^2 - \omega_p^2}{2\omega\sqrt{\omega^2 - \omega_p^2}} q_L \qquad (8)$$

also taken in the limit $\tau \to \infty$, where the longitudinal wave number can be written as $q_L = \left\{ (\omega^2 - \omega_p^2)/D \right\}^{1/2}$.

Inserting the resonance condition $q_\parallel = \omega/v$ into the local dispersion relation the following expression appears $c/v = \left[ (\omega^2 - \omega_p^2)/(2\omega^2 - \omega_p^2) \right]^{1/2}$. Note that $\omega < \omega_p/\sqrt{2}$ which means that the moving electron can excite plasmariton-like surface waves governed by the low frequency branch of the dispersion relation.

166

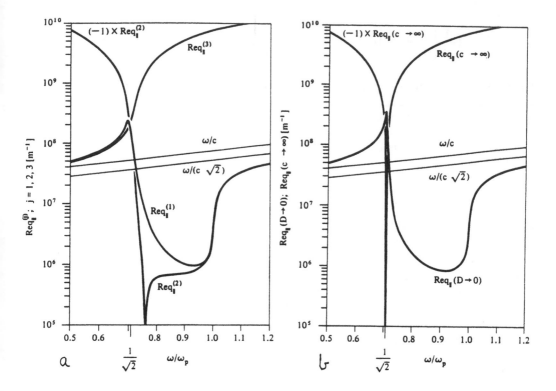

Fig 1. (a) Real parts of the non-local surface mode dispersion relation, based upon the hydrodynamic approach with a finite $\tau$. (b) Real parts of the surface mode dispersion relation in the electrostatic $(c \rightarrow \infty)$ and local $(D \rightarrow 0)$ limits, respectively. Both with a finite collision frequency.

Inserting the resonance condition $q_{\parallel} = \omega / v$ into the electrostatic dispersion relation it is realised that $\omega = (\omega_p / \sqrt{2}) / \left[1 - (\sqrt{D} / v)\right]^{1/2}$ which means that $v > \sqrt{D}$ in order to excite surface plasmons at frequencies above $\omega_p / \sqrt{2}$.

In order to give a full description of the excitation of the different non-local surface modes a more detailed study of the resonance conditions associated with the non-local surface mode dispersion relation has to be undertaken.

**REFERENCES**

Fuchs R and Kliewer K L 1971 Phys. Rev. B3 2270
Keller O and Pedersen J H 1988 Proc. Soc. Photo-Opt. Instrum. Eng. 1029 18
Kliewer K L and Fuchs R 1968 Phys. Rev. 172 607
Pedersen J H 1995 Inst. Phys. Conf. Ser. No 147 59

*Inst. Phys. Conf. Ser. No 153: Section 5*
*Paper presented at Electron Microscopy and Analysis Group Conf. EMAG97, Cambridge, 1997*

# Electron beam damage in titanium dioxide films

**M S M Saifullah, C B Boothroyd, G A Botton and C J Humphreys**

Department of Materials Science and Metallurgy, University of Cambridge, Pembroke Street, Cambridge CB2 3QZ, UK

**ABSTRACT:** Single electron transistors are destined to be one of the most attractive candidates in the future for high density integrated circuits because of their low power consumption. Recently, it has been shown that room temperature single electron transistors can be fabricated by nano-oxidation process of titanium using a scanning tunneling microscope (Matsumoto et al. 1996). In the present study, nano-reduction processes in amorphous $TiO_x$ and crystalline $TiO_2$ under an intense nanometer-sized electron probe will be discussed. It is observed that crystalline $TiO_2$ is more amenable to electron beam damage than amorphous $TiO_x$, contrary to what is observed in many crystalline oxides and fluorides.

## 1. INTRODUCTION

The finely focused nanometer-sized electron probe in a scanning transmission electron microscope (STEM) offers a means of nanometre scale processing of semiconductor devices and for the production of nanowires. In many inorganic materials, exposure to such a finely focused intense electron beam can result in chemical changes as well as direct removal in the material. It has been seen that in amorphous metal fluorides such as $AlF_3$ (Kratschmer and Isaacson, 1986), $FeF_3$ (Saifullah et al. 1995) and $CrF_3$ (Saifullah et al. 1997), a thin film of metal is left after exposure to an intense electron beam in a STEM. In the case of metal oxides, the nanometer probe either removes the material from under the beam, as in amorphous $AlO_x$, or reduces the oxide as in the case of amorphous $SiO_2$. Electron energy loss spectroscopy (EELS) studies of irradiation damage of $SiO_2$ show that a nanometer sized silicon dot is left after the removal of oxygen (Chen et al. 1993). Irradiation damage studies conducted on $TiO_2$, $Ti_2O_3$ and $TiO$ in a STEM have shown that these materials damage in the same way, that is by reduction (Berger et al. 1987). Further studies of beam damage carried out on an anatase film showed that the changes in the fine structure of the electron energy loss spectrum are consistent with changes in the interatomic distance and the shift in d band onset shows a transition from insulator to metal. (Rez et al. 1995)

In this paper, we investigate the reduction characteristics of amorphous $TiO_x$ (a-$TiO_x$) and crystalline $TiO_2$ (c-$TiO_2$, anatase) films using an intense electron probe in a STEM fitted with a Gatan imaging filter, to form a nanometer-sized metallic material which can be potentially used to fabricate single electron devices in the future.

## 2. EXPERIMENTAL

The a-$TiO_x$ thin films were deposited on single crystal rock salt substrates cooled to liquid nitrogen temperature by conventional RF reactive sputtering from a 99.9% pure Ti target in a 70% Ar + 30% $O_2$ gas atmosphere at a power of 25W. The sputtered $TiO_x$ films were amorphous and transparent. They were floated on distilled water and collected on nickel grids then heat treated at 300°C in air for one hour to produce a mixture of crystallised (anatase) and amorphous areas in the film. This facilitates a direct comparison of the irradiation damage behaviour of crystallised as well as amorphous areas under identical

168

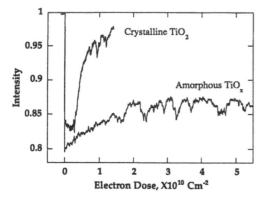

Figure 1: Drilling curves showing transmitted current intensity (incident beam = 1) as a function of electron dose for a-TiO$_x$ and c-TiO$_2$. Here both the materials show Type II drilling behaviour (Allen et al. 1993).

conditions. The electron beam damage mechanisms were studied in a VG HB501 STEM by recording both the transmitted electron intensity and energy loss spectra collected with a Gatan imaging filter as a function of irradiation time. A 100μm objective aperture was used and the EELS spectra were acquired at an energy dispersion of 0.2eV per channel. The acquisition time for time resolved EELS spectra was 5 seconds corresponding to a current density of $14.4 \times 10^8$ Cm$^{-2}$.

## 3. RESULTS AND DISCUSSION

The a-TiO$_x$ and c-TiO$_2$ films exhibit different irradiation damage behaviour when subjected to a ~2nm diameter electron beam. If the transmitted current is plotted against electron dose, as shown in Figure 1, both show a gradual increase in transmitted current with increasing dose suggesting a continuous loss of material from under the irradiated volume i.e., "Type II" behaviour (Allen et al. 1993). It is seen that c-TiO$_2$ damages much faster than a-TiO$_x$ and the latter does not appear to drill through fully. This is a very curious result because the general observation until now is that amorphous materials are more amenable to irradiation damage than crystalline materials. Also, it is usually observed that amorphous materials show "Type I" behaviour where the material suffers from an initial rapid loss of mass followed by an interval during which little mass loss occurs and finally, an abrupt transition to a complete hole. None of this is observed in the case of a-TiO$_x$. Figure 2(a,b) shows the holes produced in c-TiO$_2$ and a-TiO$_x$. In the case of c-TiO$_2$, it can be seen that surrounding each hole is a halo, dark in the bright field and bright in the annular dark field, suggesting titanium metal has migrated out from the hole during irradiation into the

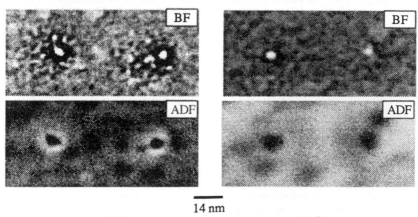

14 nm

(a)                                                    (b)

Figure 2: Holes drilled in (a) c-TiO$_2$ and (b) a-TiO$_x$ films. In the case of c-TiO$_2$ surrounding each hole is a halo, dark in the bright field and bright in the annular dark field. Note that this feature is absent in a-TiO$_x$.

surrounding crystalline matrix. This is absent in the case of a-TiO$_x$. The irradiation damage

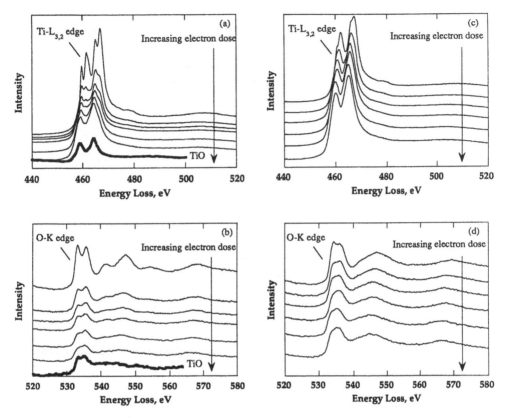

Figure 3: Time resolved electron energy loss spectra of electron beam damage in (a and b) c-TiO$_2$ and (c and d) a-TiO$_x$ showing the changes in Ti-L$_{3,2}$ and O-K edges. The doses in each case are (from top to bottom, $\times 10^{10}$ Cm$^{-2}$) 0, 0.43, 0.86, 1.3, 1.7 and 2.0.

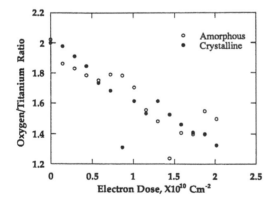

Figure 4: Ratio of oxygen to titanium ratios during irradiation damage of c-TiO$_2$ and a-TiO$_x$ as a function of electron dose. The dose of $1.5 \times 10^{10}$ Cm$^{-2}$ corresponds to the formation of a hole through c-TiO$_2$ film.

of a-TiO$_x$ is similar to that of amorphous SiO$_2$ (Chen, 1994). They both damage showing "Type II" behaviour. The material under the beam is reduced and the metal ions do not migrate out from the hole.

EELS spectra from irradiated material show that both c-TiO$_2$ and a-TiO$_x$ have similar threshold energies for the Ti-L$_{3,2}$ edges, but that these materials have different crystal field splitting. For c-TiO$_2$ four peaks are clearly visible (Figure 3a), whereas for a-TiO$_x$ the crystal field splitting is visible only as a small shoulder on the Ti-L$_{3,2}$ peaks (Figure 3c). As a function of irradiation time these spectra show the disappearance of the crystal field split as well as shift of the Ti-L$_{3,2}$ edges to lower energies, ($\sim$2eV) indicating reduction in the oxidation state (Figure 3a, c). A similar effect is observed in the irradiation damage of amorphous SiO$_2$ where both the Si-L$_{3,2}$ and the Si-L$_1$ edges move to lower values as oxygen is lost from the irradiated area (Saifullah et al. 1996). The oxygen K-edge also shows a gradual disappearance of the crystal field split peak in both cases and reduces to a shoulder for higher doses (Figure 3b, d).

The oxygen to titanium ratios were quantified using c-TiO$_2$ as the standard and plotted as a function of electron dose in figure 4. With increasing electron dose it is observed that the O/Ti ratio for both c-TiO$_2$ and a-TiO$_x$ drops steadily to ~1.3 from a similar initial composition. Although the rate of drilling is not same for c-TiO$_2$ and a-TiO$_x$ they show similar rate of oxygen loss. It appears that even after a hole has been drilled through c-TiO$_2$, it continues to lose oxygen (Figure 4). The similarity of the final spectra in both cases suggest that the final product is the same. It is interesting to see that the irradiation damage mechanisms in c-TiO$_2$, Ti$_2$O$_3$, TiO (Berger et al. 1987) and a-TiO$_x$ are different. It appears that the crystal structure of the material plays a dominant role in enhancing the irradiation damage in crystalline titanium oxides and suppressing it in a-TiO$_x$. The Ti-L$_{3,2}$ and oxygen K-edge of crystalline TiO and damaged c-TiO$_2$ and a-TiO$_x$ shows good resemblance (Figure 3 a,b).

## 4. CONCLUSIONS

Electron beam irradiation damage experiments conducted in a STEM showed that a-TiO$_x$ films were more difficult to damage than c-TiO$_2$ films. This observation is opposite to what has been found for other amorphous materials where the damage is fastest. The irradiation damage of a-TiO$_x$ resemble a-SiO$_2$. The time resolved electron energy loss spectroscopy show that both a-TiO$_x$ and c-TiO$_2$ films loose oxygen continuously during irradiation and that the Ti-L$_{3,2}$ edges shift to a lower value indicating an insulator to metal transition. The damage mechanisms in c-TiO$_2$ and a-TiO$_x$ is strongly dependent on the crystal structure. The EELS spectra suggest that the final product appears to be the same after irradiation.

## 5. REFERENCES

Allen R M, Lloyd S J and Humphreys C J 1993 Inst. Phys. Conf. Ser. No. **138**, 87.
Berger S D, Macaulay J M and Brown L M 1987 Phil. Mag. Letts., **56**, 179.
Chen G S, Boothroyd C B, Humphreys C J 1993 Inst. Phys. Conf. Ser. No **134**, 503.
Chen G S 1994 Ph.D. Thesis; University of Cambridge.
Kratschmer E and Isaacson M 1986 J. Vac. Sci. & Tech. B, **5**, 361.
Matsumoto K, Ishii M, Segawa K, Oka Y, Vartanian B J and Harris J S 1996 Appl. Phys. Lett. **68**(1), 34.
Rez P, Weiss J K, Medlin D L and Howitt D G 1995 Microsc. Microanal. Microstruct. **6**, 433.
Saifullah M S M, Boothroyd C B, Morgan C J and Humphreys C J 1995 Inst. Phys. Conf. Ser. No **147**, 325.
Saifullah M S M, Boothroyd C B, Botton G A and Humphreys C J 1996 Paper presented in European Electron Microscopy Conference - 96 (EUREM-96), Dublin.
Saifullah M S M, Boothroyd C B, Botton G A and Humphreys C J 1997 To be presented in International Centennial Symposium on the Electron, Cambridge, UK.

*Inst. Phys. Conf. Ser. No 153: Section 5*
*Paper presented at Electron Microscopy and Analysis Group Conf. EMAG97, Cambridge, 1997*
© *1997 IOP Publishing Ltd*

# Observation of the mixed dynamic form factor in the Ag $M_{4,5}$-edge

M Hölzl[a], G A Botton[d], M Nelhiebel[a,b], C J Humphreys[d], B Jouffrey[b], W Grogger[c], F Hofer[c], P Schattschneider[a,b]

[a]Institute für Angewandte und Technische Physik, TU Wien, Wiedner Hauptstr.8-10/137, A-1040 Wien
[b]Ecole Centrale Paris, F-92295 Châtenay-Malabry, France
[c]Zentrum f. Elektronenmikroskopie, A-8010 Graz
[d]Department of Material Science and Metallurgy, Cambridge University, Pembroke Street. Cambridge CB2 3QZ

**ABSTRACT:** The mixed dynamic form factor MDFF $S(\mathbf{q}, \mathbf{q}', \omega)$ is a generalization of the well known dynamic form factor (DFF) $S(\mathbf{q}, \omega)$ (e.g. Schattschneider et al). It gives rise to change in the ionization cross sections in crystals, as compared to the atomic model. In order to see whether the interference term is experimentally accessible in the intermediate energy range, we investigate the $M_{4,5}$ edge (energy loss 367 eV) in epitaxially grown Ag.

## 1. THEORY

When the incident electron is modelled in the two-beam case, and the outgoing electron is a plane wave, there are two different inelastic scattering amplitudes $f_E(\mathbf{q})$ $\mathbf{f_E}(\mathbf{q}')$ for ionization with energy loss $E$, where $\mathbf{q} = \mathbf{k}_f - \mathbf{k}_0$ is the momentum transfer between final $\mathbf{k}_f$ and initial $\mathbf{k}_i$ vacuum states of the fast electron, $\mathbf{q}$ and $\mathbf{q}'$ are related through $\mathbf{q}' = \mathbf{q} - \mathbf{g}$ with $\mathbf{g}$ a reciprocal lattice vector.

In order to calculate the inelastic scattering cross section one has to sum over the two scattering amplitudes $f_E(\mathbf{q})$ $\mathbf{f_E}(\mathbf{q}')$ weighted with the corresponding Fourier coefficients $C_{0,g}$. which are assumed to be real here, as usual, in order to simplify the derivation:

$$\frac{\partial \sigma}{\partial E \partial \Omega} = \mid C_0 f_E(\mathbf{q}) + C_g f_E(\mathbf{q}') \mid^2 = \qquad (1)$$
$$= \mid C_0 \mid^2 \mid f_E(\mathbf{q}) \mid^2 + \mid C_g \mid^2 \mid f_E(\mathbf{q}') \mid^2 + 2 C_0 C_g f_E(\mathbf{q}) f_E(\mathbf{q}')$$

In addition to the direct terms, interference produces a mixed term in the cross section.

## 2. THE MIXED DYNAMIC FORM FACTOR (MDFF)

The direct terms are given by a summation over transition matrix elements between initial and final states $|i>, |f>$ of the crystal. The perturbing potential is the Coulomb field of the passing fast electron. The direct terms can be written as

$$| f_E(\mathbf{q}) |^2 = \frac{(2\pi)^4 4\pi e^2}{a_0^2 q^4} S(\mathbf{q}, E) \tag{2}$$

with the dynamic form factor (DFF)

$$S(\mathbf{q}, E) = \sum_{i,f} |<i| e^{i\mathbf{qR}} |f>|^2 \delta(E - E_f + E_i). \tag{3}$$

In analogy

$$f_E(\mathbf{q}) f_E(\mathbf{q}') = \frac{(2\pi)^4 4\pi e^2}{a_0^2 q^2 q'^2} S(\mathbf{q}, \mathbf{q}', E), \tag{4}$$

and the MDFF within dipole approximation is

$$S(\mathbf{q}, \mathbf{q}', E) = \sum_{i,f} <i| e^{i\mathbf{qR}} |f><f| e^{i\mathbf{q'R}} |i> \delta(E - E_f + E_i) \approx \mathbf{qq}' h(E) \tag{5}$$

where $h(E)$ describes a separable energy function with the shape of an energy loss spectrum.

Under these conditions, the two cross sections for positive and negative excitation errors $w$ can be written as:

$$\sigma^- := \frac{\partial \sigma_-}{\partial E \partial \Omega} = \frac{(2\pi)^4 4\pi e^2}{a_0^2} \left[ \frac{C_0^2}{q^4} S(\mathbf{q}, E) + \frac{C_g^2}{q'^4} S(\mathbf{q}', E) + \frac{2C_0 C_g}{q^2 q'^2} S(\mathbf{q}, \mathbf{q}', E) \right] \tag{6}$$

$$\sigma^+ := \frac{\partial \sigma_+}{\partial E \partial \Omega} = \frac{(2\pi)^4 4\pi e^2}{a_0^2} \left[ \frac{C_0^2}{q^4} S(\mathbf{q}, E) + \frac{C_g^2}{q'^4} S(\mathbf{q}', E) - \frac{2C_0 C_g}{q^2 q'^2} S(\mathbf{q}, \mathbf{q}', E) \right]. \tag{7}$$

The $C_g$ depend on $w$. The terms $S(\mathbf{q}, E)$ and $S(\mathbf{q}', E)$ describe Lorentzian angular profiles centered at the zero and $\mathbf{g}$ Bragg reflection in the diffraction pattern. Eqs. 6 and 7 show how the MDFF comes into action, and gives a clue of how to e xtract it from a measurement at different $w$.

## 3. ENERGY SPECTROSCOPY DIFFRACTION PATTERNS (ESD)

In the following 5-beam simulations we assumed the outgoing electron to be a Bloch wave, a fact that complicates eqs. 6 and 7 slightly. Absorption was also included. based on potentials given by the EMS program. Figs. 1,2) show traces along the $\mathbf{g}_{220}$ direction with excitation errors of $w = +1$ and $w = -1$. The direct terms (dotted) have the form of Lorentzians centered at the $\bar{g}_{220}, 0$ and $g_{220}$ position, the mixed term resembles the Kikuchi band. Both terms are added to give the total distribution (solid line).

Subtracting ESD images Eqs. 6 and 7 will enhance the MDFF, according to the different signs. To compare experiment and theory we build the relative difference $(\sigma_- - \sigma_+)/(\sigma_- + \sigma_+) *$ 200 along the symmetry line bisecting the vector $\mathbf{g}_{220}$, perpendicular to the reciprocal lattice vector. Here, the direct terms (DFFs) cancel. The result (fig. 3) shows a central minimum as predicted by the scalar product in eq. 5, but the predicted amplitude is much too high. This could be caused by the dipole approximation, or by the use of a large selected area aperture. hence integrating over many excitation errors, thus weakening the effect.

The ESDs were taken with a Philips CM20, equipped with a Gatan imaging filter at the Zentrum für Elektronenmikroskopie in Graz.

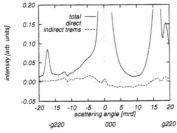

Figure 1: simulation of (220) ESD for the Ag $M_{4,5}$ edge with an excitation error of $w = -1$

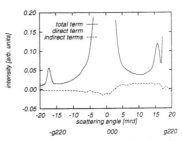

Figure 2: simulation of (220) ESD for the Ag $M_{4,5}$ edge with an excitation error of $w = +1$

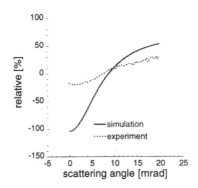

Figure 3: comparison of experimental and simulated relative difference along the bisector of $g_{220}$.

## 4. MOMENTUM DEPENDENCE OF SPECTRA

Within the dipole approximation, both the DFF and the MDFF can be factorized into an $E$- and a $\mathbf{q}$ dependent part. Intensities can then be written as $I = l(\mathbf{q})h(E)$. Energy loss spectra taken in the forward direction (0 spot position) and in the middle between the 0 and $g$ spot (referred to as $g/2$) are as follows

$$I_0 = l(0)h(E), \tag{8}$$
$$I_{g/2} = l(g/2)h(E). \tag{9}$$

Hence,

$$I_{g/2} = I_0 \frac{l(g/2)}{l(0)}. \tag{10}$$

40 energy filtered ESDs were recorded in steps of 5 eV, starting at an energy loss of 300 eV and a window width of 10 eV. Traces along the 0 spot position and position $g/2$ were then compiled.

Figure 4 shows the measured intensities $I_0$ and $I_{g/2}$ as a function of energy loss. The solid line is the intensity $I_{calculated}$ calculated from eq. 10. Up to 450 eV, it coincides with the experiment. We conclude that the dipole approximation is valid for a wide range of energy losses but seems to fail for higher energies beyond the edge. We tentatively assume that this is caused by the absence of monopole and quadrupole terms in the dipole approximation.

174

Including the monopole and the quadrupole terms into calculations will lead to better results.

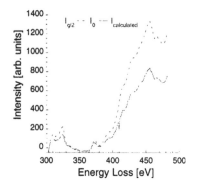

Figure 4: energy loss spectrums shows the validity of the dipole approximation

## 5.  CONCLUSION

The central minimum already seen in the MDFF of the Si K-edge (Schattschneider et al, unpublished), obtained from ESDs, is also found for Ag $M_{4,5}$. The dipole approximation predicts a much too high amplitude for this minimum. Momentum-dependent energy loss spectra also indicate that the dipole approximation fails for energy losses $\sim$ 100 eV above the Ag M-edge.

## REFERENCES

Schattschneider P, Jouffrey B, Nelhiebel M 1996 Phys. Rev. B **54**, 3861

*Inst. Phys. Conf. Ser. No 153: Section 5*
*Paper presented at Electron Microscopy and Analysis Group Conf. EMAG97, Cambridge, 1997*
© *1997 IOP Publishing Ltd*

# A New Method to Obtain Long Camera Lengths and High Momentum Resolution in ω-q Patterns

P A Midgley

Department of Materials Science and Metallurgy, University of Cambridge, Pembroke Street, Cambridge, CB2 3QZ.

**ABSTRACT:** A simple new method is described to obtain long camera lengths and high angular resolution. The specimen is raised above the image plane so that diffracted images of the convergent probe (a 'diffraction pattern') are seen in image plane. The angular resolution in this pattern depends on the probe size and specimen defocus but can be as high as 2 μrad making the new method extremely useful for studying small q features in angle-resolved energy loss spectra. Using a narrow slit at the entrance aperture to a Gatan Imaging Filter (GIF) it was possible to form ω-q patterns with very high angular resolution. Spectra from diamond are used as examples to show the potential worth of this new method.

## 1. INTRODUCTION

In the small q (large λ) limit, a beam of fast electrons passing through a thin film loses energy due to various solid state excitations. These can include surface and volume plasmons, standing waves and Cerenkov radiation. Such excitations often have a momentum dependence and in order to see their character a momentum-resolved spectrum, or ω-q pattern, must be formed. In a TEM this requires long camera lengths (~100 m) and high angular resolutions of $~10^{-5}$ - $10^{-6}$ radians (Pettit, Silcox and Vincent, 1975). In commercial machines in selected area diffraction mode camera lengths rarely exceed 5m. To achieve longer camera lengths a variety of methods have been proposed (Ferrier and Murray 1966, Yeh and Geil, 1967) but require either a modification of the specimen holder (to raise the specimen height a few cm from its original position) or a large reduction in the excitation of the objective lens (and thus a loss of spatial resolution). A simple new method is described here for achieving very long camera lengths and thus high momentum resolution ω-q patterns but still retaining a moderate spatial resolution. It is based on a field emission TEM to form a small probe (in this case an Hitachi HF-2000 FEGTEM) and a GIF.

## 2. THE METHOD

Fig. 1 shows a schematic of the new method. A convergent beam with a small convergence angle is focused on the specimen conjugate initially to the image plane. The specimen is then raised above (or below) the plane giving an array of diffracted beams. As the specimen height (defocus) is increased the distance between the diffracted spots increases, thus increasing the 'effective camera length' and angular resolution, as seen in the image plane. The spatial resolution is determined by the extent of the region illuminated by the defocused beam. Thus, the two are coupled: as the defocus increases to improve the angular resolution, the spatial resolution worsens. In addition, choice of convergence angle is important. Generally, the convergence angle should be as small as possible to reduce the effects of spherical aberration on the probe size (to maintain good angular resolution) and to keep the area illuminated to a minimum (for the best spatial resolution). For very small convergence angles the diffraction limit will dictate the ultimate angular resolution.

The method can be calibrated simply by using a sample of known planar spacing (and thus g-vector). By geometry, from the sketch in Fig. 1,

$$g = 2z\theta_B$$

where g is the distance between diffracted spots in the image plane, z is the defocus and $\theta_B$ is the Bragg angle for the known g-vector. g is calculated by measuring the distance on the viewing screen (or negative) and dividing by the microscope magnification. Once z has been determined, the region illuminated and thus the (linear) spatial resolution, $\Delta r$, can be found by

$$\Delta r = 2z\alpha$$

where $\alpha$ is the convergence semi-angle.

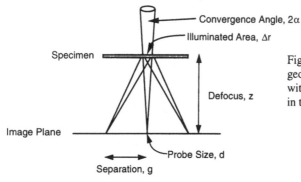

Fig. 1 Schematic showing the geometry of the new method with dimensions as discussed in the text.

The angular resolution, $\Delta\theta$, is determined by the ratio of the probe size, d, to the separation, g, as measured in the image plane. Thus

$$\Delta\theta = (d/g)\, 2\theta_B$$

As a function of defocus,

$$\Delta\theta = (d\, 2\theta_B) / 2z\theta_B = d/z$$

Momentum resolution, $\Delta q$, is given by

$$\Delta q = \Delta\theta / \lambda = d / z\lambda$$

The maximum achievable defocus on the Hitachi HF-2000 (with the specimen in a standard position) is $z \approx 1$ mm and with a 10μm condenser aperture ($2\alpha = 1.1$ mrad, probe size $\approx$ 2nm) this gives $\Delta\theta \approx 2$μrad or $\Delta q \approx 1\times10^{-4}$Å$^{-1}$ and $\Delta r \approx 1$μm. With a 'diffraction pattern' appearing in the image plane using this technique, the 'effective camera length' is given by a combination of the defocus, which separates the diffraction spots and determines the angular resolution, and the overall microscope magnification (including that of the GIF) which acts to magnify the whole pattern. Typically, a microscope magnification of 30,000x (together with a GIF magnification of 18x) will produce a pattern whose details are easily resolved on the CCD camera of the GIF. With a defocus of 1mm, this corresponds to an effective camera length of 360m (onto the CCD). By increasing the magnification to its maximum, 1,500,000 x , this leads to a maximum camera length of 18km!

## 3.    EXAMPLES AND APPLICATIONS

Although long camera lengths are useful for high resolution diffraction such as the measurement of shape transforms and dynamical fine structure, in the author's opinion the main benefit of this new technique is the ability to switch quickly from standard microanalysis mode to one with high angular resolution simply by raising the specimen height above the image plane. This can be of great importance if the 'character' of the energy loss feature is to be understood (e.g. whether there is a dispersion of $\Delta E$ with q). To obtain the clearest results, it is necessary to use a narrow angle-selecting slit, in this case composed of two sharp metal blades inserted just above the entrance aperture of the spectrometer, which allows a particular direction of scattering to be selected. The width of the slit defines the area of reciprocal space which will be dispersed to form the spectrum. The slit width can be adjusted simply by rotating the 2-blade assembly and thus changing the projected gap between the blades. To select the scattering vector relative to the sample then either the sample must be rotated or the 'diffraction pattern' rotated relative to the slit by carefully selected combinations of projector lens settings. Figure 2 illustrates how the slit can be used (a) with the 000 beam incorporated within the slit or (b) displaced so as to include components of q perpendicular to the slit.

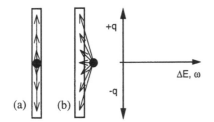

Fig. 2 Illustration of how the scattering vector is chosen by a narrow slit. In (b) the slit is displaced slightly so that the scattering vector chosen has a component of q perpendicular to the slit. This range of scattering is then integrated across the slit width and dispersed in energy to form the ω-q pattern.

To illustrate the effectiveness of this new technique, spectra from diamond will be used as examples. Fig. 3 shows a typical low loss spectrum from diamond, oriented to be near the <100> zone axis, taken in image coupled mode with a collection angle of about 3mrad. Arrows mark various features in the spectrum. Perhaps the most interesting features are the peak in the 5-12eV range, shown as B, and the shoulder at 23eV, shown as C. The former could be related to the density of states rising after the band gap (5eV) and the latter to either an interband transition or a surface plasmon. A is the zero loss peak and D is the volume plasmon. The new technique offers a way of revealing the character of such features and unambiguously identifying their origin.

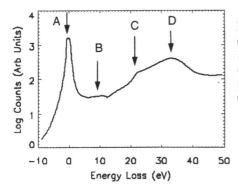

Fig. 3 The low loss EELS spectrum of diamond (shown with an intensity log scale) taken in image coupled mode with an acceptance angle of ~3mrad. Four features in the spectrum are highlighted and are discussed in the text.

Fig. 4 shows a ω-q pattern taken at the same orientation as the spectrum in Fig. 3 using the new method. There are two strongly dispersing features in the 0 to 20 eV range, a sharp non-dispersing feature at ~23 eV and a broad feature at ~34 eV. An explanation for the features in Fig. 4 are shown in Fig. 5. The lowest lying feature is that due to Cerenkov radiation seen in the region of the spectrum where $\varepsilon_1\beta^2>1$ ($\varepsilon_1$ is the real part of the dielectric function and $\beta = v/c$). For 200kV electrons, $\beta=0.695$ so $\varepsilon_1>2.07$ which, from the published dielectric constants (Phillip and Taft, 1964), gives a cut-off of about 12eV and corresponds to the sudden drop in intensity seen in the figure. The other dispersive feature is a hybrid state. Above 12eV the excitation is surface plasmon-like in character, where $\varepsilon_1<0$, and below 12eV the excitation changes to a waveguide mode (Chen, Silcox and Vincent, 1975), where $\varepsilon_1>0$. These features, when integrated over the scattering angle, give rise to the peak in the spectrum in the 5-12eV range, and is therefore not related to the density of states as might have first been thought. The lack of dispersion seen with the feature at 23eV indicates that it is not surface related but is more likely an interband transition.

An energy-selecting slit of 2eV was used to image the features in the ω-q pattern of Fig. 4 and these are shown in the montage of Fig. 6. The images are dominated by a 'double annulus' feature which corresponds to the scattering of the Cerenkov radiation (outer) and the waveguide mode (inner). The broad feature at 33eV is the volume plasmon.

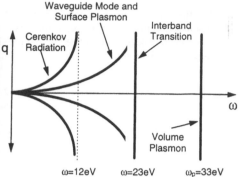

Fig. 4 ω-q pattern taken near the <100> axis of diamond. The angular range seen in the pattern is ~100 μrad.

ω=12eV    ω=23eV    $\omega_p$=33eV

Fig. 5 A schematic explanation of the features seen in Fig. 4.

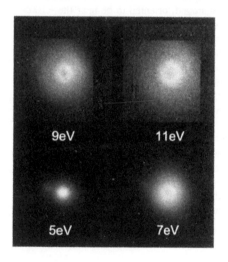

Fig. 6 A montage of images taken near the <100> zone axis of diamond with a 2eV energy slit width centred at energy loss values of 5, 7, 9 and 11 eV. The figure shows the scattering around the direct beam that corresponds to Cerenkov radiation (outer annulus) and a waveguide mode (inner annulus). The increase in annulus radius corresponds to the dispersion seen in the ω-q pattern of Fig. 4.

## 4.    CONCLUSIONS

A simple method of achieving very long camera lengths and high angular resolution has been demonstrated. By using a narrow angle-selecting slit at the entrance to a GIF it is possible to form ω-q patterns with high resolution. This enables the dispersive character of low energy loss features in the EELS spectrum to be investigated easily and thus to confirm their origin.

### REFERENCES

Chen C H, Silcox J and Vincent R 1975 Phys. Rev. B **12**, 64
Ferrier R P and Murray R T 1966 J. Royal Microsc. Soc. **85**, 323
Pettit R B, Silcox J and Vincent R 1975 Phys. Rev. B **11**, 3116
Phillip H R and Taft E A 1964 Phys. Rev. **136**, A1445
Yeh G S Y and Geil P H 1967 J. Mat. Sci. **2**, 457

### ACKNOWLEDGEMENTS

This work was performed whilst the author was at the H.H. Wills Physics Laboratory at the University of Bristol and Drs R. Vincent and S. Madras are thanked for their help and support.

*Inst. Phys. Conf. Ser. No 153: Section 5*
*Paper presented at Electron Microscopy and Analysis Group Conf. EMAG97, Cambridge, 1997*
© 1997 IOP Publishing Ltd

# *Ab initio* EELS : beyond the fingerprint

C.J.Pickard and M.C.Payne

Cavendish Laboratory, Madingley Road, Cambridge CB3 0HE

**ABSTRACT:** A new scheme is presented for the calculation of Energy Loss Near Edge Structure (ELNES) within the framework of a Density Functional Theory (DFT) total energy code. It includes a highly efficient approach to the Brillouin zone integrations, and deals with the single particle core hole effects within the supercell approximation with significant improvements on comparison to experiment for bulk Diamond.

## 1   INTRODUCTION

Electron Energy Loss Spectroscopy (EELS) probes the unoccupied electronic states of materials, transitions taking place from core or valence band states — giving rise to ELNES (and at higher energy, extended structure) or Low Loss spectra respectively (see the left hand panel in Figure 1). Knowledge of the spectral properties (i.e. variation with energy and symmetry) of these states is important in several ways. Since the local potential directly effects the energy distribution of the electronic states they are a *fingerprint* of local structure (the focussed beam of electrons in a STEM forms a local probe). The unoccupied states also give indirect access to occupied states, and hence to information as to the bonding energetics (Muller 1996). Finally, since they dictate the electronic — e.g. optical and transport — properties of the system, direct observation of the unoccupied states themselves is of interest.

With the development of computer power and theoretical tools the interpretation of ELNES can be taken beyond the fingerprint level through calculation. If these calculations are well controlled they can demonstrate clearly the degree to which the energy loss spectra is understood. If theory and experiment do not agree the understanding is then proved to be incomplete and theory must be further developed. Once a model of the spectra has been developed it may be used as an aid to intuition (suggesting further experimental possibilities) or as a quantitative tool in the interpretation of experimental results. In this paper a scheme for the *ab initio* calculation of ELNES is presented, building on earlier work (Pickard 1995), based on a DFT planewave pseudopotential total energy code — CASTEP (Payne 1992).

## 2   ENERGY LOSS NEAR EDGE STRUCTURE

In the subsequent discussion the focus is on the ELNES region of the energy loss spectrum. For a fast incident electron (treated as a planewave) which looses an energy $E$ and momentum $\mathbf{q}$ the differential cross section can be related to the imaginary part of the dielectric constant, given within the single particle approximation for a periodic system by:

$$\epsilon_2(\mathbf{q}, E) = \frac{2e^2\pi}{\Omega\epsilon_0 q^2} \sum_{\mathbf{k},c} |\langle \psi_{\mathbf{k}}^c | e^{i\mathbf{q}\cdot\mathbf{r}} | 1s \rangle|^2 \delta(E_{\mathbf{k}}^c - E_{1s} - E),$$

in the specific case of an excitation from a 1s core state, where $\Omega$ is the unit cell volume.

In this scheme the final states and energies — $|\Psi_{\mathbf{k}}^{c}\rangle$ and $E_{\mathbf{k}}^{c}$ — derive from an *ab initio* electronic structure calculation described in Section 3. The sum is over the first Brillouin zone $\mathbf{k}$ and unoccupied bands $c$, i.e. all final states. The two main elements to the calculation of ELNES within this scheme is the evaluation of the density of states (DOS) term, which is described further in Section 4, and the transition matrix elements, which are dealt with in Section 5.

## 3 THE PLANEWAVE PSEUDOPOTENTIAL APPROACH

A DFT planewave total energy code (CASTEP) is used to obtain the final electronic states and energies. It is based on DFT within the Local Density Approximation. A planewave basis set and periodic boundary conditions are used (and hence summations over all states become sums over bands and an integration over the first Brillouin zone). Non-local pseudopotentials are used to deal with the rapidly varying core regions of the wavefunctions. The technique is reviewed by Payne *et al* (1992).

## 4 BRILLOUIN ZONE INTEGRATION

A highly efficient Brillouin Zone integration scheme has been developed which uses a very low k-point sampling. It is important to use an extrapolative approach, since interpolative ones are hindered by band crossing at low sampling densities, introducing spurious singularities in the DOS. The information for the extrapolation is obtained using **k.p** perturbation theory to second order (to obtain the correct analytic behaviour at van Hove singularities) within a set of sub-volumes into which the Brillouin zone is divided (efficiently chosen to make full use of symmetry at low sampling densities) — some corrections are required for the use of non-local pseudopotentials. The resulting piecewise quadratic representation of the bandstructure is directly converted do a DOS using the analytic quadratic approach of Methfessel (1983). The right panel of Fig. 1 show the convergence properties of this scheme (which is described in greater detail in Pickard 1997).

## 5 ELNES MATRIX ELEMENTS

In the pseudopotential approximation the core states are not explicitly calculated, and the pseudowavefunctions differ from the true ones within a radius $r_c$ from the nucleus ($r_c$ is typically of the order of 0.5Å). In the evaluation of the matrix elements the initial core states are taken from all electron calculations for isolated atoms — consistent with the frozen core approximation on which the pseudopotential approximation is based. In practice the dipole approximation is taken, but it has been shown that it is straightforward to go further in this formalism (Pickard 1997). The matrix elements are then directly evaluated and corrected for the pseudopotential error using the Projector Augmented Wave approach of van de Walle and Blöchl (1993) — see the left panel of Fig. 2 (further details are given in Pickard 1997). In this way the matrix elements are evaluated quantitatively, resulting in an absolute prediction for the cross-section. This is in contrast to many previous approaches — which frequently are simply symmetry projected local DOS.

## 6 SINGLE PARTICLE CORE HOLE EFFECTS

The "single particle" core hole effects derive from the un-screening of the nuclear charge as an electron is excited from a localised core state. In a metal this potential is almost entirely re-screened and there is very little effect due to the core hole. In an insulator the screening

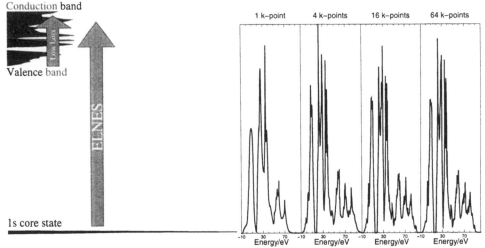

Figure 1: *Left:* Pictorial representation of the energy loss process for Diamond. *Right:* Convergence of the K-edge of Diamond (including occupied states) with number of k-points for the piecewise quadratic extrapolative method using the primitive 2 atom unit cell.

of the hole by the valence electrons is only partially achieved. Physically, the excited atom is like an impurity in the material with an associated distortion of the electronic structure (and no atomic relaxation on the time-scale of the excitation). In fact, impurity models — such as the Clogston-Wolff model — have been used to model core hole effects in systems where the screening is almost complete and hence the impurity potential differs only slightly from the unperturbed one. Weijs (1990) studied the core hole effects in the X-ray absorption spectra of transition metal silicides in this way.

For many systems the hybridisation due to core hole potential cannot be treated perturbatively, and the effects must be calculated explicitly. Within a bandstructure approach this is achieved by performing a supercell calculation in which there is a single excited potential in each cell. The spectra is then evaluated for that excited atom. The supercell must be large enough that the neighbouring excited potentials do not interact with each other. Such calculations have previously been performed for Diamond (Mauri 1995) and Graphite (Ahuja 1996). Within the pseudopotential approximation, the more common Z+1 approximation is improved upon by generating a special pseudopotential with partially occupied core states. Comparisons for Diamond show the Z+1 approximation to be generally good (Pickard 1997), but there are some small differences at the threshold.

The right-hand panel in Fig. 2 show the effects of the core hole on the Diamond K-edge. A significant improvement in the calculated spectra is observed, but the total intensity at the threshold is overestimated. According to Batson (1993) the passage of a swift electron in an EELS measurement has an effect on the excitation process, reducing the expected excitonic enhancement. It should be noted that, in contrast with the usual picture, the core hole has an effect well above the threshold. In Diamond, there is effectively a second threshold at 17eV above the first and so it is not surprising that there is also distortion there.

## 7   CONCLUSION

ELNES can be calculated within the framework of a total energy pseudopotential code. The description of the wavefunctions in terms of planewaves allows the straightforward direct evaluation of the transition matrix elements, and corrections for the pseudopotentials can be

182

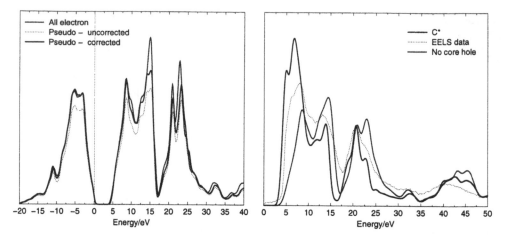

Figure 2: *Left:* The effect of correcting the ELNES matrix elements for a Diamond K-edge. Comparison is made with an all electron planewave calculation. *Right:* The inclusion of core hole effects in Diamond using a 16 atom supercell and a single excited atom. Improved agreement with experiment is observed at the threshold. Experimental data by Brian Rafferty.

included. Some of the computational expense of the technique is offset by efficient Brillouin zone integrations, and the importance of including the core hole potential is demonstrated. It is envisioned that such an scheme, in concert with the structure prediction of these *ab initio* techniques will prove to be a powerful tool in the investigation of microstructure.

## ACKNOWLEDGEMENTS

C J Pickard would like to thank the EPSRC for funding, and L M Brown, E Sandre, J Yuan and P Rez for many helpful comments in the course of this work.

## REFERENCES

Ahuja R *et al* 1996 Phys. Rev. B **54** 14396
Batson P E 1993 Phys. Rev. B **47** 6898
Mauri F and Car R 1995 Phys. Rev. Letters **75** 3166
Methfessel M S, Boon M H and Mueller F M 1983 J. Phys. C **16** 1949
Muller D A *et al* 1996 Acta Materialia **44** 1637
Payne M C *et al* 1992 Rev. Mod. Phys. **64** 1045
Pickard C J *et al* 1995 Inst. Phys. Conf. Ser. **147** 211
Pickard C J 1997 PhD Thesis
van de Walle C G and Blöchl P E 1993 Phys. Rev. B **47** 4244
Weijs P J W *et al* 1990 Phys. Rev. B **41** 11899

*Inst. Phys. Conf. Ser. No 153: Section 5*
*Paper presented at Electron Microscopy and Analysis Group Conf. EMAG97, Cambridge, 1997*
© *1997 IOP Publishing Ltd*

# Passivation of ultrathin metallic films for EELS studies

**J. Yuan, V. Stolojan, C. A. Walsh and G. D. Lian**

Cavendish Laboratory, Madingley Road, Cambridge CB3 0HE, U.K.

**ABSTRACT**: Using EELS we show that the study of the electronic structure of metals can be seriously impaired by the presence of a surface oxide layer. In this paper, we report our effort at suppressing and/or replacing the 'native' surface layer by the deposition of a thin passivation layer. This is carried out either as an integral part of thin film growth or after the electron beam transparent sample has been prepared. The latter process is carried out inside a UHV sputtering and coating system and is particularly suitable for cross-sectional samples. The effectiveness of the passivation layer approach is verified using EELS studies of ultrathin Fe films.

## 1. Introduction

Electron energy loss spectroscopy (EELS) has been used extensively for many years to characterise the chemical composition of materials at the high spatial resolutions achievable in transmission electron microscopes (TEM), because it is less affected by the beam broadening effect. Increasingly people are starting to extract, from the fine structure observed in EELS, information about the electronic structure (Pearson, Ahn and Fultz, 1993) and, in favourable cases, the magnetic structure (Yuan and Menon, 1997) of materials. Because of their sensitivity to a change in the chemical state of the atoms probed, it is vitally important that no unwanted modification of the chemical state of the solids is introduced by the experimental process. In many materials, such as polymers and oxides, the threat comes from electron beam damage, which can be very serious. In studying metallic thin film samples, the problem of oxide formation poses an even bigger challenge, as it inevitably leads to the formation of a surface layer with atoms often in a different chemical state to those in the interior (i.e. the bulk) of the sample. The formation of native oxides is often not controllable and can also lead to complex mass transport between the surface and the interior of the film. In the past, such a problem has been circumvented in the EELS studies of the metals by concentrating on thick samples, reducing the importance of the surface oxide layers. However, for EELS fine structure studies, a consideration of the signal-to-noise ratio requires the sample thickness not to be excessive. In grain boundary studies the curvature of the high angle grain boundary can also impose a more stringent limit on the thickness of the films which can be studied (Yuan *et al.*, 1995). In this paper, we first present an EELS quantification of the surface oxide layer in stainless steel. Then we describe two approaches to overcoming such problems with overlayer coating. Finally, we discuss the choice of coating materials for EELS studies and an alternative approach.

## 2. EELS quantification of the surface oxide layer

To study the effect of the surface oxide on EELS quantification, we have prepared a 316 stainless steel sample (annealed for 2 hours at 700°C) using the standard jet polishing technique

184

followed by Ar ion beam thinning. The sample was then quickly (within 10 minutes) transferred into a VG HB501 scanning transmission electron microscope (STEM) operating at 100 keV. The oxide thickness was assessed by monitoring the O 1s, Cr 2p and Fe 2p core level absorption intensities as a function of the local thickness, which were determined from the plasmon excitation probability. The result is plotted in Fig. 1. The formation of the oxide is also clearly shown through a reduction in the average 3d electron occupancy deduced from the Cr 2p absorption spectrum analysis. Assuming a constant oxide thickness at the surface, we can estimate that the oxide surface layer is of the order of 20 Å (on each surface). The optimal film thickness for EELS experiments is less than about 400 Å.

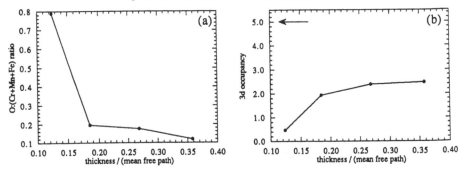

Fig. 1 (a) Oxide content and (b) estimated 3d electron occupancy
(the arrow points to that expected of bulk material).

## 3. Passivation of ultrathin film during growth

The practical solution of ion beam 'cleaning' the surface of the electron-beam transparent sample just before the EM examination clearly does not always work for EELS analyses of thin foils. In addition, the process has to be repeated if the same area has to be examined subsequently. In the rest of this paper, we report our effort at suppressing and/or replacing the 'native' surface layer by the deposition of a thin passivation layer.

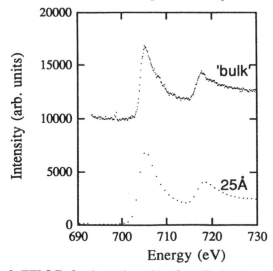

Fig. 2 EELS Fe 2p absorption edges from 'bulk' Fe and from a
thin Fe film with a Cr passivating layer.

First we show that an ultra-thin overlayer is very effective at protecting the bulk material from oxidation. Fig. 2 shows the EELS Fe 2p core edge spectrum from a 25 Å thick Fe film deposited on a GaAs substrate and protected by a 25 Å thick Cr layer. The deposition was carried out in a UHV environment. Also shown is a 'bulk' Fe 2p absorption spectrum acquired from a thick Fe film using EELS. The two spectra are very similar, confirming the effectiveness of the thin passivation layer in this case. We also point out that the substrate of this plane-view sample (fig. 3) has a GaAlAs layer specially grown near the Fe/GaAs interface so that a thin window can be chemically etched out of the back of the substrate. The thickness of the sample within the window was further reduced by ion beam thinning, however no hole was produced. If a hole was achieved in the ion beam process, we found from EELS analysis that some part of the ultrathin Fe film can become oxidised. This illustrates the importance of passivating both sides of thin films.

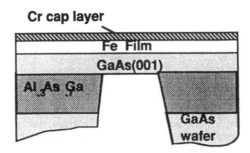

Fig. 3, Electron microscopy sample of Fe grown on GaAs with a Cr passivation layer.

## 4. Ex-situ passivation of a TEM foil

Most electron microscopy samples have to be passivated ex-situ. One of the commonest reasons is the need to produce an electron-beam transparent sample first before considering the passivation issue. One common technique used in the final stage of the production of electron-beam transparent samples is ion beam sputtering. The ideal solution for these samples would be to add a passivation layer to the sample surface as part of the ion beam sputtering process. Although some ion beam thinners have the capability of being used both to sputter and to coat, the processes cannot be completed with the same sample stage and the vacuum has to be broken for the changeover. Even if these processes can be combined together, the vacuum conditions in existing ion beam thinners is still unsatisfactory.

We have constructed a UHV system to perform the required ex-situ passivation of electron microscopy samples of environmentally sensitive materials (Fig. 4). The UHV chamber is modified from a VG electron beam evaporator with standard rotary-diffusion pumping system, supplemented by a titanium sublimation pump (TSP). The pressure of the chamber can reach $10^{-10}$ mbar even without the aid of the TSP, and hence is adequate for the UHV coating purpose. The removal of the existing surface layer of the sample is performed with a *VG AR2* ion sputtering gun operating at up to 10keV. Typical material removing rate is about 2 Å per µAmin. The passivation layer is deposited using a home made electron beam evaporator, consisting of a source material in the form of a central electrode, surrounded by a tungsten filament which is heated to produce electron emission. The source electrode is fed from a linear drive with 25mm travel. This allows the consumption of the source electrode to be compensated for.

To carry out the ex-situ passivation, the electron microscopy sample, of 3 mm diameter, is clamped at the edge in a rotary specimen holder. One of the surfaces is first cleaned by ion beam sputtering. The sample is then rotated 180° for it to be coated by the e-beam evaporator. The same process is repeated for the other surface.

186

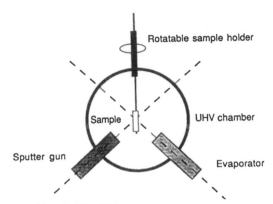

Fig. 4 The UHV sputter-coater system.

## 5. Discussion

Although the passivation of environmentally sensitive materials is routinely carried out in many areas of practical importance, its application in EELS studies requires careful consideration. As EELS is a bulk-sensitive technique, it will sample the electronic excitation of all atoms under the electron beam. Hence, increasing the relative mass-thickness of the overlayer will have a detrimental effect on the signal-to-noise ratio of the signal from the protected materials. In this respect, a coating of Al, Cr or Ni is to be preferred to one of Ag and Au. However, the possibility of a reaction between the overlayer and the protected layer is another important factor for consideration and will obviously depend on the metals involved. In both cases, a thin coating of carbon would have some advantage. With the wide availability of fullerene molecules, deposition of an overlayer of carbon could be achieved quite easily and controllably using the electron beam heating. The effectiveness of a carbon overlayer for passivation purposes, however, needs to be tested out.

Whilst the passivation process may work for EELS quantification, it could adversely affect the imaging of the microstructure of the protected films. An alternative method is to ion beam clean the sample within a microscope with a higher standard of vacuum than is the case for most of today's microscope. Compared with the undertaking to construct a UHV electron microscope, the building of a UHV evaporating system is the easier option, unless the surface layer is the object of investigation.

## 6. Summary

We have demonstrated the need for better surface control of metal foils used in high resolution EELS analyses in electron microscope. We have also shown that a relatively thin overlayer of suitable metals can be very effective in minimising the surface oxide problem. We have built a UHV sputter-coater for the passivation of real electron microscope samples.

## Acknowledgements

Funding from the EPSRC and excellent technical support from the Cavendish Workshop are gratefully acknowledged. C.A. Walsh would like to thank Magnox Electric for support and G.D. Lian is grateful to the Royal Society for a visiting Fellowship. We would also like to thank Dr. J. A. C. Bland for the provision of the Fe/GaAs sample.

## References

Pearson D H, Ahn C C, and Fultz B 1993 Phys. Rev. B, 47, 8471.
Yuan J, and Menon N K 1997 J. Appl. Phys., 81, 5087.
Yuan J, Brown L M, Walmsley J, and Fisher S B 1995, J. Microsc., 180, 313.

*Inst. Phys. Conf. Ser. No 153: Section 5*
*Paper presented at Electron Microscopy and Analysis Group Conf. EMAG97, Cambridge, 1997*
© *1997 IOP Publishing Ltd*

# The correlation between soot structure and adsorbed calcium studied by PEELS

**J T Gauntlett, P J Shuff, R Brydson***

Shell Research & Technology Centre, Thornton, P.O. Box 1, Chester CH1 3SH
* School of Materials, University of Leeds, Leeds, LS2 9JT.

**ABSTRACT:** Spatially resolved PEEL and EDX spectra of the edges of soot spherules and necks between have been taken by exploiting the nanoprobe capabilities of the VG HB601 North West STEM. The soot was extracted from engine lubricants. The sensitivity to calcium (from lubricant additives) in soot is similar for both spectroscopies. The distribution of calcium on the soot varies with samples. In some cases the calcium only adheres to the soot in areas with low graphitic nature, as evidenced by the C K-ELNES.

## 1.     SOOT IN LUBRICANTS

The combustion process in diesel engines produces a substantial amount of soot, most of which is removed by the exhaust. However, some soot is retained in the lubricant and can, in certain situations, enhance wear of sliding surfaces, alter the rheology of the lubricant or provide a reactive surface for adsorption of lubricant components. As pressure grows to increase the time between engine oil changes in all classes of vehicle, it becomes more important to understand the fundamental chemical interactions between the carbon surfaces and the different species found in lubricants. Calcium, which is sometimes added as carbonate to neutralise acids and prevent deposit formation, can adsorb onto the soot under certain conditions, apparently lowering the available concentration for deposit inhibition.

## 2.     THE STRUCTURE OF SOOT AGGREGATES

Soot consists of aggregates of spherules that vary between 30 and 100 nm in diameter. It is, therefore, thin enough to study by transmission electron microscopy. Figure 1 shows a relatively heavily sooted area of a holey carbon grid, where the lubricating oil has been gently removed by capillary de-oiling. Here the substantial structures are aggregates of many smaller structures that have simply fallen out of the liquid together onto the holey carbon. The smallest structures, which are barely visible at this magnification, represent the true, covalently bonded aggregates of soot spherules.

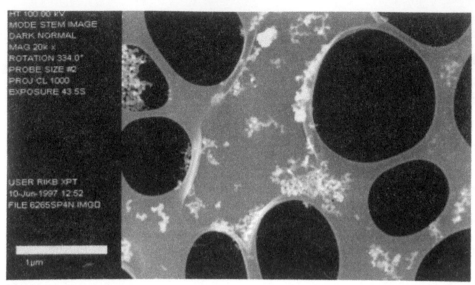

Figure 1

## 3.    ADVANTAGES OF A DEDICATED STEM

The nanoprobe capabilities of the VG HB601 North West STEM allow us to take spatially resolved PEEL and EDX spectra of many different species in this image with a few minutes acquisition time.  At high magnifications it is possible to take spectra of the edges of spherules and necks between them.  Unfortunately, we spent so much time taking spectra, we forgot to collect any high quality images at high magnifications!

The sensitivity to calcium in soot is similar for both spectroscopies (the objective aperture is optimised for EELS at 6 mrad., whereas 11 mrad is optimal for EDX) and both are employed, with data acquired in parallel.

## 4.    DISTRIBUTION OF CALCIUM ON THE SOOT

In certain samples crystals of lubricant additive decomposition products can be found both separately from the soot, such as that in the lower left of Figure 2, or associated with the soot, such as the two bright species highlighted in Figure 1, or that shown in close up in Figure 3.

However, in other cases the calcium is much more finely dispersed - and consequently harder to find.  Only soot overhanging the holes in the carbon film can be analysed, due to the carbon film background.  "Letter box" analysis areas of a few nanometres side are then used to locate locally high concentrations of calcium.  This can be laborious and would benefit immensely from imaging capabilities.

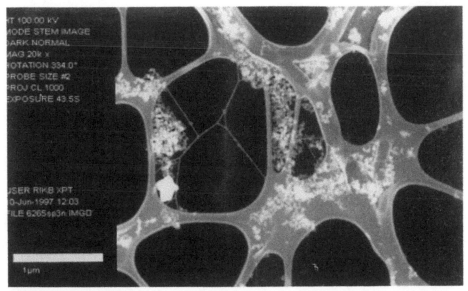

Figure 2

Figure 3

## 5. ELNES

There is an interesting correlation between the presence of calcium and the C K-ELNES. Calcium is only observed on areas of the soot where the graphitic C K-ELNES peak is very low (Figure 4). This implies that the calcium binds to the more "polymeric" carbon in the soot. This is reasonable, as these domains are likely to have more chemical functionality. Figure 5 shows the absence of calcium in an area of soot with more graphitic nature.

The Ca L-ELNES show no strong crystal field splitting of the $L_{II}$ or $L_{III}$ peaks, which implies that the Ca co-ordination is not octahedral. The presence of sulphur and phosphorus

190

in the EDX and EELS spectra implies that the calcium is present as sulphate or phosphate. There is evidence in the sulphur L-ELNES to support this, though it is not conclusive.

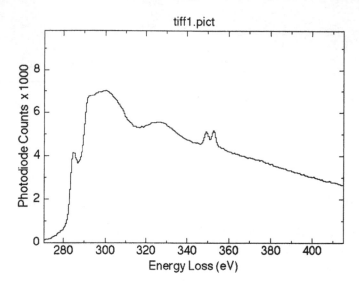

Figure 4´

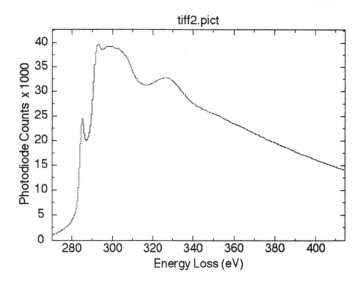

Figure 5

*Inst. Phys. Conf. Ser. No 153: Section 5*
*Paper presented at Electron Microscopy and Analysis Group Conf. EMAG97, Cambridge, 1997*
© *1997 IOP Publishing Ltd*

# Blind deconvolution of blurred images

**T D Walsh and R Vincent**

H H Wills Physics Laboratory, University of Bristol, Tyndall Avenue, Bristol,
BS8 1TL

**ABSTRACT:** We outline a method for deconvolution of an unknown point spread function from a blurred image. The algorithm has been moderately successful for simulated noise-free data but does not always find an exact solution. However, blind deconvolution techniques are developing rapidly and it is anticipated that applications to electron microscopy may become practical when the problems of finding the optimum solution within the limits set by statistical noise are better understood.

## 1.    INTRODUCTION

There are many physical situations in microscopy and astronomy where due to experimental constraints, the recorded image or diffraction pattern is blurred by some unwanted point spread function (PSF). The recorded image $g(x,y)$ is a convolution between an unknown ideal image $f(x,y)$ and the PSF $h(x,y)$, where

$$g(x,y) = f(x,y) \otimes h(x,y). \tag{1}$$

If the PSF is known, then in principle it is possible to deconvolute the PSF from the blurred image to produce the true image by a    straight-forward application of Fourier transforms, where

$$f(x,y) = FT^{-1}\left( \frac{FT(g(x,y))}{FT(h(x,y))} \right). \tag{2}$$

Evidently, eq.(2) diverges if the Fourier transform of h is zero at any point, even if h is known exactly. The addition of noise to both g and h exacerbates the problem, and the standard solution is typically some version of the Wiener filter, where the high frequencies in the transform of f are attenuated. A more general approach is based upon the maximum entropy principle, which recognises that the problem is ill-conditioned, and therefore has many solutions which equally satisfy a goodness-of-fit or $\chi^2$ statistic. By adding the further constraint that the entropy of f should be maximised, we select a solution with no structural detail that is not essential to satisfy the data and is consistent with any prior knowledge.

If the PSF is not well characterised, a related and more difficult problem is the separation of an unknown PSF and an unknown object from an experimental blurred image. This is known as the blind deconvolution problem.

Subject to the application of sufficient constraints and the absence of noise, the solution to the blind deconvolution problem is believed to be unique for functions in two or more dimensions, excluding the trivial solution with f=g and h equal to a $\delta$-function. Iterative methods developed from the Gerchberg-Saxton algorithm have been shown to converge upon estimates for f and h (Lane 1987), but suffer from the problem of trapping in local minima of the error function unless severe ab-initio constraints are repeatedly imposed at every cycle of the program. A general overview of blind deconvolution methods is given by Kundar and Hatzinakos (1996 a&b).

## 2.    METHOD

Lane (1992) suggested a new method for blind deconvolution of a blurred image by minimising an error measure between g(x,y) and the convolution of estimates for f(x,y) and h(x,y). In principle, both f and h may be complex but we consider here only the case where the functions represent intensities and are positive semi-definite. If both f and h are sampled over a field of n×n pixels, then the combined functions may be represented by a vector $\mathbf{z}$ in a space of $2n^2$ dimensions.

The error measure $\varepsilon$ is defined in L as

$$\varepsilon^2(\mathbf{z}) \equiv \varepsilon^2(h,f) = \sum_x \sum_y (g(x,y) - h(x,y) \otimes f(x,y))^2. \tag{3}$$

Lane used a conjugate gradient technique to minimise $\varepsilon(\mathbf{z})$. In order to do this, it is necessary to define gradients for the error measure $\partial\varepsilon/\partial z_i$ where $z_i$ corresponds to an element of $\mathbf{z}$. There are two different forms of the gradient depending upon whether $z_i$ corresponds to a pixel of h(x,y) or of f(x,y). The arrays of gradients $\partial\varepsilon/\partial h$ and $\partial\varepsilon/\partial f$ are calculated using Parseval's theorem and are given by

$$\partial\varepsilon/\partial h = FT^{-1}\left(2H^*(FH - G)\right), \tag{4}$$

and    $$\partial\varepsilon/\partial f = FT^{-1}\left(2F^*(FH - G)\right), \tag{5}$$

where capital and lower case characters represent a Fourier transform pair of functions.

In Lane's algorithm, constraints upon the PSF and object function are applied in a fairly ad hoc manner. An improvement suggested by Thiébaut (1995) enforces the constraints on the PSF in a strict manner. Typical constraints which must be enforced are

(1) Finite support of the object: f(x,y)=0 for a known set of pixels.

(2) Semi-definite positive: all values in h and f are positive definite or zero.

(3) Conservation of flux: $\sum h(x,y) = 1$.

Constraints (2) and (3) are obeyed by making the substitutions $f = \psi_f^2$, $h = \psi_h^2/\sum\psi_h^2$ and then minimising $\varepsilon$ with respect to $\psi_f$ and $\psi_h$. Having made these substitutions, it is necessary to redefine the gradients, where

$$\partial f/\partial\psi_f = 2\psi_f \tag{6}$$

and    $$\partial h/\partial\psi_h = 2\frac{\psi_h(1-h)}{\sum\psi_h^2} \tag{7}$$

A straight forward multiplication of eqs. (6) and (7) with eqs.(2) and (3), respectively, produces gradients of $\varepsilon$ with respect to $\psi_h$ and $\psi_f$. The first constraint is achieved in a fairly elegant manner. If the initial value of any $\psi_h$ or $\psi_f$ is set to zero, then by eqs. (6) and (7) the gradient with respect to that pixel is also zero and the minimisation routine will never move the pixel intensity away from zero.

## 3.     RESULTS AND DISCUSSION

The performance of the algorithm for simulated images is illustrated in Fig. 1, where the original image $f_0(x,y)$ is shown in (a), followed by the PSF $h_0(x,y)$ in (b). Both functions were set to zero beyond a circular cut-off which was used as the finite support for reconstruction (constraint(1)). Within its support, the PSF was chosen to have a Gaussian profile with an overall half width of 20 pixels. The convolution of (a) and (b) generated the blurred image (c) used as input data, $g(x,y)$ for the algorithm, combined with the initial guesses for the image and PSF shown in (d) and (e), respectively. From this starting point, the algorithm reduced the error function $\varepsilon$ (eq.(3)), and produced estimates for the image and PSF illustrated in Figs.1(g) and (h). These reconstructed images are evidently similar to the true image (a) and the PSF (b), but a quantitative measure of the errors showed there were significant differences between the estimated and ideal images. In general, the error, e, was defined as

$$ e = \sqrt{\sum (a_0 - a)^2 / \sum \left( a_0^2 \right)} , \qquad (8) $$

for an ideal image $a_0(x,y)$ and its estimate $a(x,y)$.

On this basis, the error in the image (Fig.1(g)) was 23%, and 2.3% for the PSF (Fig.1(h)). However, the error between the convolution of the reconstructed image and PSF and the original blurred image (Fig.1(c)) was only 0.025%, where the modulus of the error function is shown in Fig.1(f). It seems likely that the algorithm converged to a local minimum of the error $\varepsilon$, characterised by a large error in $f(x,y)$.

If noise is added to the blurred image used as a starting point for reconstruction, the noise completely swamps the small differences in $\varepsilon$ needed to select a reliable solution. Figs. 1(j) and (k) show an estimated object and PSF, respectively, reconstructed from the blurred image in Fig. 1(c) degraded by the addition of Poisson noise scaled to a maximum intensity of 4,000 counts. The reconstructed image, and PSF are very noisy and have respective errors of 53% and 89%. However, the error between the convolution of the estimated image and PSF, and the original noisy image remained small, being only 0.5%. This was lower than the difference between the blurred image (Fig. 1(c)) and the noisy blurred image (1.9%), implying that the algorithm was fitting to the noise. This is confirmed by Fig. 1($l$) which shows the modulus of the differences between the convolution of Figs. 1(j) and (k) and the noisy blurred image. It is apparent that the largest errors do not correspond to the most intense areas as expected for Poisson noise.

## 4.     CONCLUSIONS

Although the blind deconvolution problem is believed have a unique solution in two dimensions for continuous data of infinite precision (Lane and Bates 1987), it is apparent in practice that small errors between the blurred image and its reconstruction admit a multiplicity of solutions. Noise exacerbates the problem, where a blind reconstruction of a noisy blurred image may well produce a reconstructed image and reconstructed PSF which bear little resemblance to the true PSF and true image, but their convolution gives a match to the blurred image which is as good as or better than the match expected taking into account the statistics of the noise. Hence blind deconvolution of experimental images using only the constraints of a finite support and positivity on the image and PSF is not likely to be possible. In order to make further progress, it is necessary to find stricter constraints (Jefferies 1993) or make use of some type of maximum entropy criterion for selecting the 'best' solution from a the ensemble of solutions which satisfy a goodness of fit statistic.

194

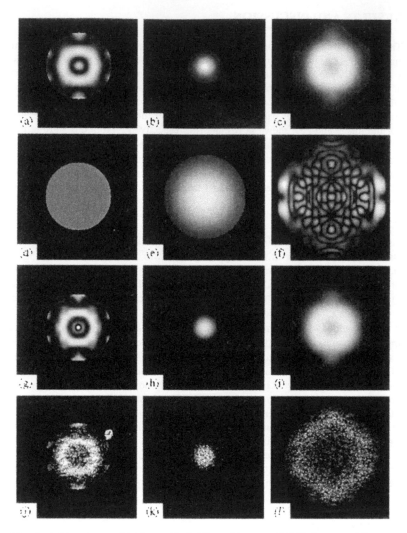

Fig. 1 a) True image, b) True psf, c) Blurred image, d) Initial guess of image, e) Initial guess of psf, f) Modulus of difference between (c) and (i), g) Reconstructed image, h) Reconstructed psf, i) Convolution of (g) and (h), j) Image reconstructed from noisy blurred data, k) PSF reconstructed from noisy blurred data, l) Modulus of differences between noisy blurred image and the convolution of (j) and (k).

## REFERENCES

Jefferies SM and Christou JC 1993 **415** 862
Kundar D and Hatzinakos D 1996a IEEE Signal Processing Magazine **13** 43
Kundar D and Hatzinakos D 1996b IEEE Signal Processing Magazine **13** 61
Lane RG and Bates RHT 1987 J. Opt. Soc. Am. A. **4** 180
Lane RG 1992 J. Opt. Soc. Am. A. **9** 1508
Thiébaut E and Conan JM 1995 J. Opt. Soc. Am. A. **12** 485

*Inst. Phys. Conf. Ser. No 153: Section 6*
*Paper presented at Electron Microscopy and Analysis Group Conf. EMAG97, Cambridge, 1997*

# Deposition of size-selected clusters and application in silicon nanofabrication

**K Seeger, S J Carroll and R E Palmer**

Nanoscale Physics Research Laboratory, School of Physics and Astronomy, The University of Birmingham, Birmingham B15 2TT, UK

**T Tada and T Kanayama**

Joint Research Center for Atom Technology (JRCAT), National Institute for Advanced Interdisciplinary Research (NAIR), 1-1-4 Higashi, Tsukuba, Japan

**ABSTRACT** Size-selected clusters from a cluster ion beam were used as nucleation sites for self-forming etching masks in silicon nanostructure production. The investigation of cluster deposition helps to understand the process of nucleation site formation. Therefore deposition experiments, performed in our laboratory, are reviewed in this article. These experiments show that it should be possible to produce ordered arrays of silicon pillars and silicon nanowires in a controlled way, as confirmed by preliminary results.

## 1. INTRODUCTION

Since Canham (1990) first reported efficient visible luminescence from porous silicon, interest in silicon as a potentially optoelectronically active material has grown dramatically. Many research groups are trying to optimize the production of efficient optically active silicon. Numerous experimental studies of porous silicon has been performed, which are reviewed, for example by Hamilton (1995). At the same time other methods for the fabrication of silicon nanostructures (e.g. pillars), compatible with standard processing techniques are being sought. A standard method for the generation of ordered arrays of small semiconductor structures is the application of lithographic techniques followed by etching. One problem with this procedure is that the resolution is limited, so that either high beam energies are needed or a further treatment is required after the etch process in order to thin the structures (Nassiopolous 1995). Another disadvantage of these lithographic techniques is that many steps have to be carried out for pattern transfer before the final product is reached.

We have discovered a way to produce silicon nanostructures by using self-forming etching masks which employ metal clusters as nucleation sites. Here we review these experiments, in which we deposited size-selected clusters and investigated the influence of cluster size on the Si pillar formation process (Tada 1997a). We also present new results demonstrating that the evaporation of copper onto silicon samples leads to the formation of rows of silicon nanopillars as a result of diffusion of the copper atoms to one dimensional surface defects. In order to understand and find efficient ways for the production of silicon nanostructures a broader range of cluster deposition experiments, performed in our group, is also reviewed in this paper. The results of these experiments show that it should be possible to produce silicon nanowires and ordered arrays of silicon pillars by using surface steps either to confine silicon clusters or to localise etching masks.

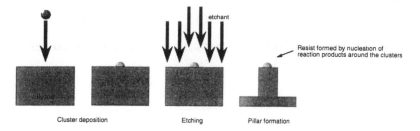

Figure 1: Schematic of the silicon pillar fabrication process. Size-selected silver clusters are deposited onto a silicon wafer. As the etching starts, reaction products of the etching gas condense around the clusters and protect the silicon underneath. The result is the creation of silicon pillars where clusters are deposited.

## 2.    PRODUCTION OF Si NANOPILLARS USING SIZE-SELECTED CLUSTERS

The process leading to the production of silicon pillars in the sub-10-nm range, reported by Tada et al. (1997a), is shown schematically in Fig. 1, and includes only a few steps. After introducing nucleation sites onto a hydrogen-terminated (100) silicon surface, the sample is Electron Cyclotron Resonance (ECR) etched with $SF_6$ as the etching gas. During the etching process gas products condense at the nucleation sites and act as a resist for the etchant. These masks protect the underlying silicon while the unprotected areas are etched away. After an etching time of around 1 minute the sample shows silicon pillars of around 100 nm height on the surface at locations attributed to the nucleation sites introduced by cluster deposition. The process is very dependent on the sample temperature. Pillar formation is only seen at -130°C while at sample temperatures of -140°C or -120°C no pillars were formed. This temperature dependence supports the mechanism proposed, i.e. condensation of etching products on the sample. When the temperature is too high, not enough resist condenses to protect the silicon, hence no pillars are formed. When the temperature is too low, condensation can take place all over the sample, so that the etching gas is not able to remove silicon at all. The nucleation sites on the sample can be of various different kinds, such as mechanical defects induced by scratching the surface, in addition to deposited clusters. However, only the latter will be discussed in this paper.

In order to investigate the influence of the size of the nucleation sites on the pillar formation process, mass-selected metal clusters were used. The clusters are formed in a gas aggregation cluster source described elsewhere (Goldby 1997). A time-of-flight mass filter with a resolution of $m/\Delta m = 20$ allows the deposition of size-selected clusters onto the silicon sample. Fig. 2 shows Scanning Electron Microscope (SEM) micrographs (taken with a tilt angel of 40°) of such silicon samples after etching. The size of the deposited clusters was varied from 200 to 600 atoms per cluster, while the deposition energy was kept constant at 500 eV. For the samples with 200 atom clusters no pillars were formed. The density of pillars increases with increasing cluster size until it is similar to the density of deposited clusters. This saturation density is reached for cluster sizes of 600 atoms. Surprisingly, the diameter of the pillars does not change with the cluster size. The average diameter of the pillars is about 17 nm. These results indicate that the clusters need to have a certain size to allow the etching products to form effective etching masks. One possible explanation could be that clusters which are too small are etched away before enough protecting material is deposited on top of them. The critical size for this process is between 200 and 300 atoms which would correspondent to a cluster diameter of 2 to 3 nm, assuming the clusters form two-dimensional islands on the surface.

The dependence of the mask formation process on the chemical nature of the cluster material was also investigated by depositing different materials as atoms onto the silicon surface. The material deposited forms islands because of diffusion of the atoms across the surface. Tada et al. (1997b) deposited AuPd by rf-sputtering and reported an average pillar diameter of 10 nm. In another experiment (Tada 1997a) both gold and silver were evaporated and then cooled in a helium stream so

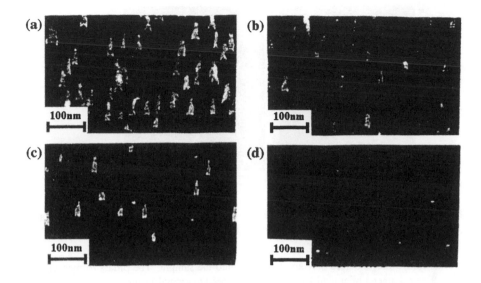

Figure 2: SEM images of silicon samples after ECR etching. Size-selected silver clusters are deposited onto the sample to act as nucleation sites for self-forming etching masks. The size of the deposited clusters is (a) 600, (b) 400, (c) 300 and (d) 200 atoms. (Tada 1997a)

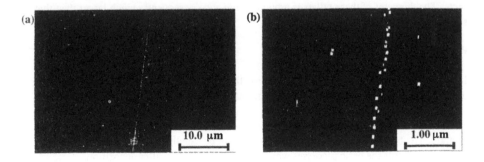

Figure 3: (a) SEM picture of silicon sample showing rows of silicon pillars. The sample was prepared by evaporating copper atoms onto the sample which diffused to line-like surface defects. Reaction products of the etching gas condense around the copper islands and act as etching masks. (b) Magnified region of the SEM micrograph in (a).

198

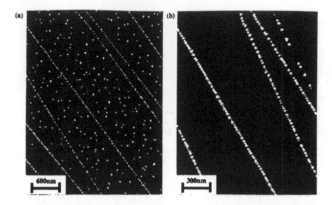

Figure 4: SEM images of HOPG after deposition of Ag at (a) 20°C, showing clusters on terraces and at steps, and (b) 165°C, showing clusters only at steps. (Francis 1996a)

that the atoms form clusters before they are deposited onto the sample. The pillars formed on the gold samples had a diameter of around 10 nm which increased to around 19 nm for the silver cluster samples. Recently we have found that the evaporation of copper (without subsequent condensation) leads to the fabrication of approximately 15 nm diameter pillars. This dependence of the diameter on the nucleation site material seems to reflex the chemical factors involved in the deposition of the etching products. Silver seems to be more chemically reactive than gold or copper. This leads to a bigger 'cap' above the clusters which acts as the etching mask so that the pillar diameter is increased.

Fig. 3 shows SEM pictures of a silicon sample which was etched after deposition of copper by evaporation. The copper atoms diffuse across the surface until they are trapped either by surface defects or other islands which are too big to move. On the sample shown in Fig. 3, the material has concentrated at line-like surface defects. This process allowed the production of rows of pillars along these defects. The diameter of these pillars is around 50 nm and consequently much larger than that of pillars at normal terrace sites on the same sample.

## 3.      PHYSICS OF CLUSTER DEPOSITION

In order to develop a better understanding of the processes involved in the formation of nucleation sites, it is profitable to consider the deposition and growth of clusters on surfaces. Cluster research has been carried out in our group both on the growth of clusters after the evaporation of atoms onto a surface and on the deposition of size-selected clusters, with emphasis on graphite as a model substrate. The behaviour of clusters on silicon will not be exactly the same as on graphite but the comparison highlights the basic processes which occur on the surface after deposition.

The formation of islands (clusters) on surfaces after evaporation of silver and gold onto graphite was studied by Francis et al. (1996a,b). They investigated the influence of the sample temperature during deposition. Fig. 4 shows SEM pictures of highly oriented pyrolytic graphite (HOPG) samples with silver evaporated at (a) 20°C and (b) 165°C. In the case of room temperature deposition clusters were formed on the terraces as well as at surface steps, while in the case of deposition at 165°C the silver atoms all diffuse across the surface until they arrive at a step. This is due to the higher diffusion rate of the atoms at increased temperatures. It is probable that this effect could be used for the *controlled* production of rows of silicon pillars. For example, we may anticipate that if copper were evaporated onto a heated, stepped surface all the copper atoms would diffuse to the steps. The copper atoms at the steps would then act as nucleation sites for the etching products, leading to the production of lines of pillars similar to those shown in figure 3 (where nucleation occurs at *random* line defects). A possible way of generating an ordered array of steps on a silicon surface is to cut crystalline silicon slightly off the principal planes. Since the steps on these 'vicinal' surfaces

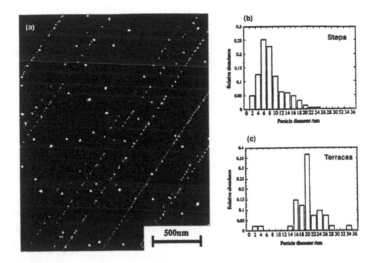

Figure 5: (a) SEM picture of a graphite sample after deposition of $Ag_{400}$ clusters with an energy of 1200 eV. Clusters are lined up at step edges, while on larger terraces the clusters also form islands. The size distribution of clusters at steps (b) is shifted to smaller sizes than at terraces (c). (Carroll 1997)

are parallel they may allow the production of ordered arrays of silicon pillars. The metal deposition rate should determine the density of the pillars.

A number of insights into the processes occurring in the deposition of size-selected metal clusters have been obtained by the application of complementary nanoscopic techniques. Goldby et al. (1996), who analyzed the diffusion and aggregation of size-selected silver clusters deposited on graphite with the SEM, found that clusters of all sizes from 50 to 250 atoms per cluster, deposited with an energy of 50 eV, are mobile on the surface. The clusters diffuse and coalesce with other clusters until they form islands of around 14 nm diameter. This preferred size has been accredited to the lattice strain between the graphite substrate and the silver islands. Further aggregation of these 14 nm islands leads to the fabrication of irregular islands and fractal structures, in which the 14 nm islands are observed as 'building blocks'.

The SEM images of size-selected Ag clusters on graphite were compared with Scanning Tunneling Microscopy (STM) images by Carroll et al. (1996). The cluster size distributions derived from these two techniques do not coincide. The mean cluster size seen in the STM images is much smaller than those of the SEM pictures. This leads to the conclusion that the tip of the STM interacts with the clusters and removes larger clusters (above 3 nm) from the field of view (e.g. by 'sweeping'). Conversely, very small clusters (smaller than 2 nm) observed with the STM cannot be resolved with the more limited resolution of the SEM. The relationship between the deposition energy and the shape and size of the clusters was also investigated in the STM study. Clusters deposited with low energy (typically 1 eV/atom) land 'softly' on the surface and have a hemispherical shape, while clusters deposited at high energy (typically 10 eV/atom) are flattened to a single atomic layer.

The interaction of size-selected clusters with surface steps was investigated by Carroll et al. (1997). Fig. 5 (a) shows an SEM micrograph of a graphite sample on which silver clusters of 400 atoms have been deposited. The picture shows clusters on terraces and at surface steps. The size distributions of the clusters on the terraces and at the steps is shown in fig. 5 (b) and (c). Clusters at the step edges have an average diameter of 6 nm while the peak for clusters on terraces is at 20 nm. Almost non of the particles on the terraces are smaller than 14 nm. The results confirm the diffusion and aggregation of clusters on the terraces but also indicate that clusters reaching a step become bound to this step, i.e. they cannot escape onto the terrace again. The aggregation of the size-selected clusters is consequently suppressed. This result is very interesting for the possible production of silicon nanostructures. One can envisage the deposition of size-selected silicon clusters onto a stepped surface and the subsequent trapping of the clusters at the steps, possibly leading to the fabrication of silicon nanowires. Indeed, the idea of cluster trapping at controlled surface defects can be extended further, perhaps to the production of 2D arrays of size-selected silicon clusters trapped at artificially prepared point defect sites.

## 4.    SUMMARY AND CONCLUSIONS

We have shown that the production of silicon pillars with diameter in the nanometre range can be achieved by the deposition of metal clusters onto a silicon surface. The clusters act as nucleation sites for the reaction products from ECR etching with $SF_6$; this forms masks which protect the silicon underneath from further eching. This leads to silicon pillar formation.

Fundamental studies of cluster deposition help to illuminate the processes included in nucleation site formation and suggest the possibility of producing silicon nanostructures by deposition and diffusion of Si atoms or clusters on stepped surfaces. Elevated sample temperatures make it possible to locate deposited atoms or clusters at preferred sites such as steps. This phenomenon may allow the production of ordered arrays of silicon pillars from vicinal silicon substrates which have a regularly spaced array of parallel steps on the surface. By evaporating material like copper onto a heated silicon sample of this type it should be possible to locate all the deposited copper atoms in clusters trapped at these steps. The clusters would act as nucleation sites for mask formation during etching. However, it would be desirable to find a way to create pillars of smaller diameter than those shown in fig. 3, produced after the growth of copper clusters at *random* line defects. The diffusion of clusters across surfaces also suggests another practical method for the production of silicon nanostructures. If silicon clusters are deposited onto a patterned (e.g. stepped) surface and diffuse to the chosen defects at appropriate surface temperatures it should be possible to produce nanometre-scale silicon wires or arrays of silicon clusters.

## 5.    ACKNOWLEDGMENTS

This project was supported in part by NEDO through the management of Angstrom Technology Partnership (ATP), by the EU and the British Council. K.S. and S.J.C. are grateful to the University of Birmingham for financial support. S.J.C. also thanks the EPSRC for the award of a studentship.

Bardotti L, Jensen P, Hoareau A, Treilleux M, Cabaud B, Perez A and Cadete Santos Aires F 1996 Surf. Sc. **367**, 276
Canham L T 1990 Appl. Phys. Lett. **57**, 1046
Carroll S J, Weibel P, v Issendorff B, Kuipers L and Palmer 1996 J. Phys.: Condens. Matter **8**, 617
Carroll S J, Seeger K and Palmer R E 1997 to be published
Francis G M, Goldby I M, Kuipers L, v Issendorff B and Palmer R E 1996a J. Chem. Soc., Dalton Trans., 665
Francis G M, Kuipers L, Cleaver J R A and Palmer R E 1996b J. Appl. Phys. **79**, 2942
Goldby I M, Kuipers L, v Issendorff B and Palmer R E 1996 Appl. Phys. Lett. **69**, 2819
Goldby I M, Kuipers L, v Issendorff B and Palmer R E 1997 Rev. Sci. Instrum., in press
Hamilton B 1995 Semicond. Sci. Technol. **10**, 1187
Nassiopoulos A G, Grigoropoulos S, Gogolides E and Papadimitrou D 1995 Appl. Phys. Lett. **66**, 1114
Tada T, Kanayama T, Koga K, Weibel P, Carroll S J, Seeger K and Palmer R E 1997a Appl. Phys. Lett., in press
Tada T, Kanayama T, Weibel P, Carroll S J, Seeger K and Palmer R E 1997b Microelectronic Engineering **35**, 293

*Inst. Phys. Conf. Ser. No 153: Section 6*
*Paper presented at Electron Microscopy and Analysis Group Conf. EMAG97, Cambridge, 1997*

# Absolute field strength determination of magnetic force microscope tip stray fields

**S McVitie, R P Ferrier and W A P Nicholson**

Department of Physics and Astronomy, University of Glasgow, Glasgow G12 8QQ, U.K.

**ABSTRACT:** Quantitative analysis of magnetic force microscope (MFM) images is only possible if the magnetic state of the sensing tip is known. In this paper we describe a method which is used to characterise the stray field produced by an MFM tip. The method utilises Lorentz microscopy techniques and tomographic field reconstruction algorithms. A specially constructed calibration specimen enables measurement of the absolute values of the tip stray field to be made.

## 1. INTRODUCTION

Since its introduction around ten years ago magnetic force microscopy (MFM) has become an increasingly important tool for investigating the stray fields at the surfaces of materials (Martin and Wickramasinghe 1987). It is based on the simple principle of scanning a sensing magnetic tip over an area of the surface of a sample and recording in some manner the interaction of the tip with the sample's stray field. The contrast obtained in MFM images is therefore dependent on the interaction of the magnetisation in the sensing tip and the stray field from the sample. Detailed knowledge of the tip magnetisation is desirable to interpret this contrast in a quantitative manner. Many of the commercially available tips are pyramidal semiconductors coated with a magnetic thin film. In such cases it is not possible to image directly the magnetic structure of the film on the tip by transmission electron microscopy (TEM) methods and hence gain knowledge of the film magnetisation.

Using methods of TEM it is possible to image the stray field in free space resulting from the magnetisation distribution in the tip. Furthermore the theory of reciprocity tells us that the interaction in MFM may be considered as between the stray field of the tip and the magnetisation in the sample. Quantitative MFM may then be considered by a deconvolution process if the field distribution of the tip is known. In this paper we describe a method whereby the absolute stray field from an MFM tip may be measured using the techniques of electron microscopy in conjunction with a specially designed calibration stage.

## 2. EXPERIMENTAL SETUP

Lorentz microscopy is the name given to a collection of imaging techniques performed in electron microscopy, which are used to reveal the magnetic structure in materials. Imaging of the stray fields from MFM tips has been demonstrated by Ferrier et al (1997) using the differential phase contrast (DPC) mode practised in a scanning transmission electron microscope (STEM). In this mode the deflection of the electron beam by magnetic induction components normal to its trajectory can be mapped using a quadrant detector placed in the far field. Difference signals from opposite segments of the detector map linearly orthogonal in-plane components of the magnetic induction integrated along the electron path and for quantitative analysis the signals are normally digitised. The instrument used in this work is a Philips CM20 which has been modified for optimum magnetic imaging performance and is also equipped with a thermally assisted field emission gun.

In the case of MFM tips the DPC images represent the integrated stray field components. Selected linescans taken from these images show the integrated field for the equivalent position of the specimen in MFM. To apply these results more usefully to MFM we require the spatial field distribution rather than the integrated field. This is achieved by acquiring DPC images over a range of orientations of the MFM tip and selected linescans are used as input data for a tomographic reconstruction algorithm. The algorithm reconstructs the 3-dimensional field in a plane and in our previous studies the field values obtained are in arbitrary units. The reason for this is that the magnetic state of the tip is not known and so no figure can be given to the absolute value of integrated field at any point.

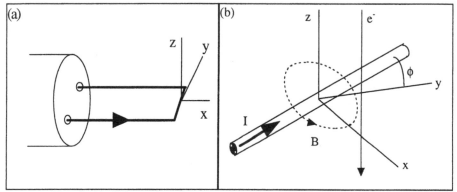

Fig. 1 (a) Schematic of current carrying wire in specimen rod for CM20 microscope. (b) Orientation of section of current carrying wire used to give known deflection to the electron beam in STEM.

For the reconstructed field to be put on an absolute basis we must calibrate the deflection of the beam using a test sample for which the integrated field is known. The field from a long straight wire carrying a current $I$ can easily be calculated from Ampère's Law. The experimental set-up is shown in Fig. 1. In Fig. 1(a) the geometry of the current carrying wire in the microscope rod is indicated along with the sense of the driving current. In Fig. 1(b) the section of wire indicated in Fig. 1(a) is shown to lie in the $yz$ plane at an angle $\phi$ to the $y$ axis. (It can easily be shown that the two other sections of the wire give no resultant deflection of the beam). For an electron beam travelling parallel to the $z$ axis such an orientation produces a finite deflection of the beam provided $\phi \neq 0°$. In this orientation deflection of the electron beam is due to the $x$ and $y$ components of field and from symmetry it can easily be seen that the net deflection due to the $x$ component is zero. The calculated deflection angle, $\beta$, is directly proportional to the integrated induction of the electron beam. This deflection is entirely due to the $y$ component and is independent of the distance of the beam from the $yz$ plane to the wire. The expression is;

$$\beta = \frac{e\lambda}{h} \int\limits_{-\infty}^{\infty} B_y \, dz = \frac{e\lambda}{h} \frac{\mu_0 I}{2} \tan \phi$$

The symbols are as discussed in the previous paragraph with $\mu_0$ being the permeability of free space, $e$ the charge of the electron, $h$ Planck's constant and $\lambda$ the wavelength of the electron radiation. As an example a magnetic thin film with a uniform saturation induction of 1.0 T and thickness 50 nm deflects an electron beam that passes through it by an angle of 30 µrad. By comparison the deflection from a wire tilted at an angle of 20° and carrying a current of 200 mA is 27 µrad. We have constructed such a wire in one of our specimen rods which has vacuum feed-throughs for connection to an external power supply. The angle $\phi$ can easily be varied by tilting the rod about its axis.

## 3. RESULTS AND DISCUSSION

The DPC images of the MFM tip show a smoothly varying integrated field and this is shown schematically in Fig. 2(a). The tip used in this study was a standard Digital Instruments tip kindly supplied by Dr Ken Babcock. Taking an intensity line trace as indicated along the direction of the dotted line in Fig. 2(a) results in a profile as shown in Fig. 2(b). This represents the integrated induction component parallel to the tip axis and is consistent with the tip being magnetised principally along its axis. Note that in general the vertical axis has arbitrary units. Also the integrated field has contributions from not just the tip but also the cantilever as can be seen from the asymmteric profile of the linescan. This has an additional effect of introducing a significant DC shift of the linescans for the different orientations imaged as previously reported by Ferrier et al (1997).

(a)                    (b)

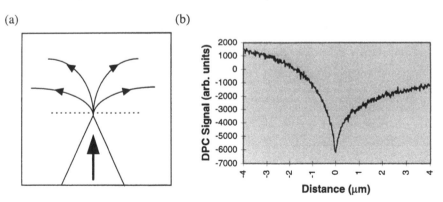

Fig. 2 (a) Schematic of field lines from a uniformly magnetised MFM tip. (b) Line trace from DPC image equivalent to (a) along direction of dotted line showing variation of integrated field component in the direction of the tip axis.

The procedure to obtain absolute values for the integrated field requires images to be taken in free space with the calibration rod. Images are acquired digitally with the wire driven with a current for a given value of tilt and additionally with no current passing through the wire. The latter images serve as a reference for zero integrated field. The former are used to assign a change in contrast to the known magnetic deflection. In the present case the calibration results indicate that the peak integrated field from the MFM tip corresponds to a deflection angle of 16.6 μrad. Reconstruction of the 3-dimensional field from the tip is made from a series of linescans taken at different orientations. As we now are able to input the absolute values of deflection angle the output from the reconstruction algorithm gives directly the absolute field distribution from the tip. The reconstructed field component parallel to the tip axis is shown in Fig. 3(a) as a grey scale image. The reconstruction plane is ~ 40 nm from the tip apex. An intensity linescan from this image represents the spatial variation of that field component and this is given in Fig. 3(b) with the absolute field values. The peak induction field from the tip for this component is ~ 40 mT. We estimate that the error in the field value using this method is at the level of 10%. The length of the tip is around 15 μm but measurements by Babcock et al (1996) suggest that the thin magnetic film may only be uniformly magnetised for ~ 0.7 μm from the apex. Initial comparison with simulations of the field indicate good agreement with the measured field values shown in Fig. 3(b).

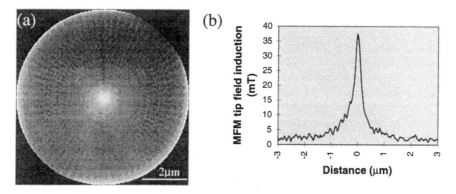

Fig. 3 (a) Grey scale image of reconstructed field (along tip axis) in a plane ~ 40 nm from the end of the tip. (b) Intensity line trace from (a) with absolute values of field plotted.

In conclusion we have demonstrated a method whereby the absolute value of integrated field for an MFM tip may be made by combining the technique of Lorentz microscopy with a suitable reconstruction algorithm and calibration sample. In future we anticipate that this method will be used to characterise MFM tips with various coatings and in different magnetic states depending on the particular application of the tip.

**REFERENCES**

Babcock K L, Elings V B, Shi J, Awschalom D D and Dugas M 1996 Appl. Phys. Lett., **69**, 705
Ferrier R P, McVitie S, Gallagher A and Nicholson W A P 1997 IEEE Trans. Mag, in press.
Martin Y, Wickramasinghe H K 1987 Appl. Phys. Lett. **50**, 1455

*Inst. Phys. Conf. Ser. No 153: Section 6*
*Paper presented at Electron Microscopy and Analysis Group Conf. EMAG97, Cambridge, 1997*
© *1997 IOP Publishing Ltd*

# Atomic force microscopy studies of the surface morphology of annealed single crystal substrates and substrate annealing effects on the $YBa_2Cu_3O_{7-x}$ thin film growth

**Y C Fan, A G Fitzgerald, H C Xu, B E Storey and A O Tooke**

Department of Applied Physics and Electronic & Mechanical Engineering, University of Dundee, Dundee DD1 4HN, UK

**P R Broussard and V C Cestone**

Naval Research Laboratory, Washington, DC 20375, USA

**ABSTRACT:** The surface morphologies of yttria-stabilised $ZrO_2$ (YSZ) and MgO single crystal substrates annealed at high temperatures in an oxygen atmosphere have been examined by AFM. Substrate annealing effects on the growth of high $T_c$ YBCO superconducting thin films have also been studied by preparing the thin films on the non-annealed and annealed substrates. STM image observations of the YBCO thin films show that the well-defined substrate surface conditions obtained after high temperature annealing play a dominant role in governing the film growth mechanism which determines the film's microstructure and surface morphology.

## 1. INTRODUCTION

In the last few years, with the increasing use of atomic force microscopy (AFM) and scanning tunneling microscopy (STM) to characterise the surface morphologies of $YBa_2Cu_3O_{7-x}$ (YBCO) superconducting thin films, the importance of substrate surface conditions in achieving of high quality YBCO thin films have been gradually recognised. Lately, much research has been carried out to improve thin film quality by suitable substrate pre-treatments such as deposition of a buffer layer (Brorsson et al 1994, Haefke et al 1992), chemical polishing (Moeckly et al 1990) and high temperature annealing (Norton et al 1990, Kim et al 1994). Among these substrate pre-treatment processes, the high temperature annealing of the substrates in oxygen has been found particularly effective in improving the thin film quality (Brorsson et al 1994, Minamikawa et al 1995, Moeckly et al 1990). Generally speaking, the high temperature annealing of substrates can effectively remove surface contamination, and at a sufficiently high annealing temperature, surface re-growth can be initiated by rearranging the surface atoms in order to reduce the surface energy (Sum et al 1995). This annealing process will usually result in well-defined atomic flat crystal surfaces with a high density of terrace steps. These terrace steps will provide favourable nuclei sites for deposited species and facilitate the epitaxial growth of YBCO thin films (Norton et al 1990). In this research work, we have systematically studied the surface morphological changes of annealed YSZ and MgO substrates by AFM. Preliminary experiments involving the deposition of YBCO thin films on non-annealed and annealed MgO and YSZ substrates have also been carried out to study substrate annealing effects on thin film growth.

206

## 2. EXPERIMENTAL

In the experiments described here, commercial YSZ (100) single crystal substrates with one side polished to an optical finish and MgO (100) substrates polished using a polishing machine with the following sequence of diamond paste grades 14, 6, 3, 1 and 0.25 μm, were used. Before annealing, the substrates were first examined by a Dimension™ 3000 atomic force microscope (Digital Instruments). The substrates were then subjected to high temperature annealing in one atm oxygen conditions at temperatures in the range of 600 to 1200 °C for periods of five and ten hours. The annealed substrates were observed again by AFM to examine the morphological changes induced by the high temperature annealing process.

Pulsed laser deposition was used to prepare the YBCO thin films on the annealed and non-annealed MgO substrates. DC off-axis sputtering was used to deposit the YBCO thin films on annealed and non-annealed YSZ substrates. The surface morphology of the thin films was studied by an STM which was composed of a Molecular Imaging PicoSTM head and a Digital Instruments NanoScope IIIa Scanning Probe Microscope Controller. The STM images of the films were acquired using a mechanically cut PtIr tip in constant current mode with a positive sample bias voltage in the range of 750 to 950 mV and a tunneling current of less than 150 pA.

## 3. RESULTS AND DISCUSSION

The substrates annealed at different temperatures exhibited distinct morphological features. For YSZ (100) substrates, when the annealing temperature was less than 600 °C, no substantial morphological changes could be observed on the substrate surfaces by comparison with the non-annealed substrates. On the surfaces of non-annealed substrates, there were many hillocks and pits as can be seen in Fig.1(a). In addition, some scratch marks and grooves caused by mechanical polishing can be observed. The typical roughness RMS values measured on the different regions were in the range 1.5 to 3 nm. When the substrates were annealed at 800 °C, the pits, hillocks and scratches became more prominent and the overall surface roughness increased (RMS values up to 4 to 6 nm) as shown in Fig.1(b). This surface roughening is believed to be caused mainly by local thermal strain released in the condensing process. The thermal evaporation of embedded impurities such as polishing materials will also contribute to this surface roughening. The substrate surfaces became smoother when the substrates were annealed at a temperature above 1000 °C for five hours, and most of the original scratches and grooves caused by mechanical polishing have lost sharpness and can only be dimly recognised. For the majority of the substrates annealed at 1200 °C, well aligned terrace steps appeared on substrate surfaces as can be seen in Fig.1(c). Atomically flat surfaces free of step structures have also been obtained for some substrates after annealing at 1200 °C. These surface morphological changes show that the surface re-growth process has been initiated by high temperature annealing through an atom migration and arrangement which results in a well defined atomically flat surface. The appearance of a high density of directional terrace steps also indicates that for part of the

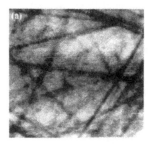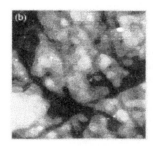

Fig.1. AFM images of (a) non-annealed YSZ substrate, (b) YSZ substrate annealed at 800 °C for five hours, (c) YSZ substrate annealed at 1200 °C for five hours showing a high density of directional terrace steps. Image scan size: 2 μm

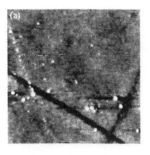  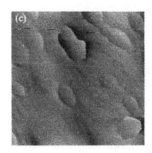

Fig.2. AFM images of (a) non-annealed MgO substrates, (b) degraded MgO substrates in air for one week, (c) MgO substrates annealed at 1200 °C for ten hours showing the appearance of high density terrace steps and condensing marks. Image scan size: 2 μm.

substrate, the surface is not exactly aligned with the nominal (100) crystal surface, and this is usually caused by miscut or vicinal polishing of the substrates. Atomically flat surfaces with or without terrace steps could be preserved after re-annealing in the range of 600 to 800 °C for long periods without further morphological changes. These thermally stable substrates are very suitable for use in the fabrication of high quality super-thin YBCO thin films.

Similar morphological changes caused by high temperature annealing have also been found to occur on MgO substrates. Fig.2 shows the AFM images of typical morphologies of chemically cleaned, degraded and annealed substrates. MgO substrates are much more sensitive to atmospheric conditions. The $H_2O$ and $CO_2$ which exist in the atmosphere are easily adsorbed by and react with the MgO substrate surface forming a layer of $Mg(HO)_2$ and $MgCO_3$ as can be seen in Fig.2(b). In order to prepare high quality YBCO thin films on MgO substrates, this contamination layer has to be removed by a high temperature annealing process. Again, after this high temperature annealing, a surface structure consisting of a high density of terrace steps was observed on the substrates as shown in Fig. 3(c). In some regions, large MgO crystallites formed in the reduction process by decomposition of the hydroxides and carbonates were frequently observed on the annealed substrates.

The non-annealed and annealed MgO and YSZ substrates have been used for YBCO thin film deposition. The films on MgO substrates were prepared by conventional pulsed laser ablation techniques under optimum deposition conditions. All of the YBCO thin films on non-annealed and annealed substrates showed superconductivity above 85K. The $J_c$ values were much higher for the films on the annealed substrates. Usually, films deposited on non-annealed MgO substrates had typical $J_c$ values around $10^3$ to $10^4$ A/cm$^2$ and sometimes $J_c$ reached $10^5$A/cm$^2$, while for films deposited on annealed substrates and particularly on the substrates annealed above 1000 °C, the $J_c$ values could easily reach $10^6$A/cm$^2$. From Fig.3 it can be seen that the films deposited on non-annealed and annealed substrates exhibit different morphologies. For the films on non-annealed substrates, the

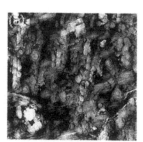  

Fig.3. STM images of YBCO thin films on (a) non-annealed MgO substrate (b) MgO substrate annealed at 800 °C, (c) MgO substrate annealed at 1200 °C showing a high density of terrace steps on the surface. Image scan size: 2 μm

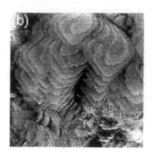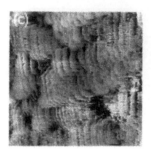

Fig.4. STM images of YBCO thin films on (a) non-annealed YSZ substrate, (b) YSZ substrate annealed at 800 °C, (c) YSZ substrate annealed at 1200 °C with high density terrace steps on the substrate surface. Image scan size: 1 μm

growth spirals were randomly distributed on the film surface and there were also some regions which did not contain spiral growth features. For the films deposited on high temperature annealed substrates, the spiral-shaped grains tended to grow along the running direction of the terrace steps and connected or overlapped with each other forming regular rows as can be see in Fig.3(c). The high $J_c$ values exhibited by these films can be attributed to the densely distributed growth spirals which are well coalesced with each other forming a good electrical path.

The YBCO films on YSZ substrates were prepared by off-axis DC sputtering from a composite YBCO target. The deposition was performed in a 100 mTorr mixed oxygen and argon atmosphere with an oxygen-argon ratio of 1:4. During deposition, the substrate temperatures were kept at 670 °C and growth rates were typically 300 Å/hour. After growth, the chamber was vented with $O_2$ to 100 Torr and the samples were furnace cooled to below 100 °C. Fig.4 shows STM images of the YBCO thin films deposited in one batch with a film thickness of 1200 Å. Well developed rectangular shaped growth spirals have been found sporadically on both non-annealed (Fig.4(a)) and low-temperature annealed (Fig.4(b)) substrates but with a low area coverage. For films prepared on substrates annealed at 1200 °C, a terrace step structure has been observed on the film surface as shown in Fig.4(c). A large number of images have been taken at several different positions on this film but no single well developed growth spiral as show in Figs 4(a) and (b) has been found. The observed directional terrace features are similar to those on the bare surfaces of high temperature annealed substrates. The film morphology is therefore to some extent a reprint of the substrate surface features. In this situation the film grows by step propagation and spiral growth is suppressed by the substrate surface conditions as also observed by Lowndes et al (1992) on miscut LaAlO$_3$ substrates.

**REFERENCES**

Brorsson G, Olsson E, Ivanov Z G, Stepantsov E A, Alarco J A, Boikov Y, Claeson T, Berastegui P, Langer V and Löfgren M, 1994 J. Appl. Phys. **75** (12), 7958

Haefke H, Lang H P, Sum R, Güntherodt H J, Berthold L and Hesse D 1992 Appl. Phys. Lett. **61** (19), 2359

Kim B I, Hong J W, Jeong G T, Moon S H, Lee D H, Shim T U and Khim Z G 1994 J. Vac. Sci. Technol. **B12** (3), 1631

Lowndes D H, Zheng X Y, Zhu S, Budai J D and Warmack R J 1992 Appl. Phys. Lett. **61** (7), 852

Minamikawa T, Suzuki T, Yonezawa Y, Segawa K, Morimoto A and Shimizu T 1995 Jpn. J. Appl. Phys. Part 1, **34** (8A), 4038

Moeckly B H, Russek S E, Lathrop D K, Buhrman R A, Li J and Mayer J W 1990 Appl. Phys. Lett. **57** (16), 1687

Norton M G, Summerfelt S R and Carter C B 1990 Appl. Phys. Lett. **56** (22), 2246

Sum R, Lang H P and Güntherodt H J 1995 Physica C **242**, 174

*Inst. Phys. Conf. Ser. No 153: Section 6*
*Paper presented at Electron Microscopy and Analysis Group Conf. EMAG97, Cambridge, 1997*
© *1997 IOP Publishing Ltd*

# Characterisation of the surface morphology and electronic structure of YBa$_2$Cu$_3$O$_{7-x}$ thin films by atomic force and scanning tunneling microscopies

**A G Fitzgerald[1], Y C Fan[1], H C Xu[1], C W An[2], B Su[2], B E Storey[1] and A O Tooke[1]**

[1]Department of Applied Physics and Electronic & Mechanical Engineering, University of Dundee, Dundee DD1 4HN, UK

[2]National Laboratory of Laser Technology, Huazhong University of Science and Technology, Wuhan 430074, P. R. China

ABSTRACT: AFM, STM and STS techniques have been used to characterise the surface morphologies and electronic structures of YBCO films. All of the films prepared under optimum laser deposition conditions exhibited dominant spiral growth features. Besides, layer growth cake-like shaped crystal grains were also frequently observed. The (*I-V*) curve analysis indicated that the spiral or layer-by-layer growth grains exhibited a metallic tunneling behaviour, while the particulates on the films showed typical semiconductor tunneling behaviour. In addition, some tiny insulator inclusions have also been revealed by CITS analysis.

## 1. INTRODUCTION

In the last few years, atomic force microscopy (AFM) and scanning tunneling microscopy (STM) have been extensively used to study YBa$_2$Cu$_3$O$_{7-x}$ (YBCO) high T$_c$ superconducting films (Hawley et al 1991, Gerber et al 1991). A wealth of information on the surface morphological features of YBCO films prepared by a variety of deposition techniques on commonly used substrates have been obtained (Schlom et al 1992 and 1994). In addition to these surface morphological characterisations, scanning tunneling microscopy has also been used to probe the surface electronic structure of the YBCO thin films by operating the STM in the scanning tunneling spectroscopy (STS) mode, and the superconducting spectra and vortex lattice images of YBCO thin films and single crystals have been directly observed by several research groups (Aprile et al 1995, Mielke et al 1995). In this paper an AFM and STM surface characterisation of YBCO thin films prepared by laser deposition is presented and the tunneling behaviour of the films at room temperature revealed by STS techniques is discussed.

## 2. EXPERIMENTAL

The YBCO films were prepared by conventional pulsed laser deposition on (100)-aligned yttria-stabilised ZrO$_2$ (YSZ), MgO and SrTiO$_3$ (STO) single crystal substrates. The well established laser deposition technique can be routinely used to fabricate c-axis oriented YBCO films with T$_c$ well above 85K and J$_c$ in the range of $10^4$ to $10^6$ A/cm$^2$. The surface morphology of the prepared films were observed by atomic force microscopy (Digital Instruments, NanoScope III ) and scanning tunneling microscopy (Molecular Imaging PicoSTM head with a Digital NanoScope IIIa Scanning Probe Microscope Controller). The AFM images of the films were taken in tapping mode with a silicon cantilever tip of beam length 125 μm and a resonance frequency of around 350 kHz. The STM image observations were carried out in the constant current mode by using a mechanically cut PtIr tip with a typical positive sample bias voltage of 800 mV and tunneling current of less than 150 pA.

The surface electronic structure of the thin films was analysed by the STS techniques at room temperature in air. The current-voltage *(I-V)* curves were obtained at a particular sample area by disabling the feedback and measuring the tunneling current as a function of sample bias voltage for a fixed tip-sample separation. The current imaging tunneling spectroscopy (CITS) technique was also used to study changes in electronic conduction properties with film location and film morphology. In the CITS measurement, at each CITS point, the *(I-V)* curve data are taken by ramping the bias voltage as described above. The current from each of the *(I-V)* curves corresponding to the *bias display* voltage was plotted as a function of the tip position, and these current images were simultaneously captured and displayed with the corresponding topographical images of the scanned area.

## 3. RESULTS AND DISCUSSION

Fig.1 shows a set of AFM images of 150 nm thick YBCO thin films (all with a $T_c$ value above 85K) on (100) aligned MgO and STO substrates. It can be seen that one of the common and prominent morphological features of films deposited by laser ablation is island growth that eventually results in

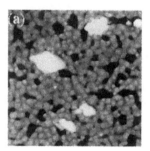

Fig.1    AFM images of the YBCO thin films grown on (a) (100) MgO at 750 °C , (b) (100) STO at 760 °C, (c) (100) STO at 810 °C. Image scan size: 5 μm

a granular structure. The films were mainly composed of hillock-shaped crystal grains with dimensions in the range of 200 to 350 nm in diameter and a few tens of nanometers in height. With increasing substrate temperatures, the growth islands tended to coalesce with each other with the obvious reduction of voids or pinholes as can be observed from Fig.1(a) to (c) . A large number of particulates were also observed on the films with dimensions of several hundreds nm up to one micrometer both in diameter and in height. These large particulates not only increased the surface roughness but also reduced the film current carrying ability because of their semiconductor-like conduction properties as will be seen later.

More detailed microstructure within the growth islands and particulates which could just be

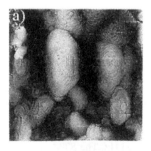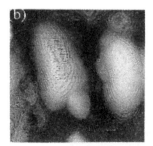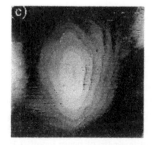

Fig.2. STM images of a YBCO thin films on a MgO substrate showing typical spiral growth features. Image scan size: (a) 1.2μm, (b) 0.88 μm and (c) 0.35 μm.

detected in the AFM images was revealed at high resolution by STM. Fig. 2 shows a set of typical STM images taken in the same region of the 2500 Å thick YBCO thin film ($T_c$=87K) grown on a MgO substrate. The majority of the islands observed in the AFM images of the films are screw dislocation generated growth spirals. Each of these grains grows by adding deposited species to a spirally expanding step on the top surface of the grain and the terrace step height corresponds to a c-axis unit cell length e.g. 1.2 nm. In our STM observations, nearly all of the c-axis oriented films on MgO substrates with high $T_c$ and $J_c$ values exhibited these spiral growth features with a density of approximately $10^8$ to $10^9$/cm$^2$.

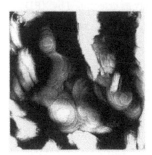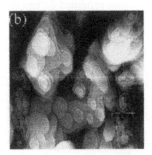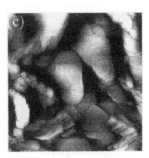

Fig.3. STM images of a YBCO thin film on a YSZ substrate showing morphological changes with location, (a) spiral growth, (b) layer-by-layer growth, (c) irregular outgrowth grains with a spiral embedded among them. Image scan size: 1.5 μm

Sometimes, the STM images taken from different regions on a film have shown distinct morphological features. Fig.3 shows the STM images acquired at different regions on a film deposited on a YSZ substrate. In the majority of the regions, the film grows by the well known spiral growth mode (see Fig.3(a)). In some regions, the observed crystal grains have a layer-cake shape as shown in Fig.3(b) which unambiguously indicates layer-by-layer growth. In a few regions, the film are predominantly composed of irregular crystal grains with a few spiral or layer-by-layer growth grains embedded among them as shown in Fig.3(c). Because all of these features have been observed on a film with a size of 1x0.5 cm$^2$, the substrate temperature and deposited species should be uniform over such a small area. The different growth features observed in this area can therefore be ascribed to different local substrate surface conditions such as local crystal defects, roughness, impurities and surface crystal orientations.

There have been some reports of changes in the spiral growth density in YBCO films with substrate temperature (Schlom et al 1994). Our recent STM observations of films grown on STO substrates have shown that not only the density but also the growth mechanism changes with the substrate temperature. Fig. 4 shows STM images of YBCO thin films on a STO substrate prepared at

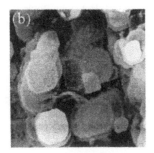

Fig.4. STM images of the films prepared on STO substrates with substrate temperature of (a) 760 °C, (b) 780 °C, (c) 810 °C. Image scan size: 0.65 μm, film thickness is roughly 1500 Å

212

temperatures of 760, 780 and 810 °C . The film prepared at 760 °C exhibited a predominant spiral growth (Fig.4(a). Both spiral growth and layer-by-layer growth were found in 780 °C films. Fig.4.(b) shows an area in the 780 °C film with layer-by-layer growth. In the film prepared at 810 °C, the majority of the crystal grains exhibited a layer-cake structure as shown in Fig.4(c). In addition, the crystal grains in the films prepared with higher substrate temperatures had a much larger dimension than that in the films prepared at low temperatures. The mobility of the deposited species in the films is strengthened by the increased substrate temperature and this facilitates the coalescence of the growth islands which in turn results in a larger number of flat layers on the top of the growth islands. The films showing these morphological features exhibited the best crystallinity as determined by XRD.

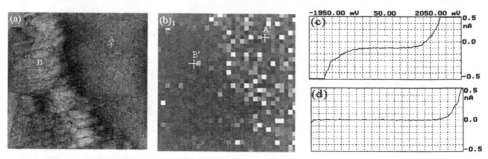

Fig.5. STS analysis of a YBCO thin film on a MgO substrate, (a) STM image of the scanned area, image scan size 3 µm, (b) tunneling current image at a *bias display* value of -500 mV, (c) I-V curve measured on a growth spiral located at position A in (a), (d) I-V curve measured on a particulate located at position B in (a)

The local surface electronic structure of the YBCO thin films has also been studied by STS techniques at room temperature in air. The typical current-voltage *(I-V)* curves obtained at position A and B (see Fig.5) which locate at a growth spiral and at the top of a particulate are shown in Fig.5(c) and (d) respectively. The *(I-V)* curves measured on the growth spirals are symmetrical about the origin, and gradually approach a straight line with the decreasing tip-sample separation showing a typical metallic tunneling behaviour. While the tunneling-voltage *(I-V)* curves acquired on the particulates and the majority of regions with no sign of spiral growth are asymmetrical to the origin e.g. no tunneling current can be detected when the sample is negatively biased. This strong current rectification behaviour shows that the large particulates on the films have a typical p-type semiconductor tunneling behaviour. The current rectification behaviour of the particulates can be presented more effectively by CITS analysis. When the sample is positively biased, the currents at all positions are approximately at the same level, and the image brightness is nearly uniform. Whereas, when the sample is negatively biased, the regions corresponding to the particulate and the non-spiral growth areas in the tunneling image Fig.5(b) become dark indicating a substantial reduction of the tunneling current. By using this CITS analytical technique, some tiny insulator inclusions in the films have also been directly detected. These tiny impurity particles are believed to be responsible for the formation of the nanometer scale holes which have been observed by Gerber et al (1991) and Schlom et al (1992) on laser ablated and DC or RF sputtered films.

**REFERENCES**

Aprile I M, Renner Ch, Erb A, Walker E and Fischer Ø 1995 Phys. Rev. Lett. **75** (14), 2754
Gerber C, Anselmetti D, Bednorz J G, Mannhart J and Schlom D G, 1991 Nature **350**, 279
Hawley M, Raistrick I D, Beery J G and Houlton R J, 1991 Science. **251**, 1587
Mielke F, Memmert U, Golubov A A and Hartmann U 1996 J. Vac. Sci. Technol. B **14** (2), 1224
Schlom D G, Anselmetti D, Bednorz J G, Broom R F, Catana A, Frey T, Gerber Ch, Güntherodt H J, Lang H P and Mannhart J, 1992 Z. Phys. B **86**, 163
Scholm D G, Anselmetti D, Bednorz J G, Gerber C and Mannhart J, 1994 J. Crystal Growth **137**, 259

Inst. Phys. Conf. Ser. No 153: Section 6
Paper presented at Electron Microscopy and Analysis Group Conf. EMAG97, Cambridge, 1997
© 1997 IOP Publishing Ltd

# An attempt at conducting AFM imaging of electron damaged aluminium trifluoride

**I R Smith and A L Bleloch**

Microstructural Physics, Cavendish Laboratory, Madingley Road, Cambridge CB3 0HE

**ABSTRACT:**   Samples of thin aluminium trifluoride films, evaporated onto a AuPd buried electrode, were electron damaged in a HB501 STEM at various exposure doses. A home made conducting AFM was then used in an attempt to study the electrical properties of these damaged regions.  EF images acquired in the STEM show widely dispersed metallic aluminium particles, believed to be protected by an aluminium oxide or aluminium fluoride coating.  Topographical AFM images of such areas were obtained and the resistivity of electron damaged $AlF_3$ was in excess of $10^{11}\Omega nm$.

## 1.    INTRODUCTION

Since its invention the atomic force microscope (AFM) has fathered many novel scanning probe techniques which, as well as imaging the surface of a sample, can also provide additional physical information (e.g. spatial variations in capacitance, magnetic force or thermal conductivity).  The ability of an AFM, with a conducting tip and cantilever, to perform localised contact conductivity measurements allows us to investigate the electronic properties of a material on the nm-scale, whilst simultaneously measuring the surface topography.  Recently this technique has been used by Dai et al (1996) to determine the resistivity of individual carbon nanotubes and by Fukuda et al (1997) to study the resonant tunnelling through a single Si quantum dot.

The aim of this study is to investigate the potential use of a conducting AFM to characterise and resistivity map regions of electron damaged aluminium trifluoride ($AlF_3$), which consist of metallic particles embedded in an insulating medium (Walsh 1989).  The damage of $AlF_3$ by an electron beam is of interest due to its potential use in nanolithography. Phahle (1977) has shown that the conduction process in amorphous, undamaged $AlF_3$ at room temperature can be explained on the basis of the Poole-Frenkel mechanism, as described by Hill (1971).  However, electron transport in e-beam damaged $AlF_3$ films has not been studied to date.

## 2.    EXPERIMENTAL DETAILS

The samples were prepared by thermally evaporating an electron transparent but electrically conducting AuPd layer onto a holey carbon film, followed immediately by the thermal evaporation of a thin 10-20nm film of $AlF_3$ at a rate of 0.7nm/s.  The $AlF_3$ was thermally heated beforehand to remove any absorbed water, at 70°C for 48 hours.

The $AlF_3$ layer was then electron beam damaged in a VG HB501 scanning transmission electron microscope (STEM) operating at 100kV, at various exposure doses such that the $AlF_3$ acts as a self-developing positive or negative resist.  Nanometre-scale holes and lines can be directly formed in amorphous $AlF_3$ by the finely focused, high current density electron beam provided by the STEM (Kratschmer and Isaacson 1986).  Lines of 1μm length and 20nm width were drawn, with a line dosage of approximately 8pC/nm.  At the edges of these lines, and at lower electron doses, the irradiated $AlF_3$ dissociates releasing the halide as a gas, leaving behind discontinuous aluminium as nm-sized metallic particles in a reduced aluminium fluoride matrix. Such a disordered conductor-insulator composite was formed by electron damaging rectangular areas (~0.4 $\mu m^2$) so that the received total area dose was approximately 1pC/$nm^2$. All samples, if stored, were kept in vacuum and the specimen was rejected if any sign of carbon contamination was observed whilst imaging in the STEM.

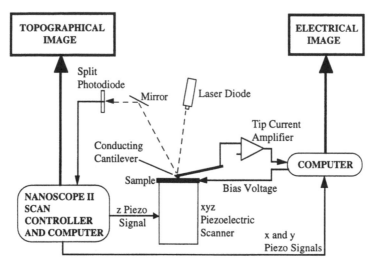

Fig. 1. Schematic diagram of our AFM for simultaneous topographical and electrical data.

A commercial, contact mode AFM (Nanoscope II, Digital Instruments), operating in air, was modified by Bond (1994) to act as a conducting AFM (Fig. 1). Topographical images were acquired in constant force mode in the usual way for the Nanoscope II. The current images were obtained by applying a constant voltage (maximum range ±10V) to the sample and measuring the current flow with a standard pre-amplifier, as well as recording the x and y piezo voltages from the Nanoscope II scan controller to locate the tip's position on the surface. The home made current pre-amplifier had a maximum gain of $10^7$V/A and system calibration showed a linear response for a wide range of resistance values (1k$\Omega$ to 10 M$\Omega$). The conducting probe was fabricated by thermally evaporating a 10nm adhesion layer of NiCr and 50nm Au layer onto commercially available $Si_3N_4$ cantilevers (200$\mu$m V-shaped cantilever with an integrated square pyramidal tip, apex radius of curvature 20-50nm). Electrical connections to the cantilever and to the AuPd layer (which acts as a buried electrode) were made with conductive silver paste and short coaxial cables, to minimise pick-up noise. The sample and cantilever holders were insulated from their surroundings and a voltmeter used to verify all electrical contacts. The current noise level for the system was approximately ±0.3nA.

## 3. RESULTS AND DISCUSSION

Fig. 2a shows a STEM annular dark field (ADF) image of part of a damaged $AlF_3$ area. Energy filtered (EF) imaging, using a ~3eV window positioned at the 15eV bulk Al plasmon energy, allows us to view the nm-sized metallic Al particles; past work by Smith (1997) is believed to show that some of the Al particles are either protected by the surrounding, reduced aluminium fluoride matrix or acquire a stable oxide coating around a metallic core, in a similar way to the Al particles grown and studied by Bagnell et al (1995).

Fig. 2b is a EF image of part of the area seen in Fig. 2a, after exposure to the laboratory atmosphere for 20 hours, still clearly indicating the presence of metallic Al particles. Fig. 2c shows a topographical AFM image of a damaged area similar to that seen in Fig. 2a. Attempts to obtain I-V characteristics or resistivity maps, showing contrast between the numerous damaged and undamaged AlF$_3$ regions, were unsuccessful in all the samples tested. The resistivity was therefore in excess of $10^{11}\Omega$nm at room temperature, assuming a contact spot radius > 10nm. The EF images show that the Al particles are widely dispersed, so perhaps it is not surprising that there appears to be no physically connected, conducting network. However, this does not explain the lack of any single tunnelling path (or weak electrical connection) between the isolated Al metal particles, the Au tip and the buried AuPd electrode. The presence of a surface aluminium oxide layer, or encapsulating aluminium fluoride matrix, may provide an explanation for the rarity of such paths, due to the increased resistance of possible tunnel junctions; further investigations will require an improvement in the sensitivity of the home made current amplifier.

Ambient conducting AFM experiments, although providing an elegant means to locate and study e-beam damaged regions, do have certain limitations. Imaging at high bias voltages can cause problems due to the magnitude of the localised electric field at the probe apex, which may lead to blunting of the tip or the degradation of both the tip and sample due to a static discharge. High quality, conducting tips are crucial for reproducible, reliable nanoprobe measurements. The Au coated cantilevers are easy to fabricate and lack any surface oxide; unfortunately the mechanically soft Au layer will suffer frictional tip wear or rupture by the high stresses present at the tip apex during AFM imaging, leaving only

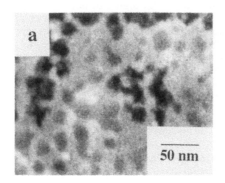

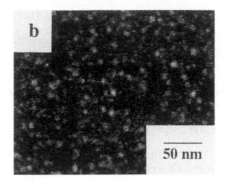

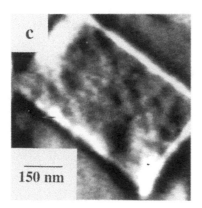

Fig. 2: Rectangular area of electron damaged AlF$_3$.
(a) ADF image obtained in the STEM immediately after electron exposure.
(b) EF image of the same area, after 20 hours exposure to the atmosphere, at the bulk Al plasmon energy (15eV).
(c) Topographical AFM topview image of a similar area to those shown above. The full z-range was 16nm.

insulating $Si_3N_4$ in contact with the sample. Testing of the tip on freshly cleaved, highly oriented pyrolytic graphite (HOPG) was carried out before and after each experiment to ensure no significant changes had occurred in the ohmic nature of the tip conductivity, otherwise the cantilever was discarded.

The possibility of an insulating surface film on the sample also needs to be considered. A carbon contamination layer, acquired when the $AlF_3$ was e-beam damaged in the STEM, may be present. The raised edges of the damaged area seen in Fig. 2c may be due to such carbonaceous contamination and this will be investigated further. Also it is important to realise that both the tip and sample will accumulate a liquid film, due to the hygroscopic nature of $AlF_3$ and the capillary condensation of vapor from the atmosphere. Beale et al (1994) has shown that a few monolayers of such material on metallic surfaces can make a dramatic difference to the contact resistance, especially with the low forces (~100nN) typically exerted by an AFM working in air.

## 4. CONCLUSIONS

It has been shown that electron damaged thin films of $AlF_3$ consist of metallic Al particles that, at best, are only weakly electrically linked. Exposure to the atmosphere is believed to cause the formation of a surface oxide layer, unless the particle is protected by a reduced layer of aluminium fluoride. It is speculated that this may account for the inability of the conducting AFM technique to acquire a nm-scale resistivity map of such damaged regions.

Although precautions were taken to counter the main problems faced by conducting scanning force microscopy, modifications to the experimental apparatus are still clearly required. Improvements to the current amplifier to help increase the signal-to-noise ratio (or the use of a lock-in amplifier) should enable the detection of pA currents. O'Shea et al (1995) has shown that tip wear may be greatly reduced if the conducting probes are fabricated by the chemical vapor deposition of boron doped diamond onto the $Si_3N_4$ cantilevers. Insulating surface films may be avoided by operating the AFM in vacuum, an ultra-dry nitrogen atmosphere or a non-polar liquid environment. However, the difficulty in not exposing the specimen at some stage to air and so forming physisorbed surface contamination layers, along with the other problems briefly mentioned above, make such low-level current measurements an experimentally demanding task.

## ACKNOWLEDGEMENTS

The authors would like to thank the Engineering and Physical Sciences Research Council and Downing College, Cambridge for their financial assistance.

## REFERENCES

Bagnall K E, Vincent R, Midgley P A, Saunders M and Steeds J W 1995 Inst. Phys. Conf. Ser. **147**,183
Beale J and Pease R F 1994 IEEE Trans. Comp. Pkg. Man. Tech. A **17**, 257
Bond S F 1994 PhD Thesis, University of Cambridge
Dai H, Wong E W and Lieber C M 1996 Science **272**, 523
Fukuda M, Nakagawa K, Miyazaki S and Hirose M 1997 Appl. Phys. Lett. **70**, 2291
Hill R M 1971 Philos. Mag. **23**, 59
Kratschmer E and Isaacson M 1986 J. Vac. Sci. Technol. B **4**, 361
O'Shea S J, Atta R M and Welland M E 1995 Rev. Sci. Instrum. **66**, 2508
Phahle A M 1977 Thin Solid Films **46**, 315
Smith I R 1997 PhD Thesis in preparation, University of Cambridge
Walsh C W 1989 PhD Thesis, University of Cambridge

*Inst. Phys. Conf. Ser. No 153: Section 6*
*Paper presented at Electron Microscopy and Analysis Group Conf. EMAG97, Cambridge, 1997*
© 1997 IOP Publishing Ltd

# Air plasma and corona discharge interactions with PVC surfaces studied by AFM and XPS

## V S  Joss and C Kiely

Department of Engineering (Materials Science Division), Liverpool University, Liverpool.

**ABSTRACT:** Tapping-mode atomic force microscopy (TMAFM) and X-ray photoelectron spectroscopy (XPS) have been used to study the effects of low power air plasma and corona discharge treatments on PVC surfaces. Despite the similarities in the chemical changes that occur, the topographical nature and structural integrity of these treated surfaces are shown to be very different. The plasma is only capable of interacting the outer surface of the polymer whereas the u.v. generated in the corona discharge penetrates into its sublayers.

## 1.     INTRODUCTION

Many  polymers require a pretreatment to improve the adhesive properties of their surface.  Such treatments range from washing the polymers in oxidising liquids to exposing the polymer to glow discharge or plasma.  The main aim of these treatments is to improve the wettability of the surface, introduce functional groups which can chemically interact with the substrate, and/or improve the structural integrity of the surface by removing weakly bound layers or crosslinking the surface.

The commodity polymer, poly (vinylchloride) (PVC), will undergo rapid degradation unless stabilisers are added to the formulation.  These low molecular weight additives prevent the discoloration of the PVC film resulting from the chemical breakdown of the polymer chain initiated by heat or u.v. light.  Unfortunately, they have a tendency to migrate to the polymer surface creating a weak boundary layer thereby reducing the adhesive properties of the surface

X-ray photoelectron spectroscopy (XPS) and tapping mode atomic force microscopy (TMAFM) have been used to monitor the effects of two types of surface treatment, plasma and corona discharge, on PVC.  The first is a novel low power plasma treatment in which the predominant reactive species are ions, mainly $O_2^+$ (O'Kell et al 1995).  The second is a corona discharge treatment, a more aggressive environment in which the polymer is exposed to significant quantities of ozone, oxygen radicals and ultraviolet radiation in addition to any ions produced.  Our experiments have included a study of the interaction of both discharges with additive free film to identify the interaction with the base polymer.  These results have then been compared to those of the plasticised film.  Ageing studies have also been carried out.

## 2.     EXPERIMENTAL

Additive free PVC samples were prepared from a ~1% solution of pure, powdered PVC resin in tetrahydrofuran.  A few drops of the PVC solution were spin-coated onto the surface of a clean glass cover slip. Plasticised PVC film, containing 3% by weight of Ba/Zn stabiliser, was supplied by Pilkington Technology Research.

Plasma treatment was carried out using a half-wave helical resonator system. The unique design of the system enabled the plasma to be ignited and sustained at very low input powers (Jones and Sammann 1990).  Samples were treated for 60 seconds in an air plasma at 0.1 Torr using a gas flow rate of 85 sccms.  Corona discharge treatments were performed using a custom made treatment rig whereby the corona was generated between the electrode bar and

218

the dielectric plate situated on a moving platform. Each sample was treated for 10 seconds at an input power of 200W.

X-ray photoelectron spectroscopy (XPS) was undertaken using a VG Microlab Mk II equipped with an aluminium Kα X-ray source (1486.6eV). Degradation of the PVC film was limited by minimising the collection time for each spectra. The surface topography of the PVC samples was investigated by Tapping-mode atomic force microscopy (TMAFM) using a Digital Instruments Nanoscope III.

## 3.    RESULTS & DISCUSSION

The surface chemistry of  the two formulations of PVC were found to be very different.  XPS analysis revealed that poly (vinylchloride) was not a main constituent of the surfaces of the plasticised polymers: the Cl2p signal intensity being 1.5% for the Ba/Zn stabilised film as compared to 35% for the spin coated film. A Ba3d signal was also detected (2.5%). The majority of the C1s and O1s signal intensity originated from the plasticiser.

Imaging the polymer surfaces using tapping mode AFM showed that spin coating the polymer onto glass produced a relatively smooth film with a peak-to-trough height of approximately 0.4nm and mean roughness of 0.2nm. Obtaining  reproducible images from the plasticised films was difficult due to the presence of *migratory* additives on the surface.

### 3.1    Air Plasma and Corona Discharge Treatment of Spin-coated PVC

Although the reactive species in the plasma and corona are different, the chemical changes induced on spin coated PVC are very similar.  Both treatments introduce small amounts of carbonyl and carboxyl functionality onto the polymer surface.  Plasma treatment, however, causes some dechlorination.  This is where the similarity ends.  TMAFM images of

Figure 1. TMAFM Images of a. Air Plasma Treated and b. Corona Discharge Treated Spin-Coated PVC   (1μm x 1μm x 100nm)

these treated surfaces are shown in Fig. 1a and b.  The plasma treated surface comprises regular conical features of approximately 8.3nm in height and 75.2nm in diameter whereas the corona treated surface contains larger, irregular globular features 13.8nm in height and 246.3nm in diameter, superimposed on a roughened surface of average peak to trough height 2.8nm.

### 3.2    Air Plasma and Corona Discharge Treatment of Plasticised PVC

For the Ba/Zn stabilised films both plasma and corona treatments resulted in a reduction of the Ba 3d signal intensity from 2.5% to 0.5%. An increase in the Cl2p signal was also detected from 1.5% to 1.9% after plasma treatment and from 1.5% to 5.8% after corona treatment. This suggests that corona treatment is more effective at removing migratory additives (plasticiser as well as stabiliser). The TMAFM images shown in Figure 2 show the extent to which the topography of the surface has changed. Plasma treatment clearly removed some of the softer migratory component of the untreated film. None of the conical features present at the surface of air plasma treated spin coated films were evident, the treatment has

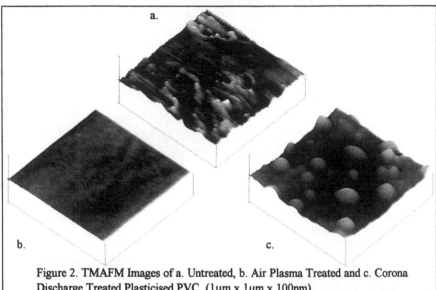

Figure 2. TMAFM Images of a. Untreated, b. Air Plasma Treated and c. Corona Discharge Treated Plasticised PVC (1μm x 1μm x 100nm)

merely smoothed the surface of the film. In contrast, the globular features of the corona treated spin coated film are also present on these treated plasticised films. It is suggested that these features are a result of chain scission caused by the interaction with ultraviolet radiation from the corona discharge and subsequent reorganisation of the polymer chains.

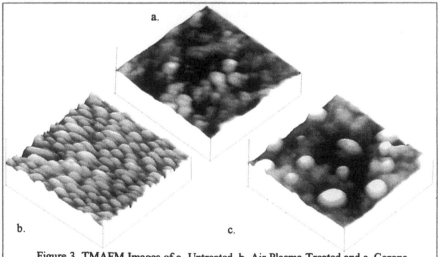

Figure 3. TMAFM Images of a. Untreated, b. Air Plasma Treated and c. Corona Discharge Treated Washed Plasticised PVC (1μm x 1μm x 100nm)

Another way of removing migratory additives is to wash the polymer surface with a suitable solvent. Soaking the polymer in trichloroethylene for 10 minutes resulted in an increase of the Cl2p signal from 1.5% to 16%, accompanied by a reduction in the Ba3d signal from 2.5% to 0.9%. The resulting washed surface, therefore, constitutes a sizable proportion of the PVC polymer. This translates to the topographical changes that occur at these washed polymer surfaces during corona and plasma treatment and can be seen in Fig. 3. The images are similar to those of the treated spin coated films. The plasma treated surfaces are prone to ageing as the additives migrate to the surface. These additives fill the valleys on the polymer surface and gradually the surface reverts to that of the untreated film which is shown in Fig. 4.

## 4.    CONCLUSIONS

In comparing the effects of two different surface treatments on a polymer it is not just the chemistry that needs to be taken into consideration. In this paper, we have shown that, although the chemical changes may be similar, great variations in the topographical nature and structural integrity of the surface can also occur. Low power plasma and corona treatment differ enormously. The ions generated in the plasma do not have enough energy to penetrate deep into the sublayers of the polymer: reaction takes place only on the surface. During corona treatment, the polymer is subjected to ultraviolet radiation

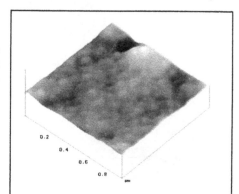

Figure 4.  Air Plasma Treated, Washed Ba/Zn Stabilised PVC (Aged 2 Days). (1μm x 1μm x 100nm)

as well as reactive ions and radicals. This u.v radiation penetrates deep into the surface of the polymer causing chain scission and subsequent rearrangement of the polymer chains which accounts for the globular features detected on all corona treated samples. Washing the surface of the plasticised polymer with trichloroethylene removes the additives and allows the plasma treatment to become more effective. However, the stability of these treated surface is poor, with additives migrating to the surface after a few days.

## ACKNOWLEDGEMENTS

The authors gratefully acknowledge Dr. Steve Pringle for his work involving plasma characterisation. This work was sponsored by the E.P.S.R.C. and Pilkington Technology Research..

## REFERENCES

Jones C and Sammann E 1990 Carbon 28(4) 515.
O'Kell S, Pringle SD, and Jones C Inst. Phys. Conf. Ser. No 147: Section 7, EMAG 95, Birmingham, 1995, 251.

Inst. Phys. Conf. Ser. No 153: Section 7
Paper presented at Electron Microscopy and Analysis Group Conf. EMAG97, Cambridge, 1997

# Probing local strain fields using electron diffraction in the scanning electron microscope

Angus J Wilkinson

Department of Materials, University of Oxford, Parks Road, Oxford, OX1 3PH, UK

ABSTRACT: This paper describes two scanning electron microscope based electron diffraction methods: electron channelling contrast imaging (ECCI) and electron back scatter diffraction (EBSD). Both methods can be used to probe local strain fields in bulk materials at the microscopic scale. ECCI allows dislocations and their collective long range strain fields to be imaged and qualitatively characterised. EBSD patterns can be used to give a quantitative measurement of elastic strain.

## 1. INTRODUCTION

Diffraction of electrons transmitted through thin foils provides the basis of a range of techniques that is an essential part of materials characterisation. Diffraction effects are of course still present in the scanning electron microscope (SEM) however the use of bulk specimens leads to significant multiple scattering and large energy losses which result in low contrast levels. Despite this SEM-based electron diffraction techniques can readily provide a wealth of information. This paper describes the two techniques of electron channelling contrast imaging (ECCI) and electron back scatter diffraction (EBSD). The emphasis will be on their use in probing local defect and strain field distributions, which will be illustrated with images and measurements from strained semiconductor epilayers on planar and patterned substrates and from fatigued metals.

## 2. ELECTRON CHANNELLING CONTRAST IMAGING

### 2.1 Mechanism and Methods

The incident electron beam can be described by a superposition of Bloch waves within the crystal and close to the Bragg condition two such waves are strongly excited. The type one wave has regions of high probability density close to the atomic sites in the lattice and is rapidly attenuated through phonon scattering processes. Large angle scattering of the incident electrons due to interaction with the heavy atomic nuclei is thus promoted and the yield of back scattered electrons (BSEs) is increased. Conversely the type two waves has high probability density centred on the channels between atomic sites and so is less rapidly attenuated. The type one wave is preferentially excited at angles less than the Bragg angle (negative deviation parameter) leading to channelling bands of raised intensity when the incident beam is rocked to form electron channelling patterns (ECPs). At larger angles (positive deviation parameter) the type two wave dominates and the intensity is decreased.

At the Bragg condition the two waves are equally excited and their excitation varies rapidly with the diffraction plane orientation. The local strain field around lattice defects such as dislocations is sufficient to tilt the diffraction planes and affect a local modification of the deviation parameter. Thus dislocations and other defects can be imaged through their modulation of the BSE signal when a bulk crystalline specimen is suitably oriented.

Although single dislocation lines can be imaged with this technique (Wilkinson & Hirsch 1995) the electron optical conditions are stringent and require a high brightness electron source. For imaging groups of dislocations and longer range strain fields discussed in this paper larger beam diameters can be used and SEMs with thermionic emitters ($LaB_6$ or even W) give good results. The images described here were all obtained from bulk specimens at beam energies from 20 keV to 40 keV, and with beam currents from 1 nA to 10 nA. A BSE detector consisting of a scintillator, light guide, and photomultiplier tube was used for most of the work. This was positioned so as to capture BSEs emitted at low take-off angles from highly tilted specimens. A conventional annular solid state diode detector mounted below the objective lens was used for the work discussed in section 2.3 where the specimen was at close to normal incidence. The diffraction conditions were established using ECPs to correctly orient the specimen using a tilt/rotate goniometer.

More details (experimental and theoretical) can be found in a recent review (Wilkinson & Hirsch 1997) and references therein.

## 2.2    Misfit Dislocations in Strained $Si_{1-x}Ge_X$ on Si

During the early stages in the relaxation of strained $Si_{1-x}Ge_x$ epilayers on (001) Si, misfit dislocations are produced in groups at the interface. The groups of dislocations can be readily imaged using ECCI even when the dislocations are at depths greater than that from which the diffraction contrast arises. In such cases the image is formed due to the long range strain field for which the group of lattice defects can be considered to be a single superdislocation with the same net Burgers vector. The form of the image contrast has been studied using two beam dynamical diffraction theory (Wilkinson & Hirsch 1995). It was found that the contrast was strongly dependent on surface stress relaxation effects. As an example fig. 1 shows intensity profiles calculated for a superdislocation 1 μm below the (001) surface of a Si specimen, with the Burgers vector **b** parallel to the diffraction vector **g** = 220. For the dotted curve displacements for a dislocation in an infinite medium were used, while for the dashed curve an image dislocation was included. These images differ considerably from the result when the surface stress relaxation is fully implemented as shown by the solid curve. The stress relaxation gives a marked increase in the contrast and even reverses the sense of the contrast compared to what would be expected based on a dislocation in an infinite medium.

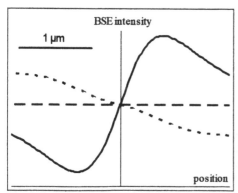

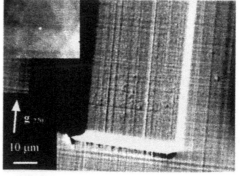

**Fig. 1:** Intensity profiles calculated for superdislocation 1 μm below surface.

**Fig. 2:** ECCI showing misfit dislocations in a $Si_{0.85}Ge_{0.15}$/Si mesa.

Examples of such defect bunches at the interface between a 1 μm thick epilayer of $Si_{0.85}Ge_{0.15}$ on Si (001) can be seen in fig. 2. In this case the Si substrate had been patterned with raised pads or mesas prior to growth of the epilayer. ECCI allows the dislocation distribution to be imaged and characterised over large areas of the as grown structures, while the mesa height of 3 μm effectively prohibits preparation of plane view TEM foils. In rectangular mesas with (110) and ($\bar{1}$ 10) sides and a 1 μm thick $Si_{0.85}Ge_{0.15}$ epilayer, no misfit dislocations were seen to propagate parallel to sides that were less than 10 μm apart. The presence of dislocations along the orthogonal <110> direction indicated that dislocation nucleation sites were present, and suggested that the stress had relaxed to a level too small to propagate dislocations along the mesa length. Further analysis of this stress relaxation using EBSD will be presented in section 3.3.

## 2.3    Dislocation Substructures in Fatigued Cu

Further examples of the use of ECCI to observe groups of dislocations are taken from recent work on fatigued Cu single crystals (Ahmed, Wilkinson & Roberts 1997). Crystals oriented along [541] for single slip were prepared with square cross-section with one pair of side faces parallel to ($\bar{1}$ 11) containing the primary [101]/2 Burgers vector. Specimens were cycled under strain control and examined in the SEM at various stages through the fatigue life. Roughening of the ($\bar{1}$ 2 $\bar{3}$) faces can be observed using secondary electrons, while ECCI observations on the ($\bar{1}$ 11) faces allow the dislocation substructure to be examined. Fig. 3 gives examples from a specimen cycled at a constant plastic strain amplitude of $\pm 10^{-3}$. After 3640 cycles the shear stress approached the saturation value of 28 MPa, and at this point ECCI revealed a structure consisting of veins of high dislocation dipole density (fig. 3a). After cycling for a further 710 cycles the stress saturated at 28 MPa, the strain localised in persistent slip bands (PSBs), and ECCI showed (fig. 3b) that the dislocations had locally rearranged into the well known 'ladder' structure within the PSBs.

Future work will investigate the stability of such structures to further cycling and how the structures influences the initiation and early growth of cracks.

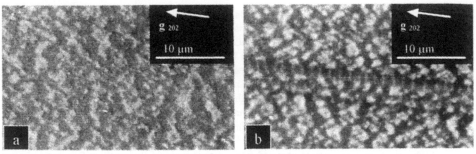

**Fig. 3:** Dislocation substructures in fatigued Cu. Regions of high dislocation dipole density appear in bright contrast. (a) initial veined matrix. (b) PSBs with ladder structure.

## 2.4    Strain Fields near a Fatigue Crack Tip

Observations have also been made of plastic zones around fatigue cracks grown in single crystal SRR99 Ni-based superalloy specimens (Wilkinson, Henderson & Martin 1996). The example given here is a fatigue crack grown in the [010] direction with an (001) crack plane, at a temperature of 650°C, a load ratio of 0.1, and a frequency of 10 Hz. The crack was grown at a constant ΔK of 23 MPa√m corresponding to the mid-Paris regime of the crack growth rate curve. After crack growth, a mid-plane section was cut and polished for the ECCI observations. Fig. 4 shows the crack tip strain field imaged using ECCI with g perpendicular to the crack plane. A pair of images is shown between which g has been reversed by tilting the specimen through a few degrees. A marked change in contrast is evident, indicating that the contrast is crystallographic in origin. The contrast inverts on

224

crossing the crack plane, and also on moving ahead of the crack tip from the crack wake. The sense of the contrast in these and images taken with other **g** vectors allows the sense of the lattice plane tilting and hence strain to be determined.

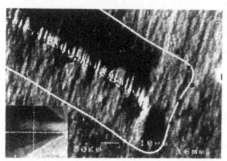

**Fig. 4:** Strain contrast from fatigue crack plastic zone inverts on reversing the sense of **g**.

## 3. ELECTRON BACK SCATTER DIFFRACTION

### 3.1 Mechanism and Methods

EBSD patterns are Kikuchi patterns formed by diffraction of the BSEs as they leave the specimen after some initial (single or cumulative) large angle scattering. They are similar in appearance to ECPs to which they are related through the reciprocity theorem. In EBSD patterns the incident beam direction is held constant and the variation of BSE flux with exiting angle is recorded, while for ECP the incident beam direction is varied and the BSE flux integrated over a fixed angular range.

In practise EBSD patterns are recorded by placing a scintillator screen close to a highly tilted sample and viewing this screen through a lead glass window with a low light level video camera mounted outside the SEM vacuum chamber. The video camera is readily interfaced with a personal computer, so that image analysis and measurements can be performed conveniently on the patterns obtained. The angular width of EBSD patterns is generally large (>60°) and is simply controlled by the size and position of the scintillator relative to the specimen.

The most widespread use of EBSD is for the measurement of crystal orientation which is having a marked impact on the study of texture as has been reviewed by Dingley & Randle (1992). Orientation measurement is not discussed further in this paper.

Elastic strain measurements have been made from EBSD patterns by Troost *et al* (1993) and Wilkinson (1996, 1997). The strain is obtained from measurement of the small shift in zone axes positions away from their expected positions in a pattern from a reference crystal at a known strain. To achieve good sensitivity the specimen to screen distance is increased so that patterns with reduced angular width but increased angular resolution can be obtained. The specimen is then oriented to bring a selected zone axis onto the screen, and a series of patterns recorded from the region of interest, including one from the reference crystal. Cross-correlation methods are then used to establish the shift of the zone axis away from its position in the pattern from the reference crystal. Repeated measurements indicated that the zone axis shift could be measured with a sensitivity of ±0.1 mrad. The specimen is then reoriented and the procedure repeated so as to determine the shift in positions of several different zone axes. The measured shifts in the zone axes positions reflect the distortion of the crystal from the shape of the reference crystal and can be used to calculate the local strain.

There are 9 degrees of freedom describing a general strain and rotation, and in principle, 8 of them can be found from this analysis, using measurements at 4 zone axes. The remaining degree of freedom is the hydrostatic dilatation of the lattice which does not affect the positions of the zone axes, (but should give a small change in the width of the Kikuchi bands).

The following two sections give results from measurements on $Si_{1-x}Ge_x$/Si heterostructures.

## 3.2 Strains in $Si_{1-x}Ge_x$ Epilayers on Plannar Si

Specimens comprising of a single $Si_{1-x}Ge_x$ epilayer on Si were prepared for 4 different nominal Ge concentrations. ECCI observations showed that misfit dislocations were only being present as isolated crosses presumably associated with a low density of coarse growth defects. The epilayer was etched away to reveal the Si substrate on part of each specimen. For each specimen pairs of patterns were recorded, one from the epilayer and the reference substrate, with the specimen in various different orientations.

The only non-zero elements of the tensor (**A**) describing the epilayer strain are on the leading diagonal and are related to a single parameter $\varepsilon$ ($a_{11} = a_{22} = \varepsilon$, and, $a_{33} = -2v\varepsilon/[1-v]$). $\varepsilon$ can be found from measurement of a single zone axis shift. EBSD measurements were made at several zone axes for each specimen and the average and the standard deviation of the strain is given in table 1 which also compares the EBSD results with those from x-ray diffraction. The EBSD measurements are in good agreement with the x-ray data and the standard deviations show a strain sensitivity of $\pm 2 \times 10^{-4}$.

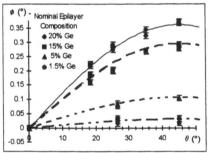

**Fig 5:** zone axis shifts ($\phi$) as a function of zone axis inclination ($\theta$).

| nominal composition | EBSD Strain (%) | EBSD Std. Dev. (%) | X-ray Strain (%) |
|---|---|---|---|
| 20% Ge | -0.73 | 0.03 | - |
| 15% Ge | -0.56 | 0.03 | -0.5936 |
| 5% Ge | -0.21 | 0.01 | -0.2178 |
| 1.5% Ge | -0.06 | 0.02 | -0.0635 |

**table 1:** strain in $Si_{1-x}Ge_x$ epilayers on Si

With this simple strain tensor the size ($\phi$) of the shift in the zone axis position is expected to increase to a maximum as the inclination ($\theta$) of zone axis direction from the interface normal is increased from 0° to ~45°. Fig. 5 shows that the data follow the expected variation indicated by the curves calculated from the strain values given by x-ray diffraction.

## 3.3 Strains in $Si_{1-x}Ge_x$ Epilayers on Patterned Si

In section 2.2 it was noted that when the sides of $Si_{0.85}Ge_{0.15}$/Si mesas were close enough together the propagation of misfit dislocation was inhibited. The finite width of the structure removes the lateral constraint and allows the epilayer to bow outward to accommodate its misfit from the substrate. This lowers the overall strain energy so that thick epilayers can be grown without the need to form misfit dislocations. Direct evidence of this elastic relaxation of the strain energy was obtained using EBSD. The shift of the positions of zone axes were measured as a function of distance across long rectangular mesas of several widths, at two epilayer thicknesses. Regions toward the edge of the mesa showed shifts in the zone axes positions that were consistent with the bowing out of the epilayer to accommodate its larger natural lattice parameter. The width of the relaxed regions was greater for the thicker epilayer. Fig. 6 gives an example for a 4 μm wide mesa with a 0.2 μm thick epilayer. In this case the relaxed region is confined to within ~1 μm of the mesa edge. Toward the centre of the mesa the position of the zone axes remained fixed showing a constant strain. As the thickness is increased the relaxed regions spread in from either side and eventually can overlap at the centre. Such a result is shown in fig. 7 for a 5 μm wide mesa with a 1 μm epilayer. For these narrower mesas the

226

relaxation spreads across the entire width of the mesa before the strain energy becomes high enough to nucleate misfit dislocations. Further growth then accumulate little additional strain energy and the epilayer can be grown to considerable thickness without the generation of misfit dislocations.

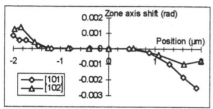

Fig. 6: evidence for bowing out of 0.2 μm thick epilayer on 4 μm wide mesa.

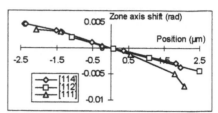

Fig 7: evidence for bowing out of 1 μm thick epilayer on 5 μm wide mesa.

## 4. CONCLUSIONS

In an SEM with a thermionic emitter the ECCI technique allows groups of lattice defects and their cumulative long range strain fields to be imaged and characterised. The technique is quick and allows large areas to be examined with minimal specimen preparation and allows the evolution of defect distributions, with for example fatigue cycles, to be followed in the same specimen. Analysis of the image contrast gives qualitative information about the sense of the local lattice plane tilting.

The EBSD method is more time consuming to apply, but gives the benefit of quantitative determination of local strains and small misorientations.

Despite the inferior spatial resolution, the ability to work with bulk specimens gives these SEM based techniques a powerful role to play at the mesoscale, at which many material properties are controlled.

### ACKNOWLEDGEMENTS

I am grateful to the Royal Society for their support through a University Research Fellowship. I thank the EPSRC for funding the development of these techniques, and the RMS for sponsoring my attendance at the EMAG'97 meeting.

### REFERENCES

Ahmed J, Wilkinson A J & Roberts S G 1997 Phil. Mag. Let. in press
Dingley D J & Randle V 1992 J. Mater. Sci. 27,4545
Wilkinson A J , Henderson M B & Hirsch P B 1996 Phil. Mag. Let. 74, 145
Wilkinson A J & Hirsch P B 1995 Phil. Mag. A 72, 81
Wilkinson A J & Hirsch P B 1997 Micron in press
Wilkinson A J 1996 Ultramicroscopy 62, 237
Wilkinson A J 1997 J. Mater. Sci. & Technol. 13, 69
Troost K Z, van der Sluis P & Gravesteijn D J 1993 Appl. Phys. Let. 62, 1110

*Inst. Phys. Conf. Ser. No 153: Section 7*
*Paper presented at Electron Microscopy and Analysis Group Conf. EMAG97, Cambridge, 1997*
© *1997 IOP Publishing Ltd*

# Strategies for achieving sub-nanometre depth resolution in secondary ion mass spectrometry (SIMS)

**D S McPhail**

The Department of Materials,
Imperial College of Science, Technology and Medicine,
Prince Consort Road,
London SW7 2BP UK.

**ABSTRACT:** The factors that limit the depth resolution in SIMS are discussed and experimental strategies for overcoming these limitations and achieving sub-nanometre depth resolution explored. **Beam induced mixing processes** are shown to be the main limitation to sub-nanometre depth resolution in some materials whereas **surface and beam induced topography** dominate in others. The development of a new generation of low energy ion beams and of novel analysis strategies such as the 'bevel-and-linescan' approach and the imaging of focused-ion-beam (FIB) milled cross-sections are described.

## 1. INTRODUCTION

Secondary Ion Mass Spectrometry is a surface analysis technique that uses a combination of sputtering and secondary ion emission to generate three-dimensional chemical maps of the distribution of elements within a solid (Benninghoven et al 1987). In-depth analysis or **depth profiling** involves the measurement of the concentration of the elements of interest as a function of depth and the depth resolution is a measure of the accuracy with which these distributions may be assessed. Typically the depth resolution is between a few nanometres and a few tens of nanometres. Measurements of the lateral distribution of the elements of interest is termed **imaging** and the development of fine-focus liquid metal ion sources has led to lateral resolutions of a few tens of nanometres. The excellent lateral and depth resolution together with a sensitivity of parts per million has led to applications in many areas of materials science. However, there is an on-going need for improvements in these figures-of-merit and new instrumentation and analysis procedures are continuously evolving. In this paper I will review the current developments in high resolution SIMS depth profiling.

The main impetus in the quest for sub-nanometre depth resolution in SIMS remains the reduction in device dimensions and the associated length metric in integrated circuit technology. Information is being sought on the shape of low energy ion implants with peaks only a few nanometres below the surface and on the spatial localisation and sheet densities of dopants confined in narrow quantum wells or even single atomic planes (delta doping). It should be emphasised, however, that sub-nanometre resolution will find many other applications, for example the study of slow corrosion, oxidation and diffusion processes in a wide range of materials

## 2. THE SIMS PROCESS

In the SIMS process the sample surface is irradiated with a beam of mono-energetic mass filtered **primary ions**, typically of energy $E_p$ in the range of 1keV to 30keV. These primary ions generate an

228

intense but short lived ($\sim 10^{-12}$s) **collision cascade (Fig. 1)** in the near surface as they are brought to rest by a series of collisions with the atoms of the matrix. Some small fraction of the atoms of the matrix in the cascade are directed upward with sufficient energy to overcome the work-function of the material and are **sputtered**. Thus SIMS is a destructive process and the sample is gradually consumed. The number of atoms of the matrix sputtered per incident primary ion is termed the **sputter yield** and is typically in the range 0.1-10 atoms per incident primary ion. The sputter yield decreases with beam energy and increases with angle of incidence $\phi$. It is different for each layer in a multi-layer structure. Some (usually small) fraction ($\alpha$) of the sputtered species are ionised, and some small fraction (**T**) of these secondary ions can be mass filtered and counted. For a species X, present at some concentration $\rho(X)$, with an ionisation probability $\alpha$ ($X^+$) then the number of ions of $X^+$ detected, $I(X^+)$, during the removal of an analytical volume V, is given by:

$$I(X^+) = \rho(X)V[\alpha (X^+)T(X^+)]\dots\dots\dots\dots\dots\dots\dots\dots\dots\dots\dots\dots\dots\dots\dots\dots\dots1$$

The product $[\alpha(X^+)T(X^+)]$ is termed the useful ion yield $Y(X^+)$ for that species and is typically in the region of $10^{-3}$ to $10^{-7}$ ions detected per atom sputtered. The useful ion yield of a species is independent of concentration in the dilute limit ($\rho<1\%$) however both sputter yield and ion yield vary from one matrix to another. In a depth profile **(Fig. 1)** the ion beam is scanned over a limited area of the sample surface so that a crater is milled down into the material. Selected ions are monitored sequentially and their intensities (viz. concentrations) recorded as a function of time (viz. depth).

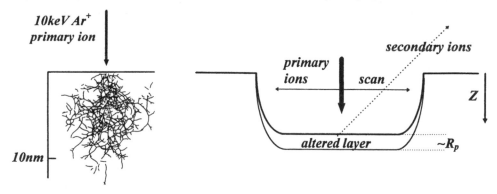

**Fig. 1** The collision cascade and the evolution of the SIMS crater in a SIMS depth profile

## 3.    FACTORS CONTROLLING THE DEPTH RESOLUTION IN SIMS

The depth resolution measures the accuracy with which an abrupt feature or interface can be resolved in a SIMS depth profile. The development of delta doping in semiconductors has produced ideal structures with which the depth resolution of the secondary ion mass spectrometer can be measured. The 'response curve' from a typical high energy SIMS measurement on a delta has a steep up-slope and a rather less pronounced down-slope **(Fig. 2)**. Often it is assumed that both the up-slope and down-slope follow an exponential dependence on depth and the depth resolution is quoted in terms of the depth, usually expressed in nanometres, over which the up-slope and down slope change by a factor of **e** (expressing the slope as $e^{\pm z/\lambda}$) or a factor of **10** (expressing the slope as $10^{\pm z/\lambda}$). It is most important to check which definition is being used as there is no consensus in the SIMS community. For the rest of this paper I will quote up-slopes and down-slopes in nanometres per decade.

A number of factors limit the depth resolution in SIMS. At the most fundamental level there are processes associated with the sputtering event that cannot be avoided. The sputtering process involves a series of discrete impacts at the sample surface, rather than the uniform removal of material one atomic layer at a time, so that some 'statistical' micro-roughness is inevitable. Oeschner in 1983 has

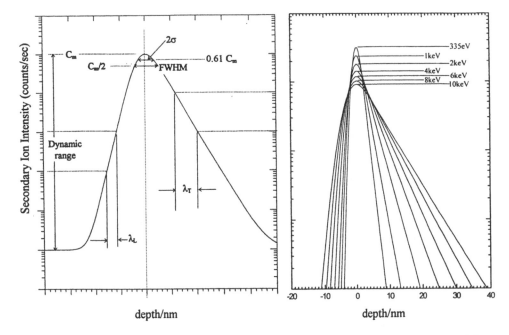

**Fig. 2.** The definitions of SIMS depth resolution. associated with the response curve from a delta doped sample

**Fig. 3.** The response curves measured from boron deltas in silicon at various primary beam energies using a FLIG gun.

estimated that the **micro-topography** will extend over five atomic planes. The information depth of the technique is less clearly understood as is the process of ion emission. However the limited number of studies that have been conducted suggest two or three atomic layers at most (see Hofmann 1994). What is clear from Eqn. 1 is that the signal intensity depends upon the analytical volume removed so that the **depth increment per data point** necessary to achieve adequate sensitivity will often impose a practical limit of several nanometres on the attainable depth resolution.

In practise, especially with semiconductors, it is often the beam induced mixing processes associated with the collision cascade that represent the dominant profile broadening mechanism. The region that is disturbed on ion impact can be regarded to a first approximation as a hemisphere of radius $R_p$, where $R_p$ is the range of the primary ion **(Fig. 1)** . The range can be expressed as (Wilson et al. 1989):

$$R_p = C\, E_p^{\,n} \cos \phi \dots\dots\dots\dots\dots\dots\dots\dots\dots\dots\dots\dots\dots\dots\dots\dots\dots\dots 2$$

where $\phi$ is the angle of incidence of the primary ion beam with respect to the surface normal. For example, for an $O_2^+$ primary ion beam in silicon C=2.15nm and n=1 so that the mixing associated with a 10keV $O_2^+$ ion is confined to a hemisphere ~22nm in diameter, whereas the mixing from a 1keV ion is confined to a hemisphere ~2.2nm in diameter. In a sputter depth profile a continuous zone of damage of thickness ~$R_p$, termed the altered layer, is established beneath the surface, as the analysis proceeds **(Fig. 1)**. The altered layer may be chemically distinct from the underlying material if the primary ion beam is reactive (for example oxygen at normal incidence in silicon generates a stoichiometric $SiO_2$ layer) with atoms from abrupt features such as deltas essentially mixed within the altered layer to an extent $R_p$. Reducing the beam energy or increasing the angle of incidence of the beam reduces $R_p$, the range of the primary ions (see Eqn. 2).

All of the discussion to date has assumed that the base of the crater is, and remains, perfectly flat and parallel to a smooth initial surface, throughout the analysis. In many materials, however, the surface topography is far greater than the range $R_p$ of the primary ions. For example, recent studies on superconducting multi-layers by Montgomery et al (1997a) have established that the resolution of the interfaces, and the shape of the signal from the interfaces, is controlled by the topography present in the etch pit as the analysis proceeds. This topography is often a combination of the initial surface topography and some beam induced topography.

## 4. STRATEGIES FOR IMPROVING THE DEPTH RESOLUTION IN SIMS

### Reduction In The Primary Beam Energy

There has been a considerable impetus from the semiconductor industry in recent years to reduce the primary beam energy in order to reduce the extent of the beam induced mixing processes ($\sim R_p$) It has been very difficult to extract useful currents (>10nA) at beam energies below 1keV because the extraction of ions from the source is limited by space charge effects. The available primary beam current fall with primary beam energy according to a power law (Child-Langmiur law), $I = kE^n$. This problem has led Dowsett and co-workers (1996) to develop a Floating Low Energy Ion Gun (FLIG). The key idea is that primary ions are transported at high energy down the ion column and are then retarded at the final lens. The FLIG is capable of delivering approximately 60nA at 250eV focused into a beam approximately 50μm in diameter. The improvement in resolution can be seen in **Fig. 3** where the best fits to the SIMS profiles from a boron delta in silicon are plotted as a function of primary beam energy. The resolution at 335eV with a normal incidence oxygen beam is 0.7nm/decade on the leading edge and 2.3 nm/ decade on the trailing edge. FLIG based analysis has been applied to semiconducting structures with excellent results (Smith et al 1996). The FLIG technology is now available on the Atomika 4500 SIMS depth profiling instrument. The main problem with low energy ion beam analysis is that sputter yields drop rather rapidly below 1keV so that it can take several hours or even tens of hours to complete an analysis, if the sample is several microns thick.

### Analysis At Large Angles Of Incidence.

The $\cos(\phi)$ term in Eqn.2 has encouraged Wittmaack (1994) and others to attempt analysis at grazing angles of incidence. Analysis at $80°$ should yield an improvement of 5.8 as compared to normal incidence analysis, due to the reduction in primary ion range. Generally, however, these experiments are very difficult to perform and unevenness in the etch pit almost invariably arises

### The Bevel And Line-Scan Approach.

An alternative approach to the high depth resolution analysis of thin films and multilayers is to synthesise high magnification bevels into the materials using appropriate chemical etchants. Chemical etching is a relatively benign way of etching the material whereas mechanical and ion beam bevelling introduce damage which is magnified on the bevel plane. Recent studies by Hsu et al (1995,1996) have shown that linear bevels can be produced up to bevel magnifications **M** of at least 10,000. These bevels can then be analysed by line-scanning the SIMS ion beam along the bevel plane. At these high bevel magnifications the lateral mixing of the ion beam (~10nm) can be ignored so that the mixing is essentially vertical (**Fig. 4**). The theory developed by McPhail et al (1997) suggests that the secondary ion signal from an abrupt feature such as a buried delta layer will be asymmetric at the highest magnifications with the trailing edge sharper than the leading edge. The broadening of the feature will be controlled by the thickness of the feature t, the beam half-width W, the analytical depth removed during sputtering $d_a$ (which will shift the feature horizontally), and a term $d_i$ which includes the secondary ion information depth and the micro-roughening associated both with the bevelling and the imaging processes. The theory suggests the FWHM of the signal F, will be:

$$F = (W/M) + (t+d_a+d_i)/2 \qquad (d_aM<W) \dotfill 3a$$
$$F = (t+d_a+d_i)/2 \qquad (d_aM>W) \dotfill 3b$$

The agreement between theory and experiment is excellent. **Fig. 5** shows the FWHM and the leading and trailing edge as a function of bevel magnification for aluminium deltas in gallium arsenide. At the highest bevel magnification (8300) the secondary ion signal has a FWHM of 2.1nm, an up-slope of 2.9nm/decade and a down-slope of 0.6nm/decade. The sharpness of the down-slope is essentially controlled by the ratio of the beam width to the bevel magnification [(5µm/8300)~0.6nm] so that improvements to the beam focus should lead to trailing edge resolutions better than 0.1nm/decade.

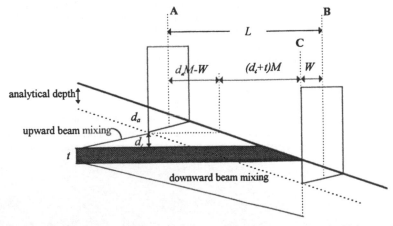

**Fig. 4** The theoretical model for the line-scan analysis of a bevelled sample in the high magnification case ($d_aM>W$). Points A and B represent the first and last positions at which information from the layer, of thickness t, is detected.

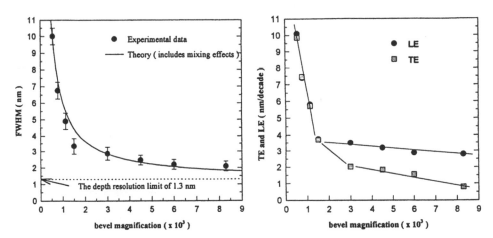

**Fig. 5** The full-width-at-half-maximum (FWHM), and the leading and trailing edge slopes as a function of bevel magnification. The fit to the data for the FWHM is from Eqn. 3a.

The bevel and linescan approach offers a number of advantages over conventional depth profiling. The analysis is rapid because the structure of interest has been opened up prior to analysis, the depth resolution is high, especially on the trailing edge, and is independent of the depth of the feature beneath the sample surface. Furthermore the bevel shape, which can be accurately assessed using a ZYGO white light scanning interferometer, is a permanent record of the depth of origin of the secondary ion signal so that depth calibration is facilitated.

Imaging Of Sections Milled With A Focused Ion Beam (FIB) system

The strategies identified above are inappropriate for a wide range of materials. Low energy SIMS depth profiling can only be used if the sample surface is flat to within a nanometre or so, the bevel and linescan approach can only be used if suitable chemical etchants can be found. An alternative is to

232

prepare and then image a bevel or cross-section through the material using a fine focused SIMS ion gun.

Recently Montgomery et al (1997b) have used such an approach in a FIB 200, focused ion beam milling system at FEI Europe (Cottenham). In this system a fine focus gallium ion beam (W~50nm) is programmed to create a trench in the material. The steep side of the trench wall can be sharpened by micro-machining, rotated and the side-wall then imaged with the same gallium probe. The image data can then be converted into a 'depth profile' by summing the image data along lines at right angles to the layers. **Fig. 6** shows the resulting images from a superconducting multi-layer. The images indicate the good lateral uniformity of the structure over several microns. The reconstructed depth profile revealed interfaces a few tens of nanometres wide. This work is at an early stage and the imaging parameters (especially the beam width and analytical depth) need to be optimised, nevertheless it has considerable potential for the analysis of rough samples and samples not amenable to chemical etching.

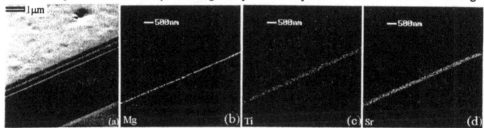

**Fig. 6.** Secondary electron (a) and ion images from the trench wall synthesised by FIB milling in a YBCO/SrTiO₃/MgO/YIG superconducting multi-layer.

## 5. CONCLUSIONS

Sub-nanometre depth resolution is now attainable during SIMS depth profiling of semiconducting structures using either low energy (FLIG based) analysis or, alternatively bevel and linescan analysis. This depth resolution has many potential applications outside the field of semiconductors, for example in the study of slow oxidation, corrosion and diffusion processes in metals, ceramics and glasses. Focused ion beam (FIB) milling and imaging is a very promising alternative to in-depth analysis if the sample is rough and/or roughens under the ion beam, although it is yet to be fully optimised.

## 6. REFERENCES

Benninghoven A, Rudenauer FG and Werner HW 1987 Secondary Ion Mass Spectrometry (John Wiley and sons) ISBN 0-471-01056-1

Dowsett MG, Smith NS, Bridgeland R, Richards D, Lovejoy AC and Pedrick P 1996 Proc. SIMS X, (John Wiley) pp 367-370

Hofmann S 1994 Surf. and Interf. Anal. **21**, 673-678

Hsu CM, Sharma VKM, Ashwin MJ and McPhail DS 1995 Surf. and Interf. Anal. **23**, 665-672

Hsu CM and McPhail DS 1995b Mikrochim Acta 1996 **13**, 419-424

McPhail DS, Hsu CM, Chater RJ, Dingle T and Walker J 1997 Proc SIMS X (John Wiley) pp 379-382

Montgomery NJ, MacManus-Driscoll JL, McPhail DS, Moeckly B, Char K 1997a J. Alloys and Compounds **251**, 355-359

Montgomery NJ 1997b PhD thesis (University of London)

Oeschner H 1983 Proc. IXth Intl. Vac. Conf., Madrid, pp 316

Smith NS, Dowsett MG, McGregor B and Philips P 1996 Proc. SIMS X (John Wiley) pp 363-366

Wilson RG, Stevie FA and Magee CW 1989. Secondary Ion Mass Spectrometry (John Wiley) ISBN 0-471-51945-6

Wittmaack K 1994 J.Vac. Sci. and Technol. B **12** pp 258-262

*Inst. Phys. Conf. Ser. No 153: Section 7*
*Paper presented at Electron Microscopy and Analysis Group Conf. EMAG97, Cambridge, 1997*

# Structural Transition of Bismuth Nano-Particles observed by Ultra High-Vacuum TEM

**Yoshifumi Oshima, Hiroyuki Hirayama and Kunio Takayanagi**

**Department of Materials and Engineering,**
**Interdisciplinary Graduate School of Science and Engineering,**
**Tokyo Institute of Technology, 4259 Nagatsuta Midori-ku, Yokohama 226, Japan**

**ABSTRACT:** Bismuth nano-particles with sizes ranging from 2 to 17 nm were found to be transformed among three different structural phases using ultra high-vacuum TEM. Below 8.5 nm, the structure fluctuated among cubic, rhombic and twin structure like a quasi-liquid phase. Between 8.5 and 11 nm, bismuth particles had a core-shell structure: the core of the cubic structure is surrounded by eight crystallites of the rhombic structure. Each rhombic crystallite has (111), (110) and (112) facets of low surface energies, and has dislocations at the boundaries between them. Above 11 nm, the particle had rhombic structure.

## 1. INTRODUCTION

For a nano-particle, the most stable structure is determined by minimizing the total surface energy, for example, Wulffs polyhedron and multiply twinned structures. But it is possible that, in order to determine the stable structure, the high inner pressure of a nano-particle is also an important physical parameter. Because the inner pressure is so large that it is about a few G Pa for a particle with size 10 nm due to the large curvature of surface. The decreased melting point of a fine particle was explained by the inner pressure (Pawlow (1909)).

In the case of bulk crystallite, Mattheiss et al (1986) reported that a structural phase transition under high pressure occurred for Bismuth (Bi). The Bi crystallite has rhombic structure which is transformed into nearly simple cubic under high pressure above 2.5 G Pa. The structure is transformed from rhombic to cubic when the rhombic angle is changed from 57 deg. to 60. Bi nano-particle is available to investigate effect of " inner pressure" on determining the most stable structure of a nano-particle (Yokozeki et al (1978)).

## 2. EXPERIMENTAL

Bismuth nano-particles with sizes ranging from 2 to 17 nm were grown on a lateral face of a holey amorphous carbon film by deposition in-situ in UHV electron microscope. The pressure around the specimen was kept at $10^{-7}$Pa by use of sputter ion pumps and a cryogenic pump. The deposition was done both at room temperature and at 400K. To avoid any contamination of the particle with the substrate, the amorphous carbon film was preheated by cleaning at 700 °C for an hour in the electron microscope before the deposition. The rate of the deposition was chosen to be 0.33 nm/s, as monitored by a quartz oscillation monitor.

High-resolution transmission electron microscope (HR-TEM) images of bismuth particles were recorded on video tapes with a time resolution of 33 msec through a TV camera attached to the electron microscope. This was because they were highly mobile and changed their orientation too rapidly to be resolved on photographic films. HR-TEM images shown in this paper were reproduced from the video tapes. They were processed digitally for noise reduction and contrast enhancement.

To clarify the size dependency of the structure of bismuth nano-particles, we measured spacings and angles of the lattice fringes in the images. We measured the angle of the crossing (110) lattice fringes to determine the rhombic angle. Sizes of particles were determined by the number of (110) lattice fringes (=0.328 nm) appearing in the particle.

## 3. RESULTS

### 3.1 Rhombic and Cubic structure

The HR-TEM images of bismuth nano-particles with sizes below 6 nm often changed their structure during observation. The particles are shown in fig.1(a) and (c), which are images seen along the [111] and [110] directions, respectively. As schematically shown in fig.1(b), the lateral facets of the particle in fig.1(a) are the (101), (112), (011) and (110) planes, and the angle between the (011) and (101) lattice planes was measured to be 86 deg. The angle is close to the corresponding 87.6 deg. of the bulk rhombic structure. The particle in fig.1(c) has, as illustrated in fig.1(d), lateral facets of (110), (111), (112) and (001) planes, and the angle between (110) and (112) is 88 deg. which is close to the corresponding 86.4 deg. of the bulk crystallite. From these HR-TEM images, the three dimensional morphology of Bi nano-particles was deduced to be a truncated rhombohedron as shown in fig.2: The eight corners of the rhombohedron are truncated by two {111} and six {100} planes, and all edges are truncated by six {112} and six {110} planes.

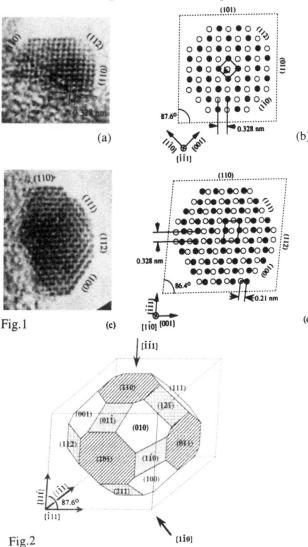

Fig.1

Fig.2

## 3.2 Core-shell structure

Particles around 8.5 - 11 nm in size often give images such as that shown in fig.3 (Oshima et al (1997)): Extra lattice planes of dislocations are introduced at positions indicated by the arrows. Due to these dislocations, the lattice fringes cross obliquely in the corner regions of the particle. The lattice fringes at the central region, on the other hand, cross orthogonally. The central region, thus, seems to have a cubic lattice, while the corner regions have a rhombic lattice. The angles between the cross fringes at the three corner regions are 84 deg., 87.6 deg. and 87.4 deg., and the lattice spacings are 0.349 nm, 0.328 nm and 0.351 nm. These lattice fringes are close to the {110} lattice fringes of the rhombic crystalline in fig.1.

Each corner region has, therefore, the rhombic structure seen from the [111] direction, and the three facets of each corner region are (101), (112) and (011). A possible structure of this particle in three dimensions is illustrated in fig. 4; The core-shell structure in fig.4(a) consists of an octahedral crystallite of cubic structure (fig.4(b)) at the center and eight outer crystallites of rhombic structure (fig.4(c)). The shaded plane in fig.4(a) is an extra plane of a dislocation which compensates for the open space between neighbouring rhombic crystallites. Each of these eight crystallites are bounded by a (111) plane of rhombic structure with the core octahedron. Thus, the core-shell structure has facets of {111}, {110} and {112} surface of low surface energies.

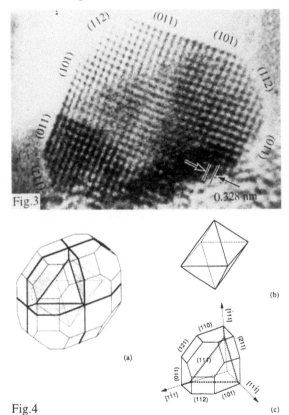

Fig.3

Fig.4

## 3.3 Rhombic structure

The structure of Bi nano-particles above 11 nm in size was almost rhombic as shown in fig.5. But, in the central region, rectangular fringes were often appeared. The rectangular fringes are not thought to be thickness fringes, because the extinction distance of the {110} lattice plane of a Bi crystallite is about 50 nm at the acceleration voltage of 200kV. These rectangular fringes may relate to lattice distortion. Concerning the morphology, the {110} and {112} planes are large, while the {1$\bar{1}$0} plane is narrow. Also, the lattice fringes near

236

the (110) and (112) planes are found to be distorted. In such a large particles, dislocations were sometimes observed (fig.5), but they almost disappeared during in-situ observation.

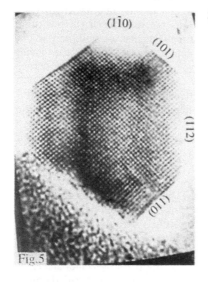

Size: 15.7 nm

Fig.5

## 4.  DISCUSSION

Bi nano-particles have three structural phases; cubic & rhombic, core-shell and rhombic structure. The cubic structure is thought to appear due to high inner pressure. Among the structural transition of Bi nano-particles, it is most interesting that the core-shell structure only exists between particle sizes of 8.5 and 11 nm. It is thought that the core is cubic due to an inner pressure from the surrounding shell blocks. The shell blocks are rhombic in structure which have almost the same rhombic angle as the bulk, but lattice fringes near the surface layer are distorted. Concerning this distortion, two phenomena are observed. One is that the {1$\bar{1}$0} surface has a tendency to change its structure into a {112} surface structure. The atomic arrangement of {112} surface planes is very similar to that of {1$\bar{1}$0}. The other is that at the {112} ({1$\bar{1}$0}) surface with a narrow region, the spacings of {1$\bar{1}$0} ({112}) lattice fringes are contracted along the <110> (<100>) direction. This contraction was not observed at the {112} surface with a wide region. Distortions like these are observed even in a large particle such as is shown in fig.5. It is thought that the distortion relates to the relaxation at the {111} surface layer. Thus, at the {111} surface layer, the spacings of {111} planes elongate slightly. High inner pressure and surface relaxation should be considered in order to understand the stability of the core-shell structure.

## REFERENCES

Mattheiss L E, Hamann D R and Weber W 1986 Phys.Rev.B**34**, 2190
Oshima Y, Hirayama H and Takayanagi K 1997 Z.Phys.D**27**, 534
Pawlow P 1909 Z.Phys.Chem.**65**, 1; 545
Yokozeki A and Stein G D 1978 J.Appl.Phys.**49**, 2224

*Inst. Phys. Conf. Ser. No 153: Section 7*
*Paper presented at Electron Microscopy and Analysis Group Conf. EMAG97, Cambridge, 1997*
© *1997 IOP Publishing Ltd*

# *In-situ* transmission electron microscopy studies of the initial oxidation stage of Cu(001)

J C Yang, M Yeadon, B Kolasa and J M Gibson

Frederick Seitz Materials Research Laboratory, University of Illinois at Urbana-Champaign, 104 South Goodwin Ave., Urbana, IL 61801, USA

**ABSTRACT:** We have examined the initial nucleation stage of $Cu_2O$ on Cu(001) due to oxidation in an *in-situ* UHV-TEM. Rapid nucleation followed by only growth of epitaxial $Cu_2O$ islands was observed, which is indicative of a surface-limited reaction. Modeling of the nucleation behavior using the heteroepitaxial concept of a zone of oxygen capture showed good agreement with the experimental data. No preferential nucleation sites were observed.

## 1. INTRODUCTION

Within the thin oxide regime, the classical theory of Cabrera and Mott (1948) has proved to be highly successful in predicting the oxidation behavior of metals, but it assumes uniform growth. Yet it is known that early stages of oxidation involve nucleation and growth of metal oxide islands (Orr 1962; Milne and Howie 1984). It is reasonable to believe that oxygen surface diffusion should play a major role in the nucleation and initial growth of the metal oxides, as has been previously suggested (Orr 1962; Holloway and Hudson 1974). The concepts of surface diffusion, nucleation and growth are similar to heteroepitaxy, such as Ge formation on Si, which has been extensively modeled recently, such as by Venables et al. (1984).

In this paper, we present our results of *in-situ* ultra-high-vacuum (UHV) transmission electron microscopy (TEM) experiments of copper oxidation and a model for the oxide nucleation, based on the heteroepitaxial concept of a zone of oxygen capture, where oxygen surface diffusion is the dominant mechanism. We have also observed that oxygen surface diffusion is the dominant mechanism for the initial growth of the $Cu_2O$ islands; these results can be found in Yang et al. (1997a).

Cu was chosen as a model metal system, since it is a simple face-centered cubic metal, with a lattice parameter $a = 3.61Å$ and a melting temperature of 1083°C. $Cu_2O$ is known to form on copper (001) due to oxidation (Roennquist and Fischmeister 1960-1961). Also, Cu is a promising metal interconnect material because of its low resistivity and good electromigration properties (Ohba 1995). However, these applications are limited by the poor oxidation behavior of copper, since Cu does not form a thin, adherent and protective oxide scale.

## 2. EXPERIMENTAL

Single crystal 99.999% pure Cu(001) films were grown on irradiated (001)NaCl in an UHV e-beam evaporator system, where the base pressure was $10^{-10}$ torr. 1000Å thick films were examined so that the film was thin enough to be examined by TEM, but thick enough for the initial oxidation behavior to be similar to that of bulk metal. The copper film was removed from the substrate by floatation in deionized water and mounted on a specially prepared Si mount. The Si mount and the microscope specimen holder allow for resistive heating of the specimen up to 1000°C.

The microscope used for this experiment is a modified JEOL200CX. The modifications permit the introduction of gases directly into the microscope column. A UHV chamber was attached to the middle of the column, where the base pressure is $< 10^{-8}$ torr, without the use of cryoshroud pump. The cryoshroud inside the microscope column can reduce the base pressure to approximately $10^{-9}$ torr when filled with liquid helium. For more details about the experimental apparatus, see McDonald et al. (1989).

After removal from the Cu film growth chamber, the copper film formed a native oxide on the surface due to air exposure. To remove the native oxide from the Cu film, the Cu film was annealed at 350°C for approximately 15 minutes (for a brief discussion of this anomalous desorption, see Yang et al. (1997b)). To oxidize the Cu film, scientific grade oxygen gas (99.999% purity) was introduced into the TEM chamber at a partial pressure of $5 \times 10^{-4}$ torr at 350°C. To remove the copper oxide formed due to *in-situ* oxidation, the specimen was annealed at 350°C and methanol gas was leaked into the TEM column, at a partial pressure of $5 \times 10^{-5}$ torr. The methanol reduces the $Cu_2O$ to Cu and forms methoxy $(CH_3O)$ species, which, upon heating, becomes *gaseous* CO and $CO_2$ (Francis et al. 1994).

## 3. OXYGEN SURFACE DIFFUSION MODEL

If oxygen surface diffusion is the dominant transport mechanism for the nucleation of copper oxides, then the probability of an oxide nucleation event is proportional to the fraction of the available surface area, where a diffusion-related "denuded" zone of area, $L^2_d$, surrounds each island.

$$dN \cong k(1 - L^2_d N)dt$$

where N is the number of nuclei, t is time, $L^2_d$ is the area of the zone of oxygen capture, where $L_d >>$ than the diameter of the oxide island, and k is a proportionality constant, which depends on the probability for Cu and O to form $Cu_2O$.

Using the boundary condition that at $t = 0$, $N = 0$, and solving the above linear differential equation, we obtain:

$$N = \frac{1}{L^2_d}(1 - e^{-kL^2_d t})$$

The above equation is fitted to the experimental data. The initial nucleation rate is equal to $kL_d$.

One obvious consequence of oxygen surface diffusion being the mechanism for nucleation is that there is a saturation island density, k. This is due to the existence of an denuded zone of oxygen capture around each oxide island long before the islands impinge on each other, where the radius of this zone is dependent on the oxygen surface diffusion coefficient. This model assumes homogeneous nucleation; hence, it is important to determine whether there exists preferential nucleation sites.

## 4. RESULTS AND DISCUSSION

Figure 1(a) is a dark field image taken from the $Cu_2O$ (110) reflection after the copper film has been annealed at 350°C. No $Cu_2O$ islands are visible in this region. After the introduction of oxygen gas, strain contrast, assumed to be the $\sqrt{2} \times 2\sqrt{2}$ R45 oxygen-induced surface reconstruction (Jensen et al. 1990), was observed. After a dwell time of a few minutes, the oxide islands were observed to nucleate rapidly, followed by growth. Figures 1(b-c) shows corresponding dark field images, taken from the $Cu_2O$ (110) diffraction spot, of the same area as shown in figure 1(a), of the copper film at successive 10 minute time increments after oxygen was leaked into the column of the microscope to a partial pressure of $5 \times 10^{-4}$ torr. These TEM images show the copper oxide islands which formed on both surfaces of the copper film. The SAD (selected area diffraction) pattern of the copper film after 30 minutes oxidation at $5 \times 10^{-4}$ torr in dry oxygen can be indexed as (001) $Cu_2O$, where the relative orientation between the copper oxide and copper film is $(00\bar{1})Cu//(001)Cu_2O$ and $[100]Cu//[100]Cu_2O$.

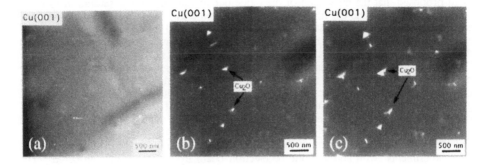

Figure 1: Cu$_2$O(110) dark field images after (a) 0 min. (b) 10 min. (c) 20 min. of oxidation at T = 350°C and P(O$_2$) = 5x10$^{-4}$ torr.

The negatives were digitized with a Leafscan™ 45. The software packages Digital Micrograph™ and NIH Image™ were used to determine the number density as a function of time. Figure 2 shows the experimental data and theoretical fit to equation (2). A good match is noted where the fit parameters, k = 1.6 μm$^{-2}$ and L =0.17 min$^{-1}$. Hence, the initial nucleation rate, kL$_d$, is 0.27 μm$^{-2}$ min$^{-1}$, and the saturation island density, k, is 1.6 μm$^{-2}$.

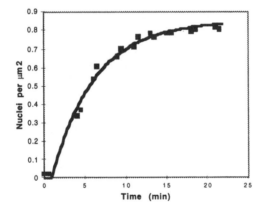

Figure 2: Nuclei density versus time.

To determine as to whether there are preferential nucleation sites, such as surface steps (Milne and Howie 1984) or dislocations (Heinemann et al. 1975), we performed repeated oxidation, reduction, followed by oxidation experiments, but no nuclei appeared at the same positions; however, the surface conditions could have changed during the cleaning process. We also examined the Cu(200) dark field images, where dislocations and surface steps are visible, before and shortly after oxidation. Some of the Cu$_2$O nuclei were at steps or dislocations and others were not. Hence, we did not observe clear evidence for preferential nucleation sites.

Other mechanisms have been previously reported in the past for the initial nucleation mechanism for metal oxides. Lawless and Mitchell (1965) have suggested that surface layer of copper is saturated with oxygen, and then nucleation of copper oxides occurs by precipitation from this oxygen-saturated surface. Heinemann et al. (1975) have investigated the initial stages of copper oxidation where the Cu was oxidized in a side chamber attached to a TEM.

They also proposed that the nucleation of the oxide occurs after there is a saturation layer of oxygen on the surface of the Cu film. We have observed a dwell time before nucleation and surface reconstruction during oxidation, proving that the Cu(001) surface has an adsorbed oxygen layer before nucleation can occur. However, the saturation layer is only ~1ML (Jensen et al. 1990); hence, the nucleation is a surface-limited process and not a precipitation process.

Orr, and Holloway and Hudson have proposed that oxygen surface diffusion is the dominant mechanism for transport, nucleation and growth for oxides of Mg (Orr 1962) and Ni (Holloway and Hudson 1974). Our experimental data on $Cu_2O$ nucleation due to oxidation of Cu agree well with their theory, although the growth model of the oxide islands required modification in order to incorporate the 3-dimensional growth of the oxide islands into the Cu (Yang et al. 1997).

The effect of the electron beam on the oxidation kinetics should be considered. Careful oxidation experiments with and without the electron beam irradiation were conducted. The effect of the electron beam is to reduce the reaction rate by approximately a factor of four, but a saturation island density was observed in both cases, indicative of a surface diffusion nucleation mechanism.

## 4 . CONCLUSION

We have observed the initial stages of copper oxidation by UHV *in-situ* transmission electron microscopy techniques. We have developed a model to describe the initial transport and nucleation of copper oxides where oxygen surface diffusion is the dominant mechanism. Neither surface steps or dislocations were clearly observed to be preferential nucleation sites. Since oxygen surface diffusion is a dominant mechanism for nucleation and initial growth (Yang et al. 1997) of the copper oxides, then factors which influence the surface structure, such as polishing and surface impurities, should alter the oxidation behavior dramatically.

We are presently conducting copper oxidation experiments at various temperatures and pressures. We are applying critical rate theory (Venables et al. 1984), which is used to describe heteroepitaxial growth, in an attempt obtain quantitative data, such as the oxygen surface diffusion coefficient and the critical nucleus size, for the initial Cu oxidation stage.

## 5 . ACKNOWLEDGMENTS

This research project is funded by the Department of Energy (DEFG02-96ER45439). M. Menezes and Professor H. Birnbaum kindly allowed the use of their UHV e-beam evaporator system. The expert assistance of M. Marshall is appreciated. Center for Microanalysis of Materials and the Materials Research Laboratory's computer facilities at the University of Illinois, Urbana-Champaign, were used for this research project.

## REFERENCES

Cabrera, N and N F Mott 1948 Reports on Progress in Physics **12**, 163
Francis, S M, F M Leibsle, S Haq and N Xiang 1994 Surface Science **315**, 284
M Bowker 1994 Surface Science **315**, 284
Heinemann, K, D B Rao and D L Douglas 1975 Oxidation of Metals **9**(4), 379
Holloway, P H and J B Hudson 1974 Surface Science **43**, 123
Jensen, F, F Besenbacher, E Laegsgaard and I Stensgaard 1990 Phys. Rev. B **42**(14), 9206
Lawless, K and D Mitchell 1965 Memoires Scientifiques Rev. Metallurg. **LXII**, 17
McDonald, M L, J M Gibson, and F C Unterwald 1989 Rev. Sci. Instrum. **60**, 700.
Milne, R H and A Howie 1984 Philosophical Magazine A **49**(5), 665
Ohba, T 1995 Applied Surface Science **91**, 1
Orr, W H 1962 Oxide Nucleation and Growth, Thesis, PhD, Cornell University.
Roennquist, A and H Fischmeister 1960-1961 Journal of the Institute of Metals **89**, 65
Venables, J A, G D T Spiller and M Hanbuecken 1984 Rep. Prog. Phys. **47**, 399
Yang, J C, M Yeadon, B Kolasa and J M Gibson 1997a Applied Physics Letters **70**(26), 3522
Yang, J C, M Yeadon D Olynick and J M Gibson 1997b Microscopy and Microanalysis **3**(2), 121.

*Inst. Phys. Conf. Ser. No 153: Section 7*
*Paper presented at Electron Microscopy and Analysis Group Conf. EMAG97, Cambridge, 1997*
© *1997 IOP Publishing Ltd*

# Newly developed UHV-FE-HR-TEM for particle surface studies

**Y Kondo[1], H Ohnishi[1], Q Ru[1], H Kimata[1] and K Takayanagi[1,2]**

[1]Takayanagi Particle Surface Project, ERATO, Japan Science and Technology Corporation, 3-1-2, Musashino, Akishima, Tokyo, 196, Japan.
[2]Department of Material Science and Engineering, Tokyo Institute of Technology, 4259 Nagatsuta, Midori-ku, Yokohama, 226, Japan

**ABSTRACT:** We developed a ultrahigh vacuum high-resolution electron microscope equipped with a field-emission gun (UHV-FE-HR TEM) to study nano-particle surfaces. The ultra-high vacuum ($2 \times 10^{-8}$ Pa) allow us to keep clean surfaces on nano-particles. The microscope is equipped with a specially designed electron biprism for electron holography to measure a particle thickness, a high speed TV camera with a time resolution of 1 ms for dynamic observation, and a miniaturized STM which gives the spectroscopic information of a particle having a definite structure determined by high-resolution electron microscopy.

## 1. INTRODUCTION

The study of nano-structures has become a key for a future material. The significance of these materials is that their physical properties are different from those of bulk material. For example, a wire in nanometer thickness shows the quantization of conductance (Agrait et al 1993, Olesen et al 1994) even at room temperature. A surface is important for the most of nano-structures, because the ratio of surface atoms to those in whole structure is the order of tens of percent.

The nano-structures can be fabricated by an intense nano-electron-beam, which is generated from a modern high-brightness field emission source coupled with adequate probe forming lenses (Humphrey et al 1990).

For the research of nano-structures, structural and spectroscopic observations should be carried out simultaneously, or spectroscopic observations should be performed on a well-defined structures. On the other hand, the dynamic observation of nano-structures is important,

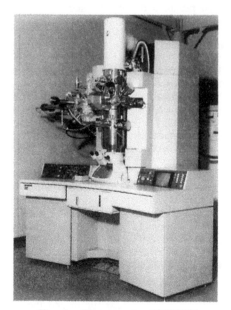

Fig. 1    Photograph of 2000VF

since some of nano-structures change their form during microscopic observation, and this phenomena can be utilized to search stable or quasi-stabilized structures of them (Iijima and

Ichihashi 1993). In addition, the surface of nano-structures should be clean to observe the native both electronic and structural states of them. Thus, an instrument to investigate these characters in one nano-structure is needed for nano-structure science.

We have developed a new ultrahigh vacuum transmission electron microscope (JEM-2000VF shown in Fig. 1) which is suitable for the nano-structure analysis. In this paper, we first report the construction of UHVTEM which is modified so as to fit with new instruments for nano-structure studies. Secondly, a result of this microscope which is formed by the nano-beam fabrication.

## 2. BASIC CONSTRUCTION AND NEW INSTRUMENTS FOR NANO-STRUCTURE ANALYSIS

The new microscope is basically a 200kV high resolution electron microscope with Cs = 0.7 mm and Cc = 1.2 mm, which enables us to see atomic image with 0.21 nm point-to-point resolution. The prototypes of this microscope are a JEM-2000FXV (Takayanagi et al 1987, Kondo et al 1991) and a JEM-2000FXVII (Kondo et al 1994a) which are UHVTEMs.

This microscope is evacuated by differential pumping system consisted of four sputter-ion pumps, two sputter-ion/titanium-sublimation combination pumps, a turbo molecular pump, a dry pump, and three rotary pumps. The ultimate vacuum level is 1 x 10$^{-8}$ Pa, and a working vacuum level is kept at below 3 x 10$^{-8}$ Pa.

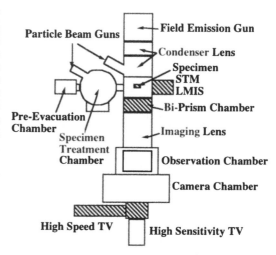

Fig. 2   Construction of UHV-FE-HR-TEM

The new microscope has been equipped with a newly developed miniature scanning tunneling microscope (STM) (Ohnishi et al 1997) to observe a STM image or a scanning tunneling spectrum of the nano-structure, a specially designed electron biprism for holography which gives a specimen thickness and an electric field around the nano-structure (Ru et al 1997), a newly developed high speed TV camera to observe fast morphological changes of the nano-structures (Kondo et al 1997), and a field emission gun which provides an electron beam having high coherency and high brightness. The details of these instruments will be shown in the individual reports. A nano-particle with clean surface will be made in-situ with a liquid metal ion source (LMIS)(Kimata et al 1997) and a arc-discharge type nano-particle gun. The construction of this TEM is schematically shown in Fig. 2. This TEM is also equipped with a commercial highly-sensitive TV camera (Gatan 622SC), a specimen treatment chamber and a UHV compatible goniometer (Kondo et al 1994b). The goniometer is equipped with an ultra-fine specimen driving system using piezo electric devices, which was modified from one reported previously (Kondo et al 1994c).

## 3. MODIFICATION

An objective lens of our UHVTEM is modified for the in-situ devices which will be attached to the TEM in near future and the STM. The pole pieces of objective lens are conventionally fixed on an adequate metallic spacer to set two of poles at axially-symmetric positions and to keep a proper distance between upper and lower poles. In conventional TEM, the spacer prevent to attach a large specimen holder which is able to receive the large STM holder, and is also an obstacle to attach an in-situ instrument such as an evaporator or a nano-particle gun. In the new UHVTEM, the spacer was eliminated. Then, the lower pole is fixed on a metallic block which support the goniometer and the upper pole is fixed on an upper yoke of the objective lens as shown in Fig. 3.

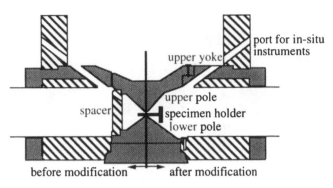

The electron biprism is attached to a chamber between an intermediate lens and the object lens. A proper distance between a image plane of objective lens and a position of a biprism is determined by an object size (particle size) and a desired resolution of reconstructed image form a hologram. We decide to see a particle of 10 nm in diameter or nano-particles in the region of 10 nm². Then, we selected two of the distances which are 10 and 30 mm under limitation of a detector size and a number of pixels the detector. The distances of 10 mm and 30 mm are proper to obtain a static high resolution hologram on a conventional film and a dynamic process of nano-particle recorded on the highly-sensitive TV, respectively. The chamber to which the biprism is attached, prepares two ports for the distances of 10 mm and 30 mm. A table of lens currents which determines the magnifications are rewritten as new to fit with a modified distance between the intermediate lens and the object lens. Thus, we are able to obtain the holograms of nano-particles by only inserting the biprism

Fig. 3   Modification of the objective lens

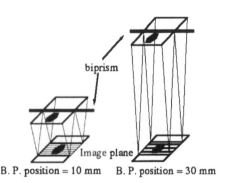

Fig. 4 Two of distances between the biprism and the image plane of the objective lens

and setting the biprism voltage. We believe that quick operation is essential for the observation of nano-particles, because they change their forms easily.

## 4. NANOWIRE AND NANO-CONTACT FABRICATED BY ELECTRON BEAM IRRADIATION

The intense electron beam irradiation is useful for nano-fabrication. By thinning down a gold film, a gold nano-wire has been obtained as a manner reproduced in Fig. 5. The gold wire has self-ordered to align its wire axis in the [110] crystallographic orientation through the thinning process. The wire has became as thin as four atomic [110] columns in the projection, while it is 10 nm long. The surface of the gold nano-wire is found to be reconstructed as descried elsewhere (Kondo and Takayanagi 1997b). The surface reconstruction is supposed to stabilizes such a wire supported in vacuum, by giving rise to a extra-surface tension. Another example of nano-beam fabrication is to study the point contact of the two tip apices. Figure 6 shows a change of the structure of a point contact during the desorption of the gold atoms at the contact. Between the two gold crystals which reveals the (200) lattice fringes of 0.2 nm separation in Fig. 6(a), a single gold atom remains, as seen in Fig. 6 (b) and (c), in the instant of splitting off the contact.

## 5. SUMMARY

UHV-FE-HR TEM allow us to study structures of nano-wires and atomic bridges at

244

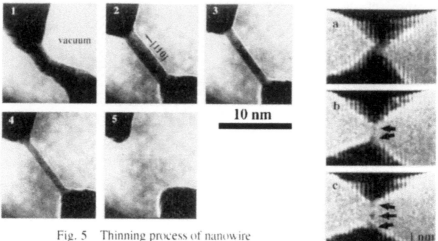

Fig. 5   Thinning process of nanowire

Fig. 6 Gold nano-contact

atomic resolution. Electron microscope technique combined with spectroscopic method would clarify properties of nano-structures in future. As we demonstrated the dynamic process of NB, the UHVTEM having an ability of dynamic observation is a powerful tool for observing nano-structures.

## REFERENCES

Agrait N, Rodrigo J G, and Vieria S 1993 Phys. Rev. B **47**, 12345

Humphrey C J, Bullough T J, Devenish R W, Maher D M and Turner P S 1990 Scanning Micros., Suppl. **4**, 185

Iijima and Ichihashi 1993 Phys. Rev. Lett., **56**, 2923

Kimata et al 1997 in preparation(will be presented at International Centennial Symposium on the Electron, Cambridge 15-17 Sept.)

Kondo Y, Ohi K, Ishibashi Y, Hirano H, Harada Y, Takayanagi K, Tanishiro Y, Kobayashi K and Yagi K 1991 Ultramicroscopy, **35**, 111

Kondo Y, Kobayashi H, Ishibashi Y, Kasai T, Nunome H, Honda T, Kersker M and Ishida Y 1994a Proc. 52nd Annual Meeting of the MSA, p496

Kondo Y, Kobayashi H, Kasai T, Nunome H, Kirkland A I and Honda T 1994b Proc. 13th Int. Cong. Electron Microscopy, Paris, p269

Kondo Y, Hosokawa F, Ohkura Y, Hamochi M, Nakagawa Y, Kirkland A I and Honda T 1994c Proc. 13th Int. Cong. Electron Microscopy, Paris, p275

Kondo et al 1997a in preparation (will be presented at International Centennial Symposium on the Electron, Cambridge 15-17 Sept.)

Kondo Y and Takayanagi K 1997b submitted

Ohnishi et al 1997 in preparation (will be presented at International Centennial Symposium on the Electron, Cambridge 15-17 Sept.)

Olesen L, Laegsgaard E, Stensgaard I, Besenbacher F, Schiotz J, Stoltz P, Jacobsen K W and Norskov K 1994 Phys. Rev. Lett., **72**, 14, 2251

Ru et al 1997 in preparation

Takayanagi K, Tanishiro Y, Kobayashi K, Akiyama K and Yagi K 1987, Jpn. J. Appl. Phys., **26**, L957

*Inst. Phys. Conf. Ser. No 153: Section 7*
*Paper presented at Electron Microscopy and Analysis Group Conf. EMAG97, Cambridge, 1997*
© *1997 IOP Publishing Ltd*

# The effect of gaseous and liquid water layers on the estimation of the composition of materials in High Pressure Scanning Electron Microscopy (HPSEM): A Monte Carlo Simulation

H R Powell K J Hollis and J S Shah

University of Bristol, H H Wills Physics Laboratory, Tyndall Avenue, Bristol BS8 1TL

**ABSTRACT:** Monte Carlo method with "single scattering approximation" and "mixed element representation" is applied to estimate the effect of electron scattering, in gaseous and liquid water layer lying on the top of the specimen, on estimation of composition. This scheme of calculations is used to discuss two particular cases: (1) To see how a copper - carbon 'step' boundary would be resolved in the presence of a layer by imaging the emerging BSE distribution and (2) To obtain characteristic (Ca) x-ray emission profile of bone in the presence of a gas/liquid layer. The results show that although the resolution can remain sharp, the contribution of the emission comes from the region extending far beyond resolution.

## 1 . INTRODUCTION

Hydrated specimens are routinely examined in High Pressure Scanning Electron Microscopy (HPSEM). In this technique the specimen can be maintained in a fully hydrated condition without water loss because it is in a saturated water vapour while being scanned by the electron beam. This variant of the technique is referred to as 'Moist Environment Ambient Temperature Scanning Electron Microscopy (MEATSEM). This can be performed in a commercially available microscope such as ESEM. During the performance of MEATSEM a focused electron beam travels through water vapour and possibly through a condensed thin layer of water before it interacts with the specimen itself. The effect of electron scattering on the actual resolution that can be obtained in HPSEM/MEATSEM can be studied by a Monte Carlo simulation.

## 2 . THE METHOD

The schemes of calculations employed are those used previously for studying BSE distribution (Shah and Weare 1995) and the characteristic x-ray emissions from the specimen for elemental microanalysis (Shah and Joyce 1996). The principal assumptions for the calculation are:

(1) The specimen contains a number of different atoms according to its chemical composition. Otherwise it is homogenous. Electrons entering in it are scattered by the individual atoms they encounter randomly, but relative to the abundance of their type. The specimen, if not elemental, and the layer on the top, was modelled by "mixed element representation" as adopted by Shah and Weare (1995) and Shah and Joyce (1996). Chemical composition of different bones has been estimated by Howell and Boyde (1994). The VFM (volume fraction mineral value) of the bone specimen in this study was taken to be 0.3. Different (liquid) water layer depths were modelled. The water vapour layer modelled here had a thickness of 5 mm at the saturated water vapour pressure of 17 torr (2261 Pa) at 20 ° C.

(2) Both elastic and inelastic collisions occur. It was assumed that in elastic collisions electrons are deflected without loss of energy and they were modelled by single scattering approximation as described by Joy (1995). On the other hand inelastic collisions were assumed to occur so that electrons lose energy without deflection. Energy loss of an electron is continuous and according to Bethe's continuous energy loss approximation (Joy 1995).

246

(4) Characteristic x-ray photons are produced due to collision events at the appropriate "inner shell ionisation" cross section:

$$\sigma_{Bethe} = 6.51 \times 10^{-21} \left( \frac{n_s b_s}{EE_c} \right) . \log \left( \frac{c_s E}{E_c} \right)$$

where $E_c$ = Critical ionisation energy of the x-ray wavelength, $n_s$ = number of electrons in the shell, $b_s = 0.9$ and $c_s = 0.65$, for the K-shell

## 3. RESULTS AND DISCUSSION

The results presented here are for incident electrons with energy 20 keV. Scattered beam profiles after the passage through the water (vapour and liquid) layers were calculated. In HPSEM, after its passage through the water vapour layer under consideration, the scattered beam profile does not alter very much Consequently, the loss of resolution is small (Fig. 1).

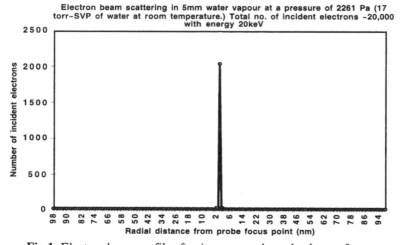

**Fig.1**: Electron beam profile after its passage through a layer of water vapour

However, it must be noted that the effect of the noise generated in the high pressure condition, which will substantially affect the resolution in practice, is not considered here. Some of the beam profiles after traversing through the water layers are shown in Fig. 2. Beam broadening due to the presence of a liquid water layer above the specimen is dependent upon its depth. At a depth of a fraction of a micron the beam is quite broad therefore the resolution will suffer appreciably. It is important to note that the central peak diminishes to zero intensity, indicating absence of electrons at the central point of incidence at the specimen, and additional maxima are generated in a circle around the centre. This circle appears at the depth of 0.4 microns and becomes dominant above the central peak at the depth of 0.6 microns. The radius of this circle increases with water depth. The existence of this phenomenon suggests that, at least theoretically, it may be possible to illuminate a point on the specimen surface by an annular beam.

'Morphological' resolution was examined by simulating the effect of scattering at a 'contrast (intensity) step function' at the surface of a specimen containing a boundary between carbon and copper. The resolution of the step for different depths of the water layer, as 'seen' by the backscattered electrons (BSE), was obtained by calculating intensity at and around the step, using summation of constituent intensities from each side of the step according to the criterion that contribution to the overall intensity is proportional to the perpendicular distance from either side of the step.

Intensity of the incident beam on the surface of a (bone) specimen after it has scattered through a water layer of different depths

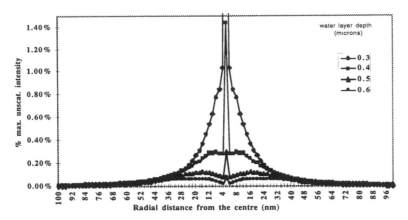

**Fig. 2**: Intensity profiles of the scattered beams after they have been through different depths of water layers.

Thus the curves shown in Fig. 3 represent the intensity variations in the BSE image of the boundary. Note that the step appears to be remarkably sharp; even though for highly scattered beams the differential contrast across the step is reduced.

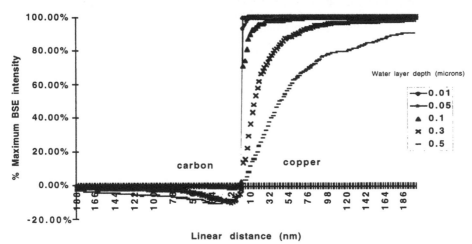

**Fig. 3**: Morphological' resolution of a boundary between copper and carbon showing modifications of the 'Contrast ( BSE intensity) Step Function' -plotted against the linear distance from the step for different depths (microns) of water on carbon-copper specimen.

In fact the image resolution of the step appears not that different to the radial resolution of the beam itself. These calculated results are broadly in line with the experimental measurements of Farley and Shah (1990). It must be born in mind that their results were on a far cruder scale than the plots in Fig. 3.

The study of the spatial distribution of the BSE from the bone specimen covered by 0.3 μm of water layer shows that the resolution limit is of the order of 50 nm.

248

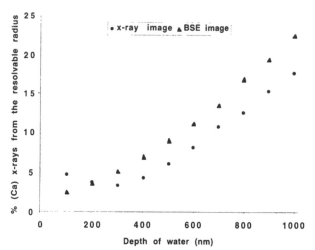

**Fig. 4** Estimation of the actual proportion of Ca within the resolvable radii for different depths of water over the bone specimen

For evaluating resolution of the (Ca) x-ray image, one can employ the Rayleigh criterion and compare it with the resolution of the BSE image by the same criterion. It was found that in both cases the resolvable radii increase non-linearly with the depth of the water layer. The % Ca x-rays coming from within the resolvable radius increases with the depth (Fig. 4) because, the resolvable radius itself and the ratio (volume of x-ray generation underneath the area of the resolution circle /that underneath the area of the 'skirt' outside this circle) increases. The reason for a small minimum in the plot for the x-ray image is not clear. In this context, one cannot compare the resolution of the secondary emitted electron (SE) image because SE emission was not modelled. The escape depth of SE emerging from the specimen is much less than that of the BSE. This has been confirmed by Monte Carlo calculations of Bache et al (1997). The effect of the depth of the water layer on the top of the bone specimen, on the estimation of an element by electron probe microanalysis (or Energy Dispersive X-ray Analysis - EDXA) can be discerned from Fig. 4. The plots therein show the extent of calcium signal present, as estimated by percentage emission of the characteristic (Ca) x-rays from the incident radius of the beam probe at the bone specimen, and the corresponding BSE resolution at different depths of water. It is clearly indicated that the quantitative estimation of the concentration from the experimental data, without correction, would be grossly inaccurate. The graphs in Fig. 4 do provide the values of the correction factors that should be applied to the experimental measurement for estimating the *actual* Ca present within the 'analysis spot'. The results indicate that the value of the correction factor will reduce as resolution worsens. In the presence of water, the resolution of elemental x-ray maps would remain relatively good but quantitative estimates will be inaccurate unless the appropriate correction factors are applied.

**ACKNOWLEDGEMENT**: *K J Hollis and J S Shah gratefully acknowledge financial grant from the BBSRC.*

**REFERENCES**
Bache et al 1997 in Proc. Microscopy and Microanalysis '97, Cleveland Ohio, MSA
Howell P G and Boyde A. 1994, Bone, **15**, 2
Joy D C 1995 Monte Carlo Modelling for Electron Microscopy and Microanalysis Oxford
    University Press
Shah J S, 1995 in Proc. ECASIA 1995, p345, John Wiley
Shah J S and Joyce D 1996 in Proc. EUREM 96 Dublin, T6
Shah J S and Weare J 1995 in ECASIA 1995, p42, John Wiley

*Inst. Phys. Conf. Ser. No 153: Section 7*
*Paper presented at Electron Microscopy and Analysis Group Conf. EMAG97, Cambridge, 1997*

# High pressure scanning electron microscopy (HPSEM): Conversion of an ordinary SEM to maintain pressures to $10^5$ Pa

**K J Hollis and J S Shah**

University of Bristol, H H Wills Physics Laboratory, Tyndall Avenue, Bristol BS8 1TL. UK

**ABSTRACT:**    The key function of a high pressure scanning electron microscope is to support an elevated pressure in the specimen chamber whilst maintaining the column of the microscope at high vacuum. Details are given of the conversion of a conventional SEM to enable operation at up to atmospheric pressure.

## 1 . INTRODUCTION

Conventional scanning electron microscopes (SEM) are designed for operation under high vacuum and primarily for electrically conducting specimens. However, specimens with high vapour constituents must be prepared in order to be viewed. This causes specimen deformation and damage. This is particularly true for delicate biological and other insulating materials. This then lead to the development of high pressure scanning electron microscopy (HPSEM) for which specially designed microscopes (e.g. ESEM) are commercially available. For performing HPSEM the specimen chamber of an SEM, operating at elevated pressures, serves to prevent water loss from hydrated specimens. This technique was first used for imaging of hydrated specimens, more closely approximating their natural state and without water loss by Shah & Beckett (1979). Since the advent of wider availability of SEM, operating at high pressures, more diversified uses have been reported, notably, dynamic experiments such as stretching of hydrated muscle tissue (Shah *et al.* 1994). The presence of a gas prevents charge accumulation on non-conducting surfaces, by virtue of the neutralising effect of ions present  in high pressure gaseous environment. The technique has been used with any ionisable gas for a variety of nonconducting specimens. (For a recent review see Shah 1995.) It is therefore desirable to extend the pressure range of HPSEM.

Although the specimen chamber operates at an elevated pressure it is necessary to maintain the microscope column at high vacuum. In order to achieve this, earlier attempts used an extremely thin, electron transparent, window to physically separate the two pressure regimes. This design was superseded during the 1970's by the use of a differential pumping arrangement (Shah 1977). This technique uses two vertically aligned pin-hole apertures to separate the column from the specimen chamber. The space between the two apertures is pumped separately resulting in three areas of high, intermediate and low vacuum. Commercial SEM are now available which normally operate up to a maximum pressure of approximately 6650 Pa (50 torr) and may be used to study hydrated material and non-conducting specimens with low vapour pressures. However, it is desirable to increase the working pressure range available in order to extend the applicability of HPSEM to other samples, for example those with high vapour pressures or at elevated sample temperatures. A JEOL JSM-35C conventional scanning electron microscope has therefore been converted to operate at up to atmospheric pressure whilst maintaining the electron optics column at below $10^{-3}$ Pa. The modified microscope may still operate in conventional mode.

## 2. Microscope Conversion

Previous designs for HPSEM conversions at Bristol University have used independent high pressure specimen chambers which may be inserted or removed from the conventional microscope as desired. The main objective of this conversion was to alter the vacuum system so that the whole of the main specimen chamber itself can be operated at a pressures much higher than that operating in commercial SEM and that in the previous Bristol designs. This gives the advantage of using the commercial sample positioning mechanisms which are already in place and also having the maximum room to add additional equipment as necessary.

The pressure differential between the microscope column and the high pressure specimen chamber is supported using the standard method of installing an additional chamber below the objective lens, containing two vertically aligned apertures. The space between these two apertures is then differentially pumped, producing an area of intermediate pressure between the column and the specimen chamber (Schumacher 1961). This chamber is mounted onto the pole-piece cover and has necessitated some modification to allow vacuum sealing of the cover and mounting of the intermediate chamber (Fig 1). The intermediate chamber is machined from low vapour pressure brass. Commercially available copper apertures are used for the vacuum apertures. The pin-holes are attached with vacuum compatible epoxy and aligned manually using an optical microscope. A commercial X-Y micro-manipulator and sliding O-ring seal assembly is used to enable alignment of the two apertures with the primary beam, with cable drives connected to rotary feedthroughs to allow aperture alignment from outside of the chamber. Although back-lash exists in a system of this kind, it is a simple task to align the apertures and once in position, drift is minimal. The intermediate chamber with its apertures can be disassembled in only a few minutes. This is extremely useful for regular cleaning of apertures in order to maintain a high image quality.

All vacuum and electrical connections are made by the replacement of two of the chamber side plates. This does not therefore effect the use of the microscope in conventional operating mode.

In operation a rotary pump is used for rough pumping the specimen chamber to the desired pressure and the inlet and pumping lines may then be used to attain a steady pressure. A rotary pump is also used to evacuate the intermediate chamber, controlling the leak rate into the microscope column. The diffusion pump to the specimen chamber may be closed off to prevent chamber pumping but this does not effect the safety interlocks on the microscope. The pumping circuit of the differential chamber is then completely independent to that of the microscope (Fig 2).

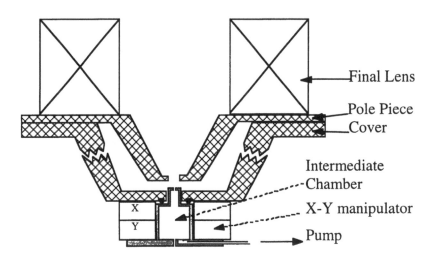

Fig 1 Configuration of the mounting arrangement of the intermediate chamber onto the pole piece cover.

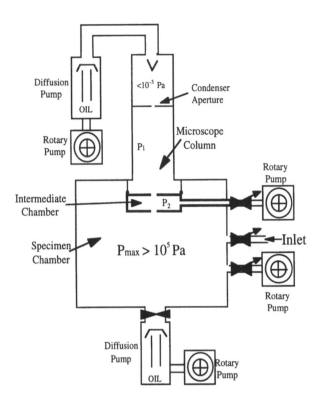

Fig 2  Revised vacuum pumping arrangement for the operation of a JSM-35C microscope at high specimen chamber pressures, showing the four pressure regions delimited by the respective apertures.

## 3. RESULTS

The maximum working pressure which may be attained in the specimen chamber ($P_{max}$) is limited by the leak rate which may be permitted into the microscope column. This is determined by the conductance's of the upper and lower apertures ($C_U$, $C_L$) and the speed at which the intermediate chamber is pumped. The conductance's of the apertures are proportional to the aperture diameter to the power of two for the molecular and to the power of four for the viscous flow regimes respectively. The relative size of these holes is therefore critical in attaining high pressures in the specimen chamber. In addition, the sizes of the apertures determine the available field of view at low magnification.

The pin-hole aperture sizes currently in use are 375 μm and 200 μm, for the top and bottom apertures respectively. The intermediate chamber is pumped using a two stage (238 Lmin$^{-1}$) rotary pump, However, the maximum pumping speed is determined by the size of the chamber outlet pipe. This arrangement of apertures allows the main chamber to reach atmospheric pressure ($10^5$ Pa) without effecting the safety interlocks in the microscope column.

Intermediate chamber pressure measurements were made using a $10^{-2}$ to 100 torr capacitance manometer. Main chamber pressures were made using both a $10^{-1}$ to 1000 torr capacitance manometer and a $10^{-4}$ to 760 torr active pirani gauge. Microscope column pressures were measured using an internally provided gauge giving a voltage output proportional to the pressure. The variations in the intermediate chamber pressure and in the nominal column pressure as the specimen chamber pressure is raised, are shown in Figures 3 and 4.

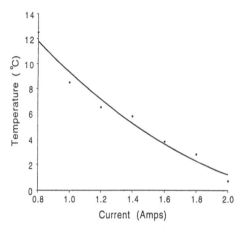

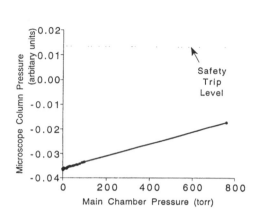

Fig 3 Graph showing the variation of the microscope column pressure as the main chamber pressure is raised. The pressure limit at which the microscope safety interlock operates is shown.

Fig 4 Graph showing the variation of the intermediate chamber pressure as the main chamber pressure is raised.

## 4. CONCLUSIONS

It is clear from the figures that the aperture sizes in use allow the pressure in the main chamber to reach $10^5$ Pa without approaching the safety pressure interlock on the microscope column. In addition, using these apertures it is possible to obtain a circular field of view of approximately 200 μm in diameter.

The investigation and optimisation of the various parameters in the system is under continued investigation. Indications are that it would be possible to increase the aperture sizes further by increasing the speed at which the intermediate chamber is pumped and this would allow a further increase in the available field of view.

It should be noted that with existing secondary electron detectors and the biased specimen current detector (BSCD) the signal to noise ratio at atmospheric pressure would deteriorate drastically. It is therefore necessary to develop a detector which can give vastly superior signal to noise ratio performance at higher pressures.

## REFERENCES

Schumacher R N 1961 Trans. 8th Vac. Symp. and 2nd Int. Congr. **2**, 1192
Shah J S 1977 British Patent 1477458
Shah J S and Beckett A 1979 Micron **10**, 13-23
Shah J S, Proc ECASIA 1995, 345, John Wiley
Shah J S, Ping G. and Durkin R. 1994 *Electron Microscopy*, **IIIB,** p775, Les Editions de Physique, Ulis, France

Inst. Phys. Conf. Ser. No 153: Section 7
Paper presented at Electron Microscopy and Analysis Group Conf. EMAG97, Cambridge, 1997
© 1997 IOP Publishing Ltd

# Comparison of the tensile properties of bovine muscle at 20°C and 3°C by *in situ* dynamic measurements and imaging in high pressure scanning electron microscopy (HPSEM)

T Tsavalos, K J Hollis and J S Shah

University of Bristol, H H Wills Physics Laboratory, Tyndall Avenue, Bristol BS8 1TL, UK

**ABSTRACT:** An improvement in image quality can be obtained by the cooling of wet samples viewed using high pressure scanning electron microscopy (HPSEM). A cooling device has therefore been developed for use with muscle samples attached to a miniature tensometer apparatus. The effect of cooling on the mechanical properties of muscle samples has been assessed. Since no variation has been found in the properties of muscle samples at 20°C and 3°C this method of cooling can be used to reduce the sample vapour pressure.

## 1. INTRODUCTION

It is well established that conventional specimen preparation techniques for scanning electron microscopy, such as critical point drying or freeze drying, cause significant shrinkage and damage to delicate structures when applied to biological specimens. For example, it has been shown that small muscle samples undergo 13% and 56% shrinkage by volume, for freeze drying and critical point drying respectively (Durkin 1994). Morever mechanical properties of muscle changes drastically on all kinds of drying including air drying (Shah *et al.* 1994a). Thus untreated, hydrated, biological samples cannot be stretched in a conventional SEM as they would undergo rapid drying in the vacuum system. Moreover charging of non-conducting surfaces would prevent imaging. The technique of high pressure scanning electron microscopy (HPSEM) was therefore developed to address these problems. It enables biological specimens to be placed in their fresh state into the microscope and tested *in situ* (Shah, 1995).

Due to the high pressure environment, the electron beam is inevitably scattered by the gas above the specimen. This has the effect of significantly reducing the intensity of the unscattered beam and increasing the level of background noise due to ionisation processes in the gas. It is therefore advantageous to keep the pressure as low as possible without causing the loss of high vapour pressure materials from the sample. One way in which this may be achieved is to reduce the temperature of the specimen, so decreasing the saturated vapour pressure of the sample components. This consideration is appropriate for the imaging of muscle specimens during dynamic measurements of their mechanical properties. This requires not only the cooling of the specimen, but an assurance that the cooling device does not interfere with viewing of the specimen in SEM or the *in situ* measurement of its mechanical properties. The aim of the present study was to find out whether tensile properties of muscle tissue significantly changes if it is cooled from room temperature to ~2°C.

## 2. APPARATUS

### 2.1 Tensometer Stage

Previous work has shown that an HPSEM stage, incorporating a miniature computerised tensometer, can be used in HPSEM for monitoring changes in the structure of specimens under tensile loading (Shah *et al.* 1994b). Due to the delicate nature of these samples a commercial tensometer could not be used and so an in-house mechanism was designed.

The stage can operate at pressures of up to ~2400 Pa, while the pressure in the microscope column remains at less than $10^{-2}$ Pa. The tensometer was developed in-house for application with muscle samples of approximately 10 mm in length. A symmetrical stretching mechanism was designed to ensure that, on stretching, the specimen area under view did not shift significantly within the viewing field. The specimen was attached on two jaws which were situated on separate slides. A computerised stepper motor drive enabled stretching by pushing the two slides apart symmetrically.

The range of the tensile force available could be varied since it was dependent on the spring constant of a specially designed force sensing spring, attached to a Linear Variable Displacement Transducer (LVDT). The accuracy of the force reading was governed by the 12 bit analogue to digital converter incorporated into the measurement system. The device was capable of measuring very low tensile stresses in the range of 1 to 30 kPa. The stretching mechanism had a full scale extension of ~3 mm and is accurate to ~2 µm, allowing a stretch of 30% with a strain resolution of $10^{-4}$ for a typical 10 mm specimen ( Shah *et al.* 1994b).

### 2.2 Cooling Device

The requirements for the cooling apparatus were to enable the cooling of an approximately 10x2x2 mm hydrated muscle specimen from ambient temperature to 0°C without interfering with viewing of the image formation or measurement of the mechanical properties of the specimen. This essentially means that there must not be any physical contact between the cooling device and the muscle specimen. Furthermore the cooling device , along with the tensometer, must fit into the modified specimen chamber.

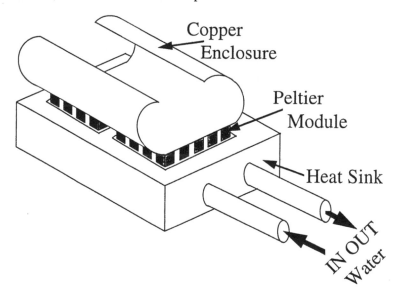

Fig. 1    Shows a schematic representation of the cooling device designed for use with an HPSEM tensometer stage. It incorporates two Peltier devices with a water cooled heat sink and copper enclosure to reduce temperature gradients in the specimen.

The approach chosen for this apparatus was the use of commercially available semiconductor heat-pumps functioning by the use of the Peltier effect. These devices have the advantages of being small and providing an easily controllable degree of cooling by the variation of the current supplied to the devices. It was found that two 1.8 Watt Peltier devices (TEC's from Melcor Thermoelectrics Ltd) coupled in series electrically and in parallel thermally were able to achieve the degree of cooling required. It was experimentally found that dissipation of the heat generated at the hot-side of these devices, due to vacuum and space constraints, could not be achieved by the use of a simple heat sink reliant on radiation and convection. It was therefore necessary to include a water cooled heat sink in the design. Additional water connections were necessary to enable water to be routed through the heat sink within the vacuum system.    Initial tests demonstrated that, although the devices provided sufficient cooling, an unacceptable temperature gradient was produced between the upper and lower surfaces of the muscle sample. In order to cool the space around the sample more evenly a copper radiation shield was constructed which wrapped almost entirely around the muscle tissue but without restricting the incidence of the electron beam on the specimen. This 'jacket' provided shielding through almost 360 degrees and so ensured that there were minimal radial and longitudinal temperature gradients in the sample. The copper was not in contact with the muscle specimen and so did not introduce friction artefacts in the measurement of the mechanical characteristics of the muscle fibre. The cooling arrangement used is shown, in schematic form, in Figure 1.

## 3.    RESULTS

### 3.1  Specimen  Cooler

Initial measurements of the cooling characteristics of the module were made using a k-type commercial thermocouple at various points on the muscle surface, at atmospheric pressure and ambient temperature. For each current setting the temperature of the muscle specimen was allowed to reach equilibrium before the temperature measurement was taken. The temperature differential across the depth of the meat was always less than 0.5°C. The results obtained are shown in Figure 2.

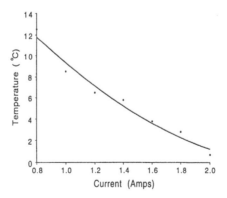

Fig. 2    Shows the variation of muscle temperature as a function of the current through the Peltier devices.

### 3.2  Mechanical  Properties  of  Bovine  Muscle

Mammalian striated muscle consists of several orders of structure from the single celled filament through fibrils, fibres, and fibre bundles to whole muscles. Although the muscle fibres themselves are intrinsically weak under tensile loading, the fibrous structure of the collagen binding the muscle fibres together results in the characteristic strength and flexibility of fresh meat. The collagenous perimysium, surrounding the muscle fibre bundles has the appearance of a fibrous sheet covering the large fibre bundles, with crimps in the sheet having

256

a mean periodicity of ~6.7 μm. How the crimps and linkages between fibre sheets react to tensile loading is currently under investigation. It has previously been found that for stresses up to 150 kPa the bovine muscle shows non-linear elastic and visco-elastic behaviour, see Figure 3.

Experiments were performed at saturated water vapour pressure over a range of temperatures on many muscle specimens.

To within the accuracy obtainable using this apparatus, no change in the mechanical properties occurs as the temperature of the specimen is reduced from the ambient temperature to 3°C.

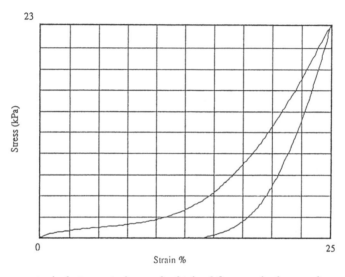

Fig. 3     Shows a typical stress-strain graph obtained from a single muscle specimen under the gradual application and then release of tensile loading

## 4.  CONCLUSIONS

Development of the cooling module has allowed the investigation of the mechanical properties of samples of bovine muscle over the temperature range of 20 to 3 °C. The data show no significant change in the behaviour of the muscle samples over this range. It will therefore be possible to perform further *in situ* tensile tests at low temperature with the confidence that the results obtained will accurately represent those at ambient temperature. The use of the specimen cooler will allow a reduction of the vapour pressure of hydrated samples and will therefore allow images to be obtained with a much higher signal to noise ratio for dynamic viewing of muscle specimen in HPSEM, consequently with improved textural and morphological details

## ACKNOWLEDGEMENTS

The authors gratefully acknowledge the BBSRC for funding this work.

## REFERENCES

Durkin R. 1994 Ph.D. Thesis University of Bristol
Shah J S 1995 Proc. ECASIA 46-49, John Wiley
Shah J S, Ping G and Durkin R. 1994a *Electron Microscopy*, **IIIB,** p575, Les Editions de Physique, Ulis, France
Shah J S, Ping G, and Durkin R 1994b *Electron Microscopy*, volume **IIIB,** p975, Les Editions de Physique, Ulis, France

*Inst. Phys. Conf. Ser. No 153: Section 7*
*Paper presented at Electron Microscopy and Analysis Group Conf. EMAG97, Cambridge, 1997*
© *1997 IOP Publishing Ltd*

# *In-situ* TEM study of the sintering of copper nanoparticles on (001) copper

**M. Yeadon, J.C. Yang, R.S. Averback, J.W. Bullard and J.M. Gibson**

Materials Research Laboratory, University of Illinois, Urbana, Illinois 61801, USA

**ABSTRACT:**   The sintering of copper nanoparticles deposited on a clean (001) copper surface has been studied in real time using a novel *in-situ* ultrahigh vacuum (UHV) transmission electron microscope with a UHV DC sputtering attachment. The particles were found to assume an initially random orientation on the foil, reorienting upon heating to assume the orientation of the substrate by a mechanism involving classical sintering and grain growth, contrary to expectation. The experiment was inspired by that of Gleiter, where rotation of macroscopic copper balls to low energy configuations was observed upon annealing. We present the results of our study and discuss the differences in reorientation mechanisms in the case of macro- and nano-scale particle sintering.

## 1.   INTRODUCTION

The current interest in nanophase materials arises primarily from the novel grain size-dependent properties which arise due to the large surface area:volume ratio. Bulk quantities of nanophase material are typically produced by compacting and sintering nanoparticulates. The nanoparticles can be produced either by evaporation, sputtering or mechanical attrition of the constituent material(s) (e.g. Edelstein and Cammarata 1996). Due to a number of experimental challenges which include reactivity with the atmosphere and imaging resolution, the structures and properties of nanophase materials are not well understood. As a consequence computer simulations are being used increasingly to predict sintering behavior (see e.g. Phillpot *et al.* 1995, Zeng *et al.* 1997). Reliable experimental data from samples of known cleanliness are essential, however in order to determine the operative mechanisms and provide a clear understanding for further investigations.

In this paper we present a study of the sintering of sputtered copper nanoparticles deposited on a clean single crystal copper foil inside an ultrahigh vacuum transmission electron microscope (UHV TEM). Our experiment is qualitatively similar to that of Gleiter (e.g. Herman *et al.*, 1976) where an array of single crystal balls of copper and silver, in the size regime 10-100µm, were sintered to single crystal copper close to the melting point in a reducing atmosphere. X-ray pole-figure analysis of the orientations of these balls showed that they rotated over many hours to form low energy grain boundaries with the substrate at their necks. In our experiments the particles are ~$10^5$ times smaller and sintered under UHV.

## 2.    EXPERIMENTAL METHOD

The present experiments were performed using a JEOL 200CX TEM modified for ultrahigh vacuum, sample heating and gas injection (McDonald *et al.* 1989). A photograph of the experimental configuration is shown in figure 1. Copper films 40nm in thickness and of predominantly (001) orientation were deposited on rocksalt by e-beam evaporation, transferred to a silicon heater block by flotation and mounted inside the microscope. The native oxide ($Cu_2O$) was removed *in-situ* by annealing in methanol vapor at 350°C.

Upon cooling the substrate, nanoparticles were generated in the UHV sputtering attachment under 1.3 Torr of ultraclean Argon gas. Immediately following generation the particles were transferred inside the microscope, in the gas phase, via a connecting pipe and deposited on the copper foil.

## 3.    RESULTS AND DISCUSSION

The morphology and orientations of the particles were examined using bright-field and dark-field imaging together with selected area electron diffraction patterns (SADPs). The particles were found to have assumed random orientations on the substrate and were stable in both orientation and position on the surface. However, upon heating to ~200°C in 10° intervals and at a rate of ~20° min$^{-1}$, the particles began to spontaneously disappear from the bright-field images. Real-time video recordings indicated the apparent diameters of the particles to shrink rapidly to zero over a time frame of approximately 0.2 seconds. Subsequent examination of dark-field images, using a $Cu_{220}$ reflection, revealed thickness changes in the film of similar dimensions to the particle diameters prior to annealing. A typical dark-field image taken immediately following annealing is shown in figure 2.

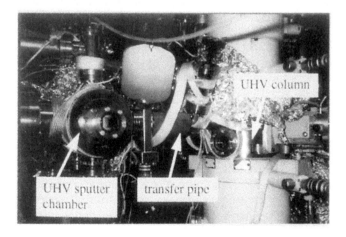

Figure 1: Photograph of the UHV DC sputtering attachment (left), UHV TEM column (right) and particle transfer pipe (center).

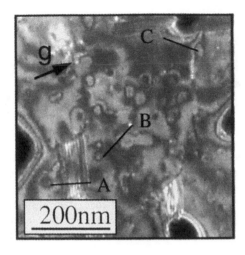

Figure 2. Dark-field image of the copper foil; arrows A, B and C indicate the presence of
thickness fringes corresponding to the positions of nanoparticles prior to sintering.

Sintering between copper nanoparticles deposited on amorphous silicon nitride substrates has been observed previously at room temperature in our system (Olynick *et al.* 1995) and it is expected that the initial contact area, or neck (figure 3(a)), between the particle and the substrate will increase towards an equilibrium configuration prescribed by a balance of forces from reductions in surface area and concomitant reductions in grain boundary area. From a simple surface and grain boundary energy balance the equilibrium contact angle between the substrate and the particle is estimated to be ~160° (see figure 3(c)). A significant driving force for neck growth will therefore exist in order to achieve this configuration. Since the equilibrium angle between the particle and substrate exceeds 90° there will come a point during neck growth (figure 3(b)) when it is favorable for the boundary to move through the particle and be eliminated.

The free energy of the particle could alternatively be reduced by particle rotation to a lower energy configuration, however our results clearly show that the particles fully reorient on the surface implying boundary elimination, in contrast to the findings of Gleiter in the case of copper particles many orders of magnitude larger. The diffraction data and image contrast are also not consistent with a rotational mechanism (Yeadon *et al.* 1997a). It was previously suggested that thermal fluctuations in the position of the grain boundary might permit migration of the boundary beyond the central (widest) region of the particle (Yeadon *et al.* 1997b). However, statistical calculations indicate this to be highly unfavorable. The dominant mechanism of mass transport in particles below ~10μm diameter and 600°C is surface diffusion (Kuczynski, 1949). Calculations of the rate of neck growth in nanoparticles, based on the modeling by Nichols and Mullins (1965) for larger particles but assuming only surface diffusion, are consistent with our experimental observations (Yeadon *et al.* 1997a). Using this model our data would indicate a 30% lower activation energy for surface self-diffusion of copper, which may be partly attributed to the ultraclean UHV

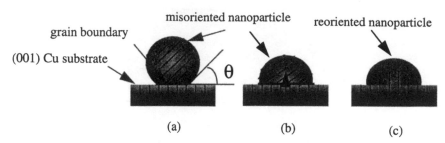

Figure 3. Schematic of the reorientation process showing (a) the contact angle between the particle and substrate, (b) neck formation, and (c) grain boundary migration.

conditions in which our sintering was performed. The difference in reorientation mechanisms is most likely a consequence of the larger surface area:volume ratio of the particles in our experiment enabling more rapid neck growth and thus reorientation by grain boundary migration.

## 4.    CONCLUSIONS

We have demonstrated that copper nanoparticles deposited on a clean copper foil reorient upon annealing by a classical process involving sintering and grain growth. This finding is in contrast to the observations of Gleiter and the difference is attributed to the large difference in the surface area:volume ratios giving more rapid neck growth in our case.

### ACKNOWLEDGEMENTS

The authors would like to thank the U.S. Department of Energy for the support of this work under contract DEFG02-96ER45439. The authors also acknowledge the use of facilities in the Center for Microanalysis at the University of Illinois, Urbana-Champaign.

### REFERENCES

Edelstein A.S. and Cammarata R.C., *"Nanomaterials: Synthesis, Properties and Applications"* (1996), IOP
Herman G., Gleiter H. and Baro G., Acta. Met. 24 (1976), p. 353
Kuczynski C., Trans. AIME 185 (1949), p. 169
Lange F.F. and Kellett B.J., J. Am. Ceram. Soc. 72 (1989), p. 735
McDonald M.L., Gibson J.M. and Unterwald F.C., Rev. Sci. Instrum. 60 (1989), p. 700
Olynick D.L., Gibson J.M. and Averback R.S., Mater. Sci. Eng. A 204 (1995), p. 54
Nichols F.A. and Mullins W.W., J.Appl. Phys. 36 (1965), p. 1826
Phillpot S.R., Wang J., Wolf D. and Gleiter H., Mater. Sci. Eng. A 204 (1995) p. 76
Yeadon M., Yang J.C., Averback R.S., Bullard J.W., Olynick D.L. and Gibson J.M., App. Phys. Lett. (1997a), *in print*
Yeadon M., Yang J.C., Ghaly M., Olynick D.L., Averback R.S. and Gibson J.M., MRS Symp. Proc. Vol 457 (1997b), p. 179
Zeng P., Zajac S., Clapp P.C. and Rifkin J.A., Mater. Sci. Eng. A, 1997, *in print*

*Inst. Phys. Conf. Ser. No 153: Section 7*
*Paper presented at Electron Microscopy and Analysis Group Conf. EMAG97, Cambridge, 1997*
© *1997 IOP Publishing Ltd*

# Monte Carlo simulation of edge artefacts in MULSAM Images

**M M ElGomati, A M D Assa'd, H Yan, J A Dell**

The Department of Electronics, University of York, Heslington, York YO1 5DD, UK.

**ABSTRACT:** A fast Monte-Carlo model to simulate electron-solid interaction in small dimension raised and buried structures has been developed. Secondary, back-scattered, Auger and low energy loss electrons and characteristic x-rays are collected. The model has been used to develop a new method of detection and correction of edge enhancement encountered in high resolution Auger imaging of topographical structures. Back-scattered electrons of energy >0.75Ep (Ep is the energy of the primary electron beam) are used to correct for Auger edge enhancement. Correction is applied to small structures, including a bevelled multi-layer sample where substrate enhancement occurs.

## 1. INTRODUCTION

Computer simulation of electron solid interaction processes, using Monte-Carlo techniques, have been established and these are utilised to investigate and quantify the artefacts which may be experienced in experimental Auger spectroscopy (AES) and multi-spectral analytical microscopy (MULSAM). This paper outlines the simulation results obtained from some typical small structures and the artefact magnitude resulting in each case. It is concluded that the simulation tools can be usefully employed in the interpretation of experimental measurements as well as evaluating the design of new experiments, without involving costly practical measurements. This is an important step towards the quantification of Auger imaging of ultra large scale integrated circuits.

## 2. SIMULATION TECHNIQUES, the single electron scattering model

To simulate the interaction of incident electrons with the materials in question a new model, which correctly applies the screening parameter and determines the Auger yield, has been developed by Asaa'd (1993) from the work of Joy (1987). It is based on a Monte-Carlo random process which is employed to establish the path taken by the electron after each collision event in the material. The accumulated results from typically 50000 electron paths are used. The electron mean free path is governed by either a polynomial fit to the Mott elastic cross-section or a modified Rutherford scattering cross-section. The energy lost by each electron as it travels between elastic scattering events is determined by a modified Bethe formula. The resulting electron path is governed by the particular characteristics of the material and the reducing energy of the electron. Simulated events which give rise to back-scattered electrons are counted and the emission of Auger electrons are estimated. It is found that rediffused primary electrons, which exit from the side of a raised structure and re-enter the substrate close to the edge, are very effective in Auger electron generation, both from the overlayer and the surrounding substrate, and are the main cause of the observed edge enhancement.

### 2.1 Experimental Configurations Employed

Two examples of sample geometry have been studied, one involves a simple metal overlayer, of Gold or Aluminium, placed on a silicon substrate and the other involves a bevelled multilayer structure with alternating platinum and cobalt layers which is representative of structures found in depth profiling experiments.

## 2.2 Simulation Experimental Results for the simple structure

Figures 1 shows typical simulated line scan results for the simple geometrical structure. Evaluation of these results reveal the large magnitude of the artefact associated with the edge of the metal layer which is mainly caused by the mechanism described above. In this example the back-scattered results for electrons with energy >0.75 Ep show the same profile as the Auger results.

It has been found that for any detector geometry and a wide range of primary beam incidence angles there is a close similarity between the back-scattered signal and the Auger profile. For this reason the back-scattered signal can be used to correct the Auger profile effectively removing the edge artefact. The result of such correction is also shown in Figure 1.

Experimental results indicate that off-normal incidence of the primary beam may give rise to an overall shift of the apparent edge position due to the skewed energy profile of the back-scattered electrons. However, the use of the correction method proposed here significantly reduces the magnitude of this effect.

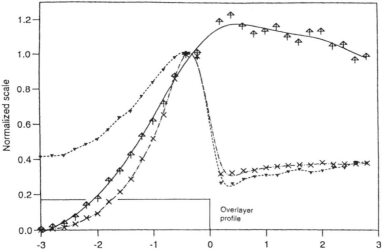

Figure 1     Electron beam position (microns) relative to the edge
(× Si Auger signal, ▼ BSE escaping between 20 and 30 degrees from the surface normal,
↑ Corrected Auger signal, Ep 20keV, Al ovelayer 140nm)

## 2.3 Substrate Enhancement

Depth profiling is a frequent requirement in surface analysis. In a multi-layer specimen electrons back-scattering from the substrate enhance the Auger signal and the extent to which this occurs depends on the sample geometry. However, the precise geometry of the specimen is frequently unknown and calculation of the enhancement cannot be performed.

The back-scattered signal from the specimen can be used to assess the Auger enhancement from an unknown specimen through the use of the equation proposed by Cazaux (1992). For normal incidence, this takes the form:

$$R_A = 1 + 4\left(\frac{\eta}{1+\eta}\right)\left(1-\frac{1}{U}\right)\left(1+\frac{\ln\left(\frac{1+\eta}{2}\right)}{\ln U}\right)$$

where $\eta$ is the back-scattered coefficient, $U$ is the reduced energy ($E_p/E_c$), $E_p$ is the primary electron energy and $E_c$ is the enegy of the electron in question.

This can be employed as a correction factor for the Auger signal which eliminates the effect of the substrate enhancement. This method has been found to give good results as shown by the example below where the corrected Auger values show only slight variation across the specimen.

### 2.4    Simulation Results for a multi-layer sample

The results for a simulated line scan across the bevelled multi-layer sample which consists of alternating 10nm layers of platinum and cobalt on a silicon substrate are shown in Figure 2 and 3. It can be seen that the back-scattered signal for normal incidence ($\theta = 0$ Figure 2) increases by a factor of two as the incident electron beam proceeds from the substrate towards the top of the bevelled structure. A similar increase is seen at other angles of incidence.

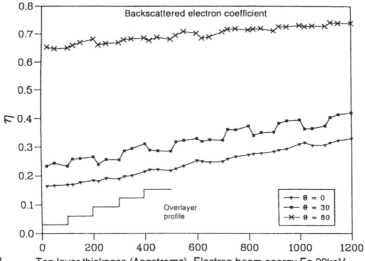

Figure 2    Top layer thickness (Angstroms), Electron beam energy Ep 20keV

The Auger yield from both layers, ($\blacktriangledown$ Platinum and $\bullet$ Cobalt Figure 3), show a general increase across the specimen although not by the same factor due to a fortunate combination of effects in this particular case. In general the increase in both factors is because the effective atomic number (Zeff) of the sample is lower when the electron beam incidence is close to the substrate and higher when the electron beam incidence is near the top of the structure. The back-scattered electron yield can thus be used to correct the Auger yield and

264

remove the substrate enhancement. The corrected Auger results, points × and⊗ for Platinum and Cobalt respectively, which are now quite uniform across the layers, are also shown in Figure 3.

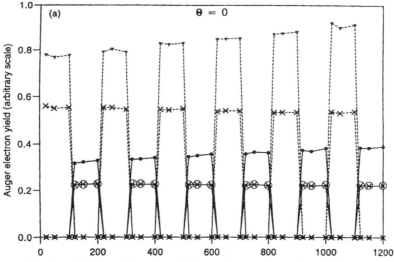

Figure 3        Top layer thickness (Angstroms), Electron beam energy Ep 20keV

## 3.    CONCLUSION

The importance of Monte-Carlo simulation can be appreciated, not only providing invaluable back-scattered electron yield results but also showing the magnitude of artefacts which can be expected in a particular experimental situation. Correction of image artefacts by the methods described above has been successfully applied to the edge enhancement artefact (ElGomati et al 1988). Reduction of the edge artefact observed in Auger imaging by an order of magnitude has been achieved and substrate effects found in Auger depth profiling have been corrected.

In the future it is hoped that the simulation tools can be usefully employed in evaluating the design of new experiments, without involving costly practical measurements, and in providing data to enable the minimisation and removal of the artefacts which are likely to be present in measurements of practical specimens.

**REFERENCES**
Assa'd A M D, El Gomati M M and Robinson J (1993), Inst. Phys. Conf. Ser. No. **130**, 403-406.
Cazaux J (1992), Microchim. Acta., **107**
ElGomati M M, Prutton M, Lamb B and Tuppen C G (1988), Surface & Interface Analysis, **11**, 251-265.
Joy D C(1987), J. Microscopy, **147**, 51.

Inst. Phys. Conf. Ser. No 153: Section 7
Paper presented at Electron Microscopy and Analysis Group Conf. EMAG97, Cambridge, 1997

# On the measurement of low energy backscattered and secondary electron coefficients

**M. M. El Gomati, A M D Assa'd, T. El Gomati\* and M. Zadrazil+**

Department of Electronics, University of York, York YO1 5DD, UK
mmg@ohm.york.ac.uk
* Department of Chemistry, University of Al-Fateh, Tripoli, Libya
+ Institute of Scientific Instruments, Brno, Czech Republic.

**ABSTRACT:** We report on measurements of the backscattered and secondary electron coefficients in the energy range 250eV-6 keV which are collected in UHV from target surfaces that were cleaned by ion bombardment. Data collected under conventional SEM pressures and with electron energies <2 keV exhibit an un-systematic behaviour with respect to the target atomic number. The results obtained from clean surfaces show that for electron energies > 600 eV, the backscattering coefficient increases with increased target atomic numbers. For lower electron energies, we see the influence of strong elastic scattering as a function of the target atomic number in a complex way. Secondary electron coefficients show similar behaviour to previously published data of a maximum yield at a given energy, albeit with different magnitudes and the energies they occur at.

## 1. INTRODUCTION

Low voltage electron microscopy (LVSEM) is becoming an important metrology tool, particularly in the microelectronics discipline but is also extensively used in the biological and material sciences because of the ability to image insulating materials. With the development and use of field emission sources in electron microscopy, some recent reports have demonstrated the power of LVSEM to detect low dopant concentrations in semiconductors (Perovic et al 1995 and Venables and Maher 1996). However, the contrast of both the backscattered and secondary electron images has been reported to depend on the incident electron energies, as well as on the type of instrument where the measurement takes place.

Low energy electrons (<5keV) have a small range within solids, typically of the order of 10-100 nanometers. It is this property which makes LVSEM attractive to use in many applications but it is also one of the reasons that makes interpretation of the results in this mode quite difficult and as yet non-quantitative. There are two contributing factors that determine the values of the backscattered ($\eta$) and secondary electron ($\delta$) coefficients, namely the composition of the sample under investigation and the incident electron energy used. At low energies the electron range in the solid and its transport properties are strong functions of the solid's electronic state configuration. Some recent reports have shown some systematic trends of these properties across the periodic table (Matthew et. al. 1997, Cumpson 1997).

The confinement of the interaction volume of the low energy incident electrons to such shallow surface regions makes the sample environment and its preparation history important parameters in determining the value of $\eta$ and $\delta$. It is therefore essential to characterise the sample environment and its surface conditions for the measurements of these parameters. The purpose of this paper is to report on measurements of η and δ from solid surfaces, which have been cleaned by energetic ions. The data were collected under ultra high vacuum (UHV) conditions in the energy range 250eV-6keV. The reported measurements are also compared with previously published data.

## 2.   EXPERIMENT

The present experimental measurements of η and δ are for a number of elements spanning the periodic table that were collected under UHV conditions $2-5 \times 10^{-10}$ mbar. Electrons of energies in the range 250eV to 6keV impinged target surfaces at normal incidence. All target surfaces were mechanically polished. These were then washed in de-ionised water followed by isopropanol alcohol for periods of between 10-20 minutes in an ultrasonic bath, prior to insertion into the vacuum system. After initial inspection of the sample surfaces, collection of η and δ data from these surfaces took place. These data are referred to as the "as inserted" or "uncleaned" samples. All the sample surfaces were then subjected to bombardment by energetic argon (Ar) ions to remove surface contaminants prior to data collection from the clean surfaces. Cleaning conditions were typically 2-5 $\mu$A cm$^{-2}$ at 2-3 keV and for periods of between 120-360 minutes. Details of the experimental set-up have been described elsewhere (El Gomati and Asaa'd 1996). Figure 1 shows a schematic diagram of the electron detector used showing a collection angle of $2\pi$ str for the present measurements. Discussion of the possible systematic errors in these measurements and an estimate of their value (of less than 3%) have been given elsewhere (Reimer and Tollkamp 1980, Asaa'd 1996).

## 3.   RESULTS AND DISCUSSION

Figure 2 shows the backscattered electron coefficient data from the Ar ion cleaned surfaces (solid symbols with a dotted line through) and those obtained from the as inserted surfaces of Bongeler et al 1993, which are very similar to ours (open symbols with solid line through). The as-inserted data in the incident energy range up to 2 keV show a complex behaviour. For example in the 1keV case, η increases as the target atomic number is increased until it reaches a maximum around targets of atomic number Z=30-35, η then appears to gradually decrease in value. Our interpretation of these results is that under conventional vacuum conditions ($\sim \times 10^{-5}$ mbar), most surfaces will grow a layer of contamination of variable thickness and composition. The most likely elements in this film are carbon, oxygen and hydrogen. Removal of the contamination film by ion cleaning reveal a similar behaviour between η and Z for all energies used down to 500-600eV incident electron energies.  These results are encouraging from the point of view of atomic number contrast interpretation in LVSEM and highlight the importance of UHV conditions in this imaging mode, as the case is for surface analysis. The η values from the clean surfaces agree in general with the data obtained by Thomas and Pattinson (1970) and by Bronshtein and Faiman (1969), which were both collected from evaporated films in UHV.

The secondary electron coefficient δ was also measured for the same set of targets used in Figure 2 and under the same experimental conditions. The results obtained from the as inserted and the Ar ion cleaned surfaces show the familiar pattern of an increase in the value of δ as the incident electron energy (E) is increased until a maximum value $\delta_m$ is reached at a given energy $E_m$. δ then gradually decreases as E is increased. The results obtained in the present study show a different value of $\delta_m$ and $E_m$ for each element studied and these also differ in value from the other published data. The discrepancy with the published data may again be due to the surface treatments of the two target surfaces. In the present measurement, $\delta$ will particularly be sensitive to the local surface chemistry, where the presence of contaminants such as oxygen or carbon is bound to have a detrimental effect on the values obtained.

Scattering of electrons inside solids is governed by elastic and inelastic events and the solid's transport properties, which together determine the size of the interaction volume of the incident electrons with a solid target. It is the ratio of the elastic to inelastic scattering and these to the solid's transport properties that determine the values of $\eta$ and $\delta$ that one obtains in a given experiment. For example, in the case of incident electron energies greater than 1-2keV, these ratios are stable and show a simple dependence of the values of these parameters on the target atomic number. However, this picture alters dramatically as the incident electron energy approaches the electron binding energies of the bombarded solids. Although this is a somewhat simplified picture, the overall pattern is the normally accepted interpretation. Even in the low energy regime (<1keV) where these parameters have strong dependence on a number of other factors, there is still perhaps a "simple" pattern of behaviour of $\eta$ and $\delta$ with respect to the target atomic number (Z) that can be uncovered. This picture can only be revealed if one performs the experiment on well characterised target surfaces and where the experimental conditions and environment preserve the surface cleanliness of the target in question.

In conclusion, the results presented in this study show that the surface condition is of paramount importance in the measurements of the backscattering and secondary electron coefficients of low energy electrons. With the increased use of field electron emitters as electron sources in modern SEMs, it is perhaps time that such instruments particularly operated in the LVSEM mode should adopt surface analysis practice of provision of sample cleaning and UHV environment.

This work was partly supported by EPSRC and the EU grant PC12283.

**References**

A M D Asaa'd, 1996, D Phil Thesis, University of York, UK.
I M Bronshtein and B S Faiman, 1969 Sec. Electron Emission, Nauka Press, Moscow.
P Cumpson, 1997, Surface & Interface Anal. **25**, 357-364.
M M El-Gomati and A M D Asaa'd, 1997, Mikrochemica Acta, In Press.
J Matthew, A Jackson and M El Gomati, 1997, J Elec. Spec. & Rel. Phen, In Press.
D Perovic et al, 1995, Ultra Microscopy, **58**, 104-113.
L Reimer and C Tollkamp, 1980, Scanning **3**, 35-39.
S Thomas and E B Pattinson, J Phys. D: Appl. Phys. **3**, (1970), 349-357.
D Venables and D M Maher, 1996 J. Vac. Sci. & Tech. **14(1)**, 421-425.

Table 1. Values of $\delta_m$ and $E_m$ obtained in the present experiment from cleaned surfaces.

| Element | C | Al | Ti | Ni | Cu | Mo | Ag | W | Pt | Au |
|---------|------|------|------|------|------|------|------|------|------|------|
| $\delta_m$ | 0.72 | 2.12 | 0.96 | 0.89 | 0.92 | 0.91 | 1.06 | 0.92 | 1.37 | 1.26 |
| $E_m$ (keV) | 0.3 | 0.35 | 0.35 | 0.45 | 0.65 | 0.35 | 0.6 | 0.55 | 0.7 | 0.8 |

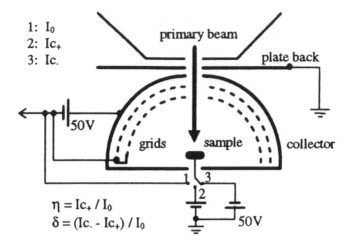

1: $I_0$
2: $Ic_+$
3: $Ic_-$

primary beam

plate back

50V

grids     sample     collector

1   3

2

$\eta = Ic_+ / I_0$
$\delta = (Ic_- - Ic_+) / I_0$

50V

Figure 1. Schematic of the electron detector used in the measurement of $\eta$ and $\delta$.

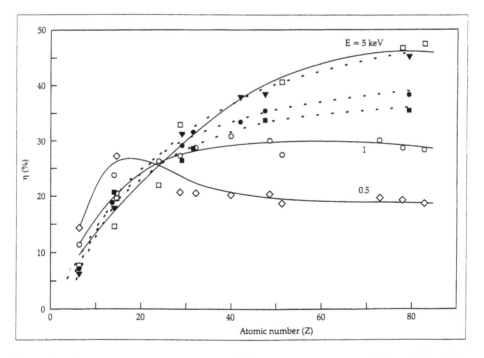

Figure 2. The backscattered electron coefficient $\eta$ as a function of the target atomic number for the as inserted data (open symbols with solid line through, data from Bongeler et al) and the surface cleaned targets (solid symbols with dotted line through).

*Inst. Phys. Conf. Ser. No 153: Section 8*
*Paper presented at Electron Microscopy and Analysis Group Conf. EMAG97, Cambridge, 1997*
© *1997 IOP Publishing Ltd*

# New Directions In Very High Resolution Analytical Electron Microscopy

**P Kruit**

Delft University of Technology, Department of Applied Physics, Lorentzweg 1, 2628 CJ Delft, The Netherlands

**ABSTRACT:** Three areas in which sub-nanometer resolution analytical microscopy can be advanced are identified: higher spatial resolution, better spectral resolution and exploration of new secondary signals. In each area, one example of an instrumental development is described. For higher spatial resolution the example is electron standing wave illumination. For spectral resolution it is monochromatization of the primary electron beam for 50 meV electron energy loss spectroscopy. For the exploration of new secondary signals the example is Auger spectroscopy in the scanning transmission electron microscope.

## 1.    INTRODUCTION

Sub-nanometer resolution analytical microscopy can be advanced by innovations in three different areas. First of all, it is still possible to improve the spatial resolution of existing methods, either by traditional means such as smaller aberrations and disturbances or by revolutionary means such as aberration correction or standing wave illumination. At these levels of resolution it is important to pay as much attention to the microscope alignment and the electronwave-specimen interaction as in high resolution transmission electron microscopy. Then, existing methods of microanalysis can be developed further, either by increasing the detection efficiency or by enhancing the spectral resolution. EELS seems to fit best with very high spatial resolution, but much can be gained in terms of energy resolution. It is time to decrease the energy spread of the primary beam such that an EELS resolution of about 50-100 meV can be contained for high resolution chemical shift measurements. Finally, as the third area of innovation, new information channels can be explored. A start has been made on adding Auger spectroscopy and Auger-EELS coincidence spectroscopy to the STEM, but much work still has to be performed before that can be as standard as X-ray analysis or EELS. The following concentrates on a discussion of necessary instrumental developments for examples in all three areas: standing wave illumination, monochromatization of the primary beam and efficient detection of Auger- and coincidence spectra.

## 2.    HIGHER SPATIAL RESOLUTION, MORE CURRENT

In recent years, the probe size in scanning transmission electron microscopes has been brought down to the resolution limit determined by a balance between diffraction and lens aberrations. At this limit, the probe current is fundamentally limited to a value of about $I = 10^{-18} \cdot B_r$, in which $B_r$ is the reduced brightness of the electron gun, expressed in A/sr m² eV. A typical $B_r$ for a Schottky source is $2 \times 10^7$, giving a probe current of 20 pA. To increase this current and/or to obtain an even better resolution, the development of aberration correctors is a promising direction. We are trying a different approach, in which the specimen is illuminated by two, mutually coherent beams. Each individual beam is limited to a narrow aperture angle, so that the effect of aberrations is under control. However, the relative angle between the beams is substantial, so that interference between the beams gives a pattern with very closely spaced fringes. By observing the changes in the intensity of secondary signals (X-ray, EELS) while moving the fringe pattern over the specimen, one obtains analytical information (Mertens et al 1996). If a biprism is used for splitting the beam into two coherent beams, the narrow fringes can be obtained but the current is not higher than in a diffraction limited probe. We are attempting to make a stable crystal beamsplitter, e.g. a Si thin foil. The instrument under construction has this foil positioned in an auxiliary objective lens. This lens is followed by a three-lens "magnification and rotation unit" which images the fringe pattern onto the specimen with adjustable fringe distance and fringe orientation. Taking all aberrations into account, we accept a current - fringe distance relation as given in fig.1, where fringe scanning electron microscopy (FSEM) is compared with STEM.

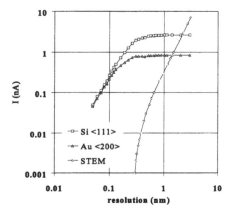

Fig.1: Current as a function of the resolution for STEM and FSEM for 10 nm field. The high tension is 100 kV, $C_s = 2.0$ mm and the FWHM defocus spread is 40 nm.

## 2.    BETTER ENERGY RESOLUTION IN EELS

All analytical methods benefit from better spectral resolution. Even if it does not supply fundamentally different information, it always helps to provide better signal to background ratio's. In EDX the expected improvement must come from new spectrometer principles. In EELS the improvement must come from a primary beam with lower energy spread. The spectrometer has intrinsically an almost unlimited resolution, only limited by power supply stabilities. The aberrations can be controlled by accepting smaller input angles. Microscopes with monochromators have been built (Terauchi et al 1991), but the combination of increased energy resolution and subnanometer spatial resolution has not yet been obtained. We have designed, and are now testing, a small monochromator directly behind the extraction electrode of a Schottky field emission gun, see fig.2 (Mook and Kruit 1997).

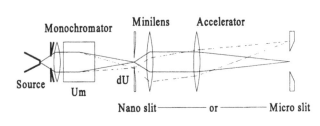

Fig.2: Schematic principle of a monochromator for a Schottky emitter

The difficulty in designing a monochromator is finding the right balance between dispersion, aberrations and stochastic Coulomb interactions. In our early designs we found that the Coulomb interactions invariably either increased the energy spread by the Boersch effect or decreased the beam brightness by the trajectory displacement effect.

A fundamentally different approach to obtaining a beam with about 50-100 meV energy spread is depicted in fig. 3 (Borgonjen et al 1997).

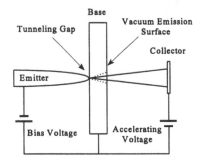
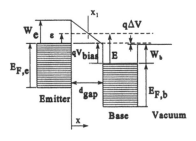

Fig. 3: Principle of Tunnel Junction Emitter. Mechanical set-up and energy levels of tip and foil.

The electron source consists of a sharp metal tip that is positioned within tunnelling range of a conducting thin foil. A tunnelling bias potential is applied across the gap such that electrons are injected into the foil. The electron beam emerging from the opposing surface of the foil has a narrow energy spread due to the energy filtering effect of the potential step at the surface. Calculations for the electron emission which include the tunnel junction, ballistic electron transport through the foil, and subsequent emission from the foil surface result in an expected foil transmission versus energy spread as given in fig.4. It is expected that the emission energy spread is 100 meV for an emission current ratio of 1%. For a typical tunnel current of 10 nA, this implies an emission current of 0.1 nA. The corresponding reduced brightness is $6.6 \times 10^7$ A m$^{-2}$ sr$^{-1}$ V$^{-1}$. For the emission to be sufficient for a practical electron source, it is necessary that the foil thickness does not exceed the mean free path for electron scattering which is approximately 5 nm at the excitation energy of interest. Polycrystalline platinum foils of 5 nm thickness have been fabricated that are mechanically stable and withstand the forces exerted by a scanning tunnelling microscope tip under ambient conditions. Further experimental effort is currently undertaken.

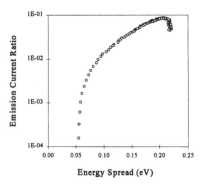

Fig.4: Calculated performance of a tunnel junction emitter at room temperature.

## 3.    DETECTION OF NEW SIGNALS

Analytical microscopy is the art of detecting the specimen's response to the bombardment with electrons. I am convinced that there is more response than what we usually listen to. We have a long running project on detecting the Auger electrons in STEM and measuring secondary electrons in coincidence with EELS, but it would be interesting to also try to detect electron stimulated desorption especially from insulators. For the Auger spectroscopy, a Philips 300 kV TEM-STEM was modified, see e.g. Kruit and Venables 1988. A schematic overview of this microscope is given in fig. 5. Some preliminary experiments have been performed.

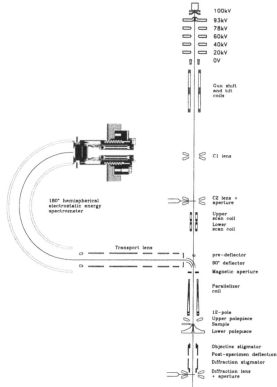

Fig. 5: Optics of the Auger TEM

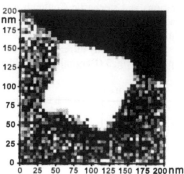

Fig.6: Oxygen (KLL) Auger map of MgO cube

We have obtained energy scans with the primary beam focused to a spot with a diameter between 3 and 10 nm. In this way we were able to measure the secondary electron signal and to identify the Auger lines of carbon (KLL-edge, 265 eV), silver (MNN-edge, 355 eV), titanium (LMM-edge, 381 and 416 eV), oxygen (KLL-edge, 503 eV) and nickel (LMM-edge, 707, 772 and 844 eV).

Fig. 6 shows a map of an MgO crystal on a carbon foil. The Auger spectrometer was tuned to the oxygen (KLL) Auger line. The background subtraction was not perfect, so only the hole in the foil is really black. the present effort is directed at building a 100 channel parallel detector for the Auger signals, because a practical use of Auger spectroscopy can only be expected if the signal rate can compete with EDX and parallel EELS. The technical challenge in the detector is maintaining a time resolution of a few nanoseconds which is necessary for parallel-parallel coincidence detection with the EELS signal.

## REFERENCES

Borgonjen EG, van Bakel GPEM, Hagen CW, Kruit P, Applied Surface Science, 1997, 165.

Kruit P and Venables JA, Ultramicroscopy **25**, 1988, 183.

Mertens BM, Buist AH and Kruit P, Proceedings of the 11th European Congress on Electron Microscopy, Dublin, Ireland, 26-30 August 1996.

Mook HW and Kruit P, 1997 These proceedings.

Terauchi M, Kuzuo R, Satoh F, Tanaka M, Tsuno K, Ohyama J, Microsc. Microanal. Microstruct., **2**, 1991, 351.

*Inst. Phys. Conf. Ser. No 153: Section 8*
*Paper presented at Electron Microscopy and Analysis Group Conf. EMAG97, Cambridge, 1997*
© *1997 IOP Publishing Ltd*

# Electron Energy Loss Spectroscopy in the 2-5 eV Regime

## U Bangert, A Harvey, R Keyse[a] and C Dieker[b]

Department of Physics, UMIST, Manchester M60 1QD, UK
[a]Department of Materials Science and Engineering, University of Liverpool, Liverpool L69 3BX, UK
[b]Institut fuer Schicht und Ionentechnik, Forschungszentrum Juelich, 52425 Juelich, Germany

**ABSTRACT.** Localized electron energy loss spectroscopy (EELS) studies in the bandgap energy regime were conducted in the GaN/3C-SiC and GaAlInP/GaAs materials systems. Requirements on experimental conditions and data analysis using the VG601UX and the possible interferences arising from instrumental factors and Cerenkov radiation losses are addressed. The bandgap of GaAs can be determined after subtraction of the zero loss peak. This procedure is, however, proned to introduce inaccuracies: GaAlInP shows deviations from the expected bandgap value of 0.5 eV. Materials not requiring subtraction of the zero loss peak fare better: the bandgap values of 3.3 and 3.1 eV obtained for GaN tie in with those obtained from Photoluminescence spectra for hexagonal and cubic GaN. Variations in the midgap states density vary with the location in the film and can be related to the amount of microstructural defects.

## 1. INTRODUCTION

EELS in the low loss regime has recently attracted attention for its prospects for localised band gap studies. One of the common approaches is to investigate the low loss region is via Kraemers-Kroening analysis [Egerton 1996], in which the real part of the dielectric function is extracted from the loss spectrum, whose first zero crossing gives the bandgap. In order for this to be successful the spectra need to be representing true energy loss processes. Common complications that can occur are the occurence of Cerenkov radiation and instrumental artifacts added into the spectra. The latter arise as a result of alignment errors and non optimized detection system parameters. We concentrate on investigating the direct low loss data taken under optimum instrumental conditions in a VG STEM with cold field emission gun. Bandgap studies have been undertaken in the GaInAlP/GaAs and the GaN/3C-SiC material systems.

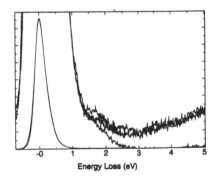

Fig.1. Low loss region of GaAs, GaInAlP together with ZLP after deconvolution of the PSF. Note the shoulder at 1.5 eV. Underlaid are the full size spectra.

## 2. EXPERIMENTAL

Measurements are carried out on cross-sectional samples in a VG601UX FEGSTEM. Millisecond acquisition times are used to minimize the effects of high voltage instabilities. In order to obtain reasonable statistics a number of spectra are taken, aligned and summed. An instrumental broadening function (PSF) is acquired for each instrumental set-up and deconvolved from the corresponding spectra. In previous work we have modelled the zero loss peak (ZLP) and subtracted it from the spectra [Bangert et al 1997]. The danger in this is that this does not take account of artifacts in the ZLP (and in the spectra) arising from alignment errors. Here we subtract an experimentally obtained ZLP, ensuring that this ZLP was acquired under identical conditions and immediately after the acquisition of a given spectrum. It was found that this indeed eradicates instrumental artifacts, such as high energy shoulders of the ZLP. An example of this can be seen in fig.1, which shows the tail section of the ZLP taken in the hole together with low loss spectra taken in GaAs (highest curve) and GaInAsP. The peak at 1.5 eV is merely due to an instrumentally induced high energy shoulder of the ZLP.

With regards to the materials systems, GaN was chosen because the bandgap is large and not influenced by the tail of the ZLP. On cubic SiC, GaN grows in the cubic and the hexagonal phase. The cubic phase is predicted to have a bandgap of 3.25 eV, the hexagonal phase of 3.4 eV [Logothetidis et al 1994]. We set out to establish, whether it is possible to distinguish the bandgaps of the different phases. GaAs/GaInAlP was investigated to see whether bandgaps as low as 1.4 eV and differences between GaAs and GaInAlP can be detected.

## 3. RESULTS AND DISCUSSION

**3.1 Radiation losses.** The dielectric constants of III-V semiconductors at bandgap energies are of the order of 10 and hence the occurence of Cerenkov radiation is to be expected. The energy and and intensity of Cerenkov radiation produced by an electron beam of a given energy decreases with deviation from rectilinear propagation. Hence the position and strength of EEL peaks due to Cerenkov radiation losses depends on the direction under which the energy loss electrons are collected. It also depends on specimen thickness. Festenberg [1969] predicted radiation losses at 2.9 and 3.3 eV at $0°$ for GaAs and GaP respectively, i.e *above* the bandgap. Tricker et al [1997] claim a peak at bandgap in GaN to be due to Cerenkov losses. The strength of the emission at a given energy and hence the strength of the associated electron loss peak decreases rapidly with increasing angle. We expect that due to the solid collection angle of 3.4 mrad for EELS the mean intensity value of the Cerenkov loss peak will be negligable. Indeed we have not observed any distinct peaks around the bandgap in any of the materials, as can be seen in the following figures. Instead the rise in all the loss spectra around bandgap can be approximated by the parabolic density of states (DOS) curve $\sqrt{(E-E_{gap})}$.

**3.2 GaInAlP/GaAs.** Fig. 2 shows residual spectra of GaAs and GaInAlP, respectively. These are the spectra from fig. 1 after subraction of the ZLP. A DOS curve is shown to intersect the zero line at 1.5 eV for GaAs, which is an acceptable value for the bandgap of GaAs. The parabola for GaInAlP intersects the zero line at 1.7 eV. The bandgap of $(Ga_{0.3}Al_{0.7})_{0.5}In_{0.5}P$ is 2.23 eV. We do not have any explanation for the discrepancy. It was, however, observed in TEM micrographs that the quaternary layer showed extraordinarily strong sharp and bands, a few nm thick, parallel to the growth surface, presumably connected with composition fluctuations as a result of the growth process. Whether this led to formation of different quaternary materials with a variation of bandgaps, the smallest of which is observed in the EEL spectrum, has yet to be clarified.

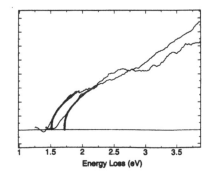

Fig.2 Residual low loss spectra of GaAs (left curve) and GaAlInAs with fitted DOS curves.

**3.3. GaN/3C-SiC.** It has been found that the GaN film used in this study exhibits grains of two phases in approximately equal proportions, a 'native' hcp phase and a 'transformed' hcp phase originating from an orderly faulted cubic phase. The more detailed epitaxial relationships, which differ from those of layers grown on other cubic substrates (e.g. GaAs), are to be published elsewhere. The remaining portion of purely cubic phase in the transformed grains is very small . Fig.3 shows a section of a native hcp grain (left side of boundary) meeting a transformed grain (right side of boundary). The left upper corner of this grain is cubic. The stacking faults become more frequent towards the bottom right corner (not shown here), until they constitute an ordered sequence, where two faultily stacked planes of the cubic structure alternate with two unfaulted planes, thus yielding the hcp structure. Fig.4 shows a succession of EEL spectra taken in native hcp grains (4a), in the cubic

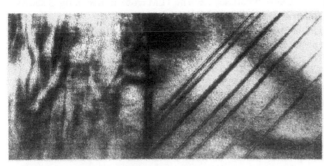

Fig.3 Cross-sectional TEM view of GaN/3C-Si, showing a native hcp grain (left) and a cubic part of a grain, which transforms into a hcp grain as the stacking faults become more frequent (towards the right, transformation not shown here).

parts of cubic grains (4b) and in the faulted (i.e. the hcp) part of cubic grains (4c). The spectra in each sequence a-c are from very localized regions close to the film surface, which constitute the most perfect parts of the film. They are normalised to each other in the rise around the bandgap energy. The spectra in 4d are from hcp and transformed cubic grains near the buffer layer interface.

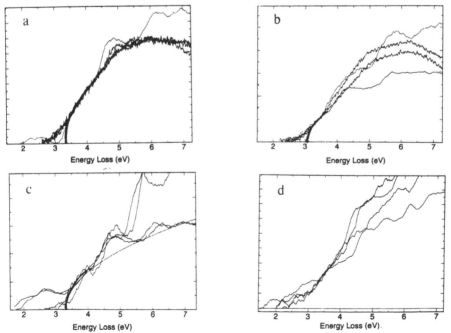

Fig. 4 Low loss spectra taken a) near the surface of native hcp grains, b) in cubic portions of cubic grains, c) in faulted (hcp) portions of cubic grains and d) near the buffer layer interface. A parabolic DOS curve (thick line) is fitted in a-c.

The spectra of the native hcp grain in fig. 4a coincide fairly well in this energy region, though genuine differences below bandgap and above 4.5 eV can be seen in the different locations. The parabolic fitted DOS curve yields a bandgap value of 3.3 eV. The energy resolution of +/- 0.2 eV allows us to conclude that there is a distribution of states between 2.8 and 3.25 eV. This is also observed in PL data (Fig.5). One curve shows a distinct peak at 2.5 eV in the region of the yellow luminescence. The TEM cross-section in fig. 3 shows a large dislocation density, which might introduce the observed midgap states.

The spectra in the small cubic parts of the cubic grains show greater variations in their shapes above 4 eV, though the rise can be fitted by a parabola with intersection at 3.1 eV. There is a smaller number of midgap states, then in the hcp grains, which could be seen in agreement with the fact that the cubic portions of the cubic grains are of great crystallographic perfection.

The spectra of the faulted regions in the cubic grains show large variations. A tentative fit yields a bandgap of 3.3 eV (as in the native hcp grain), but there are a considerable amount of midgap states in two of the spectra with a general broad peak at 2.5 eV.

EEL spectra taken closer to the interface with the AlN buffer (Fig. 4d) in hcp and transformed cubic grains do not exhibit a clear rise at the bandgap energy and the spectra differ largely from each other. There is a continuous rise dominated by midgap states starting between 1.7 and 2.0 eV in particular in the transformed cubic grains, bearing witness of the huge amount of defects, which can be observed in these parts of the film. The density of states in the native hcp grains at the interface is slightly less (lower curves) then in the transformed grains.

Fig.5 shows a PL spectrum of the GaN film, which consists of multiple peaks with the highest shoulder at 3.26 eV. This agrees with our value for the native hcp phase. Strangely neither the EELS nor the PL results give the predicted energy of 3.4 eV for the hcp phase. Further peaks occur at 3.2 and 3.156 eV, which tie in with values measured in the cubic grains. The large intensities at 2.84 and 2.37 eV are reflected in the broad EELS midgap peaks around 2.5 eV in the faulted cubic portions as well as lower down in the film.

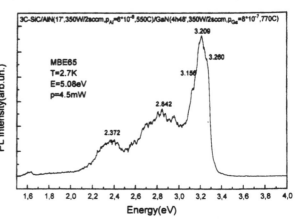

Fig.5 PL spectrum of GaN/3C-SiC

## 4. CONCLUSIONS

It is possible to give a localized account of the bandgap in GaN. Differences in the cubic and the hcp phase can be detected as well as an increase of midgap states in the regions of high defect densities. The values for the bandgap energies are confirmed by PL spectra, which exhibit multiple peaks. Furthermore the bandgap of GaAs can be detected by simple manipulation of the raw data. It was found sufficient for band gap determinations in smaller bandgap materials, following an optimized, careful data acqusition procedure, to subtract an experimental ZLP from the spectrum. Cerenkov peaks were not found to interfere with EEL spectra at bandgap, which coud be fitted by a parabolic DOS function.

## ACKNOWLEDGEMENTS

We thank the Anglo German foundation for financial support and D Freundt for supplying the samples.

## REFERENCES

Bangert U, Harvey A J and Keyse R 1997 Ultramicroscopy **68**, 173

Egerton R F 1996 in 'Electron Energy Loss Spectroscopy in the Electron Microscope', edt: Egerton, Plenum press New York and London, p 414

Festenberg C V 1969 Z. Physik **227**, 453

Logothetidis S, Petalas J, Cardona Mand Moustakas T D 1994 Phys. Rev. B **50**(24), 18017

Tricker D M, Natusch M K H, Boothroyd C B, Xin Y, Brown P D, Cheng T S, Foxon C T and Humphreys C J 1997, proc of the MSM X, to be published in Inst. Phys. Conf. Ser.

Inst. Phys. Conf. Ser. No 153: Section 8
Paper presented at Electron Microscopy and Analysis Group Conf. EMAG97, Cambridge, 1997
© 1997 IOP Publishing Ltd

# Numerical simulation of valence losses in MgO cubes

J Aizpurua*, B Rafferty[†], F J García de Abajo[§] and A Howie[†]

* Facultad de Químicas, UPV/EHU, Aptdo. 1072, 20080 San Sebastián, Spain
[†] Cavendish Laboratory, Madingley Road, Cambridge CB3 OHE, UK
[§] Facultad de Informática, UPV/EHU, Aptdo. 649, 20080 San Sebastián, Spain

ABSTRACT: Valence losses can be conveniently computed using classical dielectric excitation theory. In the non-relativistic case, the losses can be simulated using a distribution of surface and interface charges. These charges interact self-consistently with each other as well as with the incident electron. This method is used to analyse experimental STEM energy loss spectra for various incident beam positions near MgO cubes. The characteristic contributions of the cube edges are identified.

## 1. INTRODUCTION

Classical dielectric theory by Fermi (1940) deals very adequately with electron energy loss spectra (EELS) in scanning transmission electron microscopy (STEM). This approach has mainly been based on analytical solutions obtained for planar interfaces (Ritchie 1957), spheres (Ferrell and Echenique 1985) and other simple structures (Rivacoba et al 1996). In practice however, much more complex geometries are often encountered such as an interface passing normally through a thin film and allowing contributions from (possibly coupled) interface, surface and edge modes (Dobrzynski 1972, Davis 1976). Such structures can now be tackled using the boundary charge method together with numerical computation. This method is based on a distribution of interface charges which interact self-consistently with the field created by the incoming electron at the interfaces of the structure. This induced charge distribution produces a field at the fast electron position which generates the energy loss. It is convenient to place a greater number of interface charges near the beam as well as in sharp regions where the electric field can vary rapidly. The interface charge approach has already been employed by Fuchs (1975) to determine the normal modes of a cube and by Ouyang and Isaacson (1989a, 1989b) to compute energy losses for a spherical particle supported on a film. Standard packages allow simulation of many real-sample geometries observed in STEM such as cubes, truncated slabs or junctions. In particular we use the boundary charge method to deal with valence losses when the electron beam passes near an MgO cube. Marks (1982) studied the energy loss functions in this material for several electron trajectories and found new excitations associated with the edge effect. Here we simulate this effect in the cube energy loss spectrum and calculate the necessary corrections in terms of appropriate loss functions.

## 2. SIMULATIONS AND EXPERIMENTAL OBSERVATIONS

### 2.1 Parallel trajectory

As a first approximation to the edge effect in an MgO cube and to compare with observed spectra, we have simulated the energy loss probability per unit length for the case of the electron beam travelling parallel to an infinitely long, square cylinder of MgO. The dielectric response function used was derived from the bulk loss function in the same material extracted from energy loss data after deconvolution. In Fig. 1 we plot the simulated spectra for two different trajectories just outside a 100nm cube as shown in the inset and compare them with the experimental observations under the same conditions. In trajectory a the beam is travelling near the centre line of the cube face and the energy loss spectrum resembles that of a single planar interface. Three main peaks are excited at 11 eV, 14 eV and about 20 eV. The simulations agree well with the experimental observations although the latter could not be normalised because of saturation of the zero loss signal. In trajectory b, where the beam travels near and parallel to the edge of the cube, the losses are shifted down in weight. The peaks are less clear than in the planar case and the peak at 20 eV almost disappears in the simulation though remaining more visible in the observations. This could be due to an error in the beam position since the 20 eV peak is restored if the trajectory is moved 2nm along the face and away from the edge. From the simulations, we can conclude that the main effect of the edge is to increase the relative weight of the low energy excitations and even completely remove the 20 eV peak for the trajectory exactly along the edge.

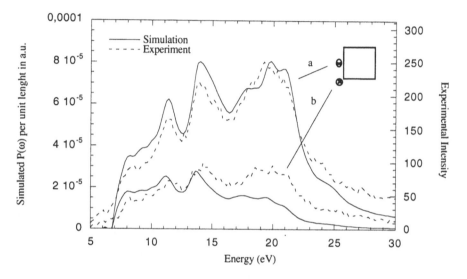

Fig 1. Energy loss probability per unit length for a 100 keV electron beam passing near an MgO cube as shown in the inset. The impact parameter is 2 nm. Solid lines correspond to the square cylinder simulations and dashed lines to the experimental observations.

## 2.2   Perpendicular trajectory

We can also study by experiment and simulation the edge effect in the loss spectra of MgO cubes when the beam travels perpendicularly to the edge. Since the exact position of the edge is not clear in all experimental situations, there can be a bulk contribution at 22 eV due to some beam penetration of the cube. In fig. 2 we present both the STEM observed energy loss spectrum and, for the identical trajectory, the total simulated energy loss probability $dP(\omega)/d\omega$ which agrees in absolute terms to a factor of about 0.7. The simulation does not reproduce the excitations observed in the band gap which must arise either from defects or from relativistic effects. In this case, the beam is travelling 2 nm just inside the edge and the contribution of the 22 eV bulk loss appears together with the other excitations from the top and bottom surfaces as well as from the edge. In the simulations we plot separately the uncorrected bulk contribution and the correction. It is clear that the main effect of the correction to the bulk is to produce an enhancement in the contribution of the low energy peaks at 11 eV and 14 eV. The planar contribution (20 eV) is not so strong in this case although it is also present in the spectrum. We cannot associate separately every peak to edge or planar excitations since they all appear mixed up, but it is possible to state that the low energy peaks are very sensitive to the edge while the higher one at 20 eV is not. This behaviour is related to the complex dielectric function which characterises MgO. A cleaner separation would be possible in the case of a free electron metal. As mentioned above, the total energy loss probability shows an important contribution in the high energy range (22 eV) due to the penetrating trajectory. The correction due to the nearby cube surfaces and edge introduces not only new characteristic peaks but also diminishes the bulk contribution via the Begenzungs effect. This bulk peak does not mask the surface and edge correction which is of the same order of magnitude for the impact parameter used.

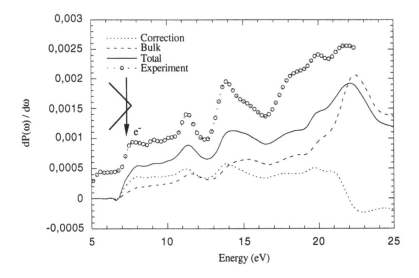

Fig 2. Energy Loss probability for a 100 keV electron impinging on an MgO edge as shown in the inset. The impact parameter is 2 nm from the edge and the different contributions to the losses are plotted together with the experimental observation.

## 3.  CHARACTERISTIC ENERGY LOSS FUNCTIONS

For situations in which the beam in the STEM passes through a complex structure constructed from two dielectric media A and B with bulk regions, planar interfaces, edges or even corners, the loss probability can (in the non-relativistic case) be described as a sum over characteristic energy loss functions.

$$\frac{d^2P}{d\omega\,dZ} = \frac{2e^2}{\pi\hbar v^2} \sum_{j=0}^{n} A_j \, \text{Im}\left[\frac{-1}{\alpha_j \varepsilon_A + (1-\alpha_j)\varepsilon_B}\right] \tag{1}$$

In this eqn. each j value represents the contribution of a different mode which, in simple situations, is recognised as a bulk mode ($\alpha_j = 1$ or 0) or a planar interface mode ($\alpha_j = 0.5$). In our analysis of MgO cubes, we can take material A to be MgO and B to be vacuum so that $\varepsilon_B = 1$. For larger cubes of dimension $L \gg v/\omega$, the cube edges do not interact significantly (except near the corners) and the value $\alpha_j = 0.3$ seems to describe the characteristic edge function. Losses in smaller cubes can still be described by eqn(1) but interactions between edges could shift the mode eigenvalues and thus the values of $\alpha_j$.

The contribution of each excitation mode is weighted with the $A_j$ factor which depends on the position of the incident beam. Thus, in the simple case of the beam travelling in vacuum at distance x parallel to and just outside an infinite planar surface, eqn(2) would consist of just one term with $\alpha_j = 0.5$ and $A_j = K_0\,(2\omega x/v)$. For the beam passing normally through a slab of thickness $L \gg v/\omega$ eqn(1), after integrating over the path length z, becomes

$$\frac{dP}{d\omega} = \frac{2e^2}{\pi\hbar v^2} L \left[ A_1 \, \text{Im}\left(\frac{-1}{\varepsilon_A}\right) + A_2 \, \text{Im}\left(\frac{-1}{0.5\varepsilon_A + 0.5}\right) \right] \tag{2}$$

where $A_1 = \ln(k_c v/\omega) - 0.5\,A_2$, $A_2 = \text{const}/L$ (Ritchie 1957). This illustrates the Begrenzung effect whereby the strength of the new surface excitation comes from a compensating reduction in the bulk excitation. More generally when the beam crosses the interface between two media, the Begrenzung reduction is shared between the bulk losses on either side.

We now return to the trajectory a of Fig. 1 where the beam is just 2nm outside the face centre for MgO cubes with $L \gg v/\omega$. The most important losses are then the planar loss and the losses from the top and bottom edge excitations. The edges on either side are too remote to be relevant and, for this beam position near an MgO cube of size L, the results are to quite a good approximation expressed by the eqn.

$$\frac{dP}{d\omega} = \frac{2e^2}{\pi\hbar v^2} L \left[ A_1 \, Im\left(\frac{-1}{0.5\varepsilon_A + 0.5}\right) + A_2 \, Im\left(\frac{-1}{0.3\varepsilon_A + 0.7}\right) \right] \tag{3}$$

where $A_1 = K_0 \, (2\omega x/v) - 3/L$ ; $A_2 = 6.75/L$ (L is expressed in nm). These impact parameter dependent coefficients are obtained for the case of MgO with an accuracy of 10 per cent.

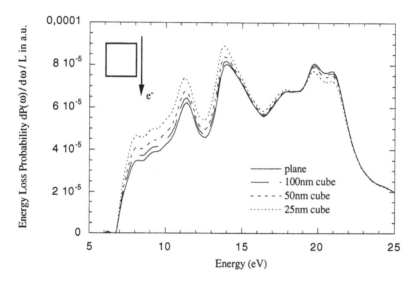

Fig 3. Energy loss probability simulated for a 100 keV electron passing near an MgO cube as shown in the inset. The probability is normalised to the length of the cube L for different sizes. The impact parameter is 2nm in all the cases.

The simulated spectra for different sizes of cube and the fixed 2nm impact parameter are shown in fig.3. Due to the presence of the edges, the planar contribution is reduced to compensate the enhancement of the low energy peaks. This effect is clearer as the cube size is smaller. In the case of a free electron material, unlike MgO, the bulk, surface and edge modes all give sharp and well separated losses. It is then possible to identify contributions from some of the other, weaker edge modes with an accuracy of 1% in the $A_j$ coefficients.

## ACKNOWLEDGMENTS

We thank Iberdrola (JA and AH), the Basque Government (JA) and de Beers (BR) for support.

## REFERENCES

Davis L C 1976 Phys. Rev. B **14**, 5523
Dobrzynski L and Maradudin A A 1972 Phys. Rev. B **6**, 3810
Fermi E 1940 Phys. R ev. **57**, 485
Ferrell T L and Echenique P M 1985 Phys. Rev. Lett. **55**, 1526
Fuchs R 1975 Phys. Rev. B **11**, 1732
Marks L D 1982 Solid State Comm **43**, 727
Ouyang F and Isaacson M 1989a Phil. Mag. B **60**, 481
Ouyang F and Isaacson M 1989b Ultramicroscopy **31**, 345
Ritchie R H 1957 Phys. Rev. **106**, 874
Rivacoba A, Apell S P and Zabala N 1996 Nucl. Instrum. Methods B **96**, 465

*Inst. Phys. Conf. Ser. No 153: Section 8*
*Paper presented at Electron Microscopy and Analysis Group Conf. EMAG97, Cambridge, 1997*
© 1997 IOP Publishing Ltd

# Identification of orientation of magnetic moments in hematite (α-Fe₂O₃) using linear dichroism in spatially resolved EELS in STEM

**N K Menon and J Yuan**

Cavendish Laboratory, Madingley Road, Cambridge,. CB3 0HE.

**ABSTRACT:** We demonstrate the effect of Magnetic Linear Dichroism (MLD) in Electron Energy Loss Spectroscopy (EELS), as applied to the antiferromagnetic compound α-Fe₂O₃. The experiments were performed in a high resolution Scanning Transmission Electron Microscope (STEM), capable of producing a nanoscale electron probe. The resultant difference spectrum compares well with theoretical simulations of MLD conducted using atomic multiplet theory, and is also consistent with known magnetic structure of Hematite. This suggests that MLD in EELS, coupled with the high spatial resolution attainable in STEM, is a useful technique in the study of magnetic microstructure.

## 1. INTRODUCTION

The study of microstructural magnetism is, at present, possible using a number of techniques that can be broadly classified into those relying on elastic scattering processes and those which are spectroscopy based. Electron energy loss spectroscopy (EELS) provides an example of the latter, and its use in studying the magnetic structure of a known antiferromagnet is investigated in this paper. The study incorporates the technique of Magnetic Linear Dichroism (MLD), defined to be the difference in absorption cross-section for incident electric fields polarised parallel and perpendicular to the magnetisation in the material. It depends on $<M^2>$ of the ions in the solid (Thole et al 1985), and can therefore be strong in any system with collinear magnetic ordering.

We chose to consider the compound hematite (α-Fe₂O₃), whose crystal structure is trigonal, with a = b = 0.5038 nm and c = 1.3772 nm. There are 6 formula units per unit cell, combining to yield the uniaxial corundum structure in which the iron atoms are coordinated by six oxygen atoms with slight deviations from octahedral (O_h) symmetry. Hematite has an antiferromagnetic ground state, with spins on specific neighbouring planes oriented in opposite directions, implying that the magnetisation $<M> = 0$. The layered structure of the ordered spins in this material provides good model system to study MLD.

## 2. MLD in EELS

It can be shown that the perturbing field due to the incident electron beam, E (**r**, t), can be decomposed in Fourier space into a superposition of components of the form

$$\mathbf{E}\,(\mathbf{q},\,\omega)\;\approx\;i\,\phi\,(\mathbf{q},\,\omega)\,\mathbf{q} \qquad (1)$$

where $i^2 = -1$, $\mathbf{q}$ ( $= \mathbf{k_i} - \mathbf{k_f}$ ) is the momentum transfer and $\phi$ is a scalar function of $\mathbf{q}$ and $\omega$. It can be seen that these component waves are longitudinally polarised parallel to the direction of momentum transfer, and are hence only capable of linear polarisation. The control of their polarisation direction relative to the magnetisation $\mathbf{M}$ makes it possible to observe MLD.

It can be seen from Eq. (1) that the control of the electric field orientation can be achieved by sampling different $\mathbf{q}$ vectors. However, the EEL spectra in STEM always contain a range of $\mathbf{q}$ vectors. If we decompose $\mathbf{q}$ into a component perpendicular to $\mathbf{k_i}$ ($\mathbf{q_\perp}$) and one that is parallel to $\mathbf{k_i}$($\mathbf{q_\parallel}$), the dichroic information can still be obtained if we can collect two different spectra containing a varying weighting of the contributions from the two $\mathbf{q}$ components. In this case, we achieve this by varying the convergence angle of the incident electrons.

## 3. EXPERIMENTAL RESULTS

The experiments were carried out at room tempearture in a dedicated VG-HB501 Scanning Transmission Electron Microscope (STEM) with an CCD based parallel EELS detector at Cambridge. The measurements were conducted on a small crystallite sample of Hematite at room temperature. The orientation of the crystallites relative to the incident beam was adjusted in order to align the c axis with the incident beam direction. This was done by observing the diffraction pattern and tilting the specimen until the characteristic hexagonal symmetry is observed. The magnetisation $\mathbf{M}$ is then perpendicular to the microscope axis.

We chose to consider the $L_{2,3}$ absorption of the $Fe^{3+}$ ion in Hematite, with the dominant excitation being $2p^6 3d^5 \rightarrow 2p^5 3d^6$, as it was expected that the dichroic effects are maximised in this case. In Fig. 1a., we have shown two spectra acquired for objective aperture sizes of 25 µm and 50 µm, corresponding to convergence angles of 3.5 and 7.0 mrad respectively for 100 keV incident electrons. In addition, we present the resultant difference spectrum. The spectra consist of two groups of peaks, corresponding approximately to transition from the spin-orbit split Fe's 2p core hole states to the crystal field split 3d states. The dichroic effect manifests itself as the small intensity difference in the fine structure within the main peaks. Since they are quite small, a large acquisition time of 20 s per single acquisition was required to collect the weak core loss signal. A larger acquisition time is not possible because of the presence of energy and specimen drifts. Further improvements to the signal-to-noise ratio are achieved by aligning and summing many similar spectra. No sign of radiation damage is observed.

## 4. SIMULATION OF MLD

The calculation of the core level EEL spectra and the corresponding polarisation dependence for the transition $2p^6 3d^5 \rightarrow 2p^5 3d^6$ of $Fe^{3+}$ in hematite uses the methods outlined by van der Laan and Thole (1991). The localised nature of the d states involved in the transition facilitates the use of atomic multiplet theory as a basis for describing a system such as hematite. In addition to the spin-orbit splitting of the 2p core hole, the multiplet description includes the p-d and d-d Coulomb and exchange interactions. The solid state effect is included through the presence of the ligand (or crystal) field associated with the octahedral arrangement of anions around the Fe atom, and also the magnetic superexchange field. The values for these parameters in addition to the Lorentz broadenings introduced to simulate lifetime and solid state effects are those given by Kuiper et al (1993).

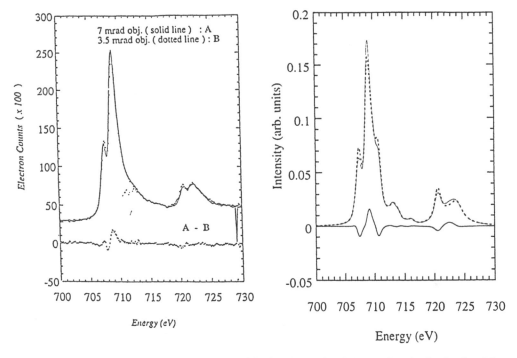

**Fig. 1.**a) (left) Normalised EELS spectra of Fe 2p absorption in Hematite obtained using 25 μm ($I_1$--dotted line) and 50 μm ($I_2$ --solid line) objective apertures. Also plotted is the difference ($I_2$-$I_1$) which is proportional to the magnetic dichroism effect (see text for details). b) Theoretical L-edges for Fe in Hematite with local symmetry corresponding to spin arrangement perpendicular to trigonal axis, incorporating weighting factors for the 25 μm (dotted line) and 50 μm (solid line) objective apertures. Also shown is the difference spectrum.

In order to simulate the core level experimental EEL spectra in STEM corresponding to the different objective aperture configurations described previously, it is necessary to take into account the distribution of momentum transfers, and hence incident electric fields present in normal imaging mode. The detected intensity is known to have the form $I = \alpha\, I_{/\!/} + (1 - \alpha)\, I_{\perp}$ where $\alpha$ is the parallel ' weighting ' fraction, dependent on the objective and collection angles, and is calculated using the model of Menon and Yuan (1997). The quantities $I_{/\!/}$ and $I_{\perp}$ are the intensities corresponding to incident polarisations purely parallel and perpendicular to the trigonal axis respectively, and their values are obtained directly from the atomic multiplet calculation. If $I_1$ and $I_2$ represent the detected intensity in the two cases considered, then the resulting difference spectrum can be shown to have the form :

$$I_1 - I_2 = (\alpha_1 - \alpha_2)(I_{/\!/} - I_{\perp}) \qquad (2)$$

In Figure 1b, we show the theoretical simulation of the Fe L-edge corresponding to objective apertures of 50 and 25 μm for which the corresponding calculated values of $\alpha$ are 0.51 and 0.71 respectively. The difference spectrum is also shown.

## 5. DISCUSSION

A comparison of the sign of the theoretical difference signal with the corresponding experimental spectrum ( Figure 1a.) suggests that the the magnetic moments are aligned perpendicular to the c axis. This is consistent with neutron diffraction studies of the spin structure of hematite (Nathans et al 1964), showing that , above a certain temperature, known as the Morin temperature $T_m$ , ($\sim -10^{\circ}C$), the moments are aligned perpendicular to the c axis of the crystal, and are always parallel within any basal plane. In addition, neighbouring basal planes are antiferromagnetically coupled. At temperatures below $T_m$, a magnetic phase transition takes place which results in the orientation of magnetic moments switching by $90^{\circ}$. The spins are then parallel to the c axis. In addition, the good agreement between theory and experiment of the amplitudes and positions of the peaks of the difference signals at energies corresponding to the two main peaks at the $L_3$ edge suggests that this system is well described by the use of atomic multiplet theory and that the above values of the weighting coefficients are reasonable.

The dichroism effect in solids can normally be attributed to either (or both) the anisotropic crystal field due to the array of oxygen anions surrounding the relevant cation or a local magnetic field **B** produced, for example, as a result of the next to nearest neighbour superexchange interaction, both of which result in a splitting of the J levels. It has been demonstrated that for hematite the local exchange field due to ordered spins has the dominant effect in MLD of the Fe's 2p absorption spectra (Kuiper et al 1993) .The different dipole selection rules for polarisation parallel ($\Delta M_J = 0$) and perpendicular ($\Delta M_J = \pm 1$) to the anisotropic axis make transitions to final states with $J' = J-1, J, J+1$ have different intensities.

## 6. CONCLUSION

We have demonstrated the phenomenon of magnetic linear dichroism in Electron Energy Loss Spectroscopy, as applied to the antiferromagnetic compound $\alpha$-$Fe_2O_3$. The form of the difference spectrum is consistent with theoretical simulations of the Fe L- edge. Since significant effects are observed with a nanosized focussed electron probe, this opens the way for high resolution study of micromagnetism with chemical specificity. It will thus be complementary the more conventional high resolution techniques such as Lorentz microscopy or holography.

## ACKNOWLEDGEMENTS

We would like to thank the EPSRC for funding this research.

## REFERENCES

Thole B T, van der Laan G and Sawatzky G A 1995 Phys. Rev. Lett. **55**, 2086
van der Laan G and Thole B T 1991 Phys. Rev. B **43**, 13401
Kuiper P, Searle B G, Rudolf P, Tjeng L H and Chen C T 1993 Phys. Rev. Lett. **70**, 1549
N K Menon and J Yuan to be submitted
Nathans R, Pickart S J, Alperin H A, Brown P J 1964 Phys. Rev. **136**, A1641

*Inst. Phys. Conf. Ser. No 153: Section 8*
*Paper presented at Electron Microscopy and Analysis Group Conf. EMAG97, Cambridge, 1997*
© *1997 IOP Publishing Ltd*

# A Study of Individual Carbon Nanotubes by Angular-Resolved EELS

**Y Murooka**[A]**, S Ando**[A]**, and M Hibino**[B]

[A]Dept. of Electronics, Nagoya University, Furo-cho, Chikusa, Nagoya, Japan
[B]CIRSE, Nagoya University, Furo-cho, Chikusa, Nagoya, Japan

**ABSTRACT :** An experimental technique has been developed in order to study individual carbon nanotubes using angular-resolved electron energy loss spectroscopy (AREELS). With this technique, AREELS spectra are acquired from individual multi-walled carbon nanotubes in two different scattering directions with respect to the tube axis. The spectra in each direction indicate the graphitic nature of multi-walled carbon nanotube. The $\pi$ plasmon peak observed in the direction perpendicular to the axis may reflect the quantization of wave function around the tube. The reduction of noise and other artefacts, to a level as low as shot noise, appears to be sufficient for quantitative analysis.

## 1 : INTRODUCTION

The discovery of carbon nanotubes (Iijima 1991) have attracted many researchers both in science and nanotechnology. Carbon nanotubes have a diameter ranging from 1 to 30nm, and can be as long as a few micrometers. They consist either of a single rolled graphite sheet (singled-walled carbon nanotube) or of several co-axial sheets (multi-walled carbon nanotube). Theoretical work on single-walled carbon nanotube (Hamada, et 1992) show that it can be either metallic or semiconducting nature depending on the arrangement of carbon atoms. Recent experiments on the conductivity (Tans et al 1997) have also shown that individual single-walled carbon nanotubes act as quantum nanowires. Despite these efforts, the study of the electronic structure of individual carbon nanotubes is still limited due to their small physical size.

A combination of EELS and electron microscopy have been used to probe the electronic structure from individual carbon nanotubes in terms of both energy and momentum transfer. Several groups (Ajayan et al 1993, Kuzuo et al 1994, Bursill et al 1994) have obtained EELS spectra from individual carbon nanotubes. A report by Cullen et al (1994) seems to be the only EELS investigation in terms of momentum-transfer dependence. They show the momentum dependence, but poor statistics and poor energy resolution hampered the investigation of $\pi$ plasmon (at 7eV) and further quantitative analysis. Here we report the development of an experimental technique which allows recording of low loss spectra from individual carbon nanotubes, in terms of momentum transfer with good statistics.

## 2 : EXPERIMENTAL APPARATUS

PEELS spectra were acquired with a Gatan PEELS system (model 666), which was attached to a Hitachi H-8000 TEM. The dispersion and the acceleration voltage were set to 0.3eV/ch and 175kV respectively. Both single-walled carbon nanotubes and multi-walled carbon nanotubes were produced by the arc-discharge method, and dispersed onto a microgrid. The specimens were heated at 250°C during observation to reduce contamination. No severe contamination or irradiation damage was observed by high resolution electron microscopy.

## 3 : DEVELOPMENT

In order to acquire AREELS from an individual carbon nanotube the specimen was illuminated by a parallel beam in selected area diffraction mode in order to have good momentum resolution. Spectra were recorded by displacing the diffraction pattern with respect to the entrance of the spectrometer. There were two practical obstacles to be tackled. One was the hysterisis of a magnetic lens which occurred while switching between imaging and diffraction modes, and the other was artefacts and noise coming into the readout from the PEELS system.

Because of its very small physical size in comparison to the selected area aperture, specimen drift can cause significant change in the signal intensity and therefore it is necessary to check its position. Drift is inevitable because of the long exposure time, e.g. a few minutes. Unfortunately switching between the imaging mode and the diffraction modes causes misalignment of the diffraction pattern with respect to the entrance. As a consequence, it is not possible to calibrate the scattering angle. We have developed a system which controls the position of the diffraction pattern digitally and allows quick calibration of the scattering angle.

Beam instability, afterglow, readout noise, and the large tail of zero loss peak initially prevented the recording of the very weak low loss signal. In order to minimise the degradation of energy resolution, a shorter exposure time was used and many spectra were summed to form a spectrum after the aligning them in terms of the zero loss. The built-in program routine for recording background was not appropriate for this work, since it does not take into account the change in the fixed read-out pattern or the broad afterglow. Therefore, the same number of background spectra were recorded with the same exposure time and then summed. This procedure also reduces exposure time to the nanotube and consequently the number of checks. Zero loss was positioned at the same pixel so that any sharp afterglow effects could be avoided.

## 4 : RESULTS and DISCUSSIONS

Fig. 1 shows the low loss spectra from a single single-walled carbon nanotube. Two spectra (Fig. 1a) are superimposed on to each other; one with the specimen and the other without the specimen. Counts in the zero loss peak are approximately 800,000. The signal in the 10-40eV region is a combination of sharp afterglow peaks (e.g. at 40eV) and the large tail of the zero loss which extends over 40eV (which is a common problem particularly on the fibre optic system). Subtraction of the two spectra (Fig. 1b) shows that sharp afterglow peaks as well as the long tail are removed e.g. in the -20- -4eV region, and the signal in the 5-60eV region is seen. The large fluctuation observed in the -4 to 4eV region is due to the change in the shape of the zero loss caused by its instability. Here the data processing we have used has succeeded in minimising the artefacts and noise down to a level of as low as that of shot noise level.

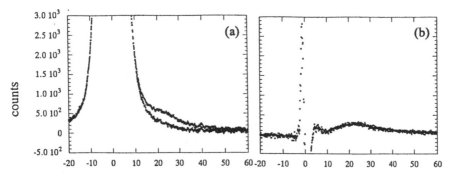

Figure 1 (a) Two low loss spectra superimposed onto to each other: one with a nanotube and the other without a nanotube; (b) subtraction between the two.

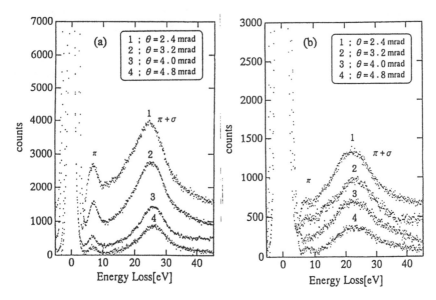

Figure 2 AREELS spectra from a single carbon nanotube: (a) the scattering direction parallel to the tube axis, and (b) the scattering direction perpendicular to the tube axis

Fig. 2 shows the AREELS spectra from a single multi-walled carbon nanotube for two different scattering directions ($\theta$=2.4-4.8mrad): the scattering direction (a) parallel and (b) perpendicular to the tube axis. Spectra in each direction are displaced vertically by a constant value for clarity.

In (a), both $\pi$ and $\pi+\sigma$ peaks are seen, and exhibit dispersions similar to those of graphite with the beam parallel to the c-axis. In (b), only the $\pi+\sigma$ peak is prominent and the relative intensity of $\pi$ peak is small. Note that the intensity of the $\pi+\sigma$ peak is lower and its width is wider than those observed in (a). These may be because excitations in the side walls as well as those in the top-bottom planes are contributing.

It is interesting to determine whether the intensity of the $\pi$ peak changes systematically in (a) but further improvement is required to investigate this. The effect of the quantisation of the wavefunctions around the carbon nanotube may then be observable.

The "surface" plasmon peak reported by Bursill et al 1994 at around 12eV is not clearly observed in (a), although energy resolution and the statistics should be good enough to detect it. Also the $\pi$ peak in (a) is about 0.5eV higher than that of graphite, whereas other work (Kuzuo 1994) reported a lower shift of the peak. These inconsistencies are still not understood and will be investigated further.

## 5 : CONCLUSIONS

An experimental technique has been developed for studying individual carbon nanotubes using AREELS. The spectra obtained in either direction indicate the graphitic nature of multi-walled carbon nanotube and, in addition, some possibility of excitation which is unique to the cylindrical geometry. AREELS spectra have been acquired from individual multi-walled carbon nanotubes in two different directions with respect to the tube axis in this work. Low loss spectra acquired from individual single-walled carbon nanotubes show that noise and artefacts can be reduced to a level as low as shot noise.

## 6 : ACKNOWLEDGEMENTS

The authors acknowledge Professor T. Matsumura of Nagoya University for producing multi-walled carbon nanotubes, and also Associate Professor Y. Saito of Mie University for providing single-walled carbon nanotubes.

## REFERENCES

Ajayan PM, Iijima S, and Ichihashi T, 1993 Phys. Rev. B47, 6859
Brusill LA, Stadelmann PA, Peng JL, and Prawer S, Phys. Rev. 1994 B49, 2882
Hamada N, Sawada S, and Ohiyama A, 1992 Phys. Rev. 68, 1579
Iijima S, Nature 1991 354, 56
Kuzuo R, Terauchi M, and Tanaka M, 1992 Jpn. J. Appl. Phys. 31, 1484
Kuzuo R, Terauchi M, Tanaka M, and Saito Y, 1994 Jpn. J. Appl. Phys. 33, 1316
Tans SJ, Devoret MH, Dai H, Thess A, Smalley RE, Geerligs LT and Dekker C 1997 Nature 386 474

Inst. Phys. Conf. Ser. No 153: Section 8
Paper presented at Electron Microscopy and Analysis Group Conf. EMAG97, Cambridge, 1997

# The segregational control of oxidation and corrosion

S B Newcomb

Department of Materials Science and Metallurgy, Pembroke Street, Cambridge, CB2 3QZ

**ABSTRACT:** The effects of different alloy additions on the way in which oxidation and corrosion can be controlled in a variety of alloys is described. In each case the form of the segregation is rather different and this necessitates the application of a range of TEM techniques which is also discussed.

## 1. INTRODUCTION

One of the features of the design of those alloys which are used in high temperature corrosive environments is the way in which either small amounts of a reactive element can be added (e.g. Stott 1987) or the thermal history of the alloy changed in ways which promote favourable degradation properties (e.g. Singer and Gessinger 1982). On the other hand, alloys designed for optimal mechanical properties and use in less aggressive environments frequently include low concentrations of elements which can turn out to be detrimental to their long term oxidation properties (e.g. Seebaruth et al 1990). In both cases it is recognised that the oxidation can be controlled by elemental segregation to the scales that are formed on their surfaces (Cotell et al 1990). It is thus critical to be able to determine the ways in which segregation can occur if the oxidation and corrosion mechanisms are to be understood and the design of new alloys improved. Here three different degradation processes are described for which various types of segregation have been found to be rate controlling.


## 2. RESULTS AND DISCUSSION

### Stress corrosion in Al-based alloys

Al-Zn-Mg-Cu alloys are both complex and subject to stress corrosion cracking (SCC) although their sensitivity to this mode of failure has been found to be dependent on the ageing treatment used prior to corrosion (e.g. Park 1988). The changes in the corrosion properties known to occur after different heat treatments not only underlines the importance of the alloy's microstructure in determining the degree of corrosion undergone but also implies that the changes occurring in the compositions of the alloy grain boundaries during corrosion need to be understood.

The microstructures of the crack tips formed in both a peak aged (T651) and overaged (T7451) alloy as exposed to 3.5%NaCl for 10 days at either 75MPa (T651) or 325MPa (T7451) have been compared. Figure 1a shows the microstructure of a region of the peak aged alloy in which a corrosion crack tip was seen. The crack is located at an alloy grain boundary rather than in the area surrounding it where multiple phase precipitation has occurred. Although the corrosion was found not to have taken place particularly uniformly along any one boundary this probably reflects the three-dimensional nature of the crack propagation. The corroded grain boundary described was found to contain $O_2$ and Cl but also to be enriched in Cu and depleted in Zn relative to the matrix. The compositional changes for the Cu and the Zn were significantly larger than those typically observed at grain boundaries in the peak aged condition when Mg depletion, rather than the enrichment found in the corroded zone here, is also observed (Rao 1981). More surprisingly, the matrix seen here was apparently overaged for distances up to 500μm from the corroded surface and did not resemble the microstructure observed in specimens which had been similarly aged but not corroded. A typical crack tip formed in the overaged 7150 alloy is shown in figure 1b from which it is apparent that the corrosion is now only weakly associated with the alloy grain boundary, which is arrowed, but is instead moving on a front which is some 1μm from it. For this ageing condition the grain boundary and matrix microstructures immediately ahead of the corroded part of the alloy were found not to exhibit any morphological features that are substantially different from those typically seen in the unexposed overaged alloy.

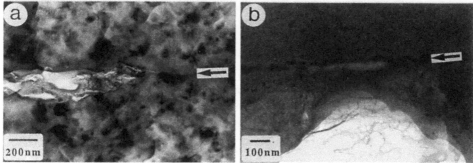

**Figure 1:** Bright field micrographs of the corrosion cracks formed in a) a peak aged and b) an overaged 7150 Al alloy.

The data summarised here is sufficient to demonstrate that there are enormous differences in the corrosion characteristics of a 7150 Al alloy as a function of its ageing treatment. For the peak aged material the observations are consistent with a process that is controlled by the microstructure in the vicinity of a grain boundary and thus with studies that demonstrate a correlation between grain boundary microstructure and susceptibility to SCC in Cu rich 7xxx alloys (Park 1988). The data indicate that corrosion is probably occurring by stress assisted anodic dissolution rather than any process involving gross hydrogen embrittlement. Nonetheless the gross changes to the microstructure of the overaged alloy ahead of the corrosion front can only be reasonably be explained on the basis of a H in-diffusion effect allowing accelerated "ageing". The observed tendency for Cu enrichment to occur at those boundaries which are undergoing corrosion in the peak aged state provides evidence for the anodic behaviour and is further consistent with the fact that overageing decreases the alloy's susceptibility to corrosion given the way in which the profiles for local Cu enrichment and Zn denudation adjacent to the boundaries are shallower after an overageing than a peak ageing treatment. While the trends described are not seen for the T7451 alloy and the material exhibits no clear tendency for its grain boundaries to be sensitive to corrosion, other data has emphasised its susceptibility to hydrogen embrittlement though only in the presence of a pre-crack (Warner et al 1994).

## Healing layer formation in FeCrSi alloys

The beneficial effects of the addition of Si to a number of metals such as Fe and FeCr alloys are well known (Wood et al 1971) and are of interest to the power generation industry for which the long term stability of FeCr alloys is required. The way in which the addition of Si to two Fe-11Cr alloys modifies the microstructures of the inner layer oxides that are formed is now described as is the way in which the overall balance of reactions which occur is delicately controlled by the formation of $SiO_2$ at the metal/oxide interface through the resultant changes in local activities in the different scales.

Dark field micrographs of the metal/oxide interface regions formed in the two alloys after 3088 hours' oxidation are shown in figure 2. Figure 2a is for an Fe-11Cr-1.10Si alloy and shows near continuous formation of a ~30nm band of $SiO_2$ at the base of the fine grained oxide lying above. Several different inner layer scales varying in thickness from 1.5-4.5μm were examined and all found to be $(FeCr)_3O_4$ based. Although $(FeCr)_2O_3$ was observed in some of the scales there was a clear distinction between its formation, the average oxide Cr content and the thickness of the scales: the thinnest layers with the highest Cr content exhibited most $(FeCr)_2O_3$ formation. The microstructure may be compared with that of the metal/oxide interface formed in an alloy of lower Si content, Fe-11Cr-0.41Si. The glassy phase is much thinner (~7.5nm) and is discontinuous (figure 2b) so that isolated $SiO_2$ layers may be observed at varying heights in the oxide (as arrowed). The scales for both Si contents exhibited significant variations in local thickness and figure 2b shows an oxide growth protruberance which extends into the metal at A. These local inner oxide thickness variations are distinct from the coarser differences seen optically and require TEM characterisation of the range of the different oxide thicknesses exhibited for the lower Si content alloy. All these scales examined contained $(FeCr)_3O_4$ and again exhibited variable formation of $(FeCr)_2O_3$ but were found to be enriched in Cr (74-78wt%) at the base of the scales even in the absence of $(FeCr)_2O_3$. Significantly, however, by comparison with the inner layer scales formed on the alloy of higher Si content, carbon was now found to be deposited in the more porous parts of the the inner layer scale, these being primarily higher up where the Fe concentration tended to be relatively high.

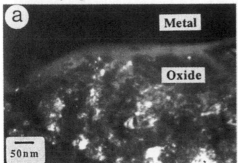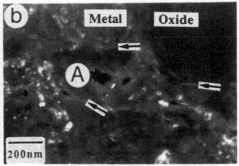

**Figure 2:** The metal/oxide interface formed in a)Fe-11Cr-1.10Si and b) Fe-11Cr-0.41Si after 3088 hours' oxidation in $3\%CO/CO_2$ at 600°C.

The reduction in the oxygen potential during the inner layer scaling process for the two alloys described here is paralleled by the formation of both $SiO_2$ and Cr-rich oxides at the metal/oxide interface after 3088 hours'exposure. The reduction in oxygen potential, however, does not occur uniformly for the oxides as is evidenced by the way the formation of $(FeCr)_2O_3$ is variable. While most of the $SiO_2$ and Cr-rich oxide formation takes place during the period of oxidation from 294 to 3088 hours, the former phase would appear to be the most critical in maintaining a low rate of oxidation. Nonetheless a $SiO_2$ layer at the alloy/oxide interface does not prevent the diffusion of Fe and Cr into the scales during the oxidation treatment and the way partial $SiO_2$ layers can be incorporated into the inward formed scale at the lower Si content demonstrates that sufficient Si is required for it to be effective in reducing the scaling rate. This correlates with the way that the Cr content of the 0.41Si alloy immediately beneath the oxide is lower (5.5wt%) than in the 1.10Si alloy (7.5wt%). The flux of Fe into the scale formed on Fe-11Cr-1.10Si is apparently sufficiently low to allow oxidant to diffuse back out of the inner scales whereas the relatively high Fe flux across the metal/oxide interface formed on the 0.41Si alloy results in its inevitable in-situ interaction with CO and the eventual deposition of carbon. The significance of the way the balance of reactions can be tipped one way or the other in the kinetically controlled reactions described is that the 0.41Si alloy has been found to go into breakaway oxidation after 5,000 hours exposure (Newcomb and Stobbs 1990) and the data described here would imply that similar instabilities will arise for the 1.10Si alloy after protracted oxidation.

## The effects of Y on alumina forming ODS Alloys

The effects of $Y_2O_3$ additions to FeCrAl are now described, the oxidation properties of the oxide dispersion strengthened (ODS) alloys being known to be significantly better than those of their ternary counterparts (Huntz 1981). However, the oxidation rate of an FeCrAl ODS alloy (MA956) has been found to be higher when the ferritic matrix contains 0.70wt% rather than 0.17wt% of dispersoid (Czyrska-Filemonowicz 1993) and this difference has prompted an investigation of the oxide grain boundaries formed on the two different alloys.

The approach taken has been to combine the use of the Fresnel Method (Ross and Stobbs 1991) with both localised probe data obtained in a STEM and energy selected images. Figure 3a shows a through focal series of bright field micrographs of an alumina grain boundary that was situated close to the scale/gas interface and figure 3b shows digitised profiles of the same boundary. It can be seen that there is a dark line in-focus and the absorption contrast seen here indicates that the boundary contains at least one relatively heavy element. The boundary was found to be decorated with a thin layer of amorphous material (Newcomb et al 1997) and this observation can be used to explain the sense of the Fresnel data where it can be seen that the boundary has a lower scattering potential than the crystalline alumina grains on either side of it although it is noted that it is unusual for a boundary with a low scattering potential to exhibit dark absorption contrast. Given that it is unlikely that the dark absorption contrast arises solely from stronger scattering from an amorphous boundary layer than from the weakly diffracting grains on either side of it, the dark absorption contrast must thus originate from scattering to high angles by heavy elements to a degree that is disproportionately high relative to their effect on $V_0$. Surprisingly, the oxides of Al, Ti and Y have similar values of $V_0$ and if it is assumed that the $V_0$ values for their amorphous phases are also similar then the lower scattering potential of the boundary is of less concern. Equally $V_0$ for yttrium is significantly higher than for its oxide and this further emphasises the fact that the yttrium present at the boundary is likely to be in its oxidised state. The chemistry of the grain boundary from which the Fresnel data were obtained was confirmed using both energy selected imaging and STEM EDX. Figures 4a-c show a zero loss image, an Y elemental map and a Ti jump ratio image respectively for three oxide grain boundaries, the Fresnel data described earlier coming from the upper of these. The Fresnel data can be used to extrapolate the fringe spacing to zero defocus from which it is possible to estimate the width of the boundary that has been described. The data indicate that the boundary is approximately 1.8nm in width while the form of the contrast changes is indicative of diffuse interfaces between the amorphous layer and the crystalline grains to its sides.

Further analyses were made of a number of grain boundaries as a function of their position within the scale formed on the 0.70% yttria alloy. Fresnel data for an alumina grain boundary seen near the middle of the scale are shown in figures 3c and 3d and the boundary can again be seen to exhibit a lower scattering potential than the alumina grains to its side as well as absorption contrast close to focus. Extrapolation of the fringe spacing to zero defocus indicated that the width of the boundary here is approximately 1.1nm while the boundary itself was found to be compositionally considerably more abrupt than in the region described earlier. Boundaries near the metal/oxide interface were similarly investigated and figures 3e and 3f show data for a boundary lying some 150nm above the alloy. Here the Fresnel images are again very different and do not show either changes in their spacing as a function of defocus nor significant absorption contrast. While the latter is simply indicative of the fact that this boundary is not significantly enriched in either Y or Ti, as is consistent with the probe analyses, it is noted that energy selected images showed a degree of Cr enrichment at this boundary. Furthermore the apparent absence of any significant spacing changes as a function of defocus shows that the boundary must be very wide as well as very diffuse. These features were reflected in the form of the plots for the spacing and contrast changes with defocus (Newcomb et al 1997) and the width of the boundry described here was found to be 4nm.

The lower yttria content was similarly examined. Figures 5a and 5b show a series of Fresnel defocus images and digitised profiles respectively of an alumina grain boundary high up in the scale. The appearance of the boundary is generally rather similar to that of the boundaries examined for the oxide formed on the higher yttria content alloy in that it exhibits both a lower potential than the crystalline $Al_2O_3$ to its sides and dark absorption contrast close to focus. The nature of the heavy element segregation implied from the Fresnel micrographs has been compared with data obtained by STEM EDX and energy filtered images, both Y and Ti again being found to enrich the boundaries but to a lower degree than was observed for the previous oxide. The differences in the boundaries formed on the two alloys is further emphasised by the Fresnel data which demonstrated that the typical widths of the boundaries, irrespective of their vertical position within the scale, are no higher than 0.5nm. In this context figures 5c and 5d show a Fresnel fringe series for a boundary situated near the metal/oxide interface, where it will be remembered rather wide and diffuse boundaries were observed in the previous alloy.

The data described above can be used to explain the differences in the oxidation behaviour of the two alloys remembering that for both alloys oxidation occurs by the grain boundary transport of oxygen (Clemens et al 1993). A simple correlation can be made

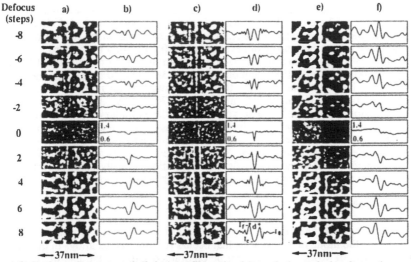

**Figure 3:** Defocus images and digital profiles of oxide grain boundaries formed on a 0.70% Y₂O₃ FeCrAl alloy near a,b) the scale surface c,d) mid-point and e,f) metal/oxide interface.

**Figure 4:** a) Zero loss image of oxide grain boundaries formed on the high Y₂O₃ alloy and b) Y elemental map and c) Ti jump ratio map for the same area.

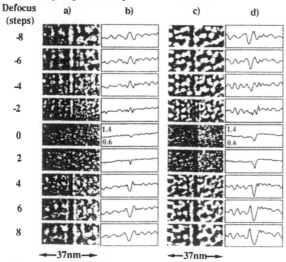

**Figure 5:** Defocus images and digital profiles for two oxide grain boundaries formed on the low Y₂O₃ alloy.

between the relatively high rate of oxidation for the 0.70% alloy and the tendency for the alumina scale formed on its surface to have rather open grain boundaries as well as the way in which the Ti and Y enrichment of the boundaries is noticeably lower on the 0.17% alloy. The origin of the differences between the oxide grain boundaries formed on the two different alloys might then lie in the greater propensity of the scale on the 0.70% alloy to form vacancies on its grain boundaries. Vacancy formation will originate from the counter diffusion of $O_2$ and Y (and Ti) (Przybylski and Yurek 1989) and once initiated should lead to an opening up of the grain boundaries. The latter should itself then be followed by an increase in the upward diffusion of Al, particularly in the oxide formed on the 0.70% alloy. The synergism between the various diffusional processes is seen very well in the boundary near the metal/oxide interface in the 0.70% yttria alloy. Here the boundary was at its coarsest and it would appear that the dilution of the Y grain boundary content is high enough to allow sufficient Al (and Cr) diffusion to increase the rate of oxidation. Such a process does not occur in the low yttria content alloy even where the Y and Ti boundary concentrations near the scale/gas interface are locally high because there are no interconnected diffusional pathways to the metal.

## 4. CONCLUSIONS

Different segregation phenomena have been described and these processes have been found to determine the mode of degradation that occurs for the various alloys in question. In many cases the type of work described can be completed without the use of particularly specialised approaches mainly because the boundary parameters are sufficiently coarse that either standard probe methods can be used to determine the local chemistry or conventional imaging used to characterise the associated changes in microstructure. In other instances modified techniques are required either to improve quantification or to examine the segregation of low concentrations of particular elements. Increasingly, methods such as PEELS and more recently energy loss filtered imaging are being used to unravel complex problems where quantitative chemically sensitive data is needed.

## ACKNOWLEDGMENTS

I am grateful to Prof.A.H.Windle for the provision of laboratory facilities as well as the Isaac Newton Trust for financial support. I am also indebted for to the late Michael Stobbs for many invaluable discussions and thank Dr R. F. Dunin-Borkowski for help with the Fresnel work.

## REFERENCES
Clemens D, Bongartz K, Speier W, Hussey R J and Quadakkers W J 1993 J. Anal. Chem. **346**, 318
Cotell C M, Yurek G J, Hussey R J, Mitchell D F and Graham M J 1990 Oxid. Met. **34**, 173
Czyrska-Filemonowicz A, Versuci R A, Clemens D and Quadakkers W J 1993 Microsc. Oxid. II ed S B Newcomb and M J Bennett (London: Inst. Mat.) pp 288-297
Huntz A M 1989 The Role of Active Elements in the Oxidation Behaviour of High Temperature Alloys (London: Elsevier Applied Science) pp 81-106
Newcomb S B and Stobbs W M 1990 Microsc. Oxid. I ed M J Bennett and G W Lorimer (London: Inst. Mat.) pp 84-90
Newcomb S B, Dunin-Borkowski R F, Boothroyd C B, Czyrska-Filemonowicz A, Clemens D and Quadakkers W J 1997 Microsc. Oxid. III In press
Przybylski K and Yurek G J 1989 Mat. Sci. Forum **43**, 1
Park J P 1988 Mat. Sci. Eng. **A103**, 223
Rao B V N 1981 Met. Trans. **12A**, 1356
Ross F M and Stobbs W M 1991 Phil. Mag. **A63**, 1 & 37
Seebaruth K, Newcomb S B and Stobbs WM 1990 Inst. Phys. Conf. Ser. No. 119 ed F J Humphreys (Bristol: IOP) pp 237-240
Singer R F and Gessinger G H 1982 Met. Trans. **13A**, 1463
Stott F H 1987 Rep. Prog. Phys. **50**, 861
Warner T J, Newcomb S B and Stobbs W M 1994 4th Int. Conf. on Al Alloys ed T H Sanders and E A Starke (Georgia: Georgia Inst. Tech.) pp 88-96
Wood G C, Richardson J A, Hobby M G and Boustead J 1971 Corros. Sci. **9**, 659

*Inst. Phys. Conf. Ser. No 153: Section 8*
*Paper presented at Electron Microscopy and Analysis Group Conf. EMAG97, Cambridge, 1997*
© *1997 IOP Publishing Ltd*

# Quantitative X-ray mapping with high resolution

**M Watanabe, D T Carpenter, K Barmak and D B Williams**

Department of Materials Science and Engineering, Lehigh University, Bethlehem PA 18015-3195

**ABSTRACT:** Quantitative X-ray mapping of thin specimens in the analytical electron microscope (AEM) is generally not very accurate because of low efficiency for X-ray generation and collection from the limited analysis volume. An intermediate-voltage scanning transmission electron microscope (STEM) VG HB603 at Lehigh, which has been carefully designed to overcome the thin-film limitations for microanalysis, has the capability of quantitative X-ray mapping with high spatial resolution. By using the HB603, sub-nanometer Gibbsian solute-element segregation at grain boundaries can be detected in mapping mode with spatial resolution approaching 2 nm (FWTM).

## 1. INTRODUCTION

X-ray energy-dispersive spectrometry (XEDS) in electron-beam instruments is a mature technique. XEDS has been used for more than 25 years to determine the composition of bulk materials in the electron probe microanalyzer (EPMA) with a spatial resolution of ~1 μm and an analytical sensitivity of <100 ppm for beam energies in the range of 20-30 kV. Quantitative X-ray mapping by the EPMA is more than 40 years old (Cosslett and Duncumb 1956) and has become a standard procedure (e.g. Newbury et al. 1991). Although EPMA maps are limited by the poor spatial resolution, this poor resolution makes quantitative mapping feasible. The large analysis volume in bulk specimens ensures that the X-ray count rate is substantial and thus sufficient counts can be gathered in a reasonable time to allow quantification.

XEDS of thin specimens in the analytical electron microscope (AEM) can be performed at individual analysis points with a spatial resolution of <5 nm and a sensitivity of <0.5 wt.% (Williams and Carter 1996). The standard way of presenting quantitative X-ray data in the AEM is the generation of line profiles by taking XEDS spectra at a series of points across the region where the concentration changes. While the line-profile approach is useful, this one-dimensional approach is both time-consuming and operator-biased.

If quantitative X-ray maps can be obtained from thin specimens while maintaining high spatial resolution, XEDS in the AEM becomes a powerful approach for characterizing materials. Unfortunately, the improved spatial resolution due to the smaller analyzed volume in thin specimens means that quantification in the AEM is less accurate than in the EPMA due to the low efficiency for X-ray generation and collection. This is why quantitative X-ray mapping is rarely attempted in the AEM and most X-ray maps obtained in the AEM are simply dot maps, which are not quantitative (Lyman 1992). For good quantitative mapping, it is generally assumed that in excess of $10^6$ counts are required (Goldstein et al 1992). Therefore, extraordinarily long times are required to collect the requisite number of counts for quantitative mapping of thin specimens because of the poor X-ray collection efficiency of typical commercial AEMs. Long counting times results in specimen drift, damage and contamination, so generally the quality of X-ray images from thin specimens does not approach that of bulk specimens in the EPMA.

These limitations of the reduced count rate can be overcome by use of intermediate-voltage instruments combined with a high-brightness field-emission gun (FEG) source to increase the beam current with maintaining the small probe size. Ultra-high vacuum stages can be used to reduce contamination and careful stage design can maximize the X-ray collection angle and peak-to-background ratio. In addition, megahertz-rate beam blanking gives large increases in the X-ray throughput and drift-correction software can be used to allow longer acquisition. All of these design requirements are incorporated in a VG HB603 AEM at Lehigh (Lyman et al. 1994). The HB603 is equipped with an Oxford windowless Si(Li) and an Oxford ultra-thin window intrinsic-Ge XEDS detector with the largest collection angles available in commercial AEMs (0.30 sr and 0.17 sr respectively). If necessary, X-ray intensities can be obtained by both detectors simultaneously. With these design features, the HB603 is capable of quantitative X-ray mapping with spatial resolutions approaching 2 nm (FWTM). Gibbsian segregation of Cu at grain boundaries in an Al-4 wt.% Cu thin specimen has been quantified using the HB603 and it has been shown that concentration line-profiles can be extracted from the AEM maps (Williams et al. 1997). The AEM maps at lower magnifications are ideal for systematic studies of grain-boundary segregation from a large number of grain boundaries in a single field of view, as long as high resolution may be maintained.

## 2. EXPERIMENTAL X-RAY MAPPING AT LOWER MAGNIFICATIONS

Although the X-ray mapping approach at high magnification is useful and allows more flexible data processing, it is still limited to showing a very local area of the specimen. For grain-boundary analysis, low magnification mapping is more useful because each image may contain many different grain boundaries. This parallel acquisition of multiple grain-boundary data is essential in order to begin making generalizations about grain boundary segregation in bulk materials. In addition to comparisons between different boundaries, it is also possible to quantify the homogeneity of single boundaries and measure the fraction of boundaries showing different levels of segregation. To ensure complete sampling of the specimen, the minimum probe diameter must be larger than the display pixel diagonals. Therefore, at lower magnifications more pixels are required for a given probe size to fully sample the field of view, and the time required to create a map with the same statistics will rise accordingly.

To obtain quantitative maps requires that the bremsstrahlung intensity should be subtracted from the characteristic peak intensity at every pixel and the background-subtracted intensity be converted into the concentration for each element by the Cliff-Lorimer k factor (Cliff and Lorimer 1975). Fig. 1a shows a low magnification (300 kX) bright-field image of an Al-4 wt.% Cu film with homogeneous thickness of 100 nm, which had been aged to cause segregation of Cu to the grain boundaries. In this specimen, the average grain size is about 100 nm and most grains are through thickness. X-ray intensity maps of the Cu K and Al K peaks were gathered at a resolution of 256 x 256 pixels, with a dwell time of 60 ms per pixel in the same field of view as Fig. 1a. Intensity maps were also gathered from background windows on either side of each peak and the net intensity maps of the Cu K and Al K lines were obtained by subtracting the background-intensity maps. Fig. 1b shows the quantitative Cu concentration map (in wt.%), calculated from the net Cu K and Al K intensity maps using a k factor experimentally

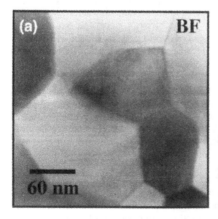
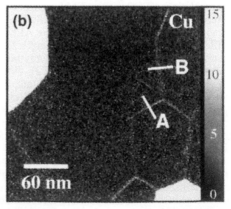

Fig.1 (a) A low magnification electron image of the grain structure in an Al-4 wt.% Cu thin-film. (b) A quantitative Cu map from a same field of view as (a), including a look-up table.

determined from stoichiometric θ-phase (Al$_2$Cu) particles which were distributed throughout the thin film. In acquiring the X-ray maps, the HB603 was operated using a probe size of ~1 nm and beam current of 0.9 nA. In addition, count throughput was increased by reducing the time constant of the pulse processor. This reduces the number of counts lost to pileup and hence reduces dead time, but this is done with some loss of energy resolution. For the Al-Cu system, energy resolution is not very important since the peaks do not overlap. Therefore, the shortest time constant was used (process time = 1 on the Oxford/Link eXL multi-channel analyzer).

The bright-field image (Fig. 1a) shows strong diffraction contrast, from which it can be seen that most of the boundaries in the region are not oriented in the ideal configuration parallel to the beam. The Cu concentration at each boundary seems different as shown in Fig. 1b. This concentration difference might be dependent on boundary inclination with respect to the beam. Fig. 2 shows the line profiles extracted from boundaries with higher (a) and lower (b) Cu concentration ,which are indicated by white lines A and B in Fig 1b. While the profile from the higher Cu boundary in Fig. 2a has a Gaussian type distribution with spatial resolution of ~4 nm (FWTM), the profile in Fig. 2b shows asymmetry and ~15 nm resolution (FWTM). Fig. 3 illustrates the orientation relationships between the incident beam and an edge-on boundary (a), and an inclined boundary (b). If the beam scans across the edge-on boundary with a projected width of ~1 nm, the line profile can be a narrow distribution with a high peak. In the case of the inclined boundary with <10 nm image resolution as shown in Fig. 3b, the profile can be abrupt on one side and broad on the other side, i.e. an asymmetric shape. A scan across such an inclined boundary also results in a lower peak height in the profile. From 120 grain boundaries, the mean Cu coverage is ~4 atoms/nm$^2$ if all the Cu atoms are assumed to be in the boundary plane.

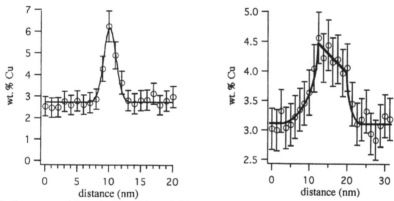

Fig. 2 Cu composition profiles (a) and (b) extracted from lines A and B in the Cu map (Fig. 1b), respectively.

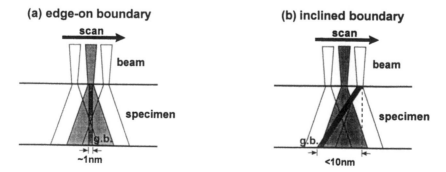

Fig. 3 Schematic diagrams of orientation relationships between the incident beam and an edge-on boundary (a), and an inclined boundary (b).

## 3.  FUTURE DIRECTIONS IN QUANTITATIVE X-RAY MAPPING

**New XEDS instrumentation**: Improving the count rate of the acquired signal, and hence the statistics, at each pixel is crucial to development of quantitative AEM mapping. The X-ray count rate can be improved using the beam blanking and the dual-detector operation in the HB603. In addition, The throughput of X-ray counts can be improved while maintaining high energy resolution by using the digital pulse-processing XEDS system (Mott and Friel 1995).

**Position-tagged spectrometry**: The energy-window approach is used for most XEDS maps including this study. However, this simple window approach is relatively poor and limited especially when the X-ray peaks of interest overlap. Currently, position-tagged spectrometry (PTS), which allows the whole spectrum to be saved at each pixel in mapping mode, is available (Mott et al. 1995). The PTS provides more flexibility in post-spectrum processing (e.g. peak deconvolution).

**Thickness map**: X-ray absorption becomes a serious problem in quantitative analysis of non-uniform thin films. Recently, the $\zeta$-factor approach has been proposed to overcome the difficulties in quantitative analysis caused by X-ray absorption (Watanabe et al. 1996). The $\zeta$-factor analysis provides quantification with absorption correction and thickness determination simultaneously. By combining the $\zeta$-factor approach with X-ray mapping, it is conceivable to extract thickness maps of non-uniform thin foils at the same time as quantitative composition maps.

## 4.  CONCLUSIONS

The Quantitative X-ray mapping of grain-boundary segregation is feasible in the VG HB603 at low magnification. Statistics of segregation measurements are improved by low magnification mapping because several boundaries can be imaged in a single map. The line profiles can be extracted from the X-ray maps across the individual boundaries and the effects of boundary tilt can be observed and modeled.

**ACKNOWLEDGEMENTS**
DTC, KB and DBW wish to acknowledge the support of NSF through the provision of a graduate research traineeship to DTC via grant #DMR 92-56322. MW wishes to thank the Materials Research Center at Lehigh University for its support.

**REFERENCES**
Cliff G and Lorimer G W 1975 J Microscopy **103**, 203
Cosslett V E and Duncumb P 1956 Nature **177**, 1172
Goldstein J I, Newbury D E, Echlin P, Joy D C, Romig, Jr. A D, Lyman C E, Fiori C E and Lifshin E
  1992 Scanning Electron Microscopy and X-Ray Microanalysis (New York: Plenum) pp 525-45
Lyman C E 1992 Microscopy: the Key Research Tool (Woods Hole: Electron Microscopy Society
  of America) pp 1-9
Lyman C E, Goldstein J I, Williams D B, Ackland D W, Von Harrach S, Nicholls A W and
  Statham P J 1994 J. Microscopy **176**, 85
Newbury D E, Marinenko R B, Myklebust R L and Bright D S 1991 Images of Materials, eds
  D B Williams, A R Pelton and R Gronsky (New York: Oxford University Press) pp 290-308
Mott R B and Friel J J 1995 X-Ray Spectrometry in Electron Beam Instruments eds D B Williams,
  J I Goldstein and D E Newbury (New York: Plenum) pp 127-57
Mott R B, Waldman C G, Batcheler R and Friel J J 1995 Proc. Microscopy and Microanalysis
  1995 eds G W Bailey, M H Ellisman, R A Hennigar and N J Zaluzec (New York: Jones and Begell
  Publishing) pp 592-3
Watanabe M, Horita Z and Nemoto M 1996 Ultramicroscopy **65**, 187
Williams D B and Carter C B 1996 Transmission Electron Microscopy (New York: Plenum)
  pp 621-35
Williams D B, Watanabe M, Carpenter D T and Barmak K 1997 Mikrochim. Acta (in press)

*Inst. Phys. Conf. Ser. No 153: Section 8*
*Paper presented at Electron Microscopy and Analysis Group Conf. EMAG97, Cambridge, 1997*

# Investigations of grain boundary embrittlement in the STEM

**V J Keast and D B Williams**

Department of Materials Science and Engineering, Lehigh University, 5 East Packer Avenue, Bethlehem PA 18015, USA

**ABSTRACT:** The problem of segregation and embrittlement of grain boundaries is ideally suited to demonstrating the capabilities of a high resolution analytical electron microscope (AEM). X-ray energy dispersive spectroscopy (XEDS) and electron energy loss spectroscopy (EELS) have been used to investigate the embrittling systems of Bi in Cu and S in Ni. Changes in chemistry and bonding at the nanometre level have been observed.

## 1. INTRODUCTION

There are many systems in which impurities segregate to grain boundaries and the subsequent effects on the material properties are as varied as the segregants themselves. One common consequence is that the material becomes brittle and fails through intergranular fracture. The temper embrittlement of steel is one such example, which is of considerable concern to industry and is still not fully understood (Bulloch (1994)).

The identification and quantification of the segregants has been the focus of many of the investigations to date, and often the distribution of segregants between boundaries is also of interest. Auger electron spectroscopy (AES) has been the most commonly used technique, however, AEM has also been applied to a number of systems (Doig and Flewitt (1977), Michael and Williams (1984)). AEM has the important advantage that the specimen does not need to be fractured in order to observe the boundary. Investigations in the electron microscope have either utilised a stationary focused probe placed directly on the grain boundary, or in a few cases, a scan across the grain boundary to obtain a profile (Hall et al. (1981), Williams and Romig (1989), Titchmarsh and Dumbell (1996)). Two dimensional X-ray mapping of grain boundary segregants has not been performed, although electron energy filtering has recently been used to map segregants (Bentley (1997), Menon et al. (1997)).

The small probe sizes, necessary for the X-ray mapping of grain boundary segregants, contain a very low current and therefore unrealistically long acquisition times are required for a two dimensional map. The VG HB603 scanning transmission electron microscope (STEM) at Lehigh University has been specifically designed to overcome these problems through improvements in X-ray collection efficiency, probe-current density and background reduction (see Lyman et al. (1994) for details). Mapping allows for the identification, quantification *and* the distribution of segregants to obtained simultaneously.

The electron energy-loss spectrum can be used to obtain information about the changes in bonding at grain boundaries associated with segregation. For example, changes in the near edge structure of Cu at a grain boundary have been observed in the presence of Bi (Bruley et al. (1996), Keast et al. (1997)). An increase in intensity just above the edge threshold, in the so-called "white line", was seen. A similar effect has also been observed $Ni_3Al$, where a reduction in white line intensity is observed upon the segregation of the ductilising B (Muller et al. (1996)). In this paper, the segregation of S in Ni is examined with XEDS and EELS. Further work on Bi doped Cu is also described.

## 2.    EXPERIMENTAL PROCEDURE

The Ni-S material was provided by Westinghouse-Bettis with a pre-melt chemistry of Ni + 0.03% S. A sequence of three rolling passes, each for a reduction of 30% and each followed by a 1000°C anneal were performed. This was followed by a final anneal of 48 h at 650°C to ensure segregation and embrittlement. Cu-Bi material was supplied by D. E. Luzzi and contains 12 at ppm Bi. It was swaged to 5 mm diameter, followed by an anneal of 1 h at 850°C. It was then further swaged to 3 mm with a further anneal of 48 h at 600°C. For both materials, TEM specimens were prepared by hand polishing 3 mm discs on SiC paper down to ~100 μm, followed by dimpling and ion milling. Electropolishing was not used so as to reduce the possibility of changes in chemistry at the grain boundaries. Specimens were immediately inserted in the microscope (within 5 minutes) to avoid growth of an oxide layer.

The VG HB603 STEM was operated at 300 kV. The probe size chosen has a FWHM of 0.8 nm and a beam current of 0.6 nA. Probe sizes have been measured by scanning the beam across MgO cubes and beam currents were measured using a Faraday cup. The microscope is equipped with both an ultra-thin window intrinsic Ge detector and a windowless Si(Li) detector. The Si(Li) detector was used for the experiments on Ni-S specimens whereas the Ge detector was used on the Cu-Bi specimens. The Ge detector was chosen for Cu-Bi because, when the grain boundary of choice was tilted parallel to the beam, the specimen was tilted toward the Ge detector and away from the Si(Li) detector. Quantification was performed using the Cliff-Lorimer equation (Cliff and Lorimer (1975)) and theoretical k factors.

X-ray profiles were obtained by stepping the beam in 0.5 nm steps along a line perpendicular to the grain boundaries, with acquisition time of 20 s for each step. At each step a full spectrum is saved and quantification performed afterwards. The X-ray maps are 64 by 64 pixels, with a dwell time for each pixel of 100-500 ms. A complete map takes ~45 minutes which is usually within the stability range of the microscope stage (< 2 nm/hour) and the electron gun (<20% decrease in probe current/hour). At each pixel in the map, only the counts in a pre-set energy window, normally about the elemental peak of interest, were stored. By judicial selection of energy windows, it is possible to obtain fully quantified maps (see Williams et al. (1997) for details).

The segregation of Bi to grain boundaries in Cu is a more challenging problem for mapping than that of S in Ni. This is because the k factor, $k_{BiCu}$ is larger by a factor of 3 than $k_{SNi}$ and, in addition, the large size of Bi means that complete coverage of a grain boundary will be acheived by fewer Bi atoms than S atoms. To ensure that a Bi map could still be obtained, the collection efficiency was further increased by a reduction in the pulse-processing time of the detector. While this does result in a degradation of the energy resolution of the detector, it is not important in mapping when only select energy windows are used and the peaks are well separated. In addition, a larger virtual objective aperture was used. This increased the probe size to 1.5 nm, thus increasing the probe current to 1.8 nA and resulted in a slightly degraded spatial resolution.

EELS experiments were performed on a VG 501 STEM fitted with a Gatan 666 parallel electron spectrometer. During the acquisition the beam was scanned over an area of 3 x5 $nm^2$ at 10 MX. In this way specimen drift can be monitored and corrected during the acquisition. The grain boundary was oriented perpendicular to the 3 nm scan. Two spectra showing the Ni $L_{23}$ edge were acquired: one where the probe was placed on the grain boundary and one with it placed nearby (within 10 nm) in the adjacent grain. The dark-current was subtracted and the channel-to-channel gain variations in the response of the photodiode array were corrected. The exponentially decreasing background under the edges was subtracted by fitting a power-law to the pre-edge region. Plural scattering was removed by Fourier deconvolution of a low-loss spectrum acquired in the same area and immediately afterwards. Comparisons between the two spectra were made to investigate grain boundary effects.

## 3.    RESULTS

Fig. 1 shows the S profile across a grain boundary in S doped Ni and Fig. 2 shows a map of S content for a similar grain boundary. The FWHM for the profile is ~1.2 nm.

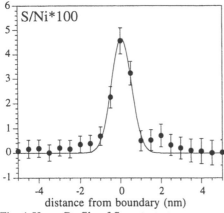

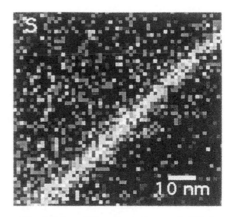

Fig. 1 X-ray Profile of S content at an embrittled grain boundary

Fig. 2 Map of S content at an embrittled grain boundary.

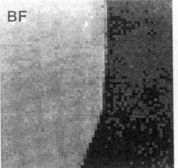
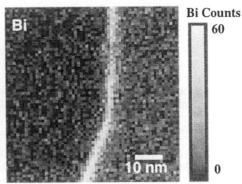

Fig. 3 Bright field image and associated Bi map at a faceted grain boundary in Bi doped Cu

In Fig 3, a similar map, but for Bi doped Cu, is seen, where the left-hand side is a simultaneously collected bright-field image and the right-hand side is the Bi map. Two facets of the grain boundary can be seen. The lower facet appears to be brighter than the upper facet and this reflects a real difference in segregation levels. When the segregation levels were checked for these two facets using a stationary, reduced-area scan and longer acquisition times for accurate quantification, the lower facet was found to contain $3.7 \pm 0.3$ atoms/nm$^2$ and the upper facet $3.0 \pm 0.2$ atoms/nm$^2$.

In Fig. 4 the Ni $L_{23}$ energy-loss spectra for a grain boundary in S-doped Ni and the nearby grain are shown. As for the case of Bi in Cu, an increase in the white-line intensity is observed at the grain boundary

## 4. DISCUSSION AND CONCLUSIONS

The profile of Fig. 1 has a very high resolution, the full width of the boundary region is around 2 nm. This comparable to, or better than, any previously obtained. Mapping of segregants is relatively routine on the VG HB603 and promises the ability to distinguish different segregation levels between facets of a grain boundary, or even along one facet.

302

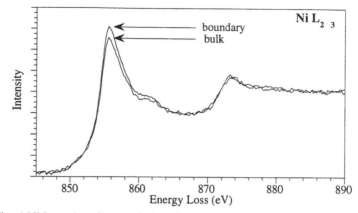

Fig. 4 Ni L$_{23}$ edges for a grain boundary and the nearby grain in S doped Ni

The spectra of Fig. 4 suggest that, in addition to identifying, quantifying and locating S segregation, we can observe changes in the bonding at the boundary that may be associated with embrittlement. Care must be taken to properly account for the thickness of the sample. A change in thickness can erroneously introduce a white line into the Ni edge. For this reason, the low-loss spectrum should be used to ensure that the grain boundary is of the same thickness as the adjacent grain. In addition, contamination should be minimal as deconvolution of an improper low-loss spectrum will also introduce a white line. The spectra in Figure 5 were collected from regions within 5 nm of one another, changes in thickness were below 3% and contamination was minimal. The difference between the spectra, although small, suggests that a similar mechanism to that observed in Bi doped Cu is occurring.

## 5.    ACKNOWLEDGMENTS

The financial assistance of the National Science Foundation is gratefully acknowledged (DMR 93 05253) as well as D. M. Symons from Westinghouse-Bettis and D. E. Luzzi from the University of Pennsylvania for supplying materials. Thanks also to M. Watanabe and D. T. Carpenter for helpful discussions and technical assistance.

## REFERENCES

Bentley J 1997 Microscopy and Microanalysis 1997, eds G W Bailey, R V W Dimlich, K B Alexander, J J McCarthy and T P Pretlow (New York: Springer) pp 533-4
Bruley J, Keast V J and Williams D B 1996 J.  Phys. D: Appl. Phys. **29,** 1730
Bulloch J M 1994 Mater. High Temp. **12,** 311
Cliff G and Lorimer G W 1975 J. Microsc. **103,** 203
Doig P and Flewitt P E J 1977 J. Microscopy **112,** 257
Hall E L, Imeson D and VanderSande J B 1981 Phil. Mag. A **43,** 1569
Keast V J, Bruley J, Rez P, Maclaren J M and Williams D B 1997 Acta Mat. in press
Lyman C E, Goldstein J I, Williams D B, Ackland D W, von Harrach S, Nicholls A W and Statham P J 1994 J. Microsc. **176,** 85
Menon E S K, Reynolds W T and Fox A G 1997 Microscopy and Microanalysis 1997, eds G W Bailey, R V W Dimlich, K B Alexander, J J McCarthy and T P Pretlow (New York: Springer) pp 553-4
Michael J R and Williams D B 1984 Met. Trans. **15A,** 99
Muller D A, Subramanian S, Batson P E, Silcox J and Sass S L 1996 Acta Mat.. **44,** 1637
Titchmarsh j M and Dumbell S 1996 J. Microscopy **184,** 195
Williams D B and Romig A D 1989 Ultramicroscopy **30,** 38
Williams D B, Watanabe M and Carpenter D T 1997 Mikrochimica Acta in press

*Inst. Phys. Conf. Ser. No 153: Section 8*
*Paper presented at Electron Microscopy and Analysis Group Conf. EMAG97, Cambridge, 1997*
© 1997 IOP Publishing Ltd

# Quantification of the composition of silicon germanium / silicon structures by high-angle annular dark field imaging

**T Walther*** and **C J Humphreys**

Department of Materials Science and Metallurgy, University of Cambridge, Pembroke Street, Cambridge CB2 3QZ, UK; *now at: CEA-Grenoble, Département de Recherche Fondamentale sur la Matière Condensée, SP2M, 17 rue des Martyrs, 38054 Grenoble, France

**ABSTRACT:** The intensity in high-angle annular dark field images of SiGe/Si heterostructures is studied as a function of the specimen thickness and orientation, the inner detection angle and the energy loss. The approximation by Rutherford scattering where the intensity is proportional to the thickness and the square of the average atomic number is only applicable within a narrow thickness range (50 to 100 nm). At small thicknesses oxide layers on the crystal surface and at large thicknesses plasmon and Si K core loss scattering reduce the image contrast and hence the apparent Ge composition. Also, strain related contrast is visible, even in off-axis conditions and for large inner detection angles.

## 1. INTRODUCTION

High-angle annular dark field imaging (HAADF) in a scanning transmission electron microscope (STEM) is a promising technique for the study of layers of materials with different average atomic numbers (Howie 1979). The HAADF signal comprises contributions from elastic, phonon, free electron and multiple scattering. The atomic number sensitivity or `Z-contrast´ is due to the dominance of phonon scattering for large angles, according to the Rutherford formula, and has been successfully used to characterise materials on an atomic scale (Treacy 1982, Shin et al. 1989). However, signal quantification as necessary for a correlation with other techniques brings with it some experimental problems. For reasons of counting statistics as well as instrumental stability an aperture is generally used which limits the inner detection angle to about 50 mrad. This allows electrons from e.g. multiple Bragg scattering, which is much more intense than phonon scattering, to contribute a background of unknown dependence on the atomic number to the image intensity. Also, the usefulness of HAADF to a quantitative characterisation of an interface between a weakly and a strongly scattering material, for which it should be ideally suited, is limited by the lack of energy filters in most STEM equipments. Light elements exhibit low-energetic core losses whereas heavy elements show corresponding transitions at higher energies. This will change the contrast of an unfiltered experimental HAADF image as compared with a simulation of elastic and thermal diffuse scattering (Wang and Cowley 1989), especially for large thicknesses (Treacy and Gibson 1993).

In this study we determine the influence of amorphous surface layers, sample orientation, inelastic scattering and inner detection angle on the HAADF contrast of SiGe/Si layers and show that a quantification of the signal necessitates a careful choice of the experimental conditions, especially the sample thickness.

## 2. EXPERIMENTAL

Specimens of SiGe/Si (001) multilayers have been prepared by cleavage on {110} planes, with occasional cleavage steps on {111}. The cleaved wedges were mounted on copper grids using a conductive silver epoxy glue and transferred into vacuum within a few minutes such that oxidation of the cleaved surfaces was minimised. The experiments were performed in a VG HB501 STEM at 100 kV with a probe size of about 0.3 nm, a set of home-built annular masks on a rotating holder in front of the dark field detector and either a Link photomultiplier (for standard HAADF) or a Gatan imaging filter with slow scan CCD (for energy-filtered HAADF and electron energy-loss spectroscopy, EELS). The nonlinear response of the detectors has been corrected for. If not mentioned otherwise, the inner detection angle was limited to about 200 mrad by the annular mask. The precise value depends on the setting of the objective lens the pre-field of which distorts, for a strong excitation, the trajectories of electrons scattered sideways. The ratio, R, of the image intensity, I, of $Si_{1-x}Ge_x$ to that of Si is given by

$$R \equiv I(Si_{1-x}Ge_x)/I(Si) = t(Si_{1-x}Ge_x)\,\overline{Z}^n(Si_{1-x}Ge_x)\Big/\Big[t(Si)\,Z^n(Si)\Big] \approx (9x/7+1)^2 \quad (1)$$

where the approximation is valid if the thickness, t, of the SiGe and the Si is comparable (as can be ensured for cleaved wedges from their geometry) and if the scattering behaves like Rutherford scattering with a dependence of power n=2 on the average atomic number, Z. Other scattering mechanisms will reduce the image contrast such that 1<n<2 (Pennycook 1989) where the precise value of n will depend on the material, the specimen thickness and orientation, the detector angle and the energy loss. Hence we obtain

$$x \equiv x(Ge) = 7/9\big[\exp(\ln R/n)-1\big] \approx 7/9\big(\sqrt{R}-1\big). \quad (2)$$

## 3. RESULTS

### 3.1 The rôle of the specimen thickness and amorphous surface layers

Line scans of the HAADF image intensity of Si and $Si_{1-x}Ge_x$ with different compositions x for a cleaved wedge are displayed in Fig. 1a as a function of the specimen thickness. The similar HAADF intensity of Si and SiGe near the specimen edge is explained by amorphous surface layers which will, to a first approximation, add a constant to the low image intensities of both crystalline materials and thus reduce the contrast. The thickness above which the influence of the surface oxide becomes negligible can be estimated: if the oxide has the thickness of typical native $SiO_2$ ($\approx$2.5 nm), is found on both top and bottom surface of the specimen (giving a factor of 2), makes up a fraction f of the number of atoms of the crystalline part and has an average atomic number similar to $SiO_2$, then pure Rutherford scattering will change R from $R_{clean}=(9x/7+1)^2$ to

$$R_{oxidised} \equiv I(Si_{1-x}Ge_x + f\,SiO_2)/I(Si + f\,SiO_2) \approx \big[9x/(7+5f)+1\big]^2. \quad (3)$$

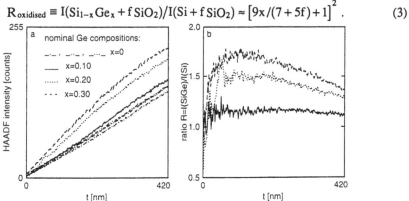

Fig. 1: Plot of the HAADF image intensity (a) and the intensity ratio of SiGe to Si (b) vs. the specimen thickness in unfiltered HAADF from a cleaved SiGe/Si wedge

For an error of $\Delta x \leq 0.03$ over the entire range $0 \leq x \leq 0.3$, $\Delta R/R=0.05$ and hence f<0.16 would be required. Taking into account that the number of atoms per volume is about 1.6 times higher in quartz than in Si, the volume fraction corresponding to f=0.16 is 0.1, i.e.: 5 nm $SiO_2$ must constitute less than 10% of the specimen. Hence, a *minimum* crystal thickness of 50 nm is required to obtain the composition from unfiltered HAADF within 10%.

## 3.2 The rôle of the inner detection angle and the sample orientation

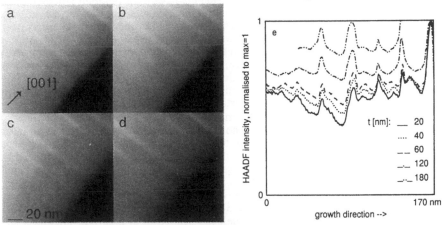

Fig. 2: HAADF images of SiGe/Si layers of different widths (x≈0.1). Specimen orientations: ‹100› zone axis orientation (a, b) and a symmetrical diffraction condition along the (040) Kikuchi band (c, d). Inner detection angles: 24 mrad (a, c) and 200 mrad (b, d). Contrast between the layers stems from strain relaxation. (e): intensity line scans from (d), demonstrating that asymmetric strain features prohibit any quantitative analysis of thin specimens. The effect decreases with increasing specimen thickness.

In thin specimens, a significant part of the strain can be relaxed by the bending of lattice planes near the surface. This strain relaxation is visible in HAADF as asymmetric light/dark line features parallel to the layers (Perovic et al. 1993). Fig. 2 indicates that this effect is slightly stronger for a zone axis orientation than for a symmetrically diffracting condition. If, however, layers are to be imaged edge-on then weakly diffracting conditions which would be favourable are difficult to establish. Channelling conditions along a zone axis should clearly be avoided for quantitative HAADF. Also, for inner detection angles ≤100 mrad the image contrast is reduced at large specimen thicknesses (Treacy and Gibson 1993), and even for scattering angles ≥200 mrad strain contrast can be seen in Fig. 2e up to about 60 nm thickness (cf. Cowley and Huang 1992, Hillyard and Silcox 1995).

## 3.3 Inelastic contributions (plasmon and core losses)

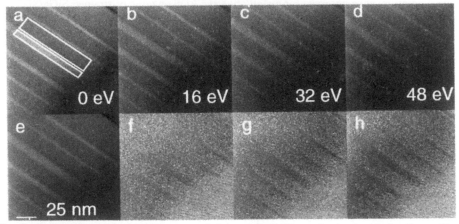

Fig. 3: Energy filtered HAADF images of cleaved SiGe/Si, as obtained with the Gatan filter; slit width: 12 eV; energy losses: 0 (a), 16 (b), 32 (c) and 48 eV (d). (e) is the weighed sum of (a) to (d). (f, g, h) are the ratios of (b, c, d) divided by (a); black=0; white=2.

Only a few energy filtered HAADF experiments have been reported (Bleloch et al. 1994). It is well-known that EELS spectra taken with large momentum transfer, i.e. at large scattering angles, yield a smaller inelastic mean free path. In other words: the specimen appears thicker at larger angles. This cannot be explained by geometry as even for a large beam convergence the longer path along which sideways scattered electrons move amounts to less than 10% after radial averaging. Hence, inelastic contributions are more important under HAADF than under bright field conditions. For cleaved Si with native oxide, the elastic & phonon component never constitutes more than 60% of the total HAADF intensity, and this contribution falls rapidly for larger crystal thicknesses (Walther 1996).

Fig. 3 shows energy filtered HAADF images of SiGe/Si layers of nominal Ge content of x=0.3 to 0.4, acquired for elastic & phonon scattering (zero loss) and for plasmon losses. As the ratio images of Figs. 3f-h show the layers as dark lines, the HAADF image contrast is reduced by plasmon as compared with elastic scattering. The intensity ratio R in Figs. 3b-d is reduced to 0.77±0.06 at a thickness of the inelastic mean free path as compared to Fig. 3a. The compositional error is enormous: $Si_{0.7}Ge_{0.3}$ could be mistaken for $Si_{0.8}Ge_{0.2}$ !

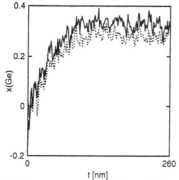

A correction for plasmon losses by using EELS spectra to determine the amount of inelastic scattering at a certain thickness and then weighing HAADF images acquired at various energy losses correspondingly halves this error according to Fig. 4. While the zero loss HAADF image of Fig. 3a yields x=0.33±0.02, Fig. 3e would give x=0.28±0.02. Constant values are approached for thicknesses t≥100 nm in both cases. This can be explained by the exclusion of the Si L contributions at 99 eV in these filtered images: at a thickness of about the inelastic mean free path of silicon (≈120 nm), this edge contributes 10% to the integrated intensity of an EELS spectrum of SiGe at 200 mrad. As contributions from the Ge L edge (at ≥ 1217 eV) are negligible, Si L losses reduce the ratio of the SiGe to the Si intensity in unfiltered HAADF for t>100 nm (Fig. 1b).

Fig. 4: Apparent composition of $Si_{0.7}Ge_{0.3}$ vs. thickness for filtered HAADF including only the zero loss, i.e. phonon scattering (solid line) or zero and plasmon losses (dotted)

## 4. CONCLUSION

Several points must be observed for quantitative HAADF imaging of strained SiGe/Si:
• At thicknesses above the mean free path the Si L contribution reduces the image contrast.
• Plasmon scattering reduces the image contrast at all thicknesses. For $Si_{0.7}Ge_{0.3}$ this leads to an underestimate of the Ge composition by about Δx=0.05 from unfiltered images.
• At small thicknesses amorphous surface layers reduce the image contrast considerably.
• A zone axis orientation and inner detector angles under 100 mrad must be avoided.
With an energy filter the first two problems are avoided. This study furthermore facilitates a correction of data acquired unfiltered which is necessary if no imaging filter is available.

## REFERENCES

Bleloch AL, Castell MR, Howie A and Walsh CA 1994 Ultramicroscopy **54**, 107
Cowley JM and Huang Y 1992 Ultramicroscopy **40**, 171
Hillyard S and Silcox J 1995 Ultramicroscopy **58**, 6
Howie A 1979 J. Microsc. **117**, 11
Pennycook SJ 1989 Ultramicroscopy **30**, 58
Perovic DD, Howie A and Rossouw CJ 1993 Philos. Mag. Lett. **67**, 261
Shin DH, Kirkland EJ and Silcox J 1989 Appl. Phys. Lett. **55**, 2456
Treacy MMJ 1982 J. Microsc. Spectrosc. Electron. **7**, 511
Treacy MMJ and Gibson JM 1993 Ultramicroscopy **52**, 31
Walther T 1996 PhD Thesis, University of Cambridge, UK
Wang ZL and Cowley JM 1989 Ultramicroscopy **31**, 437

*Inst. Phys. Conf. Ser. No 153: Section 8*
*Paper presented at Electron Microscopy and Analysis Group Conf. EMAG97, Cambridge, 1997*
© 1997 IOP Publishing Ltd

# Electron microscopic characterization of doped sintered α-Al₂O₃

**K Kaneko, T. Gemming\*, I Tanaka§**

Japan Science and Technology Corporation (JST), International Joint Research Program, Ceramics Superplasticity, c/o Japan Fine Ceramics Centre 2F, 2-4-1, Mutsuno, Atsuta-ku, Nagoya 456, JAPAN

\*Max-Planck-Insitut für Metallforschung, Seestraβe 92, D-70174 Stuttgart, Germany
§Dept. Mater. Sci. & Engg., Kyoto University, Sakyo, Kyoto, 606-01 JAPAN

**Abstract:** High-resolution transmission electron microscopy (HRTEM) and analytical electron microscopy (AEM) was employed to investigate grain boundary structures, chemical compositions and chemical bondings of Zr-doped polycrystalline α-Al₂O₃. HRTEM revealed that there was no amorphous films at any grain boundary phases as well as no triple pocket was observed. From the O-K edge ELNES, no-apparent chemical shift nor changes of the shape of EELS were observed during this experiment. However, a slight modification due to structural changes became apparent after application of spatially differential ELNES method.

## 1. INTRODUCTION

Alumina is a common ceramic material with many commercial applications, particularly polycrystalline Al₂O₃ has been technologically useful for its insulation and its chemical resistance. Grain boundary of doped and undoped α-Al₂O₃ have been intensively studied by many workers. Particularly, a special grain boundary known as Σ11 tilt boundary of bicrystal has been extensively investigated in the sense of atomic and electronic structures by a group of the Max Planck Institute and others in recent years (Bruley (1993), Brydson et al. (1995), Höche et al. (1994) and Mo et al. (1996))

In the case of polycrystalline ceramics materials, structures of grain-boundaries become more complicated than the simple Σ11 boundary for two reasons; 1) the orientations between adjacent grains are expected to be almost random when the materials are fabricated by sintering of fine powders, 2) grain boundaries in sintered α-Al₂O₃ samples are often decorated by impurities inherent to starting powders or intentionally doped as sintering agents.

In the present study, EELS work are carefully repeated for several random grain boundaries of 0.1 wt% - Zr-doped α-Al₂O₃ system. A first principles molecular orbital (MO) calculation of both the Zr-doped α-Al₂O₃ have been made in order to understand the possible influence of the presence of donors to the ELNES.

## 2. EXPERIMENTAL DETAILS

TEM specimens were prepared by the standard mechanical thinning method, dimpling, then Argon ion milling. The specimens were coated with a thin carbon layer to avoid charging-up effects.

HRTEM was conducted on a JEOL 1250 operated at 1250 kV with a point resolution of 1.3 Å. AEM was achieved with a dedicated STEM (Vacuum-Generators HB 601 UX) operated at 100 kV, with a field emission gun. An energy resolution of better than 0.4 eV was attained by the full-width-half-maximum (FWHM) of zero-loss with the spectrometer entrance aperture limited to about 16 mrad. For measuring the EELS edges, the beam current was kept around approximately 0.2 nA with a mean beam diameter of 5 Å (FWHM).

The DV-Xα method using a program code SCAT has been successful for reproduction of ELNES of various compounds by Tanaka and Adachi (1996a and 1996b). The advantages of this method compared with other theoretical methods are as follows: 1) Computations are made in a first principles manner. Therefore, no empirical parameters except for atomic numbers and coordinations are necessary as an input. 2) Since it uses flexible and numerical atomic orbital basis-functions, quantitatively reliable calculations with respect to molecular orbital energies can be made up to 20 eV above the Fermi energy only with minimal basis-set. It is greatly advantageous for understanding of the origin of spectral features in terms of chemical bondings. 3) Effects of core-hole can be rigorously taken into account, because all core electrons are included in the diagonalization of Hamiltonian.

## 3.    RESULTS AND DISCUSSION

As can be seen from the HR-TEM images (Fig. 1), the two grains usually exhibit a mismatch in orientation, and distorted near the grain boundary within approximately 0.3 nm width. No precipitates are also observed. The three grain junctions and the grain boundaries are found free of second phase. EDS results obtained showed apparent segregations of Zr at grain boundaries (Fig. 2).

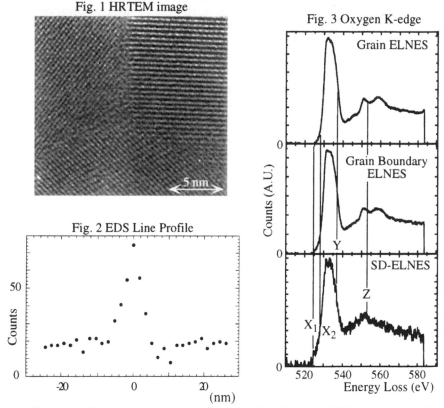

Fig. 1 HRTEM image

Fig. 3 Oxygen K-edge

Fig. 2 EDS Line Profile

Two sets of spectra were recorded essentially with 0.1 eV/channel energy dispersion from various energy ranges to observe O-K ELNES and the low-loss region. The regions of 20 to 30 nm thickness was selected as optimum area to obtain EELS. Standard method of obtaining ELNES were applied throughout the experiment (Egerton 1996). The O-K ionization edge was used since it exhibits saw teeth profile with a sharp rise at the threshold followed by a slow decay. The original spectrum is dominated by the bulk matrix component due to the finite size of the beam. The spatial difference (SD-) ELNES represents the contribution of impurities and structures to Al and O atoms at the interface which could possess a different local coordination to those in the bulk $\alpha$-$Al_2O_3$ by removing the energy-dependent background. Hereafter, O-K ELNES and theoretical considerations are discussed.

Comparisons between the one obtained from bulk $\alpha$–$Al_2O_3$ and the other obtained from grain boundaries seen in Fig. 3 showed that there were petite changes as a slight broadening on the low energy side and a small pre-peak edge near the threshold of O-K edge. SD-ELNES shown exhibits the peak-shift to the lower energy, and peaks $X_1$ and $X_2$.

Computations are made for $\alpha$–$Al_2O_3$ (Fig. 4a), tetragonal $ZrO_2$ (Fig. 4b) and $\alpha$–$Al_2O_3$ with a substitutional $Zr^{4+}$ ion using model clusters of $(Al_8O_{33})^{42-}$ and $(Zr_{13}O_{56})^{60-}$, and $(ZrAl_7O_{33})^{41-}$ (Fig. 4c), respectively. These clusters are centered by a cation, and are embedded in the Madelung potential generated by point charges outside the clusters. In the case of $(ZrAl_7O_{33})^{41-}$ cluster, $Zr^{4+}$ ion was put at the center of the cluster.

Since both $Al_2O_3$ and $ZrO_2$ are fully oxidized metal oxides, major components of valence band and conduction band are O-2p and metal-spd orbitals; They are separated by a band-gap. Although the computed band-gap is larger than the experimental one because we use a small model cluster, the computed DOS well reproduces the shape of DOS of bulk crystals for both $\alpha$–$Al_2O_3$ and t- $ZrO_2$. The LDOS of central Zr ion, its coordinated O-ions and Al-ions are combined to make the DOS of the $Zr^{4+}$ in $\alpha$–$Al_2O_3$. The Zr-4d band is found to be located in between the band-gap of $\alpha$–$Al_2O_3$. Since the Zr-4d orbitals interact with O-2p and Al-3spd orbitals, notable increase in O-2p and Al-3spd PDOSes can be seen just below the bottom of the conduction band of $\alpha$–$Al_2O_3$ when $Zr^{4+}$ is present as indicated by arrows in right middle and bottom panels of Fig. 4(c). The O-2p PDOS increased more significantly than the Al-3spd PDOSes below the conduction band-edge since Zr-O interaction is more stronger than the Al-Zr interaction. In the present calculation, no lattice relaxation around $Zr^{4+}$ was included. Formal valency of substitutional Zr was assumed to be $Zr^{4+}$. More detailed electronic calculation may change the PDOS features for some extent, however, it cannot alter the qualitative arguments of the location of Zr-4d band as well as its orbital components: For example, if formal valency of Zr is assumed to be 3+, an electron occupy the Zr-4d band and its MO energy should increase. If an oxygen vacancy is present near the $Zr^{4+}$ impurity, a vacancy level should appear and as a result the MO energy of the Zr-4d band should be altered. It could be influenced also by the inclusion of lattice relaxations. However, it is natural to expect that the Zr-O interaction is more stronger than the Al-Zr interaction in any of these cases.

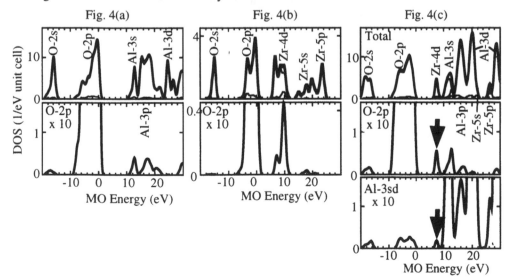

Fig. 4(a)    Fig. 4(b)    Fig. 4(c)

## 4. CONCLUSION

HRTEM and EDS/EELS studies of random grain boundaries of fine-grained 0.1 wt% Zr-doped polycrystalline $\alpha-Al_2O_3$ samples have been made. No intergranular phases can be found at any boundaries observed in the present work, and the atomic structure of the grain boundary region was not strongly distorted. At the same time, however, segregations of Zr at the grain boundaries was detected.

It is true that the newly appeared peaks in the Al-3spd and O-2p PDOS due to the presence of substitutional $Zr^{4+}$ ion in $\alpha-Al_2O_3$ coincide with either $X_1$ or $X_2$ peaks that were observed by experiments. However if it is the case, the intensity of these new peaks should be more stronger. The origin of these peaks may therefore be more likely ascribed to the mishappening of the chemical bonding of $Al_2O_3$ at the grain boundary. As discussed above, the peak $X_2$ may be ascribed to Al atoms coordinated only by four oxygen atoms. The peak $X_1$ should be due to more heavily mishappening defects. Presence of Zr ion at the boundary may simply contribute to the mishappening.

## ACKNOWLEDGMENT

We thank Ms M. Sycha and Ms S. Schulz (MPI/MF) for preparing S/TEM specimens. KK would like to acknowledge Dr. S. Tsurekawa and Dr. K. Saitoh (Univ. of Tohoku), Dr. Y. Ikuhara (Univ. of Tokyo), Dr. H. Gu, Dr. T. Nagano, Dr. F. Wakai and Mr. S. Honda (JST) for valuable discussions.

## REFERENCES

Bruley J 1993, Microsc. Microanal. Microstruct., **4**, 23.
Brydson R, Müllejans H, Bruley J, Trusty P A, Sun X, Yeomans J A and Rühle M 1995, J. Micros., **177**, 369.
Egerton R F 1996 Electron Energy Loss spectroscopy in the Electron Microscope   2nd  edition (New York: Plenum Press)
Höche T, Kenway P R , Kleebe H-J, Rühle M, and Morris P A 1994 J. Am. Ceram. Soc., **77**., 339
Mo S-D, Ching W-Y and French R H 1996, J. Am .Ceram. Soc., **79**, 627.
Tanaka I and Adachi H 1996a Phys. Rev. B., **54**, 4604
Tanaka I and Adachi H 1996b J. Phys. D: Appl. Phys., **29**, 1725

*Inst. Phys. Conf. Ser. No 153: Section 8*
*Paper presented at Electron Microscopy and Analysis Group Conf. EMAG97, Cambridge, 1997*
© *1997 IOP Publishing Ltd*

# Application of EXELFS to glasses in a dedicated STEM

**Y Ito, H Jain and D B Williams**

Department of Materials Science and Engineering, Lehigh University, Bethlehem, PA, 18015 USA

**ABSTRACT** : Extended energy-loss fine structure (EXELFS) in an electron energy-loss spectrum (EELS) has been applied to the determination of the element-specific local structure (e.g. partial radial distribution functions (RDF)) of glasses. EXELFS spectra have been acquired by using a Gatan 666 PEELS spectrometers attached to a VG HB501 (100 keV) scanning transmission electron microscope (STEM). Pure silica glass (amorphous $SiO_2$) was chosen as a reference specimen since its structure is well known. RDFs around oxygen and silicon atoms were obtained by the Fourier transform method.

## 1. INTRODUCTION

EXELFS, the electron analogue of extended X-ray absorption fine structure (EXAFS), is particularly appropriate for light elements which are the most important atoms in common oxide glass systems (e.g. oxygen, silicon, etc.). EXELFS can be obtained from nanometer-scale area in a transmission electron microscope. Recent developments of EXELFS provide comparable quality of data to that obtained from a synchrotron source (Qian et al., 1995). However, there has been only limited EXELFS study on $SiO_2$ (Yuan et al., 1995) since most of the silicate glasses are susceptible to electron beam damage in an electron microscope (Hobbs, 1984). In this paper, RDF around oxygen and silicon atoms were obtained from oxygen and silicon K-edges with minimum beam damage. RDFs around oxygen from heavily beam damaged area were compared with those from undamaged area.

## 2. EXELFS AND RDF

For amorphous materials, EXELFS is regarded as a measure of the short-range order due mainly to scattering from nearest-neighbour atoms. An incoming high-energy electron ejects a core electron from an atom in the specimen which is represented by the outgoing spherical wave. Weak oscillations can arise from interference between the outgoing spherical wave and reflected waves due to elastic backscattering of the electron wave from neighbouring atoms.

Approximating the ejected-electron wavefunction at the backscattering atom by a plane wave and assuming that multiple backscattering can be neglected, EXAFS theory gives

$$\chi(k) = \sum_j \frac{N_j}{r_j^2} \frac{f_j(k)}{k} \exp(-2r_j/\lambda_i) \exp(-2\sigma_j^2 k) \sin[2kr_j + \phi_j(k)], \tag{1}$$

where $r_j$ is the radius of a particular shell of neighbouring atoms, $N_j$ is the number of atoms in shell $j$ and $N_j/r_j^2$ (as a function of r) is the RDF. Therefore, $\overline{\chi}(k) = |RDF|$ can be obtained as the modulus of by Fourier transform of (1). $f_j(k)$ is the backscattering amplitude from the atoms in the same shell, $\lambda_i$ is the inelastic mean free path of the ejected electron, $\sigma_j^2$ is the mean square displacement of the excited atom, and $\phi_j(k)$ is the phase shift due to both the excited and to the j-th backscattering atoms.

## 3. SPECIMEN AND EXPERIMENTAL

A $SiO_2$ film was deposited onto a holey carbon film using a magnetron-sputtering machine purged with argon gas with a radio frequency (RF) gun (100 and 300 W). The target used was an amorphous $SiO_2$ disk (99.9995%) (Pure Tech Inc.). Film thickness was approximately 200 Å estimated from the low-loss EEL spectra of the specimen. The holey carbon film was used for a mechanical support as well as for preventing the film from charging.

A VG HB501 STEM with Gatan 666 PEELS spectrometer was used. However, in order to avoid electron beam damage, the probe was over-focused (approximately 7 μm overfocus) (Ito et al., 1997). Spectra were obtained with this defocused probe raster-scanned at magnification of 5 million times. Convergence semi-angle of the focused probe was 15 mrad and collection semi-angle was 35 mrad. Acquisition time for an oxygen and silicon K-edge was 5 sec and 300, respectively.

## 4. DATA PROCESSING

After the removal of dark-current noise, flat-field correction and subtraction of the pre-edge background, contribution of plural scattering was removed by applying the Fourier-ratio deconvolution (Egerton, 1996). In-house software performed extraction of oscillatory component of EXELFS ($\chi$(eV)-function) by using either n-th order polynomial or power-law fitting, conversion of the scale of $\chi$(eV) into k-space ($\chi$(k)-function), correction for k-dependence of backscattering, truncation of $\chi(k)$ by multiplying a top-hat function of a unit amplitude, and Fourier transform of $\chi(k)$ (raw radial distribution function, $\overline{\chi}(r) = |RDF|$). Figs. 1a-c are examples of the output of the software. For RDF calculation, the first two peaks of oxygen K-edge were omitted since these peaks possibly include contributions from multiple scattering and the plane wave approximation may start breaking down (Ito et al., 1997). Tentative phase-shift correction was applied. Here, the first peak of RDF was compared with the value in the literature. Assuming the phase shift is linear in k-space, the difference of the first peak position from the value in literature was added to measured values of subsequent peaks. These peak positions were also compared with values in the literature (Table 1).

## 5. RDF OF BEAM DAMAGED AREA

RDFs from oxygen K-edge were obtained by a focused STEM probe from an undamaged area and from a heavily damaged area (Fig. 1d). An area of 35x45 $nm^2$ was damaged by raster-scanning a probe for 20 min prior to the acquisition of EEL spectra. The spectrum was collected from a 7x9 $nm^2$ region of the damaged area with a focused probe.

The first peak representing the Si-O bond shifted towards shorter distance in the RDF of the beam damaged area and also the peak is sharper than that of the undamaged one. Under the stationary STEM probe, a-$SiO_2$ can be reduced to pure Si (Chen et al., 1993). Oxygen atoms are desorbed from the surface layers by a Knotek and Feibelman electron-stimulated desorption mechanism, and it is assumed that oxygen atoms in the bulk are similarly displaced. By raster-scanning a STEM probe across the $SiO_2$ film, a mixture of silicon and silica in the beam damaged area are expected to be formed.

## 6. RDF AROUND SILICON

The original silicon K-edge spectrum was smoothed by convolving with a Gaussian function (FWHM 5 eV). Parameters for the 5th order polynomial fit were chosen such that the fitted curve followed closely to the power-law fit (Fig. 2a). The obtained RDF closely resembles that obtained by XAFS (Greaves et al., 1981). The peak at 2.36 Å in Fig. 2b corresponds to the Si-Si bond distance. The RDF obtained from XAFS shows a small shoulder on the right-hand side of the peak. Although the RDF in Fig. 2b does not show a clear shoulder, an asymmetric first peak indicates the possible existence of an extra peak, which agrees with the XAFS results.

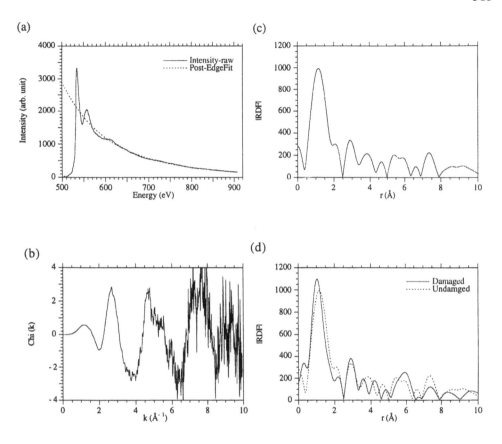

Fig. 1 EXELFS around oxygen K-edge. (a) Power-law curve fitting for extraction of oscillatory component. (b) $\chi(k)$-function. (c) RDF without phase shift correction. Fourier transform was performed for the k-range between 4.41 and 9.65 Å$^{-1}$ in Fig. 1b. (d) RDF of electron-beam damaged $SiO_2$.

Table 1 Peak position in RDF about oxygen with phase-shift correction.

| Peaks | i*<br>Peak position<br>without phase-<br>shift (Å) | ii<br>Peak position<br>from the<br>literature (Å) | iii<br>Peak position<br>with phase-<br>shift (Å)† | iv<br>Difference iv-ii (Å) |
|---|---|---|---|---|
| 1st | 1.16 | 1.61** | 1.61 | 0 |
| 2nd | 2.08 | 2.61** | 2.59 | -0.02 |
| 3rd | 2.91 | (3.43)*** | 3.42 | -0.01 |

* The relative error in peak positions in the RDFs is estimated to be ±0.01.
** Values are from (Greaves, 1990).
*** Value is inter-tetrahedra O-O of α-quartz (Ogata et al., 1987).
† Constant value 0.51 Å (ii-i) was added as a phase-shift correction.

314

(a)                                          (b)

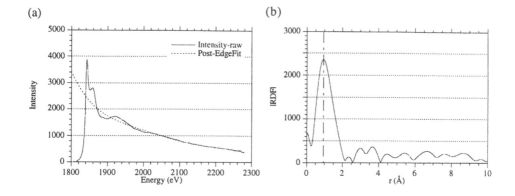

Fig. 2 EXELFS around silicon K-edge. (a) 5th order polynomial curve fitting for extraction of oscillatory component. (b) RDF without phase shift correction.

## 8. CONCLUSION AND FUTURE WORK

EXELFS of oxygen and silicon K-edges of an a-SiO$_2$ film on a holey carbon film were acquired in a VG HB501 STEM. In order to minimise electron beam damage, a defocused probe was used. Peak positions of the RDFs with phase-shift correction showed an excellent agreement with the literature values. The maximum deviation of positions of the first three peaks of RDF around oxygen was less than 0.02 Å from the literature values. The RDF around silicon also showed very good agreement with that from EXAFS, which confirms the reliability of the data analysis. the data process was sensitive enough so that An RDF of an electron-beam damaged area of the specimen showed an apparent change from that of the undamaged. The software can be improved in spline fitting and the truncation of $\chi$-function, and in noise reduction in raw EEL spectra. In future, curve fitting method will be applied for more detailed analysis (Qian et al., 1995).

## ACKNOWLEDGEMENTS

The authors would like to thank Drs. G.N. Greaves, D.J. Wallis and J. Bruley for helpful discussions. Also, Dr. K. Barmak and Mr. J. Kim are thanked for assistance with the magnetron sputtering system. This work is supported by the US Department of Energy (DE-FG02-95ER45540). The magnetron-sputtering system is supported by NSF (DMR-941116).

## REFERENCES

Chen, G.S., Boothroyd, C.B. and Humphreys, C.J. (1993) Appl. Phys. Lett. **62,** 1949.
Egerton, R.F. (1996) *Electron energy-loss spectroscopy in the electron microscope,* 2nd edn. Plenum Press, New York.
Greaves, G.N. (1990) *Glass Science and technology,* **4B,** eds. Uhlmann, D.R. and Kreidl, N.J., Academic, Boston, 1.
Greaves, G.N., Fontaine, A., Lagarde, P., Raoux, D. and Gurman, S.J. (1981) Nature **293,** 611.
Hobbs, L.W. (1984) *Quantitative electron microscopy,* eds. Chapman, J.N. and Cravan, A.J., SUSSP publications, Edinburgh, 399.
Ito, Y., Winkler, D., Jain, H. and Williams (1997) to be published in J. Non-Cryst. Solids.
Ogata, K., Takeuchi, Y. and Kudoh, Y. (1987) Z. Kristallogr. **179,** 403.
Qian, M., Sarikaya, M. and Stern, E.A. (1995) Ultramicroscopy **59,** 137.
Yuan, Z.W., Csillag, S., Tafreshi, M. and Colliex, C. (1995) Ultramicroscopy **59,** 149.

Inst. Phys. Conf. Ser. No 153: Section 8
Paper presented at Electron Microscopy and Analysis Group Conf. EMAG97, Cambridge, 1997
© 1997 IOP Publishing Ltd

# PEELS of STEELS

**E. S. K. Menon[†], A. G. Fox[†], W. T. Reynolds, Jr.,[*] and G. Spanos[§]**

[†] Center for Materials Science, Department of Mechanical Engineering, US Naval Postgraduate School, Monterey, CA 93943, USA; [*] Department of Materials Engineering, Virginia Polytechnic Institute, Blacksburg, VA 24061, USA; [§] Code 6324, Naval Research Laboratory, Washington DC 20375, USA.

**ABSTRACT:** Some of the salient results of parallel electron energy loss (PEELS) studies of the detection, distribution and quantification of carbon in various steels are presented. The feasibility of detecting C in ultra low carbon alloy steels containing as little as 0.03wt% C by PEELS is demonstrated followed by examples of electron energy loss images showing C partitioning between retained austenite and ferrite or martensite. In addition, an example of C and Mo segregation to ferrite:martensite interfaces during diffusional austenite decomposition in a Fe-C-Mo steel is presented.

## 1. INTRODUCTION

The influence of C and other alloying elements on the kinetics of austenite decomposition in steels is well documented (Aaronson 1962). Since the properties of heat treated steels are strongly influenced by the partitioning of alloying elements, especially carbon in various phases like austenite ($\gamma$) ferrite ($\alpha$) and martensite ($\alpha$'), a knowledge of the local chemistry is of paramount importance. A variety of qualitative as well as quantitative techniques including Energy (or Wavelength) Dispersive X-ray Spectroscopy, Atom Probe Field Ion Microscopy, Electron Energy Loss Spectroscopy and Imaging are extremely useful and have indeed been employed in this area of research (Bhadeshia et al 1982; Stark et al 1988&1990; Reynolds et al 1988; Bentley and Kenik 1994; Ozkaya et al 1994; Yuan et al 1995; Hofer et al 1996). Undoubtedly, the analysis of C, whether in large quantities or in small amounts remains an extremely challenging problem to the microscopist and adequate precautions must be taken as cautioned by Egerton (1996). Here, the difficulties encountered in C analysis in low carbon steels (all compositions are given in wt% unless specified otherwise) are discussed and experimental results from two specific problems are presented.

A variety of steels were examined during the course of this study. Details of experimental procedures are described by Reynolds et al 1988 and Fonda et al 1996. The thin foils were examined at room temperature in a Topcon 002B microscope with a LaB$_6$ filament operating at 200keV equipped with a Gatan Imaging Filter (GIF). The thin foils were stored in vacuum prior to TEM examination at room temperature. The conventional practice of using a cold stage to reduce sample contamination in the microscope was not followed here in order to avoid any phase transformation of the $\gamma$ phase.

## 2. RESULTS AND DISCUSSION

### 2.1 Detection of C by EELS

The detection of C in steels is very demanding and specimen contamination from hydrocarbons present in the microscope column can distort the results. Hence, as a prelude to this study, we monitored the C-K edge counts obtained from various steel samples after several minutes of continuous electron bombardment. A brief summary of the results can be obtained from an examination of the electron energy loss spectra displayed in Fig. 1(a). Two

316

sets of spectra are shown; those collected with large electron probes (~80 nm) exhibited as second derivatives while those obtained with a 6 nm electron probe could be easily displayed as intensity versus electron energy loss plots. This figure clearly illustrates the enormity of the problem as C build-up occurs on the sample surface during set up and collection (note the t = 0 curve) when a small probe is used for spectroscopy. It may be pointed out that a visible contamination spot does not appear on the sample surface until about 10 minutes of continuous exposure to a 6 nm electron probe. Thus, in all our work, we used large probes (~60-100nm) and spectra collection times were limited to less than 10 seconds. Another example of C spectra from an ultra-low carbon (ULC) steel is illustrated in Fig. 1(b).

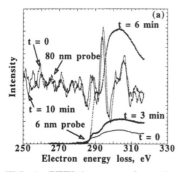
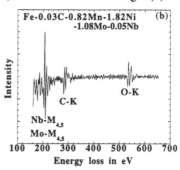

FIG. 1 : PEELS spectra from the fusion zone (FZ) of a steel weldment containing (a) Fe-0.025C-1.40Mn-2.72Ni- 0.49Mo and (b) from the heat affected zone (HAZ) of an ultra-low carbon steel weldment. The thicknesses of the electron transparent foils in these regions were less than 20% of the electron mean free path and 10 spectra were cumulatively acquired for 0.8 seconds each, with a collection angle of 100 millirad in the image coupling mode.

## 2.2 Carbon in retained austenite

Properties of weldments are strongly influenced by the distribution of alloying elements through their effect on phase stability and transformation kinetics. A continuous change in alloy composition is inherent to weldments in the regions extending from the HAZ in the base plate to the FZ due to back-melting and solid state diffusion. A large amount of retained austenite is commonly found in the resulting microstructures and we have examined the extent of C partitioning between the microconstituents in several areas of ULC steel weldments. Often, thin films of the $\gamma$ phase are retained in between martensite plates or laths formed in the HAZ as illustrated in Fig. 2(a), while large idiomorphs of the $\gamma$ phase were

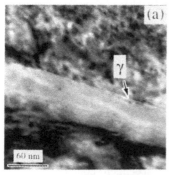
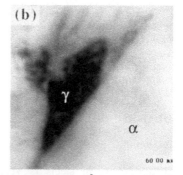

FIG. 2: (a) and (b) are filtered (slit width = 6eV) images obtained from the HAZ and FZ of Fe-0.03C-0.82Mn-1.82Ni-1.08Mo-0.05Nb and Fe-0.025C-1.40Mn-2.72Ni- 0.49Mo steel weldments, respectively. Note the thin film of retained $\gamma$ at an inter-lath boundary in (a).

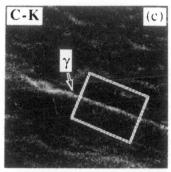
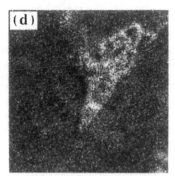

Fig. 2 (contd.): C-K jump ratio images (slit width = 10eV) corresponding to Figs. 2 (a) and 2(b) are shown in 2(c) and 2(d).

found in the FZ as depicted in Fig. 2(b). The corresponding C-K elemental image or jump ratio image extracted from the regions shown in Figs. 2(a) and 2(b) presented in Fig. 2(c) and 2(d) respectively indicate the expected C enrichment in the γ phase. The spectrum shown in Fig. 1(b) was obtained from the region marked X here using an ~80nm probe.

### 2.3 C and Mo segregation at ferrite:martensite interfaces

Elements like Mo drastically reduce the kinetics of α growth in austenite γ and, based on microstructural and kinetic evidence, the occurrence of this transformation stasis phenomenon has been ascribed to a "solute drag like effect" caused by the segregation of Mo atoms to α : γ boundaries (Shiflet and Aaronson 1990; Reynolds et al 1990). We have examined the interfacial segregation of C and Mo in an Fe-0.24wt%C-0.93wt%Mo alloy. Fig. 3 illustrates the C and Mo distribution near a typical α : α' inter-phase boundary. The segregation of the Mo and C atoms all along the α : α' boundary is clearly seen. The concentrations of the segregated elements appear quite uniform in these images. The elemental images also suggested that interfacial segregation is confined to a region ~ 6 nm thick. However, it must be borne in mind that this is an exaggerated figure since the interface in Fig. 3(a) is not exactly parallel to the incident electron beam and it was very difficult to determine the exact focusing conditions near the core loss edges.

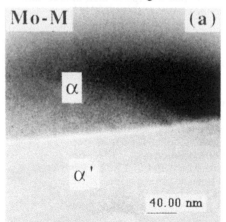
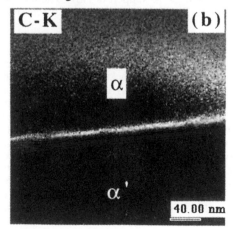

FIG. 3 : (a) Mo-M$_{4,5}$ and (b) C-K elemental images showing interfacial segregation along an α : α' boundary in Fe-0.24wt%C-0.93wt%Mo, austenitized at 1473K and isothermally reacted at 903K for 60 seconds.

318

## 2.4 Quantification of C from PEELS

From the results presented so far, it appears that provided sufficient care is exercised in collecting the spectra, quantification of C in steels is certainly possible. As an example, results from the isothermally heat treated Fe-C-Mo alloy are presented here. Electron energy loss spectra (C-K, Mo-$M_{4,5}$ and the Fe-$L_{2,3}$ core losses) were acquired from several locations within the ferrite, austenite or martensite single phase regions as well as the $\alpha : \gamma$ interfaces, and an attempt was made to quantify the results. The average atomic

TABLE I. Experimentally determined average atomic concentrations.

|            | at%C       | at%Mo      |
|------------|------------|------------|
| ferrite    | 0.14±0.02  | 0.87±0.10  |
| martensite | 1.25±0.15  | 0.83±0.10  |
| interface  | 6.75±0.13  | 2.93±0.31  |

concentrations of C and Mo obtained from 6 to 8 measurements are tabulated in Table I. The atomic concentrations reported were obtained by taking the ratio of the number of Mo or C atoms to the total number of C, Mo and Fe atoms. The values of Mo reported here appear rather high since both ferrite and martensite appear to have more than the average alloy concentration (Fe-1.11at%C-0.93at%Mo). This may be associated with errors in using the power law background model and those in the cross sections determined by the hydrogenic model. The amount of C and Mo at the interface are underestimates since large probe sizes were employed and the sample thickness is much less than the probe size. Consequently, the cross sections are calculated using the area of the electron probe rather than the area of the interface where the atoms contributing to the electron energy loss signals actually reside.

## 3. CONCLUSIONS

The experimental results presented here clearly indicate that using the GIF carefully, it is certainly possible to gather qualitative as well as quantitative information about the distribution of C in various microconstituents present in steels. Several examples of spectroscopy and energy filtered imaging have been used illustratively to gather microchemical information from steel samples.

## ACKNOWLEDGEMENTS

ESKM, AGF and GS acknowledge the support of ONR for this research and WTR acknowledges the support of NSF. The authors would also like to express their appreciation to Dr. M. Saunders for enlightening discussions during the course of this work.

## REFERENCES

Aaronson H I 1962 The Mechanism of Decomposition of Austenite by Diffusional Processes, eds V F Zackay and H I Aaronson (New York: Interscience) pp. 387
Bentley J and Kenik E A 1994 Electron Microscopy 1994, ICEM 13, eds B Jouffrey and C Colliex ( Les Ulis Cedex A: les éditions de physique) pp. 623
Bhadeshia H K DH and Waugh A R 1982 Acta Metall. **30**, 775
Egerton R F 1996 Electron Energy Loss Spectroscopy (New York: Plenum) 341
Fonda R W, Spanos G and Vandermeer R A 1994 Scripta Metall., **31**, 683
Hofer F, Warbichler P, Buchmayr B and Kleber S 1996 J. of Microscopy **184**, 163
Ozkaya D, Yuan J and Brown L M 1994 Electron Microscopy 1994, ICEM 13, eds B Jouffrey and C Colliex ( Les Ulis Cedex A: les éditions de physique)pp. 663
Reynolds W T Jr., Brenner S S and Aaronson H I 1988 Scripta Metall., **22**, 1343
Reynolds W T Jr., Li F Z, Shui C K and Aaronson H I 1990 Met. Trans. **21A**, 1433
Shiflet G J and Aaronson H I 1990 Met. Trans. **21A**, 1413
Stark I and Smith G D W 1988 Phase Transformations '87, eds G W Lorimer (London: Inst. of Metals) pp. 475
Stark I, Smith G D W and Bhadeshia H K D H 1990 Met. Trans. **21A**, 837
Yuan J, Brown L M, Walmsley J and Fisher S B 1995 J. of Microscopy **180**, 313

*Inst. Phys. Conf. Ser. No 153: Section 8*
*Paper presented at Electron Microscopy and Analysis Group Conf. EMAG97, Cambridge, 1997*
© *1997 IOP Publishing Ltd*

# Effects of the sputter cleaning process in cathodic arc evaporation

**M MacKenzie, AJ Craven, WAP Nicholson and P Hatto***

Dept. of Physics & Astronomy, University of Glasgow, Glasgow, G12 8QQ
*Multi-Arc (UK) Ltd., Medomsley Road, Consett, Durham, DH8 6TS

**ABSTRACT** This paper reports on the interface region between a CrN coating and a Ti(6%Al, 4%V) substrate after sputter cleaning with Cr. EELS on a VG HB5 STEM and diffraction showed that Cr sputter cleaning of the Ti(6%Al,4%V) resulted in penetration of Cr into the substrate giving a bcc alloy of Ti and Cr with a graded composition. Subsequent deposition of CrN resulted in an initial layer of $Cr_2N$ as the system was changed from sputter cleaning to coating followed by a {022} textured fcc CrN layer.

## 1. INTRODUCTION

Reactive cathodic arc evaporation is a technique used to deposit thin films for hard coatings. Prior to deposition of the coating, the substrate surface is conditioned by a process known as sputter cleaning. This conditioning process modifies the substrate surface in a way that results in good adhesion of the coating. Ti(6%Al, 4%V) is an $\alpha+\beta$ alloy of Ti which is widely used in the aerospace industry. Samples of Ti(6%Al, 4%V) which had been sputter cleaned with Cr and coated with CrN are investigated.

## 2. RESULTS

Fig. 1 is a high magnification XTEM image of an interface between Ti(6%Al, 4%V) and CrN showing two distinct layers which are non-flat and of non-uniform thickness. The layer closest to the substrate is of the order of 7nm wide. The second layer has a similar width of about 10nm. These are both crystalline with small grains. The CrN coating is columnar. Fig. 2 shows an EELS line profile stepping across such an interface. The line profile has been split into 4 sections each of which will be discussed below; these are substrate (S), Ti:Cr alloy (A), coating 1 (C1) and coating 2 (C2).

The first section is substrate which has not been chemically altered by the sputter cleaning or the coating processes and is therefore of little interest to this discussion.

The next region is a Ti:Cr intermixed region of graded composition. The position of the first plasmon in the EELS data was used to confirm that this was an alloy (Williams and Edington 1976) and not simply an artefact of specimen tilt resulting in overlapping layers. Fig. 3 shows the effect of summing the low-loss spectra from pure Ti and pure Cr compared to a spectrum collected from the point in the intermix where there are roughly equal amounts of Ti and Cr.

Fig. 2 shows a distinct step in the coating composition on going from region C1 to C2. Comparison of the N edge in spectra from regions C1 and C2 shows a distinct change in shape of the fine structure for the two regions. If these spectra are compared with spectra from hexagonal $Cr_2N$ and fcc CrN powder standards, as in fig. 4, then it can be seen that the ELNES on the N K edge in C1 matches that of the $Cr_2N$ powder standard indicating that they have the same structure. Similarly C2 can be seen to have the same structure as the CrN standard.

Looking at fig. 2 it can be seen that the Cr-N ratios in regions C1 and C2 are higher than in the powder standards which are close to stoichiometry. Thus the layers in the coating are sub-stoichiometric with compositions of $Cr_2N_{0.6}$ and $CrN_{0.6}$ in regions C1 and C2, respectively.

Fig. 2 also shows an interpenetration of Ti into region C1 for about 12nm. This tailing is unlikely to be an artefact due to tilt as the estimated error in the orientation of the specimen for microanalysis is an order of magnitude smaller. Thus the Ti is substituting for Cr in region C1.

Diffraction techniques were employed in order to confirm the above crystallographic structures. These were performed on a Philips CM20 microscope.

In view of the narrowness, and non-uniformity of the interface layers together with their very fine grain size and random orientation, it was not possible to isolate and index their diffraction patterns in XTEM. Diffraction patterns were obtained from a planar specimen of Ti(6%Al, 4%V) which had been sputter cleaned with Cr and then back-thinned to electron transparency. Most of these patterns, such as the one shown in fig. 5 contained more than one phase. One incomplete diffraction pattern, shown in fig. 6, was obtained from the very edge of the specimen where it is likely that only the alloy layer was present. These rings indexed as a bcc pattern with a lattice parameter of 2.97±0.06Å. This

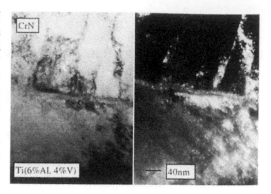

Fig. 1 XTEM bright and dark field images of CrN on Ti(6%Al, 4%V)

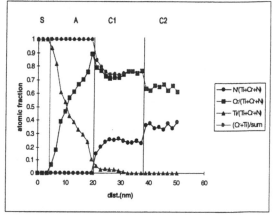

Fig. 2 EELS line profile across the interface between CrN and Ti(6%Al, 4%V)

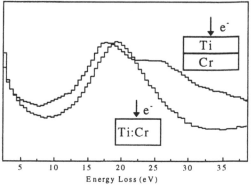

Fig. 3 First plasmons in a Ti:Cr alloy and overlapping Ti and Cr layers

value is in between that expected for either pure β-Ti(6%Al, 4%V) or pure bcc Cr and suggests a bcc alloy with a Cr-Ti ratio of 0.257±0.005. The rings in this pattern are also present in fig. 5. The other rings in fig. 5 indexed as hcp α-Ti(6%Al, 4%V) and are from substrate which had been unaltered by the sputter cleaning process. Diffraction patterns from other regions also confirmed the presence of a small amount of bcc β-Ti(6%Al, 4%V).

The addition of Cr to Ti(6%Al, 4%V) should stabilise the bcc β phase. The Ti:Cr phase diagram (Hansen 1958) shows that there is very little solubility of Cr in α-Ti. However, at higher temperatures there is solubility of Cr in β-Ti across the full range of compositions. The sputter cleaning process is known to be energetic and probably causes local migration of atoms similar to a high temperature effect. Therefore, it is quite likely that the high temperature region is the relevant part of the phase diagram.

Fig. 7 shows a diffraction pattern obtained by averaging over a relatively large region of coating in XTEM using the technique described by Craven et al (1997). This pattern indexed as fcc CrN with an {022} texture normal to the substrate surface.

To date convincing diffraction patterns have not been obtained and indexed from the $Cr_2N$ region.

In summary, diffraction confirmed the structures of the substrate, the Ti:Cr alloy and the CrN coating to be hcp α-Ti(6%Al, 4%V) with a small amount of bcc β-Ti(6%Al, 4%V), a bcc Ti:Cr alloy and an fcc CrN coating respectively.

## 4. DISCUSSION & CONCLUSIONS

All of the coatings studied exhibited growth of sub-stoichiometric CrN on an interfacial region consisting of a metal alloy of graded composition followed by a sub-stoichiometric $Cr_2N$ layer.

The fcc CrN coating grows with a columnar structure and an {022} texture. Differences in the surface finish of the coating, grain size and width of the interfacial region were observed for different deposition parameters.

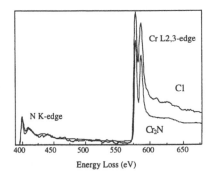

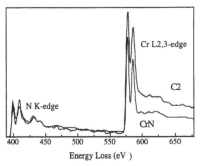

Fig. 4 EELS spectra from regions C1 & C2 compared with $Cr_2N$ and CrN powder standards

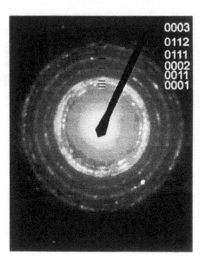

Fig. 5 Diffraction pattern from a planar sample showing a bcc and an indexed hcp pattern

The initial growth of sub-stoichiometric $Cr_2N$ is likely to be a result of switching from sputter cleaning to coating. After sputter cleaning is complete, the substrate bias is reduced and nitrogen gas introduced into the chamber. This is usually done manually and involves adjusting the pressure to the correct level while the substrate bias is varied to ensure stability. This typically takes between 30s and 2 minutes to perform. The deposition times for the coatings were about 13-14 minutes. The $Cr_2N$ accounts for roughly 5% of the total coating so it is not unreasonable to assume that it is a result of the changeover. During this changeover process, the cathode becomes poisoned (Hovsepian and Popov 1994) i.e. the surface changes from being metal to nitride as $N_2$ gas is introduced into the chamber. Such poisoning causes a decrease in the emission rate of Cr. Thus, the emission of Cr is initially higher and there is less gas in the chamber so there is not enough $N_2$ gas to form CrN and the coating grows as sub-stoichiometric $Cr_2N$. As the gas pressure increases and the rate of Cr evaporation decreases, deposition of CrN becomes possible and the crystallographic structure changes.

The presence of a chromium-titanium alloy at the interface was observed. This is important because it is known that the presence of a $TiCr_2$ phase in β alloys of Ti can cause embrittlement (Polmear 1981). The presence of this alloy shows that there is some penetration of chromium into the substrate during the sputter cleaning process with the Ti(6%Al, 4%V) transforming to the β phase.

This paper has also demonstrated the use of the ELNES to determine the structure of a material in a situation where it proved difficult to obtain unambiguous diffraction information.

Fig. 6 An indexed bcc pattern from the same sample as fig. 7

013
022
112
002
011

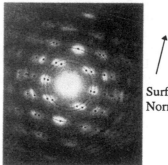

Surface Normal

Fig. 7 Diffraction pattern from the CrN coating with a simulated {022} texture pattern

## REFERENCES

Craven A J, Nicholson WAP and Lindsay R 1997 these proceedings

Hansen M 1958 Constitution of Binary Alloys. McGraw-Hill Book Company (New York)

Hovsepian Pand Popov D 1994 Vacuum **45**, 5, 603

Polmear I J 1981 Light Alloys : Metallurgy of the light metals. Edward Arnold (London)

Williams D B and Edington J W 1976 J. Micros. **108**, 113

*Inst. Phys. Conf. Ser. No 153: Section 8*
*Paper presented at Electron Microscopy and Analysis Group Conf. EMAG97, Cambridge, 1997*
© *1997 IOP Publishing Ltd*

# ELNES in metal nitrides

**M MacKenzie, AJ Craven, I Parkin\* & A Nartowski\***

Department of Physics & Astronomy, University of Glasgow, Glasgow, G12 8QQ
\*Department of Chemistry, University College London, London, WC1H 0AJ

**ABSTRACT** This paper reports on the ELNES in transition metal nitrides with the rocksalt structure. The effect of the lattice parameter on the shape of the fine structure is investigated for different binary nitrides. Titanium and vanadium nitrides of various stoichiometries were also studied. Some preliminary results are presented from an investigation of the effect of substitution on the metal sub-lattice in the ternary $(Ti_{0.9}V_{0.1})N_{0.8}$.

## 1. INTRODUCTION

Metal carbides and nitrides are found in many systems of technological interest e.g. as precipitates in steels or as hard coatings on cutting and shaping tools. Due to the extensive mutual solubility of metal carbides and nitrides with the same structure, substitution on both sub-lattices is common as are vacancies on the non-metal sub-lattice. When there is excess non-metal, it can be accommodated either by the presence of vacancies on the metal sub-lattice or by the excess atoms occupying other interstitial sites. In addition, overgrowth of one phase on another is common in precipitate systems and intermixing at interfaces is common in PVD multilayer systems. Thus new ways of characterising such materials are of interest.

One approach is to use the electron energy loss near-edge structure (ELNES) present on ionisation edges in electron energy loss spectra. The change in the ELNES with change of crystal structure is well known but there are also clear systematic changes in the ELNES in binary compounds with the same crystal structure if the elements present are changed (Craven 1995). There are changes in both absolute and relative energies of features in the ELNES. These changes reflect the lattice parameter, the deviation from stoichiometry and the elements present. Substitution on the sub-lattice also causes changes in the ELNES.

## 2. MATERIALS & PROCEDURES

The nitrides investigated in this paper were standards produced by various means. Some of the materials were produced using solid state metathesis, which is a fast, self-energetic reaction resulting in the formation of a fine ceramic powder (Parkin 1996). Some of the $VN_x$ samples investigated were made using a high temperature route involving the hot pressing of mixtures of binary compounds (Lengauer 1985). Further samples were produced by arc melting or bought as commercial powders. The samples were all crushed in propan-2-ol with a pestle and mortar and drops of the suspension were placed onto holey carbon films. The samples were investigated in a VG HB5 STEM operated at 100kV using a GATAN model 666 parallel-detection electron energy loss spectrometer. Spectra were collected with a convergence and collection semi-angles of 8mrad and 12.5mrad, respectively. The spectra were subsequently processed using EL/P software provided by GATAN.

324

## 3. RESULTS

The overall shape of the ELNES on the N K edge from neighbouring transition metal nitrides such as TiN, VN and CrN is similar as shown in fig. 1. However, in the case of ScN, the thresholds of the N K edge and the Sc $L_3$ edge are very close making it unlikely that the N edge can be distinguished. ScN is very susceptible to oxygen and fig. 2 shows the change in the ELNES on the Sc edge as the oxygen concentration changes. Even when there is no oxygen detected the Sc threshold peaks split into two, unlike the Ti, V or Cr edges with the lower energy one being much less intense. As the oxygen content increases, this splitting decreases but the lower energy peak increases relative to the higher energy one.

It is known that there is an increase in the separation of the peaks on the N K edge as the lattice parameter decreases (Craven 1995). This variation can be seen in fig. 1. VN and CrN have very similar lattice parameters, and hence very similar peak separations on their N K edges. The peaks in the TiN spectrum are more compact as a result of its larger lattice parameter. The energy scales of the spectra can be interpolated to take account of the differences in lattice parameters. This is illustrated in fig. 3 where the separations of the peaks in CrN and VN can be seen to be much more similar to that of TiN after interpolating the energy scales to match that of the TiN.

Differences in the ELNES also result from changes in the stoichiometry of the material. The lattice parameter changes with composition and hence the relative separations of the features in the ELNES should decrease. This effect has been investigated for $TiN_x$ and $VN_x$. Figs. 4 and 5 show the N K edge in various $TiN_x$ and $VN_x$ compounds. In these systems, there is only a small variation in lattice parameter with stoichiometry and hence there is little change in the separation of peaks. However, there are differences in the overall shape of the ELNES. As the N content increases in the $VN_x$ spectra the relative height of P4 to P1 increases and P5 becomes more distinct. There are also differences in the depth of the dip between P2 and P3. Similar differences can be seen in the $TiN_x$ spectra although the effects are less pronounced than in the $VN_x$ spectra.

The effect of substitution on the metal sub-lattice was investigated by examining the (Ti,V)N

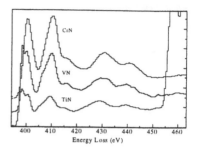

Fig. 1 N K edge in CrN, VN and TiN

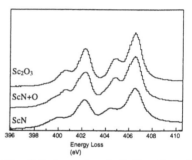

Fig. 2 Sc $L_{2,3}$ edges in ScN with and without oxygen compared to that in $Sc_2O_3$

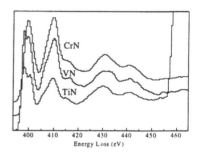

Fig. 3 N K edge in CrN, VN and TiN with energy scales interpolated for lattice parameters to match TiN

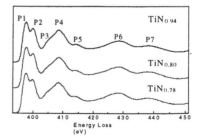

Fig. 4 N K edge in $TiN_x$

system. Fig. 6 shows the N K edge in TiN, VN and in $(Ti_{0.9}V_{0.1})N_{0.8}$. The fine structure on the N K edge in the $(Ti_{0.9}V_{0.1})N_{0.8}$ is different to that of either the TiN or the VN with differences in both the shapes and the relative positions of the peaks. It is interesting to note that the peaks shift rather than spread suggesting the presence of a virtual crystal rather than a simple averaging of the TiN and the VN structures. The composition of the $(Ti_{0.9}V_{0.1})N_{0.8}$ sample is commensurate with the separation of peaks P1 and P6.

One of the difficulties in the quantitative analysis of the EELS data is the removal of the background from under the Ti $L_{2,3}$ edge in the presence of a N K edge. The background removal routine provided in EL/P failed to fit an $AE^{-r}$ background under the Ti $L_{2,3}$ edge due to the fine structure on the N edge. It was decided to remove the background using the N K edge in a CrN spectrum to match the N K edge in TiN having corrected for the effect of the different lattice parameters as described above. This technique was successful and resulted in the satisfactory removal of the background. Fig. 7 shows an example of this technique being applied using the N edge in a CrN spectrum to remove the background from a Ti edge. It can be seen that, although the N edge in the CrN spectrum is not a perfect fit, it is perfectly adequate. The N edges in the spectra were aligned to take into account the chemical shift in the edge onsets and the counts in the spectra were matched in a 25eV window from 422eV.

Fig. 8 shows the relationship between P6-P1 and $1/a^2$ for all of the above nitrides and transition metal nitrides from the second and third transition series. The solid line is the fit to the original data in Craven (1995). The graph now includes the $TiN_x$, $VN_x$, $(Ti_{0.9}V_{0.1})N_{0.8}$ and CrN data It can be seen that the $TiN_x$ and $VN_x$ points cluster together as a result of the small change in lattice parameter with stoichiometry. This shows that the introduction of vacancies to the structure has a less pronounced lattice parameter than substitution on the metal sub-lattice.

## 3.    DISCUSSION & CONCLUSIONS

This paper has shown some of the uses of the ELNES as a tool for microanalysis. The changes

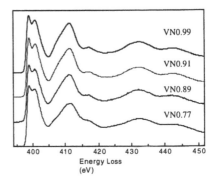

Fig. 5 N K edge in $VN_x$

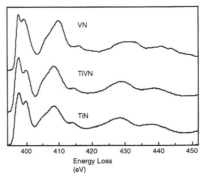

Fig. 6 N K edge in VN, $(Ti_{0.9}V_{0.1})N_{0.8}$ and TiN

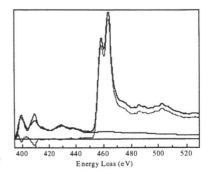

Fig. 7 N K edge in CrN being used to remove the background from the Ti $L_{2,3}$ edges in the presence of N after correcting for the difference in lattice parameter

326

in the ELNES as a result of lattice parameter have been shown for different transition metal nitrides. Results have been obtained which show the effect of the stoichiometry of a binary metal nitride on the fine structure. Although the change in separation of peaks P1 and P6 with stoichiometry of the material is small due to the smallness of the change in lattice parameter there are distinct changes in the overall shape of the ELNES. This change in shape may prove useful for determination of unknown stoichiometries. Preliminary investigations on the effect of going from binary to ternary compounds indicate that there are changes in the separation of the peaks commensurate with the change in lattice parameter. Further work is clearly required.

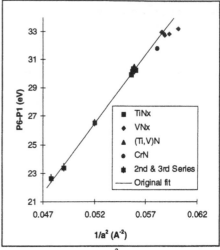

Fig. 8 P6-P1 against $1/a^2$ for various transition metal nitrides

## REFERENCES

Craven 1995 J. Micros. **180**, Pt 3, 250-262
Lengauer W & Ettmayer 1985 Journal of Less Common Metals **109**, 351-359
Parkin I P 1996 Chemical Society Reviews, 200-207

## ACKNOWLEDGEMENTS

The authors would like to thank Professor Lengauer and Dr Chatfield for the provision of samples and Engineering and Physical Sciences Research Council and Multi-Arc (UK) Ltd for the provision of financial support.

*Inst. Phys. Conf. Ser. No 153: Section 8*
*Paper presented at Electron Microscopy and Analysis Group Conf. EMAG97, Cambridge, 1997*
© *1997 IOP Publishing Ltd*

# A method of studying texture development in thin films on a nanometre scale

**A J Craven, W A P Nicholson and R Lindsay**

Department of Physics and Astronomy, University of Glasgow, Glasgow G12 8QQ, Scotland.

**ABSTRACT:** Texture is important in many thin film applications such as hard coatings, magnetic recording and semiconductor devices. The texture depends on the growth conditions and composition and develops as the film grows. A method of investigating such development on a nanometre scale is described. It uses cross-sectional specimens in the transmission electron microscope. The ultimate spatial resolution is the diffraction limit imposed by the angular resolution required to resolve the diffraction rings. This is determined by the particular system under investigation but is typically a few nanometres.

## 1. INTRODUCTION

Polycrystalline thin films deposited onto substrates are important in many technological applications. In hard coatings, the texture of the film can affect the performance. The texture depends on the method and conditions used to deposit the film as well as on its composition. In the semiconductor industry, there is great potential benefit if layers of poly-silicon can be grown with a common axis normal to the substrate. In magnetic recording, texture plays an important role in determining the performance of thin film recording media. Thus there is great interest in determining how quickly the texture in a film develops as it grows. If that development occurs within a few tens of nanometres of the substrate, it is quite difficult to investigate.

One approach is to use a cross-sectional specimen in the TEM. An electron probe with a diameter of a nanometre or less can be formed in a TEM equipped with a field emission gun (FEG). This can be scanned over the specimen along a line parallel to the substrate surface giving rise to a diffraction pattern averaged over many grains hence showing the texture. Unfortunately the probe convergence is so large that the rings overlap and the diffraction pattern is difficult to interpret. With suitable adjustment of the electron optics, the probe convergence can be traded off against spot size to achieve an angular resolution compatible with the problem. Provided that the source is sufficiently bright, there is sufficient current to work at the diffraction limit so that the product of the spatial and angular resolutions is approximately the wavelength of the electrons. Thus the low order rings can be resolved in diffraction patterns from crystals with small unit cells with a spatial resolution of a few nanometres.

## 2. ANGULAR AND SPATIAL RESOLUTION

Figure 1 shows a schematic diffraction pattern in which the adjacent rings have Bragg angles of $\vartheta_n$ and $\vartheta_{n+1}$. If the incident probe has an angular radius, $\alpha$, then the widths of the rings will be $2\alpha$ and they will just touch if $\alpha = \vartheta_{n+1} - \vartheta_n$. The Airy disc diameter, $D$, for this value of $\alpha$ is given by

$$D = 2.4 / \left( 1/d_{n+1} - 1/d_n \right)$$

where $d_n$ and $d_{n+1}$ are the crystal plane spacings giving rise to the rings.

Because $\alpha$ is so small when forming a probe under these circumstances, in principle, there are only two contributions to its size i.e. the Airy disc and the Gaussian image of the source (Craven, 1980). With the diameter of the latter equal to half that of the former, the probe has a diameter $\sim D$

and contains a current $\sim\beta\lambda^2$ where $\beta$ is the source brightness and $\lambda$ the electron wavelength. In principle, this current is independent of the accelerating voltage for a given emitter and, for a typical field emission source, has a value of $\sim10^{-10}$A. Thus, with a FEG, a diffraction limited probe can be used in practice.

In texture studies, the angular resolution around the rings is also of interest. As shown in Figure 1, the angular diameter of the incident probe corresponds to a range of azimuthal angles on each ring. On the $m^{th}$ ring, this is $2\varphi_m$ which is given in radians by $\alpha/\vartheta_m = d_m\left(1/d_{n+1} - 1/d_n\right)$ on the above criterion.

The optimum conditions are very dependent on the material under study. Table 1 one shows the effect of the diameter of the Airy disc on the diffraction patterns from four simple crystal structures for values of $1/d$ up to $\sim8$ nm$^{-1}$. Y in the columns headed S shows that this ring is separated from the previous one. The columns headed $2\varphi$ give the value in degrees for each ring at that Airy disc diameter. The greater ring separations in the diamond (Si) and bcc (Cr) structures allow all the low order rings to be separated for smaller values of $D$ than is the case for the fcc rocksalt (TiN) and especially for the hcp (Ti) structure. However, the azimuthal resolution is then rather poor on the inner rings when $D$ is small.

In practice, one might expect to be able to resolve spatial features, diffraction rings and features around diffraction rings to better than $D$, $2\alpha$ and $2\varphi$ respectively and the improvement could be up to a factor of 2 in favourable circumstances.

## 3. THE TECHNIQUE

We have used this technique with a Philips CM20 FEG TEM/STEM equipped with an objective lens with a 20mm gap designed for magnetic studies. However, it should be possible to use the same technique with the standard objective lenses and even with other FEG TEM/STEMs. The optical configuration in the STEM mode produces a demagnified image of the electron source and, for optimum resolution, this should contribute less to the probe size than diffraction and spherical aberration. In these circumstances, a probe of $\sim0.5$ nm in diameter with a value of $\alpha$ of $\sim6.5$ mrad can be obtained. The optical configuration of the objective lens is shown in Figure 2 which shows that condenser 2 forms a virtual object for the objective lens (solid rays). By strengthening condenser 2 lens

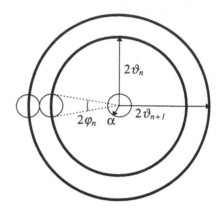

Figure 1. Schematic diffraction pattern from a probe of angular radius, $\alpha$, showing the mid-positions of two diffraction rings with Bragg angles $\vartheta_n$ and $\vartheta_{n+1}$. $2\varphi_n$ is the azimuthal angle subtended on the $n^{th}$ ring by the angular diameter of the probe.

| $D$ | | 2 nm | | 5 nm | | 10 nm | |
|-----|------|---|-----------|---|------------|---|
| | hkl | S | $2\varphi$ | S | $2\varphi$ | S | $2\varphi$ |
| TiN | 111 | - | 17 | - | 7 | - | 3 |
| | 200 | | 15 | Y | 6 | Y | 3 |
| | 220 | Y | 10 | Y | 4 | Y | 2 |
| | 311 | Y | 9 | Y | 4 | Y | 2 |
| | 222 | | 8 | | 3 | Y | 2 |
| | | | | | | | |
| Si | 111 | - | 21 | - | 9 | - | 4 |
| | 220 | Y | 13 | Y | 5 | Y | 3 |
| | 311 | | 11 | Y | 4 | Y | 2 |
| | 400 | Y | 9 | Y | 4 | Y | 2 |
| | 331 | | 9 | Y | 3 | Y | 2 |
| | | | | | | | |
| Cr | 110 | - | 14 | - | 6 | - | 3 |
| | 200 | Y | 10 | Y | 4 | Y | 2 |
| | 211 | Y | 8 | Y | 3 | Y | 2 |
| | 220 | Y | 7 | Y | 3 | Y | 1 |
| | | | | | | | |
| Ti | 100 | - | 17 | - | 7 | - | 3 |
| | 002 | | 16 | | 7 | Y | 3 |
| | 101 | | 15 | | 6 | Y | 3 |
| | 102 | Y | 12 | Y | 5 | Y | 2 |
| | 110 | | 10 | Y | 4 | Y | 2 |
| | 111 | | 10 | | 4 | Y | 2 |
| | 103 | | 9 | | 4 | Y | 2 |

Table 1. The effect of the Airy disc diameter on the separation of adjacent diffraction rings in various structures plus the corresponding value of $2\varphi$ in degrees.

and weakening the objective lens, the convergence angle at the specimen is reduced with a corresponding reduction in the demagnification of the source.

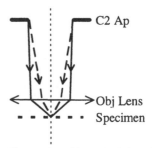

In principle, this is a very straightforward modification of the optics but, in practice, the result is very sensitive to the objective alignment, the diffraction focus and the diffraction stigmation. If these are not correctly adjusted, which is not easy with the coherent illumination provided by the field emission gun, the diffraction rings become non-circular with non-coincident centres as the objective lens strength is reduced. Further, when the rocking points of the scan system are adjusted to give a stationary diffraction pattern on the screen, the magnifications in orthogonal directions on the scanned image differ markedly and are not reproducible.

Figure 2. The principle of decreasing the probe convergence

When correctly adjusted, the probe can be scanned over an area of the specimen while a diffraction pattern is recorded. With the Philips' scanning unit, a single line scan can be used. Thus, if the specimen is a cross-section of a film deposited on a planar substrate which is oriented so that the interface is parallel to the electron beam and the scan rotation control is used to align the line scan parallel to this interface, then an average diffraction pattern from grains a given distance from the interface can be obtained. The spatial resolution normal to the interface is limited by the probe diameter. In practice, even if the interface is perfectly planar, the accuracy with which the line scan can be made parallel to the interface is limited and the line traced by the probe on the specimen may not be straight. Limiting the line length to $\sim 10^2$ times the resolution required is a sensible criterion. If the probe is scanned too far from the axis, lens aberrations cause the diffraction pattern to shift and this places a limit on the length of the scanned line for a given angular resolution. This is typically several microns. For other purposes, it can be useful to scan an area rather than a line. A rectangular area with an aspect ratio chosen by the operator is available with the standard scan unit. However, it is possible to envisage ways of averaging over selected portions of a field of view containing arbitrary shaped and separated portions of a film which has been etched away in selected place after deposition, as is common in semiconductor devices.

## 4. RESULTS AND CONCLUSIONS

X-ray diffraction on TiN/NbN multilayers deposited by reactive arc evaporation shows that a 2 μm thick coating with individual layers 10 nm thick has an (002) texture whereas the texture is (111) if the layer thickness is 30 nm. To investigate the texture development, a coating of 6 pairs of nominally 10 nm thick layers followed by 5 pairs of nominally 30 nm thick layers was deposited on a Si substrate and capped with Hf. Figure 3 shows a CTEM image of a cross-section of this specimen. Above the image are three diffraction patterns recorded by scanning a line parallel to the substrate in the 1st and 6th 10 nm TiN layers and the 5th 30 nm TiN layer respectively. The (111) and (200) rings are clearly resolved and the STEM image showed a spatial resolution ~5nm, clearly resolving the individual layers. The strong intensity in the (200) ring in the direction of the surface normal in the left hand pattern shows a well defined (100) texture even very close to the substrate with a dispersion of ~10° in the [100] direction. This texture is preserved in the second pattern although the grain size has now increased. The dispersion of the [100] direction has doubled and it is less likely to be found exactly parallel to the surface normal. The third pattern shows that the intensity of the (111) reflection in the surface normal direction has increased markedly after nine 30 nm layers of growth but that the change of texture is not complete at this stage.

Figure 4 shows a cross-section of a poly-silicon film grown on a Corning 7059 glass substrate (Kakinuma, 1995). The substrate is the darker strip on the left hand side of the image. The dark patches are the result of the carbon coating necessary to render the surface of the glass conducting. The three diffraction patterns above the image were recorded along lines parallel to the interface at the positions indicated. They show that the film is initially random polycrystalline and

330

that it develops the final (110) texture over a considerable distance. When the texture has developed, it is again less likely that the texture axis is exactly parallel to the surface normal.

In conclusion, this technique offers a way of studying texture development on a scale of a few nanometres in a range of technologically important systems.

Figure 3. An image of a cross-section of a TiN/NbN multilayer on a Si substrate thinned by a focused ion beam. It consists of 6 pairs of layers, each layer being nominally 10nm thick, followed by 5 pairs of layers, each layer being nominally 30 nm thick and capped with Hf. Diffraction patterns, obtained by scanning a probe along lines in the layers indicated, are shown above.

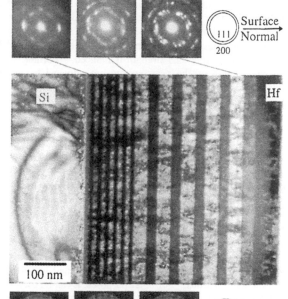

Figure 4. An image of a cross-section of a poly-silicon film on a Corning 7059 glass substrate thinned by tripod polishing. The substrate is the dark strip down the left hand side. The dark patches on the poly-silicon are the result of the carbon coating deposited to render the surface of the substrate conducting. Diffraction patterns, obtained by scanning a probe along a line parallel to the substrate at the points indicated, are shown above.

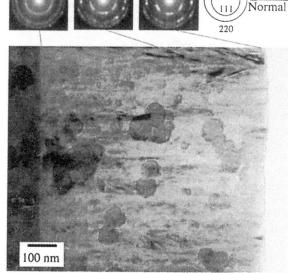

## REFERENCES

Craven A J 1980 Inst of Physics Conf Ser **52**, 47

Kakinuma H 1995 J Vac Sci Tech **A13**, 2310

## ACKNOWLEDGEMENTS

The authors would like to thank Peter Hatto and Clive Davies of Multi-Arc (UK) for depositing the TiN/NbN multilayer; John Walker of FEI for preparing the TiN/NbN cross-section using an FIB; Hiroaki Kakinuma for supplying the poly silicon material; EPSRC and Motorola for financial support.

*Inst. Phys. Conf. Ser. No 153: Section 8*
*Paper presented at Electron Microscopy and Analysis Group Conf. EMAG97, Cambridge, 1997*

331

# TEM and X-Ray Analysis of $(Al_yGa_{1-y})_xIn_{1-x}P$ Quantum Well Layers Under Compressive and Tensile Strains

**S A Hall P C Mogensen[a] U Bangert and S W Bland[b]**

Condensed Matter Physics Group, Dept. of Physics, UMIST, PO Box 88, Manchester, M60 1QD, UK.
[a]Dept. of Physics and Astronomy, University of Wales at Cardiff, PO Box 913, Cardiff CF2 3YB, UK.
[b]Epitaxial Products International Ltd. (EPI), Pascal Close, Cypress Drive, St. Mellons, Cardiff, CF3 0EG, UK.

**ABSTRACT :** Plan view and cross-sectional TEM, and X-Ray microanalysis in a STEM have been used to study the structural integrity of a series of strained $(Al_yGa_{1-y})_xIn_{1-x}P/Ga_xIn_{1-x}P$ quantum well (QW) layers ranging from -1% to +2%. In the compressive strained samples relaxation in the form of island formation has been observed before the Matthews and Blakeslee critical limit is reached. At the highest tensile strain the onset of similar 'wavy growth' is seen.

## 1. INTRODUCTION

The motivation for the development of strained AlGaInP lasers is the large direct bandgaps which may be achieved, leading to emission in the ~610-690nm range. The use of strain is essential to optimise the wavelength and improve device performance for use in such diverse applications as optical data storage, plastic fibre communications systems and 633nm HeNe replacements. The use of both tensile and compressive strain QWs, where the net mismatch is below 1% is widespread (Bour et al 1994). However at higher strains there is evidence of poorer laser performance (Valster et al 1992).

In this work a series of layers incorporating tensile and compressive strain QWs ( mismatches between -1% and +2% respectively ) have been examined, and through a study of their microstructure we have correlated the appearance of strain induced structural non-uniformities with a worsening of the laser performance. In the following sections we describe the sample structures and present results from TEM and X-Ray analysis which show a deterioration in the strained layers as the nominal compressive mismatch exceeds +1.4% and in the case of the tensile QWs as the strain approaches -1%.

## 2. LAYER STRUCTURE

All the samples have been grown at EPI by Metal-Organic Chemical Vapour Deposition on n-type GaAs substrates, mis-oriented by 10° from the [100] direction towards the [111] to limit the well documented tendency of the alloys to spontaneously order. The compressive strain samples have a single $Ga_xIn_{1-x}P$ QW positioned centrally in an

$(Al_xGa_{1-x})_{0.5}In_{0.5}P$ barrier region. The tensile strain samples have a $Ga_xIn_{1-x}P$ double quantum well (DQW) in a corresponding barrier region. Outer $(Al_{0.7}Ga_{0.3})_{0.5}In_{0.5}P$ layers 1.5μm thick on either side form the cladding for a waveguide structure. The structures and compositions of the QW and waveguide core are shown in Tables 1 and 2 respectively, these layers are undoped. Both samples were capped with a 0.25μm heavily p-doped GaAs layer.

| Layer description | Material | Thickness ( μm ) |
|---|---|---|
| Waveguide core | $(Al_{0.4}Ga_{0.6})_{0.5}In_{0.5}P$ | 0.1 |
| Quantum well | $Ga_xIn_{1-x}P$ | 0.01 |
| Waveguide core | $(Al_{0.4}Ga_{0.6})_{0.5}In_{0.5}P$ | 0.1 |

**Table 1. Compressive strained laser structure. Four samples have been grown: Sample 0(+1.0% strain) [x=0.38], Sample 1(+1.4% strain) [x=0.32], Sample 2(+1.7% strain) [x=0.28] and Sample 3(+2.0% strain) [x=0.24].**

| Layer description | Material | Thickness ( μm ) |
|---|---|---|
| Waveguide core | $(Al_{0.45}Ga_{0.55})_{0.5}In_{0.5}P$ | y |
| Quantum well | $Ga_xIn_{1-x}P$ | 0.01 |
| Inner Barrier | $(Al_{0.45}Ga_{0.55})_{0.5}In_{0.5}P$ | 0.025 |
| Quantum well | $Ga_xIn_{1-x}P$ | 0.01 |
| Waveguide core | $(Al_{0.45}Ga_{0.55})_{0.5}In_{0.5}P$ | y |

**Table 2. Tensile strained laser structure. Four samples have been grown: Sample 4(0.0% strain) [x=0.51,y=0.0975], Sample 5(-0.4% strain) [x=0.58,y=0.091], Sample 6(-0.7% strain) [x=0.61,y=0.0835] and Sample 7(-1.0% strain) [x=0.76,y=0.077].**

## 3. COMPRESSIVE STRAINED SAMPLES : ELECTRON MICROSCOPY AND X-RAY ANALYSIS

TEM studies were performed on a Philips CM200 series microscope at the Manchester Materials Science Centre and STEM analysis in a VG 601UX system at the Northwest STEM facility. The bright field cross-sectional TEM images of samples 0(+1.0%), 1(+1.4%), 2(+1.7%) and 3(+2.0%) are shown in Figures. 1a, 1b, 1c and 1d respectively. In Fig. 1a the quantum well (the central dark band running from right to left) can be seen to be intact with a width of approximately 10nm as specified in Table 1. However in both Fig. 1b and 1c the quantum well region, again running from right to left, is seen to be 'wavy' and irregular with a widely fluctuating thickness. In Fig. 1d the irregularities are seen together with several threading dislocations propagating from the supposed well region through to the cap. The intended increase in the mismatch between the well region and the surrounding material has resulted in distinctly non-planar growth, with non-uniform strain fields propagating into the material surrounding the well region. The 'bands' or 'striations' seen around the well regions in all samples are due to slight compositional fluctuations in the growth of the quaternary due to growth condition variations.

The breakdown of uniform growth is confirmed in Fig. 2, a plan view TEM image of sample 2 (+1.7%) which shows the presence of islands in the well region. These islands have a mean diameter of 150nm, which is comparable to the pitch of the undulations seen in Fig. 1c and an approximate density of $3x10^9 cm^{-2}$, representing a coverage of ~ 40%. It appears that the increase in mismatch has led to the transformation of 2D layer Franck Van-der-Merwe

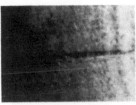

Fig. 1a  Sample 0 (+1.0%)
(360 nm by 250 nm)

Fig. 1b  Sample 1 (+1.4%)
(560 nm by 430 nm)

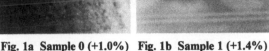

Fig. 1c  Sample 2 (+1.7%)
(630 nm by 450 nm)

Fig. 1d  Sample 3 (+2.0%)
(2.10 μm by 1.48 μm)

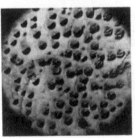

Fig. 2  Sample 2 (+1.7%)
(2.60 μm by 2.66 μm)

growth in the well region to 3D Stranski-Krastanov (S-K) growth as a probable strain relieving mechanism. S-K growth is well documented as a preferred growth mode in high mismatch systems, including Si\Ge (Eaglesham and Cerullo 1990) and InGaAs\GaAs (Snyder et al 1991).

Table 3 shows the results from X-Ray analysis performed on the plan view specimen in a region less than 40nm thin.    X-Ray spectra were measured inside the central part of an island and compared with spectra measured for the surroundings.    The composition of the islands corresponds closely to the intended well composition and the surrounding composition to that of the waveguide.  This indicates that no segregation of elements has occurred, simply that there is a 3D growth of ternary quantum well material surrounded by quarternary waveguide material.  This is further confirmed by dark field STEM imaging of a cross section (not shown) indicating clear composition contrast between the islands and waveguide region.

Therefore it appears that in these compressive strained laser structures the intended strain is accommodated through the formation of islands for mismatch > +1.4%, and in the case of the +2.0% mismatched sample this islanding is accompanied with the nucleation of threading dislocations at the island peaks, which may possibly be caused by local strain at island peaks.

**Table 3. Results of X-Ray analysis on Plan View Sample**

| Element | Nominal Well fraction (%) | Measured Island fraction (%) | Nominal Waveguide fraction (%) | Measured Surrounding fraction (%) |
|---------|---------------------------|------------------------------|--------------------------------|-----------------------------------|
| Ga | 14.25 | 15.5 | 15.00 | 19.7 |
| Al | 0.00 | 1.0 | 10.00 | 8.7 |
| In | 35.75 | 35.7 | 25.00 | 23.3 |
| P | 50.00 | 47.8 | 50.00 | 48.3 |

## 4.    TENSILE STRAINED SAMPLES : ELECTRON MICROSCOPY

Figures 3(a) and (b) show cross sectional TEM images from tensile strained samples 5(-0.4%) and sample 7(-1.0%).   Fig 3(a) shows the DQW intact with straight interfaces and with dimensions as expected. This is also the case for sample 4(0%) and sample 6(-0.7%) which are not shown. Fig. 3(b) again shows the DQW however in this case the well widths can be seen to be fluctuating from ~110nm to ~180nm with a periodicity of ~230nm. It appears that this level of tensile mismatch (-1%) is sufficient to produce 'wavy growth' in this system. This relative

334

Fig. 3a Sample 5 (-0.4%)
(360 nm by 220 nm)

Fig. 3b Sample 7 (-1.0%)
(480 nm by 280 nm)

strain level is less than that required for the compressive samples to exhibit this behaviour, and further work on more highly tensile strained specimens is required to investigate whether a similar or more dramatic continued growth phase occurs with

increasing tensile strain compared to that of compressive.

## 5.    LASER DEVICE PERFORMANCE

The detrimental effect on the laser device performance is dramatic with a twofold increase in threshold current density ($J_{th}$) between samples 1(+1.4%) and 2(+1.7%) both of which have higher threshold current densities than sample 0(+1.0%) at all temperatures. This is due primarily to increased optical losses and reduced optical confinement resulting from phenomena such as scattering in the islands region. Sample 3(+2.0%), which shows threading dislocations in addition to the islanding, fails to operate as a laser at any temperature.

The tensile strained devices do not show such an increase in $J_{th}$ with strain at all temperatures and it can be seen that if the optical losses scale with the inhomogeneity of the well region, then their impact will not be as significant for the tensile strain lasers. The main problem with the tensile devices is that of thermal leakage of carriers from the well to waveguide core. This is a direct consequence of the composition required to achieve the tensile mismatch required. This composition increases the bandgap of the well region resulting in small band offsets. This effect would possibly be enhanced in samples exhibiting fluctuating well widths which may lead to even smaller band offsets due to the possibilty of narrower QW regions.

## 6.    DISCUSSION AND CONCLUSIONS

The structural integrity of a series of strained $(Al_yGa_{1-y})_xIn_{1-x}P/Ga_xIn_{1-x}P$ QW layers has been examined and it has been found via TEM that in a high compressive strain regime 3D island formation is found to occur as a relief mechanism for the intended mismatch as opposed to misfit dislocation formation. X-Ray analysis confirms that the islands formed are made up from the ternary well material and no strain related compositional segregation occurs. Tensile strained samples begin to exhibit a similar style of 'wavy growth' at a relatively smaller strain level. Further work on greater strained tensile samples is required before a direct comparison can be made. Laser device performance can be seen to deteriorate due to the 'non-planar' growth seen in the compressive samples resulting in larger $J_{th}$.

## REFERENCES

Bour D P, Geels R S, Treat D W, Paoli T L, Ponce F, Thornton R L, Krusor B S, Brigans R D
    and Welch D F 1994 J. Quant. Electron. **30**, 593
Eaglesham D J and Cerullo M 1990 Phys. Rev Lett. **64**,1943
Matthews J W and Blakeslee A E 1974 J. Cryst. Growth **27**, 118
Snyder C W, Orr B G, Kessler D and Sander L M 1991 Phys. Rev. Lett. **66**, 3032
Valster A, Van der Poel C J, Finke M N and Boermans M J B 1992 Digest 13th Int. Semicond.
    Laser Conf. **G-1**, 152

*Inst. Phys. Conf. Ser. No 153: Section 8*
*Paper presented at Electron Microscopy and Analysis Group Conf. EMAG97, Cambridge, 1997*
© *1997 IOP Publishing Ltd*

# Study of oxygen content and dielectric function of YBCO$_{7-\delta}$ by EELS

**S Schamm, C Grigis, D Lessik, G Zanchi and J Sévely**

CEMES / CNRS, BP 4347, 31055 Toulouse cedex, France

**ABSTRACT:** We have measured by electron energy loss spectroscopy the energy and momentum dependence of electronic excitations for the compound YBCO$_{7-\delta}$ as a function of δ. For the determination of the important parameter δ, we use an approach based on the study of plasmons in addition to the modelling of the O-K edge already proposed. The methods give equivalent results but the collection of the low loss signal is however easier.

## 1. INTRODUCTION

Investigation of materials by electron energy loss spectroscopy (EELS) in the transmission electron microscope is suitable for examining electronic properties with high spatial resolution. An interesting advantage is that topological, structural and chemical information of the bulk material can be obtained at the same time.

Concerning the high temperature superconductors YBCO$_{7-\delta}$, one important point is the ability to determine the O content (δ), which controls the structural and electromagnetic properties of the material. While YBCO$_6$ (δ=1) is a tetragonal insulating, antiferromagnetic compound, YBCO$_7$ (δ≥0.6) is an orthorhombic two-dimensional superconducting (T≤91K) "strange metal".

In the EELS spectra, two kinds of signals can be used to determine δ. They correspond to the individual excitation of the O-1s core level electrons as proposed by Browning et al (1992) and to the collective excitation of the valence electrons, the plasmons. This work deals with the quantitative exploitation of the two signals. The complex dielectric function ε(**q**,ω), which gives information on the electronic structure of materials, is also considered.

## 2. METHODS

### 2.1 Modelling of the O-K edge

Important structural features in YBCO$_{7-\delta}$ are the CuO$_2$ planes perpendicular to the c direction of the crystal and the CuO chains parallel to the b direction. When going from YBCO$_6$ to YBCO$_7$, oxygen incorporation in the CuO chains is associated with charge transfer from the planes to the chains in such a way that a p-type doping occurs in the CuO$_2$ planes. Considering the electronic structure of YBCO$_{7-\delta}$ near the Fermi level, as discussed with the charge transfer model by Fink et al (1994), p-type doping implies that the Fermi level is shifted to lower energies in the valence band, which mainly has the O2p character (hybridization with Cu3d states). Thus, on doping, hole states of O2p character are created in this band.

In EELS experiments, the signal measured at the edge of a core level excitation is connected to the site-selective unoccupied density of states (DOS) being directly proportional to the imaginary part of the dielectric function ε$_2$. According to dipole selection rules, the unoccupied part of the O2p states can be reached from the O1s core level excitation. The investigation of the O-K edge directly evidences the O2p character of the holes created by O incorporation in YBCO$_{7-\delta}$. As shown Fig. 1a, in the O-K near-edge features a pre-edge appears just before a main absorption edge, the intensity of which is sensitive to δ. It decreases when δ increases and vanishes for δ~0.6. According to these observations and based on a three-gaussian simulation of the O-K edge, Browning et al (1992) proposed a method for the evaluation of the oxygen content.

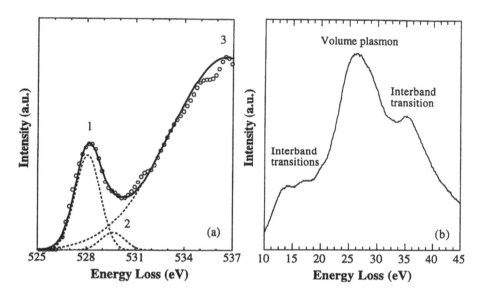

Fig. 1 :  (a) O-K near-edge features in YBCO$_{6.9}$ and their simulation by three gaussians ;
(b) corresponding spectrum in the low-loss range.

## 2.2 Energy of the volume plasmons

Another approach for determining the O content uses a more direct method based on study of the low-loss range of the EELS spectra. As illustrated on Fig. 1b, different features exist in this energy domain. The spectrum is dominated by the volume plasmon near 26 eV corresponding to the collective excitation of all the valence electrons. The additional structures around this peak at lower and higher energies are caused by interband transitions. The energy of the volume plasmon is sensitive to the variation of the number of the valence electrons when going from YBCO$_6$ to YBCO$_7$. This variation can be measured and correlated with the results of the quantitative treatment of the O-K pre-edge peak.

## 3. EXPERIMENTS AND TREATMENTS

Spectra of both the low-loss range and the O-K near-edge were acquired from samples of unknown δ value but rather O-rich or O-poor. These were Y$_1$Ba$_2$Cu$_3$O$_{7-\delta}$/Y$_2$BaCuO$_5$ composites prepared by directional solidification, kindly provided by S. Pinols from ICMAB/CSIC in Barcelona. Contributions mainly from the (a,b) plane of the YBCO structure were collected. We used a Philips CM30-ST microscope working at 100 keV, equipped with a Gatan PEELS. The energy dispersion and resolution were 0.1 eV/channel and 1 eV respectively while the size of the analysed area was around 100 nm.

We applied Browning's method to the O-K distributions. An example of the accuracy of such a modelling with gaussians is shown on Fig. 1a. For each spectrum, the energy positions of the gaussians, 528.1±0.15 eV, 529.5±0,15 eV and 536.6 eV respectively, were nearly constant. The height and the width of each gaussian are determined by a least squares fitting to the experimental data. Before modelling, the spectra were deconvolved using the corresponding low loss spectrum in order to remove the effects of multiple scattering and of changing the thickness. The relative intensity of the pre-edge to the main edge (A$_1$/A$_3$) is calculated. Then, $\delta$ is extracted by means of the calibration curve determined by Browning on YBCO samples of known composition.

Simultaneously, we measured the corresponding volume plasmon energy from the energy loss function $Im(-1/\varepsilon)$, which is evaluated from the low-loss spectra by a program we developed for this purpose.

## 4. RESULTS

### 4.1 O content : O-K edge and $E_p$

$\delta$ is evaluated from the values $A_1/A_3$ and the calibration curve. The results of the treatment for representative spectra corresponding to different O contents are shown in Table 1. In this case, the lower the value of $\delta$, the more accurate is the measurement, at worst, it is 0.1.

| $A_1/A_3$ | $A_2/A_3$ | $7-\delta$ |
|-----------|-----------|------------|
| 0.11 | 0.03 | **6.9** |
| 0.10 | 0.03 | **6.8** |
| 0.09 | 0.04 | **6.7** |
| 0.07 | 0.04 | **6.6** |
| 0.04 | 0.04 | **6.5** |
| 0.02 | 0.04 | **~6.4** |

Table 1 : Areas of the first and the second gaussians relative to the area of the third one and corresponding O contents.

The curve of Fig. 2 associates the volume plasmon energy, $E_p$, and the O content $7-\delta$. The plasmon energy varies from around 25.6 eV to nearly 26.2 eV for the lowest (~6.4) and highest (~6.9) oxygen content, respectively. A linear relationship is evidenced between $E_p$ and $\delta$. From this graph, evaluating $\delta$ from the measure of $E_p$ has the same precision as the gaussians method applied to the O-K edge, that is about 0.1.

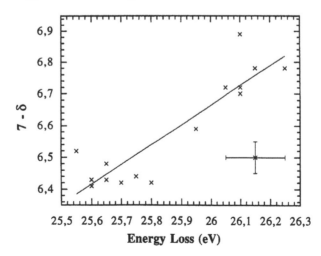

Fig. 2 : Variations of the O content, $7-\delta$, with the volume plasmon energy.

### 4.2 O content : $Im(-1/\varepsilon)$, $\varepsilon_1$ and $\varepsilon_2$

Fig. 3 compares the variations of $Im(-1/\varepsilon)$, $\varepsilon_1$ and $\sigma$, $\sigma = \omega\varepsilon_0\varepsilon_2$ being the optical conductivity, when the energy loss changes from 10 to 50 eV. The results concern two

338

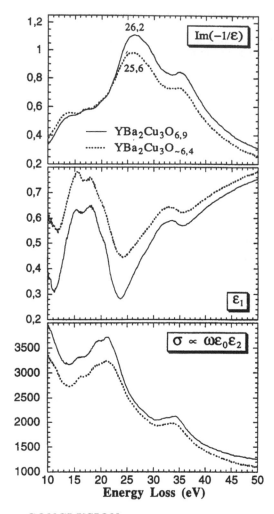

crystals corresponding to the lowest and highest O content.

The maxima in $\sigma$ are directly connected with interband transitions and can be compared with the band structure diagram determined by experiment and calculation (Romberg et al 1990). If we compare $Im(-1/\varepsilon)$ and $\sigma$, we can see that the maxima of $\sigma$ correspond to relative minima in the energy loss function. The energy position of the maxima is unchanged for the two crystals.

Commonly, zero crossing of $\varepsilon_1$ with positive slope signifies a collective excitation in a material. In $YBCO_{7-\delta}$, the situation is unusual since this crossing does not occur. This is believed to be the result of interband transitions at energies close to the plasmon energy. However, considering particular criteria defined by Ehrenreich et al (1962), it can be shown that there is effectively a plasmon for the two samples. The collective excitation is more damped for $7-\delta\sim6.4$ than for $7-\delta\sim6.9$ because the intensity of the corresponding loss function at the plasmon energy is lower.

Fig. 3 : Variations with energy of the energy loss function, $Im(-1/\varepsilon)$, of the real part of the dielectric function, $\varepsilon_1$ and of the optical conductivity, $\sigma$.

## 5. CONCLUSION

The volume plasmon energy in $YBCO_{7-\delta}$ is also a pertinent parameter for a measure of oxygen content. Its use is more direct than the modelling of the O-K edge. The plasmon signal has also the advantage of being more intense and thus easier to collect.

Differences in the dielectric functions of $YBCO_{6.4}$ and $YBCO_{6.9}$, such as the volume plasmon energy and the intensity of $\sigma$, can be related to differences in the number of valence electrons. The stronger plasmon damping for $7-\delta\sim6.4$ is probably the result of an energy of collective excitation closer to the lower interband transition energies than for $7-\delta\sim6.9$.

## REFERENCES

Browning N D, Yuan J and Brown L M 1992 Physica C **202**, 12
Ehrenreich H and Philipp H R 1962 Int. Conf. Phys. Semiconductors, ed A C Strickland (Bartholomew Press, Dorking, UK) p 367
Fink J, Nücker N, Pellegrin E, Romberg H, Alexander M and Knupfer M 1994 Journal of Electron Spectroscopy and Related Phenomena **66**, 395
Romberg H, Nücker N, Fink J, Wolf T H, Xi X X, Koch B, Geserich H P, Dürrler M, Assmus W and Gegenheimer B 1990 Z. Phys. B- Condensed Matter **78**, 367

*Inst. Phys. Conf. Ser. No 153: Section 8*
*Paper presented at Electron Microscopy and Analysis Group Conf. EMAG97, Cambridge, 1997*
© *1997 IOP Publishing Ltd*

# Modelling of electron energy-loss spectroscopy detection limits

**M K H Natusch, G A Botton and C J Humphreys**
Department of Materials Science and Metallurgy, University of Cambridge, Pembroke
Street, Cambridge, CB2 3QZ

**O L Krivanek**
Cavendish Laboratory, University of Cambridge, Madingley Road, Cambridge, CB3 0HE
and Department of Materials Science, University of Washington, Seattle, WA 98195, USA

**ABSTRACT:** We develop the methodology to predict spectral signal-to-noise ratios
(SNRs) of electron energy-loss spectroscopy (EELS) by computing spectra taking
account of valence electron excitations, inner-shell excitations and elastic scattering
together with the noise characteristics of the detection system. The consideration of
multiple scattering in our model allows us to apply it to a wide range of specimen
thicknesses and collection angles. The detectability of an element is judged by the SNR.
The calculated SNRs at the oxygen and magnesium K-edges in MgO agree well with
experiment.

## 1. INTRODUCTION

The physical and chemical properties of heterogeneous materials often depend on the
distribution of trace elements on a sub-nanometre scale. This is the case for instance for the
electronic properties of semiconductors or the mechanical properties of composites. Hence a
microanalytical technique is required that provides high spatial resolution as well as trace
element sensitivity. In a transmission electron microscope the energy distribution of
transmitted electrons can be analysed using electron energy-loss spectroscopy (EELS) which
is a technique that suits the above requirements since it has been shown in recent years that it
can detect atomic concentrations below 100 ppm on a nanometre spatial scale.

An evaluation of the detection limit, i.e. the minimum detectable concentration of a trace
element contained within a matrix or at an interface, provides an important performance
parameter for every microanalytical technique. A theoretical model that predicts the
detection limits of EELS would allow one to check whether an element present at a
suspected concentration should be detectable in an EELS spectrum acquired under particular
conditions. It would also show how to optimise acquisition and processing parameters so
that the detection limits become as low as possible.

Predicting the detection limits of EELS is made non-trivial by the complex nature of the
spectrum. We are developing a prediction model based on computing the artificial spectrum
expected from a given concentration of an element of interest in a given matrix.

## 2. MODELLING THE SPECTRUM

A characteristic ionisation edge in EELS is always superimposed on a background. This
pre-edge background results from all possible scattering events whose threshold energy is
less than the edge energy and all combinations of these events according to a Poisson
multiple scattering distribution. The background needs to be subtracted without distorting
the signal on top of it. The detection limits in EELS arise from the correct identification of
this weak signal on top of the large background in the presence of noise. Our aim is to model
the absolute intensity of the background and the noise in the spectrum.

340

## 2.1 Single Scattering Distribution

We assume that the specimen is amorphous and that dynamical diffraction effects can be neglected. These assumptions can be well approximated even in a crystal provided it has been oriented in non-diffracting condition. In this condition the elastic scattering can be approximated by Lenz cross-sections as described by Egerton (1995). Even though elastic scattering has no obvious influence on an energy-loss spectrum, multiple elastic-inelastic scattering changes the angular distribution of higher energy loss events.

The collective excitations are modelled by Drude-Lorentz type oscillators which not only describe the volume plasmon but also interband transitions. The damping parameter of the volume plasmon which varies with scattering angle is chosen in such a way that the plasmon distribution changes smoothly into that for the single-electron excitation at the cut-off angle.

The core-loss cross-sections are computed using the parametrised Hartree-Slater generalised oscillator strength tables as provided by Gatan's EL/P software version 3.0 and described by Rez (1982). The two-dimensional polynomial interpolation algorithm given by Press et al. (1992) contains a convenient monitor for the accuracy of the extrapolation. Once the accuracy leaves a specified narrow range, the core-loss cross-sections are forced to be hydrogenic. The routines provided by Egerton (1995) then allow to cover the whole spectral range.

## 2.2 Multiple Scattering

Following the procedure described by Su and Zeitler (1993), we group the various contributions to an EELS spectrum into four channels, the unscattered distribution, $P_0$, elastic scattering, $P_1$, low-loss scattering, $P_2$ and core-loss ionisation, $P_3$. The final distribution of electrons at all energy losses and scattering angles, $P$, can be expressed in terms of the Fourier transformed distributions denoted by the tilde as

$$\tilde{P} = e^{-t/\lambda}\left(1 + \tilde{P}_1 + \left[\tilde{P}_2 + \tilde{P}_3 + \tilde{P}_2\tilde{P}_3\right] + \tilde{P}_1\left[\tilde{P}_2 + \tilde{P}_3 + \tilde{P}_2\tilde{P}_3\right]\right), \quad (1)$$

where $t/\lambda$ denotes the specimen thickness in units of the total mean free path. Equation (1) shows the combination of the single scattering distribution and the multiple low loss and high loss events in the square bracket with the elastic distribution.

In our calculations a standard Fourier transform algorithm as given by Press et al. (1992) performs the energy transform. The two-dimensional angular transform reduces to a one-dimensional Hankel transform of zero order as described by Johnson (1987) on account of the scattering symmetry in the non-diffracting specimen.

The results of a calculation of this kind are shown below in figure 1 for various collection angles and for two specimen thicknesses. Figure 1(b) shows how the visibility of the magnesium L-edge at 51 eV is reduced for increasing specimen thickness. The visibilty of the oxygen K-edge at 532 e, however, is still surprisingly high.

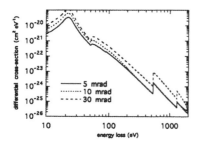
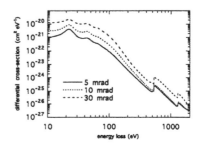

Fig. 1(a) Calculated MgO spectra for $t/\lambda = 1$.     Fig. 1(b) Calculated MgO spectra for $t/\lambda = 3$.

### 2.3 Poisson and Detector Noise

Once the total intensities in the spectrum are obtained from the above cross-section by multiplying it with the areal density in the specimen and the beam dose, Poisson noise can be added to the spectrum before it is converted into detector counts. The Gatan Imaging Filter utilises a slow-scan CCD camera with a size of 1024x1024 pixels to detect electrons. Provided we know how many pixels per energy channel are used to actually acquire the spectrum we can add the two components of the detector dark noise. These are the time-independent electronic read-out noise and the dark current noise which depends on acquisition time. The two components of the detector noise are comprised in the detective quantum efficiency (DQE) which is a measure of how much noise the detector adds to the signal. It has been measured to have a maximum value of about 0.8 in the present system.

### 3. COMPARISON WITH EXPERIMENT

To be able to compare the model described above with experimental data, momentum resolved spectra have been acquired in the ω-q mode of the Gatan Imaging Filter mounted on top of a VG STEM HB501 as described by Krivanek et al. (1994). The intensities of the calculations have been adjusted for the reduced aperture in a momentum resolved measurement with a 0.4 mrad wide slit in the image plane of the spectrometer.

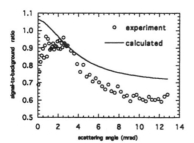

Fig. 2(a)  Signal-to-background ratio for the oxygen K-edge with a 150 eV window.

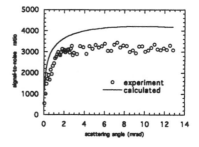

Fig. 2(b)  Signal-to-noise ratio for the oxygen K-edge with a 150 eV window.

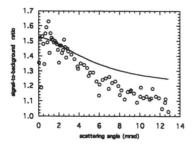

Fig. 3(a)  Signal-to-background ratio for the magnesium K-edge with a 150 eV window.

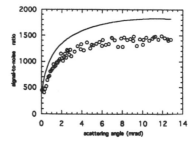

Fig. 3(b)  Signal-to-noise ratio for the magnesium K-edge with a 150 eV window.

We calculate the edge intensities in the theoretical spectra by simply subtracting a calculated background spectrum without the edge of interest from the original spectrum. For the experimental spectra we assume that the background under the edge can be removed error-free by a maximum likelihood estimation of the power-law background as described by Pun et al. (1985). While we expect that the signal-to-background ratio can be predicted well provided the multiple scattering calculation is correct, the real test for the model is the signal-to-noise ratio since we are forced to compare absolute counts.

It is apparent from the figures that both the signal-to-background ratio and the signal-to-noise ratio are overestimated in the calculation and more so for higher angles. Stray scattering in the spectrometer contributes to the background at high energy loss and this might act to reduce the experimental signal-to-background ratio. The error-free removal of the background in the experimental spectra is also questionable.

## 4. MODELLING DETECTION LIMITS

Once we have confidence in the calculated signal-to-noise ratios, it is possible to calculate the concentration of a trace element which produces a signal that is just detectable. According to the Rose criterion, this is the case when the SNR exceeds 3. Trebbia (1988) calculates the minimum detectable signal under the two risks that one concludes that an element is present whereas it actually is absent or one finds that an element is absent while it is actually present in the specimen.

A model as described above can also be used to optimise acquisition parameters so that the detection limits become as low as possible. The experimental parameters that effect the SNR are the high tension, the convergence and collection angles, the beam current, the acquisition time and the width and position of the energy windows. Starting from the model given above, one can in principle quickly calculate the set of parameters that maximises the SNR.

## 5. CONCLUSION

We have developed a model to calculate the spectral signal-to-noise ratios in an EELS spectrum. This model allows one to check whether an element present at a suspected concentration should be detectable in an EELS spectrum acquired under particular conditions. The results indicate that the model can successfully predict the measured signal-to-noise ratio, and that it can be used to predict attainable limits for the detection of trace constituents and the optimum acquisition conditions.

At present our approach is limited to minimum mass fraction (MMF) detection limits. We plan to extend this also to minimum detectable mass (MDM) limits. We will also introduce white lines into the calculated spectra from tabulated data, so that experimental first and second difference spectra can be compared with differentiated calculated spectra which have been adjusted for the instrumental broadening.

**Acknowledgements**

MKHN gratefully acknowledges financial support by Gatan Inc. and the EPSRC.

**REFERENCES**

Egerton RF 1996 Electron energy-loss spectroscopy in the electron microscope (New York: Plenum Press) 2nd ed.
Johnson HF 1987 Comput Phys Commun **43** 181
Krivanek OL, Bui DN, Ray DA, Boothroyd CB, Humphreys CJ 1993 Proc ICEM **13** 167
Press WH, Teukolsky SA, Vetterling WT, Flannery BP 1992 Numerical Recipes (Cambridge: Cambridge University Press) 2nd ed.
Pun T, Ellis JR, Eden M 1985 J Microsc **137** 93
Rez P 1982 Ultramicroscopy **9** 283
Su DS, Zeitler E 1993 Phys Rev B **47** 14734
Trebbia P 1988 Ultramicroscopy **24** 399

*Inst. Phys. Conf. Ser. No 153: Section 8*
*Paper presented at Electron Microscopy and Analysis Group Conf. EMAG97, Cambridge, 1997*
© *1997 IOP Publishing Ltd*

# Dielectric theory in the superlattice limit

J P R Bolton[†], T Neyer[§] and P Schattschneider[§]

†Department of Physics, The Open University, Walton Hall, Milton Keynes MK7 6AA, UK.
§Institut für Angewandte und Technische Physik, Technische Universität Wien, Wiedner Hauptstrasse 8-10/137, A-1040 Wien, Austria.

ABSTRACT Retarded dielectric theory gives an explicit but complicated formula for the electron energy loss probability in the case of cross-sectional incidence on a multilayer. This formula predicts suppression of longitudinal bulk modes in thin layers, but experiment suggests that even stronger finite-size effects are present. Here we obtain an energy loss formula for the limiting case of an infinite superlattice; the result is much simpler than for a large multilayer and, being expressed in terms of integrals over $k_z$, is well-suited to investigations based on a wavenumber cut-off.

## 1 Dielectric theory in multilayers

Semiclassical retarded dielectric theory allows an electron energy loss spectrum to be interpreted in terms of bulk dielectric properties. The method usually models the beam as a narrow probe and takes the dielectric response of each medium be local, characterised by a wavevector-independent dielectric function. Maxwell's equations, subject to the usual boundary conditions, are solved with source terms appropriate for an unscattered electron and the work done by the electron is interpreted semiclassically in terms of the exchange of virtual quanta.

Dielectric theory has been applied to many different media and geometries, including anisotropic bulk media, single slabs, spheres, cylindrical holes and wedges. Assuming abrupt interfaces it is possible to extend the theory to general multilayered systems. We are interested in cross-sectional incidence on a multilayered slab; for convenience, we briefly reproduce results obtained previously for this case [1, 2].

(i) *Dispersion relation of coupled interface modes*
The dispersion relation for coupled interface modes in an $n$-layered slab takes the form:

$$\left[ C_{n0} \right] \left[ \tilde{C}_{n0} \right] = 0 \tag{1}$$

where $[C_{n0}]$ is extracted from a general transfer matrix product

$$A^{(ji)} = \left( \begin{array}{cc} [C_{ji}] & [D_{ji}] \\ [E_{ji}] & [F_{ji}] \end{array} \right) = \prod_{k=i}^{j} \left( \begin{array}{cc} h_{k+1,k}^{+} f_k^2 & h_{k+1,k}^{-} \\ h_{k+1,k}^{-} f_k^2 & h_{k+1,k}^{+} \end{array} \right) \tag{2}$$

with

$$h_{ji}^{\pm} = q_j \epsilon_i \pm q_i \epsilon_j, \qquad f_k = e^{q_k d_k} \quad \text{for} \quad k \neq 0 \quad \text{and} \quad f_0 = 1.$$

Here, $d_i$ is the thickness, $\epsilon_i$ the dielectric function and $q_i = \sqrt{k_{\parallel}^2 - \epsilon_i \omega^2/c^2}$ the inverse attenuation length of layer $i$; $\hbar k_{\parallel}$ is the in-plane momentum transfer. Throughout, we use the convention that any quantity (such as $\left[\tilde{C}_{n0}\right]$) carrying a tilde accent has the same form as its relative without, but with each $h_{ji}^{\pm}$ term replaced by the corresponding $\tilde{h}_{ji}^{\pm}$, where

$$\tilde{h}_{ji}^{\pm} = q_j \pm q_i.$$

This dispersion relation factors into coupled TM modes ($[C_{n0}] = 0$), with all magnetic fields in the plane of the slab and $\mathbf{k}_{\parallel}\cdot\mathbf{E} = 0$, and coupled TE modes ($\left[\tilde{C}_{n0}\right] = 0$), with all electric fields in the plane of the slab and $\mathbf{k}_{\parallel}\times\mathbf{E} = \mathbf{0}$.

(ii) *Energy loss probability*
We take the $x$-axis to be the beam direction and the $z$-axis to be the stacking direction, with layer $i$ extending from $z_{i-1}$ to $z_i$ (and $z_{-1} = 0$). The incoming electron has charge $-e$, speed $v$ and is at position $z_b$ in layer $m$. With the normalisation chosen so that the work done per unit path length is

$$\frac{dW}{dx} = \int_0^{\infty} d(\hbar\omega)\, \hbar\omega \int_{-\infty}^{\infty} dk_y\, \frac{d^3P}{d(\hbar\omega)dk_y dx},$$

the energy loss probability density emerges as

$$\frac{d^3P}{d(\hbar\omega)dk_y dx} = \frac{e^2}{4\pi^2\varepsilon_0\hbar^2 v^2}\, \mathrm{Im}\, \chi_m^{(n)}\Big|_{k_x=\omega/v}$$

with

$$\chi_m^{(n)} = \frac{-1}{q_m \epsilon_m k_{\parallel}^2}\left(q_m^2 \frac{\gamma_{nm}^+ \zeta_{m-1,0}^-}{\left[C_{n0}\right]} - \epsilon_m k_y^2 \frac{v^2}{c^2}\frac{\tilde{\gamma}_{nm}^- \tilde{\zeta}_{m-1,0}^+}{\left[\tilde{C}_{n0}\right]}\right) \tag{3}$$

where the position-dependence of the beam enters only through the factors

$$\gamma_{nm}^{\pm} = [C_{nm}]\,e^{-q_m(z_b-z_{m-1})} \pm [D_{nm}]\,e^{+q_m(z_b-z_{m-1})}$$

$$\zeta_{m-1,0}^{\pm} = [C_{m-1,0}]\,e^{+q_m(z_b-z_{m-1})} \pm [E_{m-1,0}]\,e^{-q_m(z_b-z_{m-1})}.$$

Equation 3 agrees with known results for a bulk medium ($n = -1$), a single interface ($n = 0$) and a single slab ($n = 1$). It has the correct symmetry properties on reversal of time, $v$ or region labelling, and gives consistent results when two neighbouring regions coalesce or when the beam region becomes very thick with the beam far from all interfaces. The pole structure has a simple interpretation: the poles of $\epsilon_m^{-1}$ correspond to longitudinal bulk plasmons while the poles of $q_m^{-1}$ give the transverse bulk modes (including Cerenkov radiation). The poles in $\left[C_{n0}\right]^{-1}$ and $\left[\tilde{C}_{n0}\right]^{-1}$ correspond to TM and TE interface plasmons, and also to transverse bulk modes in all layers (because the dispersion relations $\left[C_{n0}\right] = 0$ and $\left[\tilde{C}_{n0}\right] = 0$ are satisfied by $q_i = 0$ for $i = 0\ldots n+1$). In practice, the TE modes are not strongly excited because they have $k_{\parallel} < \omega/c$, while the in-plane momentum transfer for cross-sectional incidence must be greater than $\omega/v$. All these poles occur in the lower-half complex frequency plane, corresponding to modes that are damped in time, so infinities are not encountered in the integral over real frequencies but, in general, lightly damped modes with poles close to the real frequency axis will tend to produce a greater scattering probability than heavily damped modes further away.

As the beam position crosses the interface between two media, Equation 3 changes abruptly in form. Nevertheless, explicit calculation shows that the energy loss probability is a continuous function of $z_b$. Continuity is maintained because the pole in $\epsilon_m^{-1}$ has a vanishing coefficient when $z_b$ is at an interface. There is a connection here with the Bethe sum rule, which requires the strength of the longitudinal bulk plasmons to diminish as the strength of the interface plasmons increases: with local dielectric functions and an infinitely narrow beam exactly at an interface, the interface plasmons give their maximum contribution and the longitudinal bulk plasmons are completely suppressed by confinement terms (begrenzungseffekt). The suppression of longitudinal bulk plasmons is further illustrated by integrating Equation 3 over beam position within a layer $m$. In the limit $|q_m d_m| \longrightarrow 0$ the integrated energy loss probability is found to contain no pole in $\epsilon_m^{-1}$. By contrast, transverse bulk modes are excited in all layers via zeros of $[C_{n0}]$ and $[\tilde{C}_{n0}]$.

## 2  Finite size effects in multilayers

We have previously reported [4] experimental results in W/Si (6:20)Å metallic multilayers, which show that bulk plasmons are suppressed more strongly than predicted by Equation 3. We explored this phenomenon in a heuristic way by identifying the bulk contribution to Equation 3, representing it as an integral over all $k_z$ values and cutting off contributions from wavelengths greater than the layer thickness. These long-wavelength modes were replaced by modes in an anisotropic effective medium. While this procedure was useful in suggesting the existence of a new finite-size effect, it is not entirely satisfactory because it is impossible to unravel the interface and confinement contributions, which cannot be represented as integrals over $k_z$ in any finite multilayer. This provides our motivation for examining Equations 1 and 3 in the superlattice limit: our aim is to re-establish the $k_z$ dependence so that more information can be extracted, allowing any cut-off to be imposed consistently. A subsidiary aim is to obtain simpler expressions, avoiding the need to choose an arbitrary large $n$ when discussing superlattices.

## 3  Taking the superlattice limit

A superlattice is composed of a basis extending from region $\alpha$ to region $\beta$ and repeated infinitely along the stacking direction with period $d = \sum_{i=\alpha}^{\beta} d_i$. We use Equation 2 to introduce a transfer matrix $X \equiv A^{(\beta\alpha)}$ for the basis and then perform a similarity transformation $X = R\Lambda R^{-1}$, where $\Lambda$ is a diagonal matrix, with eigenvalues

$$\lambda_{1,2} = \frac{\operatorname{Tr} X \pm \sqrt{(\operatorname{Tr} X)^2 - 4 \det X}}{2}. \tag{4}$$

The invariance properties of Tr and Det ensure that these eigenvalues are independent of the starting layer $\alpha$ chosen for the basis. Although $X$ is not hermitian the diagonalisation is always possible provided the eigenvalues are distinct. Powers of $X$ can then be expressed as

$$X^n = \frac{1}{\lambda_2 - \lambda_1} \begin{pmatrix} (\lambda_2 - X_{11})\lambda_1^n - (\lambda_1 - X_{11})\lambda_2^n & -X_{12}(\lambda_1^n - \lambda_2^n) \\ -X_{21}(\lambda_1^n - \lambda_2^n) & (\lambda_2 - X_{11})\lambda_2^n - (\lambda_1 - X_{11})\lambda_1^n \end{pmatrix}.$$

The dispersion relation for TM modes is found by setting the $(1,1)$ matrix element of $X^n$ equal to 0, giving $(\lambda_1/\lambda_2)^n = (\lambda_1 - X_{11})/(\lambda_2 - X_{11})$. In the large $n$ limit appropriate for a superlattice it follows that $\lambda_1/\lambda_2$ is a pure phase factor, which we identify as $e^{2ik_z d}$ by means of Bloch's theorem. Further simplifications, using Equations 2 and 4 lead to the final dispersion relation

$$G(k_\parallel, k_z, \omega) \equiv X_{11} + X_{22} - 2\cos(k_z d) \prod_{i=\alpha}^{\beta} (2q_i \epsilon_i e^{q_i d_i}) = 0 \quad \text{for} \quad -\pi/d < k_z \leq \pi/d, \tag{5}$$

in agreement with previous results for a binary superlattice [3]. For real materials, with complex dielectric functions, the solutions of the dispersion relation lie in the lower-half complex frequency plane, with the imaginary part of the frequency giving the inverse decay time the mode. This suggests a further simplification. Since our energy loss formula involves only real frequencies, the TM and TE dispersion relations cannot be *exactly* satisfied (infinities are avoided) in Equation 3; the two eigenvalues $\lambda_1$ and $\lambda_2$ must therefore have different moduli. In the limit of an infinite superlattice any difference is significant, allowing us to neglect the ratio $(\lambda_S/\lambda_L)^n$ compared to 1, where $\lambda_L$ and $\lambda_S$ are the eigenvalues of larger and smaller modulus respectively. In the superlattice limit we can suppose that

$$X^n \simeq \frac{\lambda_L^n}{\lambda_L - \lambda_S} \begin{pmatrix} -(\lambda_S - X_{11}) & X_{12} \\ X_{21} & \lambda_L - X_{11} \end{pmatrix}.$$

Taking the beam to be in the middle of a superlattice, and choosing the basis to start from the beam layer, we use the above matrix elements in place of $[C_{nm}]$, $[C_{m-1,0}]$ etc. This leads to a much simplified expression for the electron energy loss probability in a superlattice:

$$\chi_m^{(\infty)} = \chi_n^{(-1)} + \frac{-1}{q_m \epsilon_m k_\parallel^2} \left( q_m^2 \frac{R^+}{\lambda_L - \lambda_S} - \epsilon_m k_y^2 \frac{v^2}{c^2} \frac{\tilde{R}^-}{\tilde{\lambda}_L - \tilde{\lambda}_S} \right)$$

where

$$R^\pm = \pm(X_{12}e^{2q_m(z_b - z_{m-1})} - X_{21}e^{-2q_m(z_b - z_{m-1})}) - 2(\lambda_L - X_{11}).$$

It is easy to express the bulk loss contribution as an integral over all $k_z$:

$$\chi_m^{(-1)} = \left( \frac{v^2}{c^2} - \frac{1}{\epsilon_m} \right) \frac{1}{\pi} \int_{-\infty}^{\infty} \frac{dk_z}{k_z^2 + q_m^2}.$$

By contour integration we have shown that the remaining terms can also be expressed as an integral over $k_z$, with the reciprocal of the superlattice dispersion relation (Equation 5) integrated over the first Brillouin zone of the superlattice:

$$\frac{1}{\lambda_L - \lambda_S} \equiv \frac{\text{sign}(|\lambda_1| - |\lambda_2|)}{\lambda_1 - \lambda_2} = \frac{1}{2\pi d} \int_{-\pi/d}^{\pi/d} \frac{dk_z}{G(k_\parallel, k_z, \omega)}.$$

These formulae are much simpler than the multilayer formula for large $n$ and are well-suited to our ultimate aim of investigating finite-size effects because a low $|k_z|$ cut-off can be applied to all terms in a consistent way.

# References

[1] Bolton, J.P.R. & Chen, M. (1995) Electron energy loss in multilayered slabs: (2). Parallel incidence. *J. Phys.: Condens. Matter* **7**, 3389–3403.

[2] Bolton, J.P.R. & Chen, M. (1995) Energy loss in multilayered slabs. *Ultramicroscopy* **60**, 247–263.

[3] Bloss, W.L. (1983) Plasmon modes of a superlattice — classical vs quantum limits. *Solid State Commun.* **48**, 927–931.

[4] Neyer, T. Schattschneider, P. Bolton, J.P.R. & Botton, G.A. (1997) Plasmon coupling and finite size effects in metallic multilayers in multilayers. To be published in J. Micros.

*Inst. Phys. Conf. Ser. No 153: Section 8*
*Paper presented at Electron Microscopy and Analysis Group Conf. EMAG97, Cambridge, 1997*
© *1997 IOP Publishing Ltd*

# EELS in the region of the fundamental bandgap.

**B Rafferty and L M Brown**

Cavendish Laboratory, Madingley Road, Cambridge, CB3 0HE, UK

**ABSTRACT:** The Bethe theory for inelastic scattering has been developed to understand the detailed shape of bandgap EELS spectra. The theory shows how the matrix elements for direct and indirect transitions can be decoupled from one another and have energy dependencies of $(E - E_g)^{1/2}$ and $(E - E_g)^{3/2}$ respectively. The zero loss peak was removed via a deconvolution routine which allowed this theory to be tested on a number of crystalline samples with bandgaps as low as 0.9 eV at an energy resolution of 0.22 eV. The agreement between theory and experiment is surprisingly good.

## 1. INTRODUCTION

The ability to view the bandgap region of semiconducting and insulating materials is an ideal way to study their electronic structure. Optical techniques have the advantage of high-energy resolutions (~ 20 meV) which allows fine detail to be resolved. However, the EELS technique used here has the considerable advantage of being carried out in a STEM where the electron beam can be focused down to approximately 4 Å and positioned anywhere on the sample. The spectra presented here have energy resolutions of 0.22 eV at best. This being defined by the FWHM of the zero loss peak; it is the zero loss peak which s the main source of problems in acquiring good EELS spectra in the region of the bandgap. A simple scaled subtraction of a zero loss peak for the sample spectrum limits the user to studying materials whose bandgaps are ~ 2 eV at best. This can be overcome by deconvolving out a measured vacuum zero loss peak from the sample spectra allowing the study of bandgaps as low as 0.9 eV at energy resolutions of 0.22 eV. The detailed shape of the spectra processed in this way could not be explained using the basic *joint density of states* picture [Bruley and Brown (1987), Batson (1992)].

## 2. EXTENDED BETHE THEORY

Bethe theory describes the single electron inelastic scattering of fast charged particles. When this is applied to fast electrons impinging upon a thin foil the scattered intensity is found to be proportional to the product between the *joint density of states* (JDOS) and the matrix element which describes the transition of the crystal electron. For a bandgap of size $E_g$ and assuming parabolic bands the JDOS has an energy dependence of $(E - E_g)^{1/2}$ between critical points separated by an energy $E_g$. When the energy dependence of the matrix elements is examined it is found [Rafferty and Brown (1997)] that for indirect transitions the matrix element contributes a further factor of $(E - E_g)$. Thus the double differential scattering cross section can be written as:

$$\frac{d^2\sigma}{dEd\Omega} \propto \left(E - E_g\right)^{1/2} \quad \text{Direct}$$

$$\propto \left(E - E_g\right)^{3/2} \quad \text{Indirect}$$

The main consequence of this result is that we have decoupled the matrix elements for transitions that are direct (optical) and those that are indirect. To test this theory best the imaginary part of the dielectric function, $\varepsilon_2$, must be examined since the structure in $\varepsilon_2$ is produced by single electron scattering events.

## 3.    EXPERIMENTAL RESULTS

EELS spectra in the region of the bandgap have been acquired from five crystalline materials, MgO (direct, 7.8 eV), CVD diamond (indirect, 5.5 eV and direct 6 eV), cubic (indirect, 6.2 eV) and hexagonal (indirect, 5.2 eV) BN and GaAs (direct, 1.5 eV). All of the spectra were acquired using the same objective and collector apertures that had semi-angles of 9 and 7 mrad respectively and at an energy dispersion of 0.05 eV/channel. These apertures ensured that all excitations to the conduction band in the first Brillouin zone would be collected. Figure 1 shows the matrix-element-weighted JDOS as extracted from the spectra via a Kramers-Kronig transformation for MgO, CVD diamond, cubic and hexagonal BN. It can be seen that the onsets of the plots look very different. To test if the theory can explain the difference in the shapes of the onsets in the JDOS data a routine was developed which would provide the size of the bandgap and the nature of the transitions.

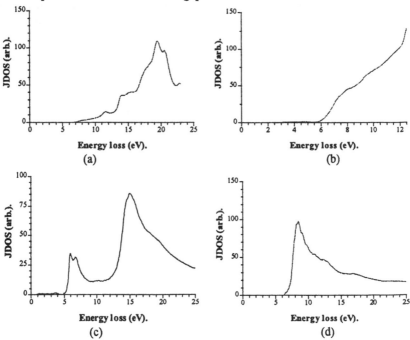

Fig. 1: The joint density of states extracted from the EELS spectra using a Kramers-Kronig transformation for (a) MgO, (b) CVD diamond, (c) hexagonal BN and (d) cubic BN.

The approximate position of the onset can be seen clearly in the JDOS data but the finite scaling of the data makes an accurate determination of the position difficult. An energy window of a few eVs is placed over the onset starting about 6 data points before the approximate position of the onset. This masked data is then plotted on a log-log plot with the start of the energy window as the origin and a trial size and position of the bandgap onset. A straight line $(Y = a + n \times X)$ is then fitted to this displaced data. The value of B and the correlation factor, $R^2$ of the fitting can then be extracted for this trial bandgap. This is then repeated for the next data point (trial bandgap) until we have gone past the approximate position of the bandgap onset. The value of 'n' and $R^2$ are then plotted as a function of the trial bandgap. The position of the maximum in $R^2$ will provide the value of the bandgap and the corresponding value of 'n' which will show if the bandgap is either direct (n = 1/2) or indirect (n = 3/2) in nature. Figure 2 shows such plots for MgO, CVD diamond, cubic and hexagonal BN. The agreement between the size of the bandgaps obtained via this technique and those in the literature is within 0.1 eV. In the case of CVD diamond it has been possible to detect both bandgaps of 5.5 eV and 6.1 eV. The values of the index 'n' for these materials have been reproduced and show the correct nature of the transitions between the critical points. The values of 'n' have been either 1/2 or 3/2 when $R^2$ has been maximised and the error in the value of 'n' is at a minimum.

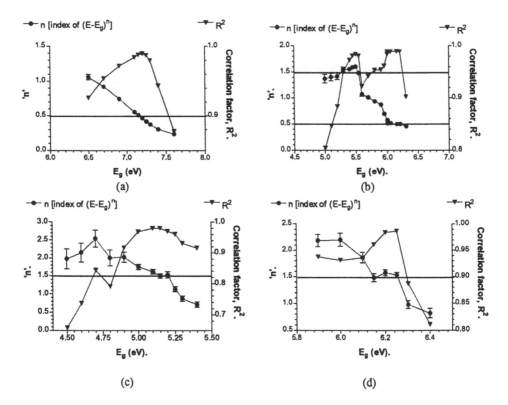

Fig. 2: Plots showing the index 'n' of the $(E-E_g)^n$ term and the correlation factor for the fitting procedure, described in the text, as a function of the position of the critical point for (a) MgO cubes, (b) CVD diamond, (c) hexagonal BN and (d) cubic BN.

350

Figure 3 shows the JDOS plot for GaAs. It was not possible to apply the above method to this data since there is a small number of data points covering the bandgap onset. Another problem with this data is that there is much structure separated by less than 1 eV. This structure can be related to the characteristic *camel back structure* of the conduction band centred on the $\Gamma$ point of the Brillouin zone.

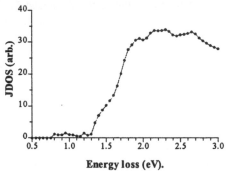

Energy loss (eV).

Fig. 3: The joint density of states for gallium arsenide. The positions of the peaks at higher energies correspond to the critical points of the camel back structure of the conduction band edge.

## 4.    CONCLUSIONS

It has been shown that to obtain the most information about the electronic structure from the bandgap region of an EELS spectrum a plot of the matrix-element-weighted JDOS must be extracted from the EELS spectrum via a Kramers-Kronig transformation. To find out about the nature of the transitions to a particular bandgap or critical point this data should be analysed using the routine set out in section 3. From the resulting plot the value of the index 'n' of the $(E - E_g)^n$ term and the value of the bandgap, $E_g$, can be obtained when the correlation factor $R^2$ is maximised. This method has been demonstrated on several materials to be robust and reliable. Thus detailed information about the electronic structure of semiconducting and insulating materials can be obtained using EELS.

## 5.    ACKNOWLEDGEMENTS

B Rafferty would like to thank De Beers Industrial Diamond Division for their financial support.

## REFERENCES

Batson P E (1992) "Transmission Electron Energy Loss Spectroscopy in Materials Science" Ed. Disko M M, Ahn C C and Fultz B, TMS publication (1992), pp 217.
Bruley J and Brown L M (1987) EMAG 1987, pp 107.
Rafferty B and Brown L M (1997) to be submitted to Physical Review B.

*Inst. Phys. Conf. Ser. No 153: Section 8*
*Paper presented at Electron Microscopy and Analysis Group Conf. EMAG97, Cambridge, 1997*
© 1997 IOP Publishing Ltd

# Application of sum rules in the interpretation of electron energy loss spectra from transition metal oxides

**N K Menon and J Yuan**

Cavendish Laboratory, Madingley Road, Cambridge CB3 OHE

**ABSTRACT**: In this paper, we model the linear dichroism of the $L_{2,3}$ absorption of Titanium in tetragonal symmetry as is relevant in the compound $TiO_2$. We develop a point charge model of the tetragonal potential, and use previous experimentally observed core level multiplet splittings to fix the values of the crystalline parameters necessary in the simulation . We then illustrate the use of the resultant difference spectra in determining ground state properties by application of a recently derived sum rule for the linear dichroism to obtain values of the electric quadrupole moment.

## 1. INTRODUCTION

Linear Dichroism (LD) is defined to be the difference in absorption cross-section for incident fields polarised parallel or perpendicular to a specific axis and this difference is related to the electronic structure of the system In antiferromagnetic materials, for example, the summed LD can be related the square of the magnetic moment of the ions in the solid.

Recently, a sum rule relating the integral of the linear dichroism from a core shell to the quadrupole moment of the ground state charge distribution has been derived. It has been shown (Carra et al (1993)), using arguments from angular momentum theory, that adding the integrated dichroic signal from two partners $J_{+,}$ of a spin orbit split edge and normalising it to the unpolarised spectrum yields

$$\frac{\int\limits_{J_{+}+J_{-}}(\mu_1+\mu_{-1}-2\mu_0)d\omega}{\int\limits_{J_{+}+J_{-}}(\mu_1+\mu_{-1}+\mu_0)d\omega} = \frac{<Q_{zz}>}{\ell(2\ell-1)h} = \frac{<\sum\limits_{i}(3\hat{\ell}_z^2-\ell(\ell+1))_i>}{\ell(2\ell-1)h} \tag{1}$$

In the above formula, $\mu$ denotes the absorption coefficient and $<Q_{zz}>$ is the expectation value of the quadrupole moment of the charge distribution of the incomplete shell, with angular momentum $\ell$ and $h = 2(2\ell+1)$ - n holes, to be probed. This is examined in this paper for a specific case.

## 2. TITANIUM IN TETRAGONAL SYMMETRY

We apply Eqn.(1) to the L- edge absorption of Titanium in tetragonal symmetry.The simulation involves the use of atomic multiplet theory (van der Laan et al (1991)), valid for this system in which the nature of the empty d-band states involved in this transition are localised. The transition probability as a function of energy loss ($\epsilon$) is, in the dipole limit, given in the Born approximation to be :

$$\Gamma(\epsilon) \propto \frac{1}{q^2}\sum_f < \Psi_f |\mathbf{q}.\mathbf{r}|\Psi_i > \delta(\epsilon - \epsilon_f + \epsilon_i) \qquad (2)$$

In the above expression, $\Psi_i / \Psi_f$ and $\epsilon_i$ , $\epsilon_f$ are the initial and final atomic wavefunctions and energies of the system respectively, found by diagonalisation of the corresponding initial and final state Hamiltonians and $\mathbf{q}$ is the momentum transfer associated with the fast incident electron.

The final state Hamiltonian includes the spin-orbit splitting of the 2p core hole and the p-d and d-d Coulomb and exchange interactions . The parameters associated with these terms are generated using a self consistent Hartree-Fock calculation. In addition, a ligand (or crystal) field is present, and is associated with the arrangement of anions around the relevant transition metal atom. In tetragonal symmetry, there exist three crystalline parameters asssociated with the crystal field: one octahedral component (Dq) and two associated with the axial distortion (Ds and Dt).

## 2.1.    Point Charge Model

A sytematic set of values for the crystalline parameters is necessary in order to produce a best fit to experiment. This is achieved here by describing the crystal field potential, V, as that due to an array of anions of charge -Ze in a distorted octahedral arrangement surrounding the central atom, with in plane interatomic distance a and out of plane distance b. By expansion of the angular part in terms of Spherical harmonics, we find that :

$$V = V_o + V_T , \text{ where}$$

$$V_T = 2Ze^2r^2\left(\frac{1}{b^3} - \frac{1}{a^3}\right)C_0^{(2)}(\theta,\varphi) + 2Ze^2r^4\left(\frac{1}{b^5} - \frac{1}{a^5}\right)C_0^{(4)}(\theta,\varphi)$$

$$V_o = \frac{7Ze^2r^4}{2a^5}\{C_0^{(4)}(\theta,\varphi) + \sqrt{\frac{5}{14}}(C_4^{(4)}(\theta,\varphi) + C_{-4}^{(4)}(\theta,\varphi)\} \qquad (3)$$

where the $C_q^{(k)}(\theta,\varphi)$ are normalised spherical harmonics of rank k and r is the distance from the origin. By consideration of the breaking of degeneracy of the 3 $t_{2g}$ levels upon distorting a perfect octahedron, we find that :

$$Dq = \frac{e^2}{6a^5}Z<r^4>$$

$$Dt = \frac{2e^2}{21}\left(\frac{1}{a^5} - \frac{1}{b^5}\right)Z<r^4> \quad \text{and} \quad Ds = \frac{2e^2}{7}\left(\frac{1}{a^3} - \frac{1}{b^3}\right)Z<r^2> \qquad (4)$$

The simulation of the L-edge spectrum is conducted by choosing the value of Dq that best fits the experimental $L_2$ edge splitting ( de Groot et al (1990) ), with both Ds and Dt set to zero. This is valid in the limit where the tetragonal distortion is small. By inspection of the formulae in (4), this constrains the term $Z < r^4 >$ and therefore also fixes Dt. The parameter Ds is then varied so as to produce a best fit to experiment. The resulting spectrum and the relevant values of the crystal parameters are shown in Fig. 1, with Lorentz broadenings introduced to simulate lifetime and solid state effects similar to those given by de Groot et al (1990).

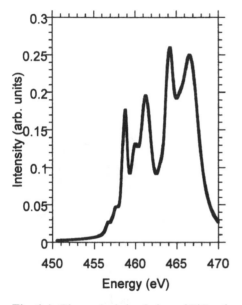

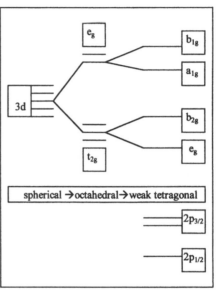

**Fig. 1a).** Theoretical simulation of Ti L-edge. The best fit values of the crystal parameters are: Dq = 0.21, Dt = 0.08 and Ds = 0.17 eV.

**Fig. 1b).** Energy level diagram for spin-orbit split 2p state , showing also crystal field split 3d levels.

## 2.2.   Linear Dichroism and quadrupole moments.

We now extend the analysis by consideration of the linear dichroism of the EEL spectra for Titanium in oxidation states 2+,3+ and 4+. The resulting theoretical linear dichroism and polarisation summed spectra are shown in Fig. 2. The (normalised) integrated dichroism is used to extract the quadrupole moment using (1) and the values were found to be $1.28 \times 10^{-5}$ , 0.06 and 0.103 for $Ti^{4+}$, $Ti^{3+}$ and $Ti^{2+}$ respectively.

The relative magnitude of the quadrupole moments can be understood by consideration of the relative ground state occupancy of the lowest lying 3d levels. When an octahedral field is distorted axially, the threefold degeneracy of the $t_{2g}$ levels is broken, with the lowest lying state being the twofold ($d_{xz}$ and $d_{yz}$) degenerate $e_g$, as illustrated in Fig. 1b. For Ti in the 4+ oxidation state, there are no 3d electrons in the ground state, and hence we expect the quadrupole moment to be zero. The non-zero theoretical value is ascribed to numerical error in the simulation. In the case of $Ti^{3+}$ and $Ti^{2+}$, the respective single and double occupancy of $e_g$ (for which $< \ell_z^2 - 2 >= -1$) would result, by considering the RHS of Eqn (1), in a ratio of moments of 0.44. The actual value from the simulation of 0.58 and this difference is ascribed primarily to the fact that in $Ti^{2+}$, the presence of the d-d electrostatic interaction means that $d_{xz}$ and $d_{yz}$ are no longer eigenstates. As a result, the true ground state is formed from a mixing of the crystal field split states, each with different values of $< \ell_z^2 >$.

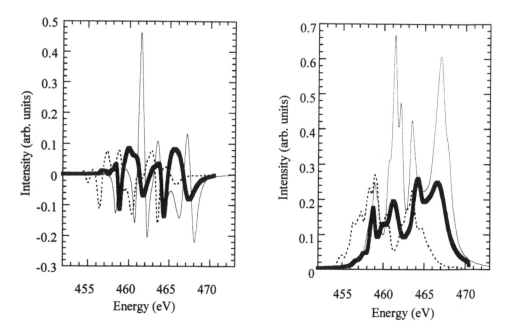

**Fig. 2.** Linear dichroism (left) and summed (right) spectra for for $Ti^{4+}$(bold line), $Ti^{3+}$(solid line) and $Ti^{2+}$(dotted line).

## 3.    CONCLUSION

We have demonstated the validity of Atomic Multiplet theory in the modelling the multiplet structure in the L-edge of Titanium compounds in tetragonal symmetry, where the agreement with experiment is good. The potential has been described by a point charge model, with the three crystalline parameters fixed by comparison with experiment. The linear dichroism is modelled, and its use in understanding the ground state electronic structure is illustrated by the use of a sum rule to extract values for the electric quadrupole moment.

## ACKNOWLEDGEMENTS

We would like to thank the EPSRC for funding this work.

## REFERENCES

Carra P et al 1993 Physica B192 182
van der Laan G and Thole B T 1991 Phys. Rev. B 43, 13401
de Groot F M F, Fuggle J C, Thole B T and Sawatzky G A 1990 Phys. Rev. B41 928

*Inst. Phys. Conf. Ser. No 153: Section 8*
*Paper presented at Electron Microscopy and Analysis Group Conf. EMAG97, Cambridge, 1997*
© *1997 IOP Publishing Ltd*

# Optimising ALCHEMI experiments

**N Jiang and I P Jones***

School of Metallurgy and Materials
* also IRC in Materials for High Performance Applications
The University of Birmingham, Birmingham B15 2TT, UK

**ABSTRACT:** ALCHEMI experiments are used to characterise sublattice occupancies in compound crystals. The accuracy of the experiments increases as the channelling of the beam electrons along special sublattice planes increases. In this paper we predict theoretically the optimum specimen orientation, beam voltage and foil thickness for systematic ALCHEMI experiments. All these predictions have been verified by experiment.

## 1. INTRODUCTION

ALCHEMI (see Spence 1992 for a review) is an experimental technique for determining the chemical occupancies of the sublattices in ordered crystals. In its systematic form the crystal is tilted about a superlattice reflection Bragg condition which alters the number of beam electrons channelled along the crystal planes of each sublattice. The ratios of the characteristic X-ray yields from the various elements present, or their EELS edges, are monitored and converted via a simple construction (Jones 1995, Hou, Jones and Fraser 1996) or formula (Krishnan 1988) into sublattice occupancies. Absolute sublattice occupancies require one additional independent measurement or assumption (e.g. Jones 1995). How good the ALCHEMI experiment is is dependent on how strong the channelling is and the differences in electron intensity on the various sublattices which are thereby achieved. In this paper we seek to maximise these differences as a function of specimen orientation, beam voltage and foil thickness. We have not considered axial channelling and delocalisation effects, which latter, because of the poor counting statistics associated with most ALCHEMI experiments, is not currently one of the more serious problems facing ALCHEMI.

## 2. METHOD OF CALCULATION

To simplify the analysis, we suppose there are only two different sublattices in an ordered compound where superlattice systematic rows consist of alternating superlattice and fundamental reflections and planes of the two sublattices alternate and are equally spaced. Neglecting delocalisation we take the degree of channelling R to be the difference in thickness averaged electron intensities ($\bar{I}$) between the two planes, A and B,

$$R = \bar{I}_A - \bar{I}_B \qquad (1)$$

To calculate $\bar{I}_A$ and $\bar{I}_B$ we have used standard n-beam diffraction theory as described by Hirsch, Howie, Nicholson, Pashley and Whelan (1977). The 'absorbed' electrons are scattered into plane waves. These excite X-rays according to the overall composition of the crystal. The subroutine ATOM (Bird and King 1990) has been used for the anomalous absorptive ratio

$\xi_g / \xi'_g$ . Normal absorption $\xi'_0$ is quite important in this context: we have set $\xi_g / \xi'_0$ to be 0.4.

To facilitate comparison between different beam voltages and atomic number elements, we have used reduced parameters (Jones 1976) in which distances parallel to the electron beam are expressed in terms of the principal extinction distance and the level of dynamical interaction along the row is represented by the dimensionless parameter $w_s = g^2/2U_g$ . $w_s$ describes the effects of beam voltage and the average atomic number of the specimen.

## 3. FEW BEAM APPROACH

According to two-beam theory (Hirsch et al. 1977), the channelling effect can be expressed as

$$R = F(w) \cdot T(w,t) \tag{2}$$

where

$$F(w) = -\frac{2w}{1+w^2} \tag{3}$$

and

$$T(w,t) = 1 - \frac{\sin(2\pi\sqrt{1+w^2}\,t/\xi_g)}{2\pi\sqrt{1+w^2}\,t/\xi_g} \tag{4}$$

in which t is foil thickness and w is the dimensionless deviation parameter.

In eqn (4) only T(w,t) involves the foil thickness and so the function T(w,t) is called the thickness factor. The variation of thickness factor with foil thickness is shown in Fig.1(a). The maximum value of the thickness factor can be obtained by a simple mathematics process and is just at the $0.715\xi_g$ ($\approx 3/4\xi_g$) when w = 0. Considering the de-channelling effects (the derivation can be found in Jiang 1997), as shown in Fig.1(b), only the first peak remains: the others are smeared out, and the degree of channelling decreases significantly.

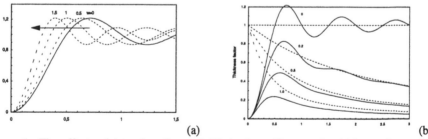

(a)                                                                                    (b)

Figure 1   The effects of (a) orientation and (b) de-channelling on the thickness factor. The thickness axis is in units of $\xi_g$. In (a) the orientation is characterised by w as indicated. In (b) the $\xi_g/\xi'_0$ values are also indicated. The full lines are the coherent two-beam approach and the dotted lines are the incoherent two-beam approach.

Fig.2 shows how the channelling effect depends on w. Overall, the channelling effect (like most diffraction phenomena) is a maximum when w ~ ±1. Many beam effects on the channelling can be obtained simply by a three-beam approach. A general solution for the three-beam eigen equation is necessary and can be obtained, but the result is too lengthy and complicated to be of much use. In ALCHEMI, the symmetry orientation is a more useful special configuration and can give some very useful results. By a derivation similar to that for two beams, the channelling at the symmetry condition using three beams has the form of:

$$R = \sqrt{2} \cdot F(w_m) \cdot T(w_m,t) \tag{5}$$

in which $w_m$ is called the "effective dynamical deviation parameter" and has the form

$$w_m = s\xi = \frac{1}{\sqrt{2}}\left(-w_s + \frac{\xi_g}{2\xi_{2g}}\right) \tag{6}$$

Since the channelling should have a maximum at $w \sim \pm 1$ as indicated by two-beam approach, the dynamical interaction parameter $w_s$ corresponding to the maximum channelling should be

$$w_s^{max} = \sqrt{2} + \xi_g/2\xi_{2g} \tag{7}$$

Hence the strongest channelling effect could be always at the symmetry condition if $w_s$ is smaller than $w_s^{max}$, as shown in Fig.3. The orientation is expressed via the transverse component of **K** along the normal to the atom planes, $K_t = \mathbf{K} \cdot \mathbf{g}/g$.

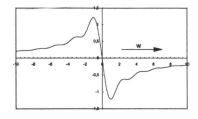

Figure 2  Channelling effect dependence on w value under two-beam approach without de-channelling. The ordinate represents the channelling R. The foil thickness is set to be $t/\xi_g = 0.5$.

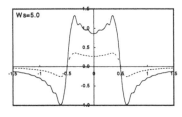
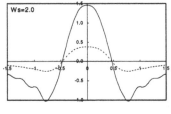

(a)                                                                    (b)

Figure 3  Three-beam approach showing the effect of the dynamical interaction parameter $w_s$ on the channelling variation. The abscissa represents the orientation, which is characterised by $K_t$. The full lines denote the results in absence of a de-channelling effect, and the dotted line denote the results in the presence of de-channelling effect $\xi_g/\xi'_0 = 0.5$. In (a) $w_s > w_s^{max}$. In (b) $w_s < w_s^{max}$.

From eqn (6), if $w_s^c = \xi_g/2\xi_g$, the channelling effect could disappear, and, if $w_s$ becomes much smaller, the channelling effect reverses. Hence at a critical beam voltage the channelling effect reverses. It should be noted that the critical voltage here is different from the usual critical voltage in electron diffraction.

So, as a result, the degree of the maximum channelling will increases as $w_s$ decreases if $w_s > w_s^{max}$ and it will decrease if $w_s^{max} > w_s > w^c$.

## 4.    EXPERIMENTAL RESULTS AND DISCUSSIONS

To examine the theoretical predictions, experimental measurements of the channelling dependence on the experimental parameters were carried out on $\gamma$-Ti$_{44}$Al$_{56}$ (the TEM specimen was kindly supplied by Dr Lee). In the gamma phase, several superlattice reflections can separate the distinguished sublattice planes, but only (001) and (110) are examined in this study.

Both the measured and calculated dependence of channelling on orientation for the (110) at 200kV are shown in Fig.4(a). All these calculated results are qualitatively in agreement with the measured ones. The strongest channelling occurs at the symmetry orientation. However, for the (001) systematic row, the channelling disappears around the symmetry

358

orientation, as shown in Fig.4(b). This is because $w_s$ (=1.51) for (001) in $\gamma$-Ti$_{44}$Al$_{56}$ at 200kV is quite near to the critical voltage ($w_s{}^c = 1.45$).

Fig.5(a) shows both the measured and calculated dependence of the channelling on foil thickness. The variation of the channelling can be divided into three fairly distinct thickness regions. In very thin foils, the channelling is very weak, and it almost disappears when the thickness tends to zero. In medium thickness foil, the largest channelling can be obtained. When the foil becomes thick, the channelling decreases slightly with foil thickness increase. The corresponding three thickness regions can also be observed at 400kV, as shown in Fig.5(b). Generally, the channelling at 400kV is stronger than that at 200kV. This is because $w_s{}^{200kV} > w_s{}^{400kV} > w_s{}^{max}$.

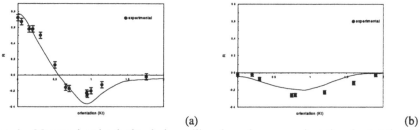

(a)                                                                 (b)

Figure 4.   Measured and calculated channelling dependence on orientation for (a) the (110) systematic row and (b) the (001) row in $\gamma$-TiAl at 200kV. The specimen thickness is about 123.2nm. The normal absorption coefficient ratio, $\xi_g/\xi'_0 = 0.40$.

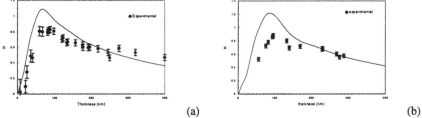

(a)                                                                 (b)

Figure 5.   Measured and calculated channelling dependence on foil thickness for the (110) systematic row in $\gamma$-TiAl at (a) 200kV and (b) 400kV. The orientation is at the symmetry condition. The normal absorption coefficient ratio $\xi_g/\xi'_0 = 0.40$.

## REFERENCES

Bird D and King Q A 1990 Acta Cryst. A **46**, 202
Hirsch P B, Howie A, Nicholson R B, Pashley D W and Whelan M J 1977 Electron Microscopy of Thin Crystals (2nd edn, New York: Krieger)
Hou D H, Jones I P and Fraser H L 1996 Phil. Mag. A **74**, 741.
Jiang N 1997 PhD Thesis, The University of Birmingham.
Jones I P 1976 Phil. Mag. **33**, 311
Jones I P 1995 "How ordered are intermetallics?" celebration for R E Smallman (published by IOP)
Krishnan K M 1988 Ultramicroscopy **24**, 125
Spence J C H 1992 Electron Diffraction Techniques **vol 1**, 465.

*Inst. Phys. Conf. Ser. No 153: Section 8*
*Paper presented at Electron Microscopy and Analysis Group Conf. EMAG97, Cambridge, 1997*
© *1997 IOP Publishing Ltd*

# New aspects in the theory of energy loss by channelled electrons

**M Nelhiebel**[†*], **P Schattschneider**[*†], **B Jouffrey**[†]

[†]Laboratoire MSSMat, Ecole Centrale Paris, Gde Voie des Vignes, F-92295 Châtenay-Malabry
[*]Institut für Angewandte und Technische Physik, Technische Universität Wien, Wiedner Hauptstrasse 8-10/137, A-1040 Wien.

**ABSTRACT:** We discuss electron channelling using the concept of the mixed dynamic form factor. The site specific information of an inelastic scattering experiment is shown to be related to interference terms. On this theoretical foundation we may simulate the expected effects and present first experimental results.

## 1.    INTRODUCTION

Electron channelling experiments are used to determine the atomic structure of crystals. In EDX-analysis, the method is known as ALCHEMI and rather commonly used (e.g. Nüchter and Sigle 1995). In electron spectroscopic techniques, however, ELCE investigations are less frequently reported in literature, despite their great potential (Taftø and Krivanek 1982). This may be due to the complex theoretical background inhibiting easy quantitative modelling of inelastic intensity under various channelling conditions.

In the following we introduce the concept of the mixed dynamic form factor (MDFF) to describe the inelastic scattering cross sections in crystals, using analogies to the known channelling behaviour of the fast electron. We shall see that site specific information is retained in the so called indirect terms of the MDFF. Using the dipole approximation, evaluation of the scattering cross section is rather simple, and one may find out experimental conditions where the indirect terms of the MDFF (and thus the site specific effects) dominate.

## 2.    THEORY

Dynamical theory of diffraction gives the correct description of a fast electron passing a crystal as a superposition of $N$ Bloch waves, each weighted with a coefficient $\varepsilon^j$, where $N$ is the number of reciprocal lattice vectors

$$\psi_{\mathbf{k}}(\mathbf{r}) = \sum_j \varepsilon^j \sum_{\mathbf{g}} C_{\mathbf{g}}^j e^{i\mathbf{k}_{\mathbf{g}}^j \mathbf{r}} . \tag{1}$$

Each Bloch wave is itself a coherent superposition of plane waves travelling in direction $\mathbf{k} + \mathbf{g}$, weighted with the coefficients $C_{\mathbf{g}}^j$.

We consider a two-beam case in special geometry where one Bloch wave, say type 1, dominates, i.e. $\psi(\mathbf{r}) \approx \varepsilon^1 \psi_1(\mathbf{r}) = \varepsilon^1 [C_0^1 \exp(i\mathbf{k}^1\mathbf{r}) + C_{\mathbf{g}}^1 \exp(i(\mathbf{k}^1 + \mathbf{g})\mathbf{r})]$. Then, the probability of presence of the fast electron as a function of $\mathbf{r}$ is ($|\varepsilon^1|^2 \approx 1$)

$$I(\mathbf{r}) \approx |\psi_1(\mathbf{r})|^2 = |C_0^1|^2 + |C_{\mathbf{g}}^1|^2 + 2\mathrm{Re}[C_0^1 C_{\mathbf{g}}^{1*} e^{-i\mathbf{g}\mathbf{r}}] \tag{2}$$

360

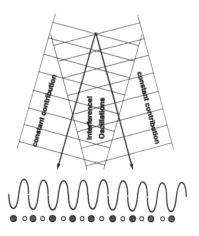

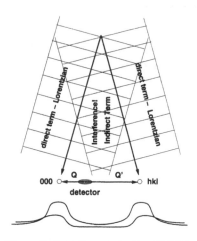

Figure 1: Channelling in the two beam case.     Figure 2: Interference in the MDFF.

The interference between the two coherent plane waves $C_0^1 \exp(i\mathbf{k}^1\mathbf{r})$ and $C_{\mathbf{g}}^1 \exp(i(\mathbf{k}^1+\mathbf{g})\mathbf{r})$, introduced in evaluating the module squared, induces intensity oscillations across the atomic planes normal to $\mathbf{g}$, see Fig. 1. This phenomenon is called channelling.

We now include inelastic scattering in this picture. Classically, the fast electron passes from a plane-wave state $|\mathbf{k}_i, E_0\rangle$ to an other plane-wave state $|\mathbf{k}_f, E_0 - E\rangle$. This may be expressed by defining an inelastic scattering amplitude $f_E(\mathbf{q})$ for ionization processes with energy loss $E$ and momentum transfer $\hbar\mathbf{q}$, $\mathbf{q} = \mathbf{k}_i - \mathbf{k}_f$, whose square is the scattering cross section. Now we add a second incident plane wave $C_{\mathbf{g}}^1 \exp(i(\mathbf{k}^1+\mathbf{g})\mathbf{r})$ coherently, letting the final state of the fast electron remain a single plane wave. Then there are *two* transitions from each of the two incident waves to the final one, represented by scattering amplitudes $f_E(\mathbf{Q})$, $f_E(\mathbf{Q}')$, cmp. Fig 2. The cross section will be the square of their sum,

$$\frac{\partial^2\sigma}{\partial\Omega\partial E} \approx \left|C_0^1 f_E(\mathbf{Q}) + C_{\mathbf{g}}^1 f_E(\mathbf{Q}')\right|^2 = |C_0^1|^2|f_E(\mathbf{Q})|^2 + |C_{\mathbf{g}}^1|^2|f_E(\mathbf{Q}')|^2 + 2\mathrm{Re}[C_0^1 C_{\mathbf{g}}^{1*} f_E(\mathbf{Q}) f_E(\mathbf{Q}')^*] \tag{3}$$

Formally, this is equivalent to Eq. (2). The inelastic intensity is composed of two direct terms (the usual Lorentzians, centered at the Bragg spots) and one indirect term, stemming from the coherent superposition of the two incident waves. By the analogy it is clear that site specific information will be retained in the interference term of the cross section.

If the electron is described by a superposition of plane waves Eq. (1), we have to consider transitions from any $\mathbf{k}_{\mathbf{g},i}^j$ before interaction to any $\mathbf{k}_{\mathbf{h},f}^l$ after interaction. As in the example before, we define the scattering amplitude $f_E(\mathbf{Q})$, $\mathbf{Q} = \mathbf{k}_{\mathbf{g},i}^j - \mathbf{k}_{\mathbf{h},f}^l = \mathbf{q} + \mathbf{g} - \mathbf{h} + \delta\gamma^{jl}$, sum up all scattering amplitudes for the various plane-waves transitions labeled by $\mathbf{Q}$ and take the square to obtain

$$\left(\frac{\partial^2\sigma}{\partial\Omega\partial E}\right)_{\mathrm{cryst}} = \left|\sum_{\mathbf{Q}} u_{\mathbf{k}_i\mathbf{k}_f}(\mathbf{Q}) f_E(\mathbf{Q})\right|^2 = \sum_{\mathbf{Q}\mathbf{Q}'} u_{\mathbf{k}_i\mathbf{k}_f}(\mathbf{Q}) u_{\mathbf{k}_i\mathbf{k}_f}(\mathbf{Q}')^* f_E(\mathbf{Q}) f_E(\mathbf{Q}')^*, \tag{4}$$

where the weight $u_{\mathbf{k}_i\mathbf{k}_f}(\mathbf{Q})$ stands for the excitation strengths of the two plane waves linked by $\mathbf{Q}$. It depends on the orientations of incident and outgoing wave to the crystal and is calculated using the dynamical theory of diffraction. We define

$$f_E(\mathbf{Q}) f_E(\mathbf{Q}')^* = \frac{(2\pi)^4}{a_0^2} \frac{4\pi e^2}{Q^2 Q'^2} S(\mathbf{Q}, \mathbf{Q}', E) \tag{5}$$

with the MDFF (Kohl and Rose 1985)

$$S(\mathbf{Q}, \mathbf{Q}', E) = \sum_f \langle i|e^{i\mathbf{QR}}|f\rangle\langle f|e^{-i\mathbf{Q}'^*\mathbf{R}}|i\rangle \delta(E - E_f + E_i), \qquad (6)$$

expressing the probability that by excitation of the target the fast electron transits from $|\mathbf{k}_{gi}^j, E_0\rangle$ to $|\mathbf{k}_{hf}^l, E_0 - E\rangle$, whereas the target state is changed from $|i\rangle$ at energy $E_i$ to any $|f\rangle$ at energy $E_f$. The inelastic cross section in crystals Eq. (4) is thus a weighted sum over direct ($\mathbf{Q} = \mathbf{Q}'$) and indirect ($\mathbf{Q} \neq \mathbf{Q}'$) terms of the MDFF (Schattschneider et al 1996), in generalization of Eq. (3).

Considering inner-shell excitations we approximate the crystal by a symmetric arrangement of independent atoms, one of which will be ionized in the interaction. That is, we substitute the target states $|i\rangle$, $|f\rangle$ by atomic wave functions $|\varphi\rangle$, $|\varphi'\rangle$ centered about atomic positions $\mathbf{a}$, and sum over all interacting atoms (Nelhiebel et al 1996). The crystal sum may be split up into a lattice sum and a sum over the unit cell, and one eventually obtains

$$\left(\frac{\partial^2 \sigma}{\partial\Omega\partial E}\right)_{\text{cryst}} = \sum_{\mathbf{QQ}'} w_{\mathbf{k}_i\mathbf{k}_f}(\mathbf{Q}, \mathbf{Q}') K_{\mathbf{QQ}'} \frac{1}{Q^2 Q'^2} S_{\text{atom}}(\mathbf{Q}, \mathbf{Q}', E). \qquad (7)$$

The weights $w_{\mathbf{k}_i\mathbf{k}_f}(\mathbf{Q}, \mathbf{Q}')$ differ from those in Eq. (4), including now thickness and absorption effects. In the three other terms we neglect the small differences between branches of the dispersion surfaces $\delta\gamma^{jl}$, i.e. $\mathbf{Q}$ and $\mathbf{Q}'$ may just differ by a reciprocal vector $\mathbf{G}$. $S_{\text{atom}}(\mathbf{Q}, \mathbf{Q}', E)$ is Eq. (6) evaluated with atomic wave functions and yields in the small angle (dipole) approximation

$$S_{\text{atom}}(\mathbf{Q}, \mathbf{Q}', E) = \mathbf{QQ}'h(E). \qquad (8)$$

Note that in this approximation angular part and energy-dependent part may be separated. $K_{\mathbf{QQ}'}$ stems from the sum over interacting atoms in the unit cell

$$K_{\mathbf{QQ}'} = \frac{1}{N_u} \sum_{\mathbf{u}}^{N_u} e^{i(\mathbf{Q}-\mathbf{Q}')\mathbf{u}}. \qquad (9)$$

In the weighted sum over atomic MDFFs Eq. (7) the indirect terms are thus multiplied with a structure dependent factor, whereas for the direct terms $K_{\mathbf{QQ}} = 1$. Site specific information is indeed restricted to the indirect terms, as expected from the analogy to electron channelling.

## 3.  SIMULATIONS AND EXPERIMENT

As a test specimen for ELCE experiments we have chosen GaAs in the 111 reflexion. With this reciprocal lattice vector the structure factor Eq. (9) changes sign for the indirect terms of the Ga and the As edge, respectively.

We are interested in experimental setups where the indirect terms of the MDFF are dominant. This will be the case in the middle between the two Bragg spots, where the direct Lorentzian terms are already attenuated. Using Eqs. (7) and (8) we simulate the *angular* part of the cross section. Fig. 3 shows a plot of the scattered intensity as a function of the scattering angle orthogonal to $\mathbf{g}$ for an excitation error $w = -0.75$, with and without contribution of the indirect terms of the MDFF. As indicated by the arrows, the indirect terms lead to an increase of Ga intensity. For As the contrary is the case. Recording the edge intensities in diffraction mode under orientations with $w = -0.75$ and $w = +0.75$, and denoting them Ga$_+$, As$_+$ for positive, Ga$_-$, As$_-$ for negative excitation error, we may build, experimentally and theoretically,

$$\frac{\text{Ga}_-}{\text{As}_-} : \frac{\text{Ga}_+}{\text{As}_+} = \frac{\text{Ga}_-}{\text{Ga}_+} \cdot \frac{\text{As}_+}{\text{As}_-}, \qquad \frac{I_-(\text{Ga})}{I_+(\text{Ga})} \cdot \frac{I_+(\text{As})}{I_-(\text{As})} = \frac{A_-(\text{Ga})h(\text{Ga})}{A_+(\text{Ga})h(\text{Ga})} \cdot \frac{A_+(\text{As})h(\text{As})}{A_-(\text{As})h(\text{As})} =: R$$

362

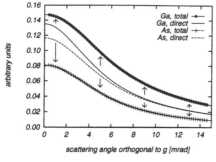

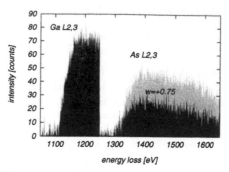

Figure 3: Simulation of the inelastic intensity as a function of scattering angle $\theta_y$ orthogonal to **g**. The scattering angle $\theta_x$ along
**g** is fixed to $\theta_{Bragg}$, and the excitation error
of the (111) spot is $w = -0.75$. Specimen
thickness 45 nm, primary energy 200 keV.

Figure 4: EELS spectra recorded in diffraction mode and normalized to the Ga edge in-
tensity. $\beta = 1.7$ mrad, $\alpha = 3$ mrad. $\theta_y = 9.5$
mrad, $\theta_x(+0.75) = 4.7$ mrad, $\theta_x(-0.75) =$
3.0 mrad.

Here, the (unknown) incident intensities in the two measurements cancel, and on the theoretical side the energy loss dependence cancels due to the separability of angular and energy-loss
part of the cross section in the dipole approximation. Thus the calculated product of elemental
ratios is determined exclusively by the angular parts $A_{+,-}$. If the spectra a normalized to the
Ga edge intensity, the measured ratio of As edges should be the calculated $R$, thereby proving
that the crystal structure assumed for simulations was correct. Fig. 4 shows one of some first
PEELS measurements for experimental parameters that would suggest $R = 1.7$, compared to
measured $R = 1.8$.

## 3.    CONCLUSION

We have shown that site specific information about crystals is retained in the interference terms of the MDFF, and pointed out the analogy to the classical interpretation of
channelling experiments. Making use of the dipole approximation, channelling experiments
may be simulated. First experimental tests suggest that a quantitative interpretation of ELCE
measurements is possible.

## REFERENCES

Kohl H and Rose H 1985 Adv. Electron. Electron Phys. **65**, 173-227
Nelhiebel M, Schenner M and Schattschneider P 1996 Ultramicroscopy **66**, 173-181
Nüchter W and Sigle W 1995 Philos. Mag. **71**, 165-186
Schattschneider P, Jouffrey B and Nelhiebel M 1996 Phys. Rev. B**54**, 3861-3868
Taftø J and Krivanek O L 1982 Phys. Rev. Let. **48**, 560-563

## ACKNOWLEDGEMENTS

This work was supported by the European Union under TMR contract No. ERBFMBICT961416

Inst. Phys. Conf. Ser. No 153: Section 8
Paper presented at Electron Microscopy and Analysis Group Conf. EMAG97, Cambridge, 1997
© 1997 IOP Publishing Ltd

# Evidence of a plasmon band in metallic multilayers

**T Neyer[†], GA Botton[‡], D Su[§], P Schattschneider[†], JPR Bolton[*], E Ziegler[+]**

[†]Institut für Angewandte und Technische Physik, TU-Wien, Wiedner Hauptstrasse 8-10
[‡]Department of Material Science and Metallurgy, Cambridge University, Cambridge CB2 3QZ
[§]Bereich Physikalische Chemie, Hahn-Meitner-Institut, Glienickerstr. 100, D–14109 Berlin
[*]Department of Physics, The Open University, Walton Hall, Milton Keynes MK7 6AA
[+]European Synchrotron Radiation Facility, BP 220, F-38043 Grenoble

**ABSTRACT:** We apply transmission electron microscopy in connection with electron energy loss spectrometry and imaging filtering in order to explore metallic multilayers in the low energy loss region.

## 1. INTRODUCTION

Metallic multilayers are widely used in modern technological applications and exhibit many novel features. In particular collective interface and surface modes may alter optical and electrical properties of the stacked device. Another essential parameter of the multilayer is the degree of interface perfection and diffusion length. We present an experimental investigation of multilayer collective modes, which has been performed on a dedicated scanning transmission facility (STEM) combined with an imaging filter device (GIF). We further estimate the steepness of the interface concentration profile using sideband analysis on a conventional transmission electron microscope (CTEM) combined with an electron energy loss spectrometer (EELS).

## 2. SIDEBAND ANALYSIS

Sidebands contain information of the modulation in wavelength of the multilayer, the quality of perfection and the coherency strain imposed by the epitaxial growth on a substrate with a different lattice parameter. For the metallic multilayers under consideration we have applied the useful model of an imperfect x-ray monochromator developed by McWhan (1988).

We shall employ the model of a sequence of $n_a$ monolayers of tungsten with a thickness $d_a$ interspersed by layers of silicon with thickness spacing of $c_0 + c_j$. $c_0$ is the average thickness and $c_j$ is the deviation of the $j$th layer from the average. This bilayer is then repeated $n$ times. For simplicity it is assumed that the scattering from silicon regions is negligible relative to that of tungsten. If we further assume that $c_j$ has a continuous Gaussian distribution with standard deviation $\sigma$ McWhan gives the total scattering intensity as a function of the wave number component $k_z$ parallel to the stacking direction ($F$ is the scattering power as a function of the wave number perpendicular to the interfaces)

$$I(k_z) = F^2(k_z)\frac{\sin^2(n_a d_a k_z/2)}{\sin^2(d_a k_z/2)}\left\{ n + 2\sum_{j=1}^{n-1}(n-j)\cos k_z j(n_a d_a + c_0) \times e^{-k_z^2\sigma^2/4}\right\}. \quad (1)$$

Fig. 1 shows the measured sideband maximum intensities of our sample up to the 7th order

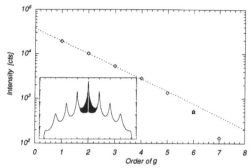

Figure 1: Logarithmic/linear plot of the measured elastic intensities of the superlattice satellites up the the seventh order, indicating an exponential decay (dashed). Diamonds correspond to measurement (triangle represents a thermal stability check). The inset shows the simulation result following Eq. 1 for the system W/Si $10\times$ (6:20 Å) with $\sigma = 0.9 \times d_W$ reproducing the experiment reasonably well.

decaying exponentially (dashed line) as proposed by McWhan's formula by setting $\sigma = 0.9 \cdot d_W$. Thus, the introduction of a Gaussian uncertainty distribution in the interface positions introduces a term equivalent to the Debye-Waller factor with $\sigma$ giving a statistical measure of the interface roughness and diffusion length.

## 3.  THEORY

Within a classical approach the specimen can be modelled as a stack of two alternating media characterized by their local dielectric functions $\varepsilon_1$ and $\varepsilon_2$. The dispersion relation for coupled interface modes in multilayers and superlattices can be found by solving Maxwell's equations.

According to the number of different regions available we end up with a large number of equations which have to fulfill conditions on each interface i.e. continuity of $E_y$ and $\varepsilon_{1,\,2}E_z$ ($z$ being the stacking direction and the $y$-axis lying within the interface). Bands of transverse magnetic (TM) and transverse electric (TE) modes exist. However, because our experimental arrangement is unable to resolve momenta to the left of the light line, the TE modes are not excited and the TM modes could be treated in the non-retarded limit. The dispersion relation $\omega = \omega(k_{\parallel}, k_z)$ for the TM modes in a periodic superlattice is then determined implicitly by

$$\frac{(\varepsilon_1 - \varepsilon_2)^2}{2\varepsilon_1\varepsilon_2} + \cosh(k_{\parallel}d) = \cos(k_z d), \tag{2}$$

with $d = d_1 + d_2$. Solutions of Eq. 2 for a W/Si superlattice with layer thicknesses 6:20 Å are shown in Fig. 2. The dielectric functions for this simulation are taken from Weaver et al (1981) and Palik (1985), carefully extended into the lower-half complex plane by expanding the smooth function $(\varepsilon + 1)/(\varepsilon - 1)$, as explained in Chen et al (1991).

From Eq. 2 we note that the bandwidth is largest at $k_{\parallel} = 0$, suggesting the band of interface plasmons may be most easily detected by fixing $k_{\parallel} = 0$ and varying $k_z$. The corresponding simulation for the plasmon band limits as a function of $k_z$ is shown in Fig. 3.

## 4.  EXPERIMENT

The experiment was performed using a dedicated VG HB501 scanning transmission electron microscope (STEM) combined with a Gatan imaging filter (GIF). The specimen was aligned so that the superlattice reflection vectors in the diffraction pattern entered a momentum selecting slit (therefore in the $k_z$ direction) which acted as a collector aperture for the spectrometer. The electrons entering the spectrometer through the slit were then energy analyzed while preserving the $k_z$ momentum distribution.

The STEM was used as a conventional TEM with parallel illumination (in selected area mode) with a stationary beam illuminating the specimen region of interest of diameter 200 nm. This results in an excellent momentum resolution limited by the small convergence of the beam

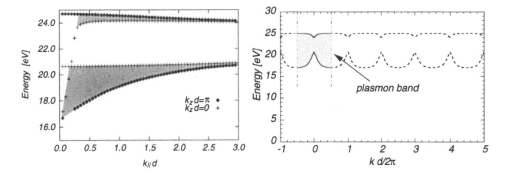

Figure 2: Plasmon bands as proposed by Eq. 2. The optical branch is defined at $k_z = 0$. The maximum splitting is given at $k_z = \pi/d$ (at $k_{\|} = 0$).

Figure 3: Interface plasmon band vs $k_z$ at $k_{\|} = 0$. The minimum band width appears at $k_z = 0$ and the maximum splitting is observed at $\pm\pi/2d$).

($< 0.1\,\mathrm{nm}^{-1}$). Additionally, as the instrument is equipped with a cold field emission gun, the energy resolution is about $0.5\,\mathrm{eV}$. Fig. 4 compares the $\omega$–$q$ plot of a W/Si $300 \times 6{:}20\,\text{Å}$ multilayer

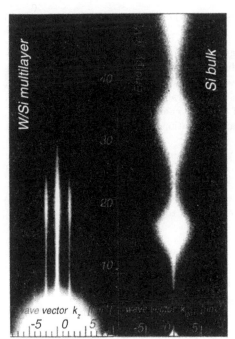

Figure 4: Comparison between $\omega$–$q$ plots of a periodic W/Si $300 \times (6{:}20\text{Å})$ multilayer (left) and bulk silicon (right). In the multilayer case the dynamical intensity range is dominated by the tail of the Lorentzians centered at the multilayer reflection spots.

Figure 5: Experimental evidence of the interface plasmon bands derived by normalizing the intensity map at the mid point of the band. The inset shows the result after the removal of the quasielastic plural scattering.

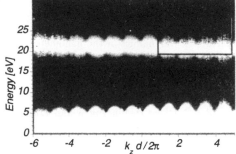

(left-hand image) with that for bulk silicon (right-hand image). In pure silicon one clearly sees the bulk plasmon at $17\,\mathrm{eV}$ which obeys a parabolic dispersion in $k_{001}$, together with multiple plasmon losses at $34\,\mathrm{eV}$ and $51\,\mathrm{eV}$. The W/Si multilayer shows very different behaviour: the stripes represent the high energy shoulder of the multilayer sidebands on which is superimposed a less intense plasmon shifted to approximately $20.5\,\mathrm{eV}$.

Inspection of the raw data showed that the dynamic range in the vicinity of the expected plasmon band, governed by the Lorentzian intensity distribution, was so high that the band was not visible. We therefore normalized the $\omega$–$q$ plane to a momentum trace parallel to $k_z$ in the middle of the band. This operation removes the dominant Lorentzian variation in $k_z$, unveiling a band structure (Fig. 5).

## 5. RESULTS

The comparison between experiment Fig. 5 and simulation Fig. 3 shows convincing similarities. The maximum splitting is obtained halfway between integers of $2\pi/d$ at $\pi/d$. Moreover, even the width of the gap at the $G$ reflections is well reproduced. This is surprising at the first glance, since the information contained in dispersion relations does not necessarily comprise excitation probabilities. In fact, this uncertainty, together with superposition of quasi-elastic plural scattering, results in a minor discrepancy between the simulation (which has a greater curvature at the lower limit of the band) and the experimental result (which has a greater curvature at the upper limit).

Similar to phonon bands, the plasmon band is restricted to the first Brillouin zone. Momentum transfers $|k_z| > \pi/d$ originate from Umklapp processes. Thus, the measured plasmon intensity in higher order Brillouin zones should be mainly given by quasi-elastic plural scattering. This contribution may be removed under the assumption that the measured intensity is the sum of shifted intensity maps weighted with the respective superlattice reflection intensity at zero energy loss, i.e.

$$I_\omega(k_z) \simeq I_\omega^{\mathrm{meas}}(k_z) - \frac{1}{I_0(0)} \sum_{n=-2}^{2} I_0(k_z + nG) \cdot I_\omega^{\mathrm{meas}}(k_z + nG), \tag{3}$$

with $I_\omega(k_z)$ describing the single scattering intensity map and $G$ being the reciprocal superlattice vector $G = 2\pi/d$. This subtraction procedure including multilayer sidebands up to the second order eliminates most of the intensity present in the $k_z$-range $|k_z| \geq \pi/d$. Nevertheless, as is shown in the inset of Fig. 5 there remains a measurable signal at values $|k_z| \geq \pi/d$. Possible explanations could be: a) nonlinearity of the CCD camera, b) aberrations in the GIF, c) inclination of the slit vs the satellites and d) approximations involved in the linear Ansatz of the correction procedure.

The experimental confirmation of interface plasmon bands in the eV-regime of metallic multilayers, together with the measurement of acoustic plasmons in the meV-regime (Gossard et al 1982) of semiconductor quantum well structures could have significant implications for nanosized multilayered structures. Furthermore, the importance of using electron microscopy as a routine probe of multilayer structures has been shown together with the applicability of classical electrodynamics in the nanometer scale.

## REFERENCES

Chen M, Bolton JPR and McGibbon AJ 1991 in EMAG91, Inst Phys Conf Ser **119** (Bristol)
Gossard AC, Olego D, Pinczuk A and Wiegmann W 1982 Phys. Rev. B **25**, 7867
McWhan DB 1988 Physics, Fabrication and applications of multilayered structures, eds P Dhez and C Weisbuch (New York: Plenum Press) pp 67–92
Palik ED 1985 Handbook of optical constants (New York: Academic Press)
Weaver JH, Krafka C, Lynch DW and Koch EE 1981 Physics Data, ed 18-2, (Karlsruhe: Fachinformationszentrum EPM Gmbh)

## ACKNOWLEDGEMENTS

The author gratefully acknowledged financial assistance by the Austrian Fonds zur Förderung wissenschaftlicher Forschung, project P10408-PHY and the British Council (Austria).

*Inst. Phys. Conf. Ser. No 153: Section 8*
*Paper presented at Electron Microscopy and Analysis Group Conf. EMAG97, Cambridge, 1997*
© *1997 IOP Publishing Ltd*

# In situ observation of small onion formation under electron irradiation of turbostratic BC$_2$N and turbostratic BN

**O Stéphan[#][*], Y Bando[#], C Dussarrat[#], K Kurashima[#] and T Tamiya[#]**

[#]National Institute for Research in Inorganic Materials, Namiki 1-1, Tsukuba, Ibaraki 305, Japan
[*]Laboratoire de Physique des Solides, Université Paris-Sud, 91405 Orsay, France

ABSTRACT:    Turbostratic BC$_2$N and turbostratic BN samples were exposed to intense irradiation regimes in a high resolution electron microscope equipped with a field emission gun. EELS analysis was used to check the sp$^2$-bonding type and the composition of the BC$_2$N starting materials. In both experiments, we observed a tendency for the basal planes to curl and form onion-like features. Based on the hypothesis that the onion formation occurs at a constant composition, a structural model is proposed for the innermost shell of a BC$_2$N onion. The overall polyhedral shape of the BN cages is explained within the frame of the octahedral model previously proposed for BN analogs to fullerenes.

## 1.    INTRODUCTION

The discovery of C fullerenes (Kroto 1985) and C nanotubes (Iijima 1991) grown in the vapor phase rose the revolutionary idea that graphite and diamond might not be the most stable allotrope form for atom assemblies of limited size. The formation of carbon onions (Ugarte 1992) in the condensed phase from the irradiation of graphitic polyhedral particles with an intense electron beam gave further evidence that spherical carbon network can be favored under high temperature and strong irradiation regimes. Recently, the synthesis of BN (Goldberg 1996), BCN (Stéphan 1996) nanotubes and closed shell nanoparticles pointed out that the idea already admitted for carbon that for a small assembly of atoms the network bents and curls into a close structure in order to eliminate the highly unfavorable dangling bonds had to be generalized to other layered materials. But in spite of theoretical predictions, so far there has been no experimental evidence for the stability of B-N and B-C-N analogs of buckminster fullerenes. Strong irradiation in a high resolution electron microscope allows structural fluidity. An atomic rearrangement of the structure is caused by the momentum transfer of high energy-electrons to the nuclei (knock-on collisions). We exposed turbostratic BC$_2$N samples and turbostratic BN to intense irradiation to study the ability of the hybrid network to include non hexagonal member rings and form curved structures.

## 2.    EXPERIMENTAL

In the experiments described here, two starting materials were used : a novel turbostratic BN sample with wide and ordered interlayer spacing (β-tBN) and a turbostratic BC$_2$N material submitted to high pressure and high temperature conditions. The observations were carried out in a field emission 300 kV analytical electron microscope (Jeol 3000F) and electron energy-loss data from both starting materials and onion features were recorded using a Gatan Parallel EELS detector. The energy resolution was estimated to be 0.8-0.9 eV.

## 3.    RESULTS AND DISCUSSION

### 3.1    BC$_2$N irradiation experiments

Local EELS analysis performed on the BC$_2$N starting materials show the simultaneous presence of boron, carbon and nitrogen. Fig.1a shows a typical EELS spectrum in the range of energy of B, C and N K core-edges from a turbostratic area, the onsets are respectively located at 188 eV, 284 eV and 401 eV. Deviations from BC$_2$N composition were found for amorphous areas while a constant BC$_2$N stoichiometry was worked out for turbostratic areas in both samples. Each core edge fine structure consists of a sharp π* peak and a well resolved σ* band characteristic of graphitic structures. This attests that all three atomic species are arranged in planes of graphite-like hexagonal rings.

368

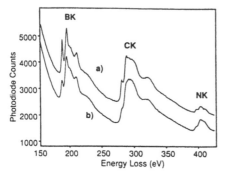

Fig. 1, electron energy-loss spectrum in the range of B, C and N -K core-edges from a) a turbostratic area (HT-HP sample) before irradiation, b) a large and onion with incomplete closed shells.

After 15 minutes of irradiation of thin turbostratic regions, curled onion like structures appear occasionally in the irradiated areas. Very rarely, we observed the formation of and isolated nested fullerenes. Fig. 2 is a time sequence of 2 pictures taken after 2 minutes of irradiation intervals showing the formation of a small nested fullerene. First the formation of an elongated structure with 7 layers and a rather large cavity is formed at the surface of curved graphitized area (first picture). Two minutes later (second picture), this feature was displaced to the edge of the specimen, which indicates that an isolated object that adheres at the surface by van der Waals interaction was formed. The closing of the very internal shell of this 6 layer nested fullerene is seen and its dimension fits approximately with that of a $C_{60}$.

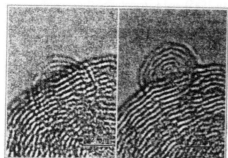

Fig. 2, formation of a small nested fullerene at the graphitized surface of a large onion. Both pictures were taken after 2 minutes of irradiation interval. Scale bar is 2 nm.

EELS analysis on large onion like structures (with incomplete closed shells) reveals that no change in the composition occurs during the irradiation process. Broadening of the $\pi^*$ peak and partial loss of the well resolved $\sigma^*$ fine structure (see Fig. 1b) accounts for the reduced ordering of the irradiated structure. No analysis could be performed on small nested fullerenes since they happened to be destroyed when exposed to the hostile analysis operating conditions. However, in the following discussion, we will make the hypothesis that the formation of small nested fullerenes is driven at a constant composition.

The transposition of carbon onion formation to that of B-C-N hybrid analogs is a puzzling case since structural stress imposed by the different length of the bonding, bonding energies (B-N and C-C are the most energetically favorable bonds whereas N-N and B-B bonds are very weak), composition are as many additional parameters that determine the stability of the closed structures. Since the innermost shell displays a spherical-like image with a diameter close to that of a $C_{60}$, we focused on a model based on the structure of that of a $C_{60}$ (20 hexagons and 12 pentagons disjoint) with the formula $B_{15}N_{30}N_{15}$. More, this model arise directly from the closing of a $BC_2N$ hexagonal network as proposed by Liu and co-workers (1989) within the frame of a ordered model. Among the three ordered structures proposed for $BC_2N$, we chose the most favorable one, according to these authors, thanks to its low structural stress and optimized number of C-C and B-N bonds. The resulting structure is composed of "lines" of B-N and C-C bonds (see Fig. 4). When introducing pentagons in the hexagonal network, one has to face the problem of chemical frustration (N-N and B-B bonds are highly unfavorable energetically and are very unlikely to occur in the closed structure). $C_{60}$ is a highly symmetrical molecule ($I_H$). Most of the symmetry elements are lost when all equivalent carbon atoms are replaced by non equivalent B, C and N atoms. An attractive model with low structural stress, optimized number of B-N and C-C bonds and high symmetry (presence of a 5-fold axis) is presented in Fig. 3.

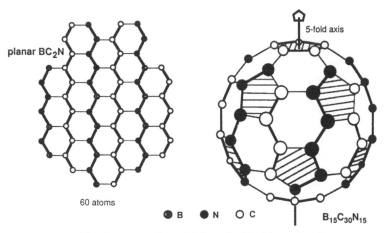

planar BC$_2$N

60 atoms

● B    ● N    ○ C

5-fold axis

B$_{15}$C$_{30}$N$_{15}$

Fig. 3, structural model for a B$_{15}$N$_{30}$N$_{15}$ molecule

## 3.2    BN irradiation experiments

Fig. 1 shows a typical sequence of High Resolution Electron Microscopy (HREM) images taken at successive time intervals during intense irradiation. In the initial stage, stripes of the surface layer - weakly bound with adjacent layer by van der Walls forces - pull off and scrolls into spirals. Figure 1a shows the simultaneous formation of 2 spirals made of 4 and 5 revolutions respectively. Let us call these objects I and II respectively. The next sequence of images shows the induced fluidity structural-transformation of both unstable objects up to a final and stable configuration (Fig. 1d). This transformation occurs by a decrease of the internal hollow. At this stage the structure is believed to have completely evolved towards nested 3D closed-cages in order to eliminate the energetically highly unfavorable dangling bonds.

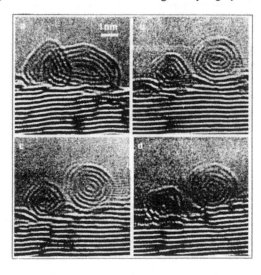

Fig. 4, time sequence showing the in-situ formation of 2 small nested cage like clusters under strong irradiation of turbostratic BN. (a) formation of 2 spirals from the scrolling of the surface plane after ≈ 20 s of strong irradiation; (b) ≈ 120 s; (c) ≈ 140 s; (d) ≈ 200 s. Note the hexagonal-like shape of the small onion 2D -image (labeled I in text) in c and the square-like shape of both onion 2D-image in their final configuration (d).

In their final configurations, the architecture of particles I and II results in the nesting of cages whose 2-D HREM images are square-like in shape. The square-like shape is all the more pronounced as the cages decrease in size. The sides of the inner most squares in onions I and II are 0.48 nm and 0.67 nm respectively.

A remarkable feature of the images shown in Fig. 1 is the polygonal shape with a small number of sides (typically 4 to 6) of the fringe patterns, which contrasts with the carbon-onion images which are either quasi-circular or polygonal with typically 10 sides. In order to interpret the observed morphology one has to look into the problem of the closure on itself of a BN hexagonal network. Due to the chemical bonding requirement that B and N atoms prefer to alternate, it is was suggested that BN-analogs to carbon fullerenes are formed from combinations of odd member rings. According to Euler's theorem, 6 squares (instead of 12 pentagons in carbon fullerenes) are needed to close the structure. The resulting structure is a octahedron whose 8 equilateral triangular faces are cut from the hexagonal network and the 6 apices are truncated by squares.

Depending on the orientation of the viewing direction of the octahedron, polygonal shapes with either 4 or 6 sides are observed. The observed shape is either square-like, hexagonal-like or rather elongated. All three types of geometry are experimentally observed as the onions evolve towards a stable configuration and rotate, setting therefore successive crystallographic orientations along the viewing direction (distorted hexagonal shape: onion I Fig. 1c; elongated shape: onion II Fig. 1b; square-like shape: onions I and II Fig. 1d) From the respective side length of the observed innermost squares, we suggest that the innermost cages of onions I and II in Fig. 1c and 1d are respectively close to $B_{12}N_{12}$ molecule and a $B_{16}N_{16}$ molecule. We note that the $B_{12}N_{12}$ cluster is the smallest BN cage which can be composed of 4-and 6-member rings disjoint, and was therefore suggested to be the analog to $C_{60}$ (Jensen 1993). More generally, octahedral polyhedra $B_{12}N_{12}$, $B_{16}N_{16}$ and $B_{28}N_{28}$ were found to have particularly high stability from electronic structure calculations (Seifert 1997).

4.    CONCLUSION

Intense electron irradiation experiments were carried on two kinds of samples, a novel turbostratic BN material and a turbostratic $BC_2N$ material. As for carbon, irradiation provokes a tendency towards curling and forming shells. In both cases, we observed the scare appearance of small nested fullerenes with a number of layers less than 10. A structural model based for a 60 atom cage was proposed based on the hypothesis that the formation process is driven at a constant composition. The competition between a growth process conserving the starting material composition and one where atoms are rearranging to lead to a more symmetrical structure with subsequent change in the layer composition is a fascinating question. EELS analysis, if made possible on such weak features, would constitute a very appropriate tool to give experimental evidence for the structure of such B-C-N hybrid fullerenes. The overall shape of the BN cages was explained within the frame of the octahedral model previously proposed for BN analogs to carbon fullerenes. The diameter of the innermost shell of the onions was estimated to be 0.4 - 0.7 nm. This range in diameter corresponds to that of the octahedral polyhedra $B_{12}N_{12}$, $B_{16}N_{16}$ and $B_{28}N_{28}$ which were predicted to be magic BN-clusters from electronic structure calculations.

ACKNOWLEDGMENTS

This work was supported by the Center Of Excellence Project at National Institute for Research in Inorganic Materials, Tsukuba, Japan.

REFERENCES

Kroto H W, Health J R, O' Brien SC, Curl R F, Smalley R E 1985 Nature **318**, 162
Iijima S, Nature 1991 **354**, 56
Ugarte D, Nature 1992 **359**, 707
Goldberg D, Bando Y, Eremets M, Takemura K, Kurashima K and Yusa H 1996 Appl. Phys. Lett. **14**, 2045
Stéphan O, Ajayan P M, Colliex C, Redlich Ph, Lambert J M, Bernier P and Lefin P 1994 Science **266**, 1683
Stéphan O, Bando Y, Dussarrat C, Kurashima K, Sasaki T, Tamiya and Akaishi M 1997 Appl. Phys. Lett. **70**, 2383
Liu A Y, Wentzcovitch R M, Cohen M L (1989) Phys. Rev. B 39, 1760
Jensen F and Toflund H 1993 Chem. Phys. Lett. **201**, 89
Seifert G, Fowler P W, Mitchell D, Porezag D and Frauenheim Th 1997 Chem. Phys. Lett. **268**, 352.

*Inst. Phys. Conf. Ser. No 153: Section 8*
*Paper presented at Electron Microscopy and Analysis Group Conf. EMAG97, Cambridge, 1997*
© *1997 IOP Publishing Ltd*

# Modelling thin-foil EDX spectra for investigating information extraction by chemometric analysis

**M W Knowles and J M Titchmarsh**

Materials Research Institute, Sheffield Hallam University, City Campus, Pond Street, Sheffield S1 1WB, UK

**ABSTRACT:** A fundamental parameters programme has been developed to generate artificial EDX spectra modelling grain boundary segregation and variations in foil thickness. Series were subjected to linear multivariate statistical analysis (MSA) to explore the form of the information components. Segregation information was confined to one component, confirming recent experiments. Self-absorption generated a series of components but suggested that just a single component would be observed in typical experiments, validating the use of MSA for EDX in many practical situations.

## 1. INTRODUCTION

MSA has recently been used to analyse compositional information from diffusion and segregation profiles at grain boundaries (Titchmarsh and Dumbill, 1996, 1997). The method reveals information components within a whole spectral data set assuming only that the components are mathematically orthogonal, i.e. arise from physically independent processes. Spectra are resolved into components (eigenspectra) which explicitly contain the variations in composition, artefacts and random noise, which subsequently can be filtered from the spectra prior to conventional processing, yielding improved sensitivity. However, there is no prior knowledge about the information sources present in any spectral series, or the form of the spectral features in the eigenspectra. A necessary and important requirement following MSA, therefore, is the interpretation of the physical origin of the eigenspectra, based on eigenspectral features and their spatial variations. To facilitate MSA eigenspectral interpretation, a fundamental parameters program has been developed to generate artificial EDX spectral series in which parameters can be varied in a controlled manner. Hence, the artificial spectra contain no artefacts and there is full knowledge of the information content. The form of the components can be unambiguously revealed and related to the chosen parameter, i.e. foil thickness, surface films, random noise, and real concentration variations. Results from controlled, artificial spectral series analysis can be correlated with components revealed by MSA of real experimental spectral series to interpret the sources of information.

## 2. THE SPECTRUM-GENERATING PROGRAM

The programme generates characteristic K and L lines using the Bethe formula (Powell, 1976) and the continuum using a modified Bethe-Heitler formula (Chapman et al, 1983) for a selected incident electron energy and detector position. Contributions from the constituent elements are added in proportion to the atomic fractions. Absorption in the specimen and detector and are incorporated through parametric expressions (Heinrich, 1987) for mass absorption coefficients. Beam current, acquisition time, specimen/detector geometry, and foil thickness are independently selected. Random noise is added using Poisson statistics which is essential in order to investigate small signal extraction. An example of an artifical spectrum from an Fe-25%P 'alloy' is shown in Fig.1.

## 3.    MSA OF SIMULATED SPECTRA

MSA of simulated spectra is being used to explore a number of problems, including: the assumption of the orthogonality of the information content of FEG-STEM spectral series containing equilibrium segregation at a grain boundary; the extent to which self-absorption can be analysed using MSA; the potential improvement in analytical sensitivity of segregation using MSA; the shapes of eigenspectra present in diffusion profiles and combinations of segregation with diffusion. Only the first two are considered here.

Measurement of boundary segregation by FEG-(S)TEM generates a profile which is roughly Gaussian in shape, reflecting the convolution of the probe current distribution with the atomic-scale width of the segregant. Hence, a series of 11 spectra was generated assuming different probe positions close to a P-segregated grain boundary in a 100nm thick foil, 100seconds acquisition time and a Gaussian probe of standard deviation 1.0nm containing 0.5nA. The P segregation was 10% of the single atom plane at the boundary. MSA of the spectra revealed 11 eigenvalues, one of these, E0, associated with the average content of the 11 spectra. The percentage information associated with the remaining 10 eigenvalues (Table 1) showed that E1 contained almost all the information; the other 9 had much smaller, similar values. The P-K counts profile of the E1 components replicated the shape of the probe current distribution used to generate the artificial spectra (Fig.2) indicating that segregation information was explicitly extracted into E1. The E1 component of the spectrum 'acquired' with the probe positioned symmetrically on the boundary plane showed a positive P peak and anti-correlated Fe, while the E2 component contained only random noise (Fig.3), as did all the other components (not shown). The results supported the assumption that the segregation information was orthogonal to all other components (which related only to noise in this ideal 'experiment') because only component E1 exhibited a P-K peak.

Self-absorption, present in all spectra, is a non-linear function of foil thickness, mass absorption coefficients and geometry. It is uncertain whether self-absorption generates a single eigenspectral component consistent with the MSA assumption of othogonality. An eigenspectrum revealed by Titchmarsh and Dumbill, 1996 has been, tentatively, ascribed to self-absorption. Two series of 12 spectra, with and without random noise, were generated for pure Fe for foil thicknesses of 10nm to 120nm in steps of 10nm. The information in the 11 eigenvalues (excluding the average, E0, in both cases) is listed in Table 1. Without noise, E1 was much larger than the others, which gradually reduced in value. With random noise, E1 was again much greater than all the others, while E2 was slightly larger than the remainder. The two E1 eigenspectral components for the thickest foil (Fig.4) had similar characteristics; the Fe-K lines were positive and the Fe-L lines were negative, consistent with the L lines being preferentially absorbed relative to the K lines when strong self-absorption was present. Adding a component such as that in Fig.4 to the average spectrum component (E0) of the series, increased K/L and generated the spectra from thicker foils, while adding the inverse spectrum reduced K/L, generating spectra from thinner foils.

Self-absorption increases as x-ray energy falls, except close to absorption edges. The second largest eigenvalues, E2, in the two series (Fig.5) both contained similar features: a K line correlated with a combination of the L line and low-energy continuum of complex shape. This was interpreted as a first order correction to the major self-absorption correction in E1 and indicated a small inconsistency in the assumption of MSA information orthogonality. Whether or not such eigenspectra are revealed in experimental data will depend upon the range of thicknesses used and the quality of the data (level of random noise). It appeared from Table 1 that, with noise-free data, the MSA attempted to fit the non-orthogonal self-absorption effect with a series of successively smaller corrections. When Poisson noise was present these corrections were smaller than the noise components and were no longer isolated as separate components. Fewer eigenvalues above the limiting noise level were, therefore, revealed, and the precise shapes of the significant eigenspectra were changed compared with the corresponding noise-free data.

The potential for non-linear effects such as self-absorption to destroy the use of MSA for thin foil EDX analysis will depend upon the relative accuracy which can be tolerated in specific situations, e.g. where a large change in foil thickness occurs across a segregated grain boundary. Comparison of the relative sizes of the eigenvalues identified with the different parameters will give some estimate of the likely limit on accuracy. When the largest self-absorption eigenvalue is only a small fraction of the segregation eigenvalue then there is unlikely to be a major influence. Further examination of this situation can be attemped now that the modelling software is available.

## 4.    CONCLUSIONS

A programme has been developed to allow ideal EDX spectral series to be generated, free from artefacts, containing selected parametric variations. The orthogonality of the EDX information has been examined using MSA. It has been shown that equilibrium segregation is explicitly extracted by MSA, confirming previous experimental results. Self-absorption limits the assumption of orthogonality which might affect the application of MSA but this is only likely in extreme experimental situations.

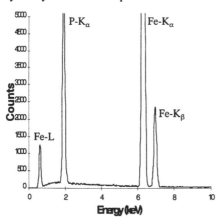

Fig.1  Artificial spectrum from an
Fe-P alloy

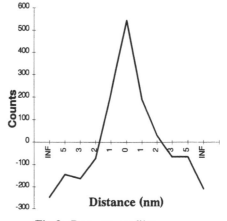

Fig.2  P counts profile:
component E1

374

Table 1 – Information in eigenvalues
(per cent)

| Eigen-value | P Segregation Profile | Pure Fe (with noise) | Pure Fe (noise-free) *$10^{-6}$ |
|---|---|---|---|
| E1 | 13.7 | 95.75 | 9.9E+7 |
| E2 | 11.4 | 1.02 | 7.E+5 |
| E3 | 10.6 | 0.44 | 5.E+4 |
| E4 | 10.2 | 0.42 | 2.E+3 |
| E5 | 9.8 | 0.39 | 2.E+3 |
| E6 | 9.6 | 0.38 | 6.E+2 |
| E7 | 9.4 | 0.35 | 6.E+2 |
| E8 | 8.8 | 0.34 | 6.E+1 |
| E9 | 8.5 | 0.32 | 2.E+1 |
| E10 | 8 | 0.3 | 2.E+1 |
| E11 | - | 0.29 | 2.E+1 |

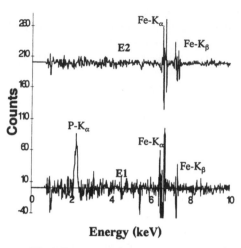

Fig.3 E1 and E2 (offset by 200 counts) components of boundary spectrum

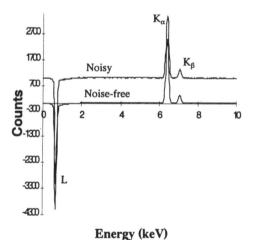

Fig.4 E1, noise-free and noisy (offset by 1000 counts) data: 120nm-thick Fe

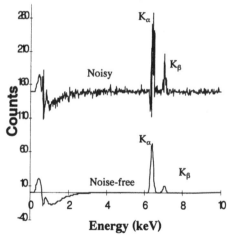

Fig.5 E2, noise-free and noisy (offset by 150 counts) data: 120nm-thick Fe

**REFERENCES**

Powell C J 1976 Rev Mod Phys A **48**, 33
Chapman J N, Gray C C, Robertson B W and Nicholson WAP 1983 X-ray Spectrom.**12**, 153
Heinrich K F J 1987 Proc.11th ICXOM, eds J D Brown & and R H Packwood (Univ.Western Ontario, London, Canada) 67
Titchmarsh J M and Dumbill S 1996 J Microscopy **184:3**, 195
Titchmarsh J M and Dumbill S 1997 J Microscopy, accepted for publication

*Inst. Phys. Conf. Ser. No 153: Section 8*
*Paper presented at Electron Microscopy and Analysis Group Conf. EMAG97, Cambridge, 1997*
© 1997 IOP Publishing Ltd

# Cap structure of the coaxial BCN nanotubes investigated by nano-EELS

**K Suenaga[1], O Stephan[1], C Colliex[1,2], F Willaime[3], H Pascard[4], N Demoncy[4,5] and A Loiseau[5]**

[1]Laboratoire de Physique des Solides, Université Paris-Sud, 91405 Orsay, France
[2]Laboratoire Aimé Cotton, Campus d'Orsay, 91405 Orsay, France
[3]Section de Recherches de Métallurgie Physique, Centre d'Etudes de Saclay, 91191 Gif-sur-Yvette, France
[4]Laboratoire des Solides Irradiés, Ecole Polytechnique, 91128 Palaiseau, France
[5]Office National d'Etudes et de Recherches Aérospatiales, 92322 Chatillon, France

**ABSTRACT:** Spatially resolved electron energy loss spectroscopy (EELS) is applied to characterize BCN nanotubes. First a quantitative procedure of elemental mapping with the help of associated calculations is presented to reveal stacking structure of graphitic carbon and BN layers along the radial direction of the nanotubes. Next the electronic structure identification, using the edge shapes, is achieved at the subnanometer level. Lastly the mapping of the $\pi^*$ weight ratio along the axial direction of the tube is the support for the investigation of the coaxial nanotube termination.

## 1. INTRODUCTION

An exclusive advantage of EELS, besides its element selectivity, is to be a probe of the electronic states at nanometer-level (Colliex 1996), e.g. the fine structure analysis of the C $K$ edge is generally used to distinguish graphitic, amorphous, fullerene-like carbon and diamond. This feature can be extended to make a selective mapping of $sp^2$ and $sp^3$ bonding by taking advantage of the bonding energy difference on the carbon $K$ edge (Muller et al. 1993). Here we demonstrate how the spontaneous mapping of $\pi^*$ weight ratio to ($\pi^*+\sigma^*$) weight, which is directly connected to the orientation of the graphitic plane with respect to the electron beam, can be used to identify which kind of planes (BN layers or C layers) cap the coaxial BCN nanotubes, which we have recently produced (Suenaga et al. 1997).

For this study a dedicated scanning transmission electron microscope (VG HB501) equipped with a parallel EELS detector (Gatan PEELS666) is operated at 100 kV in a «Spectrum-Image» mode (Jeanguillame and Colliex 1989) to record the element selective and symmetry selective profiles along and across the nanotubes.

## 2. CHEMICAL MAPPING ALONG THE RADIAL DIRECTION

The «Spectrum-Image» mode is used to obtain a series of spectra when scanning the incident electron probe across the BCN nanotubes with a subnanometer step. Each spectrum covers the three $K$ edges of B, C and N with 0.3 eV energy dispersion. The acquisition time is reduced to 1 sec for each spectrum in order to prevent beam induced effects (irradiation damage and contamination) at the cost of a reduction of signal to noise ratio for the nitrogen $K$ edge counts in some cases. Another line-scan for the low-loss region is also required as close as possible to the $K$ edge profile for further quantification.

The quantification of each spectrum is done to obtain the number of selected atom per unit area ($N_k$:) as follows (Egerton 1996);

$$N_k = I_k(\Delta)/(\sigma_k I_0)$$

where $I_k(\Delta)$: the integrated intensity of the chosen edge within the energy window $(\Delta)$, $\sigma_k$ : the scattering cross section and $I_0$ : the zero-loss intensity. Here we use the Hydrogenic cross section for each $\sigma_k$ with convergence angle correction ($\alpha = 7.5$ mrad and $\beta = 16$ mrad), which is provided by Gatan EL/P 3.0 software. With a help of the power-law background each $K$ edge weight is counted in the energy window $(\Delta)$, which is set to 35 eV for all the elements (184 - 219 eV for B, 282 - 317 eV for C and 405 - 439 eV for N).

Fig. 1(a) shows thus obtained elemental profiles of B, C and N across one of the BCN tubes. Four maxima in the C profile and two minima in coincidence with the two maxima of the B (or N) profile indicate the anti-correlation between C and B (or N).

An associated calculation is made to simulate these profiles for various distributions of layers with different chemical compositions (Fig. 1(b)). The best agreement with experiment appears in this case for a 14 layer tube, 12 nm in external diameter, formed with 3 inner layers of graphitic carbon, covered by 6 $(BN)_x C_{2y}$ layers with $y/x \approx 0.05\pm0.05$, and again 5 carbon layers. (see Fig. 1(c).) The intermediate BN layers are likely made of pure BN, but carbon substitution into the BN layers (upto 10 at% at maximum) cannot be ruled out when taking the uncertainty of the Hydrogenic cross sections into account.

Fig. 2 shows the fine structure changes in the B and C $K$ edges when the incident probe goes from the edge position to the center position along a line-scan across the coaxial nanotube. Their sharp $\pi^*$ and $\sigma^*$ peaks confirm the graphitic network with $sp^2$ type configuration. Moreover, the evolution of the weight ratio between the $\pi^*$ and $\sigma^*$ peaks is characteristic of a cylindrical structure, where the orientation of the electron beam with respect to the graphitic layers varies continuously from parallel to perpendicular (Stephan et al. 1996). The shadowed boxes in the figure show the energy windows used for the symmetry selective profiling.

## 3. SYMMETRY SELECTIVE PROFILE ALONG THE AXIAL DIRECTION

A cap structure of one of the coaxial tubes is shown in a bright field image (Fig. 3), in which one can see the tube terminating in two steps. This tube is of ~ 1μm length and ~12 nm in diameter and has the tri-shell structure (C/BN/C layers) like the former one shown in fig. 1. The particles in darkest contrast correspond to metallic hafnium boride that has been used as a catalyser to produce these coaxial tubes. Three spectra also shown in the figure correspond to

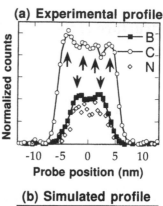

**(a) Experimental profile**

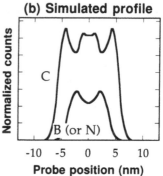

**(b) Simulated profile**

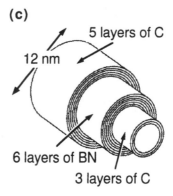

**(c)**

**Fig. 1** (a) Elemental profiles measured across one of the coaxial BCN nanotubes. (b) Simulated profiles for C and B (or N) distributions for the coaxial hetero-structure shown in (c).

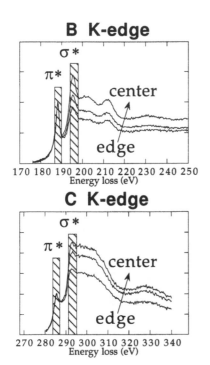

**B K-edge**

σ*

π*

center

edge

170 180 190 200 210 220 230 240 250
Energy loss (eV)

**C K-edge**

σ*

π*

center

edge

270 280 290 300 310 320 330 340
Energy loss (eV)

**Fig. 2**. The fine structures evolution of B and C K edges. The energy windows shown in the figure are for the symmetry selective profiles.

the marks in the micrograph. At the point A (within the tri-shell structure), we confirm the graphitic C and BN layers by all the $K$ edges fine structures of B, C and N correspond to their sp$^2$ configurations. Between the two terminating caps (point B) there appears little carbon content but only the BN layers. The carbon content comes back at the top of the tube (point C) but the $K$ edge shape of carbon is no longer indicating graphitic network but rather look like that of amorphous carbon.

Along the indicated line in the previous figure the π* peak weights of B and C $K$ edges are profiled as well as all the weights within the ($\Delta$) of three edges (Fig. 3 left). The intensity of the π* weight relative to the whole edge weight increases at the two terminating caps where the graphitic layers are parallel to the incident beam. The first cap (I) contains both the BN layers and C layers, (otherwise maybe BCN layers), judging from the clear indications of π* peak enhancement for both B and C profiles. While at the second cap (II) the BN layers are dominant and most of carbon amount would be amorphous or substituted into the BN layers guessing from the C $K$ edge there described above. This situation is schematized in the figure. The detailed interpretations and the forming mechanism of these cap structures will be discussed elsewhere.

## 4. CONCLUSION

We show here how the spatially resolved EELS can be applied to characterize BCN nanotubes and their cap structure. These results demonstrate the possibilities of EELS, such as the chemical analysis, the bonding state identification, and the symmetry selective mapping at nanometer-level, as a promising approach to investigate the local chemistry and the local electronic structure of a nanometrically inhomogeneous material.

## REFERENCES

Colliex C 1996 J. Electron Microscopy **45**, 44
Egerton R F 1996 «EELS in transmission electron microscope», 2nd ed., (Plenum, New York)
Jeanguillame C and Colliex C 1989 Ultramicroscopy **28**, 252
Muller D A, Tzou Y, Raj R and Silcox J 1993 Nature **366**, 725
Stephan O, Ajayan P M, Colliex C, Cyrot-Lackmann F and Sandre E 1996 Phys. Rev. B **53**, 13824
Suenaga K, Colliex C, Demoncy N, Loiseau A, Pascard H and Willaime F 1997 (submitted for publication)

A support from the NEDO International Joint Research Grant «Novel nanotubular materials: production, characterization and properties» is acknowledged.

378

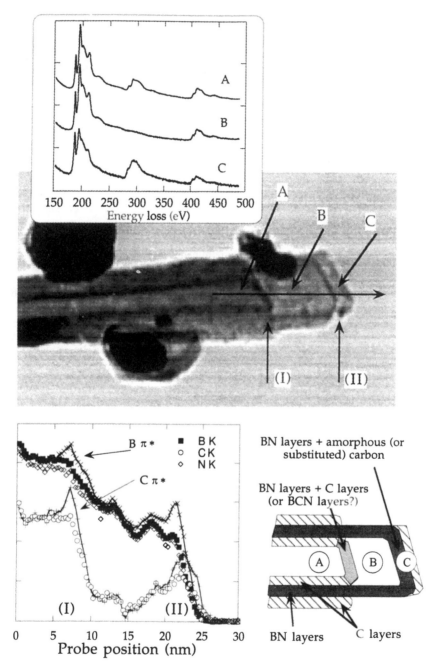

**Fig. 3** A cap structure of one of the BCN nanotubes. A bright field (center) shows two steps of terminating layers (I) and (II). Point EELS spectra (inset), corresponding to the marks in the micrograph, show the $sp^2$ structure except the C $K$ edge at the point C. Changes in the $\pi^*$ weight ratio relative to the $(\pi^*+\sigma^*)$ profiles indicate the terminating layers of the BN and C. This situation is schematized in the figure (right-bottom).

*Inst. Phys. Conf. Ser. No 153: Section 9*
*Paper presented at Electron Microscopy and Analysis Group Conf. EMAG97, Cambridge, 1997*
© *1997 IOP Publishing Ltd*

# Structural characterisation of the VMgO catalyst system

**A Burrows, A Pantazidis\*, C Mirodatos\* and C J Kiely**

Department of Materials Science and Engineering, The University of Liverpool, UK

\*Institut de Recherches sur la Catalyse, CNRS, Villeurbanne, France

**ABSTRACT:** A VMgO catalyst used in the oxidative dehydrogenation of propane (ODHP) reaction has been examined in detail by HREM. As prepared material is shown to consist mainly of MgO crystallites exhibiting an unusual {111} facet termination stabilised by a disordered vanadium containing overlayer. The structural state of the overlayer has been found to vary depending on the gaseous environment to which the catalyst has been subjected. Normal reaction conditions gives rise to a weakly ordered structure whilst strong reducing conditions induce a pronounced structural ordering of the overlayer. Subsequent reoxidation treatments cause the overlayer to revert back to a disordered state. These changes can be correlated with changes in the electrical conductivity of the material.

## 1. INTRODUCTION

The vanadium-magnesium-oxide (VMgO) binary oxide system has been shown to be a promising catalyst for the oxidative dehydrogenation of propane (ODHP) to propene (Chaar *et al* 1988). More recently, Pantazidis and Mirodatos, 1996, demonstrated for a series of VMgO catalysts that the highest propene yields were obtained from a catalyst containing 14 wt% of vanadium after calcination at 800°C. The aim of the present study is to achieve a better understanding of the specific interactions taking place between the reactants and the VMgO catalyst in ODHP. We have applied a diverse range of techniques such as *in-situ* XRD, *in-situ* EXAFS, XPS, electrical conductivity measurements as well as TEM to analyse this complex materials system. In this paper, we present results from our electron microscopy studies along with some electrical conductivity measurements which reveal the sensitivity of the active surface of this catalyst to its gaseous environment.

## 2. EXPERIMENTAL

A 14 wt% vanadium VMgO catalyst was prepared from brucite $(Mg(OH)_2)$ according to the recipe described by Pantazidis *et al* 1997. The catalyst was then tested for the ODHP reaction at 500°C in a $C_3H_8/O_2/He$ mixture $(P(C_3H_8) = 2.4$ kPa, $P(O_2) = 2.0$ kPa) for 50 hours (*used* sample). Reduction/reoxidation (redox) sequences were also carried out at 500°C in the presence of i) $O_2$, ii) $C_3H_8$ (*used-P*) and finally iii) $O_2$ (*reoxidised*). All these samples, including the *fresh* (unused) material, were examined in a JEOL 2000EX high resolution electron microscope operating at 200 kV.

## 3.    RESULTS AND DISCUSSION

HREM studies of the *fresh* VMgO catalyst have shown it to be a mixture of three phases. Discrete magnesium orthovanadate ($Mg_3V_2O_8$) and magnesium pyrovanadate ($Mg_2V_2O_7$) crystallites comprise about 10 and 1 vol% of the catalyst respectively. The dominant morphology, however, was of rather featureless MgO platelets, up to 200nm in size, which are often stacked together. Edge-on HREM imaging of these MgO platelets surprisingly shows them to be terminated by polar {111}-type planes (Fig. 1) which are expected to be energetically unfavourable in relation to the non-polar {100} type surfaces

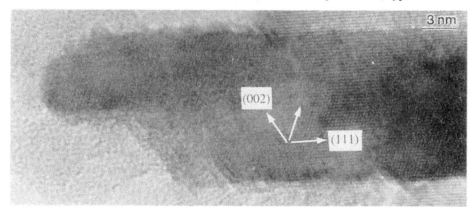

**Fig. 1** An MgO platelet showing {111} facet termination

Further HREM analysis, in conjunction with EDX studies, showed that the platelets were covered by a thin disordered V containing layer (Fig. 2a). Prolonged irradiation of the platelets was found to disrupt the overlayer and the freshly exposed {111} MgO surfaces were seen to rapidly reconstruct into a series of lower energy {100}-type facets. It is highly likely that in the unperturbed state, the thin overlayer is stabilising the unusual {111} facet termination of MgO which we have shown derives from the hexagonal crystal structure of the Mg(OH)$_2$ precursor (Burrows *et al* 1997). Examination of the *used* sample revealed that the MgO still retained the {111} platelet morphology which implies that the overlayer remains intact during the reaction (the bulk VMgO phases showed no obvious morphological changes). However, detailed inspection of lattice fringe patterns indicates that the overlayer now exhibits a weakly ordered structure (Fig. 2b).

When the catalyst was exposed to a strongly reducing propane reaction mixture above 500°C, the *used-p* sample, the minority bulk $Mg_3V_2O_8$ and $Mg_2V_2O_7$ phases remain relatively unaltered. However, the surface overlayer on the [111] MgO platelets is shown by HREM to re-organise into a strongly ordered structure. (Fig. 3). The lattice fringe spacings and/or intersection angles of the overlayer structure do not match to any reported bulk VMgO phase, nor does it correspond to $V_2O_5$, $V_2O_4$ or $V_2O_3$. *In-situ* EXAFS studies have shown that ordering of the overlayer is accompanied by a reduction of constituent V ions from the +5 to the +3 oxidation state (Vorbeck *et al* 1997). Our analysis along with *in-situ* XRD evidence leads us to suspect the formation of a new cubic spinel type phase with composition $Mg_{1.5}VO_4$. This disorder/order transformation is reversible, since subsequent exposure of the catalyst to oxygen (*reoxidised* sample) at 500°C restores a disordered overlayer similar in appearance to the *fresh* sample shown in Fig. 2a.

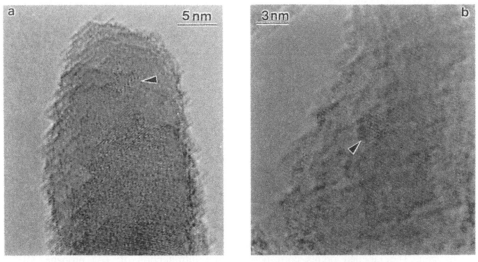

**Fig. 2** a) Disordered V containing layer on a *fresh* MgO platelet; b) a weakly ordered overlayer on the *used* catalyst after 50h of use at 500°C in $C_3H_8/O_2/He$

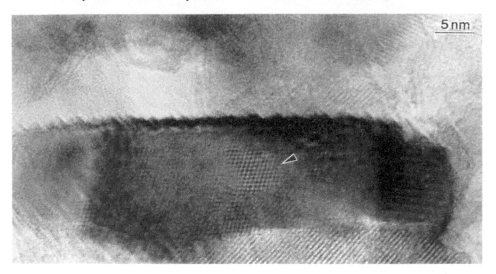

**Fig. 3** The strongly ordered overlayer structure after exposure to pure propane at 500°C

Structural and electronic changes taking place during redox cycles can also be followed using *in-situ* electrical conductivity measurements as shown in Fig. 4. At time 0, the catalyst was under an oxygen atmosphere at 500°C (*fresh* sample). When pure propane is introduced, point A, the electrical conductivity increases by several orders of magnitude indicating a severe reduction of the catalyst. This effect can be accounted for by assuming that electrons arising from a $V^{5+}$ to $V^{3+}$ reduction become available for conduction via associated vacancies. The re-introduction of oxygen at point B results in the electrical conductivity returning to its original value. Thus, the process is reversible. When a propane/oxygen mixture is introduced at point C (corresponding to the conditions for the *used* sample), the conductivity increases but to a lower value than in the presence of pure

382

propane. The catalyst is therefore reduced to a lesser extent (relative to that in pure propane) under what would be normal reaction conditions.

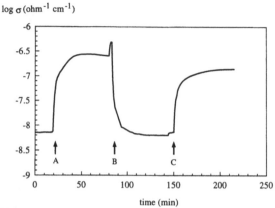

**Fig 4.** The variation of electrical conductivity during redox cycles at 500°C, initial atmosphere oxygen. The labelled points correspond to the introduction of: (A) propane, (B) vacuum, then oxygen, (C) reaction mixture

## 4.    CONCLUSIONS

HREM observations of the 14V/VMgO catalyst show that a disordered thin V containing overlayer stabilises an unusual {111} platelet morphology of MgO. This layer has been shown to be structure sensitive to the gaseous environment to which it is it is subjected. The structural changes that do occur are reversible and can be correlated with electrical conductivity measurements. Reduction in propane causes strong overlayer ordering (high conductivity) whereas oxidation treatments induce a transformation back to a more disordered structure (low conductivity). Normal reaction conditions causes a partial ordering of the overlayer characterised by intermediate conductivity. The exact nature of this overlayer and its role in the catalytic activity displayed by this material are key questions which are currently being investigated in more depth.

## ACKNOWLEDGEMENTS

This  work was funded by the EEC under the JOULE programme, contract number JOE3CT950022.

## REFERENCES

Burrows A, Kiely CJ, Pantazidis A, Mirodatos C and Derouane E, 1997, Cat. Lett. (submitted)
Chaar M A, Patel D and Kung H H, 1988, J. Catal. **109** p463
Pantazidis A and Mirodatos C, 1996, Heterogeneous Hydrocarbon Oxidation, eds B Warren and S T Oyama, ACS Symposium Series **638** p207
Pantazidis A, Mirodatos C, Burrows A and Kiely C J, 1997, J. Catal, in press
Vorbeck G, Balic Zunic T, Türnqvist, Villadsen J, Perregaard J, Burrows A, Kiely C J and Højlund-Nielsen P E, 1997, EUROPACAT III (in press)

*Inst. Phys. Conf. Ser. No 153: Section 9*
*Paper presented at Electron Microscopy and Analysis Group Conf. EMAG97, Cambridge, 1997*
© *1997 IOP Publishing Ltd*

# Nano-sized metal clusters investigated by a STEM-based mass spectroscopic technique

J C Yang, S Bradley*, M Yeadon and J M Gibson

Frederick Seitz Materials Research Laboratory, University of Illinois at Urbana-Champaign
Urbana, IL 61801, USA

* UOP, 50 East Algonquin Road, Des Plaines, IL 60017, USA

**ABSTRACT:** We have examined carbon-supported PtRu$_5$ specimens by a transmission electron microscopy, including a novel STEM-based mass-spectroscopic technique, which gives the approximate number of atoms of individual nano-sized clusters. The combination of these techniques demonstrate that the PtRu$_5$ clusters are oblate on the carbon support.

## 1. INTRODUCTION

Knowledge of catalysts' sizes and shapes on their support material is crucial in understanding catalytic properties. With increasing interest in nanosized catalytic materials, it is vital to obtain structural information at the nanometer level. Transmission electron microscopy (TEM) has proven very useful in obtaining structural information. In particular, Z-contrast is a powerful technique for gaining unique insights into the structure of the supported metal clusters (Nellist and Pennycook 1996). Crewe et al. (1975) have demonstrated that single atoms with high atomic number on a support can be visualized in annular dark field (ADF) images in a scanning transmission electron microscope (STEM). To obtain the number of atoms in a cluster from the absolute measured intensity of the ADF image, it is important that the detected contrast is due to incoherent scattering. Treacy and Gibson (1993) have investigated the issues of coherence in Z-contrast to demonstrate that very high inner detector angles are necessary for the detected electrons to be due to predominantly incoherent scattering. We have shown experimentally that very high angle (~100mrad) annular dark-field (HAADF) images in a dedicated STEM is dominated by incoherent scattering for metal clusters. Hence, the absolute intensity can be used to measure the number of atoms of individual nano-sized clusters on a support material. We demonstrated the robustness of this technique with Re$_6$ clusters supported on graphite, where the measured number of atoms for individual clusters was 6±2 Re atoms (Singhal et al. 1997). Our method builds on the earlier work of Treacy and Rice (1989), but in contrast to their relative intensity measurements, absolute measurements are required for the conclusions of our study.

In this proceedings, we will present our results from a supported bimetallic catalyst, PtRu$_5$. PtRu$_5$ is of interest for methanol oxidation in batteries (Radmilovic et al. 1995). Our data demonstrates that the shape of the PtRu$_5$ particle is, surprisingly, oblate on the carbon substrate.

## 2. EXPERIMENTAL

PtRu$_5$ compounds were produced by a molecular precursor method (Nashner et al. 1997). EDS (energy dispersive X-ray spectroscopy) on a VG HB601 STEM was performed on these specimens, where the electron beam was rastered over the entire particle. The particle was imaged during the EDS data acquisition such that specimen drift could be corrected.

Imaging for the STEM-based mass spectroscopic technique was performed on a Field Emission Gun (FEG) Vacuum Generators HB501 STEM operated at 100kV. The microscope was equipped with the digital scanning acquisition package Gatan Digiscan™. The inner angle of the annular dark field detector was masked to 96mrad, and the outer angle was ~343mrad. The incident beam convergence half-angle was 12 mrad.

Image analysis was performed on the unprocessed digital images. The intensity for each particle was integrated over a circular region sufficiently large to enclose the particle completely. The background is subtracted from the integrated signal by fitting the background to a linear function on an annulus two pixels wider than the integration radius. In order to convert the results into absolute cross-sections ($\mathring{A}^2$) we use the formula:

$$\sigma = \frac{T}{\mu_{DF} E N_0}$$

where T is the integrated signal : $T = \sum_p I_p \delta_p$

$I_p$ is the intensity of pixel p and $d_p$ is its area, $\mu_{DF}$ is the ADF detector efficiency, E is the image exposure time (seconds) and $N_0$ is the electron number flux density per second. The product $\mu_{DF} N_0$ is the number of counts detected by the annular detector for an unscattered beam in one second. The detector efficiency can be determined by tilting the electron beam onto the detector and measuring the detected signal. A known degree of attenuation is required to avoid saturation, and this is set by Faraday cup measurements. Another method is to use the carbon as an internal standard. The data presented here was obtained by using the carbon as an internal standard. The tilted beam method was also performed in order to confirm the measurements.

## 3. RESULTS AND DISCUSSION

Approximately 20 clusters were examined by EDS, where the relative concentrations of the Ru and Pt were: Ru mean = 78 ± 10% and Pt mean = 22±5%. EDS spectra taken from the entire region at 1 million magnification gave: Ru mean = 77±10% and Pt mean = 23±5%. Hence, we conclude that the relative chemical concentration of Pt to Ru between each PtRu₅ particle is reasonably constant. It is important to determine the chemistry of the metal clusters since this STEM-based mass-spectroscopic technique gives the scattering cross-section, but not the identity of the atom which is scattering the electrons. For example, 1 Pt atom will scatter as many electrons as 2 Ru atoms. EDS data is necessary to distinguish between these two scenarios.

Simultaneous BF and HAADF images were taken (Figure 1(a) and 1(b)). The distribution of the PtRu₅ clusters was not uniform; there existed regions with virtually no clusters and regions with large number of clusters. The clusters were also observed to be preferentially located at ledges. A wide spread in intensity is visually observable, which correlates to a wide spread in the measured scattered cross-sections. The theoretical cross-section for Pt is $0.019\mathring{A}^2$, and Ru is $0.00895\mathring{A}^2$ by the partial wave method, so that the theoretical cross-section for PtRu₅ is $0.063\mathring{A}^2$.

Figure 2 is a histogram of 27 particles, where 1 cluster corresponds to 1 group of PtRu₅. The average experimentally measured cross-section ($0.245$ $\mathring{A}^2$) corresponded to 4 PtRu₅ clusters or 24 atoms. The diameters of these particles were also measured from the STEM, and the average was 15.6Å. The sizes of the clusters were confirmed on a high resolution electron microscope (Hitachi H9000). Using the number density for Ru atoms, if the shape of a particle were spherical, then, for a diameter of 15Å, the particle would contain 21 PtRu₅ clusters, whereas for a hemispherical shape, the particle would contain 11 PtRu₅ clusters. Hence, since the average number of PtRu₅ clusters is 4 and the average diameter is 15.6Å, the particles must be oblate on the carbon black substrate. No sintering of these clusters was observed during the course of the experiments.

Figure 3 is a plot of the measured diameter vs. the number of atoms. The data fits better to $d^3$ instead of $d^2$, demonstrating that the clusters maintain their 3-D (dimensional) more than 2-D aspect ratio.

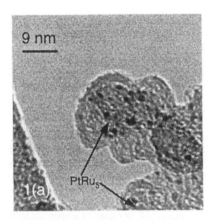
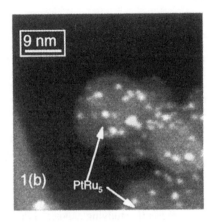

Fig. 1. (a) Bright field and (b) High angle annular dark field image of $PtRu_5$ clusters on Carbon black.

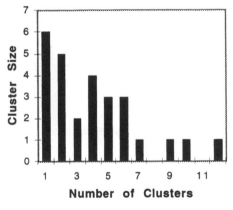
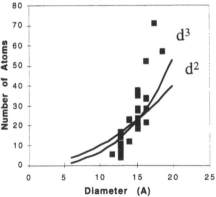

Fig.2. Histogram of Cluster Sizes for $PtRu_5$

Fig. 3. Diameter vs. number of atoms per cluster. A better fit for $d^3$ (3-D aspect ratio) than $d^2$ (2-D) is observed.

A few investigators have suggested the existence of raft-like structures of metal clusters on support materials. Nellist and Pennycook (1996) have also reported raft-like shapes of Ru on $\gamma$-$Al_2O_3$, based on Z-contrast imaging.

Nashner et al. (1997) have also examined $PtRu_5$ clusters supported on carbon using EXAFS, which gives averaged information about the coordination of the Pt and Ru atoms. They proposed that the $PtRu_5$ are semi-hemispheres on the carbon black, based on the average coordination numbers of the Pt-Pt, Pt-Ru and Ru-Ru bonds. Hence, there is reasonable agreement with the our data and the Nashner et al. proposed model, except that our model is

386

that the clusters are flatter than the Nashner et al. model, i.e. the contact angle of the spherical cap is << 90°.

A possible reason for the oblate structure of the $PtRu_5$ on carbon is that Ru forms bonds with C. Hence, it could be energetically favorable for the Ru to maximize the Ru-C bonds.

## 4. CONCLUSION

It is vital to understand the structure and chemistry of catalytic materials on their support material, since this will clearly affect their catalytic activity. We have developed recently an unique mass spectroscopic technique based on very high angle annular dark field imaging in a dedicated scanning transmission electron microscope, which gives the approximate number of atoms for individual supported metal clusters(Singhal et al. 1997). The critical aspect of this STEM-based technique is that the absolute measured intensities are used to determine the number of atoms of individual supported nano-sized clusters. The comparison of the measured diameters and the number of atoms detected provides unique insights into the structure of the clusters on the support material.

Using this technique in combination with other electron microscopy techniques, we have clearly shown that the $PtRu_5$ nano-sized clusters are oblate on carbon black. EDS demonstrated that the Pt to Ru ratio is 1 to 5. The STEM-based mass spectroscopic technique demonstrated a wide spread in the number of atoms in each cluster, where the average number of atoms is 24. The comparison of the measured diameters and the number of atoms demonstrate the oblate structure of the supported $PtRu_5$.

We are presently studying nanosized Pt clusters on C and $Al_2O_3$ in order to experimentally probe the possible existence of "magic numbers" for supported metallic clusters.

## 5. ACKNOWLEDGMENTS

We thank J. R. Shapley (Department of Chemistry), for providing the organometallic compounds investigated in this research. Useful discussions with M. Nashner (Materials Science and Engineering Department), R. Nuzzo (Department of Chemistry) and R. I. Masel (Dept. of Chemical Engineering) are thankfully acknowledged. This research was supported by the Department of Energy, No. DEFG02-96ER45439, and involved extensive use of the facilities within the Center for Microanalysis of Materials of the Frederick Seitz Materials Research Laboratory at the University of Illinois at Urbana-Champaign.

## REFERENCES

Crewe, A. V., J. P. Langmore and M. S. Isaacson 1975 Physical Aspects of Electron Microscopy and Microbeam Analysis (New York: Wiley) pp 47

Nashner, M. S., A. I. Frenkel. D. L. Adler, J. R. Shapley and R. G. Nuzzo 1997 J. of Am. Cer. Soc. **in press**.

Nellist, P. D. and S. J. Pennycook 1996 Science **274**, 413

Radmilovic, V., H. A. Gasteiger and P. N. Ross Jr. 1995 Journal of Catalysis **154**, 98

Singhal, A., J. C. Yang and J. M. Gibson 1997 Ultramicroscopy **67**, 191

Treacy, M. M. J. and J. M. Gibson Ultramicroscopy **52**, 31

Treacy, M. M. J. and S. B. Rice 1989 Journal of Microscopy **156**(2), 211.

*Inst. Phys. Conf. Ser. No 153: Section 9*
*Paper presented at Electron Microscopy and Analysis Group Conf. EMAG97, Cambridge, 1997*
© 1997 IOP Publishing Ltd

# HREM study of the reoxidation of heavily reduced Ce-La mixed oxides of interest in environmental catalysis.

**G Blanco, J J Calvino, C López-Cartes, J A Pérez-Omil, J M Pintado and J M Rodríguez-Izquierdo**

Departamento de Ciencia de los Materiales e Ingeniería Metalúrgica y Quimica Inorgánica. Facultad de Ciencias. Universidad de Cádiz. Polígono Rio San Pedro s/n. Apdo 40 Puerto Real, Cádiz 11510. Spain

**ABSTRACT:** HREM, in combination with structural modelling, image processing and image simulation, has been applied to characterise in detail the nanostructure of different Ce-La mixed oxides reduced in hydrogen at 1223K and further reoxidised in oxygen at room temperature. The formation of a thin, protective, surface layer on top of hexagonal-$Ln_2O_3$ (Ln: Ce,La) mixed oxide crystals has been evidenced after this treatment. The chemical nature and orientation relationships between the surface and bulk phases are discussed on the basis of the reactivity of these mixed oxides against oxidation and hydration/ carbonation processes.

## 1. INTRODUCTION

Mixed oxides based on cerium, such as Ce-Zr (Fornasiero et al 1995) or Ce-Pr (Logan et al. 1994), are being intensively investigated as possible substitutes for $CeO_2$ in the so called Three Way Catalysts (TWC's), currently employed to remove contaminant components from the exhaust gases of cars. In this context, according to preliminary data obtained in our lab, Ce-La mixed oxides also present interesting textural and chemical properties that allow them to be considered good alternatives to pure ceria in TWC catalytic cartridges (Bernal et al. 1997). This work, which forms part of a more extensive research project focused on the characterisation of both the nanostructure and redox behaviour of different Ce-Lanthanide mixed oxides, reports specifically the results of a detailed HREM investigation of a series of Ce-La mixed oxides. The structure of these mixed oxides was determined after reduction in hydrogen at 1223 K and further reoxidation in oxygen at room temperature. The information gained from this study helps us to understand the behaviour of these new materials under cycling oxidising/reducing atmospheres, a condition established during the usual operation conditions of TWC catalysts, and thought to be responsible for their deactivation.

## 2. EXPERIMENTAL

Ce-La mixed oxides with 20% at., 40% at. and 57% at. lanthanum contents were prepared as described elsewhere (Bernal et al. 1997). Reduction treatments were performed at 1223 K in pure hydrogen flowing at 60 cc.min$^{-1}$. After reduction the samples were flushed with helium at 1223 K and cooled, also under helium, down to 200K. A treatment in flowing oxygen from 200K up to room temperature was performed to reoxidise the reduced mixed oxides.

Experimental HREM images were recorded using a JEOL2000-EX microscope with a point to point resolution of 0.21 nm. The structural models employed as supercells for image calculations were built using the RHODIUS program developed in our lab (Botana et al. 1994). HREM simulated images were obtained using the EMS program (Stadelmann 1987). Image processing was achieved using the SEMPER 6+ software.

## 3. RESULTS

Figure 1 shows a HREM micrograph representative of the Ce-40%La mixed oxide, after the reduction/reoxidation treatment described above. According to the digital diffraction pattern (DDP), shown inset in this figure, the fringe contrast pattern observed in this image can be interpreted as due to {100} planes of a hexagonal $Ln_2O_3$ phase. It can be observed than, in addition to the {100} reflections of the mixed sesquioxide structure, two other spots labelled as $\mu_1$ and $\mu_2$, at 1/4{100} and 3/4{100} respectively, are also present in the DDP. These spots are those which give rise to the low frequency contrast modulation observed in the experimental image.

The presence of these additional diffraction spots in this sample has been confirmed by SAED. Figure 2 shows an experimental electron diffraction pattern, which can be indexed as due to the [001] zone axis of a $Ln_2O_3$ crystal , where the $\mu_1$ spots are clearly identified.

These low frequency modulation contrasts can be interpreted as Moiré fringes due to the growth of a thin surface layer on top of hexagonal-$Ln_2O_3$ cores. Taking into account that after reduction, the mixed oxide was treated in oxygen , this surface layer could be the result of the partial oxidation of the sesquioxide to a fluorite-type $LnO_{2-x}$ phase.

To confirm this hypothesis the structural model shown in figure 3 was built using the RHODIUS program, and the corresponding simulated images were obtained for different defoci. The result of this

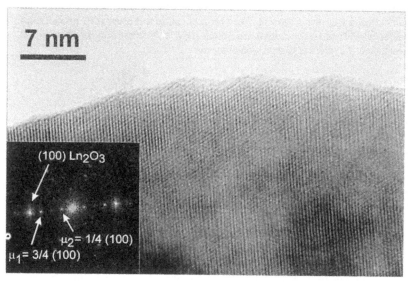

**Figure 1.-** Experimental HREM image recorded on the Ce-40% La mixed oxide after reduction at 1223 K for 5 hours and subsequent reoxidation at room temperature. A DDP of this area is included

calculation is shown in figure 4(a). As can be observed, the contrasts seen in the simulated image match fairly well with those observed in the experimental image, figure 1, and contain the low frequency modulation features. Moreover, in the DDP corresponding to this calculation, figure 4(b), both the $\mu_1$ and the $\mu_2$ reflections are also reproduced.

Taking into account that lanthanide sesquioxides suffer intense hydration/carbonation processes as a consequence of their reaction, at room temperature, with air (Bernal et al 1987), a second possibility to explain figure 1 could be that the surface layer grown on the sesquioxide cores was made of a hydrated/carbonated mixed sesquioxide phase. This reaction could have taken place during the transfer of the sample from the microreactor where it was prepared into the microscope.

After searching for different phases described in the literature for the $Ln_2O_3-H_2O-CO_2$ system a model consisting of an Ancylite type $Ln(OH)CO_3$ surface layer on top of the sesquioxide was built and its HREM image calculated, figure 4(c). As can be observed, the calculated image reproduces the basic contrasts features and the modulations observed in the experimental image. In this case the DDP, figure 4(d), only contains the $\mu_1$ spots, while the $\mu_2$ are very weak. These $\mu_1$ diffraction effects should

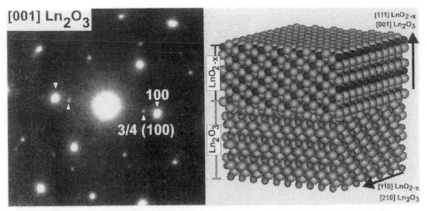

Figure 2.- SAED pattern corresponding to the Ce-40%La sample after the reduction/reoxidation treatment

Figure 3.- Structural model of the $LnO_{2-x}/Ln_2O_3$ supercell employed to calulate the image shown in figure 4(a).

be assigned in this model to the {020} planes of Ancylite instead of being due to double diffraction, Moiré, effects

From these two models, the former is the one that explains better complementary chemical characterisation data obtained from these samples, as it is the absence in the corresponding IR spectra of hydroxyl or carbonate bands with significant intensity, or the presence in the XRD patterns of small, wide, fluorite-type peaks (Bernal 1997). If these data and the fact that the freshly prepared samples were exposed to air during only a few minutes are considered simultaneously, it seems most likely that the crystalline shell formed upon the sesquioxide cores consists of a partially reduced mixed fluorite structure. This fluorite shell could prevent the inner sesquioxide crystal being transformed further into hydroxy-carbonated phases by reaction with air and the complete oxidation of the sesquioxide.

## 4. CONCLUSIONS

HREM is revealed to be a very powerful technique for detecting surface crystalline phases grown on top of a hexagonal-$Ln_2O_3$ structure. The interpretation of the Moiré contrast features, using image simulation and image processing, indicates the likely formation of a fluorite mixed oxide phase as a consequence of the surface reoxidation of the heavily reduced Ce-La mixed oxides. The precise structural relationships between the surface and bulk phases have been established on the basis of experimental and calculated images.

## REFERENCES

Bernal S, Botana F J, Garcia R and Rodríguez-Izquierdo J M, React. Solids **4** (1987) 23

Bernal S, Blanco G, Cifredo G, Pérez-Omil J A, Pintado J M, Rodríguez-Izquierdo J M, J. Alloys and Compounds **250** (1997) 449

Botana F J , Calvino J J, Blanco G, Marcos M and Pérez-Omil J A, "Electron Microscopy 1994", **2B** (1994) 1085

Fornasiero P, di Monte R., Rao G R., Kaspar J, Meriani S, Trovarelli A, Graziani M, J. Catal.., **151** (1995) 168

Logan A D and Shelef M, J. Mater. Res., **9** (1994) 468

Stadelmann P, Ultramicroscopy **21** (1987) 131

**ACKNOWLEDGEMENTS:** This work has received financial support from DGICYT under projects PB94-1305 and PB95-1257, and from CICYT, under project MAT96-0931.

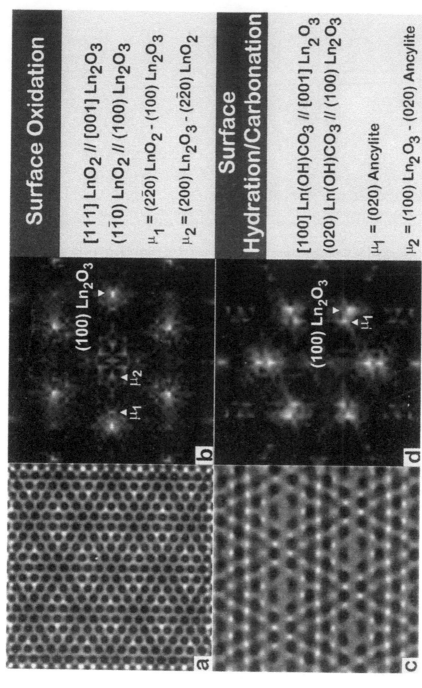

**Surface Oxidation**

[111] $LnO_2$ // [001] $Ln_2O_3$

$(1\bar{1}0)$ $LnO_2$ // $(100)$ $Ln_2O_3$

$\mu_1 = (2\bar{2}0)$ $LnO_2 - (100)$ $Ln_2O_3$

$\mu_2 = (200)$ $Ln_2O_3 - (2\bar{2}0)$ $LnO_2$

**Surface Hydration/Carbonation**

[100] $Ln(OH)CO_3$ // [001] $Ln_2O_3$

$(020)$ $Ln(OH)CO_3$ // $(100)$ $Ln_2O_3$

$\mu_1 = (020)$ Ancylite

$\mu_2 = (100)$ $Ln_2O_3 - (020)$ Ancylite

Figure 4.- (a) Image calculated from the model shown in figure 3 ; (b) DDP of (a); (c) Image calculated from a model consisting of an ancylite, $Ln(OH)(CO_3)$, shell grown on a $Ln_2O_3$ core; (d) DDP of (c). The simulation parameters were the following: HV=200 kV, df= 50 nm, Cs=0.7 mm, ds=10 nm, t=1.2 mrad. The crystallographic details concerning the orientation relationships established in each case between the surface and bulk phases, and the assignment of the moire reflections are indicated at the right boxes.

*Inst. Phys. Conf. Ser. No 153: Section 9*
*Paper presented at Electron Microscopy and Analysis Group Conf. EMAG97, Cambridge, 1997*
© *1997 IOP Publishing Ltd*

# Secondary electron imaging in characterisation of heterogeneous supported metal catalyst systems.

**R Darji and A Howie.**

Cavendish Laboratory, Madingley Road, Cambridge, CB3 0HE.

**ABSTRACT :** In recent years, high resolution secondary electron imaging has emerged as a useful tool in supported metal catalyst characterisation. However comparison of images taken in a high resolution SEM and a STEM suggest that the contrast mechanisms are rather complicated and here further studies (including Monte Carlo simulations) of these effects are presented.

## 1. INTRODUCTION

Prior to 1985 the low resolution of the SEM seriously limited its use for the SE image characterisation of supported metal catalysts. However in the scanning transmission electron microscope (STEM) it proved possible not only to get good SE images of the support but also, by comparison with annular dark field images, to identify which particles lay on the support surface in the case of Pt on C catalysts (Imeson et al. 1985). More recently several authors (Liu and Spinnler 1993, Smith et al. 1995 and Darji and Howie 1997) have employed SE imaging in the high resolution SEM (HRSEM) to image small particles in bulk catalyst samples.

Detailed comparison of the images in the STEM and in the HRSEM suggested (Darji and Howie 1997) however, that the SE contrast mechanisms are more complicated and rather different in the two cases. The SE II contribution, electron backscattered (or scattered through large angles), is more significant in the HRSEM, making particles visible at or below the surface, even in cases when they are quite invisible (though located on the surface) in the STEM. Further studies of these effects, including Monte Carlo simulations are reported here.

## 2. EXPERIMENTAL IMAGES

Comparison of SE images taken at 100kV in a VG HB501 dedicated STEM (Fig 1) with those taken at 10kV in JEOL 6000F HRSEM (Fig 2), shows interpretation of images requires care. A standard resolution test sample of small gold particles supported on thin carbon film was used as a model system. In Fig 1, the STEM SE image reveals the presence of the particles only when they are facing the detector. However, in the HRSEM (Fig 2), the particles are visible irrespective of which side of the support they sit. These results are explained by the SE II contribution noted above and by the contribution to the SE cascade arising from the extra stopping power of the particle. If the particle raises the mean energy loss by an amount comparable to that given by one escape depth thickness of support, the effect on the SE signal will be detectable (Darji and Howie 1997).

Although the SE signal is still useful in imaging surface topography, these images show that the HRSEM cannot be reliably used to identify particles sitting on the support surfaces. Such a result could have implications in the context of interpreting SE images, particularly in an environmental SEM where the SE signal is often the only one available.

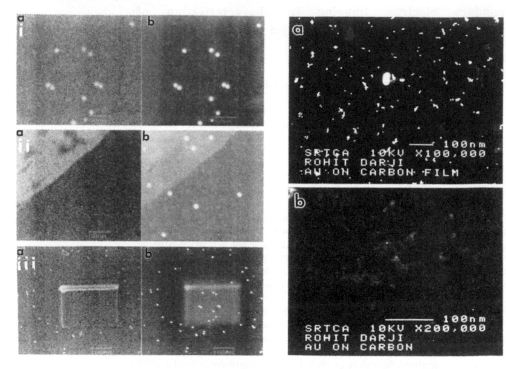

Fig 1 : SE (a) and ADF (b) images taken in STEM with Au particles facing detector (i), facing away from detector (ii), and embedded (iii).

Fig 2 : SE images of Au particles facing detector (a) and facing away from detector (b) in a HRSEM.

## 3 :    MONTE CARLO SIMULATIONS

Monte Carlo modelling of the above case allows a better understanding of the contribution of the two mechanisms to particle visibility.  The model used is similar to that of Joy (1995).  However, in this case all electrons are tracked down to a cut-off of 50 eV (a lower cut-off although desirable leads to an enormous increase in computational time) and each inelastic collision is modelled using inelastic cross sections for plasmon, single-electron and core electron excitations as opposed to the stopping power method used by Joy (1995).  This approach (originally coded by Rocca (1986)) allows the cascade to be fully simulated down to the cut-off energy. Further modifying the code allows a variety of information to be extracted from a simulation.  The code and results obtained will be discussed in more detail elsewhere. Here however, some of the more pertinent results are presented in outline.

(a)                                    (b)

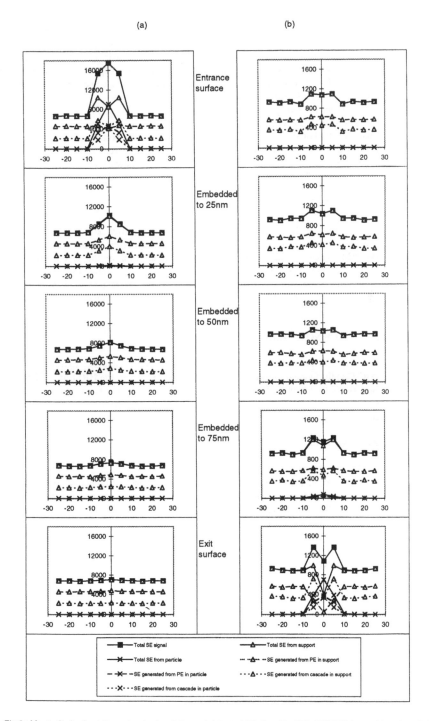

Fig 3 : Monte Carlo simulations showing breakdown of detected SE signal for 10 keV HRSEM case (a) and 100keV STEM case (b).

The geometry studied was that of a 10nm cubic particle of gold embedded in a thin film of carbon to a variety of depths. Fig 3 shows the relative contribution to the emitted SE signal depending on how the SE was generated. Essentially splitting the signal into whether it was generated in the particle or in the support and whether it was generated directly from primary electron or from a cascade electron. The case for both the HRSEM and STEM are presented.

From the two possible mechanisms given to account for particle visibility, in the STEM only the stopping power mechanism can be seen to produce the contrast seen. In the HRSEM, although this mechanism is also present, an increased SE II contribution (interactions with backscattered primaries) is seen to be more significant, especially when the particle is embedded to depths beyond a few times the mean escape depths. These simulations confirm the qualitative analysis of the images.

## 4 : CONCLUSIONS

Large angle scattering and extra stopping power are mechanisms which adequately explain the different SE contrast of supported particles in STEM and HRSEM images. The HRSEM can be a useful tool in the characterisation of supported metal catalysts.

More sophisticated Monte Carlo simulations would be useful, taking proper account of the observed band structure effects in the SE transport and escape process (Shih et al. 1997)

## 5 : ACKNOWLEDGEMENTS

RD acknowledges the EPSRC and Shell Research and Technology Centre Amsterdam (SRTCA) for funding through a CASE award and access to the HRSEM. RD thanks J Buglass, N Goesbeek, P Corbett and A Knoester for useful discussions and assistance with the HRSEM. Thanks are due to F Rocca and J Bhalla for their Monte Carlo code.

## REFERENCES

Darji R and Howie A 1997, Micron **28**, 95
Imeson D, Milne R H, Berger S D and McMullan D 1985 Ultramicroscopy **17**, 243
Joy D C 1995, Monte Carlo modelling for Electron Microscopy and Microanalysis, eds M Lapp, J-I Nishizawa, B Snavely, H Stark, A C Tam and T Wilson (Oxford University Press, Oxford) Chapters 6 and 8
Liu J and Spinnler G E 1993 Proc. 51st Ann. Meet. Microsc. Soc. Amer. pp 784-785
Rocca F 1986 PhD Thesis University of Cambridge
Shih A, Yater J, Pehrsson P, Butler J, Hor C and Abrams R 1997, J Appl Phys **82** (4), p1860
Smith D J, Yao M H, Allard L F and Datye A K 1995 Catalysis Letters **31**, 57

*Inst. Phys. Conf. Ser. No 153: Section 9*
*Paper presented at Electron Microscopy and Analysis Group Conf. EMAG97, Cambridge, 1997*
© 1997 IOP Publishing Ltd

# Structure of neodymia clusters supported on magnesia surfaces

**A Burrows, C J Kiely, J A Perez Omil[*], J J Calvino[*] and R W Joyner[#]**

Department of Materials Science and Engineering, The University of Liverpool, UK

[*]Departmento de Quimica Inorganica, Facultad de Ciencias, Universidad de Cádiz, Spain

[#]Catalysis Research Centre, Nottingham Trent University, Nottingham, UK

**ABSTRACT:** MgO doped with $Nd_2O_3$ can be used to catalyse the oxidative coupling of methane reaction. Epitaxial microclusters of neodymia have been found to exist on the MgO (100) surface when the doping level of $Nd_2O_3$ is less than 3 at%. To assist in the interpretation of high resolution micrographs of these features, image simulations using a number of candidate model structures have been carried out. Using two plausible models for the epitaxial relationship between $Nd_2O_3$ and MgO, our simulations show that the most likely model is based on an incoherent parallel interface.

## 1. INTRODUCTION

The oxidative coupling of methane (OCM) to form $C_2$ hydrocarbons is a very important area of study in heterogeneous catalysis. The huge reserves of natural gas that exist represent an untapped fuel source that could conceivably replace our current reliance on fuels derived from crude oil. Many simple oxides will catalyse the OCM reaction (Kalenik and Wolf 1993) though usually the products produced are in such small quantities that the process is not economically viable. However, synergistic effects have been demonstrated by doping one oxide with another which can lead to a significant improvement in catalytic performance. MgO doped with $Nd_2O_3$ is one such system and has been studied in detail by TEM for a number of years (Burrows *et al* 1997). It was found that six different $Nd_2O_3$ morphologies can exist depending on the $Nd_2O_3$ loading and catalyst preparation route. One of these morphologies, the epitaxial microcluster, is the subject of a detailed HREM investigation in this paper. Observed in samples doped with <3 at% $Nd_2O_3$, epitaxial microclusters are commonly found on MgO (100) surfaces and extend laterally for no more than 3.5 nm. The fact they are difficult to observe in profile suggests that the cluster thickness is only one or two monolayers thick. We postulate of two viable orientation relationships between the microcluster and the support. Image simulation techniques have allowed us to determine which model gives the best match to experimentally obtained images and also to deduce the most likely shape of the clusters.

## 2. EXPERIMENTAL

Details of the catalyst preparation procedure used have been published elsewhere (Sinev *et al* 1991). The experimental image used for detailed analysis here was obtained using a JEOL 2000EX high resolution electron microscope operating at 200 kV. The point to point resolution of this instrument is 0.21 nm. Simulations were carried out using the EMS suite of programs (Stadelmann 1987) based on supercells created using the Rhodius fortran program developed at Cadiz (Botana *et al* 1994).

## 3.    RESULTS AND DISCUSSION

Figure 1 shows an experimental image of three neodymia epitaxial microclusters supported on the MgO (100) surface viewed along the [011] MgO direction. Lattice fringe measurements indicate that the clusters bear a close structural resemblance to the cubic crystal form of $Nd_2O_3$ for which $a = 1.1056$ nm (the other bulk form of neodymia has a hexagonal unit cell) although the match is not precise due to pseudomorphic strain effects.

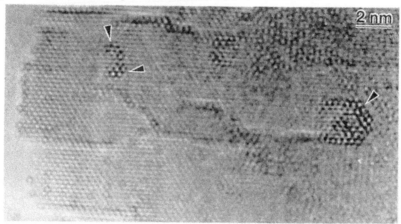

**Fig. 1.** Experimental image showing three clusters of $Nd_2O_3$ (arrowed) epitaxially oriented on the MgO (100) surface. The viewing direction **B** = [011] MgO.

For the image simulations of these clusters, we postulate two models. The first, Model A, is based on parallel epitaxy between the two crystal structures thus:

$$Nd_2O_3[001]//MgO[001] \text{ and } Nd_2O_3[100]//MgO[100]$$

There are two possibilities for this model i) pseudomorphically strained (400) $Nd_2O_3$ planes matching to (200) MgO crystal planes or ii) the unstrained case in which there is near coincidence of every third $Nd_2O_3$ (400) plane with every fourth (200) MgO plane, i.e. an incoherent interface. In the former case, a mismatch of 32% would have to be accommodated which is physically unrealistic whilst in the latter possibility the 3:4 mismatch is only 1.6%. Therefore, for Model A we have assumed an incoherent interface.

The second viable epitaxial model, Model B, takes the form:

$$Nd_2O_3[001]//MgO[001] \text{ and } Nd_2O_3[110]//MgO[100]$$

which is based on a 45° rotation of the neodymia unit cell about the coincident [001] direction. This is a coherent interface where the (440) planes of $Nd_2O_3$ are almost matched with the (200) planes of MgO. The lattice mismatch for Model B is 7.2%. Both models are shown as supercells edge on in Figures 2a and b respectively where the epitaxial plane is (001) MgO. Simulations based on these supercells were carried out for a range of microscope defoci assuming a hemi spherical cluster of 3.2 nm diameter (which approaches the maximum lateral extent that was observed for these features) and supporting MgO thickness of 3 nm.

The simulations obtained from the two models from calculations along the z-axis of the supercell, i.e. parallel to the MgO [011] zone axis, are presented in Figure 3. It is apparent that the contrast observed in the series of simulated images for Model A matches well with the experimental image, far better than Model B in fact. Whilst the hemi spherical model is a good first approximation to the cluster shape, we are experimenting with other cluster morphologies, i.e. square rafts with either {100} or {110} terminating faces, a truncated hemisphere and circular monolayers. Furthermore, the effects of varying support thickness and introducing pseudomorphic strain to the cluster for Model B are also under investigation. A more detailed version of this work will be published elsewhere (Burrows *et al* 1997).

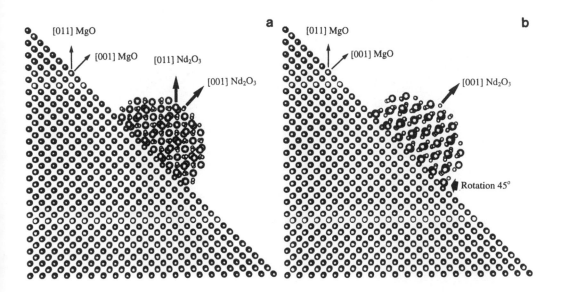

**Fig. 2** Supercells of the two models considered for the epitaxial microclusters, a) parallel epitaxy - incoherent interface and b) 45° rotation epitaxy.

## 4.    CONCLUSIONS

Computer image simulation techniques have been used to investigate the contrast observed in an experimental image of epitaxial microclusters of $Nd_2O_3$ on the (100) surface of MgO. We have demonstrated that a parallel epitaxy model matches experimentally obtained images better than one based on a 45° rotation of the $Nd_2O_3$ unit cell with respect to the MgO substrate. The simulations also indicate that to a first approximation, the clusters can be thought of as hemi-spherical in shape. Further refinements of our modelling procedure are currently underway.

**REFERENCES**

Botana F.J., Calvino J.J., Blanco G., Marcos M., and Pérez-Omil J.A.,"13th International Congress on Electron Microscopy (ICEM13), Paris, France" abstracts, p1085, 1994.
Burrows A, Kiely C J, Hutchings G J, Joyner R W and Sinev M Y, 1997, J. Catal. **167** p77
Burrows A, Kiely C J, Perez Omil J A, Calvino J J and Joyner R W, 1997, Ultramicroscopy (submitted)
Kalenik K and Wolf E E, 1993 Catal. R. Soc. Chem. **10** p154
Sinev M Y, Tyulenin Y P, and Rozentuller B V, 1991 Kinet. Catal. **32** p807
Stadelmann P, 1987 Ultramicroscopy **21** p131

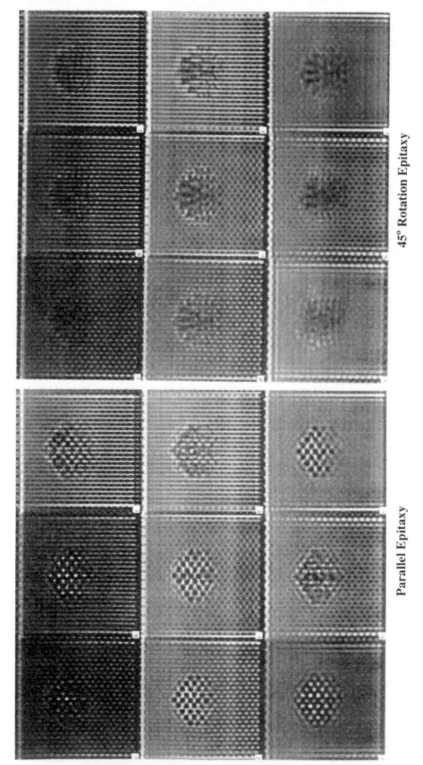

**Parallel Epitaxy**

**45° Rotation Epitaxy**

**Fig. 3** Computer image simulations based on the supercells in Figure 2. For both sets, the microscope defocus varies from 10 to 90 nm in steps of 10 nm from *a* to *i*.

*Inst. Phys. Conf. Ser. No 153: Section 9*
*Paper presented at Electron Microscopy and Analysis Group Conf. EMAG97, Cambridge, 1997*
© 1997 IOP Publishing Ltd

# Atomic resolution Z-contrast imaging of ultradispersed catalysts

**P D Nellist and S J Pennycook***

Cavendish Laboratory, University of Cambridge, Madingley Road., Cambridge CB3 0HE, UK.
*Oak Ridge National Laboratory, Solid State Division, PO Box 2008, Oak Ridge, TN 37831-6031, USA.

**ABSTRACT:** Using Z-contrast imaging in a 300 kV scanning transmission electron microscope, direct images of single metal atoms on a support material in industrial catalyst systems are formed. The simultaneously collected conventional bright-field images reveal orientational information about the support, which allows possible configurations of Pt atoms on a $\gamma$-Al$_2$O$_3$ support surface to be deduced. Image intensities also allow the three-dimensional configuration of a raft of Rh atoms on a $\gamma$-Al$_2$O$_3$ support to be determined.

## 1. INTRODUCTION

In the study of the atomic dispersion of a catalytic metal and its configuration relative to the support material, transmission electron microscopy (TEM) is an invaluable tool. Generally, the rough, insulating nature of the support precludes the use of any other direct imaging technique, such as scanning probe microscopy. Although single atom imaging has been demonstrated using conventional TEM (Iijima, 1977), a specially thin and smooth single crystal support was used. In the case of catalysts, the contrast that is observed from the support usually prevents the imaging of metal clusters smaller than about 1 nm in diameter (Datye and Smith, 1992). Using Z-contrast imaging in a high-resolution scanning transmission electron microscope (STEM), however, allows individual atoms to observed on industrially relevant supports, such as SiO$_2$ and $\gamma$-Al$_2$O$_3$.

## 2. ATOMIC RESOLUTION Z-CONTRAST IMAGING

A Z-contrast image is formed in a scanning transmission electron microscope (STEM) by detecting the electrons that are scattered out to a large, high-angle annular dark-field (HAADF) detector as a focussed electron probe is scanned over the specimen. Because of the large size of the detector, it is an incoherent imaging mode (Jesson and Pennycook, 1995) that forms a direct map of atomic locations with intensities that are strongly dependent on the atomic number (Z) of the observed atom. This ability to distinguish between heavy and light atoms makes it a powerful method for the observation of supported metal atoms, and it is therefore extremely useful in the study of catalyst materials. Using a 100 kV STEM, clusters as small as three atoms have been detected (Treacy and Rice, 1989), but the individual atoms were not resolved.

The VG Microscopes HB603U 300 kV STEM ($C_S$=1 mm) at Oak Ridge National Laboratory is capable of forming a probe only 0.13 nm in diameter. Such a small probe enhances the signal-to-background ratio for supported metal atoms, so that single atom sensitivity is now possible for tiny clusters on real catalyst supports. Using a small axial detector in a STEM also allows the conventional bright-field (BF) image to be formed simultaneously with the Z-contrast image. Because it is formed coherently, the BF image is dominated by contrast due to scattering from the support material. When not too far from a

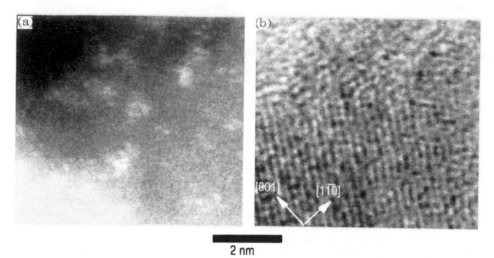

2 nm

Fig.1. Simultaneously collected (a) Z-contrast and (b) BF images from a sample of 3 wt% Pt on $\gamma$-Al$_2$O$_3$.

major zone axis, the BF contrast allows the orientation of the support material to be determined.

## 3.   IMAGING OF PLATINUM ON $\gamma$-ALUMINA

Simultaneously collected Z-contrast and BF images from a sample of 3 wt% Pt on $\gamma$-Al$_2$O$_3$ are shown in Fig. 1.  In Fig. 1a some of the clusters can be seen to be resolved into single atoms, whilst Fig. 1b is dominated by the support contrast.  The strong {222} $\gamma$-Al$_2$O$_3$ fringes visible in Fig. 1b allow the orientation of the support to be determined.  The beam direction is close to [110], and the orthogonal directions are indicated on Fig. 1b.  Performing a bandpass filter on Fig. 1a results in Fig. 2.  Trimers of Pt atoms were often observed in this sample, and Fig. 2 shows the possible surface sites that were deduced (Nellist and Pennycook, 1996).  Near the bottom of Fig. 2, dimers can be seen that have configurations of one side of the trimer.

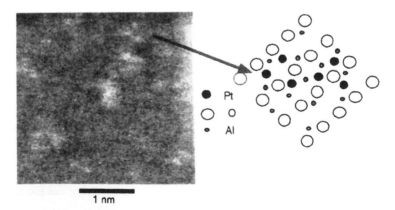

1 nm

Fig. 2.  A bandpass filtered version of Fig. 1a.  From the orientational information in Fig. 1b possible $\gamma$-Al$_2$O$_3$ surface sites for a Pt trimer are indicated.

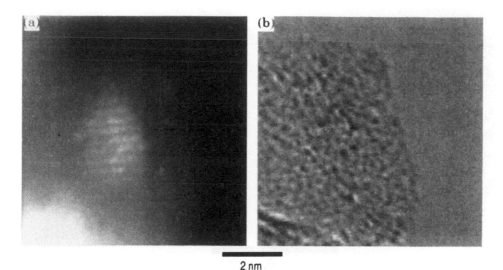

Fig. 3. Simultaneously collected (a) Z-contrast and (b) BF images from a sample of Rh on $\gamma$-Al$_2$O$_3$.

## 4. IMAGING OF RHODIUM ON $\gamma$-ALUMINA

Fig. 3 shows a simultaneous Z-contrast and BF pair of images from a sample of 1.2 wt% Rh on $\gamma$-Al$_2$O$_3$. The rows of Rh atoms in Fig. 3a are almost parallel to the facets of the $\gamma$-Al$_2$O$_3$, again suggesting an interaction with the support material. A bandpass filtered version of Fig. 3a is shown in Fig. 4. Most of the atom-like feature intensities in Fig. 4 have a similar intensity. Since part of the cluster has become obviously disordered, we may associate this intensity with the presence of one atom, and therefore the cluster is mainly a monolayer raft. Some of the features are more intense than the majority, with an intensity 2.5 times greater than the single atoms. Because of partial coherence along the beam direction, this intensity is consistent with these features being two atoms arranged in a column (Nellist and Pennycook, 1996). The position of some of these columns at the edge of the raft, or even isolated from it, suggests that something is providing the support for such a column. One possibility is that the Rh raft has become partially incorporated into the first few layers of the $\gamma$-Al$_2$O$_3$, which is providing the necessary support.

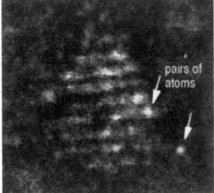

pairs of atoms

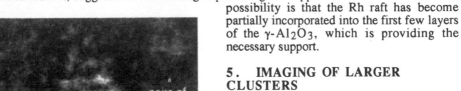

2 nm

Fig. 4. A bandpass filtered image of a region of Fig. 3a.

## 5. IMAGING OF LARGER CLUSTERS

We have demonstrated how Z-contrast is able to image single metal atoms on a support. However, it is also useful in the study of larger metal clusters. In Fig. 5, the support is so thick that the conventional BF image barely shows any contrast arising from the Rh cluster. The Z-contrast image shows the cluster clearly, and it is possible to identify a twin near the centre of the particle.

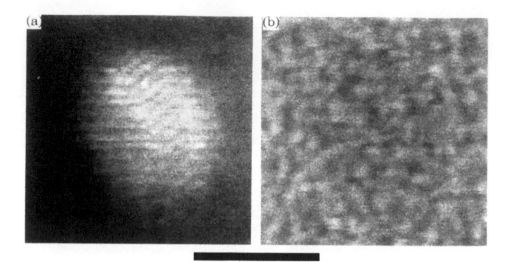

2 nm

Fig. 5. A simultaneously recorded (a) Z-contrast and (b) BF image of a Rh cluster supported on γ-Al$_2$O$_3$.

## 6. CONCLUSIONS

The small illuminating electron probe now available in a 300 kV STEM allows direct imaging of the atomic configuration of supported metal atoms in industrial catalyst systems. The ability to simultaneously record the conventional BF image can allow the metal atom configurations to be related to the orientation of the support material, and possible surface sites deduced. Image intensities also reveal information on the three-dimensional configuration of such clusters.

With this information available, it should now be possible to relate the cluster configurations to the catalytic activity experimentally observed, and to study the changes in the cluster structures as the catalyst degrades.

## ACKNOWLEDGEMENTS

We thank D. R. Liu and E. Lowenthal for the provision of the catalyst samples. This work was performed at Oak Ridge National Laboratory, managed by Lockheed Martin Energy Research for the U.S. Department of Energy under contract DE-AC05-96OR22464, and through an appointment to the ORNL Postdoctoral Research Associates Program administered by ORISE.

Nellist P D and Pennycook S J 1996 Science **274**, 413
Iijima S 1977 Optik **48**, 193
Datye A K and Smith D J 1992 Catal. Rev. Sci. Eng. **34**, 129
Treacy M M J and Rice S B 1989 **156**, 211
Jesson D E and Pennycook S J 1995 Proc. Roy Soc. Lond. A **449**, 273

*Inst. Phys. Conf. Ser. No 153: Section 9*
*Paper presented at Electron Microscopy and Analysis Group Conf. EMAG97, Cambridge, 1997*
© *1997 IOP Publishing Ltd*

# STEM Analysis of Bi-Metallic Catalysts in Mesoporous MCM-41

D Özkaya[a]   JM Thomas[ac]   DS Shephard[b]    T Maschmeyer[bc]   BFG Johnson[b]   (
Sankar[c] R Oldroyd[c]
[a]Department of Materials Science and Metallurgy, Pembroke St., Cambridge CB2 3QZ, UK
[b]Department of Chemistry, Lensfield Road, Cambridge, CB2 1EW, UK
[c] Davy-Faraday Research Laboratories, The Royal institution of Great Britain, 21 Albemarle Stree
London W1X 4BS

**ABSTRACT**:   Electron microscopy of Ag-Ru and Cu-Ru bi-metallic catalysts in MCM-41 i
carried out in a dedicated field emission gun scanning transmission electron microscope (VC
HB501).  The distinctive advantage is that the beam damage is limited to scanned area since onl
the part of the sample that the investigation is being carried out is exposed to electrons; this i
not necessarily the case in a conventional TEM. This negates the more conventional wisdom tha
the intense probe will damage the material before it can be investigated. Other advantages c
using STEM are more conventional: high resolution analytical analysis of the catalyst particles i
possible using the sub-nanometer probe and high angle annular dark field (Z-contrast) imagin
allows the analysis of the distribution and homogeneity of the particles within channels mor
readily then any other method.  Examples of the application are shown for Ag-Ru and Cu-Ru bi
metallic particles in MCM-41.  Inferences are also drawn in terms of the number of particle
superposed or clumped together inside the channels.

## 1-    Introduction

MCM-41 is a mesoporous silicate material with a honeycomb like structure composed c
approximately 3nm diameter channels. Figures 1a and 1b show STEM bright field images from bot
parallel [001] and perpendicular to pore axis [100] respectively. It is possible to vary the size of thes
channels by changing the preparation parameters (Beck and Vartuli, 1996). The channel widt
determines the size of the molecules that can be loaded as catalysts, and to a lesser extent the size c
the molecules that it will be selective towards if used as a catalyst support.
Bi-metallic nano-clusters have been investigated as catalysts since their superior catalytic propertie
were realised by Sinfelt (1983) compared to that of their solid solution bulk metal counterparts. Th
nano-particles are introduced into the channels using specific techniques which use mixed-meta
carbonyl clusters as precursors (see Thomas and Thomas, 1996 for general information), furthe
details are published elsewhere (Shephard et al, 1997). Transmission electron microscopy has th
ideal resolution to investigate MCM-41. However the classic problem of beam damage i
mesoporous materials bedevils MCM-41 as well (although to a lesser extent compared to zeolites du
to lack of de-stabilising (AlO$_4$)$^-$ complexes).
The investigation of mesoporous materials by dedicated STEM has previously been restricted to th
investigation of beam damage effects (see McComb, 1990) rather than for problem solving purpose
or testing existing models on electron scattering from hollow crystals (Walsh, 1991).
In this paper we demonstrate how useful a tool dedicated STEM has proved to be in th
investigations of use of MCM-41 as a support for bi-metallic particles.  High angle annular dark fiel
imaging has revealed the relative distribution and homogeneity of Ru-Ag and Ru-Cu bi-metalli
particles within MCM-41 channels and the EDX has confirmed the relative atomic proportions of b
metallic particles. It is clear from the investigation of MCM-41 that the ability of the STEM to onl
irradiate the area that is being investigated is a very distinct advantage and the intense probe does nc
prevent these materials from being properly investigated by STEM (or at least at a beam dose c
6x10$^{25}$electrons/m$^2$)

404

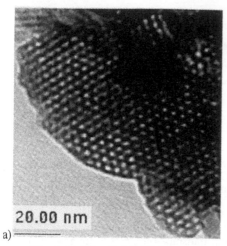
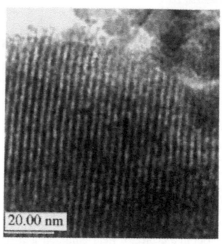

a) 20.00 nm    b) 20.00 nm

Fig 1 Bright field STEM images of MCM-41 from both a) pore axis [001] and b) perpendicular to pore axis [100]respectively.

## 2- Experimental

Experiments are conducted using a field emission dedicated scanning transmission electron microscope (VG HB501), with a windowless EDX detector. MCM-41 samples have been prepared by crushing the particles between two glass slides and spreading them on a holey carbon film supported on a Cu grid for Ag-Ru MCM-41 and Ti grid for Cu-Ru MCM-41. The samples were briefly heated under the light bulb in the specimen preparation chamber of STEM prior to putting them in the column. The details of preparations of bi-metallic MCM-41 can be found in Shephard *et al.* (1997). Both the $Ag_3Ru_{10}$ (Ag-Ru) and $Cu_4Ru_{12}$ (Cu-Ru) organometallic carbonyl cluster precursors have been structurally and spectroscopically analysed before and after insertion into the MCM-41channels and after thermal heat treatment in vacuo (200°C, 0.01torr) by several techniques including single crystal X-ray diffraction, EXAFS, XANES, NMR and FT-IR (solution and solid state) together with elemental analysis for C, H and N (Shephard *et al.,* 1997).

## 3- Ag-Ru MCM-41

Figure 2a and b show bright field and high angle annular dark field images of $Ag_3Ru_{10}$ (confirmed using EDX, at%Ru/at%Ag=3.32) particles within MCM-41 channels (.

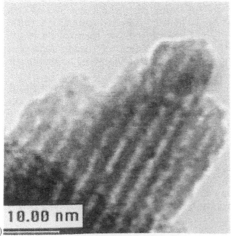
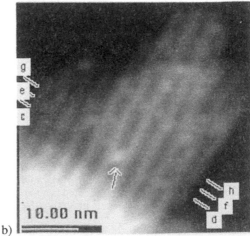

a) 10.00 nm    b) 10.00 nm

Fig 2 Bright field (a) and high angle annular dark field (b) images of $Ru_{10}Ag_3$ particles (bright ) within MCM-41 channels. Channel contrast is bright in a) and dark in b).

It is possible to clearly see the MCM channels on both bright field and annular dark field image
The Ag-Ru particles are extra bright in high angle annular dark field image (fig2b) due to Z-contra:
The particles can be seen to be smaller than the channel size and anchored to the walls of the MCM
41 from only one side apart from one cluster (arrowed in fig2b) which is likely to be two particl·
superposed or clumped together. Figure 3 shows the analysis of the intensity profiles from thr·
different lines marked in figure 2b. cd is from part of the sample which contains the suspected doub
particle in the middle and a single particle on the right, ef is from a part with no particle for referen·
and gh is from a part which contains only one particle on the left. From the contrast envelope of tl
three line profiles, it is clear that the sample has near cylindrical cross section, thinner towards tl
edges and thicker in the middle.  It is possible to eliminate the contrast effect from the MCM-41 t
subtracting the reference line profile ef from the other profiles. This is shown in the bottom ·
figure 3 marked as (cd-ef) for the double particle (longer arrow) in the middle as well as a sing
particle on the right and (gh-ef) for the single particle (shorter arrow) on the left. From the ratio of tl
intensity peaks it can be confirmed that the initial presumption of two superposed or clump·
particles is right for the middle particle. This leads to the more important inference that the MCM-∠
channels are decorated with single particles anchored to the side walls of the MCM-41 which in tui
makes the bi-metallic particles effective catalysts.

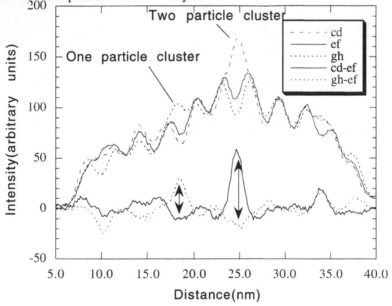

Figure3 The intensity profiles from three different areas marked in figure 2b. cd contains
suspected superposed particles in the middle and a single particle on the right, ef is from a pa
with no particle for reference and gh is from a part which contains only one particle on the lei
The two profiles at the bottom (cd-ef) and (gh-ef) show the intensity profiles due to particl·
alone. Note the differences in the intensity at the middle due to superposed particles and sing
particles in either side

## 4-    Cu-Ru MCM-41

Figures 4a and b show high angle annular dark field images from the $Cu_4Ru_{12}$ bi-metallic particl·
(confirmed by EDX, at%Ru/at%Cu=2.9) before and after an electron exposure ·
$12x10^{25}$ electrons/$m^2$ respectively. Figure 4a show an MCM-41 cylindrical bundle of 4 channe
across and the bundle is curved at either end four particles inside a channel have been marked. ]
figure 4b the structure due to MCM-41 is completely lost and due to beam damage and vitrifie·
However it is still possible to notice that the 4 particles that were marked in the earlier image are sti
there and completely intact and there are no sign of coagulation in spite of the fact that they are vei
close to each other under an intense beam. Comparing the two images shows this also holds for tl·
other particles.

is observation shows the strength of the bond between the MCM-41 structure and the Cu-Ru
isters. If this material is to have some re-usability as a catalyst, this kind of strong anchoring is
finitely a desirable quality.

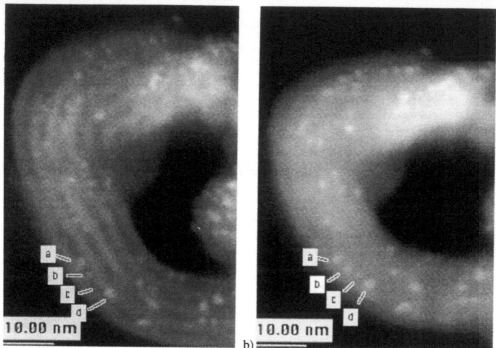

b)

Figure 4 High angle annular dark field images of $Cu_4Ru_{12}$ bi-metallic particles inside
MCM-41 before and after vitrification due to electron beam damage. Note that the particles
marked have kept their position exactly before and after beam damage (no coagulation
amongst particles). This shows the strength of anchoring between the Cu-Ru cluster and the
MCM-41.

## Summary

o examples of investigation of bi-metallic clusters inside the MCM-41 mesoporous silicate
erial by STEM are presented. High angle annular dark field imaging has helped to analyse the
ribution and homogeneity of the particles inside the channels and the behaviour of the Ru-Cu
icles after beam damage helped to draw inferences about the strength of the bond formed between
bi-metallic clusters and the walls of the MCM-41.
clear from the investigation of MCM-41 that the ability of the STEM to only irradiate the area that
eing investigated is a very distinct advantage.

ferences
felt J H *Bi-metallic Catalysts* 1983 J Wiley New York
mas J M and Thomas W J 1996 *Principles and Practice of Heteregeneous Catalysis* VCH,
inheim
k J S and Vartuli J C 1996 Current Opinion in Solid State and Mat. Sci. **1** 76
phard DS Mashmeyer T Johnson BFG Thomas JM Sankar S Ozkaya D Zhou W Oldroyd RD
7 Angew. Chemie Int.Ed.Engl. (in press)
Comb D W (1990) PhD thesis University of Cambridge
lsh C A Phil.Mag. 1991 **63** 1063-1078

*Inst. Phys. Conf. Ser. No 153: Section 10*
*Paper presented at Electron Microscopy and Analysis Group Conf. EMAG97, Cambridge, 1997*
© *1997 IOP Publishing Ltd*

# Combined microscopic studies of wide-gap materials

## J. W. Steeds

H H Wills Physics Laboratory, University of Bristol, Tyndall Avenue, Bristol BS8 1TL

**ABSTRACT**: The information provided by a wide variety of microscopies and related spectroscopies may be combined to give a more complete understanding of the complex growth processes involved in fabricating wide-gap materials. Two examples will be illustrated. One is the growth of thick oriented diamond films by CVD on silicon substrates: the other is the growth of epitaxial layers of GaN. The microscopies that have been used are optical microscopy (including interference, Raman and photoluminescence spectroscopy), scanning electron microscopy (including electron back-scattered diffraction and cathodoluminescence) and transmission electron microscopy (including CBED, cathodoluminescence, and energy-selected imaging). Information has been deduced about the growth processes, the stresses induced and point and extended defect incorporation.

## 1. INTRODUCTION

There has been a remarkable recent growth of interest in wide-gap semiconductors such as diamond, gallium nitride, aluminium nitride and silicon carbide. These materials have high breakdown fields, they can operate at high powers, high temperatures, high frequencies and are radiation hard. They are also of interest for a wide variety of electronic applications, as sensors and, in the case of the nitrides, as light emitting diodes and lasers.

While the potential advantages of these materials have been realized for a long time it is the recent advances in crystal growth, impurity, stoichiometry and doping control, defect incorporation and subsequent processing that have generated realistic prospects for marketable devices. Electron microscopy and analysis is playing an important role in this development and I shall attempt to demonstrate that these materials are particularly well suited to study by electron-based techniques by highlighting recent work in this laboratory on diamond and GaN as examples of what can be achieved.

## 2. SPECIMEN PREPARATION

Diamond is frequently grown on Si substrates by CVD processes that involve microwave plasmas, hot filaments or plasma torches. As the diamond films are very hard and also much harder than their substrates, specimen preparation for TEM observation is time consuming and difficult. Special mechanical polishing wheels have been developed for diamond but, even so, randomly oriented polycrystalline samples are the most difficult to polish. Ion thinning rates are also slow and much slower than the rates for the Si substrates. As an attempt to reduce these problems we have recently explored the use of a finely focused scanned laser beam to cut wedge-shaped cross-sections through relatively thick films (≤100μm in thickness). A wedge angle of about 5° is used and the specimens are subsequently reduced in thickness by unidirectional ion thinning at a low angle to one of the wedge surfaces.

An alternative approach is to use a focused ion beam to prepare electron transparent sections from chosen regions on a specimen. Widely used in silicon technology (Young 1993) this technique has also been used for the preparation of electron transparent specimens of both diamond (Dennig et al 1996) and GaN. GaN, frequently grown on sapphire substrates, also presents something of a challenge to specimen preparation techniques because in this case the substrate is considerably harder than the epitaxial layer grown on it.

Once prepared, the samples of wide gap materials are rather resistant to electron radiation damage and do not deteriorate noticeably with age. This is a very attractive feature that facilitates the performance of a variety of experiments in areas of interest.

## 3. GROWTH OF ORIENTED DIAMOND ON Si(100)

It is now well established that oriented diamond can be grown on (100) Si substrates by bias-enhanced nucleation (Yugo et al 1991). These films may be with random misorientation about the [100] growth direction, or highly oriented with respect to the Si surface lattice plane. There is a large mismatch between the lattice constants of Si and diamond but as a result of the larger expansion coefficient of Si there is an exact 3:2 ratio between the two values at a given temperature within the diamond growth range. Some have argued that this accounts for the highly oriented growth, others argue that a thin intervening layer of cubic SiC is created with a lattice constant intermediate between that of Si and diamond. We have therefore performed experiments on a cross-section through the Si/diamond interface of a highly oriented growth. The specimen was mounted in a Hitachi HF2000 with a cold field-emission source and a Gatan imaging PEELS system to form high resolution plasmon-loss images. It is important to be aware of the possible presence of amorphous carbon layers at the interface. These can certainly be produced under particular growth conditions but, unfortunately, they have features of their low-loss spectra that closely resemble those of SiC. The major electron energy loss of graphitic amorphous carbon occurs at about 22eV as compared with about 20eV for SiC. Both peaks are very broad. However, the first plasmon loss of Si occurs at about 17eV and amorphous graphitic carbon has a $\pi^*$ loss at around 6eV. The way to reveal the nature of the interface is, therefore, to form images at the 6eV and 20eV losses. At 20eV both SiC and amorphous graphite carbon will be revealed but the 6eV image distinguishes the two. Figs 1 and 2 illustrate this procedure. In Fig 1 the 17eV image reveals the Si in a cross-section through the interface with a oriented diamond film. The 21.5eV image shows an essentially continuous line of either SiC or graphitic a-C; the absence of any part of this line is an image taken at 6eV revealed no evidence of interfacial a-C indicating that SiC has been formed everywhere. By contrast, Fig. 2 shows a region where there is clear evidence of graphitic a-C at one place along the interface.

Alternative methods exist for investigating the interface region. Using a 2nm-sized probe, convergent beam electron diffraction patterns were obtained at various positions across the interface of a <110> oriented region. Three similar patterns were generated with lattice constants equal to those of Si, cubic SiC and diamond respectively. In addition high resolution lattice images were generated at the same zone axis in the interface regions. Fourier transformation of the images revealed that each of the Bragg peaks of the <110> diffraction pattern were tripled. By masking off the inner peaks of each triad and forming an image with the outer peaks a lattice image restricted to the Si side of the interface was generated. By selecting the central peak of the triad the lattice image was restricted to a narrow strip along the diamond/Si interface; the innermost peaks gave a lattice image of the diamond. While completely conclusive results were thus obtained by both diffraction and high resolution experiments about the existence of an intervening SiC layer, these two techniques are both quite challenging to perform and limited to a very small region of the interface. Imaging PEELS gave results that were not only easy to obtain but also covered very much greater distances along the interface.

It is well-known that diamond films are often in a highly stressed state. The differential contraction on cooling from the growth temperature would account for a compressive stress in the film but the stresses deduced often exceed the likely values originating in this way. We found that most of the oriented diamond films that we studied had well defined square facets. By performing cathodoluminescence (CL) experiments on these square facets we frequently found nitrogen-vacancy related emission at approximately 532nm and 575nm. The second of these two emission peaks has been carefully studied under the influence of uniaxial stress applied along different crystallographic directions (Davies 1979). Splittings of the 575 nm peak occur that are similar to the splittings observed in our CL

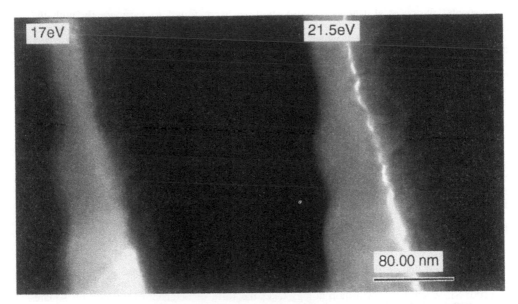

Figure 1     Cross-sectional TEM image of the interface between highly oriented (100) diamond (right) on silicon (left), without graphitic regions at the interface but with a narrow continuous layer of SiC revealed in the 21.5eV loss (courtesy of J. A. Wilson and A. Gilmore).

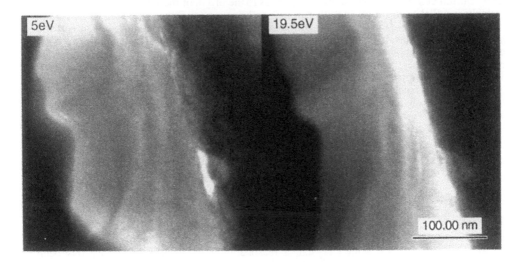

Figure 2     Cross-sectional TEM image of the interface between highly oriented (100) diamond (right) on silicon (left). The 5eV loss image reveals graphitic regions at the interface and the 19.5eV loss image reveals both SiC and amorphous carbon at the interface. (Courtesy of J. A. Wilson and A. Gilmore).

experiments (Burton et al 1995a). These splittings were found to vary greatly with the position of the electron beam creating the excitation in the diamond and the accelerating voltage used that determines the depth at which carriers were created. By assuming that the stress at the centres of the square facets is biaxial in the plane with an independent stress in the direction normal to the surface, the analysis of Davies could be used to make stress determinations in the material (Steeds et al 1996). The results revealed a biaxial tensile stress near the surface, accounting for the fractures that were sometimes observed, that changed to compression at greater depths. An alternative approach is to use confocal micro-Raman spectroscopy to deduce the stress values. We have shown that the results of the two different methods are broadly compatible, but the region of overlap is not very great because the depth (and lateral) resolution of Raman spectroscopy is much less than that of CL experiments. However, the back-scattered Raman technique can be used to probe much greater specimen depths. The results were then compared with finite element model calculations (Pickard et al 1997). Nomarksi interference optical microscopy, with associated back scattered diffraction or electron channelling measurements in a SEM, gave confirmatory evidence of the stresses that had been determined (Burton et al 1995b).

## 4. DEFECTS AND THE ORIGIN OF COMPRESSIVE STRESS IN DIAMOND

It is well known that {111} growth sectors of CVD diamond tend to have a much higher density of extended defects than {100} growth sectors. In the case of oriented growth on (100) Si, when the growth morphology takes the form of octahedra truncated by square facets, TEM samples show almost defect free (100) growth sectors but the surrounding octahedral regions have enormous defect densities. Careful diffraction contrast experiments performed by tilting by ±45° about the perpendicular <010> axes, lead us to conclude that each of four separate faces of the octahedra had inward pointing tetrahedral faults created by stacking faults or nano-twins on the three {111} planes inclined to the growth surface in that sector (Burton et al 1995a). The question arises as to the origin of these open tetrahedral faults (no faults parallel to the local growth surface) and so various forms of spectroscopy were carried out in the different regions. For sufficiently large crystallites it was possible to perform micro-Raman experiments in low defect density and high defect density regions of an electron transparent TEM sample. These experiments clearly indicated the presence of non-diamond carbon in the highly defective regions. Confocal micro-Raman mapping of the material allowed us to perform depth resolved information about the ratio of non-diamond to diamond carbon in the individual crystallites that confirmed these conclusions (Pickard et al 1998). An alternative approach that we adopted was to perform PEELS experiments in the Hitachi HF2000. These experiments revealed that the highly defective regions had a small but definite $\pi^*$ pre peak before the main diamond absorption edge, while the defect-free regions did not. We conclude that the inclusion of non-diamond carbon in {111} growth sectors is responsible for the high defect densities observed and is the probably origin of the high compressive stress that can occur in many CVD diamond films.

## 5. DEFECTS IN GaN

Wurtzite-structured GaN layers grown on sapphire substrates normally contain large numbers of extended defects. Even when grown by homo-epitaxy on the nitrogen-terminated face of a bulk wurtzite-structured crystal, a reasonably high dislocation density has been observed. A very effective method for determining the Burgers vectors of dislocations has recently been developed that relies on the splitting of Bragg lines in large angle convergent beam electron diffraction (LACBED) patterns (Chou et al 1992). We applied this method to N-face homo-epitaxial layers where the dislocations were reasonably widely separated and found, in close proximity, dislocations with **a**, **c** and **a** + **c** Burgers vectors (Ponce et all 1996a).

In addition to dislocations, tubular inversion domains were observed in many specimens that were aligned along the growth direction. Polarity determination was performed by CBED and used to show that inversion occurred within the tubes (Ponce et al 1996a, Ponce et al 1997). Diffraction contrast experiments were performed and the results interpreted in terms of an atomic model for the inversion domain boundary structure (Cherns et al 1997a).

Nanopipes were yet another defect found in the various specimens studied. At first sight similar in dimensions to the tubes of inversion domain parallel to the c axis, these were in fact hollow and generally hexagonal in cross section (Ponce et al 1997a)). The existence of a c screw dislocation was demonstrated in association with the majority of these nanopipes by the use of LACBED but occasionally nanopipes were found that lacked any associated dislocations (Cherns et al 1997b).

## 6. DOPANT DISTRIBUTION IN DIAMOND AND GaN

One of the key issues with wide-gap semiconductors is that of dopant incorporation. Most are relatively easy to dope in one sense but not the other. For example, diamond is easily doped p-type but not n-type; for GaN the opposite is the case. B dopant distribution in CVD diamond is of some considerable interest but no really reliable technique has been established for studying it, although some promising results have been obtained by CL (Graham 1994). At the high doping levels often of interest the CL intensity is greatly reduced and spectrally very broad and unspecific. N donors, although not easily incorporated after growth and very slow to diffuse at temperatures of interest, are readily identified when associated with vacancies on account of certain well-defined CL and PL (photoluminescence) peaks. Single substitutional N atoms in the diamond lattice can be associated with vacancies by electron irradiation and hence rendered detectable by CL or PL microscopy. Two N+ vacancy and three N+ vacancy optical centres are also well-known (Mainwood 1994). Interstitial N does not have a known optical signature.

In the case of GaN, p-doping is normally achieved by addition of Mg but it often requires an activation process to make it effective and only relatively low doping concentrations and hole mobilities have been achieved. As-grown undoped GaN is normally n-type but the nature of the donor has yet to be determined conclusively. Si can be added to achieve controlled n-type doping in addition to and distinct from the natural intrinsic n-type behaviour. One major difference between GaN and diamond is the polar nature of GaN. This both leads to the possibility of non-stoichiometry in GaN and also to a difference between the zone centre LO and TO phonons in the polar material. LO phonons have a long range coulomb field that both raises them in energy relative to TO phonons near the zone centre but also leads to electronic coupling effects that do not exist in the case of phonons in diamond. As electrons are added to the conduction band of n-type material, the plasmon frequency ($\omega_p$) increases in energy according to the relation

$$\omega_p^2 = \frac{ne^2}{\varepsilon_o \varepsilon_\infty m^*}$$

where n is the donor concentration ($cm^{-3}$), m* is the electron effective mass and $\varepsilon_\infty$ is the high frequency dielectric permittivity.

When the electron concentration is high enough for the plasma frequency to come close to the LO phonon frequency, strong intermixing occurs and intensity is taken out of the LO mode and redistributed into hybridized coupled LO phonon-plasmon modes. The intensity of the LO phonon mode is then a measure of the donor concentration (Kozawa et al 1994).

On account of its hexagonal structure the phonon modes in GaN involve both non polar modes ($E_2$ symmetry) and polar LO phonon modes with both $A_1$ and $E_1$ symmetry. However, for back-scattered Raman spectroscopy with normal incidence on the c face of the GaN, only the $A_1$ (LO) modes are detected. These are found to vary considerably in intensity depending on the n doping level of specimen and also to vary from place to place within a specimen. In particular, a MOCVD GaN film grown by homo-epitaxy on a nitrogen terminated surface was studied that exhibited typical large hexagonal facets on its growth surface. These hexagonal hillocks had typical dimensions of 20-50μm, some were very shallow peaks others much higher. We concentrated on a very shallow peak with a feature at its apex that is known to be a column of inversion domain with opposite polarity to the adjoining matrix. Raman spectra were generated at the centre of the hexagon and in surrounding regions. The centre of the hexagon gave a strong $A_1$ (LO) peak indicating a low

412

donor concentration while the edges of the facets had particular low $A_1$ (LO) peak intensities indicating high donor concentrations. The centres of the triangular faces making up the nexagonal pyramids had intermediate values of $A_1$(LO) intensity (Ponce et al 1996c, Ponce et al 1997b). Such charges of $A_1$ (LO) intensity can be converted to n donor leads in a quantitative fashion (Wetzel et al 1997)).

## 7. SUMMARY

A number of examples have been given where the combined application of microscopy and spectroscopy to wide-gap materials has led to important information about crystal growth, the origin of internal stresses, defects impurities and dopant distributions. It has been argued that this approach is of considerable importance in the development of high quality materials for specific applications.

## ACKNOWLEDGEMENTS

The author wishes to acknowledge collaboration with a wide variety of colleagues in his research group at Bristol University in the work that has been described. Partial support from the EPSRC, Hitachi and NATO is gratefully recorded.

## REFERENCES

Burton N C, Butler J E, Lang A R and Steeds J W 1995b Proc. Roy. Soc. Lond. **A449**, 555
Burton N C, Steeds J W, Meaden G M, Shreter Y G and Butler J E 1995a Diamond & Rel. Mat. **4**, 1222.
Cherns D, Young W T, Saunders M, Steeds J W and Ponce F A 1997a Phil. Mag. in the press
Cherns D, Young W T, Steeds J W, Ponce F A and Nakamura S 1997b J. Cryst. Growth **178**, 201.
Chou C T, Preston A R and Steeds J W 1992 Phil. Mag. **A65**, 863
Davies G 1979 J. Phys. C. Solid State Physics **12**, 2551
Dennig P A, Koizumi S, Liu H I and Sato Y 1996 Private Communication
Graham R J, Shaapur F, Kato Y and Stoner B R 1994 Appl. Phys. Lett. **65**, 292
Kozawa T, Kachi T, Kano H, Yaga Y, Hashimoto M, Koide N and Manase K 1994 Appl. Phys. Lett. **75**, 1098
Mainwood A, 1994 Phys. Rev. **B49**, 7934
Pickard C D O, Davis T J, Gilmore A, Wang W N and Steeds J W 1998 Diamond Conference Edinburgh, in the press.
Pickard C D O, Davis T J, Wang W N and Steeds J W 1997 Diamond & Rel. Mat. **6**, 1062
Ponce F A, Bour D P, Young W T, Saunders M and Steeds J W 1996b Appl. Phys. Lett. **69**, 337
Ponce F A, Cherns D, Young W T and Steeds J W 1996a Appl. Phys. Lett. **69**, 770
Ponce F A, Cherns D, Young W T, Steeds J W and Nakamura S 1997a Mat. Res. Symp. Proc. Vol 449 "III-V Nitrides" eds. Ponce F A, Monstakas T D, Akasaki I and Monemar B A p405.
Ponce F A, Steeds J W, Dyer C D and Pitt D 1996c Appl. Phys. Lett. **69**, 2650
Ponce F A, Steeds J W, Dyer C D and Pitt D 1997b Mat. Res. Symp. Proc. Vol 449 "III-V Nitrides" eds. Ponce F A, Monstakas T D, Akasaki I and Monemar B A, p731.
Steeds J W, Burton N C, Lang A R, Pickard D, Shreter Y G and Butler J E 1996 in Polycrystalline Semiconductors IV, eds Pizzini S, Strunk H P and Werner J H, Solid State Phenomena Vols 51-52, Scitec, Switzerland p271
Wetzel C, Walukiewicz W, Ager J W 1997 Mat. Res. Symp. Proc. Vol 449 "III-V Nitrides" eds. Ponce F A, Monstakas T D, Akasaki I and Monemar B A, p567.
Young R J 1993 Vacuum **44**, 353
Yugo S, Kanai T, Kimura T and Muto T 1991 Appl. Phys. Lett. **58**, 1036

*Inst. Phys. Conf. Ser. No 153: Section 10*
*Paper presented at Electron Microscopy and Analysis Group Conf. EMAG97, Cambridge, 1997*
© *1997 IOP Publishing Ltd*

413

# Investigation of SiGe quantum dot structures by Large Angle CBED and Finite Element Analysis.

**A Hovsepian D Cherns and W Jäger\***

H H Wills Physics Laboratory, University of Bristol, England
\* Centre for Microanalysis, Faculty of Engineering, University of Kiel, Germany

**ABSTRACT:** LACBED and two beam dynamical calculations are used to profile quantum well structures of Si/SiGe/Si containing self assembled Ge rich quantum dots. It is shown that a simple measurement of the asymmetry of two-beam rocking curves can give the Ge content of the buried layer. Finite element analysis is used to examine bulk strain relaxation effects in the Si cladding layers, resulting in a small (5 - 30 %) correction to the LACBED results.

## 1.   INTRODUCTION

Osbourn (1982) first suggested that lattice mismatched epitaxy could be used to create structures with new electronic properties, and subsequently a range of strained layer structures have been developed. Quantum dots are an important class of such structures and are expected to have particular uses in optoelectronic devices, where optical efficiency is enhanced by quantum confinement of charge carriers within the islands.

Attempts to characterise these materials must include the examination of the strain, composition and size of the islands, all of which affect the electronic properties. The large angle convergent beam electron diffraction (LACBED) technique has previously been used to measure strain and layer thickness in strained semiconductor multilayer structures (Grigorieff et al, 1993). Structures were profiled using low order reflections by quantitatively matching rocking curves over a wide range of deviation parameter s. In this paper we show that single quantum well structures of SiGe in Si containing self assembled quantum dots can be profiled using only the rocking curve behaviour at small s. In addition finite element analysis is used to determine the limits of the technique.

## 2.   SAMPLE PREPARATION

Two samples were grown by molecular beam epitaxy (MBE), at a substrate temperature of 700°C. A Si buffer layer 100 nm thick was deposited onto a thermally cleaned Si (001) substrate, followed by a pure Ge layer which was capped with another 100 nm Si layer. Enough Ge was deposited to produce a uniform 6 monolayer (sample 1) or 12 monolayer (sample 2) pure Ge quantum well (1 monolayer = 0.14 nm). In reality Stranski-Krastanow 3D growth occurs, resulting in self-assembled islands typically 200 nm in diameter. Plan view TEM samples were prepared by mechanical backthinning using a South Bay Technology tripod polisher producing a very clean, flat surface.

## 3.    PRINCIPLES OF THE LACBED METHOD.

The geometry and advantages of the LACBED technique are well documented (Cherns 1993). Here we make particular use of the small probe size (~1 nm) of an Hitachi HF2000 FEG TEM to combine a small defocus (and thus high spatial magnification) with excellent filtering of thermal diffuse scattering. In addition, by examining the rocking curve over a limited range of s, smaller structures can be investigated for a given defocus.

To interpret LACBED images we first consider the sample morphology. Assuming the Ge layer is, in fact, a SiGe layer pseudomorphic with the substrate, there is a normal displacement R(z) across the layer that arises for two reasons; (1) a displacement due to the different lattice spacing of Si and SiGe and (2) an additional displacement due to the Poisson expansion of the layer in the growth direction produced by compression in the growth plane to match the substrate. It is easily shown that for isotropic elasticity the total displacement across a SiGe layer of thickness t is

$$R = \xi t \left( \frac{1+\upsilon}{1-\upsilon} \right)$$

where $\xi$ is the natural mismatch, and v is Poisson's ratio. If we examine planes inclined to the growth direction there is a phase shift of the planes across the SiGe layer given by $2\pi G \cdot R$ where $G$ is the reciprocal lattice vector.

A consideration of two beam dynamical diffraction (Whelan 1978) suggests that the SiGe layer influences the rocking curve by a change in extinction length (or structure factor) across the layer, as well as by a phase term involving $2\pi G \cdot R$. Calculations show that $G \cdot R$ introduces an asymmetry to the rocking curves which can be measured by $I_2/I_1$ (Fig. 1). This relationship (Fig. 2) suggests that the effect of the structure factor on the rocking curve at small s is negligible for thin SiGe layers containing less than 32 monolayers of Ge. Therefore the layer can be treated like a shear in pure Si and R(z) approximates to a step function.

Profiling the layer is very simple because a LACBED disc combines real and reciprocal space information (Fig. 3), and because the ratio $I_2/I_1$ measured along a LACBED contour can be directly related to the Ge content using Fig. 2a. To reduce the number of degrees of freedom in the model the position of the layer in the sample is not allowed to vary, but is assumed to lie between equally thick Si layers (such regions can easily be found experimentally).

## 4.    FINITE ELEMENT ANALYSIS OF STRAIN RELAXATION.

Implicit in the above analysis is the assumption that the thickness and composition of the quantum well are invariant in the growth plane, in which case the strain in the sample is only a function of the z. However the strain around an island is not uniform. Finite element analysis has been used to calculate the strain around islands using the Abaqus software package, for the model in Fig. 4 (a cross section of a pure Ge buried layer with dimensions in table 2).

Due to the larger lattice spacing of Ge the Si planes are bent around an island as illustrated in Fig. 4. These plane rotations depend principally on $T_1$, $T_2$ and the island density. Strain relaxation in the Si cladding layers causes the average Si lattice parameter in the growth direction to be reduced above islands and increased between them. Fig. 5 shows the variation of R(z) normal to the growth plane in the centre of an island, and half-way between islands. If we assume that, to a reasonable approximation, LACBED actually measures the average integrated displacement $R_{ave}$ shown in Fig. 5. Then to correct for relaxation the measured values of $R$ must be multiplied by the ratio of $R_{QW}/R_{ave}$.

## 5.    RESULTS

An Hitachi HF2000 FEG TEM with a Gatan Imaging Filter attachment has been used to acquire LACBED images using only elastically scattered electrons. Patterns were recorded using the (7,3,1) reflection (extinction length $\xi_g = 700$ nm for 200 keV electrons) which minimises dynamical effects and simplifies interpretation of the images (if scattering is extremely dynamical the main peak becomes a broad double peak which makes automated image analysis unnecessarily complex). A profile of $I_2/I_1$ along each LACBED contour is recorded and converted to an uncorrected Ge content using fig 2a. This value is then corrected for relaxation to determine the actual Ge content of the islands and the wetting layer. Several islands and wetting layer regions were examined and the typical Ge content in monolayers can be seen in the table below

Table 1

| | Island thickness / ml | Wetting layer thickness / ml |
|---|---|---|
| Sample 1 | 19 $^+$/. 1 | 4 $^+$/. 1 |
| Sample 2 | 26 $^+$/. 3 | 5 $^+$/. 1 |

## 6. CONCLUSION

The results in table 1 confirm that LACBED is a highly effective technique for the examination of the non-uniform morphology of single quantum well structures. Sensitivity to **R** can be increased by the use of higher order reflections, and the spatial resolution of LACBED with a field emission source should allow profiling of islands as small as 50 nm in diameter for the cladding layer thicknesses used here. An important advantage of the plan view geometry is the ease of sample preparation and the large region of interest available for examination, allowing quick and easy examination of many samples. Characterisation can be completed by the use of LACBED in conjunction with thickness profiling of the layers in cross section, thereby revealing the average composition across a layer (Hovsepian et al, to be published).

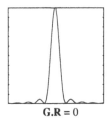 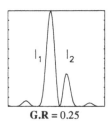

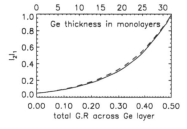

Fig. 1. Electron rocking curves showing the effect of the phase term $2\pi$**G.R** on symmetry

Fig. 2. $I_2/I_1$ vs. **R**. Solid line (a) treats the layer as a shear in a pure Si crystal, dashed line (b) includes the effect of the structure factor.

 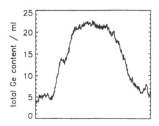

Fig. 3. LACBED contour along an island showing a change in symmetry, and the profile of the island showing total Ge content.

416

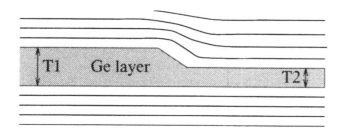

Fig. 4. Bending of Si planes
around a buried Ge island

|  | T1 | T2 | Thickness of Si layers | Separation of islands |
|---|---|---|---|---|
| sample 1 | 2.2 nm | 0.55 nm | 100 nm | 400 nm |
| sample 2 | 2.2 nm | 0.55 nm | 100 nm | 800 nm |

Table 2. Dimension of Finite Element Model.

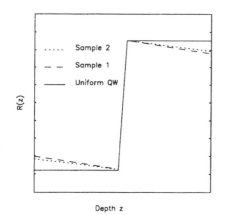

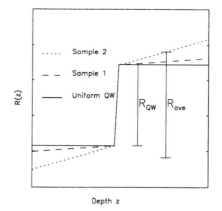

Fig. 5a. The displacement R(z) across
an island (FEM calculation)

Fig. 5b. The displacement R(z) midway
between islands (FEM calculation)

REFERENCES

Qsbourn G C (1982) J. Appl. Phys. **53**, 1586
Grigorieff N and Cherns D. (1993) Phil. Mag. A **68 no1** pp 121-136
Cherns D (1993) J. de Physique **4**, Colloque C7, supplement au J. de Physique **3**, 2113-2122
Whelan M J (1978) Diffraction and Imaging Techniques in Material Science, eds S Amelinckx
   R Gevers and J van Landuyt (North-Holland Publishing Company) pp 43-106
Hovsepian et al. To be published.

© *1997 IOP Publishing Ltd*

# TEM Study of Facet Junctions in YBa$_2$Cu$_3$O$_{7-\delta}$/PrBa$_2$Cu$_3$O$_{7-\delta}$ Thin Film Patterned Multilayer Structures

**M Barnett, M Aindow, JS Abell, PJ Hirst\*, NG Chew\* and RG Humphreys\***

School of Metallurgy and Materials, The University of Birmingham, Birmingham, B15 2TT.
\* DERA Electronics Division, St Andrews Road, Great Malvern, Worcs., WR14 3PS.

**ABSTRACT:** Transmission Electron Microscopy has been used to examine the growth of an YBa$_2$Cu$_3$O$_{7-\delta}$ thin film over a patterned step in an epitaxial PrBa$_2$Cu$_3$O$_{7-\delta}$ thin film on (001) MgO. The YBa$_2$Cu$_3$O$_{7-\delta}$ grows epitaxially over the step, but in cross-section strong strain contrast features were observed running from the facet junction at the step to the film surface. Selected Area Diffraction Patterns from such regions show that there is a small misorientation (<0.3°) across the feature which can be accommodated by dislocations as in a low angle grain boundary or elastically as the strain field of an interfacial wedge disclination at the facet junction.

## 1. INTRODUCTION

There has been a large number of studies in which the growth of YBa$_2$Cu$_3$O$_{7-\delta}$ (YBCO) thin films onto planar substrates has been examined: see for example the review by Shinde and Rudman (1994). In devices such as Josephson junctions which are fabricated from High Temperature Superconducting (HTS) films, however, non-planar structures are required and the presence of any defects in such regions is critical to device performance. Despite their importance, there have been very few studies published of the microstructures in such regions e.g. Verbist et al (1996).

In our work we have used cross-sectional Transmission Electron Microscopy (TEM) techniques to examine the microstructure of YBCO thin films, deposited by co-evaporation over photolithographically patterned steps in epitaxial PrBa$_2$Cu$_3$O$_{7-\delta}$ (PrBCO) thin films on (001) MgO substrates. In this paper we present preliminary observations obtained from these multilayer structures, and discuss the possible origins and consequences of an unusual strain contrast feature observed in the step area.

## 2. EXPERIMENTAL

The facet junction geometry studied in this work is part of the ramp junction structure shown schematically in Fig 1. A planar epitaxial bilayer, consisting of 0.2µm PrBCO and 0.35µm YBCO was grown onto a nominal (001) MgO substrate using electron-beam co-evaporation, as described by Chew et al (1988). A 45° step was then produced by photolithographic patterning and Ar-ion beam etching. The etching was stopped when the thickness of the bottom PrBCO layer is reduced by half, this being detected by Secondary Ion Mass Spectrometry (SIMS). The base PrBCO layer is a sacrificial buffer used to prevent etching of the MgO substrate

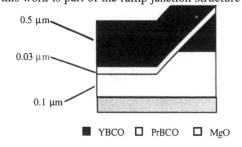

0.5 µm

0.03 µm

0.1 µm

■ YBCO  □ PrBCO  □ MgO

Fig 1: Schematic of the junction structure.

418

which can lead to grain-boundary formation in the overlying films (Tanaka et al (1993)). A thin (0.03μm) PrBCO barrier was then grown over the etched step, followed by a 0.5μm YBCO layer. Cross-sectional TEM specimens were prepared by mechanical polishing and Ar-ion beam milling to perforation. Specimens were examined in a Philips CM20 TEM, operating at 200kV with typical observations being made in bright-field, using **g**=005 with s large and positive.

## 3. RESULTS AND DISCUSSION

### 3.1 Misorientation at Step Structures

A bright-field TEM micrograph of the facet junction structure is shown in Fig. 2a, with the SADP obtained from this region parallel to [010]MgO shown in Fig. 2b. The SADP shows a superposition of the diffraction information from the YBCO, PrBCO and MgO, and indicates that the orientation relationship between these materials is:

$$(001)_{YBCO} \parallel (001)_{PrBCO} \parallel (001)_{MgO} \tag{1}$$
$$[010]_{YBCO} \parallel [010]_{PrBCO} \parallel [010]_{MgO}$$

This is the preferred epitaxial orientation for all layers in this multilayer structure, and appears to have been maintained despite the deposition of YBCO over a steep patterned slope in the PrBCO. Upon closer examination of the higher-order 00n diffraction spots (i.e. parallel to c* arrowed in Fig. 2b), however, some subtle spot splitting can be detected, indicating that there is a crystallographic misorientation of the upper YBCO layer with respect to the PrBCO layer immediately below it about the [010] direction in the vicinity of the step. Due to the finite width of the diffraction maxima, an accurate measurement of the misorientation from the SADP is difficult, but we estimate that the magnitude is ≈0.3°. The origin of this misorientation can be understood if we consider the growth of the upper YBCO layer over the large step in the previous PrBCO layer and take into account the difference in the lattice parameters between the crystal structures of these two materials.

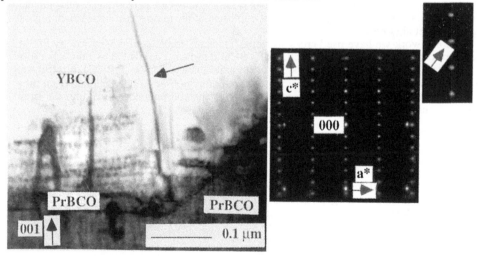

Figure 2a. Bright-Field TEM micrograph of the facet junction structure, 2b. with accompanying diffraction pattern.

The HTS oxides YBCO and PrBCO are nominally isostructural orthorhombic defect perovskites whose lattice parameters depend on the exact cation stoichiometry and oxygenation state. Moreover, whilst both materials exhibit deoxygenated tetragonal crystal structures at the growth temperatures, there are no accurate data available for the lattice parameters. In the following discussion we have, therefore, used the room-temperatures values for these structures in the orthorhombic, fully stoichiometric and fully oxygenated state at room temperature. These values for the lattice parameters are: YBCO, a=0.3821nm, b=0.3888nm, c=1.1693nm; PrBCO, a=0.3823nm, b=0.3877nm, c=1.1793nm (JCPDS-ICDD 1995, records 39-486 & 45-216). At the (001) interfaces between the steps the YBCO deposit will adopt orientation relationship (1) under these conditions and there will be a misfit of ≈0.04% between the YBCO and PrBCO lattices at this interface. The step risers are, however, sufficiently large that, in the initial stages of growth, deposition on these facets can be considered separately. For the step angle of ≈45° shown in Fig. 2, the PrBCO surface exposed on the riser is approximately parallel to (103) which is one of the more closely packed planes in this crystal structure. Thus, one might expect that an epitaxial YBCO deposit on this surface would adopt the orientation relationship:

$$(103)_{YBCO} \parallel (103)_{PrBCO} \tag{2}$$
$$[010]_{YBCO} \parallel [010]_{PrBCO}$$

There is a subtle difference between orientation relationships (1) and (2) since the angle between (103) and (001) is ≈45.80° for PrBCO and ≈45.56° for YBCO. Thus YBCO grown on the step risers exhibiting orientation (2) would be misoriented by ≈0.24° about [010] with respect to that exhibiting orientation (1) on the (001) PrBCO regions between the steps. The sense and magnitude of this misorientation is consistent with that observed in the experimental SADPs such as Fig. 2b obtained from the region of the step.

## 3.2 Accommodation of the misorientation

We next consider the way in which the misorientation is accommodated at the step structure. Diffraction contrast images such as Fig. 2a show strong strain contrast features which appear to emanate from the facet junction and extend to the YBCO deposit surface. These features do not appear in images obtained away from the steps and are presumably associated with the misorientation between the regions which exhibit orientation (1) and (2). There are two possible ways in which this orientation could be accommodated. The first is an array of crystal dislocations, i.e. a low-angle grain boundary (LAGB), extending from the facet junction to the free surface of the YBCO deposit. The second is as an elastic distortion between the two regions and this can be described more formally as the strain field of an interfacial line defect with wedge disclination character (i.e. a disclination or dispiration - Harris (1977)) at the facet junction on the YBCO/PrBCO interface.

For a 0.24° LAGB we would expect to resolve the individual dislocations in the array for diffraction contrast images obtained parallel to the tilt axis such as Fig. 2a. No such defects were observed and the contrast is instead more continuous and reminiscent of a narrow bend contour. This would be consistent with the strain field of a disclination or dispiration at the interfacial facet junction but there have been very few observations of such defects in crystalline materials because their elastic strain energies are usually unreasonably large (e.g. Pond, 1989). In the present case, however, both the wedge angle and the thickness of the crystal (i.e. the YBCO film) are small and the energy will be more modest. In an attempt to compare the two possibilities we have estimated the total elastic strain energies for both configurations. Such an estimate will neglect any contribution from core energies but will still be valid since the core energy of dislocation is usually much less than the long range elastic strain energy. Much less is known about the core energies of disclinations and dispirations but since the core corresponds to the distortion in the immediate vicinity of the interfacial facet junction, this contribution would be the same for both configurations and can be neglected in any comparison.

For the simplest low angle tilt grain boundary in which an array of pure edge dislocations with the same Burgers vectors, **b**, produces a misorientation of $\omega$ radians, then the spacing, d, of the dislocation array is:

$$d = |\mathbf{b}|/2\sin(\omega/2) \tag{1}$$

Since the strain contrast features lie roughly parallel to (100), an appropriate value for **b** would be [100], giving |**b**|=0.382nm. For the step structure considered here $\omega \approx 0.005$ radians (0.3°) giving d≈0.07μm. Thus for the 0.5μm thick YBCO film there would be seven dislocations in the array. If the interactions between edge dislocations are neglected then the elastic energy per unit length of a parallel dislocation array in an isotropic medium is simply the sum of the individual elastic energies, i.e.:

$$E = n \; G|b|^2 \ln(L/r)/4\pi(1-\nu) \qquad\qquad [2]$$

where G is the shear modulus, $\nu$ is Poisson's ratio, r is the core cut-off radius, n is the number of dislocations in the array, and L is the width of the dislocation array. Using r=|b|/2, n=7, L=0.5$\mu$m, $\nu$=0.33 and G=40GPa (Wang et al (1995)) gives E$\approx$0.04$\mu$Jm$^{-1}$.

We can obtain the equivalent value for the other configuration by assuming that the defect at the facet junction would have no displacement component, i.e. it is a pure wedge disclination. The elastic energy of per unit length of a perfect wedge disclination in an isotropic medium was given by Li (1972) as:

$$E = G\omega^2 L^2/16\pi(1-\nu) \qquad\qquad [3]$$

Using the same values for the parameters G, $\omega$, L and $\nu$ as before gives E$\approx$0.007$\mu$Jm$^{-1}$. So in this case the elastic energy of the disclination would be of the same order of magnitude as, but lower than, the equivalent low angle grain boundary. Thus it is reasonable to interpret the contrast feature in Fig. 1 as the strain field of an interfacial defect with wedge disclination character.

If the value of L were increased to >3.55$\mu$m, however, then equation [3] would give a larger value of E than equation [2]. Although this critical value of L will not be very accurate because of the assumptions made, we would expect that in the earliest stages of epitaxial growth over such step structures the misorientation would be accommodated elastically as interfacial defects with disclination character. At some later stage in the growth the elastic energy will rise to the point where dislocation introduction is favoured and a low angle grain boundary structure would develop.

## 4. CONCLUSIONS

TEM observations of multilayer YBCO/PrBCO step structures on (001) MgO substrates have revealed regions misoriented by 0.3° in the uppermost YBCO layer at the steps with strong contrast features running from the interface facet junction to the free surface between these regions. It has been shown that the misorientation can be accommodated by an interfacial wedge disclination or a low-angle boundary with the former being favoured for thinner deposits.

**ACKNOWLEDGMENTS:** We are grateful to Professors I.R. Harris and J.F.Knott for the provision of laboratory facilities, and to EPSRC and DERA for financial support.

## REFERENCES

N.G.Chew, S.W.Goodyear, J.A.Edwards, J.S.Satchell, S.E.Blenkinsop and
    R.G.Humphreys 1988 Appl. Phys. Lett. **53**, 2683
J.C.M.Li 1972 Surface Science **31**, 12
W.F. Harris 1977 Scientific American **273**, 130
R.C.Pond 1989 Dislocations In Solids, ed. F.R.N. Nabarro, (North Holland),
    **8**, ch. 38.
S.L.Shinde and D.A.Rudman (editors) 1994 Interfaces in High-$T_C$ Superconducting Systems
    (New York: Springer)
S.Tanaka, H.Kado, T.Matsuura and H.Itozaki 1993 IEEE Trans. Appl. Supercond.
    **3**, 2365
K.Verbist, O.I.Lebedev, G.Vantendeloo, M.A.J.Verhoeven, A.J.H.M.Rijnders and
    D.H.A.Blank 1996 Supercond. Sci. Tech. **9**, 978
Q.Wang, G.A.Saunders, D.P.Almond, M.Cankurturan, K.C.Goretta,
    1995 Phys. Rev. B **52**, 3711

*Inst. Phys. Conf. Ser. No 153: Section 10*
*Paper presented at Electron Microscopy and Analysis Group Conf. EMAG97, Cambridge, 1997*
© *1997 IOP Publishing Ltd*

# Charge- and Orbital ordering in LaSr2Mn2O7 examined by HRTEM and low-temperature electron diffraction

J Q Li[1], Y Matsui[1], T. Kimura[2], Y. Moritomo[2] and Y. Tokura[2]

[1]National Institute for Research in Inorganic Materials, 1-1 Namiki, Tsukuba 305, Japan.
[2]Joint Research Center for Atomic Technology, Higashi, Tsukuba 305, Japan.

**ABSTRACT:** New layered manganite LaSr2Mn2O7 is prepared, and examined by HRTEM and by low-temperature electron diffraction to examine the possibility of charge- and orbital-ordered states occuring below 200K. The crystal structure of this layered manganites is almost perfect and no planar or intergrowth defects has been detected by HRTEM. At low temperature of 200K-110K, *hk0* electron diffraction patterns clearly exhibit the appearances of characteristic super-reflection spots along the $\mathbf{a^*+b^*}$ direction. The superstructure is interpreted in terms of the $Mn^{3+}/Mn^{4+}$ charge-ordering, associated with $d_z^2$ orbital-ordering to induce Jahn-Teller distortion around the $Mn^{3+}$ ions.

## 1. INTRODUCTION

Discovery of the so-called colossal magnetoresistance (CMR) in hole-doped manganites, $Re_{1-x}Ae_xMnO_3$ (Re: trivalent rare-earth, Ae: divalent alkaline-earth elements, x: hole concentration), has stimulated considerable interest in the related systems (Chahara et al 1993; Tokura et al 1994). Interplay of charge, orbital and spin distributions at the Mn-sites governs the insulator-to-metal transition and the related CMR properties. One of the most striking phenomena due to such interplay is the "charge-ordering", as first reported by part of the present authors for $La_{0.5}Sr_{1.5}MnO_4$ with $K_2NiF_4$ structure (Moritomo et al., 1995). Recently, the authors prepared a new manganite phases, $La_{2-x}Sr_{1+x}Mn_2O_7$, and examined their magnetic and transport properties in detail. Quite large CMR effect was successfully obtained for the x=0.4 compound (Moritomo et al 1996). For the x=0.5 one, LaSr2Mn2O7, however, no evidence of the insulator-to-metal transition has been detected, and this was considered to be due to charge-ordering at the Mn-sites to keep the phase insulating even at low temperature. In the present study, we examined the perfections of crystal structure of LaSr2Mn2O7 by HRTEM, and also examined the possibility of charge-ordering by low-temperature electron diffraction technique.

## 2. EXPERIMENTAL

Single crystalline samples of LaSr2Mn2O7 were melt-grown by the floating-zone (FZ) method. The detailed process of the sample preparation is described by Kimura et al (1996). The resistivity and magnetic susceptibility data of this compound strongly suggested that the phase transition due to charge-ordering takes places below 200K. X-ray diffraction measurements at room temperature indicate that the specimen have tetragonal structures with a space group I4/mmm. For low-temperature electron diffraction and high resolution electron microscopy studies, the sample was ground under $CCl_4$, dispersed on Cu grids coated with holy-carbon support films, and subsequently examined with a Hitachi H-1500 type of high-voltage, high resolution microscope (Matsui et al., 1991), equipped with Gatan's liq.$N_2$ specimen cooling holder.

## 3. RESULTS AND DISCUSSIONS

### 3.1 HRTEM Observations

First, the $LaSr_2Mn_2O_7$ sample was examined by high resolution electron microscopy to characterize its structural perfection as the 327-type of layered structure. Figure 1 shows example of the resultant HRTEM image taken along the [100] zone-axis of the tetragonal lattice. The micrograph was obtained from a thin region of the crystal under the defocus value around the Scherzer defocus (-60nm), and so, it represents the metal atom positions as dark dots. It is clearly observed that the perovskite slabs with two $MnO_2$ planes are sandwiched by the $[La(Sr)O]_2$ slabs. Such arrangements of the dark dots are quite consistent with those expected from the 327-structure. Example of the calculated image for a defocus value of -60nm and a thickness of 2.5nm is also displayed at the right-bottom corner of the micrograph. The most important conclusions obtained through these HRTEM experiments are that the present specimen has a quite perfect structure without any planar and intergrowth defects.

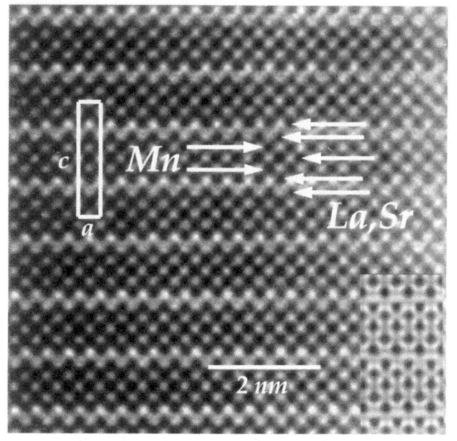

Fig. 1 HRTEM image of $LaSr_2Mn_2O_7$ taken along the [100] direction (800kV). Simulated image is also shown at the right-bottom corner.

### 2.2 Low-Temperature Electron Diffraction

Low temperature electron diffraction experiments are performed to examine the possibility of charge-order (CO) transition in $LaSr_2Mn_2O_7$. Electron diffraction observations are made between the room temperature and 110K. The evidence for a CO

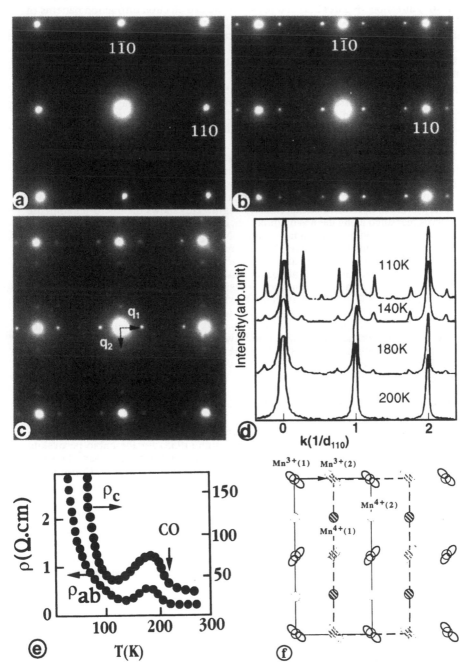

Fig 2. Electron diffraction patterns of LaSr$_2$Mn$_2$O$_7$ obtained at (a) 300K and (b)110K, respectively. (c) The diffraction pattern at 110K, exhibiting the presence of 90$^o$ twin domains. (d) The micro-photometric density curves showing the temperature variation of the superlattice reflections. (e) The temperature dependence of resistivity for x=0.5, showing CO transition at ~210K. (f) The schematic representations of the possible charge and orbital ordered state.

transition was detected as the appearances of extra superstructure reflections along the $\mathbf{a^*}$+$\mathbf{b^*}$ (or the $\mathbf{a^*}$-$\mathbf{b^*}$) direction. Figures 2(a) and (b) show the $hk0$ electron diffraction patterns of LaSr$_2$Mn$_2$O$_7$ obtained at room temperature and 110K, respectively. The diffraction pattern at room temperature can be indexed with the **I4/mmm** tetragonal structure with lattice parameters a=b=0.385nm and c=1.98nm. The most striking feature revealed in Fig. 2(b) is that a series of sharp satellite spots appear along the $\mathbf{a^*}$+$\mathbf{b^*}$ direction in addition to the fundamental Bragg reflections. The wave vector of this structural modulation was found to be nearly commensurate and can be written as $\mathbf{q}$=$\mathbf{a^*}$[1/4, 1/4, 0] ($\mathbf{a \cdot a^*}$=1). It is also worth noting that two sets of superstructure reflections around each basic Bragg spot frequently appear, which is considered to originate from different areas where the superstructure variants are related by twinning, and the twin-domains are rotated by 90 degree with respect to one another. Fig. 2(c) presents a typical electron diffraction pattern obtained at 110K, in which two sets of superstructure reflections are indicated as $\mathbf{q}_1$ and $\mathbf{q}_2$. Temperature dependence of the superstructure reflections has also been investigated, from the room temperature (300K) down to 110K. In most cases, the weak satellite reflections become visible at around 200K, and then the intensities of the superstructure spots increase progressively with the decrease of temperature. Figure 2(d) shows the micro-photometric density curves measured along the $\mathbf{a^*}$+$\mathbf{b^*}$ direction, exhibiting the changes of the intensities of superstructure peaks with varying temperature. No significant change of the periodicity of the structural modulation has, however, been found in a range of our present experiments. From the data of electrical transport and magnetic susceptibility, the CO transition was proposed at around 210K, and this is almost consistent with our electron diffraction data. Typical result of the temperature dependence of resistivity obtained from the single crystal sample, in which the CO transition is clearly exhibited at around 210K. The basic characteristics of the CO modulation has also been investigated extensively. It is found that the satellite spots at (h$\pm$ 1/4, h$\pm$ 1/4, 0) become very weak in some cases and their intensities decrease very rapidly as the sample was tilted away from the [001] zone axis. This indicates that the superlattice reflections at (h, h, 0)$\pm$ $m\mathbf{q}$ positions are caused by the double reflection effect. This fact suggests that a pure transverse lattice distortion has been induced in the LaSr$_2$Mn$_2$O$_7$ crystal lattice by CO transition.

### 2.3 Possible Model of Charge-Ordering

In LaSr$_2$Mn$_2$O$_7$ the average ionicity of Mn ion is Mn$^{3.5+}$, and therefore the charge-order (CO) model, originally proposed by Goodenough (1955) for the cubic perovskite La$_{0.5}$M(II)$_{0.5}$MnO$_3$, can be used as a starting model to explain our experimental results. LaSr$_2$Mn$_2$O$_7$ has the layered structure with the body-centered space group **I4/mmm**, and therefore the CO patterns have a relative glide-component between the neighboring layers. In Fig. 2(f), we show a proposed model of charge-ordering of Mn$^{4+}$(3d$^3$) and Mn$^{3+}$(3d$^4$) in agreement with the geometry of satellite reflections observed at low temperature. First, Mn$^{4+}$-Mn$^{3+}$ charge-ordering is induced to form double periodicity along the 110 direction. Then, it is considered that d$_{z2}$ orbital-ordering at the Mn$^{3+}$ sites takes places to produce additional double periodicity due to Jahn-Teller distortion, and finally the four-times periodicity of **L**=4**d**$_{110}$ along the [110] direction has been achieved. More detailed data and results of the present study will be published in the near future (Li et al., 1997).

### REFERENCES

Chahara K, Ohno T and Kozono Y 1993 Appl. Phys. Lett. **63**, 1990
Goodenough J. B 1955 Phys. Rev. **100**, 564
Kimura T, Tomioka Y, Kuwahara H and Y. Tokura 1996 Science **274**, 1698.
Li J Q, Matsui Y, Kimura T and Tokura Y 1997 Phys. Rev. submitted.
Matsui Y, Horiuchi S, Bando Y, Kitami Y, Yokoyama M, Suehara S, Matsui I and Katsuta T 1991 Ultramicroscopy **39** 8.
Moritomo Y, Tomioka Y, Asamitsu A, Tokura Y and Matsui Y 1995 Phys. Rev.B 51,
Moritomo Y, Asamitsu A, and Tokura Y 1996 Nature **380**, 141
Tokura Y, Urushibara A and Furukawa N 1994 J. Phys. Soc. Jpn. **63**, 3931.

*Inst. Phys. Conf. Ser. No 153: Section 10*
*Paper presented at Electron Microscopy and Analysis Group Conf. EMAG97, Cambridge, 1997*
© 1997 IOP Publishing Ltd

# Profiles of cracks formed in tensile Epilayers of III - V compound semi-conductors

**R T Murray   M Hopkinson\* and C J Kiely**

Department of Materials Science and Engineering. The University of Liverpool, L69 3BX.

\* EPSRC III-V central Facility, The University of Sheffield, S1 4DU, UK.

**ABSTRACT:** Cracks are formed in tensile epi layers grown beyond their critical thickness. Cross-section TEM has demonstrated "U" and "V" profiles, cracks with {111} tails, as well as exfoliation along (001) planes in the substrate. High resolution TEM and STEM/EDX have revealed further features of the crack formation mechanism.

## 1.   INTRODUCTION

If an epilayer is deposited pseudomorphically on to a single crystal substrate, whose lattice parameter is longer than that of the layer, elastic strain energy is stored in the structure, principally within the epilayer. As growth proceeds the stored energy increases and beyond a critical thickness $t_c$, cracks can form so as to release some of this energy as shown by Matthews and Klockholm (1972) for garnets. To predict a critical thickness for crack induction it is necessary to postulate that an incipient crack can only harvest energy within a finite range of itself. Equating the stored elastic energy within a range t of the crack line to the energy of the new crack surfaces led Murray et al (1995) to deduce that for In Ga Al arsenide layers on InP $t_c$ = 8.6 x $10^{-12}$ $f^2$ meters for misfit f.

We have shown (Murray et al 1996) remarkable agreement between this equation and the behaviour of III-V epilayers despite the observation that the cracks always penetrate into the substrate thus acquiring additional surface energy (Murray & Kiely 1997).

Where the epilayer thickness exceeds $t_c$ an array of cracks form. For thickness below $4t_c$ it has been shown by Olsen et al (1974), Aragon et al (1993) and Murray et al (1995) that the cracks are bounded only by (110) habit planes but for thickness in excess of $6t_c$ the (1$\bar{1}$0) family will nucleate. Prior to the formation of (110) cracks some degree of energy relief is provided by the formation of surface ridges which are often bounded by {115} planes. In this paper we will report on characteristics of these cracks when seen in cross-section.

## 2.   EXPERIMENTAL

In Ga Al As layers were grown epitaxially on [001] InP substrates held at 500 ± 20°C by MBE and their physical parameters are listed in table 1. M944 and M1382 are populated only by the (1$\bar{1}$0) array whereas layer M530 is sufficiently thick to support both (110) and (1$\bar{1}$0) families.

Specimens were prepared by argon milling in either a GATAN duo mill with angle of 14° or in a PIPS mill with an angle of 4° and energy of 4 keV. HREM was performed at 200 keV in a JEOL 2000Ex with the section area orientated to the [110] zone axis.

**Table 1.**

| Wafer No. | M530 | M944 | M1382 |
|---|---|---|---|
| Crack Plane | (110) and (110) | (110) | (110) |
| Thickness (nm);spacing (μm) | 150 ; 1.0 | 70 ; 1.15 | 80 + 20 ; 5.0 |
| Composition | $In_{25}Ga_{19}Al_{56}As/InP$ | $In_{25}Ga_{75}As/InP$ | $In_{24}G_{76}As/In_{24}Al_{76}As/InP$ |

## 3.    "V" AND "U" PROFILE CRACKS

As illustrated in fig 1 the simplest profile for cracks observed in [110] cross-sections is a V in which the sides of the crack appear to hinge about the crack tip. Since the substate and epilayer differ in modulus and in the sign of their initial strain it is not surprising that the crack faces which penetrate into the substrate are not exactly flat. Dislocations can frequently be seen to grow from the crack tip.

In thick layers $>4t_c$ there is a possibility that the continuing flux of epi materials will close the crack mouth, The resulting bridge (fig 1) maybe coherent with the existing epilayer, is heavily faulted, nor does it lead to a planar surface. Perhaps more interestingly, the tips of such cracks are always rounded and the infill is crystalline (fig 3). EDX analysis in a Vacuum Generators HB601 scanning transmission electron microscope indicates that the infill is InGaAlAs of a similar composition to the epilayer. We believe that subsequent to capping, material already present on the walls of the crack migrates to the energetically more stable position at the crack tip.

## 4.    DEVIATION ONTO {111} AND (001) PLANES

For epilayers thicker than $3t_c$ most cracks will deviate from $(1\bar{1}0)$ onto either $(1\bar{1}1)$ or $(1\bar{1}\bar{1})$ planes finally terminating in a sharp tip. In any field of cracks there is no preference between the two planes. However, one might expect that a growing crack could be influenced by the relative distance to its neighbours and the resultant asymmetry in the rate of change of strain. To check this 18 crack triplets have been assessed. Eight deviated towards their nearer neighbour eight away from it, and two did not deviate so there is no evidence for such an interaction.

When a deviated crack is formed it effectively hinges open about the new tip. There is then a vertical component of displacement to the lip above the {111} segment (fig 2). A simple geometrical analysis then predicts

$$L_{111} = \frac{D}{\cos\theta}\left(\frac{\tan\theta}{\tan\phi} - 1\right)^{-1} \qquad \text{eqn (1)}$$

where $\theta$, $\phi$, w, x, z and $\sigma z$ are defined in figure 2, which also displays the stepped width in multi layers as found in M1382.

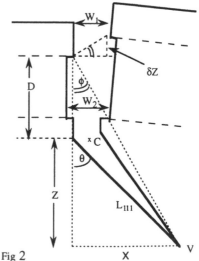

Fig 2

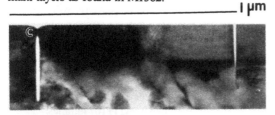

Fig 1. V crack with incipient {111}segment and a U shaped crack with imperfect cap C. Layer M530.

Fig 2.  Cross section of a crack as in Fig 4. δz i the height difference due to rotation about V at the tip of the {111}segment --- epilayer interfaces, ......... construction lines____ crack profile. C marks the crank point (r marks angles $\phi$.

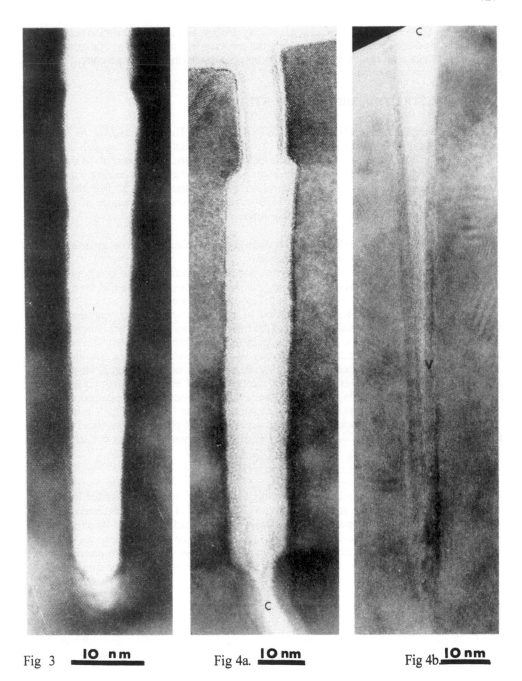

Fig 3   **10 nm**      Fig 4a. **10 nm**      Fig 4b. **10 nm**

Lattice image of a U crack in M530 showing the coherent crystalline deposit at the tip.

Fig 4. Lattice image of the upper portion of a crack in wafer M1382. Note the coherent overhang and $\delta z$ displacement of the upper interface. (b) The lower half of the crack rotated about the crack point C. Using eqn 1 predicts the crack tip to be at V despite the plastic deformation beneath it.

428

In layers thicker than $6t_c$ the {111} segment may undergo an additional deviation onto (001) and for sufficiently thick layers a link to a neighbouring crack can lead to exfoliation of a rectangular flake. Thermal shock such as immersion in liquid nitrogen promotes this process generating micron sized chips.

## 5.    HIGH RESOLUTION MICROSCOPY STUDIES

Fig 4 shows a profile of M1382 which has two epilayers. The initial $In_{24}Ga_{76}$ As layer is approximately $4t_c$ and has generated a field of cracks which are subsequently coated with 20nm of $In_{24}Al_{76}$ As. The ambitious aim of this study was to examine the atomic arrangement about the crack tip and at the interfaces. To achieve adequate resolution and interpretability it is necessary to obtain cross-sections which are thin even at the crack tip which is about 200nm in from the free surface and to orient the feature studied accurately to the [110] zone axis despite the presence of complex relaxation modes in the wedged section. We have found

(a)     The substrate/In Ga As interface is poorly defined and its coherency can not be determined. However, there is a step in the crack width at this level.
(b)     The InGaAs/In Al As interface is strain free and highly coherent. There is a symmetrical overhang of the Al rich layer. Since the interface is coherent we cannot explain this overhang by delamination. It may however be due to preferential oxidation under ambient.
(c)     As predicted in fig 2 there is a z component of any line which connects equivalent (001) planes across the gap. The indeterminate nature of the free surface close to the crack means that tan $\phi$ can be more accurately measured at the In GaAs/InAl As interface. However, there are two choices for W, the crack width, either $W_1$ just above or $W_2$ just below this interface. Defining D as the depth from the interface, to the crank point we can then predict L using equation 2. In all cases when $L_{111}$ is calculated using $W_1$ it is within 10% of the experimental length, but using $W_2$ it is up to 50% in error. We conclude that $W_1$ is the true crack width and that $W_2$ has enlarged after the growth of the layer. A refinement of this analysis enables one to conclude that point V is the tip of the crack and that the strain field locked into this region is the result of plastic deformation at the time of formation of the crack, possibly due to the high kinetic energy involved (Murray & Kiely 1997).

## 6.    CONCLUSIONS

Cracks formed in III V epilayers grow under tension may readily be profiled by cross-sectional TEM. Low magnification studies show both "U" "V" and (111) deviation as common crack forms, but have not so far elucidated the mechanism whereby a crack determines its direction of deviation. In thick layers further deviation onto a (001) plane within the substrate occurs and this can lead to the spallation of micro flakes. High resolution studies have enabled us to determine the position of the crack tip with its surrounding plastic deformation but the mechanism whereby overhangs are formed in complex layers has not been determined.

## REFERENCES

Aragon G, De Castro M J, Molina Sl, Gonzalez L, Briones F and Garcia R, (1993) MRS Fall Meeting.
Matthews JW, and Klokholm E 1972 Mat. Res. Bul 7, 213.
Murray R T and Kiely CJ (1997) to be published
Murray RT, Kiely CJ, Hopkinson M and Goodhew PJ (1995) Ins, Phys Conf Series 146, 207.
Murray R T , Kiely C J and Hopkinson M (1996) Phil.Mag. A74, 383.
Olsen G H, Abrahams MS and Zamerowski T J, (1974) J Electrochem Soc, 121, 1650.

*Inst. Phys. Conf. Ser. No 153: Section 10*
*Paper presented at Electron Microscopy and Analysis Group Conf. EMAG97, Cambridge, 1997*
© 1997 IOP Publishing Ltd

# The relationship between epitaxial growth, defect microstructure and luminescence in GaN

D M Tricker, P D Brown, Y Xin§, T S Cheng*, C T Foxon* and C J Humphreys

Department of Materials Science and Metallurgy, University of Cambridge, Pembroke Street, Cambridge CB2 3QZ, UK
*Department of Physics, University of Nottingham, University Park, Nottingham, NG7 2RD
§Now at ORNL, Solid State Division, Oak Ridge, TN 37831-6031, USA

ABSTRACT: MBE grown epitaxial GaN has been characterised by a variety of microscopies. Large scale features such as zincblende inclusions in wurtzite GaN are revealed by a combination of RHEED, SEM/CL and conventional TEM. Substrate pitting is best characterised by optical microscopy coupled with plan view TEM of samples prepared through the interface plane. Plan view HREM combined with g.R or g.b analysis of cross-sectional samples clarifies the displacements associated with prismatic stacking faults and threading dislocations. High spatial resolution EELS and CL provide insight into the effect of defects on material spatial or spectral purity.

## 1. INTRODUCTION

The wide band-gap semiconductor gallium nitride has recently become the focus of intense interest as the basis of high brightness blue/green LEDs (Nakamura et al 1994) and room temperature continuous wave blue lasers (Nakamura et al 1997). Surprisingly, such device structures contain very high densities of defects and yet still emit intense blue light. Gaining understanding of the relationship between defect microstructure and luminescence is primarily of fundamental interest; i.e. if the defects do not affect the light output, then why so? However, there is also need for the optimisation of growth and device fabrication procedures; e.g. a reduction in the defect density is still desirable in view of concerns over light scattering in laser structures (Liau et al 1996).

We are studying GaN grown by molecular beam epitaxy (MBE) on a range of substrates: {001} and {$\bar{1}\bar{1}\bar{1}$}B GaAs and GaP, {0001} sapphire and {0001} 6H SiC (Cheng et al 1995). The various large and fine scale defect microstructures which develop as a function of substrate, substrate orientation and growth condition are being characterised using a range of microscopical techniques, with a view to understanding the mechanisms of defect nucleation and development and thereby improve the growth. For example, the combination of reflection high energy electron diffraction (RHEED), cathodoluminescence (CL) in an SEM and conventional TEM can provide useful information on the distribution of cubic inclusions within epitaxial wurtzite GaN. Conversely, atomic models for individual dislocations and domain boundaries may be deduced from lattice images combined with diffraction contrast analysis (Xin et al 1997b) and these serve as the basis for band structure calculation, while the effect of such defects on band structure (and hence luminescence) can be investigated using high spatial resolution electron energy loss spectroscopy (EELS) (Tricker et al 1997a).

The purpose of this paper is to draw together results from these diverse parts of the project, in order to summarise our present state of understanding of the relationship between growth, defect microstructure and luminescence.

## 2. EXPERIMENTAL RESULTS

Results were obtained using various microscopes: a JEOL 4000EX-II for high resolution imaging, a JEOL 4000FX with post-column Gatan imaging filter for chemical mapping, a Philips CM30 and a JEOL 2000FX for conventional microscopy and energy dispersive X-ray analysis, a Philips 400ST modified to take a RHEED stage and a VG HB501 scanning transmission electron microscope (STEM) equipped with a Gatan imaging filter for electron energy loss spectroscopy (EELS). CL studies were performed using an Oxford Instruments MonoCL2 system.

## 2.1 Large scale inhomogeneities

It is important to be aware of any gross inhomogeneities in the as-deposited GaN thin films that might influence the spatial or spectral uniformity of the luminescence. Thus, the initial examination of samples using a combination of optical microscopy, SEM, X-ray or RHEED techniques is advisable. Indeed, we now perform routine assessment of the large scale structural integrity and uniformity of as-deposited GaN films using RHEED in a modified Philips 400ST. Unlike RHEED systems fitted to MBE growth chambers and primarily used to monitor surface constructions during growth, the higher electron energies (120keV) associated with TEM/RHEED facilitate much higher penetration depths for characterisation of the sub-surface material, combined with a full tilt, translate and rotate facility in order to access a variety of in-plane zone axes for a more complete structural analysis.

Thus, not only does RHEED enable rapid feedback in the event of changes in growth having a dramatic consequence on the structural integrity of the deposited films, it also ensures that time is not wasted making TEM foils of mixed phase or highly faulted structures. Here we simply show how from one RHEED pattern of a GaN/{1̄1̄1̄}B GaAs (fig. 1) it is apparent that the epilayer is not single phase wurtzite and that cubic inclusions are present. Such inclusions affect both the spatial and spectral uniformity of the luminescence of this film as illustrated by the CL data of fig. 2 acquired from a plan view TEM foil examined in SEM.

When imaging at 386nm (corresponding to a band gap of 3.21eV) localised spots of luminescence can be distinguished. The length scale of their distribution corresponds to that of cubic inclusions as viewed in TEM (fig. 3). Cubic inclusions are considered to initiate at the interface region (Xin et al 1996). Accordingly, control of the initial stages of MBE epitaxial growth remains an important issue for the subsequent development of the layer microstructure.

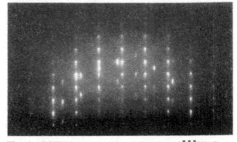

Fig. 1: RHEED pattern from GaN on {1̄1̄1̄}B GaAs indicating the presence of a small amount of zincblende phase embedded in the (0001̄) oriented wurtzite film.

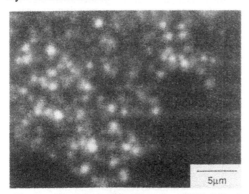

Fig. 2: CL image of a plan view TEM foil of (0001̄) GaN on {1̄1̄1̄}B GaAs, taken at 3.21 eV, showing luminescence from cubic inclusions.

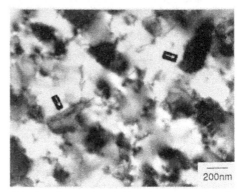

Fig. 3: Conventional TEM image showing the distribution of cubic inclusions.

Other large scale features which might have an influence on light scattering are pits in the substrate. These are particularly prevalent for GaN grown on GaAs or GaP substrates, as illustrated by the cross-sectional GaN/(001)GaAs TEM image of fig. 4. Such pits are not artefacts of specimen preparation since they are also observed at the interface when an as-deposited film is viewed in the optical microscope (fig. 5). These same features are more clearly seen to be pyramidal pits in plan view TEM of a foil containing the epilayer/substrate interface (fig. 6). In certain cases, chemical mapping across interface cross-sections shows some interdiffusion, particularly of N, arising as a consequence of substrate nitridation prior to growth (Xin et al 1997b), but many of the features at the interface are true voids. The GaN epilayer

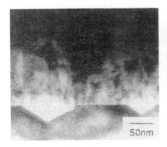

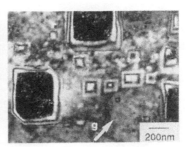

Fig. 4: Cross-sectional specimen of GaN/(001)GaAs showing pits in the GaAs substrate.

Fig. 5: Optical micrograph of a film of Be-doped GaN on (001) GaAs showing a high density of pits at the interface.

Fig. 6: Plan view dark field image of a specimen containing the GaN on (001) GaAs interface from the same film as in fig. 5, showing gross pitting. $g = 220$.

defines the line of the initial epilayer/substrate interface but does not in-fill the pit created. Hence, we have argued (Tricker et al 1997b) that such pits form during the initial stages of growth as Ga is sourced from the GaAs or GaP substrate to feed the GaN growth, with displaced As and P presumably escaping as dimers to the vacuum by pipe diffusion along the high density of threading defects.

## 2.2 Fine scale inhomogeneities

Films of single crystal zincblende GaN grown on (001) oriented GaAs and GaP substrates exhibit a very high density of inclined {111} stacking faults (Xin et al 1997a). Conversely, a high density of threading dislocations, typically greater than $10^{11}$ cm$^{-2}$, is observed in $(000\bar{1})$ wurtzite films grown on $\{\bar{1}\bar{1}\bar{1}\}$B GaAs and GaP. These dislocations often delineate low angle subgrain boundaries forming a mosaic cell structure separating columns of GaN growth. They are predominantly of mixed type (**b** = $1/3<2\bar{2}03>$) although edge (**b** = $1/3<11\bar{2}0>$) and screw type dislocations (**b** = $<0001>$) are also observed. They thus correspond to relative tilt and twist misorientations of adjacent columns, and so are considered to arise due to the coalescence of misoriented 3D islands during the initial stages of growth, as illustrated schematically in fig. 7, which in turn is related to roughening of the substrate during the process of nitridation prior to growth. Fig. 8 is a lattice image showing emergent mixed character threading dislocations. Micropipes or coreless screw dislocations have never been observed in our MBE grown GaN, in contrast to their occurrence in MOCVD GaN (Cherns et al 1997).

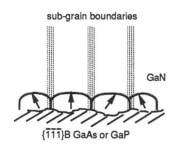

sub-grain boundaries

GaN

$\{\bar{1}\bar{1}\bar{1}\}$B GaAs or GaP

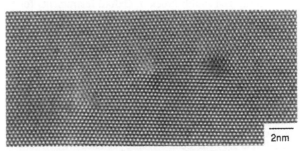

2nm

Fig. 7: Schematic illustrating the origin of threading dislocations in $(000\bar{1})$ GaN on $\{\bar{1}\bar{1}\bar{1}\}$B GaP or GaAs.

Fig. 8: Lattice image of $(000\bar{1})$ GaN on $\{\bar{1}\bar{1}\bar{1}\}$B GaAs, plan view specimen, showing threading dislocations imaged end on.

No inversion domains are observed within wurtzite GaN grown on the polar $\{\bar{1}\bar{1}\bar{1}\}$B GaAs and GaP surfaces by MBE and this differs from MOCVD material grown on non polar (0001) sapphire substrates (Rouvière et al 1997). It is considered that the polarity of the GaN growth is dictated by the polarity of the substrate in the former case. However, prismatic stacking faults are commonly observed lying on $\{1\bar{2}10\}$, with the less frequent observation of domain boundaries on $\{10\bar{1}0\}$. The $\{10\bar{1}0\}$ boundaries run through the complete film thickness and are

432

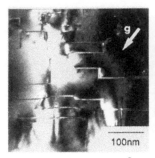
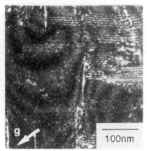
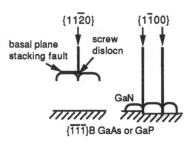

Fig. 9: Basal plane and {1$\bar{2}$10} prism plane stacking faults in a cross-sectional specimen of Mg-doped GaN on {$\bar{1}\bar{1}\bar{1}$}B GaAs, dark field image, **g** = 10$\bar{1}$2.

Fig. 10: Same specimen as fig. 9 tilted to reveal basal plane stacking faults. Dark field image, **g** = 10$\bar{1}$1.

Fig. 11: Schematic illustrating the origin of planar defects during growth of (000$\bar{1}$) GaN on {$\bar{1}\bar{1}\bar{1}$}B GaP or GaAs.

thought to arise from surface irregularities at the substrate surface. Conversely, the prismatic stacking faults arise from the coalescence of neighbouring growth columns, one of which contains a basal plane stacking fault. The {1$\bar{2}$10} boundaries when characterised by lattice imaging and **g.R** analysis both in plan view and cross-section (figs. 9 and 10) are found to exhibit a displacement vector of 1/2<10$\bar{1}$1> (Xin et al 1997b; as observed in AlN by Drum, 1964). The presence of a basal plane stacking fault within an individual column of GaN leads initially to a relative displacement of 1/6<20$\bar{2}$3> between adjacent columns. The extra 1/6<10$\bar{1}$1> displacement needed to bring two columns into a 1/2<10$\bar{1}$1> relative displacement on the prism plane is provided by a stair-rod screw dislocation at the intersection of the basal and prism plane faults (fig.11).

Such prismatic stacking faults are associated with an extra line at 3.42 eV in photoluminescence (PL) spectra (Salviati et al 1997). By contrast, we have found by high spatial resolution EELS that the band gap is reduced in the vicinity of these stacking faults from 3.4 ± 0.3 eV to 2.8 ± 0.2 eV (Tricker et al 1997a). Thus, some correlation between prismatic stacking faults and a lowering in luminescence energy can tentatively be made. For threading dislocations on the other hand, no change in band structure has been observed by EELS, and they are therefore considered to have little effect on luminescence.

**Acknowledgements** With thanks to Robin Taylor and Michael Brand of the University of Cambridge for development of the RHEED stage; and to David Evans of DTC Maidenhead, and Simon Galloway and Judith Brock of Oxford Instruments for SEM/CL data.

## REFERENCES

Cheng TS, Jenkins LC, Hooper SE, Foxon CT, Orton JW and Lacklison D 1995 Appl. Phys. Lett. **66** 1509
Cherns D, Young WT, Steeds JW, Ponce FA and Nakamura S 1997 J. Crystal Growth **178** 201
Drum CM 1964 Phil. Mag. **11** 313
Liau ZL, Aggarwal RL, Maki PA, Molnar RJ, Walpole JN, Williamson RC and Melngailis I 1996 Appl. Phys. Lett **69** 1665
Nakamura S, Mukai T and Senoh M 1994 Appl. Phys. Lett. **64** 1687
Nakamura S, Senoh M and Nagahama S 1997 Appl. Phys. Lett. **70** 868
Natusch MKH, Botton GA and Humphreys CJ 1997 IOP Conf. Ser. to be published
Rouvière J-L et al 1997 IOP Conf. Ser. to be published
Salviati G, Zanotti-Fregonara C, Albrecht M, Christiansen S, Strunk HP, Mayer M, Pelzmann A, Kamp M, Ebeling KJ, Bremser MD, Davis RF and Shreter YG 1997 IOP Conf. Ser. to be published
Tricker DM, Natusch MKH, Boothroyd CB, Xin Y, Brown PD, Cheng TS, Foxon CT and Humphreys CJ 1997a IOP Conf. Ser. to be published
Tricker DM, Brown PD, Cheng TS, Foxon CT and Humphreys CJ 1997b Appl. Surf. Sci. to be published
Xin Y, Brown PD, Boothroyd CB, Preston AR, Humphreys CJ, Cheng TS, Foxon CT, Andrianov AV and Orton JW 1996 Mat. Res. Soc. Symp. Proc. Vol. **423** 311
Xin Y, Brown PD, Dunin-Borkowski RE, Humphreys CJ, Cheng TS and Foxon CT 1997a J. Crystal Growth **171** 321
Xin Y, Brown PD, Humphreys CJ, Cheng TS and Foxon CT 1997b Appl. Phys. Lett. **70** 1308

*Inst. Phys. Conf. Ser. No 153: Section 10*
*Paper presented at Electron Microscopy and Analysis Group Conf. EMAG97, Cambridge, 1997*

# Statistics of partially coherent dark-field images of amorphous materials

M M J Treacy and J M Gibson*

NEC Research Institute, Inc., 4 Independence Way, Princeton, NJ 08540, USA
*University of Illinois, Department of Physics, 1110 W. Green St., Urbana, IL 61801, USA

ABSTRACT: This paper presents a kinematical scattering model that describes the variance of the speckle intensity observed in dark-field images of amorphous materials. A simple expression is given for the limiting case of uncorrelated atoms, that takes into account partial spatial coherence of the electron illumination.

## 1. INTRODUCTION

It has long been suspected that the speckle observed in dark field images of amorphous materials can reveal underlying structural order (Rudee & Howie 1972; Gibson & Howie 1978; Howie 1978). Until recently, effective application of the dark-field speckle technique was hampered by the meaningless speckle that is also produced by chance alignments in randomly arranged atoms (Rudee & Howie 1972; Krivanek 1975; Gibson 1978). Consequently, the mere presence of speckle says nothing about short-range or medium-range order in amorphous materials. However, statistical measurements are discriminating.

Early models of speckle predicted that coherent scattering from a random assembly of atoms should generate a coherent dark-field image intensity distribution of the type (Dainty 1975; Goodman 1975)

$$P(I) = \frac{1}{\bar{I}} \exp(-I/\bar{I}) \tag{1}$$

where $\bar{I}$ is the mean image intensity. The second moment of this distribution equals $2\bar{I}$, giving a normalized variance $V = 1$. However, experimental tilted dark field images of amorphous materials invariably have $V < 1$. For example, the tilted dark-field image in Fig. 1, of a 20 nm thick Ge film, has $V = 0.1$.

A phenomenological refinement of the above model leads to the gamma distribution (Goodman 1975)

$$P(I) = \frac{m^m}{(m-1)!} \frac{I^{m-1}}{\bar{I}^m} \exp(-Im/\bar{I}). \tag{2}$$

$m$ provides a crude measure of the partial coherence, and is equivalent to the number of independent partial illumination sources irradiating the specimen. $m = 1$ renders equation (1). As $m \to \infty$, $V \to 0$, as expected when a large number of speckle images are superimposed. The value $V = 0.1$ for Fig. 1, implies $m \approx 10$, indicating that Fig. 1 is equivalent to the incoherent superposition of about ten coherent dark-field images. Despite its simplicity, the principle difficulty with the gamma distribution is that it is difficult to

Figure 1: Tilted dark field image of amorphous germanium showing speckle.

understand, in experimental terms, the meaning of $m$ as an independent variable. Furthermore, in the incoherent limit ($m \to \infty$), the gamma distribution fails to take into account the statistics of the atom distribution in thin disordered specimens.

In previous work, we derived a simple kinematical scattering model for the normalized image intensity variance expected from a given structure in the presence of partially coherent illumination (Treacy & Gibson 1996). Using this model, we are able to simulate dark-field speckle variance of any structural model using partial coherence as an experimental parameter. This model appears to be robust, and recently we used it to demonstrate that evaporated amorphous germanium and silicon thin films contain 1 – 3 nm paracrystalline domains which can be transformed into a continuous random network on annealing (Gibson & Treacy 1997).

In this paper we outline part of our theory for the speckle variance, and discuss its predictions in the limit of uncorrelated atoms.

## 2.    THEORY

The normalized intensity variance $V(\kappa)$ of a dark-field image formed with scattering vector $\kappa$ is (Treacy & Gibson 1996)

$$V(\kappa) = \frac{\langle I^2(\kappa) \rangle - \langle I(\kappa) \rangle^2}{\langle I(\kappa) \rangle^2} \qquad (3)$$

Assuming kinematical scattering, and diffraction-limited imaging, the average tilted dark-field image intensity $\langle I(\kappa) \rangle$ from an elemental assembly of atoms $j$ is given by

$$\begin{aligned}
\langle I(\kappa) \rangle &= \frac{\lambda^2 f^2(|\kappa|)}{\mathcal{A}} \sum_j \sum_l e^{-2\pi i \kappa \cdot r_{jl}} \iint_0^{Q_{ap}} e^{2\pi i q \cdot r_{jl}} d^2q \\
&= \frac{\lambda^2 f^2(|\kappa|)}{\mathcal{A}} \pi Q_{ap}^2 \sum_j \sum_l F_{jl} A_{jl} \qquad (4)
\end{aligned}$$

$\kappa$ is the scattered wavevector into the center of the objective aperture, $\lambda$ the electron wavelength, $f(|\kappa|)$ the atomic scattering factor, $\mathcal{A}$ is the image area, and $Q_{ap}$ is the objective aperture radius in reciprocal space. $A_{jl}$ is the value of the Airy disk amplitude function centered on atom $j$ as sampled by atom $l$, and is obtained from the integral over $\boldsymbol{q}$. $A_{jl}$ can also be expressed in terms of the Bessel function $J_1$ as

$$A_{jl} = \frac{2J_1(Q_{ap}|\boldsymbol{r}_{jl}|)}{Q_{ap}|\boldsymbol{r}_{jl}|}. \tag{5}$$

$F_{jl}$ is the strength of the coherence volume associated with atom $j$ at atom $l$. In the coherent limit, $F_{jl} = e^{-2\pi i \boldsymbol{\kappa} \cdot \boldsymbol{r}_{jl}}$. For uniform hollow-cone illumination $F_{jl}$ takes the form of a narrow volume that is elongated and oscillatory along the optic axis (Treacy & Gibson 1993).

The mean square intensity is

$$\langle I^2(\boldsymbol{\kappa}) \rangle = \frac{\lambda^4 f^4(|\boldsymbol{\kappa}|)}{\mathcal{A}} \sum_j \sum_l \sum_m \sum_n e^{-2\pi i \boldsymbol{\kappa} \cdot (\boldsymbol{r}_{jl} + \boldsymbol{r}_{mn})}$$

$$\times \iint d^2\boldsymbol{q}_1 \iint d^2\boldsymbol{q}_2 \iint d^2\boldsymbol{q}_3 \; e^{2\pi i(\boldsymbol{q}_1 \cdot \boldsymbol{r}_{jn} + \boldsymbol{q}_2 \cdot \boldsymbol{r}_{nl} + \boldsymbol{q}_3 \cdot \boldsymbol{r}_{mn})} \tag{6}$$

$$\langle I^2(\boldsymbol{\kappa}) \rangle \approx \frac{\lambda^4 f^4(|\boldsymbol{\kappa}|)}{\mathcal{A}} \pi^3 Q_{ap}^6 \sum_j \sum_l \sum_m \sum_n F_{jl} F_{mn} A_{jn} A_{nl} A_{mn} \tag{7}$$

Equation (6) is subject to the joint condition $|\boldsymbol{q}_1 - \boldsymbol{q}_2 + \boldsymbol{q}_3| \leq Q_{ap}$. Replacing the integrals over $\boldsymbol{q}_1$, $\boldsymbol{q}_2$ and $\boldsymbol{q}_3$ by the product $A_{jn} A_{nl} A_{mn}$ is strictly valid only when atom pairs $jn$, $nl$ and $mn$ each align in columns parallel to the optic axis. This approximation is valid for hollow cone illumination where the $F_{jl}$ emphasizes those $j$ and $l$ that align in columns.

Inserting equations (4) and (7) into equation (3) yields the variance as

$$V = \frac{\sum\limits_{j,l,m,n} \left[N_0 A_{jn} A_{nl} - A_{jl}\right] A_{mn} F_{jl} F_{mn}}{\sum\limits_{p,q,r,s} A_{pq} A_{rs} F_{pq} F_{rs}} \tag{8}$$

where $N_0 = \mathcal{A}\pi Q_{ap}^2$, the number of pixels in the image. The 3-body component $N_0 A_{jn} A_{nl} - A_{jl}$ is the core speckle term, and essentially describes the amount of projected overlap between atoms $j$ and $l$, as mediated by a third atom $n$.

## 3. DISCUSSION

To treat the uncorrelated atom limit, we decompose $\sum_{j,l,m,n}$ into 15 distinct subcomponents, within each of which no atom is counted twice. These subcomponents are represented by the domains

$$
\begin{array}{lllll}
j = l = n = m & (j = n) \neq (l = m) & (j = m) \neq (l = n) & (j = l) \neq (m = n) & (j = m = n) \neq l \\
(l = m = n) \neq j & (j = l = n) \neq m & (j = l = m) \neq n & (m = n) \neq j \neq l & (j = n) \neq l \neq m \\
(l = n) \neq j \neq m & (j = m) \neq l \neq n & (l = m) \neq j \neq n & (j = l) \neq m \neq n & j \neq l \neq m \neq n
\end{array}
$$

Assuming that atom positions are fully random, and that there is a large number of atoms in the specimen, terms of the type $F_{jl}$, $F_{jl} F_{mn}$ average to zero, whereas terms of the type $F_{jl}^2$ are non-zero. Gathering terms in this way, equation (8) can be written in the uncorrelated limit as

$$
\begin{aligned}
V_{random}(\boldsymbol{\kappa}) &= \frac{1}{\langle n \rangle} - \frac{1}{N} + 2(1 - 1/N)(1 - 1/N_0)\overline{F^2}(\boldsymbol{\kappa}) \\
&\approx 2\overline{F^2}(\boldsymbol{\kappa}) + \frac{1}{\langle n \rangle} - \frac{1}{N} \tag{9}
\end{aligned}
$$

Here, $N \gg 1$ is the total number of atoms in the image, and $< n >= N/N_0$ is the average number of atoms per pixel column. $\overline{F^2}(\kappa)$ represents the average value of $F_{jl}^2(\kappa)$ ($j \neq l$). When atom positions are fully random, $\overline{F^2}(\kappa)$ is determined entirely by the illumination geometry. However, when there is any short- or medium-range order present, equation (9) breaks down, and the full form, equation (8), must be used.

Under conditions of perfect spatially coherent illumination, we can set $\overline{F^2} = 0.5$. With $N, N_0 \gg 1$, we get

$$V_{random}^{coh} \approx 1 + \frac{1}{\langle n \rangle} - \frac{1}{N} \qquad (10)$$

The dominant term is the $V = 1$ term, which is the result expected for negative exponential statistics when $N \to \infty$ in an uncorrelated sample. The $1/\langle n \rangle$ term is the Poisson statistics contribution coming from the variation in the number of atoms per pixel, and the $1/N$ term is from the sample statistics. The columnar $1/\langle n \rangle$ "shot noise" term is not considered in the negative exponential or gamma-distribution models, which assume infinite thickness where $\langle n \rangle \to \infty$.

In practice, the illumination is never fully coherent, and $\overline{F^2} \ll 0.5$, giving values of $V \ll 1$, a result that is consistent with experimental data.

In the limit of perfectly incoherent illumination, such as that obtained by a very high angle annular detector, we have $\overline{F^2} = 0$, and thus

$$V_{random}^{incoh} \approx \frac{1}{\langle n \rangle} - \frac{1}{N} \qquad (11)$$

which is simply a measure of the statistics of the projected atom images.

In our treatment we concern ourselves with the first and second moments of the intensity distribution only, expressed in terms of the normalized variance $V$. This is the simplest measure of the image speckle. The above treatment could, in principle, be extended to the higher order moments, which are sensitive to higher order atom correlations. However, at this level of higher-order analysis, it may be more fruitful to directly simulate the dark field images for the given models and to compare simulation with experiment. An important facet of speckle analysis is how to invert the data to extract a model. This problem has yet to be solved.

## REFERENCES

Dainty J C 1975 Laser speckle and related phenomena, ed J C Dainty
   (New York:Springer-Verlag) p 5.
Gibson J M and Howie A 1978 Chemica Scripta, **14**, 109
Gibson J M 1978 PhD thesis Cambridge University
Gibson J M and Treacy M M J 1997 Phys. Rev. Letts. **78**, 1074
Goodman J W 1975 Laser speckle and related phenomena, ed J C Dainty
   (New York:Springer-Verlag) pp 60–68.
Howie A 1978 J. Non-Cryst. Solids **31**, 41
Krivanek O L 1975 PhD thesis Cambridge University
Rudee M L and Howie A 1972 Philos. Mag. **25**, 1001
Treacy M M J and Gibson J M 1993 Ultramicroscopy **52**, 31
Treacy M M J and Gibson J M 1996 Acta Cryst. A **52**, 212

*Inst. Phys. Conf. Ser. No 153: Section 10*
*Paper presented at Electron Microscopy and Analysis Group Conf. EMAG97, Cambridge, 1997*
© *1997 IOP Publishing Ltd*

# STM surface modification of $YBa_2Cu_3O_{7-\delta}$ thin films

**S E Johnson[1], M Yeadon[1#], M Aindow[1], P Woodall[2], F Wellhöfer[2] and J S Abell[1]**

[1]School of Metallurgy and Materials; [2]Interdisciplinary Laser Ablation Facility; University of Birmingham, Edgbaston, Birmingham B15 2TT..

**ABSTRACT:** During imaging by STM, we have observed surface modification of $YBa_2Cu_3O_{7-\delta}$ high temperature superconducting thin films (grown by PLD) on single crystal (001) MgO and $SrTiO_3$ substrates. Repeated scanning of the same area led to gradual removal of material from the edge of mesas and nucleation and growth of holes. Removal was not necessarily restricted to the top layer as etching occurred at all exposed surfaces. Scanning is finally interrupted by the influence of the insulating substrate. The absence of debris at the edge of the scanned area implies that the loss of material was due to electric field induced effects rather than milling.

## 1. INTRODUCTION

STM is regularly used to image the surface topography of single crystals and thin films of high Tc superconducting oxides. The surface morphology of $YBa_2Cu_3O_{7-\delta}$ thin films typically consists of spiralled protrusions thought to form at the emergent threading segments of dislocation half loops caused by the lattice misfit between film and substrate (Yeadon 1997). STM has also been used as a tool for nanoscale modification of surfaces such as movement of clusters, atoms and molecules on surfaces and mechanical milling (Virtanen 1991, Harmer 1991). Surface modification of $YBa_2Cu_3O_{7-\delta}$ thin films during imaging by STM has been attributed to two main types of mechanism: milling, caused by tip sample interaction, and electric field induced effects (etching). Milling would be expected to occur when the bias voltage is set too low (or the set point current too high) such that the tip approaches the surface until it is in physical contact. As the STM performs a scan, the tip rasters across the surface of the sample removing material and depositing it at the perimeter of the scan area. Milling has also been achieved by setting very low gain values (Virtanen 1991) to limit the response time of the tip to sudden changes in height of the sample surface. Field induced effects would be expected to occur at high bias voltages where the electric field between the tip and the sample is greater. Heyvaert (1992) used a stable set point current and increased the bias voltage. He concluded that etching was controlled by the bias and the scanning time. Bertsche (1996) varied both the bias voltage and set point current and attributed the etching to either field enhanced chemical decomposition or field induced evaporation.

In this paper we report initial observations of the STM surface modification of $YBa_2Cu_3O_{7-\delta}$ on MgO and $SrTiO_3$ and consider the possible origins of this process.

---

[#] Now at University of Illinois, USA.

## 2    EXPERIMENTAL

A series of $YBa_2Cu_3O_{7-\delta}$ thin films (thickness $\approx$15nm) were deposited by pulsed laser deposition onto single crystal (001) MgO and $SrTiO_3$ substrates at deposition temperatures of 780°C and 820°C respectively, and a dynamic oxygen pressure of approximately 53Pa. Samples were then cooled at 30°Cmin$^{-1}$ to room temperature under 70kPa oxygen. To limit the extent of degradation due to exposure to the atmosphere, the samples were examined immediately after removal from the deposition chamber. A Digital Instruments Nanoscope II STM was used in constant current mode and the tips were prepared mechanically from 250μm Pt(80%)Ir(20%) wire. Scans were taken at a sample bias of 800mV and with a tunnelling current of 0.05nA at an estimated tip-sample distance of 9.6Å giving a field strength of approximately 0.8GVm$^{-1}$.

## 3    RESULTS

The series of $YBa_2Cu_3O_{7-\delta}$ thin films on MgO and $SrTiO_3$ substrates were examined using STM. Their surfaces consisted of layered structures with step heights of approximately one unit cell similar to those shown in scan 1 of figs. 1 and 2 for films on MgO and $SrTiO_3$ respectively.

The surface of films on MgO deteriorated with each successive scan as shown in the sequence of images in fig.1. The modification of the surface was rapid with extensive erosion by the sixth scan. Holes were already present in the first image but in subsequent scans more holes appeared and the mesa edges began to erode. By the fourth scan the two uppermost mesas had been removed entirely. Holes continued to appear with further scans and occurred in short rows of three or four. These holes grew in size while the mesa edges eroded towards them until the local area was completely etched leading to the break up of the mesa structure.

The surface structure of films deposited onto $SrTiO_3$ was more well-defined than that of films on MgO as shown in scan 1 of fig.2. The surface coverage, prior to modification, was more uniform with distinct tiered mesas (B, A, C). Surface modification occurred almost immediately with the loss of the small mesa D and the significant erosion of mesa C. Subsequent erosion of mesa A, however, occurred at a much slower rate. Mesa A in scan 6 still retained a semblance of its original structure which was in stark contrast to the structures that remained in the $YBa_2Cu_3O_{7-\delta}$ film on MgO in scan 6 of figure 1. Two holes can also be seen in mesa $A$ in the sixth scan, and were located in the area beneath the site of the uppermost mesa (C). With further scans these holes remained the same size whilst the edges of the mesas receded towards them. The preferential edge erosion towards the holes continued until the mesa structure was bisected in the 16th scan. The much larger platelet ($B$) also suffered edge erosion (as shown in scan 14) and several shallow holes appeared during the etching experiment. A deeper hole was visible at the site of the small mesa, ($D$), which was eroded within the first two scans. Scan 18 is 300 x 300nm showing the location of these holes and the extent of the modification. There is no sign of debris at the perimeter of the modified area.

## 4    DISCUSSION

The surface modification of $YBa_2Cu_3O_{7-\delta}$ on MgO was predominantly by the formation and expansion of holes which broke up the mesa structure as compared to the surface modification of $YBa_2Cu_3O_{7-\delta}$ on $SrTiO_3$ which was mainly by the steady erosion of the mesa edges. Neither modification was by the displacement of material which would have been visible

scan 1          scan 2          scan 3

scan 4          scan 5          scan 6

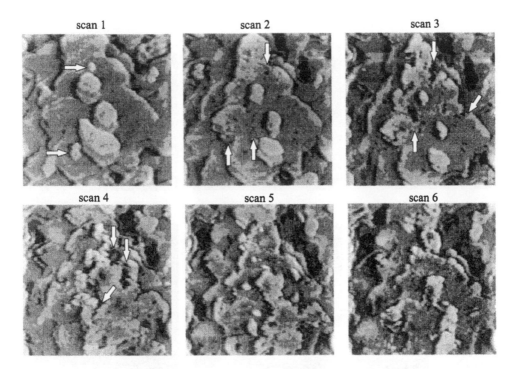

Figure 1: STM modification sequence of $YBa_2Cu_3O_{7-\delta}$ on MgO (400x400nm)

as debris at the perimeter of the scans. This would indicate that, in each case, modification was by an electric field or localised heating effect such as evaporation, ablation or decomposition. Large electric fields intrinsic to STM imaging may cause evaporation of the sample surface if the material being imaged has a lower evaporation threshold than the material used for the STM tip (Tsong 1978). A concurrent effect may also have been the localised heating of the sample by the electrons impinging onto its surface. A sample with low electrical and/or thermal conductivity would prevent the dissipation of these electrons leading to the build up of both charge and heat (Staufer 1994). The creation of holes in the $YBa_2Cu_3O_{7-\delta}$ surfaces of both samples would suggest that these areas were the sites of defects in the $YBa_2Cu_3O_{7-\delta}$ structure where the energy barrier for evaporation from the surface may be lowered or where the $YBa_2Cu_3O_{7-\delta}$ layer may be oxygen deficient and, possibly, more easily decomposed. The higher density of holes created in the $YBa_2Cu_3O_{7-\delta}$ thin film on MgO may be due to the presence of a larger number of defects, caused by the greater lattice mismatch.

## 5    CONCLUSION

$YBa_2Cu_3O_{7-\delta}$ thin films were deposited by pulsed laser deposition on MgO and $SrTiO_3$ and were imaged by scanning tunnelling microscopy. Both samples suffered surface modification during long term imaging. $YBa_2Cu_3O_{7-\delta}$ surfaces on MgO were eroded much more severely and more rapidly than that of $YBa_2Cu_3O_{7-\delta}$ on $SrTiO_3$. As no debris was observed at the perimeter of the scans the modifications were attributed to electric field evaporation and localised heating effects.

440

scan 1       scan 3       scan 6

scan 14       scan 16       scan 18

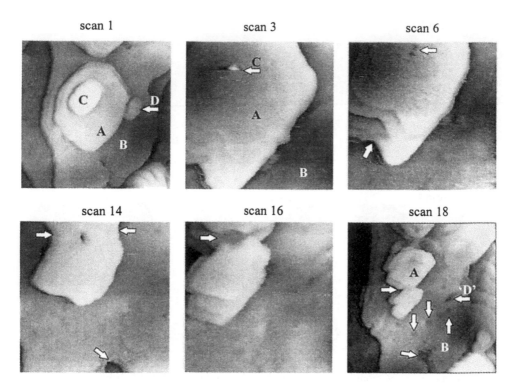

Figure 2: STM modification of $YBa_2Cu_3O_{7-\delta}$ on $SrTiO_3$. (1&18: 300x300nm, 3-16: 140x140nm)

It may be advisable to consider the possibility of surface modification when imaging specimens, over long periods of time, which have a low thermal or electrical conductivity, and/or field evaporation values lower than that of the material used for the STM.

## REFERENCES

Bertsche G, Clauss W, Kern D P (1996) Appl Phys Lett **68** (25) pp3632-3634
Harmer M A, Fincher C R and Parkinson B A 1991 Jnl Appl Phys **70** (5) pp2760-2763
Heyvaert I, Osquiguil E, Haesendonck C Van and Bruynseraede Y1992 App Phys Lets **61** (1) pp111-113
Staufer U 1994 Chapter 8, Scanning Tunnelling Microscopy II, 2nd Ed, Springer Series in Surface Science **28**
Tsong TT 1978 Surface Science **70** pp211-233
Virtanen J A, Suketu P, Huth G C and Cho Z H 1991 Jnl App Phys **70** (6) pp3376-3378
Yeadon M, Aindow M, Wellhöfer F and Abell J S 1997 Jnl Crystal Growth **172** pp145-155

*Inst. Phys. Conf. Ser. No 153: Section 10*
*Paper presented at Electron Microscopy and Analysis Group Conf. EMAG97, Cambridge, 1997*
© *1997 IOP Publishing Ltd*

# A TEM study of secondary epitaxial orientations in laser-ablated YBa₂Cu₃O₇₋δ thin films

**DJ Norris and M Aindow**

School of Metallurgy and Materials, University of Birmingham, Elms Road, Edgbaston, Birmingham, B15 2TT, UK.

**ABSTRACT:** A TEM study of laser ablated $YBa_2Cu_3O_{7-\delta}$ thin films grown below the usual temperature has revealed a variety of misoriented grains embedded within a standard epitaxial YBCO matrix. They exhibited regular shapes and, in many cases, adopted well-defined crystallographic orientations with respect to the substrate which could be explained in terms of a 'Near Coincident Site Lattice' model.

## 1. INTRODUCTION

The properties of $YBa_2Cu_3O_{7-\delta}$ (YBCO) films are known to be affected by the presence of microstructural defects. Grain boundaries in particular cause significant degradation of the critical current density ($J_c$); whereas threading dislocations and precipitates are thought to act as localised flux pinning centres which enhance the $J_c$. Since the defect microstructures of such films are known to depend critically on the process parameters it should be possible to adjust these parameters to avoid those features which degrade device performance. For most films the grain boundaries formed do not have random misorientations but instead correspond to interfaces between regions with different epitaxial orientation relationships. The deposition conditions are usually adjusted to give only

$$(001)_{YBCO} // (001)_{substrate}$$
$$[100]_{YBCO} // [100]_{substrate}$$

and any other orientations which occur thus appear as embedded grains or precipitates in this matrix.

In our work we have deposited YBCO films onto (001) MgO substrates by Laser Ablation using a wide range of deposition parameters to determine the optimum conditions for our apparatus. In films grown away from these optimum conditions we found many different embedded grains each of which adopted a distinctive regular shape and a well-defined crystallographic orientation with respect to the substrate. The range of these orientations was much wider than that reported previously (Ramesh *et al.* 1990, Ravi *et al.* 1990 and Chan 1994). In this paper we present TEM observations of such grains with particular emphasis on their orientation and morphology.

## 2. EXPERIMENTAL DETAILS

A 1100nm YBCO film was deposited onto an (001)MgO substrate by Pulsed Laser Deposition using a Questek KrF excimer laser ($\lambda$=248nm) at a fluence of ~1.5Jcm⁻² and a pulse repetition rate of 10Hz. The (001) MgO substrate was heated to 750°C which was lower than the "optimum" temperature of 780°C which gives the highest quality films in our apparatus. The substrate was positioned 7cm from a polycrystalline YBCO pellet which had been conditioned for 1min by the laser prior to deposition. Deposition was carried out under an oxygen pressure of ~53Pa (400mTorr). After deposition the oxygen pressure was increased to ~66kPa (~500Torr) for *in-situ* oxygenation for 10min. The sample was then cooled at a rate of 30°Cmin⁻¹ to 400°C and then allowed to cool naturally to room temperature. Plan-view TEM specimens was prepared in the usual manner by mechanical dimpling followed by ion-milling to perforation at liquid nitrogen temperature to minimise beam damage and examined using a Philips CM20 electron microscope operating at 200kV.

## 3.    RESULTS

The YBCO film consisted of a matrix with the (001) orientation relationship plus a wide variety of embedded grains. An area which contains a large number of such grains is shown in fig.1 - on average these constituted about 1% of the total film volume. These grains could be classified into three main types according to their shape as shown in fig. 2 .

The square-shaped grains, labelled A in figs.1 and 2, were oriented such that $[001]_g$ // $[001]_m$ (and thus // $[001]_{MgO}$), where g and m refer to the grain and matrix, respectively. The rotation, $\theta$, of these grains about the common [001], was measured from selected area diffraction (SAD) patterns. It was found that $\theta$ adopts certain well-defined values and examples of representative SAD patterns are shown in fig.3. Of these orientations, the one which was observed most frequently was $\theta = \sim 23.5°$. The size of these grains was fairly constant at $\sim 0.8\mu m$ across, irrespective of the value of $\theta$. The boundaries which delineate the grains were all [001]/$\theta$ assymetric tilt boundaries with the boundary plane parallel to $(100)_g$ or $(010)_g$ in each case.

The triangular-shaped grains, labelled B in figs. 1 and 2, were all oriented in a similar manner such that $<110>_g$//$<1\bar{1}0>_m$ with $[001]_g$ misoriented with respect to $[001]_m$ by $\sim 13°$. These grains were slightly smaller than the square grains, ranging from $\sim 0.3$ to $0.8\mu m$ across. For each of these grains, there are two mixed tilt/twist boundaries parallel to $(100)_m$ and $(010)_m$ and an assymetric tilt boundary parallel to $(110)_m$.

The grains which had a trapezoidal or more irregular shape, labelled C in figs.1 and 2, were invariably found to have no simple orientation relationship with respect to the surrounding matrix. They did, however, appear to be oriented in a similar manner to the square grains, but with an additional tilting of $[001]_g$ with respect to the $[001]_m$.

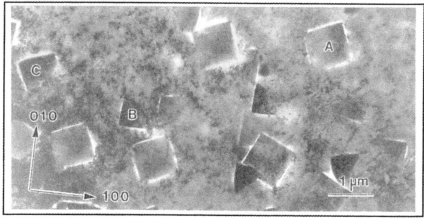

Fig 1.  Bright-field TEM image of an epitaxial (001)YBCO film with embedded grains

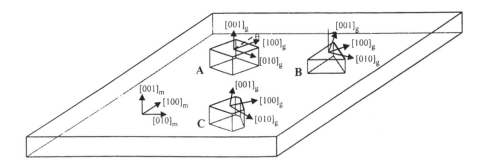

Fig 2.  Schematic representation of the morphology and orientation of the types of grain.

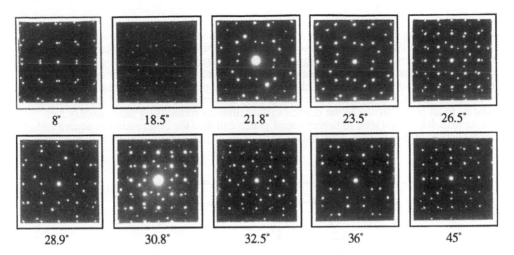

Fig 3. SAD Patterns showing the various misorientations of the square grains.

## 4. DISCUSSION

Previous studies of grain misorientations have dealt mainly with polycrystalline textured films where the microstructure consisted largely of interconnected grains which were misoriented with respect to each other (Ramesh *et al.* 1990, Ravi *et al.* 1990 and Suzuki *et al.* 1993). Such microstructures were thought to arise because of the grapho-epitaxial nature of the YBCO/MgO interface whereby the deposit adheres weakly to the substrate surface (Norton and Carter 1991). More recent studies have, however, shown that the YBCO/MgO interface can exhibit a more conventional semi-coherent interface (Aindow *et al.* 1996) which is not consistent with a graphoepitaxial model.

For the film considered in the present study, the microstructure consisted mainly of a matrix with the usual epitaxial orientation with a variety of embedded grains which adopt well-defined orientations. Here again, this is not consistent with the grapho-epitaxial model but may instead result from local surface or interface reconstructions which could favour certain secondary orientations. One such model which has been considered for this system previously is the 'Near Coincident Site Lattice' (NCSL) approach whereby the grains might adopt orientations which give a high density of near-coincident lattice sites in the interface between YBCO and MgO (Ramesh *et al.* 1990, Ravi *et al.* 1990 and Suzuki *et al.* 1993). For our material we have tested this hypothesis for the square grains by comparing those values of $\theta$ which would be expected on this basis with those observed experimentally. At certain values of $\theta$ the translation vector $[u\ v\ 0]_g$ lies parallel to $[100]_{MgO}$ giving a modest misfit, $\delta$, between the vectors $[u\ v\ 0]_g$ and $[h\ 0\ 0]_{MgO}$. If $\delta=0$ then these would correspond to $\Gamma h$ 2D CSL (coincident site lattice) orientations where $\Gamma$ is defined as the reciprocal density of coincident MgO lattice sites in the YBCO/MgO interface. In each case the grain orientations in our samples correspond to reasonable values of $\Gamma$ and/or $\delta$ and the data from the analysis are summarised in Table 1. The most commonly observed grain orientation (~23.5°) was close to several 2D NCSLs, i.e. $\theta$ = 23.9°, 23.6°, 23.2° and 22.6°, each of which has $\Gamma <16$ and $\delta$ <0.65%. Thus the experimental observations presented here are consistent with an NCSL model for interface structures in these secondary epitaxial orientations. What is less clear is why such a wide variety of NCSL orientations should be exhibited in this particular epitaxial system.

All of the triangular grains with the matrix were oriented with $[001]_g$ rotated by $\approx 13°$ with respect to $[001]_m$ about the common <110>. This corresponds approximately to an orientation relationship:

$$(\bar{1}\ 1\ 18)_g // (001)_{MgO}$$
$$[110]_g // [110]_{MgO}$$

The lattice mismatch at this orientation is $\approx 8.8\%$ along the common <110>. The misfit in the orthogonal direction has a modest value only if long translation vectors such as $[9\bar{9}1]_g$ and $[9\bar{9}0]_{MgO}$ are considered, which differ by $\approx 5.8\%$. Thus the occurrence of this orientation could not be explained easily on the basis of an NCSL model. Nevertheless, the fact that more of the grains adopt this specific orientation relationship than any other suggests that it must correspond to a relatively low energy interface structure.

The orientation of the boundaries delineating the grains is presumably related to the boundary energies and/or mobilities. If they are free to adopt low energy configurations (i.e. if the boundary mobility is high enough at the growth temperature) then one might expect the boundaries to lie perpendicular to the deposit surface, thereby minimising the boundary area. The same principle would tend to favour round grains and thus the observation of square and triangular grains implies that the habit planes correspond to particularly low energy structures for these orientations.

| $\theta$ | $\delta$ (%) | h / $\Gamma$ | u | v |
|----------|--------------|--------------|-----|-----|
| 8.13° | 0.24 | 13 | 14 | 2 |
| 18.4° | 3.39 | 3 | 3 | 1 |
| 21.8° | 1.2 | 5 | 5 | 2 |
| 22.6° | 0.65 | 12 | 12 | 5 |
| 23.2° | 0.23 | 7 | 7 | 3 |
| 23.6° | -0.099 | 16 | 16 | 7 |
| 23.9° | -0.36 | 9 | 9 | 4 |
| 26.26° | -2.5 | 2 | 2 | 1 |
| 28.8° | 0.32 | 21 | 20 | 11 |
| 30.96° | -0.26 | 16 | 15 | 9 |
| 31.6° | 0.01 | 14 | 13 | 8 |
| 32.5° | 0.36 | 12 | 11 | 7 |
| 35.5° | 1.4 | 8 | 7 | 5 |
| 36.87° ʳ | 1.7 | 9 | 8 | 6 |
| 45° | 2.77 | 4 | 3 | 3 |

r - the misfit varies with multiples of the smallest coincident translation vector

Table 1. Geometric parameters for NCSLs which occur at orientations close to those for the embedded grains.

## 5.    CONCLUSIONS

A TEM study of an YBCO film deposited onto (001) MgO by pulsed laser deposition at a lower than normal deposition temperature, has revealed a variety of misoriented sub-grains which were individually embedded within the film matrix. The number of orientations observed was far wider than those reported previously and the origin of some of these could be explained in terms of the 'Near Coincident Site Lattice' model.

**ACKNOWLEDGEMENTS**

We are grateful to Professors IR Harris and JF Knott for the provision of the laboratory facilities, Drs F Wellhöfer and P Woodall for the use of the laser ablation facility, and to EPSRC for financial support.

**REFERENCES**

Aindow M, Norris DJ and Cheng TT, 1996 Phil. Mag. Lett. **74** 267
Chan SW, 1994 J. Phys. Chem. Solids **55** 1415
Norton MG and Carter CB, 1991 J. Crystal Growth **110** 641
Ramesh R, Hwang D, Ravi TS, Inam A, Barner JB, Nazar L, Chan SW, Chen CY, Dutta B, Venkatesan T and Wu XD, 1990 Appl. Phys. Lett. **56** 2243
Ravi TS, Hwang DW, Ramesh R, Chan SW, Nazar L, Chen CY, Inam A and Venkatesan T, 1990 Phys. Rev. B **42** 10141
Suzuki H, Fujiwara Y, Hirotsu Y, Yamashita T and Oikawa T, 1993 Jpn. J. Appl. Phys. **32** 1601

*Inst. Phys. Conf. Ser. No 153: Section 10*
*Paper presented at Electron Microscopy and Analysis Group Conf. EMAG97, Cambridge, 1997*
© 1997 IOP Publishing Ltd

# Structure / Property Relationship in Negative Temperature Coefficient Thermistors

**G. D. C. Csete de Györgyfalva, I. M. Reaney, R. D. Short, A. N. Nolte\*, K. Meacham\* and R. Twiney\***

Department of Engineering Materials, University of Sheffield, Mappin Street, Sheffield, S1 3JD, UK.
\*Bowthorpe Thermometrics, Crown Industrial Estate, Taunton, Somerset, TA2 8QY, UK.

**ABSTRACT:**
Scanning and transmission electron microscopy have been used to investigate the reaction mechanism and microstructure of negative temperature coefficient, $Ni_{1-x}Mn_{2+x}O_4$ based thermistors. Evidence of unreacted NiO was found which contained florets of spinel with a cube//cube relationship to the rock salt matrix. In ferroelastically distorted samples, lamellar and herring bone domain structures were observed. Electron diffraction patterns from all samples examined exhibited streaking parallel to <110> directions which was associated with the appearance of sinusoidal, 10nm wide fringes in TEM images.

## 1    INTRODUCTION

The $Ni_{1-x}Mn_{2+x}O_4$ solid solution series can be used to form spinel-structured, semiconducting ceramics that exhibit high, negative values for the temperature coefficient of resistance (TCR). The wide range of resistivities and TCRs available in this system make it ideal for the manufacture of temperature sensors. $NiMn_2O_4$ is cubic ($Fd\bar{3}m$) with the inverse spinel structure but as $x$ increases a distorted tetragonal ($I4_1/amd$), spinel structure is stabilised (D.G. Wickham 1964). The tetragonal distortion which is present in the system not only depends on composition but on the processing route. The tetragonality can be accounted for by $Mn^{3+}$ octahedral site occupancy. This gives rise to a distortion, brought about by the non-symmetrical nature of some ions with partially occupied 'd' or 'f' orbitals. This is often referred to as the Jahn-Teller effect (D. F. Shriver *et al*, 1992). Rousset *et al*, 1992 in their investigation of $Ni_{1-x}Mn_{2+x}O_4$ based ceramics using transmission electron microscopy, observed lamellar and tweed-like domain structures in tetragonal ceramics. In addition, a finer fringe contrast was observed with a separation of only a few nanometers. Although cubic ceramics did not exhibit the lamellar and tweed-like domain structures, the fine fringe contrast was still present.

Small polaron conduction, sometimes termed as 'electron hopping', has been suggested by Brabers and Terhell (1982) as the main conduction mechanism in the $Ni_{1-x}Mn_{2+x}O_4$ system at room temperature. The above workers postulated that mixed valence manganese cations ($Mn^{3+}$, $Mn^{4+}$) present on the octahedral sites give rise to these small polaron pathways. It has also been suggested that the concentration of the $Mn^{4+}$ is subject to changes in processing temperature (Dorris and Mason, 1988).

## 2    EXPERIMENTAL PROCEDURE

### 2.1    Ceramic Production

75g batches of the raw materials, NiO and $Mn_2O_3$, were weighed on a balance accurate to 0.01g. The powder was then transferred to a polypropylene pot with zirconia milling media containing 100 cm³ of de-ionised water (contaminants from the poly-propylene pot burn off during the calcination) and was milled several hours. After milling, the mixture was suction-filtered using a Buchner funnel through grade 3 filter paper. The slurry which was left on the filter paper was dried overnight. The now dry powder was transferred to a refractory crucible and placed in a furnace where it was pre-reacted. The calcined powder was returned to the milling pot with the same amount of water, and milled and dried as described above.

About 0.5g of dry powder was placed in a 1 cm diameter press and compressed using a hydraulic jack to approximately 15 K.Pa., and held at this pressure for 1-2 minutes. However, cup and cone fractures of the green pellets were frequently observed, seemingly irrespective of the applied pressure. Therefore, a hand die was used and the powder compacted manually. The reduction in green density which occurred as a result of this method did not translate to the sintered discs. The compacted discs were placed on a refractory tray and sintered under the required conditions.

## 2.2    Analysis

Electrical analysis was carried out by taking resistance measurements at precisely controlled temperatures (+/- 0.001°C) with a Hewlett Packard Precision Resistometer. Microanalysis was carried out using scanning and transmission electron microscopy. For scanning electron microscopy (SEM), samples were mounted onto aluminium stubs using silver paste. Fractured and sintered surfaces were examined using a CAMSCAN 600 SEM operating at 20kV. TEM samples were prepared by grinding ceramics to 20 μm and ion milling to perforation using a Gatan Duomill operating at 6kV and 0.6 mA. The samples were analysed using a Phillips 400 (120kV) and a JEOL 3010 (300kV) TEMs. The JEOL 3010 was equipped with a Link energy dispersive (EDS) x-ray detector and the associated computer hardware.

## 3      RESULTS AND DISCUSSION

### 3.1    Electrical Analysis

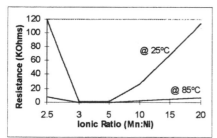

*Figure 1: Composition versus resistance for the solid solution series $Ni_{1-x}Mn_{2+x}O_4$*

Figure 1 shows a plot of composition versus resistance for the $Ni_{1-x}Mn_{2+x}O_4$ solid solution series at 25°C and 85°C. 25°C and 85°C have been chosen since these temperatures are 'industry standard' to evaluate properties over the temperature range of operation. A region of low resistance can be seen at an ionic ratio of around 3.5Mn:Ni which is particularly evident at room temperature. At 85°C, the resistivity is generally much lower and the minimum is not so clearly observed.

### 3.2    Scanning Electron Microscopy

Figures 2 and 3 are SEM images from fractured and sintered surfaces of samples. Second phases are clearly evident in the micrographs (arrowed in Fig. 2) which were confirmed by EDS as nickel rich regions. Some dendritic features were also observed on the surface of certain grains (Fig. 3). The secondary phases could be either nickel oxide which has not fully reacted, or material produced by decomposition of the spinel phase during sintering. Wickham *et al.* (1964) suggested that, although the reaction between NiO and $Mn_2O_3$ can start as low as 750°C, above 950°C the product begins to decompose. Therefore as temperature is increased above 950°C, regions of NiO form according to:

$$NiMn_2O_4 \rightarrow NiO + Mn_2O_3$$

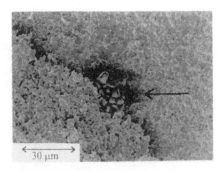

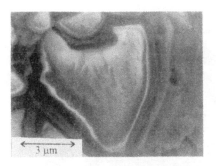

Figure 2: SEM image showing a secondary nickel rich phase (x=0)

Figure 3: Scanning electron micrograph of dendritic features on grain surfaces (x=0)

### 3.3    Transmission Electron Microscopy

Figure 4 is a TEM image of a nickel rich region commonly observed in all samples examined. The micrograph shows a NiO grain containing 'florets' of spinel nucleating and growing in a cube//cube relationship with the matrix. Such a microstructure is unlikely to represent a decomposition reaction and it is thought to arise because of incomplete reaction in the system. The dendritic structures observed by SEM are likely to be related to this transformation.

Preliminary studies using x-ray diffraction easily distinguished samples which were predominantly cubic from those which were distorted. Distorted samples showed peak splitting and multiple spinel phases thought to arise from an inhomogeneous distribution of Ni and Mn. In distorted samples, TEM revealed the presence of planar defects consistent in appearance with ferroelastic domain walls often in a herring-bone configuration, (Fig. 5). It is assumed that these domains are associated with a phase transition which occurs on cooling.

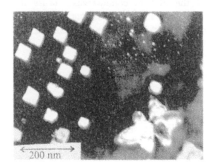

Figure 4:  Dark field (DF) TEM image showing florets of spinel in a NiO grain.

Figure 5: Brightfield (BF) TEM image showing a grain with a herring bone domain structure.

The distortion away from cubic is evidenced in electron diffraction patterns by significant spot-splitting along the <110> directions (Fig 6). In addition to spot-splitting, the diffraction pattern exhibits streaking parallel with <110> directions. Dark field images obtained in two beam conditions, so as to enhance any contrast arising from the streaking, revealed the presence of sinusoidal fringes lying approximately parallel with {110} planes and separated by a few nanometres (Fig. 7). Similar fringes were observed by Rousset et al., 1992. Careful examination of cubic samples revealed identical fringes, Fig. 8. The presence of streaking in electron diffraction patterns and fringe contrast in TEM images strongly suggests an incommensurate modulation in the structure. An incommensurate modulation can be viewed in simple terms as an electron density distribution of a long wavelength, compared to unit cell dimensions, superposed onto that associated with the conventional lattice.

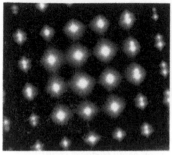  

Figure 6: Electron DP of a <001> pseudocubic zone axis show spot splitting and streaking, x=0.3.

Figure 7: DF TEM image showing the sinusoidal fringe contrast in distorted cubic sample.

Figure 8: BF TEM image of the same feature as in Fig. 7 but in cubic samples.

The structure of spinel is based upon close packed oxygens with some of the tetrahedral and octahedral sites occupied by cations. The tetrahedral occupation describes an NaCl type face centered arrangement whereas cations occupying octahedra form chains along <110> directions. According to Brabers and Terhell (1982) the electron hopping conduction mechanism is associated with $Mn^{4+}$ and $Mn^{3+}$ valence states. The fine fringes observed along <110> directions described above are separated by ~10nm or about 10 unit cell thicknesses. It is possible that the incommensurate modulation is related to the distribution of cations along <110> directions. For most compositions in $Ni_{1-x}Mn_{2+x}O_4$ solid solution series, a combination of $Ni^{2+}$, $Mn^{3+}$ and $Mn^{4+}$ could all be in close proximity to one another. Without further analysis it is impossible to say whether this is the case, but initial electronic measurements indicate that the highest conductivity occurs in compositions where $x=^1/_3$. At this stoichiometry, there is a ratio of two manganese cations to one nickel cation in the octahedral interstices. This may provide the required pathways to maximise the small polaron conduction mechanism.

## 4    CONCLUSION

1) NiO second phase is often found in samples. The internal structure of the NiO grains contains florets of spinel which have a cube//cube relationship with the rock salt matrix.
2) Distorted spinel structures contain lamellar and herring-bone domain structures arising from a phase transition.
3) Fringes separated by a few nanometres have been found in all samples examined. These are thought to be related to the distribution of different valence state cations on octahedral sites.

## REFERENCES

D.G. Wickham, J. Inorg. Chem. 26, 1369-1377, (1964)
E.G. Larson et al -J. Phys Chem. Solids - Pergamon Press, Vol 23 1771-1781, (1962)
D. F. Shriver, P. W. Atkins,C. H. Langford; Oxford Univ. Press (1992)
J. Jung, J. Topfer, J. Murbe, A. Feltz; J. Europ. Ceram. Soc., 6, 351-359 (1990)
A. Rousset et al Journ. de Phys. III, No. 4, 833-845 (1992)
V.A.M. Brabers; J. Terhell; Phys. Stat. Sol. (a) 69, 325-332, (1982)
S.E. Dorris, T.O. Mason; J. Am. Ceram. Soc. 71 [5] 379-385, (1988)

*Inst. Phys. Conf. Ser. No 153: Section 10*
*Paper presented at Electron Microscopy and Analysis Group Conf. EMAG97, Cambridge, 1997*
© *1997 IOP Publishing Ltd*

# The effect of substrate vicinal offcut on the morphology and defect microstructure of YBa$_2$Cu$_3$O$_{7-x}$ thin films on (001) SrTiO$_3$

D Vassiloyannis[1], DJ Norris[1], SE Johnson[1], M Aindow[1], P Woodall[2] and F Wellhöfer[2]

1. School of Metallurgy and Materials, 2. Interdisciplinary Laser Ablation Facility, The University of Birmingham, Edgbaston, Birmingham B15 2TT, U.K.

ABSTRACT: STM and TEM studies are presented of epitaxial YBa$_2$Cu$_3$O$_{7-x}$ on SrTiO$_3$ offcut from (001) by 0-6° about [100]. For a 1° offcut the growth spiral density is lower and the film is smoother than on 0° substrates. For higher offcuts the spirals do not form, the films become rougher and the threading dislocation density is much lower. It is proposed that these effects are related to the additional steps on the YBCO surface.

## 1. INTRODUCTION

Pulsed laser ablation is a highly reliable deposition technique which is used widely to produce heteroepitaxial thin films and multilayers of high-T$_c$ superconducting ceramics but the commercial exploitation of these films is inhibited by the presence of defects which arise due to mismatch between the substrate and the deposit. Features such as 3-D growth spirals, steps, precipitates, impurity phases etc. affect dramatically the effective thickness and the surface quality of the films. These factors are particularly important when other layers are deposited subsequently onto the YBCO such as in the fabrication of high quality Josephson-junctions. It has been observed recently (Haage *et al.*, 1997) that films deposited on substrates with vicinal offcut are more homogeneous across usefully large areas and contain fewer defects. It has been suggested (Zama *et al.*, 1996) that introducing steps on the substrate surface changes the growth mechanism of the films from 3D-spiral to layer by layer growth and that this improves the coherence across the substrate-film interface leading to the development of smoother and better quality films.

In our laboratory a systematic study of the effects of offcut angle on the microstructure and properties of YBCO thin films grown on SrTiO$_3$ vicinal substrates is being performed. Films have been deposited on a series of substrates offcut intentionally from (001) and analysed using a wide variety of characterisation techniques. In this paper we present preliminary STM and TEM data from these films and discuss the way in which the offcut affects the morphological and microstructural development.

## 2. EXPERIMENTAL DETAILS

A series of seven SrTiO$_3$ substrates offcut intentionally from (001) in 1° increments from 0° to 6° were used in our experiments. YBCO films 230 (±10) nm thick were grown by pulsed laser ablation under 400 mTorr oxygen with the substrates being heated to 820 °C during the deposition (for details see Wellhöfer *et al.*,1997). A Digital Instruments Nanoscope II STM operating in constant current mode at a bias of 800 mV and a tunnelling current of 50 pA was used to characterise the surface morphology of the films. Cross-sectional TEM specimens were prepared such that the specimen surface normal was parallel to the offcut axis. The specimens were polished and dimpled mechanically in the usual manner and Ar ion-milled to perforation using liquid nitrogen to avoid heating damage. The TEM study was performed on Philips CM20 and JEOL 4000FX microscopes operating at 200 kV.

## 3. RESULTS

### 3.1 STM data

The STM data obtained from the samples showed that the morphology of each film was fairly uniform from area to area but that there were dramatic differences from sample to sample indicating a strong dependence on the substrate offcut angle. Data from representative areas (1.2μm square) of the 0°, 1°, 2° and 6° specimens are presented in Fig. 1 as "topview" plots with the greyscale indicating the height above the lowest point in the scan. The 0° sample (Fig. 1a) is typical of epitaxial films on nominal substrates - the roughness has an amplitude, $R_a$ of ≈6nm and is characterised by atomically flat terraces separated by steps 1.3nm in height (i.e. one unit cell). The steps are arranged into distinctive spiral growth features which are usually associated with threading dislocations emerging at the YBCO deposit surface (Schlom et al. 1992; Yeadon et al. 1997). The density of these features is ≈$10^9$ cm$^{-2}$. For the 1° sample (Fig. 1b) the morphology is rather different and is dominated by 1.3nm steps whose average orientation is roughly parallel to the offcut axis. Some spiral features can still be distinguished on this surface but they are much less pronounced and their density is lower at ≈ $3·10^8$ cm$^{-2}$. The overall roughness is also significantly lower with $R_a$ ≈3.5nm. Images obtained from the 2° sample (Fig 1c) also exhibited the step and terrace structure parallel to the offcut axis but in this case no growth spirals were observed and the deposit was somewhat rougher with $R_a$ ≈6nm. This tend was also reflected in the remaining samples with the roughness increasing to $R_a$ ≈15.5nm for the 6° sample (Fig. 1d).

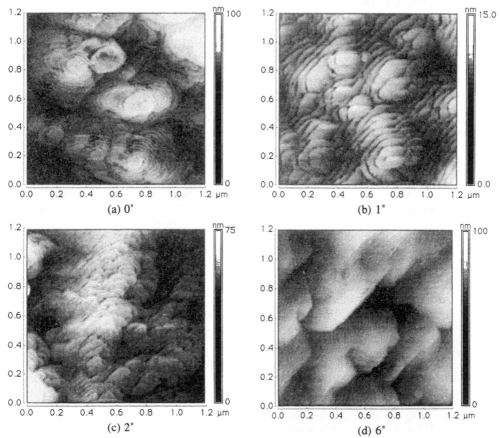

**Figure 1.** STM images YBCO films on vicinal (001) SrTiO$_3$ substrates.

These data are in good agreement with observations reported in literature for the growth of YBCO (e.g. Schlom 1993) on vicinally offcut substrates.

## 3.2    TEM Observations

Preliminary TEM observations obtained from cross-sectional samples of these films revealed that the change in surface morphology with vicinal offcut is accompanied by a change in the defect microstructure. A weak-beam dark field (WBDF) image and a selected area diffraction pattern obtained from the 0° sample are presented in Figures 2a and b, respectively. We should emphasise that this is *not* a representative area but rather one which has been chosen because it has a higher defect density than the average and includes examples of all the features seen in the deposit. Only occasional dislocations were observed in the substrate and none are present in this area. The SADP shows that the deposit is epitaxial with no discernible misorientation. There are dislocations (e.g. at A) threading from the YBCO/SrTiO$_3$ interface to the deposit surface and their density in this area is $\approx 5 \cdot 10^9$ cm$^{-2}$. There is also an embedded grain of another orientation at B. There were very few such features in our film but these are observed commonly in YBCO deposits on (001)SrTiO$_3$.

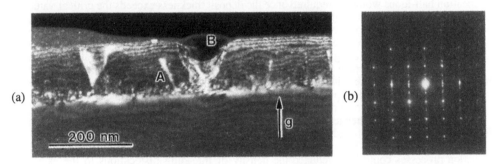

**Figure 2.** TEM cross-section of the 0° sample; (a) WBDF image obtained using $\mathbf{g} = 005_{YBCO}$; (b) SADP obtained with $\mathbf{B}$ // [010].

The corresponding WBDF image and SADP obtained from a representative area of the 2° sample (Fig. 3) showed a very different microstructure. Here again the deposit was epitaxial with no measurable misorientation but the $00n_{YBCO}$ diffraction peaks were more diffuse and streaked than for the 0° specimen (Fig. 3a). This could be due to disorder or (001) stacking faults in the YBCO but we cannot exclude the possibility that this is an ion-milling artefact. In this sample there were significant numbers of dislocations in the SrTiO$_3$ substrate and these appeared to be concentrated in the region near the original surface suggesting that they may be introduced during vicinal polishing. In contrast, no threading dislocations were observed in the YBCO deposit for this or any of the other areas examined.

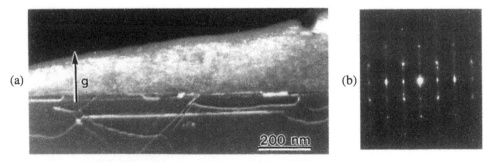

**Figure 3.** TEM cross-section of the 2° sample; (a) WBDF image obtained using $\mathbf{g} = 005_{YBCO}$; (b) SADP obtained with $\mathbf{B}$ // [010].

452

## 4. DISCUSSION

We consider firstly the morphology and microstructure of the 0° sample. This is consistent with the model of Yeadon *et al.* 1997 whereby the deposit is initially pseudomorphic and then relaxes at some critical thickness by the introduction of dislocation half-loops from the deposit surface or some heterogeneous source. The emergent threading segments of the half-loops have surface steps associated with them; these act as preferential growth sites and wind up into growth spirals during subsequent deposition since they are pinned at the dislocations (e.g. Schlom *et al.* 1992). Once the misfit has been relieved completely the density of the threading dislocations (and thus of the growth spirals) will reduce with increasing thickness as segments with dissimilar Burgers vectors interact, reacting to form one dislocation from two or even annihilating one another. Since our films are well above the critical thickness for this system ($\approx$20nm, Yeadon *et al.* 1997) we would expect to observe a high density of spirals together with a similar density of threading dislocations as shown in figures 1a and 2a.

If we now consider the 1° sample then one might expect that the microstructure would be similar since the steps on the substrate would not influence the growth mode, i.e. we would still expect 2D, layer-by-layer growth with an initially pseudomorphic deposit. The formation of dislocation half-loops would be required once the deposit thickness exceeds the critical value and thus significant numbers of threading dislocations should be present. The offcut could, however, affect the microstructure through the surface morphology. Before half-loops are introduced the average surface orientation will be maintained by a high density of steps parallel to the offcut axis and separating (001) terraces, even on an otherwise atomically smooth surface. Growth on this surface would, therefore, be much more uniform than for the 0° sample. Moreover, above the critical thickness only a small proportion of the deposit will be incorporated at the additional steps which arise at threading dislocations. There will thus be less of a tendency for spirals to form giving smoother films as observed experimentally.

For larger offcuts such as the 2° sample spiral formation is almost totally suppressed. This will also mean that the threading dislocations themselves are more mobile since they will not be pinned at the spirals. For much thicker deposits such as ours, therefore, reaction and annihilation processes could have reduced the density of threading dislocations to the extent where they cannot be observed by cross-sectional TEM (i.e. $<10^8$ cm$^{-2}$). The increase in the $R_a$ with offcut angle for offcuts of 2° or higher may be due to step bunching which can occur when the steps become closely spaced.

## 5. CONCLUSIONS

Films of YBCO on vicinal SrTiO$_3$ exhibit different morphologies and microstructures from those on nominal (001) substrates. For 1° offcut the film surface is smoother and terraced with fewer growth spirals and for 2° offcut no spiral features or threading dislocations were observed. These effects are probably related to the terraced structure of the surface. For offcuts $>1$° the surfaces become steadily rougher and this is probably due to step bunching.

### ACKNOWLEDGEMENTS

The authors would like to thank Prof. I.R. Harris for provision of laboratory facilities and EPSRC for financial support.

### REFERENCES

Haage T, Habermeier H -U and Zegenhagen J 1997 Surface Science **370** pp L156-L162
Schlom D G, Ansalmetti D, Bednorz J G, Broom R, Catana A Frey T, Gerber C, Güntherodt H -J, Lang H P, Mannhart J and Müller K A 1992 Z. Phys. B Condens. Matt. **86** pp 169
Schlom D G, Ansalmetti D, Bednorz J G, Gerber C and Mannhart J 1993 Mat. Res. Soc. Symp. Proc. **280** pp 341
Wellhöfer F, Woodall P, Norris D J, Johnson S, Vassiloyannis D, Aindow M, Slaski M and Muirhead C M 1997 proceedings of COLA'97 (in press)
Yeadon M, Aindow M, Wellhöfer F and Abell JS 1997 J. Crystal Growth **172** 145
Zama H 1996 Jpn. J. Appl. Phys. **35** pp 3388-339

*Inst. Phys. Conf. Ser. No 153: Section 10*
*Paper presented at Electron Microscopy and Analysis Group Conf. EMAG97, Cambridge, 1997*
© 1997 IOP Publishing Ltd

# Study of bevelled InP-based heterostructures by low energy SEM and AES

**R Srnanek, A Satka, J Liday, P Vogrincic, J Kovac,
M Zadrazil[*], L Frank[*], and M El Gomati[#]**

Microelectronics Department, Slovak Technical University, Bratislava, Slovak Rep.
[*] Institute of Scientific Instruments, Brno, Czech Republic
[#] Department of Electronics, University of York, York, United Kingdom

ABSTRACT: We report on low energy scanning electron microscopy characterisation of chemically bevelled InP/InGaAs/InGaAlAs and InP/InGaAsP heterostructures. A thickness of 3 nm for InP/InGaAlAs, 2 nm for InGaAs/InGaAlAs and 4-6 nm for InP/InGaAsP interfaces were obtained. These were compared with measurements by Auger electron spectroscopy (AES).

## 1. INTRODUCTION

InP-based heterostructures are widely used in optoelectronic and microelectronic devices. The position, width and quality of heterointerfaces play important roles in the performance of these devices. Such parameters had been previously estimated from photoluminiscence (Abraham et al. 1993), AES (Bresse 1987), and secondary ion mass spectroscopy (Hsu 1995). To our knowledge, a study on bevelled surfaces by low energy SEM has not been published yet. In this paper, we present a low energy and optical interference microscopy study of InP/InGaAs/InGaAlAs and InP/InGaAsP chemically bevelled heterostructures to determine the position, width and quality of interfaces.

## 2. EXPERIMENTAL

Two types of heterostructures, prepared on a (001)-oriented InP substrate were used. Type A (sample A) consisted of 10 pairs of alternating (130 nm thick) InP and $In_{0.73}Ga_{0.27}As_{0.6}P_{0.4}$ layers. Type B (sample B and C) is a structure of Resonant Cavity Enhanced PIN photodiode. This structure consists of a Bragg mirror (20 pairs of InP and $In_{0.52}Ga_{0.27}Al_{0.21}As$, 101 and 93 nm thick, respectively) followed by an undoped InP layer (600 nm thick), undoped $In_{0.52}Ga_{0.27}Al_{0.21}As$ (93 nm thick), undoped $In_{0.53}Ga_{0.47}As$ layer (200 nm thick), $p$-doped $In_{0.52}Ga_{0.27}Al_{0.21}As$ layer (470 nm thick) and p-doped $In_{0.53}Ga_{0.47}As$ layer (5 nm thick).

The apparatus used for bevel etching is that described by Srnanek et al. (1997). It consists of two tanks, the first is filled with BPK-221 etchant (Adachi 1982) while the second is partly filled with deionized water. Bevel formation is a result of a linear movement of the water/etchant interface over the sample. Bevels with angles 0.00018 rad (sample A), 0.00014 rad (sample B) and 0.0003 rad (sample C) were prepared. $HCl:H_3PO_4$ were used for InP and $KOH:K_3Fe(CN)_6$ etchants for other layers. Bevel surfaces were examined by optical interference microscopy.

Precise determination of the layer thickness, interface positions and quality has been accomplished by the Scanning Low Energy Electron Microscope (SLEEM) and a Tesla BS300 SEM. The SLEEM has been adapted from a Tesla BS 343 SEM, Mullerova and Frank (1993). The adaptation consists of placing the cathode lens assembly below the objective pole-piece and in

454

making provisions for a high-voltage specimen bias to govern the incident electron energy. In this configuration, the high energy electron beam is low-aberration focused and afterwards the final probe is decelerated within the axial field of the cathode lens. Lenc (1992), and Mullerova (1994) have shown that the SLEEM image resolution depends little on the incidence electron energy, deteriorating only by few units between 10-30 keV energy and a near-to-zero energy. The observations of the bevelled multilayer structures were made in the incidence energy range from 100 to 1000 eV.

The observations of the bevelled samples by SEM Tesla BS 300 were carried out at 2 keV electron energy, at which the contrast and signal /noise ratio are in compromise. Signal from the SEM has been electronically enhanced by dynamic DC level suppression, images were finally processed by the median filtering technique.

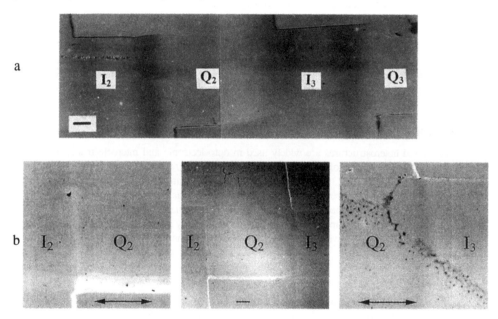

Fig. 1:  The surface of a bevelled InP/InGaAsP structure observed by optical interference microscopy (a) and by SLEEM (500 eV) (b). The bars indicate the distance of 100 μm and capitals I, Q denote InP and InGaAsP layers, respectively.


## 3. RESULTS AND DISCUSSION

The bevelled surface of sample A observed by optical interference microscopy and by low energy SEM is shown in Fig. 1. The layers of InP (denoted as $I_2$, $I_3$) and InGaAsP (denoted as $Q_2, Q_3$) were visualised by selective etching. Due to different refractive indices it is possible to distinguish InP as black stripes from InGaAsP (white-grey stripes). A part of the bevelled surface observed by SLEEM is shown in Fig. 1 (b, middle). The marks ($Q_2$, $I_3$) made by selective etching are clearly seen. On an unetched part of the bevel (horizontal middle band) InP and InGaAsP layers are slightly distinguishable (different colours). Interface layers are more clearly visible: black coloured at $I_3/Q_2$, and white coloured at $Q_2/I_2$. The interfaces are better visible at a higher magnification (Fig. 1(b) left and right). From the width of the interface in Fig. 1(b) and from the knowledge of the value of the bevel angle, the actual interface thickness was estimated: 6 nm for InP/InGaAsP interface, 4 nm for InGaAsP/InP interface. These values are comparable with Bresse's (1986) measurements by AES.

The bevelled surface of sample B and sample C observed by SLEEM and SEM are shown in Fig. 2 and Fig.3. Each individual layer is clearly distinguished at low magnification ~ X100 (Fig.2a, Fig.3): InGaAlAs (Q) and InP layers of the Bragg mirror, followed by InP bottom contact layer, InGaAlAs ($Q_N$) 'grading', InGaAs (T) absorption and InGaAlAs ($Q_P$) top contact layer.

The thickness and quality of interfaces were studied at magnifications approximately 4000 (Fig. 2b). The cellular structure of InP/$Q_N$ and T/$Q_P$ interfaces (denoted X and Z, respectively) were observed and thickness of 3 nm and 2 nm, respectively, were estimated. This might be caused by different compositions of the interface. Probably it is $InAs_xP_{1-x}$ as it was measured by Abraham et al. (1993). $Q_N$/T interface is the sharpest one. Its thickness was estimated as high as 1 nm. Small protrusions on InGaAs layer are probably caused by bevel etching.

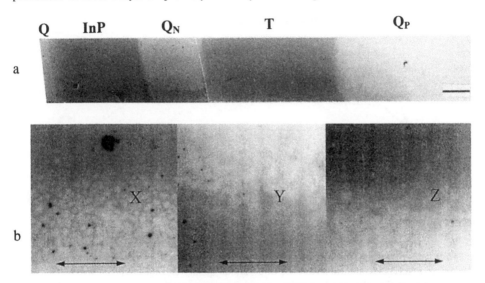

Fig. 2: The surface of a bevelled InP/InGaAs/InGaAlAs structure observed by SLEEM (500 eV). The bars denote 200 μm (a) and 10 μm (b). X, Y, Z denote interfaces InP/InGaAlAs, InGaAlAs/InGaAs and InGaAs/InGaAlAs respectively.

Fig. 3. Surface of the bevelled InP/InGaAs/InGaAlAs RCE PIN photodiode structure as observed in secondary electron mode of the SEM.

The bevelled etched surface was analysed also by AES, point-by-point, across the uncovered bevelled layers. The observed distribution of principal constituents in layer $Q_2$ and at interfaces with layers $I_2$ and $I_3$ is shown in Fig. 4 as a dependence of elemental concentrations on the position $l$. The obtained graph reveals that the distribution of elements is homogeneous and, at the same time, that the interfaces are equally wide ($\Delta l_1 = \Delta l_2$). Following Honig 1976, the width of

456

the interface is considered to be the difference between the 16 and 84 % of the Auger signal of a given element (in our case of As) inside the layer. Then the widths of the two interfaces, upon transforming the bevelled lengths $\Delta l$ into layer thicknesses by considering the magnification of 5 600 reached through bevelling, were approx. 9 nm. Naturally, this value has to be corrected for the diameter of the primary electron beam, and the escape depth of Auger electrons in the sample. The beam diameter used was approx. 10 μm, which corresponds to approx. 2 nm at the present magnification. A similar value belongs also to the escape depth of Auger electrons of As (1228 eV). The overall contribution of these two factors is roughly 3 nm (Hofmann 1977). Hence, the actual widths of the two interfaces shown in Fig. 4 are approximately 6 nm.

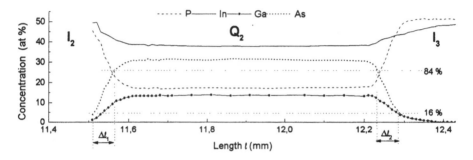

Fig. 4: AES distribution map of constituents in sample A as a dependence of elemental concentrations upon the manipulator position, denoted as length $l$.

### 4. CONCLUSION

A study of SLEEM, SEM, AES and optical interference microscopy of bevelled InP/InGaAsP and InP/InGaAs/InGaAlAs structures grown by MOCVD is presented. It is shown that the different interfaces of structure A are of a various quality and different thickness (6 or 4 nm). AES studies, on the other hand, revealed the same interface widths (approximately 6 nm) for the interfaces. On structure B, three interfaces were studied. Cellular structure has been found at InP/InGaAlAs and InGaAs/InGaAlAs heterointerfaces with corresponding thickness of 3 nm and 2 nm.

### ACKNOWLEDGEMENT

This work was supported by Copernicus Grant No. DEMACOMINT 12283, the Slovak Grant Agency, No.4219/1997, and by EPSRC.

### REFERENCES

Abraham P et al. 1993 Appl. Surf. Science **65/66**, 777.
Adachi S 1982 J. Electrochem.Soc. **129**, 609.
Bresse J F 1986 J. Appl. Phys.**59**, 2026.
Hofmann S In: Proc. 7th Int. Vac. Congr. & 3rd Int. Conf. Solid Surf., Vienna 1977, Vol. **3**, 2613.
Honig R E 1976 Thin Solid Films **31**, 89.
Hsu M et al. 1995 Surf. Interface Analysis **23**, 665.
Lenc M and Mullerova I 1992 Ultramicroscopy **45**, 159.
Mullerova I and Frank L 1993 Scanning **15**, 193.
Mullerova I and Frank L 1994, Mikrochimica Acta **114/115**, 389.
Srnanek R, El Gomati M, Novotny I and Pudis D, 1997, J. Crystal Growth (In Press).

*Inst. Phys. Conf. Ser. No 153: Section 11*
*Paper presented at Electron Microscopy and Analysis Group Conf. EMAG97, Cambridge, 1997*
© *1997 IOP Publishing Ltd*

# Analysis of interface structures by quantitative HRTEM

O. Kienzle and F. Ernst

Max-Planck-Institut für Metallforschung, Seestraße 92, D-70174 Stuttgart, Germany

**ABSTRACT:** This paper reviews the method of 'quantitative' HRTEM (high-resolution transmission electron microscopy) we have developed to determine the atomistic structure of interfaces. This method includes quantification of error limits for the atom column positions determined by iterative digital image matching. Extending previous work, we introduce a procedure to discriminate between the reliabilities of different types of atom columns in the refined structure. As an example for the application of our method we present results we have obtained analyzing the structure of a $\Sigma = 3$, (111) grain boundary in $SrTiO_3$.

## 1. QUANTITATIVE HRTEM

HRTEM (high-resolution transmission electron microscopy) constitutes a powerful technique to investigate the atomistic structure of interfaces. However, straight-forward interpretation of HRTEM images of interfaces in terms of „projected atom columns" may lead to errors: Close to an interface, the contrast pattern of the atom columns often differs from the corresponding pattern in regions of unfaulted crystal. Moreover, dynamical electron diffraction and aberrations of the electron optics may *displace* the contrast pattern against the actual (projected) positions of the columns. Nevertheless, it is possible to interpret HRTEM images of interfaces safely by *comparing* them with simulated images of model structures. This method becomes powerful if one compares experimental and simulated images not merely by visual inspection but *quantitatively*, by means of digital image processing (Ernst et al 1996b).

For a given HRTEM image (a noise-reduced supercell image) we *iterate* such quantitative image comparisons to perform an automatic *structure refinement* (Fig. 1), which determines the structure *r* that yields the best-matching simulated image to the experimental image[†]. For the quantitative comparisons between a simulated image $\underline{A}$ and an experimental image $B$ we evaluate the normalized Euclidean distance (NED)

$$D[\underline{A}, B] := \frac{\|\underline{A} - B\|}{\sqrt{\|\underline{A}\| \cdot \|B\|}}, \tag{1}$$

---

[†] In this analysis we regard structures as vectors of column coordinates and images as vectors of pixel intensities. We denote structures by small vector symbols, images by capital vector symbols, and simulated images by underlined vector symbols.

where $\|...\|$ denotes the Euclidean length.

While we can refine the positions of atom columns to any numerical precision, the coordinates only have a limited *reliability*. This arises because it is impossible to simulate HRTEM images perfectly – even if one *knew* the structure of the specimen. Amorphous layers on the specimen, residual shot noise, uncertainties in the electron optical parameters characterizing the microscope, and approximations in the algorithm used for image simulation always cause a *residual discrepancy* $D^*$ between the best-matching simulated image and the experimental image. Therefore, we can only safely distinguish between two structures whose image vectors differ by more than $D^*$. In other words, we cannot distinguish between those structures whose image vectors lie within the hypersphere with radius $D^*$ around the image $\underline{R}$ of the refined structure $r$. Thus, the average error limit for the column coordinates corresponds to the 'typical' displacement that one needs to apply to every column of the supercell in order to change the image vector by $D^*$.

We obtain the length of this typical displacement in the following way: Starting out from the refined structure $r$, we create new structures $r_\sigma$ by displacing the atom columns into random directions by distances drawn randomly from a Gauß distribution with a pre-set standard deviation $\sigma$. For each new structure $r_\sigma$ we simulate the HRTEM image $\underline{R}_\sigma$ and calculate $D[\underline{R}_\sigma, \underline{R}]$, the NED to the simulated image $\underline{R}$ of the refined structure. We perform these calculations for a variety of different standard deviations $\sigma$, and for each $\sigma$ we test several different sets of randomly chosen column displacements.

In this way we establish a correlation between image discrepancy $D$ and *average* column displacement $\sigma$. The 'reliability function' $D[\sigma]$ translates image discrepancy (in terms of NED) to structural discrepancy (in terms of average column displacement). The average reliability of the column positions in the refined structure $r$ corresponds to the standard deviation $\sigma^*$ that belongs to the residual image discrepancy $D^*$ under the given imaging conditions.

If the interface structure under study features atom columns of $M$ different types along the viewing direction, one would like to specify seperate reliabilities for the columns belonging to each type. Provided that simultaneous shifting of several columns results in an image mutation that consists of *orthogonal*, non-interfering contributions of the individual columns, we can express the reliability function by *partial* reliability functions $D_i[\sigma]$:

$$D[\sigma] = \sqrt{\sum_{i=1}^{M} D_i^2[\sigma]} . \tag{2}$$

Each partial reliability function $D_i[\sigma]$ describes how the image changes when displacing only the $N_i$ columns of type $i$, while keeping all other column positions fixed. Assuming that each one of the type $i$ columns contributes to $D_i[\sigma]$ with the same weight, and describing the contribution of an individual column by the 'single column reliability function' $d_i[\sigma]$, we have

$$D_i[\sigma] = \sqrt{\sum_{k=1}^{N_i} d_i^2[\sigma]} = \sqrt{N_i\, d_i^2[\sigma]} , \tag{3}$$

implying

$$d_i[\sigma] = D_i[\sigma]/\sqrt{N_i} . \tag{4}$$

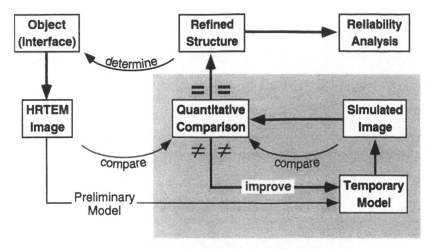

Fig. 1: Flow diagram of structure determination by quantitative HRTEM.

Evaluating the graphs of the $d_i[\sigma]$ at $\sigma^*$ indicates the *individual* reliabilities of the $M$ different column types. In section 2 we show an example where the above assumption of non-interfering column contributions actually constitutes a valid approximation.

We have applied our method of 'quantitative' HRTEM to analyze the structures of various 'special' grain boundaries in materials with various types of interatomic bonding. These studies include grain boundaries in Cu (Hofmann and Ernst 1994a; Hofmann and Ernst 1994b; Ernst et al 1996a), NiAl (Nadarzinski and Ernst 1996), $PbTiO_3$ (Stemmer et al 1995), and $SrTiO_3$ (Kienzle and Ernst 1997; Kienzle et al 1997). The interface structures obtained by these studies serve to test model structures obtained by computer simulations and to understand the under-lying physics. Moreover, our method of assessing the reliability of atom coordinates obtained by iterative digital image matching allows us to quantify the damage that electron beam irradiation introduces in the specimen: Evaluating HRTEM images recorded during systematic 'beam damaging experiments' one can estimate the maximum tolerable observation time for structure determination at given reliability (Dehm et al 1996; Nadarzinski and Ernst 1996).

## 2.  EXAMPLE: THE $\Sigma = 3$, (111) GRAIN BOUNDARY IN $SrTiO_3$

Polycrystalline $SrTiO_3$ (strontium titanate) serves as electroceramics, for example in varistors. The unique electron transport properties of this material arise from space charge layers forming at the grain boundaries owing to segregation of charged point defects. Because the extent to which point defects segregate depends on the atomistic structure of the grain boundaries, it is important to understand the building principles of grain boundaries in this material. For these reasons, we have analyzed the atomistic structure of the $\Sigma = 3$, (111) grain boundary by quantitative HRTEM.

Under standard conditions, $SrTiO_3$ adopts the perovskite structure with a lattice parameter of $a = 0.3905$ nm. The starting material for our study was a bicrystal grown by the Verneuil method and doped with Fe to 0.06 wt %. HRTEM studies were performed with a JEM 4000 EX (JEOL), operating at an acceleration voltage of 400 kV. For image recording we employed a Gatan slow-scan CCD camera with

Fig. 2: Experimental HRTEM image of the $\Sigma$=3, (111) grain boundary in SrTiO$_3$.

1024 × 1024 pixels and a dynamic range of $2^{14}$. Fig. 2 presents an experimental HRTEM image of the grain boundary in <110> projection. Image simulations indicate that in regions of unfaulted perovskite structure the bright spots of the contrast pattern correspond to the positions of Sr-O columns. The boundary, however, exhibits a different contrast pattern (compare the inset at the top of Fig. 2, which shows an enlargement of the frame marked at the bottom). This pattern hardly permits 'intuitive' interpretation in terms of atom column types and their positions.

To prepare the experimental image for the structure refinement we first adjusted the sampling rate and the average image intensity to match the corresponding settings of simulated images. The signal-to-noise ratio was improved by spatial averaging. The *imaging conditions* under which the experimental image has been recorded, including the local sample thickness $t$, the defocus setting $f$ of the objective lens, and the beam tilt, were determined by matching the simulated image of an unfaulted perovskite crystal to the experimental image at some distance away from the grain boundary. For this purpose we employed a simulated-evolution optimization algorithm (Möbus 1996).

The subsequent procedure of structure refinement requires a preliminary model structure as input (Fig. 1). In the present case we have *constructed* this starting structure by assuming that the boundary plane constitutes a mirror plane of the bicrystal and that the spacings of {111} atom layers parallel to the boundary plane correspond to those of the perovskite structure. This leads to two alternatives, (i) a model with Sr-O and O columns on the boundary plane or (ii) a model with Ti columns on the boundary plane. Image simulations immediately rule out the latter. Thus, we have started the structure refinement with model (i), depicted in Fig. 5b. The refinement procedure optimizes the structure model by iteratively matching the simulated image to the experimental image. In each iteration loop, the algorithm (i) modifies the structure model by shifting the atom columns, (ii) tests the actual model by calculating the normalized Euclidean distance $D$ to the experimental image and (iii) improves the model to reduce $D$ (Fig. 1). The structure to which this algorithm converges yields the best-matching simulated image and thus constitutes

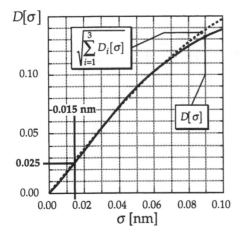

Fig. 3: Reliability function. The residual discrepancy $D^* = 0.025$ corresponds to an average reliability of 0.015 nm.

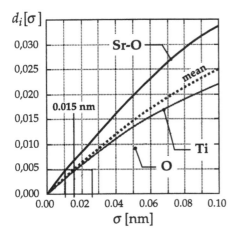

Fig.4: Single column reliability functions for the different column types of the <110> projection of SrTiO₃. The label 'mean' marks the mean single column reliability defined by Eq. (5).

the 'refined' structure, id est the best possible model structure under the boundary conditions of the refinement procedure (column model of the boundary, noise, etc.).

To analyze the reliability of the column positions we determined the reliability function $D[\sigma]$ (by shifting all columns in the supercell) and the partial reliability functions $D_{SrO}[\sigma]$, $D_{Ti}[\sigma]$, and $D_O[\sigma]$ by shifting only the Sr-O, Ti, and O columns, respectively. Fig. 3 depicts $D[\sigma]$ (bold line, the shaded area describes the scattering obtained for different runs with the same $\sigma$). From this graph we read that the residual discrepancy of $D^* = 0.025$ between the simulated image of the refined structure and the experimental image corresponds to an average reliability of 0.015 nm. The dotted line corresponds to the right side of Eq. (2). The coincidence of this graph with the graph $D[\sigma]$ up to $\sigma^*$ and beyond indicates that the hypothesis of non-interfering contributions of the individual column shifts constitutes a valid approximation under the present conditions, and thus Eqs. (2) and (3) actually apply. Fig. 4 presents the 'single column reliability functions' of the different column types according to equation (4), together with the mean 'single column reliability function' $\underline{d}[\sigma]$, which we define by

$$\sqrt{\sum_{i=1}^{M} N_i \, \underline{d}^2[\sigma]} := \sqrt{\sum_{i=1}^{M} N_i \, d_i^2[\sigma]} \,. \qquad (5)$$

Analyzing the abscissae of the graphs in Fig. 4 at the ordinate of $\sigma^* = 0.015$ nm we obtain a reliabilities of 0.010 nm for the Sr-O columns, 0.015 nm for the Ti columns, and 0.025 nm for the O columns.

Fig. 5 a shows the refined structure. Compared to the preliminary model in Fig. 5 b, the Ti-Ti spacing $\rho$ has grown from 0.23 nm to (0.32 ± 0.02 nm). This causes the bicrystal to expand by (0.06 ± 0.01) nm normal to the boundary plane. Like the preliminary model the refined structure exhibits mirror symmetry with respect to the boundary plane.

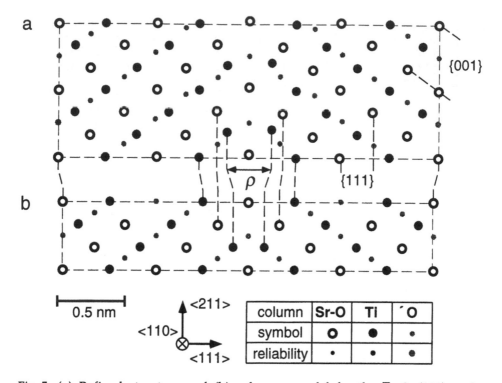

a

{001}

ρ    {111}

b

0.5 nm

<211>

<110>⊗

<111>

| column | Sr-O | Ti | ´O |
|---|---|---|---|
| symbol | ○ | ● | • |
| reliability | • | • | ● |

Fig. 5: (a) Refined structure and (b) reference model for the $\Sigma = 3$, (111) grain boundary in SrTiO$_3$.

## REFERENCES

Dehm G, Nadarzinski K, Ernst F and Rühle M 1996 Ultramicroscopy **63**, 49

Ernst F, Finnis M W, Gust W, Koch A, Schmidt C and Straumal B 1996a Z. Metallkde. **87**, 911

Ernst F, Hofmann D, Nadarzinski K, Stemmer S and Streiffer S K 1996b Intergranular and Interphase Boundaries in Materials, eds A C Ferro, E P Conde and E A Fortes (Zürich: Trans Tech Publications) pp 23

Hofmann D and Ernst F 1994a Ultramicroscopy **53**, 205

Hofmann D and Ernst F 1994b Interf. Sci. **2**, 201

Kienzle O and Ernst F 1997 J. Am. Ceram. Soc. **80**, 1639

Kienzle O, Exner M and Ernst F 1997 Atomic Resolution Microscopy of Surfaces and Interfaces, eds R Hamers and D J Smith (Pittsburgh: Materials Research Society) in press

Möbus G 1996 Ultramicroscopy **65**, 205

Nadarzinski K and Ernst F 1996 Phil. Mag. A **74**, 641

Stemmer S, Streiffer S K, Ernst F and Rühle M 1995 Phil. Mag. A **71**, 713

*Inst. Phys. Conf. Ser. No 153: Section 11*
*Paper presented at Electron Microscopy and Analysis Group Conf. EMAG97, Cambridge, 1997*

# Faceting at ZnO-Bi$_2$O$_3$ Interphase Boundaries in ZnO-based Varistor Ceramics

**H H Hng and K M Knowles**

University of Cambridge, Department of Materials Science and Metallurgy, Pembroke Street, Cambridge, CB2 3QZ, U.K.

**ABSTRACT:** ZnO-based varistor ceramics containing Bi$_2$O$_3$, Sb$_2$O$_3$, MnO and Co$_3$O$_4$ obtained by water quenching from the sintering temperature of 1250 °C show extensive faceting at ZnO – δ-Bi$_2$O$_3$ interfaces. Trace analysis showed that facets are the ZnO basal plane, ZnO prism planes or low index pyramidal planes. The most prominent facets were generally found to be the ZnO prism planes which are populated by equal proportions of zinc and oxygen atoms and are thus relatively favoured low energy planes.

## 1. INTRODUCTION

Commercial ZnO varistors are made from ZnO powders doped with oxides such as Bi$_2$O$_3$, Sb$_2$O$_3$ and Co$_3$O$_4$ and subjected to controlled sintering and post-sintering heat treatments. Generally, microstructures of varistors show ZnO grains, with Bi$_2$O$_3$ at triple junctions and grain boundaries, and a variety of second phases such as Zn$_2$Sb$_3$Bi$_3$O$_{14}$ pyrochlore and Zn$_7$Sb$_2$O$_{12}$ spinel. Varistor behaviour is enabled by the formation of suitable highly resistive electrical barriers at ZnO – ZnO interfaces (Gupta 1990). The network of Bi-rich phases provides a skeleton in the microstructure which is believed to be responsible for low-voltage leakage currents (Olsson et al 1993).

Cooling from the sintering temperature is a critical variable in the attainment of the characteristic varistor non-linear current-voltage characteristics, with faster cooling rates able to suppress varistor behaviour unless the material has a post-sintering heat treatment (Olsson et al 1989). During relatively fast cooling, faceting of ZnO – Bi$_2$O$_3$ interphase boundaries is observed (Olsson et al 1989, Sung et al 1992 and Olsson et al 1993). Sung et al (1992) reported pronounced faceting parallel to the (0001) ZnO basal planes at ZnO – β-Bi$_2$O$_3$ interphase boundaries in an air-quenched ZnO – Bi$_2$O$_3$ – MnO – TiO$_2$ varistor material, with smaller facets on either the {1$\bar{1}$00} or the {11$\bar{2}$0} ZnO prism planes.

Here we report observations of faceting at ZnO – δ-Bi$_2$O$_3$ interphase boundaries in a ZnO-based varistor ceramic containing Bi$_2$O$_3$, Sb$_2$O$_3$, MnO and Co$_3$O$_4$ obtained by water quenching from a sintering temperature of 1250 °C. Our observations contrast with those of Sung et al (1992), but can be rationalised straightforwardly from a consideration of the wurtzite crystal structure adopted by ZnO.

## 2. EXPERIMENTAL PROCEDURE

ZnO varistor samples were prepared by a conventional mixed-oxide ceramic technology. Powders containing 95 mol% ZnO, 2 mol% Sb$_2$O$_3$, 1 mol% Bi$_2$O$_3$, 1 mol% MnO and

464

1 mol% $Co_3O_4$ were mixed and ball-milled for 24 hours using zirconia beads in deionised water. The powders were dried and pressed into discs 20 mm in diameter and 5 mm thick. The green compacts were sintered for 2 hr in an atmosphere of ambient air at 1250 °C and then water quenched. Specimens for transmission electron microscopy were prepared using standard ion beam thinning methods and examined using a JEOL 2000FX at 200 kV.

## 3. RESULTS

The general microstructure of the sample was predominantly ZnO grains interspersed with smaller $Zn_7Sb_2O_{12}$ spinel grains. Bi-rich phases occurred primarily at triple and multiple grain junctions, and as thin intergranular layers. Selected area diffraction patterns of triple junction regions confirmed that the Bi-rich phase was $\delta$-$Bi_2O_3$ (Fig. 1), although it should be noted that this electron diffraction pattern appears to be consistent with an $Fm3m$ space group for this structure, rather than the reported $Pn3m$ space group (Medernach and Snyder 1978) because of the absence of weak $\{110\}$ and $\{001\}$ reflections. The presence of $\delta$-$Bi_2O_3$ observed in this water quenched sample is consistent with the fact that the $\delta$-phase is the high-temperature polymorph of the $Bi_2O_3$ phase (Medernach and Snyder 1978).

Faceting at ZnO – $Bi_2O_3$ interphase boundaries was a frequent observation. A typical example of a faceting behaviour is shown in Fig. 2, in which a thin layer of $\delta$-$Bi_2O_3$ is sandwiched between two ZnO grains. A general observation of such interfaces was that faceting occurred on one of the ZnO – $Bi_2O_3$ interfaces, but not the other. Faceting also occurred at ZnO – $Bi_2O_3$ interfaces where $Bi_2O_3$ was in contact with spinel, whereas there was no indication of faceting at $Bi_2O_3$ – spinel interfaces (Fig. 3). Occasional examples were also found of curved facets (Fig. 4), suggestive of faceting on a much finer scale than is evident in Fig. 2.

Trace analysis was able to determine straightforwardly the crystallography of facets relative to the ZnO grain with which they were in contact, but not with respect to the $\delta$-$Bi_2O_3$ because of the difficulty in obtaining low index zones from the thin $\delta$-$Bi_2O_3$ intergranular regions. Several faceting regions were analysed, the results for which are summarised in Table 1. It is evident that there is preferential faceting parallel to certain ZnO planes. The long facets

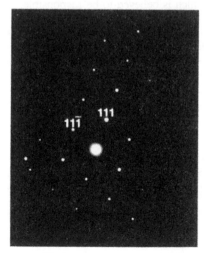

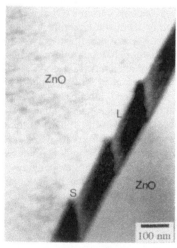

Fig. 1  [$1\bar{1}0$] zone selected area diffraction pattern of $\delta$-$Bi_2O_3$.

Fig. 2  Faceting at a ZnO – $\delta$-$Bi_2O_3$ interface between two ZnO grains (L - long facet, S - short facet).

tend to be parallel to the ZnO prism planes (either $\{1\bar{1}00\}$ or $\{11\bar{2}0\}$) or pyramidal planes ($\{1\bar{1}01\}$ or $\{11\bar{2}1\}$), whereas the short facets are either the ZnO (0001) basal plane or pyramidal planes ($\{1\bar{1}02\}$ or $\{11\bar{2}2\}$).

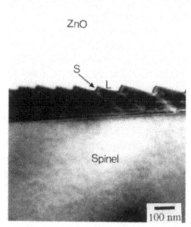

ZnO

Spinel

100 nm

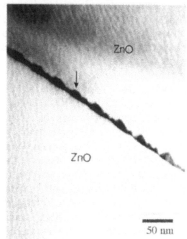

ZnO

ZnO

50 nm

Fig. 3   Faceting at a ZnO – $\delta$-Bi$_2$O$_3$ interface between a ZnO grain and a spinel grain (L - long facet, S - short facet).

Fig. 4   Examples of curved facets (arrowed) seen occasionally at ZnO – $\delta$-Bi$_2$O$_3$ interfaces.

Table 1. ZnO facet planes

| Boundary | Long Facet | Short Facet |
|---|---|---|
| 1 | $(11\bar{2}1)$ | $(10\bar{1}2)$ |
| 2 | $(10\bar{1}0)$ | $(01\bar{1}2)$ |
| 3 | $(01\bar{1}0)$ | $(0001)$ |
| 4 | $(10\bar{1}0)$ | $(0\bar{1}12)$ |
| 5 | $(11\bar{2}1)$ | $(2\bar{1}\bar{1}2)$ |
| 6 | $(10\bar{1}1)$ | $(0001)$ |
| 7 | $(2\bar{1}\bar{1}0)$ | $(0001)$ |

## 4. DISCUSSION

Our results confirm the faceting at ZnO – $\delta$-Bi$_2$O$_3$ interphase boundaries in quenched samples reported by Olsson et al (1989, 1993), although they did not comment on the crystallography of the facets, nor did they provide a rationale for their occurrence in quenched samples, but not in samples cooled at slower rates. Sung et al (1992) showed evidence of pronounced faceting parallel to (0001) ZnO planes at ZnO – $\beta$-Bi$_2$O$_3$ interphase boundaries which they interpreted in terms of the anisotropic growth of ZnO grains, known to be favoured by the addition of TiO$_2$ to the system ZnO – Bi$_2$O$_3$ – MnO (Sung and Kim 1988). Sung and Kim hypothesised that without addition of TiO$_2$, interfacial energies of ZnO were markedly less anisotropic, to the extent that anisotropic grain growth is no longer observed. It is also relevant in this context that, in his study of the location of bismuth-rich intergranular phase in a zinc oxide varistor, Clarke (1978) found that the grain boundary was parallel to the basal plane in each of the four cases in which a continuous intergranular phase was found to exist.

Our trace analysis of general grain boundaries here in the quenched $ZnO$ – $Bi_2O_3$ – $Sb_2O_3$ – $MnO$ – $Co_3O_4$ material shows that a continuous intergranular $Bi_2O_3$ phase is observed when the macroscopic interface is far from (0001) $ZnO$ and that faceting is a general phenomenon at $ZnO$ – $Bi_2O_3$ interfaces under conditions of rapid cooling.

In each of the cases examined, faceting was found on relatively low index $ZnO$ planes. A consideration of the morphology of $ZnO$ grown by anodic electrocrystallisation in conditions of relatively low supersaturation, in which hexagonal plates with pronounced (0001) habit planes and narrow prism plane edges (either $\{1\bar{1}00\}$ or $\{11\bar{2}0\}$)) were observed by Perkins (1977) shows that the broad (0001) basal plane surfaces will contain either all by zinc atoms or all by oxygen atoms, whereas the $\{1\bar{1}00\}$ and $\{11\bar{2}0\}$ prism planes are terminated by equal proportions of zinc and oxygen atoms. Thus, as a rule, growth perpendicular to the (0001) planes is more difficult than growth perpendicular either to the $\{1\bar{1}00\}$ or $\{11\bar{2}0\}$ prism planes and so the (0001) planes are planes of lowest energy, with the prism planes second lowest.

Our results can be rationalised in terms of the reduced growth anisotropy relative to the systems studied by Perkins (1977), Sung and Kim (1988) and Sung et al (1992), while still favouring relatively low index $ZnO$ planes both as preferred sites of dissolution of $ZnO$ into the surrounding $Bi_2O_3$-rich liquid and also as subsequent reprecipitation of $ZnO$ during sintering. In the examples we have studied there was no clear tendency for grain boundaries to be close to, or parallel to, basal planes, which is why (0001), while still a 'preferred' plane in that it was observed as a short facet, was not seen as a long facet. Slower cooling enables growth at facets to take place, as a result of which the $ZnO$ – $Bi_2O_3$ interfaces will become smoothed out in shape.

## 4. CONCLUSIONS

Faceting at $ZnO$ – $\delta$-$Bi_2O_3$ interphase boundaries in water-quenched varistor samples is shown to be a general observation. In all the cases examined, the facets are found to be low index $ZnO$ planes. A simple model to account for these observations has been proposed.

## ACKNOWLEDGEMENTS

We would like to thank Prof. A H Windle for the provision of laboratory facilities and Nanyang Technological University, Singapore for financial support for HHH.

## REFERENCES

Clarke D R 1978 J. Appl. Phys. **49**, 2407
Gupta T K 1990 J. Am. Ceram. Soc. **73**, 1817
Medernach J W and Snyder R L 1978 J. Am. Ceram. Soc. **61**, 494
Olsson E, Dunlop G L and Österlund R 1989 J. Appl. Phys. **66**, 5072
Olsson E, Dunlop G L and Österlund R 1993 J. Am. Ceram. Soc. **76**, 65
Perkins J 1977 J. Cryst. Growth **40**, 152
Sung G Y and Kim C H 1988 Advanced Ceram. Mater. **3**, 604
Sung G Y, McKernan S and Carter C B 1992 J. Mater. Res. **7**, 474

*Inst. Phys. Conf. Ser. No 153: Section 11*
*Paper presented at Electron Microscopy and Analysis Group Conf. EMAG97, Cambridge, 1997*
© *1997 IOP Publishing Ltd*

# Analysis of planar defects in $Nb_2O_5$- and $Bi_2O_3$-doped $BaTiO_3$ ceramics.

**W E Lee, M A McCoy, R Keyse\* and R W Grimes\*\***

Department of Engineering Materials, University of Sheffield, Mappin St., Sheffield, S1 3JD.
\*Department of Materials Science and Engineering, University of Liverpool, L69 3BX.
\*\*Department of Materials, Imperial College, London, SW7 2BP.

**ABSTRACT:** The structure and chemistry of planar defects in $Nb_2O_5$- and $Bi_2O_3$-doped $BaTiO_3$ grains exhibiting "core-shell" microstructures has been examined. In addition to 90° ferroelectric domain boundaries within the core, twins and stacking faults were observed both with interfaces lying along {111}. While the twins bisected grains completely, stacking faults were observed only in the shell region and the latter are enriched in Nb and Bi relative to the surrounding matrix. A stacking fault structure is proposed based on a double $BiO_3^{3-}$ layer with partial cation site occupancy by $Ba^{2+}$ and charge compensation by $Nb^{5+}$ substitution in adjacent octahedral sites.

## 1. INTRODUCTION

The temperature-stable dielectric response of X7R barium titanate ($BaTiO_3$) dielectrics is based on the formation of a core-shell microstructure (Chiang 1988) in which individual grains contain a tetragonal, ferroelectric core surrounded by a cubic, paraelectric shell. During sintering of a model X7R system Nb and Bi dopants react with the $BaTiO_3$ grains to form a number of intermediate Aurivillius compounds, such as $BaBi_4Ti_4O_{15}$ and $Bi_3TiNbO_9$, and a tungsten bronze phase $Ba_4Bi_2Ti_4Nb_6O_{30}$ believed to act as the dopant source (Pathumarak et al 1994, Pathumarak and Lee 1996). The dopants diffuse rapidly along grain boundaries and surfaces of the $BaTiO_3$ grains and more slowly into the grain interiors by volume diffusion. By controlling the sintering conditions and starting chemistry the grains exhibit a dopant concentration gradient from the grain edge to the core. The resulting microstructure of a typical commercial X7R capacitor consists of grains with a core containing a low dopant concentration surrounded by a dopant-rich shell (Alsaffar et al 1993). The undoped core is ferroelectric, containing characteristic 90° ferroelectric domain boundaries. Since the dopant concentration in the paraelectric shell controls the Curie temperature, the variation of dopant levels in the various shells causes a distribution in the Curie temperature with a corresponding flat response of dielectric constant with temperature. As well as ferroelectric domain boundaries, other planar defects have frequently been observed within the microstructure, consisting either of twin boundaries (Recnik et al 1994) or stacking faults (Chiang 1988). In the present work the structure and chemistry of stacking faults in a model X7R Nb/Bi-doped $BaTiO_3$ ceramic have been characterised.

468

## 2. EXPERIMENTAL

Polycrystalline barium titanate ceramics containing approximately 4.5 mol% $Nb_2O_5$ and 4.5 mol% $Bi_2O_3$ were prepared using conventional mixed oxide powder processing with uniaxial pressing at 125MPa and sintering for 1h at 1100°C. Standard ceramographic grinding and polishing and TEM sample preparation procedures were followed ending with Ar ion milling at 5kV and carbon sputter coating. Specimens were examined in a JEOL 200CX TEM and a JEOL 3010 HREM operating at 200 and 300keV respectively. The stacking fault displacement vector **R** was determined using the **g.R** fault contrast analysis technique described by Amelinckx (1993) and they were examined in HREM in <110> projection. For each region a series of images at well-defined defocus steps was recorded. Image simulation was performed using a multislice code in the EMS V3.30 software package developed by Stadelmann (1987). EDS analysis was performed using a VG-601UX dedicated STEM operated at 100keV (at the University of Liverpool).

## 3. RESULTS AND DISCUSSION

Fig. 1 shows typical $BaTiO_3$ grains containing a ferroelectric core whose contrast is dominated by ferroelectric domain boundaries surrounded by a paraelectric shell. As well as the ferroelectric domain boundaries two other types of planar defect were common: twin boundaries bisecting the grains (Fig. 1a) and stacking faults located exclusively in the shell (Fig. 1b). Trace analysis showed that both the twins and stacking faults lie along {111} while **g.R** contrast analysis of the stacking faults (McCoy et al 1998) revealed the displacement vectors were normal to the interface plane along <111>.

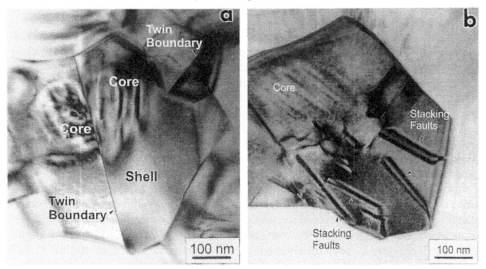

Figure 1. Bright-field TEM images of core-shell microstructure showing a) twin boundaries extending right across a grain and b) stacking faults within the shell.

A stacking fault viewed edge-on near Scherzer defocus is shown in <110> projection in Fig. 2. The change in stacking sequence across the boundary can be seen by noting the discontinuity in the stacking of the Ba-O columns which appear as white dots in the micrograph. HREM confirmed the {111} interface plane and the <111> displacement vector direction of the TEM analysis. Detailed measurements across the boundaries showed the magnitude of **R** is 1/2<111> plus a rigid body translation of approximately 0.1<111>. HREM images also show a high level of noise at the interface consistent with a high degree of disorder along the fault.

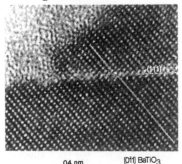

04 nm          [011] BaTiO₃

Figure 2. HREM image (near Scherzer defocus) of the shell region in <110> projection showing a stacking fault edge-on. Under these imaging conditions the white dots correspond to Ba-O columns.

Compositional analysis of stacking faults within the shell revealed much higher dopant concentrations compared to the matrix immediately adjacent to the fault (Fig. 3). Quantitative analysis from 5 different regions consistently showed a higher level of both Nb (~3.9atom% cf. ~2.9%) and Bi (~3.6% cf. ~2.3%).

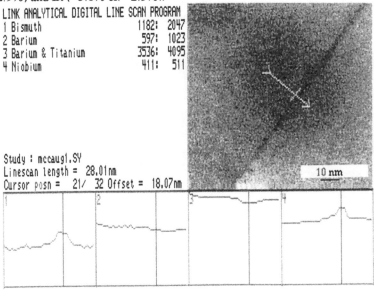

Figure 3. EDS linescans for (1) Bi, (2) Ba, (3) Ti (as determined by Ba+Ti/Ba) and (4) Nb taken across a stacking fault. The position of the fault is noted by the vertical line in the scans.

Examining the stacking sequence along [111] for the unfaulted $ABO_3$ perovskite structure reveals that it consists of alternating layers of $AO_3^{4-}$ and $B^{4+}$ separated by 1/6[111]. Various ways of introducing a (111) stacking fault have been examined by McCoy et al (1998) which, when accounting for the defect chemistry associated with the presence of Nb and Bi and charge neutrality requirements, lead to the model proposed in Fig. 4 for the stacking fault structure. Note the double $AO_3$ layer. The model includes the rigid body translation measured from HREM micrographs as well as a slight translation of $Bi^{3+}$ ions to be consistent with expected Bi-O bond lengths in this ionic environment.

470

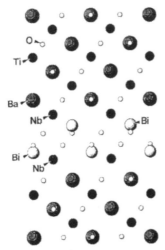

Figure 4. Stacking fault model projected along [01$\bar{1}$], incorporating both a change in stacking sequence and an in-plane shear, with substitution of $Bi^{3+}$ for $Ba^{2+}$ along the interface and $Nb^{5+}$ in adjacent $Ti^{4+}$ sites.

A simulated HREM image using this model is superimposed on an image taken at Scherzer defocus (-42nm) in Fig. 5. The agreement suggests the model is reasonable although the high level of noise and residual contrast within the interface region (unaccounted for in the simulations) suggests a high level of local disorder, possibly due to local site relaxation due to A-site occupancy by both Bi and Ba.

Figure 5. Comparison of HREM and simulated images (inset). Under these conditions the white dots correspond to Ti columns.

## REFERENCES

Alsaffar R, Azough F and Freer R 1993 Inst. Phys. Conf. Series **130**, 379
Amelinckx S 1993 in Materials Science and Technology - A Comprehensive Treatment, eds R W Cahn, P Haasen and E J Kramer (VCH Weinheim) Vol 2A pp 40-48
Chiang S-K 1988 PhD thesis, The Ohio State University, Columbus, Ohio, USA
McCoy M A, Keyse R, Lee W E, and Grimes R W 1998 submitted to J. Mater. Sci.
Pathumarak S, Al-Khafaji M and Lee W E 1994 Brit. Ceram. Trans. **93**, 114
Pathumarak S and Lee W E 1996 J. Mater. Sci.: Mater. in Electr. **7**, 161
Recnik A, Bruley J, Mader W, Kolar D and Ruhle M 1994 Phil. Mag. **B70**, 1021
Stadelmann P 1987 Ultramicroscopy **21**, 131

*Inst. Phys. Conf. Ser. No 153: Section 11*
*Paper presented at Electron Microscopy and Analysis Group Conf. EMAG97, Cambridge, 1997*
© *1997 IOP Publishing Ltd*

# Microstructure and chemistry of intergranular glassy films in liquid phase sintered alumina and relationship to wear rate

R Brydson[*], P Twigg[*], S Chen[*], F L Riley[*], X Pan[+] and M Rühle[+]

[*]School of Process, Environmental and Materials Engineering, University of Leeds, Leeds LS2 9JT, U.K.
[+]Max-Planck-Institut für Metallforschung, Institut für Werkstoffwissenschaft, D-70174 Stuttgart, Germany.

ABSTRACT: PEELS, EDX and HREM have been used to study the two-grain boundaries and three grain junctions in polycrystalline aluminas sintered with various metal oxide/silica additions of between 0 and 10 wt% to promote densification. The additions resulted in a continuous glassy film at grain boundaries of a thickness which was found to be independent of additive level. The chemistry of the intergranular glass was remarkably different from that found at triple pockets and depended on the molar ratios of the additives which profoundly affected the material's erosive wear rate.

## 1. INTRODUCTION

Polycrystalline alumina microstructures and their related properties often depend critically on both the presence of dopants and of residual impurities such as Mg, Ca and Si. The usefulness of dopants in promoting sintering and controlling grain growth in nominally pure alumina materials is fully recognized. Additives forming liquid silicates have been used for more than fifty years in the large scale production of many grades of commercial sintered aluminas. Among these the $Al_2O_3$-CaO-$SiO_2$ system has achieved considerable importance because it contains a number of eutectics with temperatures in the range 1200 to 1400 °C which generates several volume percent of a wetting liquid and sintering of the alumina to full density is readily achieved through processes assumed to involve the solution and recrystallization of alumina particles in the liquid. The characteristic microstructures of such sintered materials consist of large, plate- or rod-like grains of α-alumina markedly elongated along the c-axis, with intergranular aluminosilicate glass and crystalline phases such as calcium hexaluminate ($CA_6$), anorthite ($CAS_2$) or gehlenite ($C_2AS$). The presence of an intergranular glassy phase is expected ultimately to control both morphology and many physical properties such as strength, toughness, creep and wear rate and dielectric properties (Davidge and Riley 1995; Twigg et al. 1996). In this work we have studied the microstructures of a range of tailored aluminas, doped with differing amounts of $CaSiO_3$ (CS) and sintered at 1400 °C. Attention was paid to the grain boundary film widths, compositions and chemistries relative to the triple point regions to see if further insight into the exact mechanism of liquid phase sintering could be achieved, and to consider the implications of the resultant microstructure for mechanical properties such as wear.

## 2. EXPERIMENT

The alumina powder was of high purity and fine grain size (ca. 400 nm AL160SG-1, Showa, Japan). The sintering additive (CaO/BaO and $SiO_2$ in varying molar ratios) was added by heterogeneous precipitation of metal hydroxide and hydrated silica on to the alumina particles. The powder was dried, calcined at 900 °C for 30 minutes, crushed, sieved and uniaxially pressed into discs, isostatically pressed at 300 MPa to a green density of 53% of theoretical and sintered at 1400 °C for times up to 10 hours. Full density was reached with 2% of additive in less than 2 hours. As an indication of the effectiveness of the calcium silicate system, undoped alumina sintered at 1400 °C for 10 hours was still only 92 to 93 % dense. Full densification was achieved when the undoped alumina sample was sintered at 1600 °C for 3 hours. Analyses of the sintered material microstructures were carried out using SEM and TEM. Grain size was measured using a linear intercept method on SEM images of polished, thermally etched surfaces. TEM samples were prepared by Ar ion milling of mechanically thinned densified material. HRTEM was performed on a JEOL 4000EX operated at 400 keV with a point resolution of 0.17 nm. Analytical TEM was conducted on VG dedicated STEMs operating at 100 keV, both energy dispersive X-ray (EDX) and parallel electron energy loss (PEELS) were used for chemical analysis. Imaging and analysis was performed on grain boundaries aligned edge on. Care was taken to minimize recording times so as to avoid radiation damage and the preferential driving of atomic species (e.g. Ca) by the electron beam when conducting concentration line profiles. Highly focused probes were found to drive Ca and Al away from the irradiated area. Radiation sensitivity appeared to be strongly associated with the presence of Si.

## 3. HREM RESULTS

Apart from the undoped alumina sample, where no amorphous film was observed at grain boundaries or triple junctions (Figure 1), all alumina samples containing CS sintering additive exhibited an amorphous film at the majority of two-grain interfaces (Figure 2 - alumina - 10 wt% CS). Generally this glassy film was continuous, however, special low angle grain boundaries in the latter samples were observed to be clean, while a small number of boundaries exhibited an apparent crystalline phase. No grain boundary orientation relationship was apparent from the data. Generally, for all samples the grain boundary film thickness measured from HREM micrographs was between 1.2 and 2 nm. The exact value in a particular sample was observed to vary from boundary to boundary. A statistical analysis on the sample containing 10 wt% CS addition gave a figure of 1.6 ± 0.25 nm for the film thickness at ca. 80% of the boundaries. Significantly there appeared to be no systematic dependence of grain boundary film thickness on the amount of sintering additive used.

Figure 1      Figure 2

## 4. STEM/EDX RESULTS

The samples containing CS additions showed segregation of both Si and Ca to the amorphous triple pockets (Figure 3 - Bright field image and Ca and Si Kα EDX maps)as well as the large grain boundary facets, while segregation of predominantly Ca to the grain boundary films was observed (Figure 4). Typical compositions for triple pockets in all three samples containing CS glass were [Si]/[Al] = 0.8 and [Ca]/[Al] = 0.2. EDX linescans across triple pockets showed a difference in the spatial extent of the Si and Ca signals suggesting a degree of phase separation.

Two-grain boundaries in all three doped samples showed strong segregation of Ca. Reduced area (6 x 8 nm$^2$) analysis at the grain boundaries and in the nearby bulk grain were used to calculate the excess Ca concentration at the grain boundary. This gave values of around 0.5 monolayers Ca for all levels of CS addition. For comparison, the undoped alumina sample showed small amounts of Ca segregation to grain boundaries at a level of around 0.1 monolayers. Evidence of smaller amounts of silicon was observed at some but not all grain boundaries.

EDX linescans across the grain boundaries in the doped samples also showed Ca segregation. Fitting a 1D Gaussian to the measured line profile and expressing the maximum counts in the profile as a fraction of monolayer of segregant relative to Al gave values of between 0.65 and 1.0 monolayers for all levels of CS addition.

Figure 3

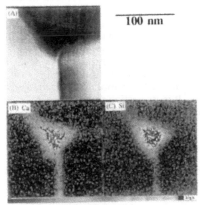

100 nm

Figure 4

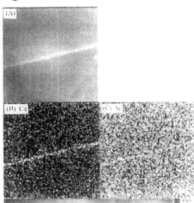

## 5. STEM/PEELS

PEELS provides an improved spatial resolution for analysis over EDX and a more accurate quantification of oxygen. Similar values to those obtained by EDX were obtained. Point analyses from triple pocket regions showed them to be Si rich and typically lying within the primary phase field of anorthite. EELS linescans across grain/triple pocket interfaces showed the segregation of Ca to the grain face and the silicon-rich nature of the centre of the triple pocket.

Point and line analyses at grain boundaries confirmed an enrichment of Ca. Analysis of the 10 wt% sample gave an excess Ca concentration of around 0.5 monolayers. PEELS linescans clearly revealed the segregation of Ca to the grain boundaries (Figure 5). The FWHM of the Ca distribution of ca. 4 nm and the integrated area gave a segregation of 0.7 monolayers. Generally there was no detectable change in Si concentration across the boundary; any residual level was attributed to a silicon oxide surface film.

Figure 5

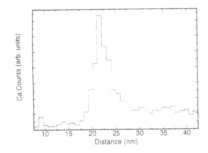

Figure 6

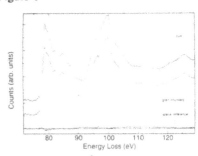

Generally, PEELS and EDX results were in agreement and appeared to be reasonably invarient for the three sintering additions. Linescans appear preferable to point analyses for the determination of grain boundary segregation levels because of the uncertainty of the exact area of analysis due to beam broadening effects. Generally the level of Ca segregation observed (between 0.5 and 1 monolayer) spread over a grain boundary thickness of 2 nm derived from HREM measurements (6 - 7 cation planes), gives an average [Ca]/[Al] atomic ratio of between 0.07 and 0.14. Assuming that the level of Si segregation is small, as evidenced by the linescans, this would put the grain boundary film composition nominally in the primary phase field of $CA_6$.

Further information using EELS can be derived from studies of the electron loss near-edge structure (ELNES) associated with each inner shell ionization edge. The form of both the Al K-, the Al $L_{2,3}$- and the Si $L_{2,3}$- ELNES can provide information on the local atomic coordinations. ELNES measurements suggest that in the triple point glass, Al is substituting in tetrahedral $SiO_4$

474

sites as would be expected for a general CAS; whereas in the Ca -rich grain boundary film Al is predominantly in octahedral, $AlO_6$, sites as in bulk $\alpha$-alumina. EELS spectra were recorded from an area of 6 x 8 $nm^2$ both on and off the grain boundary and, by subtracting a normalized version of the bulk $\alpha$-alumina spectrum from the interface spectrum, we obtain the Al $L_{2,3}$-ELNES spatial difference spectrum, which is essentially zero over the whole energy range (Figure 6 - Al $L_{2,3}$-ELNES spectra from a) bulk alumina, b) interface and c) spatial difference between b and a). As the basic form of the Al $L_{2,3}$-ELNES is sensitive to the nearest neighbour coordination, this implies that the Al coordination remains octahedral in the glassy grain boundary film. Similar measurements on the Ca $L_{2,3}$- and O K-ELNES were indicative of Ca in eightfold coordination (or higher) and a reduction in medium range order in the glass. It is interesting to note that in crystalline $CA_6$ the Al sites are predominantly octahedral and the Ca sites twelvefold coordinated to oxygen.

## 6. OTHER SYSTEMS AND RELATIONSHIP TO WEAR RATE

Other molar ratios and combinations of sintering additives have recently been investigated. Decreasing the $CaO:SiO_2$ molar ratio from 1:1 to 1:10 for a given additive level, results in the presence of Si in the grain boundary film which causes Al to become tetrahedrally coordinated to oxygen. Replacing CaO with BaO ($BaO:SiO_2$ = 1:1), gave similar levels of Ba segregation at grain boundaries as that measured for Ca, together with an absence of Si.

Figure 7 shows the effect of Si concentration [$Al_2O_3$ + 10 wt%($xSiO_2$ + CaO)] and Figure 8 cation species in the additive on the wet erosive wear rate normalised for the effect of matrix grain size. This reveals the important influence of the exact grain boundary chemistry on the initiation and propagation of microcracks.

Figure 7

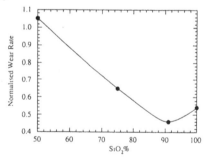

Figure 8

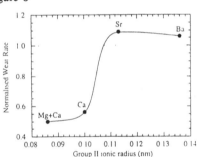

## 7. CONCLUSIONS

The relatively high doping levels of the CS additive used to make these dense sintered alumina materials led to the development of very thin Ca-rich amorphous films (of nominal composition $CA_6$)at the majority of two-grain interfaces, with the surplus CAS material being accomodated in the grain edge triple pockets. The equilibrium phases at 1400 $^{\circ}$C are alumina, $CA_6$ and $CAS_2$. Thus we propose a model during sintering which consists of alumina grains in contact with a thin film of $CA_6$, which is in turn in contact with an anorthite glass. The average two-grain boundary film thickness and composition in the densified material did not appear to be related to the amount of additive used (2 to 10 wt%, and the range used for normal commercial sintering of alumina) and thus presumably to the volume of liquid generated during sintering. Changing the molar ratios of CaO and $SiO_2$, or replacing Ca with Ba, changes the grain boundary chemistry which is reflected in the wear rate. This is believed to be because grain boundary microcrack initiation and propagation, leading to crack linking and loss of surface grains, are rate controlling.

## REFERENCES

[1] Davidge R.W. and Riley F.L. 1995 Wear **186-187**, 45-49.
[2] Twigg P.C. et al. 1996 Phil. Mag. A **74**, 1245-1252.
[3] Brydson R. et al. 1997 J. Am. Cer. Soc. - in press.

*Inst. Phys. Conf. Ser. No 153: Section 11*
*Paper presented at Electron Microscopy and Analysis Group Conf. EMAG97, Cambridge, 1997*
© *1997 IOP Publishing Ltd*

# A superspace group of the modulated structure in homologous compounds $In(In_xM_{1-x})O_3(ZnO)_m$

**Chunfei Li, Yoshio Bando, Masaki Nakamura and Noboru Kimizuka\***

**National Institute for Research in Inorganic Materials, 1-Namiki, Tsukuba, Ibaraki 305, Japan**
**\*Universidad de Sonora, CIPM, Hermosilo, Sonora, C.P.83000, Mexico**

**ABSTRACT:** The modulated structures of $In(In_xM_{1-x})O_3(ZnO)_m$ have been studied by HRTEM. Superspace groups of the modulated structure for $In(In_{0.5}Al_{0.5})O_3(ZnO)_{13}$ and $In(In_{0.5}Fe_{0.5})O_3(ZnO)_{13}$ are assigned as $_B^1B m$ and $_P^1 P\bar{1}$ , respectively. It is shown that the periodicities of the modulation waves are about 7.4 nm and 5.9 nm for $In(In_{0.5}Al_{0.5})O_3(ZnO)_{13}$ and $In(In_{0.5}Fe_{0.5})O_3(ZnO)_{13}$, respectively, where these values are located between those for $In_2O_3(ZnO)_m$ and $InMO_3(ZnO)_m$. The electron diffraction patterns are well interpreted based on a superspace group description.

## 1. INTRODUCTION

A series of homologous compounds having a general chemical formula of $InMO_3(ZnO)_m$ (M=In, Fe, Ga and Al, m=integer) was first synthesized by one of the present authors (Kimizuka 1988). From X-ray analysis, the crystal symmetries were determined either as rhombohedral $R\bar{3}m$ for m=odd numbers or hexagonal $P6_3/mmc$ for m=even numbers. The structure consists of $InO_2^{1-}$ layers (In-O layer) interleaved with $MZn_mO_{m+1}^{1+}$ (M/Zn-O layer) layers along the *c*-axis. The distribution of M and Zn ions in the M/Zn-O layer was supposed to be random.

However, a modulated structure of $InFeO_3(ZnO)_m$ has been found by analytical electron microscopy, where a sinusoidal modulated contrast was caused by the ordering of Fe ions in the Fe/Zn-O layer (Uchida 1994, Bando 1994). A Further study of the homologous compounds $In_2O_3(ZnO)_m$ revealed that the shape and the periodicity of the modulated waves strongly depend on the kinds of trivalent metal atoms M in the M/Zn-O layer (Li 1997).

In the present paper, the modulated structures of $In(In_{0.5}Al_{0.5})O_3(ZnO)_{13}$ and $In(In_{0.5}Fe_{0.5})O_3(ZnO)_{13}$ have been studied by HRTEM, where the M sites are occupied by two different atoms with respect to $InMO_3(ZnO)_m$. Based on the analysis of electron diffraction patterns showing the satellite spots, superspace groups of the modulated structures have been determined. The features of the present modulated structure in terms of its shape and periodicity are also discussed with those of $InMO_3(ZnO)_m$.

## 2. EXPERIMENT

The samples were prepared by heating powders of $In_2O_3$, $M_2O_3$ and ZnO in Pt-sealed tube at 1350 °C for about 3 days. Electron diffraction patterns and high-resolution images were taken by using a high-resolution transmission electron microscope (JEM-2000EX), operated at 200 kV.

## 3. RESULTS AND DISCUSSION

Fig.1(a) and (b) show the electron diffraction patterns for $In(In_{0.5}Al_{0.5})O_3(ZnO)_{13}$ and $In(In_{0.5}Fe_{0.5})O_3(ZnO)_{13}$, respectively, in which their schematic drawings showing the arrangements of the satellite spots are indicated. In order to explain the satellite spots, it is necessary to describe a relation between the space group of the subcell and the superspace group as follows.

In the previous study, no satellite spots were found for the compounds $InMO_3(ZnO)_m$ (M=In, Fe, Ga and Al, m=integer), where m are smaller than 6. However, for the compounds having m larger than 6, satellite spots were observed, suggesting the formation of the modulated structure.

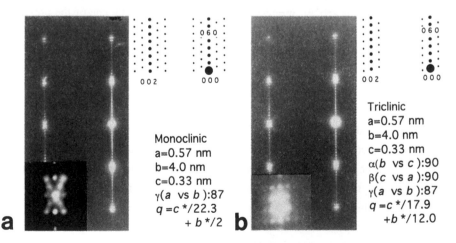

Monoclinic
a=0.57 nm
b=4.0 nm
c=0.33 nm
$\gamma(a$ vs $b)$:87
$q = c^*/22.3$
    $+ b^*/2$

Triclinic
a=0.57 nm
b=4.0 nm
c=0.33 nm
$\alpha(b$ vs $c)$:90
$\beta(c$ vs $a)$:90
$\gamma(a$ vs $b)$:87
$q = c^*/17.9$
    $+b^*/12.0$

Fig.1 Electron diffraction patterns for (a) $In(In_{0.5}Al_{0.5})O_3(ZnO)_{13}$ and (b) $In(In_{0.5}Fe_{0.5})O_3(ZnO)_{13}$ and their schematic drawing showing the arrangements of the satellite spots.

It should be also pointed out that the symmetry of the subcell structure for m larger than 6 was found to be neither $R\bar{3}m$ nor $P6_3/mmc$. The lacks of $\bar{3}$ symmetry in $R\bar{3}m$ and $6_3$ in $P6_3/mmc$ imply that the crystal symmetry is either monoclinic for m=odd numbers or orthorhombic for m=even numbers. The lattice relations between a new monoclinic unit cell and the hexagonal (rhombohedral) are expressed as $a_m = \sqrt{3}a_r$, $b_m = \sqrt{(c_r/3)^2 + (\sqrt{3}a_r/3)^2}$, $c_m = a_r$, where its lattice geometry is shown in Fig.2(a). Based on the new subcell, the satellite spots can be expressed as $ha^*+kb^*+lc^*+mq$, where q is a modulation vector with $c^*/T$. Since the satellite spots were indexed by one dimensional modulation vector, the present modulated structure is described by a four dimensional superspace group. It is then derived that the space groups of the subcell are Bm and $A2_1am$ for m=odd and even numbers, respectively, while the corresponding superspace groups are $_P\overset{Bm}{\underset{1}{}}$ and $_P\overset{A2_1am}{\underset{1\bar{1}\bar{1}}{}}$, respectively.

In Fig.1(a), the main spots are indexed by the monoclinic unit cell with a space group of Bm. The extinction rules for the satellite spots are k+m=2n. This implies that the superspace group is $_B\overset{Bm}{\underset{1}{}}$. While in Fig.1(b), the arrangements of satellite spots are different from that of Fig.1(a), where they are not arranged in a rectangular shape. Therefore, the corresponding subcell must be described as a triclinic cell instead of the monoclinic cell, where the result is indicated into the figure. The space and corresponding superspace groups are then $p\bar{1}$ and $_P\overset{P\bar{1}}{\underset{1}{}}$, respectively. The periodicities observed are 7.4 nm in (a) and 5.9 nm in (b). These values are located between that of $In_2O_3(ZnO)_{13}$ (4.3 nm) and $InFeO_3(ZnO)_{13}$ (8.6 nm).

A high-resolution lattice image of $In(In_{0.5}Fe_{0.5})O_3(ZnO)_{13}$ is shown in Fig.3, where the corresponding diffraction pattern is shown in Fig.1(b). The image contrast is almost the same as those of $InMO_3(ZnO)_m$. The In-O layers appear as darker contrast, while that of M/Zn-O as gray contrast. The zig-zag contrast appearing in the M/Zn-O layer suggests the formation of the modulated structure. The angle of the zig-zag shape normal to the c axis is about 52 ° and this angle is located between that of

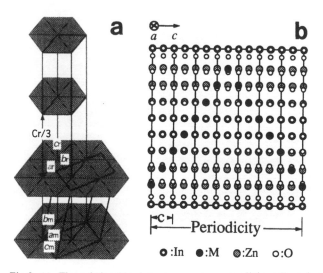

Fig.2 (a) The relationship between a new monoclinic cell and a hexagonal (rhombohedral) cell (b) One of the possible model for the modulated structure where M atoms are ordered to form a zig-zag shape.

$InFeO_3(ZnO)_{13}$ ($40°$ ) and $In_2O_3(ZnO)_{13}$ ( $60°$ ). The corresponding structure model for the modulation is shown in Fig.2(b), where M atoms are ordered within the M/Zn-O layer.

## 4.    CONCLUSION

The modulated structures in $In(In_xM_{1-x})O_3(ZnO)_m$ are observed by high-resolution electron microscopy. The superspace groups of the modulated structures in $In(In_{0.5}Al_{0.5})O_3(ZnO)_{13}$ and $In(In_{0.5}Fe_{0.5})O_3(ZnO)_{13}$ are described as $_B Bm$ and $P P\bar{1}$ . The modulated structures appear as zig-zag shapes within the M/Zn-O layer, whose periodicities are located between that of $In_2O_3(ZnO)_{13}$ and $InMO_3(ZnO)_{13}$. It is considered that the modulated structure is formed by the ordering of $In_xM_{1-x}$ atoms.

## REFERENCES

Bando Y, Kitami Y, Kurashima K, Tomita T, Honda T and Ishida Y 1994, Microbeam Analysis **3**, 279

Kimizuka N, Mohri T, Matsui Y, and Siratori K 1988, J. Solid State Chem. **74**, 98

Li C, Bando Y, Nakamura M and Kimizuka N 1997, J. Electron Microsc. **46**, 119

Uchida N, Bando Y, Nakamura M, and Kimizuka N 1994, J. Electron Microsc.**43**, 146

Fig.3 A high-resolution lattice image of $In(In_{0.5}Fe_{0.5})O_3(ZnO)_{13}$. The zig-zag contrast with periodicity of about 5.9 nm is found.

*Inst. Phys. Conf. Ser. No 153: Section 11*
*Paper presented at Electron Microscopy and Analysis Group Conf. EMAG97, Cambridge, 1997*
© *1997 IOP Publishing Ltd*

# Interfacial analysis of MMCs by dedicated FEGSTEM

**AJ Papworth and P Fox**

Department of Material Science, The University of Liverpool, Liverpool, L69 3BX.

**ABSTRACT:** An investigation has been carried out into the formation of the spinel $MgAl_2O_4$ phase on the surface of fibres in aluminium alloy MMCs. Interface analysis was performed by a FEGSTEM using a windowless EDS detector and PEELS. It was found that the $MgAl_2O_4$ phase was formed initially during the casting process by a reaction between the oxygen within the air spaces of the preform and the alloy. The $SiO_2$ binder was not reduced during casting nor during cooling even though diffusion of magnesium and aluminium atoms has been reported to occur.

## 1. INTRODUCTION.

Over the last twenty years, there have been many interface analyses carried out on metal matrix composite (MMC) systems. The one system that has drawn the most attention is the aluminium/alumina fibre composite. This type of composite is usually made by the squeeze-casting method, where an alumina preform is infiltrated with an aluminium alloy by the application of high-pressure. In order to maintain structural integrity of the preform a binder is required. In this system, the binder is nearly always silica Mortensen (1993). Any interfacial analysis of this system, must include this binder layer, which is on the surfaces of the fibres. Most interface analyses have shown that; a reaction layer ($MgAl_2O_4$) is formed at the interface between the fibre/binder and the matrix metal. The reaction $Mg + 2Al + SiO_2 \rightarrow MgAl_2O_4 + 2Si$, is commonly thought to be the cause of this reaction layer, by the reduction of the $SiO_2$ layer during T6 heat treatment as shown by Dudek et al (1993), and Molins et al (1991).

The analysis reported here was carried out using a VG HB 601UX dedicated FEGSTEM optimised for light element detection. This analysis found that $MgAl_2O_4$ had formed at the interface between Saffil™ fibre and matrix metal even when the $SiO_2$ binder was not present Two further experiments were performed, the aim of which was to identify the formation mechanism of the $MgAl_2O_4$ layer. In the first, oxygen was removed from the preform and the surface of the fibres, by a roasting technique, while for the second the $SiO_2$ binder was replaced by $CaF_2$ binder. EDX spectroscopy was used for the elemental analysis and PEEL spectroscopy was used to find the nature of the bonds between the elements. The analysis showed that $MgAl_2O_4$ was formed on all the surfaces of the Saffil™ reinforcement in all experiments.

## 2. RESULTS/DISCUSSION.

All the analyses reported here have been of as-cast composite material, which has been slow cooled in the casting die. The alloy used in these experiments was a common piston alloy, which contained 1 wt% Mg, amongst other elements. The reinforcement was Saffil™ or aluminosilicate fibres. Fig. 1 shows, that although it is not visible on the micrograph; there is a reaction phase 30 nm thick on the surface of the silica binder, as indicated by the Mg and Si traces. This reaction layer has also been identified on the surface of the Saffil™ fibre, where there was no binder present, as can be seen in Fig. 1. Analysis of a casting, where the preform was made using a CaF$_2$ binder showed that the MgAl$_2$O$_4$ reaction layer was still forming on the surface of the binder as indicated by Fig. 2. Therefore, the reaction layer is not solely reliant on the reaction with SiO$_2$. Pai (1995) reviewed of the role of magnesium in aluminium alloys, and identified an oxygen gas layer, which is always present on the surface of ceramic materials as the cause of the MgAl$_2$O$_4$ reaction layer.

A casting was made with a preform that had the oxygen gas layer removed by a magnesium-getting technique. Fig. 3 is an analysis of this casting. It can be seen from the magnesium trace that the MgAl$_2$O$_4$ reaction layer is still present. Measurements taken from the magnesium trace show that the reaction phase is 8 nm thick. The thickness of the reaction layers shown in Fig. 3 and 4 are consistent with oxide thickness measurements taken from sessile drop experiments performed in air and high vacuum Ning Wang (1992).

Measurements of oxide layers of this kind have to be considered with caution, as the high energy probe of the FEGSTEM can damage the specimen. As the oxide of the reaction layer and the fibre degrade under the intense probe, the aluminium and magnesium atoms move to the outer surface of the void as it forms. This can give the impression that there is a greater concentration of metal than there actually is, or in the case of magnesium, magnesium appears to have diffused into the fibre, Fig. 5. Fig. 6 is a line-scan of the same region as Fig. 5, but the acquisition time has been reduced from 5000 ms to 2000 ms. It can be seen from the magnesium trace that diffusion into the fibre has not taken place. A more dramatic example of this can be seen in the aluminium X-ray map of Fig. 7. The micrograph was taken just after the first line-scan. The X-ray maps were made after the second line-scan. The aluminium map clearly shows the affects of the high energy probe on the aluminosilicate fibre, where the aluminium has increased in density around the void caused by the beam.

The problem of damage caused by the probe, means that EDX measurements of the reaction layer are not reliable. Therefore, PEELS analysis was carried out on the reaction layer. The aim of this analysis was to determine the bonding of the silicon found within the reaction layer. The analysis showed, by the PEELS finger printing technique, that the silicon had the bonding characteristics of silicon-dioxide and not silicon Fig. 8. Therefore, the SiO$_2$ binder was not reduced by the reaction Mg + 2Al + SiO$_2$ → MgAl$_2$O$_4$ + 2Si.

## 3. CONCLUSION.

From the results obtained from this investigation the following conclusions can be drawn.
1. During the casting process an oxide film MgAl$_2$O$_4$ was always present on the surface of the liquid alloy.
2. The film was formed from oxygen within the air spaces of the preform and had a nominal thickness of 30 nm.
3. During the casting process aluminium or magnesium did not reduce SiO$_2$ to produce MgAl$_2$O$_4$.
4. The MgAl$_2$O$_4$ layer on the infiltration front was continuously deposited on the fibre surfaces during infiltration.
5. The high energy probe of the FEGSTEM damages oxides, causing the metal elements to concentrate in the region around the void.

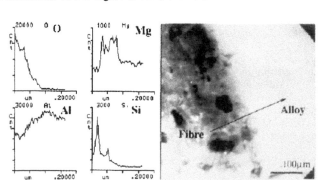

Fig.1. The magnesium trace clearly shows the reaction layer on the surface of the fibre, although the silica binder is not present.

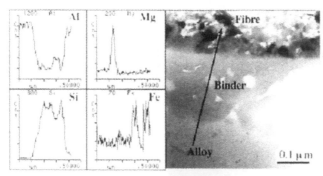

Fig. 2. The magnesium trace clearly shows the reaction layer on the surface of the silica binder.

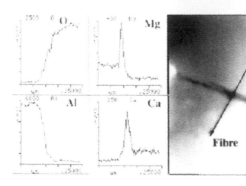

Fig. 3. The calcium trace shows that magnesium has not penetrated the $CaF_2$ binder. Therefore the 35 nm layer of $MgAl_2O_4$ would have had to have formed before coming in contact with the fibre/binder.

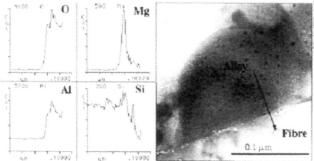

Fig. 4. The magnesium peak shows a 6 nm layer of $MgAl_2O_4$ between the fibre and the intermetallic.

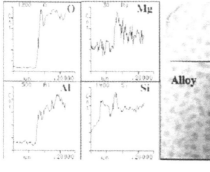
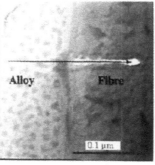

Fig. 5. The magnesium trace gives the impression that magnesium has defused into the aluminosilicate fibre.

482

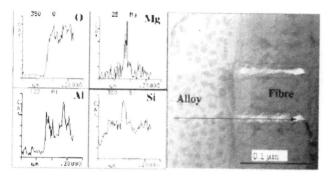

Fig. 6. The reduction of the acquisition time from 5 seconds to 2 seconds shows that magnesium diffusion into the fibre was an artefact and in reality magnesium had not diffused into the fibre.

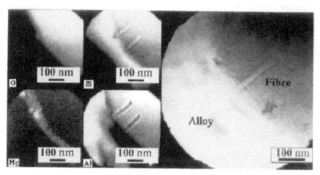

Fig. 7. The aluminium map shows a higher concentration of aluminium around the beam damaged region.

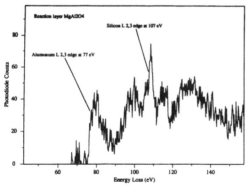

Fig. 8. EELS showing both the aluminium L 2,3 edge of MgAl$_2$O$_4$ and the silicon L 2,3 edge of silicon dioxide.

## 4.    ACKNOWLEDGEMENT

The authors wish to thank Vernaware Ltd, Bolton, UK. for all their assistance in the making of the preforms, and one of the authors (AP) is indebted to Verna Group for their financial support

## 5.    REFERENCES

Dudek H. J., Kleine A., Borath R.,  Mat. Sci. Eng. A167 (1993) 129-137
Molins R. Bartout J.D., Bienvenu Y., Mat. Sci. Eng. A135 (1991) 111-117
Mortensen A., Michaud V.J., Flemings M.C., J.O.M. January (1993) 36.
Ning Wang, Wang Z., Weatherly G.C., Met. Trans. A 23A May (1992) 1423-1430
Pai B.C., Ramani G., Pillai R.M., Satyanarayana K.G., J. Mat. Sci. 30 (1995) 1902-1911

*Inst. Phys. Conf. Ser. No 153: Section 11*
*Paper presented at Electron Microscopy and Analysis Group Conf. EMAG97, Cambridge, 1997*
© 1997 IOP Publishing Ltd

# Orientation-dependent equilibrium film thickness at interphase boundaries in ceramic-ceramic composites

## S Turan* and K M Knowles‡

* Anadolu University, Department of Ceramics Engineering, Yunusemre Campus, 26470 Eskisehir, Turkey.

‡ University of Cambridge, Department of Materials Science and Metallurgy, Pembroke Street, Cambridge, CB2 3QZ, U.K.

**ABSTRACT:** Theoretical considerations of the Hamaker constant for the interaction of two media across a third suggest that the highly anisotropic nature of the refractive indices of a medium such as $h$-BN will produce an orientation dependence on equilibrium film thickness through their effect on the dominant dispersion energy term in the Hamaker constant. High resolution transmission electron microscope observations of $h$-BN – 3C SiC interphase boundaries confirm the expected trend in orientation dependence on equilibrium film thickness.

## 1. INTRODUCTION

The existence of an equilibrium film thickness at general high angle grain boundaries in ceramics is now established experimentally (Kleebe et al 1992 and Turan and Knowles 1995) and theoretically (Clarke 1985, Clarke 1987 and Clarke et al 1993). Theoretical considerations show that an equilibrium film thickness arises from the competition between attractive dispersion forces determined by the dielectric properties of the grains and repulsive disjoining forces (steric and double-layer forces). Experimental results in the literature are in accord with this approach. There is a general perception that the equilibrium film thickness does not depend on the misorientation across an interface, unless the interface is recognised to be one of low energy, such as a twin boundary, in which case films are expected to be absent, as observed experimentally (Schmid and Rühle 1982 and Knowles and Turan 1996).

In this paper, we briefly review the theory for equilibrium film thicknesses at grain boundaries before considering its extension to interphase boundaries and possible circumstances where an orientation-dependent equilibrium film thickness might be expected. High resolution transmission electron microscope observations of $h$-BN – 3C SiC interphase boundaries where the thickness of the amorphous intergranular film is shown to be dependent on the orientation of the interphase boundary with respect to the $h$-BN (0001) planes can thus be understood qualitatively.

## 2. THEORETICAL BACKGROUND

In the absence of any applied and capillary stresses, the equilibrium film thickness at a grain boundary can be calculated from the force balance between the attractive dispersion forces ($\Pi_{DISP}$) and the repulsive disjoining forces ($\Pi_{ST} + \Pi_{EDL}$):

$$\Pi_{DISP} = \Pi_{ST} + \Pi_{EDL} \tag{1}$$

This can be written more explicitly in the form:

$$\frac{H_{\alpha\beta\alpha}}{6\pi h^3} = 4a\eta^2 \exp\left(-h/\xi\right) + \frac{16k_BT}{z^2\pi b_L h^2} \tanh^2\left[\frac{ze\Psi_s}{4k_BT}\right] \kappa^2 h^2 \exp\left(-\kappa h\right) \qquad (2)$$

(Clarke et al 1993), where $h$ is the film thickness, $\xi$ is a molecular correlation distance, $a\eta^2$ is a constant which is the free energy difference between ordered and disordered states of the film, $z$ is the ion charge, $b_L$ is the Bjerrum length, $\Psi_s$ is the electrostatic potential on the surface of the grains, $\kappa^{-1}$ is the Debye screening length and $H_{\alpha\beta\alpha}$ is the Hamaker constant of the interface containing the amorphous film. As Clarke (1987) has discussed, a suitable estimate for $H_{\alpha\beta\alpha}$ can be made by using the Tabor-Winterton approximation (Tabor and Winterton 1969):

$$H_{\alpha\beta\alpha} = \frac{3}{4} k_BT \left(\frac{\varepsilon_\alpha - \varepsilon_\beta}{\varepsilon_\alpha + \varepsilon_\beta}\right)^2 + \frac{3h\nu_e}{16\sqrt{2}} \frac{\left(n_\alpha^2 - n_\beta^2\right)^2}{\left(n_\alpha^2 + n_\beta^2\right)^{3/2}} \qquad (3)$$

where $\varepsilon_\alpha$ and $\varepsilon_\beta$ are the zero frequency dielectric constants of the grains and the intergranular film respectively and $n_\alpha$ and $n_\beta$ are the refractive indices in the visible of the respective phases and $\nu_e$ is the characteristic absorption frequency, assumed to be the same for all three materials, with a value $\approx 3 \times 10^{15}$ s$^{-1}$. For a given interface, the trend which arises from a consideration of equation (2) is that the lower the Hamaker constant, the higher the equilibrium film thickness (Clarke 1987).

For the more general problem of films at interphase boundaries, where two isotropic media $\alpha$ and $\gamma$ interact across a third isotropic medium $\beta$, the appropriate approximate expression for the Hamaker constant $H_{\alpha\beta\gamma}$ is given by Israelachvili (1992) as:

$$H_{\alpha\beta\gamma} = \frac{3}{4} k_BT \left(\frac{\varepsilon_\alpha - \varepsilon_\beta}{\varepsilon_\alpha + \varepsilon_\beta}\right)\left(\frac{\varepsilon_\gamma - \varepsilon_\beta}{\varepsilon_\gamma + \varepsilon_\beta}\right)$$

$$+ \frac{3h\nu_e}{8\sqrt{2}} \frac{\left(n_\alpha^2 - n_\beta^2\right)\left(n_\gamma^2 - n_\beta^2\right)}{\left(n_\alpha^2 + n_\beta^2\right)^{1/2}\left(n_\gamma^2 + n_\beta^2\right)^{1/2}\left\{\left(n_\alpha^2 + n_\beta^2\right)^{1/2} + \left(n_\gamma^2 + n_\beta^2\right)^{1/2}\right\}} \qquad (4)$$

from which it follows that the equilibrium film thickness will not depend on the misorientation across the interphase boundary, but is determined instead solely by the relative dielectric properties. In practice for ceramics, the dispersion energy term containing the refractive indices in the visible of the respective phases dominates the right hand side of equation (4), so that the dielectric properties at optical frequencies are the most important.

To generalise equation (4) to include anisotropic media, a sufficiently good estimate for $H_{\alpha\beta\gamma}$ can be made by substituting into the equation the values of the static dielectric constants and refractive indices in the visible normal to the interface under consideration (Turan 1995). While this is at best a first approximation to model the effects of crystal anisotropy (and note that this is different from the geometry considered by Parsegian and Weiss (1972) for anisotropic HgCl crystals), the dominance of the magnitude of the square of the optical refractive index normal to the interface in any calculation of the van der Waals dispersion forces ensures that this approximation is not unreasonable.

Hamaker constants for various interfaces calculated using this approach indicate that while the predicted difference in the equilibrium film thickness with a change in interface orientation may well be inside experimental error for anisotropic ceramics with relatively similar principal static dielectric constants and refractive indices such as $Al_2O_3$ and $Si_3N_4$, an orientation dependence on equilibrium film thickness would be expected where there is quite noticeable anisotropy in one of the media, e.g. for $h$-BN, where the principal refractive indices are $n_e = 1.65$ parallel to [0001] and $n_o = 2.13$ perpendicular to [0001] (Ishii and Sato 1983).

For the specific example of an interphase boundary between *h*-BN and 3C SiC containing an amorphous SiO$_2$ film, four possible interphase boundary configurations that might arise between *h*-BN and 3C SiC and for which there is experimental evidence (Turan and Knowles 1997) are sketched in Fig 1. In (a) and (b) the interface is parallel to (0001) *h*-BN, but whereas in (a) there is a set of {111} SiC planes perpendicular to the (0001) *h*-BN planes, in (b) they are parallel. The significant difference in (c) and (d) is that the interface plane is no longer parallel to (0001) *h*-BN.

Taking the refractive indices of SiC and SiO$_2$ to be 2.654 and 1.448 respectively (Clarke 1987), calculated Hamaker constants for these interphase boundaries are: (a) 50 × 10$^{-21}$ J, (b) 50 × 10$^{-21}$ J, (c) ≈ 140 × 10$^{-21}$ J and (d) 150 × 10$^{-21}$ J. Without further, more detailed, calculations taking into account for example the degree of epitaxial alignment assumed to exist in the silica layer (Clarke 1987), the equilibrium film thickness would be expected to be the same, relatively high value, for (a) and (b) and to be lowest in (d).

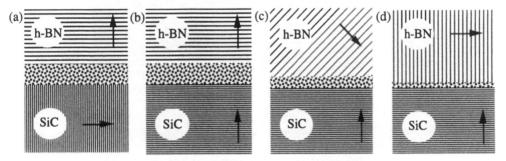

Fig. 1    Schematics of possible interphase boundaries between *h*-BN and 3C SiC containing amorphous silica films. The arrows indicate the stacking direction of (0001) and (111) planes of *h*-BN and SiC respectively.

## 3. EXPERIMENTAL RESULTS

Interphase boundaries oriented similar to the cases illustrated in Figs. 1(a) – (c) have been observed at *h*-BN – 3C SiC interfaces in hot isostatically pressed Si$_3$N$_4$ particulate-reinforced SiC composites in which *h*-BN inclusions arose indirectly during the densification process from boron oxide present on the surface of fine boron nitride particles used to prevent chemical reactions between SiC and the canning medium (Turan and Knowles 1995). Examples from HRTEM using a JEOL 4000EX TEM of such interphase boundaries are shown in Fig. 2. The film thicknesses are ≈ 12 Å for both the *h*-BN – 3C SiC interphase boundaries in Figs. 2(a) and (b), whereas in Fig. 2(c), where the (0001) *h*-BN planes make an angle of 70° with respect to the interface plane, the film thickness is ≈ 6.6 Å, lower than in (a) or (b).

## 4. DISCUSSION AND CONCLUSIONS

The trend seen in the experimental HRTEM images of *h*-BN – 3C SiC interfaces shown in Fig. 2 is consistent with the trend in the magnitudes of the Hamaker constants calculated using the Tabor-Winterton approximation and suggests that the dependence of the film thickness on the orientation of the interphase boundary with respect to the *h*-BN (0001) planes observed in Fig. 2 can be understood at least qualitatively. This indicates that it would be worthwhile making further, more detailed, calculations of the Hamaker constants for the *h*-BN – SiO$_2$ – 3C SiC system using as full a set of dielectric data as possible, from which it would be possible to make quantitative predictions of the equilibrium film thickness by solving equation (2).

## ACKNOWLEDGEMENTS

We would like to thank Profs. C J Humphreys and A H Windle for the provision of laboratory facilities.

486

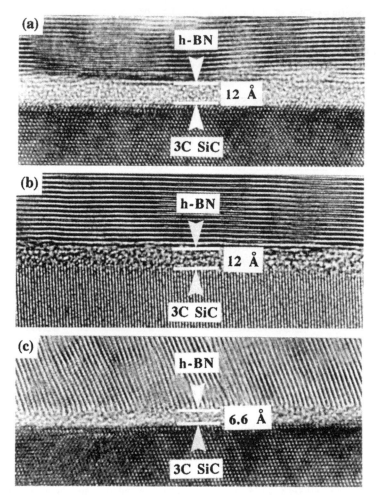

Fig. 2    HRTEM observations of three differently misoriented interphase boundaries between
$h$-BN and 3C SiC grains in hot isostatically pressed $Si_3N_4$ particulate – SiC composites
made using as-received starting powders showing orientation dependence on
equilibrium film thickness.

## REFERENCES

Clarke D R  1985  J. de Physique **46** Coll. C4,  51
Clarke D R  1987  J. Am. Ceram. Soc. **70**,  15
Clarke D R, Shaw T M, Philipse A P and Horn R G  1993  J. Am. Ceram. Soc.  **76**,  1201
Ishii T and Sato T  1983  J. Cryst. Growth  **61**,  689
Israelachvili J N  1992  Intermolecular and Surface Forces  (London: Academic Press)
Kleebe H-J, Hoffmann M J and Rühle M  1992  Z. Metallkünde  **83**, 610
Knowles K M and Turan S  1996  Materials Science Forum  **207-209**,  353
Parsegian, V A and Weiss G H  1972  J. Adhesion  **3**,  259
Schmid H and Rühle M  1984  J. Mat. Sci.  **19**,  615
Tabor D and Winterton R H S  1969  Proc. Roy. Soc.  **A312**,  435
Turan S  1995  Ph.D. Thesis  University of Cambridge
Turan S and Knowles K M  1995  J. Am. Ceram. Soc.  **78**,  680
Turan S and Knowles K M  1997  submitted to Phil. Mag. A

*Inst. Phys. Conf. Ser. No 153: Section 11*
*Paper presented at Electron Microscopy and Analysis Group Conf. EMAG97, Cambridge, 1997*
© 1997 IOP Publishing Ltd

# Microscopy of Calcium Phosphate Glass Ceramics

## A.J. McDermott, I. M. Reaney, W.E. Lee and P. F. James

Department of Engineering Materials, Sir Robert Hadfield Building, University of Sheffield, Sheffield, S1 3JD

**ABSTRACT:** Crystallisable glasses for biomedical applications have been developed in the $CaO-P_2O_5-SiO_2-Al_2O_3-TiO_2$ system. The phase evolution during heat treatment has been studied using a combination of x-ray diffraction and transmission electron microscopy with energy dispersive spectroscopy. At lower temperatures (800°C - 900°C), $CaO-P_2O_5$ rich 'florets' were observed. While at higher temperatures (1000°C - 1200°C) the floret structure was no longer apparent and coarsening of the nanodendrites occurred. This was accompanied by the appearance of anatase laths. At 1200°C, the $TiO_2$ laths were present as rutile.

## 1    INTRODUCTION

In recent years there has been significant research into calcium phosphate based glasses and glass ceramics for possible use as biomedical materials (Liao *et al.* (1987)). Hubert *et al.* (1983) classified biomaterials into two subgroups, bioinert and bioactive, based on their chemical reactivity in the physiological environment. Bioinert materials, such as alumina and stainless steel, undergo little or no change during long-term exposure to the physiological environment. These implant materials have to be physically bonded to the bone. Bioactive materials, such as bioglass and hydroxyapatite, are designed to undergo selected surface reactivity, resulting in a chemical bond between the tissue and the implant surface. This results in a strong, fibrous bond with the bone.

Since bone is mainly composed of calcium, phosphorus and oxygen, calcium phosphate glass ceramics are usually considered to be bioactive. In certain calcium phosphate systems, apatitic phases can be nucleated (Kokubo *et al.* (1986)) such as hydroxyapatite ($Ca_{10}(PO_4)_6(OH)_2$). As bone is mainly composed of natural hydroxyapatite, this artificial compound can act as scaffolding or filler to aid bone growth.

Hill *et al.* (1995) suggested that the ideal glass ceramic bone substitute material would need to: (i) crystallise readily into a calcium phosphate phase, (ii) be easily castable as a liquid into complex shapes, (iii) undergo bulk crystallisation, resulting in a fine grained material with high strength and fracture toughness, and (iv) be bioactive, stimulating bone growth by the release of calcium and phosphate ions.

Abe *et al.* (1976) investigated crystallisable calcium phosphate glass ceramics with a 1:1 $CaO:P_2O_5$ molar ratio. Ideally, the glass ceramic matrix should have a $CaO:P_2O_5$ molar ratio similar to natural bone, about 1.67:1 (Van Raemdonk *et al.* 1984). The purpose of this study is to demonstrate glass formability in compositions which contain close to this molar ratio (2:1) and also to investigate the phase evolution that occurs upon heat treatment, with the overall aim to produce a high strength bioactive glass ceramic for dental crowns or bone substitution.

## 2    EXPERIMENTAL PROCEDURE

Glasses were prepared from standard laboratory reagent (SLR) grade $CaCO_3$, $NH_4H_2PO_4$ , $Al(OH)_3$ and $TiO_2$ along with Loch Aline sand (99.5% $SiO_2$). The premixed glass batches were placed in alumina crucibles and sintered at 1000°C for 24 hours. The crucibles were then transferred to an electrically heated glass melting furnace at 1450°C for 30 minutes. The glass was poured and allowed to cool. If undissolved batch was observed the glass was crushed and remelted. The annealing treatment was at 600°C for 1hr.

X-ray diffractometry (XRD) samples were cut through the bulk and ground to a flat surface. XRD was carried out using CoKα radiation at a scanning rate of 0.33/2θ and step size of 0.2° with a Philips goniometer. Samples for transmission electron microscopy (TEM) were prepared by conventional ceramographical techniques. Bright-field (BF) imaging was undertaken using Philips 400 and 420 electron microscopes, operating at 100kV. The Philips 420 electron microscope was equipped with a LINK Energy Dispersive Spectroscopy (EDS) system.

# 3    RESULTS AND DISCUSSION

## 3.1    Glass Melting

Based on previous on work by Reaney *et al.* (1996), a series of glass compositions was produced in the $CaO-P_2O_5-TiO_2-Al_2O_3-SiO_2$ system. As simple binary composition of $2CaO:P_2O_5$ does not readily form a glass, $SiO_2$ is used in the composition to increase the molar fraction of network former. $Al_2O_3$ and $TiO_2$ are present to act as agents to promote volume nucleation, essential if control over the crystalline particle size in the glass ceramic is to be maintained (Nan *et al.* (1992)). The $TiO_2$ has the added benefit of decreasing the glass viscosity and lowering the liquidus temperature. However, it promotes phase separation of the $P_2O_5$ and $SiO_2$ networks (Reaney *et al.* (1996)). Additions of $Al_2O_3$ inhibit the phase separation, presumably by allowing the formation an $AlPO_4$ structural unit which is similar to that of the $SiO_2$ tetrahedra. The most promising composition produced was glass AM4, (mol%) CaO 39%, $P_2O_5$ 20%, $TiO_2$ 16%, $Al_2O_3$ 10% and $SiO_2$ 15%. The following section discusses the phase evolution of glass AM4 as a function of increasing temperature.

## 3.2    Glass Crystallisation

Figure 1 shows the XRD spectra obtained at various temperatures. Glass AM4 heated at 800°C for 10 hours displayed only a small volume of devitrication. Two phases are apparent at this temperature, IP and titania in the form of anatase. For the IP phase, similar peak positions and relative intensities were observed by Reaney *et al.* (1996). These authors identified IP by EDS analysis as a $CaO-P_2O_5$ rich phase and by electron diffraction to have an hexagonal Bravais lattice. Hydroxyapatite has a hexagonal Bravais lattice with space group $P6_3/m$. The difference in d-spacing from pure apatite to those in the AM4 composition could be due to the substitution of Ti for Ca in the unit cell, producing an apatitic compound based on a $(Ca,Ti)_{10}(PO_4)_6(O,OH)_2$. This however remains to be proven.

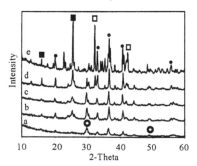

Figure 1:   XRD Spectra for samples held at (a) 800°C, (b) 900°C, (c) 1000°C, (d) 1100°C and (e) 1200°C for 10 hours.
Where • = IP, $CaO-P_2O_5$ rich phase identified by Reaney *et al.* (1996), ■ = UP, unknown phase, O = Anatase and □ = Rutile

When held at 800°C for 10 hours (Figure 2(a)) small 'florets' of a $CaO-P_2O_5$ rich phase are observed throughout the material. The distinct morphology of this phase has arisen, because it has grown with nano-dendritic arms, 10-20 nm wide, nucleating from a central point. It was found by EDS that residual glass at this temperature comprised of all the original oxides.

Figure 2(b) depicts glass AM4 held at 900°C, now the florets have grown into spherulites of the same composition and have impinged upon one another. This produces a microstructure with 'grain boundaries' containing residual glass. Although it was difficult to detect discrete anatase particles, Ti enriched regions (particles) could be found in the $CaO-P_2O_5$ rich matrix. Such data correlates with the concept of $TiO_2$ particles embedded in a $CaO-P_2O_5$ rich phase.

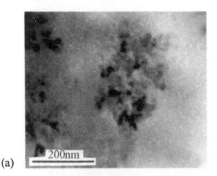
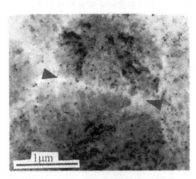

(a)    200nm          (b)    1μm

Figure 2: BF TEM images of glass AM4 held at (a) 800°C for 10 hours and (b) 900°C for 10 hours. The arrows in figure 2(b) denote the 'grain boundary' of residual glass'.

At 1000°C there was a distinct change in the microstructure; the florets were no longer apparent (figure 3(a)). Instead, coarser and coherent $CaO-P_2O_5$ rich particles, up to 0.25 μm in diameter had formed. This was accompanied by the appearance of laths of anatase up to 1μm in length.

Samples heated for 10 hours at 1100°C displayed the same phases in XRD as were observed at 900°C and 1000°C, Figure 1. However TEM revealed that the microstructure was coarser than at lower temperatures. EDS of the residual glass demonstrated that it was becoming enriched in Al and Si, compared to the composition of the residual glass at 800°C.

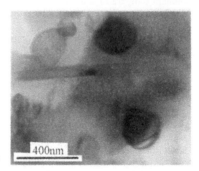

(a)    200nm          (b)    400nm

Figure 3: BF TEM images of glass AM4 held at (a) 1000°C for 10 hours and (b) 1100°C for 10 hours. There is a distinct difference in $TiO_2$ lath length and the size of the $CaO-P_2O_5$ coherent particles with temperature.

A comparison between the XRD spectra at 1100°C and 1200°C reveals the peaks to be sharper and better defined with anatase no longer present. Instead titania is observed in the form of rutile. It was at this temperature that the various phases in the crystallised glass could best be distinguished by EDS analysis as the phases were no longer overlapping. Five distinct phases were found. Figure 4(a) shows the $CaO-P_2O_5$ phase, whereas Figure 4(b) reveals the presence of $AlPO_4$ characterised by its extreme beam sensitivity and heavy faulting. This was also observed by Nan *et al.* (1992) and is surrounded by rutile laths in the image. In addition there was evidence of a CaO-$SiO_2$-$Al_2O_3$-$P_2O_5$ phase. This is thought to be based on wollastonite ($CaSiO_3$), as this is often crystallised in CaO-$P_2O_5$ glass ceramics (Kokubo *et al.* (1986)). In this phase the aluminium phosphate could substitute into the calcium silicate network and its presence could explain the peaks designated UP in Figure 1. The final phase identified by EDS analysis was $TiO_2$-$P_2O_5$ rich and has so far not been fully investigated as it is infrequently observed.

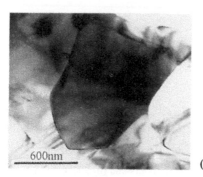 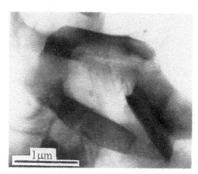

(a) 600nm          (b) 1 μm

Figure 4: BF TEM images of glass AM4 held at 1200°C for 10 hours. 4(a) denotes a CaO-$P_2O_5$ rich particle, whereas 4(b) depicts $AlPO_4$ phase surrounded by rutile laths.

## 4    CONCLUSIONS

In this study we have found that appropriate additions of $Al_2O_3$, $SiO_2$ and $TiO_2$ to 60 mol % $2CaO$:$P_2O_5$ allow the formation of a castable glass. In this system a calcium phosphate rich phase volume nucleates at low temperature (800°C) along with interspersed particles of anatase. Samples held at 900°C show CaO-$P_2O_5$ rich spherulites which impinge upon another. At temperatures above 1000°C, laths of $TiO_2$ are formed. They are anatase between 1000-1100°C and rutile at 1200°C. As the heat treatment temperatures are increased the CaO-$P_2O_5$ phase becomes more coherent and a CaO-$SiO_2$-$Al_2O_3$-$P_2O_5$ phase becomes apparent.

## 5    REFERENCES

Abe Y, Arahori T and Naruse A 1976 J. Am. Ceram. Soc. **59**, 487

Aoki H 1991 Science and Medical Applications of HA (Tokyo: Takayama Press System Centre) p165

Hill R and Wood D 1995 J. Mater. Sci: Mat. in Med. **6**, 311

Hubert S F, Hench L L, Forbes D and Bowman L S 1983 Ceramics in Surgery (Amsterdam: Elsevier Science Publishers B V) p4

Kokubo T, Ito S, Sakka S and Yamamuro T 1986 J. Mater. Sci. **21**, 536

Kokubo T, Kushitani H, Ohtsuki C, Sakka S and Yamamuro T J 1992 J. Mater. Sci: Mat. in Med. **3**, 79

Liao Y, Huang Z and Zhen A J 1987 Non-cryst. Sol. **95&96**, 1087

Nan Y, Lee W E and James P F 1992 J. Am. Ceram. Soc. **75**, 1641

Reaney I M, James P F and Lee W E 1996 J. Am. Ceram. Soc **79**, 1934

Van Raemdonk W, Ducheyne P and De Messer P 1984 Metal and Ceramic Biomaterials, Vol II, Strength and Surface (FL: Boca Raten) p143

*Inst. Phys. Conf. Ser. No 153: Section 11*
*Paper presented at Electron Microscopy and Analysis Group Conf. EMAG97, Cambridge, 1997*
© *1997 IOP Publishing Ltd*

# Electron Microscopy of TiO$_2$-based Thick Films for Gas Sensors

## P. I. Gouma, and M. J. Mills

Center for Industrial Sensors and Measurements, The Ohio State University, Columbus, OH, 43210, USA

**ABSTRACT:** The structure and chemistry of powders and thick films of TiO$_2$ (anatase)-based systems to be used in gas sensing applications have been characterised using electron microscopy techniques (SEM, TEM,EDX, X-ray mapping). Valuable information has been obtained about the relative morphological, structural and chemical features of the various components of the materials examined, which were either loose,heat-treated powders or films screen-printed on alumina substrates. This information is being used to guide the design of improved gas-sensing systems.

## 1. INTRODUCTION

In the last two decades extensive research in the area of gas sensors has been focussed on oxide semiconductors. TiO$_2$-based oxides have been identified as potential gas-sensing devices. Their electrical behaviour is sensitive to differences in the concentration of gases, such as CO (Azad et al 1994). Titania has three naturally occurring polymorphs: rutile, anatase, brookite, with rutile being the stable and most widely used form of titania. Earlier research at the Center for Industrial Sensors and Measurements (CISM) has revealed that the gas detecting and measuring (i.e. sensing) properties of the anatase form of titania are significant. Furthermore, they can be enhanced by the addition of an insulating oxide.

Characterisation studies of the anatase-based systems are still lacking. Therefore, the purpose of this project was to study two anatase-based systems doped with various components, each of which is having a different effect on the physical / chemical behaviour of the system: (i) rare earth oxide additions (Y$_2$O$_3$ and La$_2$O$_3$, respectively) for enhanced sensitivity and microstructural stability of the system-(rare earth oxides are known to inhibit the anatase to rutile transformation (Hishita et al 1983), and (ii) catalysts such as Pd and CuO, for faster detection of (or higher sensitivity to) changes in the gas concentration (Larsson et al 1996). These powder materials have been screen-printed on alumina substrates to form thick films (i.e. films of thickness >10μm). The structure, the morphology, and the distribution of each constituent of a given system were the main microstructural aspects investigated. The thickness and uniformity of the films were also studied. The analysis was performed on loose powders and thick films prior to sensing tests.

## 2. EXPERIMENTAL METHOD

Commercial anatase powder was mixed with rare earth oxides and catalysts to produce two different systems: (i) TiO$_2$-7 wt% Y$_2$O$_3$- 5wt% Pd, heat-treated at 700°C; and (ii) TiO$_2$- 10wt%La$_2$O$_3$-2wt%CuO, heat-treated at 500°C–the material preparation and processing steps (including the heat-treatments) were performed by the Synthesis Group of CISM. Thick films were prepared by the Device Fabrication Group of CISM, using the screen-printing technique. Loose, heat-treated powder material was deposited on carbon tape and was subsequently carbon-coated for SEM examination. Small samples of the thick films (2.5mm x 2.5mm) were also used for SEM imaging. Cross-sectional samples were also prepared for measurement of film thickness. Both a JEOL 820 and a Philips XL30 FEG SEM have been used for this investigation. The operation voltage used was 10kV at maximum. Both were equipped with EDX analytical systems. X-ray mapping was also performed using a EDAX system.

For the TEM examination, thin foils were initially prepared using a "sandwich" technique, which involved the mixing of the powder or the film with a resin. Unfortunately, long milling times resulted in artifacts that did not allow the analysis of these foils. Instead, the powder (or films) were suspended in ethanol and were subsequently deposited on holey carbon copper grids. The samples were studied in a Philips CM200 microscope operating at 200kV equipped with an EDS detector. HRTEM was also performed using a HITACHI H9000 NAR at 300kV.

## 3. RESULTS

### 3.1 $TiO_2$-7 wt% $Y_2O_3$- 5wt% Pd (HT at 700°C)

This is the first of the two complex systems studied, having three components. Secondary electron images of these thick films revealed a non-uniform structure across the sample, in terms of its thickness and density. Regions of very high density could be identified, 150-200µm long, as well as smaller areas, typically 50µm long, which contained long powder aggregates (see figure 1). Back-scattered electron imaging revealed atomic number contrast effects, so that fine white particles appeared in a grey matrix. Some of the larger particles were identified by EDX analysis to be $Y_2O_3$ (see fig 2a, b: SE image and EDX spectrum). This large, 2µm long, flakey yttria particle was surrounded by numerous, fine (200nm thick), spherical, $TiO_2$ particles. There were long channels of pores all over the sample.

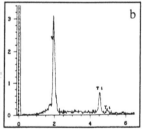

Fig. 1: Secondary Electron SEM micrograph showing the surface of a $TiO_2$-based thick film

Fig. 2: (a) Secondary Electron SEM micrograph showing a particle of yttria surrounded by fine $TiO_2$ powder; (b) EDX spectrum showing the yttrium peak and Ti from the matrix

X-ray mapping was used to differentiate between the yttria and palladium particles, since both are "heavier" than $TiO_2$. The results are shown in figure 3. Chemical mapping allowed the identification of the different phases that were present in the system, and manifested their distribution. Thus, fine

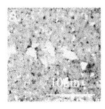
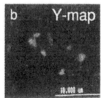
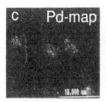
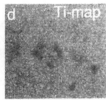

Fig. 3: (a) Back-Scattered Electron SEM micrograph showing a typical area of the $TiO_2$-based thick film; (b), (c), and (d) are the X-ray maps for the elements Y, Pd, and Ti, respectively.

clusters of Pd particles were found dispersed in the matrix, whereas the yttria flakes were larger and could be easily observed.

Another feature examined was the cross-sectional morphology of the film. Polishing of the film was very difficult, since the film tended to peel off from the substrate. The film's thickness was not uniform and varied between 18-24 μm.

## 3.2 TiO₂- 10wt%La₂O₃-2wt%CuO (HT at 500°C)

The first microstructural aspect of the system that has been investigated was the morphology of the material and the relative distribution of its constituent components. The straightforward approach was to use backscattered electron imaging, thus taking advantage of the different atomic numbers of Ti, La, and Cu, respectively. Figure 4a is a low magnification BEI micrograph which illustrates the contrast of each component– the background grey phase is the $TiO_2$ matrix. Imaging alone was not sufficient to differentiate between the secondary components, so EDX analysis was used to identify the location of $La_2O_3$ and CuO particles, respectively. Figure 4b typically shows large $La_2O_3$ particles with "bright" contrast (sizes ≥0.5μm long) and clusters of very fine CuO particles located at different sited on the $TiO_2$ matrix.

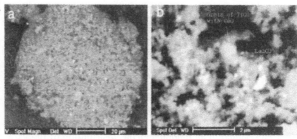

Fig. 4: (a) Back-scattered Electron SEM micrograph showing the distribution of secondary and ternary phases (regions of light contrast) in the $TiO_2$ powder; (b) micrograph of the same area taken at higher magnification, in which the various components of the system have been labelled with help from EDX analysis.

Characterisation of the material by TEM showed the $TiO_2$ phase particles to vary in size between 100-200nm on average (see figure 5a). Figure 5b shows a electron diffraction pattern of the [111] zone axis of anatase. Computer simulations using the EMS program verified the orientations. Further analysis of the $TiO_2$ particles has revealed the presence of fine clusters (fig 6a). EDX analysis performed on these clusters has shown the presence of Cu-see figure 6b. (This Cu peak had a significantly higher intensity than the background Cu signal from the support grid). This finding might be related to a dispersion of CuO particles or to reduced Cu particles.

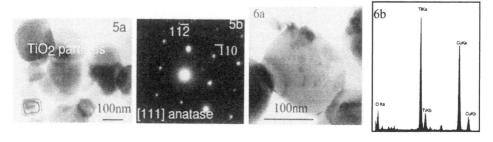

Fig. 5: (a) many beam TEM micrograph showing the matrix particles; (b) SAD of the [111] pole of anatase.

Fig.6: (a) Many beam TEM micrograph showing fine particles (dark contrast) on a $TiO_2$ powder particle; (b) EDX spectrum from this area revealing Cu peaks along with the Ti and Oxygen ones, suggesting that the particles are Cu-rich dispersions.

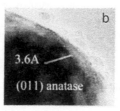

Fig. 7: (a) HRTEM micrograph showing part of an anatase-type particle (on zone axis [100] anatase) covered by a continuous amorphous film; (b) HRTEM micrograph of another anatase particle surrounded by platelet-like structures.

High Resolution electron microscopy was employed to study the distribution of secondary particles on $TiO_2$ in more detail. The only titania phase detected was anatase. Most of the anatase particles were covered by a continuous amorphous layer, as shown in figure 7a. This is not an unusual surface related phenomenon. What was striking though was the presence of a few particles having irregular features protruding from their surfaces. One such particle is shown in figure 7b. The surface layer covering this particular titania particle could have been a broken amorphous film, or fine catalyst platelets.

## 4. DISCUSSION

The systems examined consisted of fine powders of several different materials. The preparation method used resulted in an inhomogeneous distribution of the various constituents of a particular system. The rare earth oxide additions had much larger particles than the average matrix powder. The effectiveness of these additions on increasing the sensitivity and stability of the system depends on the dissolution of these oxides in the anatase lattice (Hishita et al 1983). A finer particle size and a more uniform dispersion of these oxides is recommended for more efficient interaction with a larger number of anatase particles.

A similar trend has been noticed for the catalysts particles. Although they were found to form clusters, these also had a non-uniform spatial distribution, thus making some sites in the films more active than others. The mechanism of the catalytic action of these additions is expected to be related to surface phenomena (Larsson et al 1996). Therefore, the catalyst particles should be distributed as dense layers on the $TiO_2$ support surface. In the case of the CuO containing system, metallic Cu is expected to form on the surfaces of the titania particles. The present study could not draw definite conclusions as far as the chemical state of the Cu-based particles observed on some matrix particles. Electron Energy Loss Spectroscopy is planned in an attempt to address this issue.

## 5. CONCLUSIONS

The present work has identified the morphological and structural characteristics of two gas-sensing systems. It has revealed an inhomogeneous distribution of the secondary components within the stucture of semiconducting oxides of these systems that might limit their sensitivity to gas detection. Further characterisation is required to understand the sensing mechanisms of these materials.

## 6. REFERENCES

Azad AM, Younkman LB and Akbar SA 1994 J. Am. Ceram. Soc. **77** [2], 481
Hishita S, Mutoh I, Kumoto K and Yanagida H 1983 Ceramics Int. **9** [2], 61
Larsson P-O, Andersson A, Wallenberg LR and Svensson B 1996 J. Catal. **163**, 279

*Inst. Phys. Conf. Ser. No 153: Section 11*
*Paper presented at Electron Microscopy and Analysis Group Conf. EMAG97, Cambridge, 1997*
© 1997 IOP Publishing Ltd

# The study of copper-alumina interfaces by EELS

**R Y Hashimoto, E S K Menon, M Saunders and A G Fox**

Center for Materials Science and Engineering, Department of Mechanical Engineering, US
Naval Postgraduate School, Monterey, CA 93943, USA

**ABSTRACT:**    A diffusion bonded copper-alumina interface has been studied using
electron energy loss spectroscopy (EELS). Investigations of the bulk alumina
(polycrystalline) show that silica, a common commercial impurity, is present at the triple
junctions in the form of fine mullite crystals in a glassy, silicon rich phase. EDX studies of
the interface revealed that silicon is present in varying concentrations from 1at% to 10at%.
Spatial difference EELS analysis of the interface shows two distinct residuals: a
coordination similar to an aluminosilicate for silicon rich areas and a coordination similar to
that of metal-to-metal bonding for low silicon regions.

## 1. INTRODUCTION

Mechanical testing of a diffusion bonded copper foil sandwiched between polycrystalline
alumina substrates shows that failure occurs predominantly in the ceramic (Dalgleish et al
1994). Hence, the metal-ceramic interface must have low flaw populations and/or strong
chemical bonding. In order to understand this behavior, the interface structure and chemistry
must be investigated. Here, we concentrate on the interfacial chemistry.

Of particular interest is the possibility that silica, which is the major impurity in commercial
alumina, has some influence on the interface properties. Previous studies suggest that silica
enhances the bonding process by providing diffusion paths for mass transport and by filling
pores at the interface (Dalgleish et al 1994).

A $Cu-Al_2O_3$ interface was prepared under vacuum by diffusion bonding of 100μm copper
foils (99.999% purity) pressed between polished, polycrystalline alumina substrates (~99.5%
purity) for several hours at ~90% of the melting temperature of copper. Graphite dies were
used to minimize the oxygen partial pressure, thus reducing oxidation products at the metal-
ceramic interface.

We have used a Topcon 002B TEM, operating at 200kV, equipped with an EDAX EDX
system and a Gatan Imaging Filter to examine the interfaces. EDX spectra, taken with 16nm
probes, have been used to establish the distribution of silica in the bulk alumina and at the
interface. EELS spectra, taken with 6nm probes, have been used to characterize coordination
effects at the triple junction and at the $Cu-Al_2O_3$ interface. The interface EELS spectra have
been further analyzed using the spatial difference technique (Müllejans and Bruley 1995)
which has been used widely in the analysis of interfaces. In this case, appropriate proportions
of bulk copper and bulk alumina spectra are subtracted from the interface spectra because the
probe size used, 6nm, encompassed the interface plus neighboring bulk regions. Energy-loss
near edge structure (ELNES) and selected area diffraction (SAD) have been used for phase
analysis.

## 2. RESULTS AND DISCUSSION

Fig. 1 is a bright-field TEM micrograph of the metal-ceramic interface. In this work, grain
boundaries and triple junctions in the alumina as well as the copper-alumina interface are
examined in detail and the salient features are briefly discussed in the following sections.

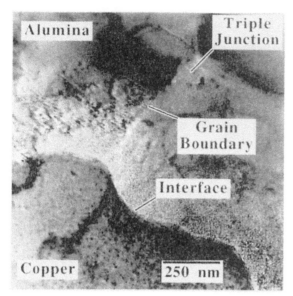

Fig. 1. Bright-field TEM micrograph of the Cu-$Al_2O_3$ interface with a triple junction nearby which is joined to the metal-ceramic interface via a grain boundary.

## 2.1 Triple Junctions

Fig. 2a is a bright-field TEM micrograph of a triple junction in the alumina. Close examination of the triple junction suggests that it contains a mixture of fine particles in a glassy matrix. This phase mixture wets the grain boundaries in the vicinity of the triple junction. A SAD pattern (Fig. 2b) acquired from the center of the triple junction indicates the particles to be crystallites of mullite, an aluminosilicate. Silica is known to segregate to the triple junctions and grain boundaries in alumina during sintering (Kingery et al 1976) and presumably reacted with the alumina to produce aluminosilicate phases.

EDX has been used to determine the distribution of silicon at the triple junction. Silicon is found to be in high concentrations near the center of the triple junction and drops below detectable limits (~0.9at%) at ~0.5μm along the grain boundary. The Si:Al ratio varies from 2.5 to 0.06, the higher silicon concentrations being measured near the center of the triple junction. There is no evidence of silicon in the interior of the surrounding alumina grains.

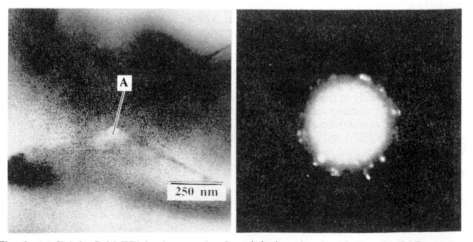

Fig. 2. (a) Bright-field TEM micrograph of a triple junction in $Al_2O_3$. (b) SAD pattern from point A in (a).

Quantitative EDX analysis for oxides from the center of the triple junction shows the composition to be $26.0\pm0.5$wt%$Al_2O_3$ and $74.0\pm0.5$wt% $SiO_2$. Error analysis for the percent concentrations is performed by considering cumulative errors due to sample thickness and density. The phase diagram shows that mullite and a silica rich solid solution phase can coexist when the alumina content is below ~72.0wt% (Davis and Pask 1972). This is consistent with the SAD results.

Electron energy loss spectra from the center of the triple junction have been acquired at 0.1eV energy resolution in the 50eV to 150eV range including the Al-$L_{2,3}$ and Si-$L_{2,3}$ edges. The spectra reveal a peak at ~107eV which could be attributed to silicon, see Fig. 3. Compared to previously published data, the spectra show good agreement with mullite (Kleebe et al 1996).

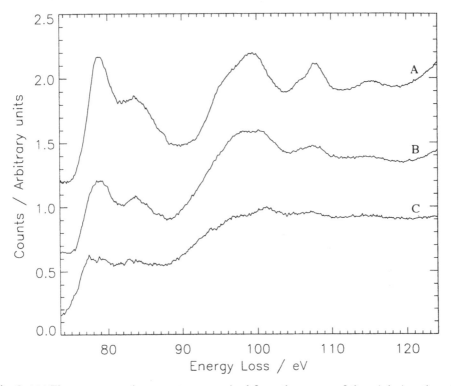

Fig. 3. (A) Electron energy loss spectrum acquired from the center of the triple junction, (B) Spatial difference residual, i.e. interface - references (Cu & $Al_2O_3$), high silicon interface region, (C) Spatial difference residual, low silicon interface region.

Combining the TEM, EDX and EELS results with the phase diagram, we conclude that the triple junction is composed of a mixture of mullite and a glassy silicon-rich phase. Diffusion along grain boundaries may provide a path for silicon to reach the metal-ceramic interface, for example, from a nearby triple junction (Fig. 1).

## 2.2 The Cu-$Al_2O_3$ Interface

Attempts to image a distinct interface phase proved inconclusive. EDX analysis shows the silicon concentration to vary from $1.0\pm0.5$at% to $10.0\pm0.5$at% along the interface. Detectable amounts of silicon are found to exist over large distances. No silicon is found in nearby regions of the bulk copper or alumina.

Electron energy loss spectra have been acquired at 0.1eV energy resolution in the 50eV to 150eV range including the $Al-L_{2,3}$, Cu-M and $Si-L_{2,3}$ edges. Two distinct spectra types are obtained from the interface. In areas shown by EDX to have a high silicon concentration, the spectra are characterized by a significant feature at ~107eV. The other type of spectra, from low silicon regions, do not appear to have this feature.

In order to determine the coordination effects at the interface, the spatial difference technique is applied to the two spectra types. In both cases the Cu-M edge (from pure Cu) and $Al-L_{2,3}$ edge (from pure $Al_2O_3$ ) are subtracted from the acquired spectra. The low silicon region residual spectrum, C in Fig. 3, shows features similar to those reported by Brydson et al (1995) and Müllejans and Bruley (1995). Following their arguments, these features can be attributed to metal to metal coordination effects at the interface. In Fig. 3 residual spectrum, B, from the high silicon region shows the possible presence of an aluminosilicate coordination (Hansen et al 1994; Kleebe et al 1996). Comparison of the spatial difference spectrum, B, from the silicon rich interface to spectrum, A, from the center of the triple junction shows similar feature locations (see Fig. 3). Further analysis of spectra acquired at higher energy losses (Si-K edge) and of Si-K elemental maps proved inconclusive because of low signal to noise ratios.

Similarities in the spatial difference spectrum from the silicon rich interface and the center of the triple junction suggest that similar coordination effects exist in both regions. Currently, we have insufficient information to unambiguously identify the exact nature of the chemical bonding since similar atomic coordinations could arise in many ways.

## 3. CONCLUSIONS

A mixture of crystalline mullite and a glassy, silicon rich phase has been found at triple junctions in bulk alumina. In addition, EDX analysis indicates low concentrations of silicon along alumina grain boundaries near triple junctions while alumina grain interiors are free of silicon.

The metal-ceramic interface proved more difficult to characterize. EDX analysis at the interface shows that silicon is present in concentrations from $1.0\pm0.5$at% to $10.0\pm0.5$at%. High silicon concentrations produce a significant EELS peak at ~107eV, whereas low silicon concentrations gave no significant feature at ~107eV. When the spatial difference technique is applied to these two types of spectra, it indicates that in the absence of the ~107eV feature the residual spectra show coordinations similar to metal-to-metal bonding and in the presence of the feature, coordinations similar to an aluminosilicate. The residual spectra from the interface areas with high silicon concentration are similar to that obtained from the center of the triple junction.

Further investigations with Auger spectroscopy and XPS are planned which could provide more conclusive chemical bonding information.

## ACKNOWLEDGEMENTS

The authors would like to thank Dr. Rowland Cannon, Dr. Brian Dalgleish, and Dr. Toni Tomsia for providing the samples and for useful discussions.

## REFERENCES

Brydson R, Müllejans H, Bruley J, Trusty P A, Sun X, Yeomans J A and Rühle M 1995 J. Miscosc. **177**, 369
Dalgleish B J, Saiz E, Tomsia A  P, Cannon R M and Ritchie R O 1994  Scripta Metall. Mater. **31**, 1190
Davis R F and Pask J A 1972  J. Am. Ceramic Soc **55**, 525
Müllejans H and Bruley J 1995 J. Microsc. **180**, 12
Hansen P L, Brydson R, McComb D W and Richardson I 1994 Microsc. Microanal. Microstruct. **5** , 173
Kingery W D, Bowen H K and Uhlmann D R 1976 Introduction to Ceramics, 2nd edition, (New York: Wiley) p 200

*Inst. Phys. Conf. Ser. No 153: Section 11*
*Paper presented at Electron Microscopy and Analysis Group Conf. EMAG97, Cambridge, 1997*
© *1997 IOP Publishing Ltd*

# Observations of high strain deformation at the surface of a worn zirconia

**JDO Barceinas-Sánchez and WM Rainforth**

Department of Engineering Materials, University of Sheffield, Mappin St., Sheffield, S1 3JD

**ABSTRACT**: The microstructure generated by the dry sliding wear of a zirconia ceramic (3Y-TZP) has been examined by TEM. High wear rates were associated with dramatic microstructural changes. The outer surface consisted of fine (~6nm) tetragonal crystallites. Below this the tetragonal grains had become elongated, corresponding to a 95% reduction in thickness. The first monoclinic zirconia was found at a depth of ~2.5µm, some of which was plastically deformed. Some evidence of dislocation flow was found in both zirconia phases. The mechanisms of high strain deformation are discussed.

## 1. INTRODUCTION

The potential of ceramics as wear-resistant materials is well recognised. Zirconia ceramics offer a combination of enhanced toughness, high hardness and good chemical inertness and should therefore offer excellent wear resistance. However, the published literature provides conflicting information. Some reports indicate that zirconia ceramics have good wear resistance (Hannink et al. 1984), while others workers have found poor wear resistance, even under mild sliding conditions (Birkby et al. 1989).

The role of transformation toughening in the wear of zirconia ceramics remains controversial. Many authors suggest that transformation of the tetragonal to monoclinic phase dominates the wear process (e.g. Birkby et al. 1989). However, the driving force for this transformation is strongly temperature dependent, its effectiveness decreasing as the temperature increases (Rainforth 1996). Rainforth & Stevens (1993) demonstrated that temperature rises generated by sliding of an Mg-PSZ against a steel counterface under mild conditions (0.24m/s speed, 6-50N load) were sufficient to limit transformation to within 200-400nm of the surface. Similarly, Rainforth & Stevens (1997) demonstrated dramatic surface changes for the dry sliding of a 3Y-TZP, including grain size refinement and gross plastic deformation of the tetragonal phase, but very little evidence of transformation from tetragonal to monoclinic zirconia.

The current study reports on cross-sectional TEM of a worn 3Y-TZP surface, tested under mild sliding conditions, in order to determine the depth and type of microstructural change and thereby to determine the wear mechanisms.

## 2. EXPERIMENTAL

The 3mol% yttria-TZP used in this study was manufactured using commercial Dai-ichi powders and was supplied in finished form by Dynamic Ceramic Ltd (pins, 10mm diameter with 30° truncated cone giving a 3mm diameter contact face). The contacting surface was diamond lapped to a high polish to remove residual stresses. X-ray diffraction confirmed that the surface was predominantly tetragonal zirconia, with no detectable monoclinic zirconia. Wear testing was undertaken on a tri-pin-on-disc machine, the full details of which are given elsewhere (Barceinas, 1997). The load of 10N/pin was applied as a dead weight, directly to the head, which was placed on top of the disc. The disc (a magnesia-partially stabilised zirconia, Mg-PSZ) was rotated at a speed of 0.24m/s. Samples for transmission electron microscopy (TEM) were prepared by cross-section (Barceinas, 1997). Prior to cross-section preparation, the surface was labelled with a sputter deposited layer of gold. All samples were examined in a Jeol 200CX microscope operating at 200keV.

## 3. RESULTS

The average wear coefficient was found to be $6.8 \times 10^{-4}$ mm$^3$/Nm, which is considered to be in the severe wear regime. SEM of the worn surface indicated extensive roughening, but no evidence of surface cracking.

A bright field image from a cross-section sample is given in Fig. 1. Three regions were present: a fine grained surface layer; a region of elongated tetragonal zirconia grains (grains 1-6 in Fig. 1) and finally a layer of monoclinic zirconia (grains 10-12).

The layer of monoclinic zirconia (grains 10-12) became progressively distorted as the surface was approached. Grain 8 contained monoclinic zirconia in the lower region, but was fully tetragonal in the upper region, Fig. 2. The monoclinic zirconia contained a high residual dislocation density. However, the diffraction conditions changed over very small distances due to sample buckling making analysis difficult, while techniques such as weak beam dark field proved impossible because of the specimen thickness.

In the outer ~2μm of the surface the tetragonal zirconia became elongated in the direction of sliding, with aspect ratio of grains increasing as the surface was approached. The extent of the elongation varied along the surface. Grains 1-6 in Fig. 1 exhibited relatively little elongation. However, several areas were identified where the elongation was much more severe, for example in Fig. 3. The average thickness of grains was 72nm (compared to the starting thickness of ~500nm). The elongated grains frequently contained Moiré fringes. A high dislocation density was found in the elongated tetragonal, although the change in diffraction conditions over a short distance and the thickness of the specimen precluded a thorough analysis. These regions exhibited a distinct texture with the $(111)_t$ planes oriented at an average of 6° to the sliding direction (indexed relative to the cubic fluorite unit cell, by convention). Extensive microcracking was present along grain boundaries.

The outer surface region, Fig. 1, contained fine (~6nm), randomly oriented tetragonal or cubic zirconia crystallites (selected area diffraction cannot differentiate between these two polymorphs).

## 4. DISCUSSION

The discussion will consider the microstructural changes in the order in which they occurred, i.e. progressing from below toward the outer surface. Analysis of the twin plane in the monoclinic zirconia ($1\bar{1}0$) indicated that it had been formed as a result of stress induced transformation from the tetragonal rather than spontaneous transformation on cooling. Thus, the stress at this depth must have been sufficient, and the temperature low enough, to drive the transformation. This indicates a temperature of less than about 200°C.

The monoclinic zirconia at this depth was distorted in the sliding direction by plastic deformation. The mechanism of deformation was difficult to determine, but must have been mechanically based, since the temperature was too low for diffusion based mechanisms. In grain No. 8, transformation of monoclinic to tetragonal zirconia had occurred, indicating a temperature at this depth of ~560°C, assuming equilibrium heating conditions.

The gross shape change in tetragonal zirconia had initiated at the depth of the monoclinic to tetragonal zirconia transformation. Thus, it is also unlikely that this shape change occurred entirely as a result of diffusional mechanisms since the temperature at this depth was too low to allow sufficient diffusion in the time available (for the wear rate of $6.8 \times 10^{-4}$ mm$^3$/Nm, the microstructural changes observed will have occurred in a time of ~17s). Moreover, the grain boundaries were smooth and free from microcracking, indicating conservation of volume. Therefore, dislocation flow is the only plausible explanation.

The outer surface consisted of a layer of fine (~6nm) zirconia grains. There was a sharply defined interface between the structure found in Fig. 3 to that in the upper region of Fig. 1 implying that this was wear debris, which had undergone further attrition between the two sliding surfaces after detachment as a wear particle, prior to re-attachment. The mechanisms of such grain size reduction are complex, but are believed to be similar to those observed in mechanical alloying (Rainforth & Stevens, 1997).

## 5.  CONCLUSIONS

Wear of the 3Y-TZP occurred by a complex mechanism. Transformation of the tetragonal to monoclinic zirconia only occurred at a depth of ~2.5μm below the surface. Evidence of plastic deformation within the monoclinic zirconia was found. Re-transformation of the monoclinic to the tetragonal phase was found at a depth of 1.8-2.5μm, indicating a temperature of ~560°C at this point. Tetragonal zirconia grains were found to have undergone extensive plastic deformation in the outer ~2μm of the surface, associated with a $(111)_t$ texture component. Some evidence of dislocation lines was found. The outer surface consisted of fine (~6nm), randomly oriented tetragonal or cubic zirconia crystallites, believed to be re-attached wear debris.

## REFERENCES

Barceinas-Sánchez JDO 1997 PhD Thesis, University of Sheffield
Birkby I Harrison P and Stevens R 1989 J. Europ. Ceram. Soc. **5,** 37
Hannink R Murray M and Scott H 1984 Wear **100,** 355
Rainforth W M 1996 Ceramics International, **22** 365-372.
Rainforth W M and Stevens R 1993 Wear **162-164,** 322
Rainforth W M and Stevens R 1997 J. Mater. Res. In press.

502

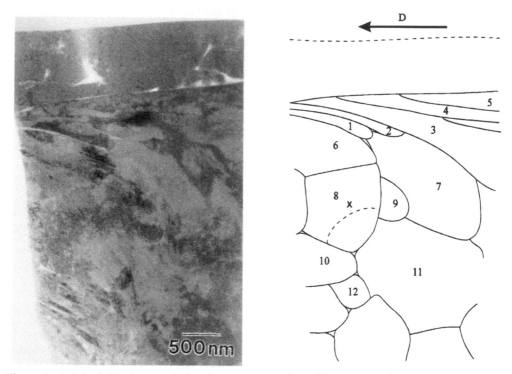

Figure 1. Bright field micrograph showing a cross-section of the worn surface (worn surface at the top) with schematic showing the grain numbers used for reference in the text.

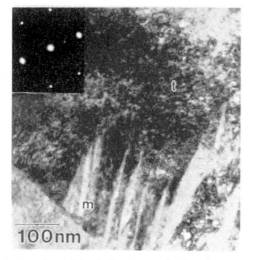

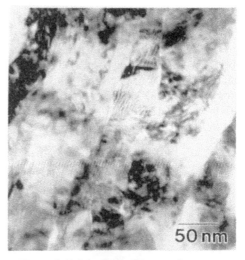

Figure 2. Bright field microgaph of grain No. 8 (from Fig. 1) showing distorted monoclinic (m) and tetragonal (t) zirconia within a single grain.

Figure 3. Bright field micrograph showing elongated tetragonal zirconia grains with grain boundary cracking.

*Inst. Phys. Conf. Ser. No 153: Section 11*
*Paper presented at Electron Microscopy and Analysis Group Conf. EMAG97, Cambridge, 1997*
© *1997 IOP Publishing Ltd*

503

# Dislocation generation during the sliding wear of ceramics

**X. Miao, WM Rainforth and WE Lee**

Department of Engineering Materials, University of Sheffield, Mappin St., Sheffield, S1 3JD

ABSTRACT: The microstructure generated by the dry sliding wear of zirconia-alumina-silicon carbide composites has been examined by TEM. Basal slip, with b=1/3<$11\bar{2}0$>, was identified in the SiC. Slip in the alumina was confined to shallow grooves, but the damage did not cross phase or grain boundaries. Basal and pyramidal slip was dominant, but no evidence of prism slip was found. Widespread stress assisted transformation of tetragonal to monoclinic zirconia occurred. Under abrasive grooves the tetragonal had not transformed but had been extensively deformed, with a subgrain size down to ~50nm.

## 1.    INTRODUCTION

The potential of ceramics as wear-resistant materials is well recognised. Zirconia ceramics offer a good combination of toughness, hardness and chemical inertness and should therefore offer excellent wear resistance. The mechanisms by which damage accumulation leads to the formation of a wear particle remains controversial, in particular, the extent to which dislocation flow is involved in this process is not known and little direct evidence has been presented.

Rainforth and Stevens (1997) and Barceinas and Rainforth (1997a) have identified extensive dislocation flow in the wear of high toughness zirconia ceramics. Similarly, Hockey (1971) has undertaken investigations of the response of both sapphire and polycrystalline alumina to indentation and abrasion. Damage mechanisms were found to depend strongly on crystallographic orientation. Hockey identified several deformation modes, including basal and pyramidal slip and basal and rhombohedral twinning. However, there remains little direct experimental evidence for dislocation flow during the sliding wear of ceramics, and where it occurs, what effect it has on the wear mechanisms.

## 2.    EXPERIMENTAL

The ceramic composites used in this work were manufactured by hot pressing (Miao et al. 1996). Two composites were investigated: a) 3mol%-yttria stabilised tetragonal zirconia (3Y-TZP) matrix with 5-30 vol% SiC platelets and b) a fine-grained alumina matrix containing 10vol% $ZrO_2$ and 20 & 30 vol% SiC platelets. The contacting surfaces were diamond lapped to a high polish to remove residual stresses. X-ray diffraction confirmed that the starting zirconia surface had tetragonal symmetry. Wear testing was undertaken on a crossed cylinder machine with a WC counterface at a speed of 0.24m/s and loads in the range 6-30N. Samples for TEM were prepared by back-thinning, with the worn surface protected from ion milling or re-deposition throughout. All samples were examined in a Jeol 200CX microscope.

## 3.    RESULTS

The microstructure and mechanical properties of the starting materials are described in detail elsewhere [5]. The wear rates obtained ($2-8 \times 10^{-7}$ mm$^3$/Nm) are typical for the unlubricated sliding wear of ceramics. Extensive cracking was observed at all SiC/matrix interfaces. Transgranular cracking was also present within many SiC platelets. The surface of the SiC contained numerous fine abrasive grooves. These were equally obvious in the alumina matrix, but not in the zirconia matrix. However, the latter exhibited evidence of local scuffing.

TEM of the zirconia composite worn surface indicated that widespread stress assisted transformation of tetragonal to monoclinic zirconia had occurred. These regions contained extensive microcracking. Surface grooving was associated with a substantial change in microstructure, Fig. 1. These regions had tetragonal symmetry, unlike the surrounding grains which were monoclinic zirconia. Dark field micrographs and selected area diffraction indicated a subgrain grain structure with an average subgrain size of ~50nm and a total rotation of 12° within a pre-existing 600nm grain. The residual strain was too great to image individual dislocations within the structure. Moreover, regions sufficiently thin to allow weak beam microscopy tended to disintegrate prior to examination of the microscope, precluding this technique.

Extensive cracking was observed in the SiC, Fig. 2, and was always associated with a locally high dislocation density. Analysis of individual dislocation lines indicated slip on the basal plane with b=1/3<1 1 2 0>. Numerous basal stacking faults were also observed. Although only basal slip was identified, the dislocation density in some areas was too high to determine the structure and therefore other slip planes may have been active (again, techniques such as weak beam were not possible because of the thickness of the specimen). Dislocations in the SiC were not always associated with surface cracks. Locally high densities were observed and these were usually found at the base of abrasive grooves (as seen in SEM images and identified by the reduced thickness of the thin sample), Fig. 3. Again, only basal slip was observed, although locally dislocation densities were too high to allow a full analysis.

The damage in the SiC platelets in the alumina composites was similar to that found in the zirconia matrix composites. Extensive dislocation damage was observed in the alumina matrix. This was exclusively associated with the base of abrasive grooves, Fig. 4. Interestingly, the damage associated with an abrasive groove changed across a grain boundary; some grains exhibited severe cracking and a high dislocation density while the adjacent grain only exhibited minimal damage. Similarly, an abrasive groove crossing from an alumina grain to a SiC platelet always exhibited more extensive damage in alumina than the SiC. Limited analysis of the slip system suggested that pyramidal slip was dominant, with occasional basal slip also present. The majority of zirconia grains in the alumina matrix had transformed to monoclinic symmetry, frequently containing cracking. Cracking was also prevalent along the SiC/matrix interfaces.

## 4.    DISCUSSION

The damage mechanisms were significantly different in the three phases examined. The most common damage mechanism in the zirconia matrix composite was the stress assisted transformation of tetragonal to monoclinic zirconia, which should promote surface compressive stresses and therefore reduce the wear rate (Rainforth 1996). Abrasive grooving had resulted in extensive deformation of the tetragonal phase. This structure was difficult to investigate in plan view. A similar structure was found in plan view by Rainforth & Stevens (1997) and Barceinas & Rainforth (1997a) working on zirconia monoliths. When viewed in cross-section, extensive dislocation glide was identified by these workers. Interestingly, the regions shown in Fig. 1 had not transformed to monoclinic symmetry, indicating local temperature rises sufficient to remove the driving force for the stress assisted transformation.

The SiC was extensively cracked, with a high dislocation density local to the cracks. However, there was also extensive dislocation activity below abrasive grooves. While dislocation glide can dissipate the surface stresses and therefore reduce the possibility of crack initiation, dislocation pile-ups can also be a source of time dependent damage accumulation leading to crack initiation. It was not clear whether the dislocation activity close to the cracks in the SiC arose from crack propagation or had been part of the crack initiation process.

The damage in the alumina grains was similar to that observed by Barceinas and Rainforth (1997b) in the water lubricated sliding wear of an alumina monolith. The dislocation damage was largely confined to the abrasive grooves, but the extent of damage changed dramatically from one grain to another. Thus, the damage induced by an abrasive strongly depended on the crystallographic orientation of the alumina grain. Grain boundaries pile-ups of dislocations produced in this way have been shown to initiate grain boundary cracking, which can ultimately promote a transition to catastrophic wear (Barceinas & Rainforth 1997b).

The observation of pyramidal slip in the alumina is at variance with classical deformation studies of alumina which suggest that prism slip is favoured at low temperature, basal at high temperature, but that pyramidal slip is the least favoured system at all temperatures. The observation is believed to be associated with one or more of: a) the higher strain rates found in wear compared to classical deformation studies, b) free surface effects and c) absorption of water which alters the critically resolved shear stress for that system.

## 5.    CONCLUSIONS

Damage in the tetragonal zirconia matrix consisted of two forms: transformation to monoclinic symmetry; extensive deformation of the tetragonal phase under abrasive grooves, considered to result from extensive dislocation flow. Extensive cracking was found in the SiC, usually associated with basal slip and basal stacking faults. Dislocation flow was found at the base of abrasive grooves, but there was no evidence to suggest that this was associated with crack initiation. Dislocation flow in the $\alpha$-alumina was associated with the formation of abrasive grooves. The extent of damage from abrasive grooves changed significantly across a grain or phase boundary, indicating the importance of crystal orientation as well as crystal structure. Slip was predominantly on pyramidal planes, with some basal slip, but no prism slip.

## REFERENCES

Barceinas-Sánchez JDO and Rainforth WM 1997a Submitted to J. Amer. Ceram. Soc.
Barceinas-Sánchez JDO and Rainforth W M 1997b Submitted to Acta Mater.
Hockey, BJ 1971 J. Am. Ceram. Soc. 5 223
Miao X Rainforth WM and Lee WE 1997 J. European Ceram. Soc. 17 913-920
Rainforth W M 1996 Ceramics International, 22 365-372.
Rainforth W M and Stevens R 1997 J. Mater. Res. In press.

Figure 1. Bright field micrograph showing deformation of tetragonal zirconia below an abrasive groove.

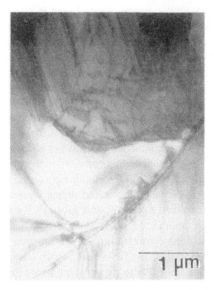

Figure 2. Bright field micrograph showing a transgranular crack and associated basal slip

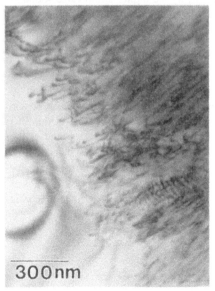

Figure 3. Basal slip beneath an abrasive groove in a SiC platelet.

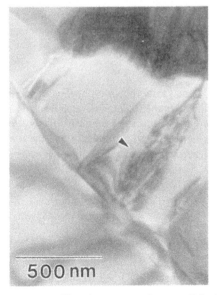

Figure 4. Abrasive groove (arrowed) in the alumina with a local high dislocation density. There is no damage in the SiC.

Inst. Phys. Conf. Ser. No 153: Section 11
Paper presented at Electron Microscopy and Analysis Group Conf. EMAG97, Cambridge, 1997
© 1997 IOP Publishing Ltd

# Grain-boundary segregation of Ho in BaTiO$_3$

## A Woonbumroong, P D Brown and C B Boothroyd

Department of Materials Science and Metallurgy, University of Cambridge, Pembroke Street, Cambridge, CB2 3QZ, UK

**ABSTRACT:** Positive temperature coefficient of resistance BaTiO$_3$ ceramics containing different donor dopant concentrations of holmium have been studied. An increase in donor dopant concentration from 0.1 to 0.8 at. % Ho results in a reduction in mean grain diameter of an order of magnitude. The grain-boundary segregation of donors has been observed and analysed using energy dispersive X-ray spectroscopy and energy filtered imaging. The potential at the boundary as determined by Fresnel contrast analysis was found to decrease for the higher Ho content sample.

## 1. INTRODUCTION

The perovskite-structure compound BaTiO$_3$ is a technologically important electroceramic widely used as dielectrics, thermistors and ferroelectrics. Donor doped BaTiO$_3$ ceramics exhibit varistor and positive temperature coefficient of resistor properties when sintered in air. Its non-linear electrical property is attributed to the existence of a potential barrier at the grain boundaries (Heywang 1961). Thus grain boundary properties are of great interest since in applications, the titanates depend on highly engineered grain boundary electrical barriers. An understanding of the nature of dopant segregation phenomena and the related changes in the electrical activity and electronic structure at the boundaries in this material is therefore crucial. In this paper we use EDX, energy filtered imaging and Fresnel contrast analysis to investigate the grain boundaries in Ho doped BaTiO$_3$.

## 2. EXPERIMENTAL DETAILS

Polycrystalline BaTiO$_3$ specimens doped with 0.1 and 0.8 at. % Ho were examined. The specimens were prepared by standard mixed oxide techniques and sintered in air with the formulation Ba$_{1-x}$Ho$_x$TiO$_3$. Commercial BaTiO$_3$ was used as the precursor material, with Ho, added in the form of Ho$_2$O$_3$, used as the donor dopant. Sufficient TiO$_2$ was added in order to ensure liquid-phase sintering. A small quantity of SiO$_2$ was also added as a sintering aid. Sintering procedures have been described in greater details elsewhere (Al-Allak et al 1988a). Samples were prepared for transmission electron microscopy using standard ion beam milling procedures and examined using a Philips CM30, VG HB501 scanning transmission electron microscope and JEOL 4000FX equipped with a Gatan imaging filter.

## 3. RESULTS AND DISCUSSION

An increase in donor dopant concentration from 0.1 to 0.8 at. % results in a reduction in the mean grain diameter from 40 μm to 2 μm. It has been shown that at high dopant concentrations, there is a considerable increase in low-temperature resistivity which dielectric measurements have demonstrated to be purely a grain boundary phenomenon (Al-Allak et al 1988a).

In order to obtain information about the changes in the concentration of the constituent elements across the boundaries, EDX analysis was carried out at 100 kV in a VG HB501 STEM by stepping a probe of size ~2 nm across ten grain boundaries of vertical but otherwise random orientation. The concentration profiles obtained from EDX shown in Fig. 1 depict preferential Ho segregation at the grain boundaries examined and provide estimates of the boundary widths to be 2.2 and 1.9 nm (± 0.2 nm) for specimens doped with 0.1 and 0.8 at. % Ho respectively.

Diffuse dark field images from both samples were obtained with the objective aperture placed well away from any diffraction spots. Fig. 2 illustrates clearly in the form of bright contrast that

508

the boundary does indeed contain a layer of amorphous material. The widths of the boundaries estimated from ten such observations are 3 nm and 2 nm (± 0.2 nm) for samples doped with 0.1 and 0.8 at. % Ho respectively.

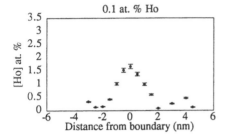
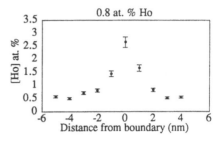

Fig. 1 STEM grain-boundary concentration profiles of Ho-doped BaTiO3 showing segregation of Ho at the boundaries.

Energy filtered images were obtained using the Ho edge at 1334 eV. Because the Ho edge has a strong initial peak three images were captured, one before the ionisation edge, one on the sharp white line peaks and another one after the edge. The spectra were quantified by averaging the intensity from the images before and after the edge then subtracting this from the window of interest. Fig. 3 shows the Ho map from the grain boundary of the specimen doped with 0.8 at. % Ho, confirming that the segregation of donor dopants to the boundary is uniform along the length of the boundary. Unfortunately, in the 0.1 at. % Ho specimen the Ho was below the limit of detectability at the boundary.

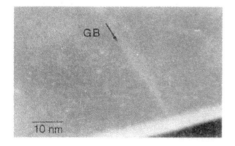
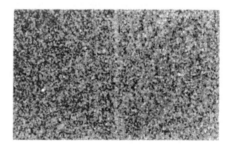

Fig. 2 Diffuse dark field image of a grain boundary in the 0.8 at. % Ho specimen.

Fig. 3 Ho energy filtered map for the 0.8 at. % Ho specimen.

Energy-filtered Fresnel series of the boundaries in each specimen were obtained at 397 kV with an energy selecting slit width of 10 eV, and the point spread function was deconvoluted from each image. A selection of the defoci analysed is shown in Fig. 4 after projecting parallel to the boundaries, together with the best fitting simulated profiles calculated using the multislice fitting procedure of Dunin-Borkowski (1997). The boundaries in both specimens exhibit a bright central fringe underfocus and a dark central fringe overfocus which is a characteristic of a lower scattering potential at the boundary relative to the surrounding bulk material. The contrast from the specimen with the higher amount of dopant is stronger than that from the specimen with less dopant. The best fitting potential profiles to the experimental data are shown in Fig. 5. The corresponding parameters describing the potential well widths and depths are given in Table 1. It is apparent that the magnitude of the real part of the potential at the boundary of the 0.1 at. % Ho specimen is smaller than that of the 0.8 at. % Ho specimen and reassuringly the potential profiles obtained from two different boundaries in the 0.8 at. % Ho sample are similar, despite the difference in Fresnel fringes of Fig. 4 (b) and (c).

The form of the imaginary part of the potential describes scattering of electrons to large angles. It seems that both specimens scatter to a similar extent, as would be expected if this scattering was due to the presence of an amorphous layer whose width is similar in both specimens. The well width was not found to vary significantly with the change in dopant concentration. Both fitted potential had a well width of $2 \pm 0.1$ nm. This is in agreement with the estimated widths obtained from the EDX concentration profiles.

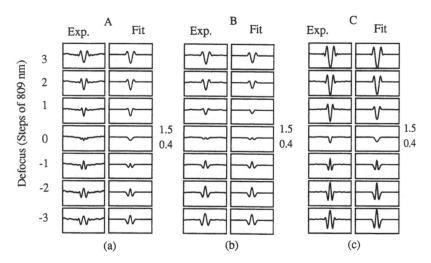

Fig. 4 One-dimensional projected Fresnel fringe profiles for specimens doped with (a) 0.1 at. % Ho, (b) and (c) 0.8 at. % Ho.

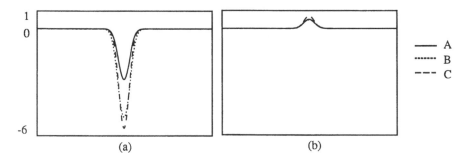

Fig. 5 (a) Variation in the real part and (b) the imaginary part of the potential.

Table 1 Parameters describing the best-fitting potential profiles

| [Ho] at. % | Set of profiles | $\Delta V$ (V) | fwhm (nm) | Absorption (fraction) |
|---|---|---|---|---|
| 0.1 | A | -2.84 | 1.79 | -0.18 |
| 0.8 | B | -4.96 | 1.97 | -0.10 |
| 0.8 | C | -5.57 | 1.75 | -0.13 |

There are many possible origins for the change in the real part of the scattering potential at an interface and several factors may be present at once. Three factors which could contribute to the change in potential considered are the change in scattering factor, the density and the effect of space charge. The effect of a change in composition on the potential will be considered as the two extreme cases where the material is fully neutral and fully ionic. As shown in Table 2, the calculated values of $V_0$ for fully neutral and fully ionic for pure $BaTiO_3$ with a tetragonal unit cell of side $a = 0.3994$ nm and $c = 0.4038$ nm are 24.37 V and 16.31 V respectively. Thus an increase in ionic character at the grain boundary without any change in composition leads to a decrease in scattering potential. The effect of composition on the electron scattering factor was considered in two cases where Ho substitutes for Ba and where Ho substitutes for both Ba and Ti (Takada et al). The values of $V_0$ assuming crystalline $BaTiO_3$ tabulated in Table 2 indicate that the effect of composition cannot solely be the cause of the observed difference in the magnitude of the real part of the potential between specimens doped with 0.1 and 0.8 at. % Ho obtained from the simulated profiles.

Table 2 Values of $V_0$ as affected by the composition, assuming crystalline $BaTiO_3$

|  | [Ho] at. % nominal | [Ho] at. % at boundary | $V_0$ (V) Ho substitutes for Ba | Ho substitutes for Ba and Ti |
|---|---|---|---|---|
| Fully neutral | pure $BaTiO_3$ |  | 24.37 | 24.37 |
|  | 0.1 | 1.7 | 24.32 | 24.21 |
|  | 0.8 | 2.7 | 24.29 | 24.12 |
| Fully ionic | pure $BaTiO_3$ |  | 16.31 | 16.31 |
|  | 0.1 | 1.7 | 16.29 | 16.26 |
|  | 0.8 | 2.7 | 16.25 | 16.23 |

Given that the differences in densities of fully crystalline and fully amorphous ceramics of identical compositions are usually of the order of 10% (Kara et al 1996), the values of $V_0$ when taking into account the amorphous layer at the boundary would be 10% less than $V_0$ of crystalline $BaTiO_3$. It is apparent that the effect of the amorphous layer does indeed explain the decrease in $V_0$ but predicts the same magnitude for the decrease in $V_0$ for both samples.

While no single contribution of the above can account for the experimentally measured values of $\Delta V$, a combination of changes in the scattering factor and a decrease in density could only do so in the unlikely event that the density changed significantly with Ho content.

One possible reason for the difference in potential between the two specimens is the presence of a space charge at the boundary. However, the relative permittivity of doped $BaTiO_3$ is large and varies enormously with Ho concentration from $4 \times 10^4$ for 0.1 at. % Ho to $1.5 \times 10^5$ for 0.8 at. % Ho (Al-Allak et al 1988b). This means that if the potential is solely affected by the contribution from space charge, an enormous amount of space charge would be needed to affect the potential at all. Also, if a boundary is treated as a strong depletion, slight inversion, back-to-back Schottky barrier then the Debye length, $L_D$, can be used as a measure of the boundary width (Desu and Payne 1990). Values of $2L_D$ for specimens doped with 0.1 and 0.8 at. % Ho calculated for temperature of 300 K assuming a value of the effective charge of +1 are 120 nm and 84 nm respectively, much larger than the observed potential well width of 2 nm. Thus it appears that no realistic space charge can account for the difference in boundary potential between the two specimens.

## 4. CONCLUSIONS

We can conclude that Ho in $BaTiO_3$ segregated strongly to the grain boundaries, which are at least partly amorphous and have a width of about 2 nm. The measured $V_0$ at the boundary decreases in proportion to the measured Ho segregation but this change in $V_0$ can only be explained if the added Ho is affecting the bonding at the boundary by e.g. making it more metallic.

### Acknowledgements
We would like to thank H M Al-Allak, University of Durham for the provision of the samples and Prof A H Windle for the provision of laboratory facilities. We are indebted to the late W M Stobbs for many useful discussions.

### References

Al-Allak H M, Brinkman A W, Russell G J, Roberts A W and Woods J 1988a J. Phys. D:Appl. Phys. **21** 1226
Al-Allak H M, Illingsworth J, Brinkman A W, Russell G J and Woods J 1988b J. Appl. Phys. **64** (11) 6477
Desu S B and Payne D A 1990 J. Am. Ceram. Soc. **73** (11) 3391
Dunin-Borkowski R E, Mao Z, Davis C A and Knowles K M 1997 in preparation
Heywang W 1961 Solid State Electron. **3** 51
Kara F, Dunin-Borkowski R E, Boothroyd C B, Little J A and Stobbs W M 1996 Ultramicroscopy **66** (1-2) 59
Takada K, Chang E and Smyth D N Advances in Ceramics, eds J B Blum and W R Cannon (Amer. Ceram. Soc.) 2ed.

*Inst. Phys. Conf. Ser. No 153: Section 11*
*Paper presented at Electron Microscopy and Analysis Group Conf. EMAG97, Cambridge, 1997*
© *1997 IOP Publishing Ltd*

# Electron microscopy studies of petroleum contaminated marine clay

**A Tuncan**

Faculty of Engineering and Architecture, Department of Environmental Engineering, Anadolu University, Yunusemre Kampusu, 26470, Eskisehir, Turkey

ABSTRACT: The effects of crude oil on physico-chemical and microstructure properties of laboratory prepared marine clays were studied. Crude oil significantly influenced clay microstructure and physico-chemical properties such as specific surface area and cation exchange capacity. Experiments were conducted on artificially prepared marine sediments to investigate this phenomenon. Studying the micrographs of oil contaminated marine clay, two types of adsorption mechanisms were observed. One was the adsorption of clay particles by the oil and the other one was the adsorption of oil by the clay particles. As a result of these mechanisms three different interactions were observed between oil and clay particles. These were the spherical agglomeration of clay particles, clay coating of the oil drops and adhesion of oil layers or sheet structures to clay surfaces.

## 1.    INTRODUCTION

The impact of progressive petroleum hydrocarbon contamination on the world oceans has become one of the most important problems for scientists to solve. Petroleum pollution in the ocean waters may adversely affect the marine environment and the ocean floor. Petroleum hydrocarbon may influence the development of clay microstructure and thus physico-chemical properties of the marine sediments. Petroleum hydrocarbon contamination of ocean sediments may occur through a variety of sources such as the following: tanker accidents, discharge from coastal facilities, offshore petroleum production facilities and natural seepage (Geyer 1980). The environmental effects of a petroleum spill in the ocean depends not only on the chemical and physical characteristics of the sediment but also on the characteristics of the petroleum product.

Petroleum is mostly composed of hydrocarbons which are immiscible with water and are chemically inactive compounds. Hydrocarbons are basically compounds of carbon and hydrogen. When a hydrocarbon such as crude oil is released into water, it spreads and floats on the surface to form an oil slick due to its low specific gravity and low solubility. Oil particles are dispersed in the water column and come into contact with suspended clay and colloidal particles. The suspended clay particles and sea water act as transfer agents for the oil to the bottom sediments. Clay particles can adsorb large quantities of hydrocarbons. This adsorption increases with salinity. If the quantity of suspended particles is high, so is the extent of oil transfer to the bottom sediments. Oil wets the clay particle surfaces and also agglomerates the particles which then subsequently sink to the bottom with increased rate of sedimentation. Suspended clay particles enter the marine environment by rivers, coastal erosion, aeolian transport and submarine volcanic eruptions. The physico-chemistry of these particles is important in understanding the behavior of marine sediments which cover the sea floor.

Sediment microstructure is a function of the clay fabric (orientation and arrangement of particles) and also the physico-chemistry of the clay-water system. Physico-chemical interactions are electrostatic in character. These interactions are responsible for repulsion and van der Waals' attractive forces between the clay particles in suspension. Changing environmental conditions in the marine water column may have significant effects on the microstructures of marine clays. If these effects are understood in any given case, the sediment behavior can be estimated with a sufficient degree of consistency. Such estimations may be important in preventing slope instability in the nearshore, environmental damage to harbor structures, foundations of offshore oil drilling platforms, oil storage reservoirs and underwater structures.

## 2. INVESTIGATION

### 2.1 Materials

Marine clay was prepared using a clay mixture of 50% illite, 21% Ca-Montmorillonite, 16% kaolinite and 13% chlorite, by weight. This mixture was called the "marine clay". Crude oil is a form of hydrocarbon which is immiscible with water. The marine clay was premixed with saline water to make a slurry. A commercially available salt mixture called "Instant Ocean" was added to tap water to simulate ocean water with a typical salt concentration of 38g/l.

### 2.2 Experimental Procedure

The experimental set up consisted of 3 rectangular sedimentation tanks. The sedimentation tanks were made up of translucent plastic sheets of 0.5 cm thickness. They were 0.3 m by 0.3 m in cross section and 0.6 m in height. The amount of the contaminating agent was measured as weight percentage of the dry clay and premixed into the clay with salt water to make a slurry mixture. The proportion of these contaminants were selected as 5%, 10% and 15% by dry weight of the clay.

The undisturbed samples were used to prepare specimens for scanning electron microscope (ETEC SEM) analysis. The SEM specimens were prepared by critical point drying technique as outlined by Bennett et al (1977). They were initially treated with acetone and a critical point drying apparatus was utilized to replace the acetone with $CO_2$. The specimens were then kept in a high vacuum desiccator for extended periods of time to ensure the removal of all the water from the pore space.

## 3. EXPERIMENTAL RESULTS

### 3.1 Physico-Chemical Properties

Physico-chemical properties of clay particles are dependent on particle surface characteristics, such as specific surface area (SSA) and cation exchange capacity (CEC). Specific surface area of clay particles is an important property which influences the particles' behavior. Surface area was determined using a Monosorb Surface Area Analyzer. The theoretical basis upon which the Monosorb operates is the Brunauer, Emmett and Teller theory (Brunauer, Emmett and Teller 1938). Cation exchange capacity is an important fundamental property useful in understanding the development of electrical charges between the clay particles and also between clay particles and other substances. Cations are attracted and held onto the clay and colloidal surfaces and also to the edges to preserve the electrical neutrality of the system. These cations are exchangeable cations because they can be replaced by cations of another type. This substitution gives a net negative charge to the clay particles. CEC was determined by the sodium saturation method (Chapman 1965).

Fig. 1 shows the relationships between the specific surface area and cation exchange capacity of the marine clay mixture and the percentage of crude oil added. Specific surface area decreases with increasing percentage of crude oil. Organic matter is assumed to coat and agglomerate the clay particles. This brings small particles together creating larger particles. The specific surface area of 15% crude oil mixed marine clay is around 1 $m^2$/g. This is similar to the specific surface area of fine sand. It was observed that the color of crude oil mixed marine clay was brownish and the dried clay flocs could not be crushed easily probably due to the "gluing" effect of the oil. This phenomena may be similar to the "spherical agglomeration" discussed by Puddington and Sparks (1975). CEC decreases with increasing percentage of the crude oil. Crude oil is a petroleum hydrocarbon, and the adsorption of these hydrocarbon chains onto the clay surfaces causes wettability to change from water-wet to oil-wet. The hydrocarbon chains attach themselves to the clay surface and displace the previously adsorbed water molecules. The surfaces of the clay particles are then coated by the hydrocarbon and become wet by oil. The adsorption of these components creates a layer around the clay particles which is not water soluble and is not displaced by water. Thus most of the exchange sites become coated by large hydrocarbon molecules.

### 3.2 Scanning Electron Microscope (SEM) Results

Spherical agglomeration observed in the sedimented specimens of marine clay micrograph can be seen in Fig. 2. The clay particles in water immiscible liquids are believed to be randomly arranged and the agglomerations are more or less spherical in shape. When oil drops and clay particles are mechanically agitated as in the case of preparing oil mixed water suspensions of clay in the settling columns, the total energy of interaction between the double layers together with the forces due to particle and oil drop motion is responsible for bridging the oil drops and clay particles into sufficiently close proximity. An interesting product of this process is observed in Fig.2, where the inside of a sherical shell of clay is exposed when the oil drop was removed by some means. The spherical agglomeration of the clay particles along with adhesion and coating of these particles on an oil drop surface is evident in this feature. Different interactions were also observed. One is the clay particles coating the surfaces of oil drops as shown in Fig. 3. The average diameter of these spherical drops ranged from a fraction of a micron to several microns. The second interaction is the adhesion of clay particles to oil layers. The coating of oil surfaces with clay particles is evident in the micrograph in Fig. 4.

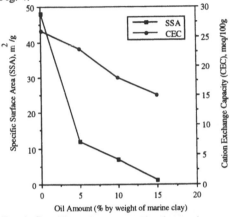

Fig. 1 Specific surface area and cation exchange capacity versus oil amount for salt water mixed marine clay.

Fig. 2 The SEM of salt water and crude oil mixed marine clay sedimented specimens. (S: spherical agglomeration)

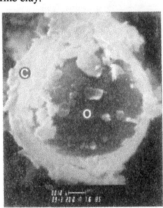

Fig. 3 The SEM of salt water and crude oil mixed marine clay sedimented specimens. (O: oil drop, C: clay particles)

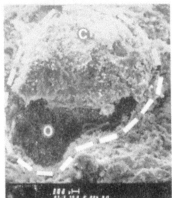

Fig. 4 The SEM of salt water and crude oil mixed marine clay sedimented specimens. (O: oil surface, C: clay surface)

Figures 5 and 7 show the x-ray diffraction spectrums (EDS) of clay particles and oil surfaces, respectively. It should be noted here that the x-ray diffraction spectrum records of the so called oil surfaces indicate no dominant presence of a mineral and therefore are interpreted as oil. Another mechanism is when the oil is incorporated into the sediment as an irregular shaped mass or a thin layer of material, as shown in Fig. 6. There is adhesion between these oil surfaces and the clay particles surrounding them. Such oil surfaces act as blockades and prevent water flowing through the soil voids, this therefore results in the net effect of lowered permeability. Crude oil mixed marine clay appeared to have a "spongy" structure in general as observed in Fig. 8. This is probably due to the agglomeration caused by oil. Finally, visual observation and handling of the crude oil mixed marine clay specimens indicated that crude oil reduces cohesion of the clay which is most probably due to the agglomeration of clay into flocs and significant reduction of the effective specific surface area of the material.

514

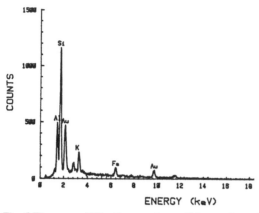

Fig. 5 The x-ray diffraction spectrum of clay surface. (refer to Fig. 4)

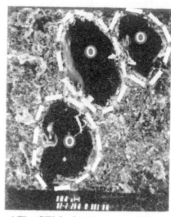

Fig. 6 The SEM of salt water and crude oil mixed marine clay sedimented specimens. (O: oil surface)

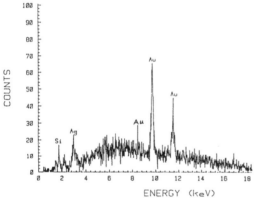

Fig. 7 The x-ray diffraction spectrum of oil surface. (refer to Fig. 4)

Fig. 8 The SEM of salt water and crude oil mixed marine clay sedimented specimens. (O: oil surface, S: spongy structure)

## 4. CONCLUSIONS

This study shows that crude oil has a significant influence on physico-chemical properties and microstructure of marine sediments. Emphasis was given to the impact of crude oil on the formation of clay microstructure and its variation which would in turn influence its physico-chemical properties. The crude oil reduced the specific surface area, cation exchange capacity and double layer thickness of the clay mixtures. Two types of adsorption mechanisms were observed. One is the adsorption of clay particles by the oil and the other one is the adsorption of oil by the clay particles. As a result of these mechanisms three different interactions were observed between oil and clay particles. These were the spherical agglomeration of clay particles, clay coating of the oil drops and adhesion of oil layers or sheet structures to clay surfaces.

## REFERENCES

Brunauer S, Emmett P H and Teller E 1938 Adsorption of Gases in Multimolecular Layers, Journal of American Chemical Society, Vol 60, pp 309-319
Chapman H D 1965 Cation Exchange Capacity, in Methods of Soil Analysis, ed C A Black Part 2 No 9, in the Series Agronomy; American Institute of Agronomy pp 891-901
Geyer R A 1980 Marine Environmental Pollution, 1:Hydrocarbons, Elsevier Scientific Publication Co
Puddington I E. and Sparks B D 1975 Spherical Agglomeration Processes, Minerals Science Engineering, Vol 7, No 3, pp 282-288

*Inst. Phys. Conf. Ser. No 153: Section 11*
*Paper presented at Electron Microscopy and Analysis Group Conf. EMAG97, Cambridge, 1997*
© *1997 IOP Publishing Ltd*

# Carbon-based materials studied by PEELS

R Brydson, X Jiang, A Westwood, S Collins, S Lu and B Rand

School of Process, Environmental and Materials Engineering, Univ. of Leeds, Leeds LS2 9JT.

ABSTRACT: PEELS has been used to characterize boron-carbon-nitrogen alloys, carbon microtubes and interfacial reaction products in a range of C/SiC composites. Particular attention was paid to the compositional homogeneity over nanometre lengthscales and to the extraction of bonding information which are both of prime importance for properties such as strength and oxidation resistance as well as for subsequent materials processing.

## 1. INTRODUCTION

Carbon-based materials offer extremely high strength to weight ratios for structural applications. Graphite is intrinsically anisotropic and a major feature in the fabrication of synthetic materials is the control of anisotropy in the various polycrystalline forms that can be produced. Manipulation of the discotic nematic liquid crystal phase that develops in polyaromatic liquids such as pitch, allows the utilisation of the exceptional mechanical, electrical and thermal properties of the graphite basal plane in oriented products such as fibres. However, a number of problems still remain such as the protection of these materials against oxidation at elevated service temperatures and the control of the bonding at fibre-matrix interfaces in carbon fibre composite materials. The former may be achieved via the incorporation of glass-forming elements into the carbon structure to form a carbon-ceramic alloy which has self-healing characteristics, while the latter is often critically affected by impurities present either within the fibre or introduced during consolidation of the composite which segregate to the fibre/matrix interface. The light element sensitivity of PEELS and electron spectropic imaging (ESI) combined with their inherent spatial resolution can provide invaluable information on elemental distributions in such materials at the required scale of nanometres. In addition, the electron loss near-edge structure (ELNES) can highlight local chemistries, bonding mechanisms and ordering at a similar level.

## 2. CARBON CERAMIC ALLOYS

These materials are new ceramics with inherent oxidation resistance and potentially attractive mechanical and thermal properties, when compared with those of analogous carbon-only materials. In CBN alloys the overall aim is to replace C-C links in processible carbonaceous pitch with isostructural and isoelectronic B-N units. The inclusion of B-N units imparts oxidation resistance to CBN pitch/cokes and thus mechanical properties are expected to be retained in products formed from these materials such as fibres. CBN alloys were made by pyrolysis of novel precursors containing $sp^2$-bonded B, $sp^2$-N and $sp^3$-C, ethyl borazines (containing BN rings) formed from the reaction of of $B_2H_6$ with acetonitrile. Further CBN alloys were formed from borazarene-based precursors (containing isolated BN units) derived from the 2,2 diaminophenyl/ $BCl_3$ adduct which additionally contained $sp^2$-bonded C. Pyrolysis was performed under dry $N_2$ at temperatures ranging between 773 K and 1273 K.

### 2.1 Borazine-based pitch

PEELS analysis gave an average composition of 33% B, 25% C, 28% N and 10% O (the latter resulting from hydrolysis). The slightly lower N content, relative to B, is believed to be as a result of preferential N loss as volatile amines during pyrolysis. Variations in boron, carbon and nitrogen levels within crystallites were observed over a scale of tens of nanometres. These were visible as bright and dark regions in the HAADF STEM image.

B K-, C K- and N K-ELNES from the alloy (Figure 1) all show a sharp leading $\pi^*$ peak corresponding to transitions to a p antibonding MO which is indicative of the presence of $sp^2$ bonding. The degree of $sp^2$ bonding for a particular element may be determined via measurement of the normalized $\pi^*$ peak area [2]. This normalized $\pi^*$ peak area is then compared with that obtained from a reference compound known to contain completely $\pi$-bonding, such as graphite or hexagonal BN. This procedure gives values close to 100 % for all the various elements in the CBN alloy indicating a fully $sp^2$-bonded structure.

The B K- and N K-ELNES of the CBN alloy prepared by this route are similar to that found in hexagonal BN, although they exhibit a slightly broader and less detailed structure. Similarly the C K-ELNES is much less detailed than that found in oriented graphite (Figure 1)and may be likened to that of turbostratic graphite. This is as expected since the samples had only been heat-treated to relatively low temperatures. Full ordering (graphitization), if it occurs at all, would only take place at temperatures above 2773 K.

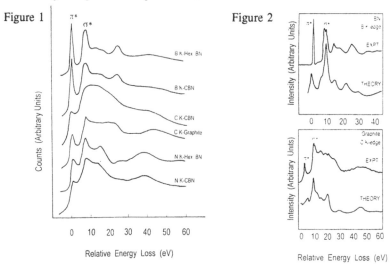

Figure 1

Figure 2

Relative Energy Loss (eV)

Relative Energy Loss (eV)

Figure 2 shows the B K- and C K-ELNES from hexagonal BN and highly oriented pyrolytic graphite together with the theoretical results of large, cluster-based MS calculations of the ELNES [3]. The clusters employed consisted of 120 and 145 atoms for hex. BN and C graphite respectively. By investigating the results obtained from smaller clusters and restricting the number of scattering events undergone by the excited electron, it is possible to identify the origin of specific ELNES features. These additional calculations show that, although the majority of the features are determined by the atomic arrangement of atoms in a single graphene or BN layer, fine spectral details are determined by inter-layer correlations. This is especially true of the sharp $\sigma^*$ feature ca. 6 eV above the $\pi^*$ peak in graphite. Furthermore the spectra all show a feature some 40-50 eV above the edge onset (arrowed in figure 2) which may be identified with scattering from the second shell (in graphite this consists of 6 next nearest neighbours in the a-b plane at a distance of 0.246 nm (0.250 nm in hex. BN)). Such a feature is known as a $\sigma^*$ scattering resonance and has been shown to be sensitive to the bond length, the energy of such a feature above the edge onset varying as the inverse square of the bond length. The positions of these scattering resonances for the B K- and N K-edges of hex. BN and the borazine-based CBN alloy are similar suggesting similar next nearest neighbour distances (NNND), whereas the C K-edge scattering resonance is at a lower energy for the CBN alloy as compared to graphite suggesting the presence of longer NNND in the alloy.

Comparing the B K- and N K-ELNES from hex. BN reveals a complementary aspect to the relative intensities of the various features which appears to be indicative of B-N bonding. Naively this may be understood as arising from the relative contributions of the two atomic species to the unoccupied MOs formed from a linear combination of atomic orbitals (LCAO). It is relatively clear that the B K- and N K-ELNES of the CBN alloy are complementary and this, together with the information on the NNND obtained from the position of the scattering resonance, suggests the presence of regions containing an sp²-bonded BN network separated by sp²-bonded C-rich regions. This is to be expected considering that the the borazine-based precursor used to form this alloy contains B-N ring structures. However, there is a lack of registry between the sp²-bonded layers which leads to a smearing out of the detailed ELNES. The latter is confirmed by relatively broad (0002) and (1000) reflections from a hexagonal graphitic-like structure in measured XRD patterns as well as arcing in measured electron diffraction patterns .

The tendency for 'phase' separation of C-C bonding from B-N bonding observed in CBN alloys has been eluded by others. The basis for this separation essentially follows from thermodynamic considerations.

## 2.2 Borazarene-based pitch

Pyrolysis of borazarene-based precursors gave CBN alloys with a much lower B and N content. Analysis of a pitch pyrolysed at 773 K gave an average composition 4.8 % B, 87.6 % C, 4.0 % N and 3.6 % O. Generally the relative N content was again lower than the relative B content, however compositions varied somewhat from region to region by around a factor of 2. Some isolated graphitic regions were observed as well as regions containing solely C and N. Generally the B and N contents were correlated as is revealed in the EELS linescan in figure 3. The level of oxygen was constant across the sample suggesting that was a surface oxide/hydrolysis product.

Figure 3 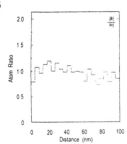  Figure 4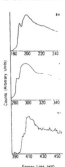

The B K-, C K- and N K-ELNES obtained from this alloy are shown in figure 4. All show a π* feature and again the respective π*/σ* ratios at the various edges suggested a completely sp²-bonded network. However, comparison of figures 1 and 4 reveals clear differences especially in the B K- and N K-ELNES which show very little structure in the case of the borazarene-based alloy. Furthermore, the scattering resonances are less distinct at the B K- and N K-edges, suggesting a poorly defined NNND. However, the energy positions suggest a longer NNND for boron and a shorter NNND for nitrogen as compared to that in hex. BN.

All this information points to the existence of isolated B-N units in an sp²-bonded C network with little registry between the graphitic layers in the borazarene-based CBN alloy which would follow from the nature of the precursor (in that it contains isolated B-N units).

## 3. VAPOUR GROWN CARBON MICROTUBES

Fibres produced by a catalysed continuous vapour growth technique were examined by TEM/STEM. The majority of the fibres consisted of a tubular structure, some 150-200 nm in diameter with a hollow centre some 20-50 nm in diameter (figure 5). The walls of the tube consisted of a turbostratic carbon structure, i.e. graphitic domains,with the c-axes generally parallel to the length of the tube, containing extremely low levels of Fe and Si impurities (< 1%).

C K-ELNES linescans across the fibre diameter, shown in figure 5, reveal the orientation and registry of the graphite layers. Figure 5, curve B shows the normalized intensity of the π* peak at 285 eV. The observed decrease is related to the changing orientation of the graphite planes relative to the scattering vector for a tubular morphology.

Figure 5, curve A shows the normalized intensity of the sharp σ* feature at 291 eV in the C K-ELNES both normalized to the higher lying σ* intensity between 294 and 309 eV. This feature is highly sensitive to the degree of layer registry. At the edges of the fibre there exists a large radius of curvature of the graphite layers giving generally good registry, while towards the centre of the fibre the registry decreases as the radius of curvature increases. These findings are essentially confirmed by HREM imaging of (0002) planes (figure 5).

518

Figure 5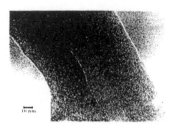

Figure 6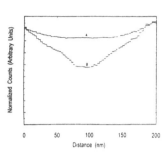

## 4. INTERFACE CHARACTERIZATION

In C fibre composite materials an important concern is the chemical interaction between the individual components during processing which may have a significant impact on overall properties. C/SiC composites offer a number of benefits over conventional C/C composites including significant improvements in oxidation resistance. The C/SiC composites studied here were prepared from commercial AS4 standard modulus surface treated (SMS) fibres wound with pitch solution to give a unidirectional green lamina which was pyrolysed at 1273 K. This was then repeatedly treated with polycarbosilane (PCS) ceramic precursor and pyrolysed at 1223 K until < 10% porosity was achieved. The final material was then heat treated at various temperatures between 1223 K and 1723 K. An important concern for overall mechanical properties and, in particular, load transfer are the reactions which occur at the fibre/matrix interface.

XPS studies of SMS fibre surfaces revealed appreciable levels of O and N, and figure 7 shows EELS linescans across the C/SiC interface in the composite processed at 1473 K revealing a discrete layer some 40 nm in width rich in oxygen and nitrogen, presumably indicating a silicon oxy-nitro-carbide compound. Further into the matrix, the O and N levels decrease and there exists solely SiC. Studies of the Si $L_{2,3}$-ELNES confirm a change in local coordination and bonding of Si between these two regions. This segregation of oxygen and nitrogen was found to be a strong function of the final heat treatment temperature.

In contrast, thin regions of SiC matrix sandwiched between two C fibres revealed an absence of oxygen which appeared to be associated with segregation of N to the C/SiC interface shown in figure 8. This difference in behaviour is related to the ease with which gaseous SiO and CO (produced during the reaction of excess C and $SiO_2$ in PCS-based SiC matrix at 1200 °C) escapes via the fibre-matrix interface. If the matrix layer is thin, the small amount of gaseous species can easily escape while a thicker matrix region may lead to the accumulation of SiO and CO which will disproportionate into $SiO_2$ and form a Si-O-C-N segregation layer (figure 7).

Figure 7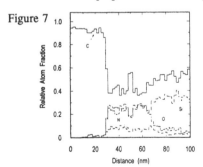

Figure 8

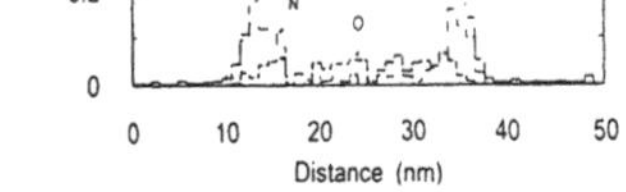

## CONCLUSIONS

This review demonstrates the power of EELS and ESI in the characterization of carbon materials at high spatial resolution, highlighting the composition, distribution, degree of homogeneity and bonding of light elements incorporated into the carbon structure.

## REFERENCES

[1] Lu S 1997 Ph.D. Thesis, University of Leeds.
[2] Bruley J, Williams D B, Cuomo J J and Pappas D P 1995 J. Microscopy **180**, 22.
[3] McCulloch D G and Brydson R 1996 J. Phys. Condensed Matter **8**, 3835.

*Inst. Phys. Conf. Ser. No 153: Section 11*
*Paper presented at Electron Microscopy and Analysis Group Conf. EMAG97, Cambridge, 1997*
© *1997 IOP Publishing Ltd*

# Crystallography of faceted boundaries in SrTiO₃ electronic ceramics

**Z Mao and K M Knowles**

University of Cambridge, Department of Materials Science and Metallurgy, Pembroke Street, Cambridge, CB2 3QZ, U.K.

**ABSTRACT:** Faceted boundaries in SrTiO₃ electronic ceramics have been studied using transmission electron microscopy. A number of orientation relationships were observed in both SrTiO₃ internal boundary layer capacitor ceramics (IBLCs) doped with either sodium or lithium as an acceptor and niobium as a donor and a SrTiO₃ semiconducting ceramic doped only with niobium. Dislocation contrast appearing in such boundaries suggested that faceting was most evident in boundaries free of intergranular films.

## 1. INTRODUCTION

Grain boundaries and grain boundary regions have a controlling influence for a number of properties in ceramic materials, including mechanical properties such as fracture strength, toughness, plastic deformation, and high temperature creep, and electrical properties such as conductivity and dielectric loss. Faceting of boundaries provides a mechanism for interface migration (Morrissey and Carter 1980) and for lowering of the energy of an initially flat interface (Sutton and Balluffi 1995), and so observations of faceting are of particular interest. While faceting of both low-angle and high-angle grain boundaries in alumina is well documented (Carter et al 1980, Morrissey and Carter 1984), it is less so in other ceramic systems. The purpose of this paper is to report and discuss observations of faceting of grain boundaries in SrTiO₃ electronic ceramics.

## 2. EXPERIMENTAL DETAILS

Sintered pellets from three different starting powder mixtures of doped strontium titanate were produced. All three mixtures contained niobium as a donor dopant. In addition, one contained sodium as an acceptor dopant and a second contained lithium as an acceptor dopant. Full details of specimen preparation and heat treatment schedules are given elsewhere (Mao and Knowles 1997).

The sintered samples were cut, mechanically ground and polished to less than 100 μm thick and ion-milled to electron transparency using a GATAN 600 disc mill. The thin-foils were subsequently examined uncoated in a JEOL 2000FX at 200 kV.

## 3. RESULTS AND DISCUSSION

Faceted boundaries were a common observation in all three types of specimen. An example of a faceted grain boundary in the sample free from either lithium or sodium is shown in Fig. 1.

Here, the two grains are oriented so that there is a near-common $\mathbf{g} = \overline{2}00$, with the [001] zone in grain 1 about 4.5° away from the [011] zone of grain 2, i.e. an orientation relationship close to $\Sigma = 29$, 43.6° about [$\overline{1}00$]. The weak beam dark field (WBDF) micrograph in Fig. 1(a) using the nearly common $\mathbf{g} = \overline{2}00$ reflection and the WBDF micrograph from grain 2 formed in $\mathbf{g}/4\mathbf{g}$ conditions with $\mathbf{g} = 01\overline{1}$ in Fig. 1(b) show the faceting behaviour well, together with the dislocation-like contrast at the steps. No second phase is evident at this particular high-angle grain boundary.

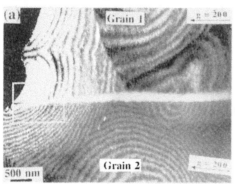 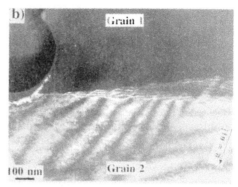

Fig. 1 (a) WBDF image taken in $\mathbf{g} = \overline{2}00$ and (b) WBDF image taken in $\mathbf{g} = 01\overline{1}$ from the boxed area in (a) of a faceted grain boundary in a donor-doped sample of $SrTiO_3$.

An example of a faceted grain boundary in a lithium-doped sample is shown in Fig. 2. Fig. 2(a) is a bright field (BF) image in which the faceted grain boundary is seen almost edge-on and Fig. 2(b) is the corresponding electron diffraction pattern from this boundary region. The approximate orientation of the grain boundary is $(\overline{1}42)_A \parallel (031)_B$. The arrowed set of spots in Fig. 2(b) is from grain A in Fig. 2(a) and is from a <211> zone. The diffraction spots in Fig. 2(b) arising from grain B in Fig. 2(a) are from a <100> zone.

It is apparent that the two zones from grains A and B are almost parallel (to within 0.5°) and that the orientation relationship between the two grains can be described to a good approximation as $[211]_A \parallel [100]_B$, $(01\overline{1})_A \parallel (010)_B$ and $(\overline{1}11)_A \parallel (001)_B$, which describes a rotation of 56.60° about a unit vector $[-0.769, -0.590, 0.245]$. If the orientation relationship evident in Fig. 2(b) is described more carefully by recognising the slight rotation of about 1° about the electron beam normal seen between $(01\overline{1})_A$ and $(010)_B$, the angle/axis description is virtually unchanged: 56.83° about a unit vector $[-0.769, -0.596, 0.253]$. These angle/axis descriptions are the ones with the lowest angles and can be compared with the tabulation of coincident-site lattice orientations given by Grimmer, Bollmann and Warrington (1974). The closest coincident-site lattice orientation is $\Sigma = 39$, 50.13° about [$\overline{3}21$], 6.34° away from the observed orientation about a rotation axis $[0.469, 0.876, 0.112]$. Fig. 2(c) is a dark field (DF) image of the boundary from diffraction spot 'a' in Fig. 2(b) and Fig. 2(d) is a dark field image from diffraction spot 'b' in Fig. (b), both showing contrast behaviour consistent with faceting.

Two micrographs from a through-focal Fresnel fringe series of micrographs taken with the boundary accurately edge-on to the electron beam are shown in Fig. 3, together with a diffuse dark field image of the boundary and a plot of Fresnel fringe spacing as a function of defocus. These suggest that any interface layer at this boundary is less than 0.5 nm in thickness, implying that it is unlikely that there is a discrete second phase at this boundary. Unfortunately, we were unable to obtain an HREM image from this particular boundary, and so no definite conclusion can be made about what contributed to the Fresnel contrast and the bright contrast in the diffuse DF image, although ion beam thinning damage and preferential grain boundary etching and subsequent sputter deposition are both possible explanations.

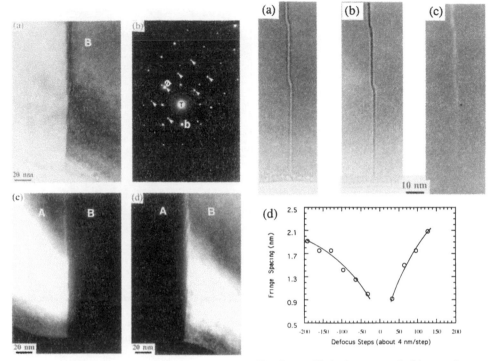

Fig. 2 (a) BF of a faceted grain boundary in a lithium-doped SrTiO₃ sample, (b) electron diffraction pattern from (a), with the arrowed spots from grain A, (c) DF from spot 'a' in (b), (d) DF from spot 'b' in (b).

Fig. 3 (a) Underfocus, and (b) overfocus Fresnel images of the boundary in Fig. 2 with the electron beam parallel to the boundary, showing the facets clearly, (c) a diffuse DF image of the boundary and (d) a plot of the distance between fringe maxima as a function of defocus.

Fig. 2 (a) BF of a faceted grain boundary in a lithium-doped $SrTiO_3$ sample, (b) electron diffraction pattern from (a), with the arrowed spots from grain A, (c) DF from spot 'a' in (b), (d) DF from spot 'b' in (b).

Fig. 3 (a) Underfocus, and (b) overfocus Fresnel images of the boundary in Fig. 2 with the electron beam parallel to the boundary, showing the facets clearly, (c) a diffuse DF image of the boundary and (d) a plot of the distance between fringe maxima as a function of defocus.

An example of a faceted low-angle boundary in a sodium-doped sample is shown in Fig. 4. The boundary is oriented parallel to the electron beam (which is travelling along [001]) and can be seen to facet on $(0\bar{1}0)$ and $(\bar{1}10)$ planes. A series of WBDF images of this boundary are shown in Fig. 5 taken with the approximate beam normals (a) $[\bar{1}21]$, (b) $[\bar{2}25]$, (c) $[\bar{2}05]$ and (d) $[\bar{1}11]$. These show dislocation contrast consistent with most dislocations having $\mathbf{b} = [100]$, given their near-invisibility in (d) and the known character of dislocations in $SrTiO_3$ (Mao and Knowles 1996). The line direction of these dislocations is [001] and their spacing on the $(0\bar{1}0)$ planes is about 21.5 nm. Therefore, using the equation $\theta = |b|/D$ where $D$ is the spacing of the dislocation array at the boundary, the tilt angle of this particular boundary can be calculated to be about 1°. The dislocations confined to the $(\bar{1}10)$ facet plane appear to be

Fig. 4 Low-angle grain boundary in sodium-doped $SrTiO_3$ faceted on $(0\bar{1}0)$ and $(\bar{1}10)$.

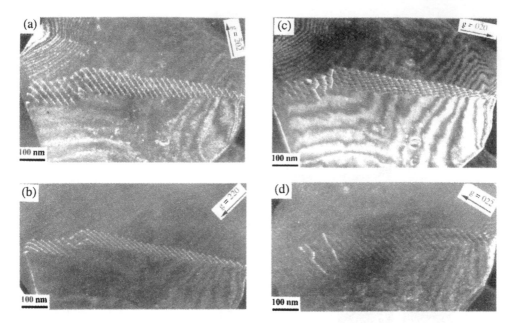

Fig. 5 WBDF images of the low-angle grain boundary in Fig. 4. (a) $\mathbf{g} = \overline{2}0\overline{2}$, (b) $\mathbf{g} = 220$, (c) $\mathbf{g} = 0\overline{2}0$ and (d) $\mathbf{g} = 02\overline{2}$. (a), (b) and (d) are in $\mathbf{g}/3\mathbf{g}$ conditions, while (d) is in a $\mathbf{g}/4\mathbf{g}$ condition.

$\mathbf{b} = [010]$, given their clear visibility in (c) and (d) and their absence in (a). Note that as in the previous examples, there is no indication of second phases either at this faceted boundary.

## 4. CONCLUSIONS

Faceting appears to be common in grain boundaries in SrTiO$_3$ electronic ceramics in cases where there is no clear evidence for a second phase. Our analysis suggests that, as in alumina, both high-angle and low-angle grain boundaries free of second phases are able to facet, thereby helping to provide a mechanism for interface migration during sintering through the lateral movement of such facets.

## ACKNOWLEDGEMENTS

ZM would like to thank the Cambridge Overseas Trust for the award of a Schlumberger Cambridge Scholarship during the course of this work.

## REFERENCES

Carter C B, Kohlstedt, D L and Sass S L 1980 J. Am. Ceram. Soc. **63**, 623
Grimmer H, Bollmann W, and Warrington D H 1974 Acta Cryst. A **30**, 197
Mao Z and Knowles K M 1996 Phil. Mag. A **73**, 699
Mao Z and Knowles K M 1997 submitted to J. Am. Ceram. Soc.
Morrissey K L and Carter C B 1984 J. Am. Ceram. Soc. **67**, 292
Sutton A P and Balluffi Interfaces in Crystalline Materials (Oxford: Clarendon Press)

*Inst. Phys. Conf. Ser. No 153: Section 11*
*Paper presented at Electron Microscopy and Analysis Group Conf. EMAG97, Cambridge, 1997*
© 1997 IOP Publishing Ltd

# Characterisation of interfacial effects in SrTiO$_3$ electronic ceramics

**Z Mao and K M Knowles**

University of Cambridge, Department of Materials Science and Metallurgy, Pembroke Street, Cambridge, CB2 3QZ, U.K.

**ABSTRACT:** The microstructures and electrical behaviour of two doped polycrystalline strontium titanate ceramics have been compared. One sample was made semiconducting through the addition of niobium as a donor dopant, while the other exhibited internal boundary layer capacitor behaviour through the additions of niobium as a donor dopant and sodium as an acceptor dopant, with a novel processing procedure in which an uninterrupted single firing process was used. Clear differences are seen in the Fresnel fringe profiles of grain boundaries and the second phases found in the two materials which relate directly to the differences seen in *I-V* curves and are consistent with a simple boundary barrier model we have established.

## 1. INTRODUCTION

The microstructure, and in particular the interfacial structure, plays a vital role in the properties of SrTiO$_3$ internal boundary layer capacitor ceramics. Sintering and post-sintering heat treatments are aimed at achieving certain interfacial electronic properties and then optimising the performance of the ceramic components. There are two approaches used to characterise the interfacial effects in these ceramics. The first approach is to measure the electrical properties of the bulk components, such as the dielectric constant as a function of frequency, current-voltage relationships and capacitance against voltage. The second approach is characterisation of the microstructure by microscopical techniques.

SrTiO$_3$ internal boundary layer capacitors prepared by an uninterrupted single firing process using lithium or sodium as an acceptor species in the starting powders are a recent innovation (e.g. Zhou et al 1991, Mao and Knowles 1995). This process technology has attractions for multilayer capacitor technology. Here, we summarise transmission electron microscopy (TEM) results from the characterisation of the interfacial structure of a SrTiO$_3$ internal boundary layer capacitor free of acceptor dopant and compare these observations with those we have made on a sodium-doped sample. *I-V* data from the two bulk sample types are also compared. The electrical properties of the two materials are then discussed in terms of the interfacial microstructure we have established here and in our previous work on sodium-doped samples (Mao and Knowles 1995).

## 2. EXPERIMENTAL DETAILS

Two powder mixtures of doped strontium titanate were prepared. Both contained 1.0 mole SrCO$_3$, 1.0 mole TiO$_2$, 0.007 mole Nb$_2$O$_5$, 0.003 mole Al$_2$O$_3$ and 0.004 mole SiO$_2$. In addition, one mixture contained 0.015 mole NaCl. After wet milling, the powders were calcined at 1150 °C for one hour in air and pressed into 3 mm thick pellets under a pressure of 0.7 MPa. The pellets free from sodium were sintered at 1360 °C for four hours in a reducing atmosphere of methane and argon to consolidate the pellets and make the $SrTiO_3$ grains semiconducting, after which they were cooled to room temperature. The pellets containing sodium were sintered at 1360 °C for four hours in a reducing atmosphere of methane and argon, after which they were cooled to 1000 °C and held at this temperature for thirty minutes in air to encourage the diffusion of sodium along the grain boundaries. Finally, these pellets were furnace cooled to room temperature.

Samples for TEM were prepared using conventional procedures for ion beam milling and subsequently examined uncoated in both a JEOL 2000FX at 200 kV and a JEOL 4000EX HREM at 300 kV. EDX spectra were collected using a Philips CM30 with a nano-probe at 200 kV equipped with a Link Analytical AN10000 EDX system. $I$-$V$ relationships were measured with a Philips PM2521 automatic multimeter and a Heathkit regulated H.V. power supply (Model IP-17) as a $DC$ voltage source whose output was checked by a Philips PM2521 automatic multimeter to ensure that the accuracy of the applied $DC$ voltage on the samples was within $V \pm \Delta V$, where $\Delta V \leq 0.05$ V. Measurements covered a current range of 2 μA – 20 A and an applied $DC$ voltage range of 0 – 50 V.

## 3.  RESULTS AND DISCUSSION

General transmission electron microscope observations of the sample doped with sodium as an acceptor showed that the average grain size of the sample was 0.5 – 1 μm (Mao and Knowles 1995), whereas that of the semiconducting sample free from sodium was 30 – 40 μm (Fig. 1).

Examination of high angle grain boundaries in the sodium-containing specimens by diffuse dark field imaging indicated the presence of amorphous phases at the grain boundaries. Through-focal series of Fresnel images from such grain boundaries aligned parallel to the electron beam confirmed the presence of thin ≤ 1 nm to 2 nm interface layers at the boundaries with a potential lower than the grains (Mao and Knowles 1995). By comparison, similar interfaces in the sample free from sodium showed markedly weaker Fresnel contrast, but still in general appeared to contain an amorphous film (Fig. 2). Only boundaries which faceted appeared to be free of second phase, in that they exhibited dislocation contrast (Mao and Knowles 1997a).

Not surprisingly, X-ray diffraction analysis of both samples failed to find evidence for crystalline second phases. However, differences were found in TEM. Sodium-doped samples contained occasional grains of $Sr_2TiO_4$ (Mao and Knowles 1995). By comparison, the occasional second-phase particles in the samples free of acceptor dopant were richer in titanium than $SrTiO_3$. An example is shown in Fig. 3. Fig. 3(a) is a dark field image of a boundary phase formed from the diffraction spot arrowed in Fig. 3(b). The size of such boundary particles and the difficulty in obtaining any recognisable electron diffraction pattern at all from low index zones precluded further, more detailed, crystallographic analysis.

$I$-$V$ data from the sample containing sodium and the sample free from sodium show clearly that the current-voltage relationships of both samples are non-linear within the current and voltage range studied, whereas the plots of ln $I$ against voltage $V$ can be fitted both to a linear relationship (Fig. 4). For the sodium-doped sample, it is apparent that there is a threshold voltage of about 20 V, above which the non-linear current-voltage relationship is evident, but for the sample free from acceptor dopant, it is not clear whether or not there is any clear threshold voltage within the accuracy of our measurements. Furthermore, the slope of ln $I$ against $V$ plot for the sodium-doped sample shown in Fig. 4(b) is significantly smaller than that of the sample free from sodium in Fig. 4(d). Specifically, the slopes are 0.23 for the

Fig. 1 Scanning electron micrograph of an etched sample of SrTiO₃ free of acceptor dopant.

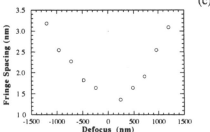

sodium-doped sample and 0.79 for the sample free from sodium. These electrical measurements confirm the semiconducting nature of the sample free from acceptor dopant and the IBLC behaviour of the sodium-doped sample.

To rationalise these electrical measurements, a simple boundary barrier model based on $p$-$n$ junction theory can be used (Mao and Knowles 1997b). With this model, the current-voltage characteristics of both types of sample can be expressed in the general form

$$\ln I - \ln I_0 = K(V - V_T)$$

where $V$ is the applied voltage, $I$ is the current, $I_0$ is determined by the band gap and

Fig. 2 Fresnel contrast from a typical boundary in a specimen free from acceptor dopant: (a) underfocus, (b) overfocus and (c) a plot of the distance between fringe maxima as a function of defocus.

Fermi level of the bulk ceramics and $K$ is a constant describing how much the applied voltage $V$ is distributed across each boundary. $V_T$ gives a relative measurement of the electrical resistance of the boundary area. The non-linear current-voltage relationship in the samples can only be observed when the applied voltage $V$ is significantly larger than $V_T$.

Fig. 3 An example of a crystalline interface phase found in the semiconducting SrTiO₃ sample free from sodium: (a) dark field image of the interface phase formed using the arrowed diffraction spot shown in the electron diffraction pattern (b).

526

The data in Fig 4(d) can then be rationalised in terms of a relatively low value of $V_T$ indicating semiconducting behaviour, whereas in the acceptor-doped sample there is a much higher value of $V_T$. $K$ is also lower for the sodium-doped sample, not only because of the semiconducting behaviour of the sample free of sodium, but also because of the much smaller grain size of the sodium-doped sample.

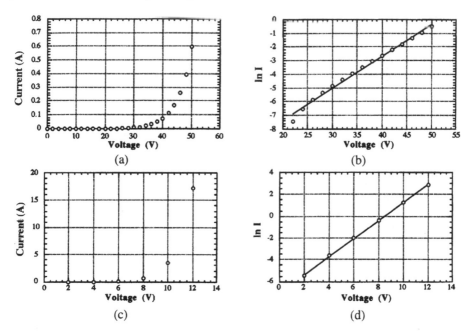

Fig. 4   *I-V* Data. (a) and (b) are from the sodium-doped sample, (c) and (d) are from the sample free from sodium. (a) and (c) are plots of current (A) against voltage (V). (b) and (d) are plots of ln *I* against voltage.

Microstructurally, the differences between the two samples can be understood in terms of the sodium addition in the IBLC sample (Mao and Knowles 1997b): the appearance of $Sr_2TiO_4$ and the high resistance boundary layers in the sodium-containing sample can both be attributed directly to the segregation of acceptor negatively-charged boundary states. In the specimen free from sodium, the titanium-rich phase is formed by the substitution of niobium ions for titanium ions in the bulk and the material remains semiconducting.

**ACKNOWLEDGEMENTS**

ZM would like to thank the Cambridge Commonwealth Trust for the award of a Schlumberger Cambridge Scholarship during the course of this work.

**REFERENCES**

Mao Z and Knowles K M 1995 Inst. Phys. Conf. Ser. **147**, 563
Mao Z and Knowles K M 1997a this conference
Mao Z and Knowles K M 1997b submitted to J. Am. Ceram. Soc.
Zhou L, Jiang Z and Zhang S 1991 J. Am. Ceram. Soc. **74**, 2925

*Inst. Phys. Conf. Ser. No 153: Section 11*
*Paper presented at Electron Microscopy and Analysis Group Conf. EMAG97, Cambridge, 1997*
© 1997 IOP Publishing Ltd

# Porous ceramics for affinity chromatography applications.

**J. Brooks\*, I. M. Reaney\*, P.F. James\*, K. Beyzavi†.**

\*Department of Engineering Materials, Sir Robert Hadfield Building, Mappin Street, University of Sheffield, Sheffield, S1 3JD, UK.

†Bioprocessing Ltd., Medomsley Rd., Consett, Co. Durham, DH8 6TJ, UK.

**ABSTRACT:** Porous ceramics for use as biological support materials have been made by casting a melt in the $Na_2O-CaO-P_2O_5-TiO_2-SiO_2$ system followed by leaching in HCl. The melt crystallised upon cooling forming a three phase material comprised of a sodium-calcium-phosphate phase, titania (rutile) and amorphous silica. Acid leaching removed the sodium-calcium-phosphate phase leaving a continuous porous network of titania and silica. Average pore diameters are of the order of 2-3 μm with a corresponding surface area of 90 $m^2/g$.

## 1. INTRODUCTION

Porous materials based on phase separated borosilicate glasses were first developed by Hood and Nordberg (1934) and have long been used as supports in the processing of biological products. These compositions due to their high silica content are susceptible to attack by alkali solutions, and since NaOH is the most widespread and effective cleaning agent of biological supports these materials will have a finite lifespan. For a material to be an effective biological support it must possess certain attributes, namely

a) chemical durability
b) mechanical stability
c) pore size compatible with proteins
d) narrow pore size distribution, and
e) modifiable surface chemistry

The pore size and distribution of the material need to be carefully controlled as the most effective pore size is twice the major axis of the immobilised protein as shown by Weetal (1969) and later by Messing (1970). Obviously if the pore size can be controlled then all manner of proteins can be purified regardless of size.

New compositions based on materials in the $Na_2O-CaO-TiO_2-P_2O_5-SiO_2$ system have been developed by Hosono et al (1990) that have greater alkali resistance and which may be promising biological support materials. SEM and TEM techniques have been applied to similar compositions to those suggested by Hosono et al. (1990) with a view to assessing their potential for use in affinity chromatography.

## 2. EXPERIMENTAL PROCEDURE

Batch compositions of $15Na_2O-32CaO-19TiO_2-15P_2O_5-19SiO_2$ (mol %) were prepared from standard laboratory reagent grade starting materials and Loch Aline sand (99.5% $SiO_2$), which were weighed (to 0.01g) then mixed by hand prior to melting in Pt crucibles. Melting was carried out in an electric furnace under an oxidizing atmosphere at 1400°C and the melt was stirred for 3 hours to promote homogeneity. Melts were cast into a steel mould then annealed at 600°C for 1 hour to relieve any internal stresses.

Solid and powdered samples were leached in 1M HCl at 100°C under reflux for 18 hours. Powdered samples were stirred throughout to prevent sedimentation.

Fracture surfaces of the leached material were examined using an SEM (Camscan) and unleached material was subjected to TEM (Phillips EM 400, Jeol 3010) and XRD analysis. For XRD analysis a scan rate of 0.33°/2θ and 0.2° step size was used. JCPDS powder diffraction cards were used as a means of identifying the phases present. Surface areas and pore size distributions of the powdered porous material were evaluated using a Micromeritics Gemini analyzer (sample degas at 200°C for 4 hours) and a Micromeritics Poresizer mercury porosimeter.

## 3. RESULTS AND DISCUSSION

According to Hosono et al (1990), casting a melt in this system produces a three phase monolithic ceramic comprised of a $NaCaPO_4$ (NCP) type phase, $TiO_2$ and amorphous $SiO_2$. The structure of the cast material is dendritic in nature as shown in Fig. 1. Grain sizes are of the order of 500 μm, and it is thought the phases evolve through spinodal decomposition during casting prior to crystallization (James, 1975). After leaching in HCl a weight loss of 57 % is observed. This is thought to be associated with the removal of the NCP phase leaving an amorphous $SiO_2$ matrix containing rutile laths which surround an interconnecting network of pores. Peaks associated with the NCP phase can be identified by comparing XRD patterns from leached and unleached samples, Figures 2a and b. All the peaks present after leaching correspond to $TiO_2$ (rutile).

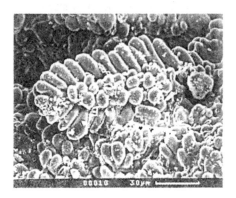 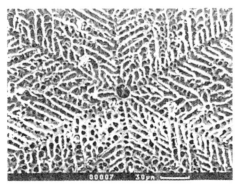

a         b

Fig. 1 - SEM micrographs of (a) as cast material and (b) material leached in HCl for 18 hours.

The evolution of the NCP phase is expected to be in accordance with the equation :

$$15Na2O+32CaO+19TiO2+15P2O5+19SiO2 \rightarrow 30NaCaPO4+19TiO2+19SiO2+2CaO$$

$$4740g \quad\quad 1520g \quad 1140g \;\; 112g$$

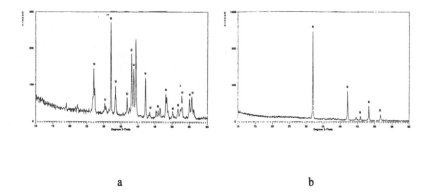

|          |          |
|----------|----------|
|    a     |    b     |

Fig. 2 - XRD traces of (a) unleached material and (b) leached material showing complete
removal of the phosphate phase. (R=rutile, U=unidentified phase).

The weight loss associated with the complete dissolution of NaCaPO$_4$ should
theoretically be 63 %. The observed weight loss of 57 % agreed well with published work by
Hosono et al and the theoretical value of 63 % assuming the phosphate phase is NaCaPO$_4$ and
is completely dissolved by the acid treatment. The slight anomalies can be attributed to
volatilization in the melt, P$_2$O$_5$ being particularly susceptible. Surface area analysis yielded
BET values of 96 m$^2$/g for 100 μm powders and 148 m$^2$/g for powders less than 45 μm. Pore
distributions determined by mercury porosimetry gave median pore diameters of 2.4 μm for the
100 μm powder and 1.7 μm for the <45 μm powder.

The structure of the NCP phase could not be conclusively identified through XRD
analysis, indicating that it may be metastable. TEM analysis was carried out in order to further
investigate the crystal structure associated with this phase. Figure 3 is a bright field TEM
image showing the NCP phase (left), rutile laths (right) and residual glass (centre). Figures
3a, b and c are electron diffraction patterns obtained from this phase. The presence of a zone
axis which has 6-fold symmetry suggests that the Bravais lattice is either cubic, hexagonal,
rhombohedral or trigonal. The ratios and angles associated with the diffraction patterns are
not exactly consistent with a cubic phase , however they do show strong similarities , although
slightly distorted, to those from face centred cubic (fcc) structures eg, Figures 4b and c look
similar to <110> and <122> zones in fcc structures. It is possible therefore that the phase has
a rhombohedrally distorted fcc lattice. In addition superlattice reflections are observed in all
patterns at either a $\frac{1}{2}$ or a $\frac{1}{3}$ {hkl} spacings. The origin of these spots is so far unknown
and will be investigated in future work.

530

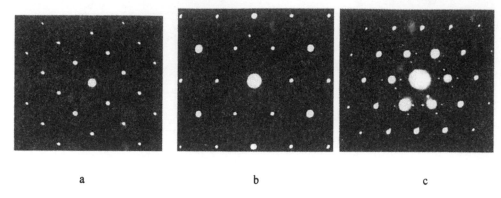

a                                    b                                    c

Fig. 3 - Selected area diffraction patterns from the NCP phase.

,

Fig. 4 - Bright field TEM image showing
NCP phase (left), residual glass
(centre) and rutile laths (right)

500 nm

Since the melt devitrifies upon cooling there can be no control over the final pore size
unless the glassy structure can be retained by rapid quenching. Subsequent carefully controlled
heat treatments could be applied to quenched samples to control the phase evolution and thus
pore size. The high $SiO_2$ content of the porous body (~50 mol %) should enable
aminosilanization chemistry to be applied quite readily, thus making the material suitable for
affinity chromatography. The presence of $TiO_2$ should reduce the rate of chemical attack
during cleaning in NaOH.

4.      CONCLUSIONS

- Casting a melt in the $Na_2O-CaO-TiO_2-P_2O_5-SiO_2$ system produces a 3 phase monolithic
  ceramic comprised of $NaCaPO_4$, $TiO_2$ and amorphous $SiO_2$.
- Acid leaching removes the $NaCaPO_4$ phase resulting in a highly porous material with an
  average pore size of 2-3 μm.
- The high surface area combined with the alkali resistance afforded by the $TiO_2$ may make
  this a suitable material for biological processing.

**REFERENCES**

Hood H P and Nordberg M E 1934 US Patent 2,106,744,
Weetal H H 1969 Nature 233, 959
Messing R A 1970 Enzymologia 38, 39
Hosono H Y.Sakai, M.Fasano and Y.Abe 1990 Jrnl. Am. Ceram. Soc. 73, 2536
James P F 1975 Jrnl.of Mater. Sci. 10, pp 1802-1825

*Inst. Phys. Conf. Ser. No 153: Section 11*
*Paper presented at Electron Microscopy and Analysis Group Conf. EMAG97, Cambridge, 1997*
© *1997 IOP Publishing Ltd*

# TEM investigation of hydrothermally synthesised Ba(Mg$_{1/3}$Ta$_{2/3}$)O$_3$ powders

**I MacLaren and C B Ponton**

IRC in Materials for High Performance Applications and School of Metallurgy and Materials, The University of Birmingham, Birmingham B15 2TT.

**ABSTRACT:** Perovskite Ba(Mg$_{1/3}$Ta$_{2/3}$)O$_3$ powders have been produced by hydrothermal synthesis at $\leq 200°$C. It was shown that the perovskite particles are irregular in shape and magnesium deficient, and that the excess magnesium was precipitated separately as the hydroxide. An improved method was developed and the resulting powder had a high proportion of rounded near-stoichiometric particles, although some irregular shaped magnesium-deficient particles were also present. Thus, a link has been found between the stoichiometry and morphology of particles, which suggests that the mode of growth is affected strongly by particle stoichiometry.

## 1. INTRODUCTION

Pure or doped Ba(Mg$_{1/3}$Ta$_{2/3}$)O$_3$ ceramics (BMT) hold great promise for use as dielectric materials in microwave resonators since they exhibit a moderate dielectric constant, an ultra-low loss at microwave frequencies, and a near-zero temperature coefficient of the resonant frequency (Matsumoto *et al.*, 1986; Matsumoto *et al.*, 1991; Chen *et al.*, 1994; Furaya and Ochi, 1994). These materials are, however, currently restricted in application, because when produced by conventional mixed oxide routes they are difficult to sinter, requiring very high firing temperatures (> 1600°C) (Matsumoto *et al.*, 1991; Chen *et al.*, 1994; Furaya and Ochi, 1994) or long sintering times ($\approx$ 100 hours at 1450°C) (Matsumoto *et al.*, 1986) to produce dense ceramics with the best microwave properties.

Chemically synthesised ceramic powders often display better sinterability than mixed-oxide derived powders on account of their better chemical homogeneity, finer particle size and controlled particle morphology. Thus, a number of workers have formed BMT ceramics from sol-gel produced powders (Renoult *et al.*, 1992; Katayama *et al.*, 1996) and such powders were found to display very high sinterability, with dense ceramics being formed at 1300°C - 1400°C. Moreover, Renoult *et al.*, (1992) found a high degree of atomic ordering after sintering at 1400°C for just 5 hours but both studies measured disappointing microwave properties at around 10 GHz; the reasons for this effect were not elucidated.

An alternative method for the production of ceramic powders which shows particular promise is hydrothermal synthesis. In this method, ceramic sols are produced by chemical reactions in an aqueous or organo-aqueous solution under the simultaneous application of heat and pressure. This has been used for the preparation of a wide variety of ceramic materials including other barium-based ceramics such as barium titanate (Vivekanandan *et al.*, 1986) and barium hexaferrite (Ataie *et al.*, 1995). In this paper, we describe the hydrothermal synthesis of perovskite barium magnesium tantalate powders and the characterisation of these powders using transmission electron microscopy (TEM) and X-ray diffraction (XRD). In particular, a link is established in this paper between the stoichiometry and morphology of perovskite particles. Some of the work set out in this paper has been reported more fully elsewhere (MacLaren and Ponton, 1997) with a greater emphasis on the chemical synthesis aspects of this work.

532

## 2. EXPERIMENTAL PROCEDURE

Precursor sols for the hydrothermal synthesis were produced using barium and magnesium acetates and hydrated tantalum oxide (produced from tantalum oxalate solution) in aqueous-based, alkaline solutions. These alkaline conditions were provided by the strong organic base, tetramethylammonium hydroxide (TMAH). TMAH was preferred to other strong bases such as sodium or potassium hydroxide, since these may result in Na or K doping of the ceramic, which is likely to affect the dielectric properties of the ceramic.

The precursor sols were treated hydrothermally in a 250 cm$^3$ PTFE-lined autoclave at temperatures of 160°C - 200°C for 2 hours with stirring. The synthesised sols were vacuum filtered and the resulting powder cake was then washed with water, followed by acetone. The powder cake was then dried at room temperature for several days and powdered using a porcelain pestle and mortar. These powders were then investigated using TEM and XRD. TEM characterisation was carried out using bright field imaging, selected area diffraction (SAD) and energy dispersive X-ray (EDX) analysis. Identification of crystalline phases in powders was performed by comparison of XRD traces with JCPDS (now ICDD) standards.

## 3. RESULTS AND DISCUSSION

Fig. 1 shows the XRD trace for a powder produced at 200°C using TMAH, with the main perovskite peaks indicated. This shows clearly that the desired perovskite phase has been formed at a temperature as low as 200°C. The peaks are somewhat broadened, however, suggesting that the particle size is extremely fine (i.e. of the order of a few nm). A pair of peaks is also noted at $2\theta \approx 24°$; these are the largest peaks for BaCO$_3$ (JCPDS 44-1487), and

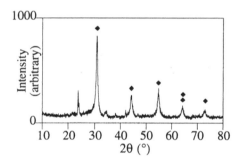

Figure 1: XRD trace for a powder synthesised hydrothermally at 200°C.

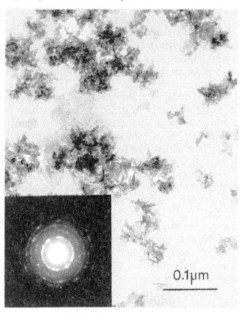

Figure 2: TEM image of particles in a powder synthesised at 160°C; inset: SAD pattern taken from these particles

some other peaks for this phase may be noted in the region of $2\theta = 34°$. A TEM image of a similar powder (produced at 160°C) is shown in Fig. 2 and the reason for the XRD peak-broadening may be seen clearly: most particles are clusters of irregular shaped crystallites with extremely small dimensions (smallest dimensions typically of the order of 10 nm). A SAD pattern taken from the particles shown in this image is shown in the inset in Fig. 2 and all the expected rings for the perovskite phase are present. Quantitative EDX analysis on the TEM showed that these particles were usually deficient in magnesium (typically 30-50 % less than is required for the stoichiometric composition); barium deficiencies were sometimes noted as well. Some thin hexagonal platelets may also be observed in Fig. 2; EDX analysis of these platelets showed peaks for magnesium and oxygen. These platelets were thought to be magnesium hydroxide since it has a hexagonal crystal structure (JCPDS standard 44-1482) and is often found as hexagonal platelets (Itatani *et al.*, 1988, 1989). It seems likely, therefore, that at least some of the magnesium not incorporated within the perovskite particles was precipitated out as the hydroxide. It is probable that any unreacted barium has remained in solution in a similar manner to that observed for the hydrothermal synthesis of $BaTiO_3$ (MacLaren *et al.*, 1997).

Thus, it is clear that this material, whilst it can be formed readily by hydrothermal synthesis, displays a marked tendency towards non-stoichiometry with the formation of Mg-deficient perovskite particles and the separate precipitation of magnesium hydroxide. For this reason, a modified method was developed using TMAH at temperatures close to

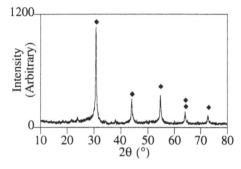

Figure 3: XRD trace for a powder produced at 190°C using an organo-aqueous solution.

200°C in an organo-aqueous solution in an attempt to rectify this situation. A XRD trace for one of the resulting powders is shown in Fig. 3, with the main peaks for the perovskite phase indicated. All the principal peaks on this trace match the perovskite phase well, and the only indication of any impurity is a small $BaCO_3$ peak at $2\theta \approx 23.9°$. The peaks are also sharper than those for materials produced at 200°C by the standard TMAH method, suggesting that the powder has a larger particle size. This is confirmed by TEM as shown in Fig. 4a, and it was found that the powder consisted mainly of rounded particles with dimensions of about 30 nm.

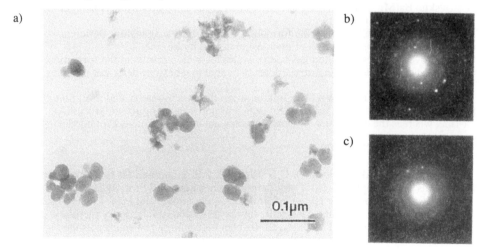

Figure 4: a) TEM image of particles in a powder synthesised at 190°C using an organo-aqueous solution; b) SAD pattern from some rounded particles; c) SAD pattern from a cluster of irregular shaped particles.

In addition to these particles, however, there are clusters of irregular shaped particles similar to those produced by the standard TMAH method, as well as a few magnesium hydroxide platelets. Fig. 4b shows a SAD pattern for a group of the rounded particles and Fig. 4c shows a SAD pattern for a cluster of irregular shaped particles. Whilst the spots in Fig. 4b are much sharper than those in Fig. 4c, they coincide well with the expected ring radii for the perovskite phase in both cases. The compositions of the two types of particles were determined using quantitative EDX; the average compositions from 4 groups of rounded particles and 3 clusters of irregular particles are shown in Table 1 overleaf (the deviations shown are simply the statistical standard deviations from the spread of measured values and do not reflect instrumental or statistical errors within the TEM/EDX system):

534

| Particle type | Ba (at. %) | Mg (at. %) | Ta (at. %) |
|---|---|---|---|
| Rounded | $52.3 \pm 2.1$ | $14.1 \pm 1.9$ | $33.6 \pm 1.1$ |
| Cluster of irregular particles | $53.2 \pm 2.5$ | $7.6 \pm 2.0$ | $39.2 \pm 4.2$ |
| Stoichiometric composition | 50 | 16.7 | 33.3 |

Table 1: Compositions of rounded particles and clusters of irregular particles from EDX

Whilst there are usually inaccuracies in measurements made using standardless quantitative EDX, a clear trend may be seen: rounded particles have near-stoichiometric compositions, whereas the clusters of irregular particles were significantly magnesium-deficient.

Thus, it seems that the way in which the perovskite particles grow and the resulting morphology is affected strongly by their composition, with magnesium-deficient compositions resulting invariably in irregular shaped, ultra-fine particles, whereas near-stoichiometric compositions result in the formation of larger, more rounded particles. If technologically useful powders are to be produced by this method, a way has to be found to alter the reaction conditions, so that the formation of stoichiometric particles is more favourable than the formation of magnesium-deficient particles. Progress has been made in this direction by the use of organo-aqueous reaction media but further work is required in order to achieve this aim fully.

## 4. CONCLUSIONS

Perovskite $Ba(Mg_{1/3}Ta_{2/3})O_3$ powders have been produced by hydrothermal synthesis at low temperatures of $200^{\circ}C$ or less.

A tendency has been found for the formation of ultra-fine, magnesium-deficient particles with the separate precipitation of magnesium as the hydroxide.

An improved synthesis method has been developed, which results in the formation of a large proportion of near-stoichiometric particles but some magnesium-deficient particles are still found.

TEM has been used to show a clear link between the stoichiometry and morphology of the perovskite particles with the magnesium-deficient particles being irregular in shape and ultra-fine, whereas the near-stoichiometric particles are larger and have a rounded shape.

## 5. ACKNOWLEDGEMENTS

The authors would like to thank Dr C.N. Elgy, Mr P.R. Knott and Dr I. Alexander of Morgan Materials Technology Ltd. ($M^2T$) for helpful discussions, and Prof. M.H. Loretto, Director of the IRC in Materials for High Performance Applications, and Prof. I.R. Harris, Head of the School of Metallurgy and Materials, for the provision of laboratory facilities. This work was done as part of a DTI/EPSRC LINK Nanotechnology project in partnership with $M^2T$.

**REFERENCES**
Ataie A, Piramoon M R, Harris I R and Ponton C B 1995 J. Mater. Sci. **30**, 5600
Chen X M, Suzuki Y and Sato N 1994 J. Mater. Sci. - Mater. Electronics **5**, 244
Furuya M and Ochi A 1994 Jpn. J. Appl. Phys. **33**, 5482
Itatani K, Koizumi K, Howell F S, Kishioka A and Kinoshita M 1988 J. Mater. Sci. **23**, 3405
Itatani K, Koizumi K, Howell F S, Kishioka A and Kinoshita M 1989 J. Mater. Sci. **24**, 2603
Katayama S, Yoshinaga I, Yamada N and Nagai T 1996 J. Am Ceram. Soc. **79**, 2059
MacLaren I and Ponton C B, 1997, submitted to J. Mater. Sci.
MacLaren I, Ponton C B, Elgy C N and Knott P R 1997 LINK Nanotechnology Project 38: "Preparation and Characterisation of Ultra-Fine Electroceramic Powders", Internal Progress Report
Matsumoto K, Hiuga T, Takada K and Ichimura H 1986 IEEE Trans. UFFC **33**, 802
Matsumoto H, Tamura H and Wakino K 1991 Jpn. J. Appl. Phys. **30**, 2347
Renoult O, Boilot J P, Chaput F, Papiernik R, Hubert-Pfalzgraf L G and Lejeune M 1992 J. Am. Ceram. Soc. **75**, 3337
Vivekanandan R, Philip S and Kutty T R N 1986 Mater. Res. Bull. **22**, 99

*Inst. Phys. Conf. Ser. No 153: Section 12*
*Paper presented at Electron Microscopy and Analysis Group Conf. EMAG97, Cambridge, 1997*
© *1997 IOP Publishing Ltd*

# Phase stability, defects and deformation mechanisms in the "exotic" intermetallic compounds $Nb_3Al$ and $Cr_2Nb$

**M Aindow, AV Kazantzis and LS Smith**

School of Metallurgy and Materials & IRC in Materials for High Performance Applications,
The University of Birmingham, Edgbaston, Birmingham, B15 2TT, UK

**ABSTRACT:** A review is presented of recent TEM investigations on two of the "exotic" intermetallics $Nb_3Al$ and $Cr_2Nb$. In Nb-Al alloys diffraction data indicate that additional long range order is present in both the A15 and A2 phases. In A15 $Nb_3Al$, planar faults on $\{001\}$ in have a displacement $1/4<021>$ and cannot move conservatively whilst extended screw dislocations with Burgers vectors $<100>$ are dissociated on $\{012\}$ cross-slip planes and are locked. In C15 $Cr_2Nb$ grains deform by either dislocation glide or twinning. The dislocations glide on $<110>\{11\bar{1}\}$ and are composed of widely separated Shockley partials due to the low stacking fault energy. Twinning occurs on only one $<112>\{11\bar{1}\}$ system in any one grain probably due to autocatalytic nucleation. Both deformation modes require synchroshear but the individual components have not been observed directly.

## 1. INTRODUCTION

Intermetallic compounds have become the subject of intense research activity in the metallurgical community, mainly because of the demand from the aerospace sector for materials which exhibit good structural properties at elevated temperatures. Most of this work has been focused on the aluminides of nickel and titanium which are ordered intermetallic compounds, i.e. their structures correspond to simple packing arrangements but with the two atomic species located on particular sites in the unit cell. Thus, TiAl and $Ni_3Al$ exhibit the $L1_0$ and $L1_2$ structures respectively which are both derived from the fcc structure. Similarly, $Ti_3Al$ and NiAl exhibits the $DO_{19}$ (ordered hcp) and B2 (ordered bcc) structures, respectively. Whilst these compounds have been exploited with some success, there is a need for materials that can withstand yet higher temperatures. One possibility is the so-called "exotic" intermetallic compounds whose structures are not ordered but instead have complex bases imposed on simple lattices. There are many such compounds including aluminides of refractory metals, silicides, and those formed between two or more transition metals. The bonding in these structures is significantly more covalent than that in the ordered intermetallics and this, in conjunction with the structure gives high strengths and melting points but very low ductilities and fracture toughnesses. At present these materials are too brittle to be used as monolithic structural materials and can only be incorporated as strengthening components in composite structures or two-phase alloys. If monolithic materials with such structures are to be developed then a much better understanding of the physical metallurgy will be required.

In this paper a brief review is presented of some recent work which has been performed in our laboratory on two of these "exotic" intermetallics: $Nb_3Al$ which exhibits the A15 structure and $Cr_2Nb$ which is a Laves phase that undergoes a polymorphic transformation from the C14 to the C15 structure on cooling. The main emphasis of this review is the contribution which detailed transmission electron microscopy (TEM) observations have made to our understanding of the phase stability, defects and deformation mechanisms in these compounds.

## 2. Nb₃Al

### 2.1 Phase Stability

In common with many intermetallic compounds of the general formula $A_3B$, $Nb_3Al$ exhibits the A15 (βW-type) structure with the space group Pm3n. The structure has a simple cubic lattice with a complex eight atom basis as shown in Fig. 1; there is a bcc arrangement of Al atoms and the Nb atoms lie in chains parallel to the cube axes. The A15 $Nb_3Al$ phase forms by a peritectic reaction between a primary bcc (A2) solid solution of Al in Nb and the liquid on solidification. In recent Nb-Al equilibrium diagrams (e.g. Jorda *et al.* 1980), the compositional ranges given for the phase fields are: 0 to 12%Al - A2 only; 12 to 18%Al - A2+A15; and 18 to 25%Al - A15 only. It has, however, been shown using x-ray diffraction that in rapidly solidified material the peritectic reaction is suppressed giving retained A2 at higher Al contents (Schulze *et al.* 1990).

Electron diffraction studies have, however, revealed more complex behaviour. Firstly there have been several observations of "forbidden" 100-type reflections in selected area diffraction patterns (SADPs) obtained from the A15 phase in binary Nb-Al alloys (e.g. Sudareva *et al.* 1986). Under most circumstances these reflections can be explained on the basis of multiple diffraction events but this cannot account for their presence in SADPs such as Fig. 2 which was obtained with the beam direction, **B**, parallel to [110], since there are no routes to these reflections at this zone axis (Aindow *et al.* 1994). It was shown that the 100-type reflections at this zone axis cannot be due to second phases or anisotropic distributions of bonding electrons and must, therefore correspond to the presence of long range order. It was proposed that this could correspond to vacancies or substitutional Nb atoms on *one* of the Al sublattice sites only. This would give a new cubic structure with the space group Pm3̄.

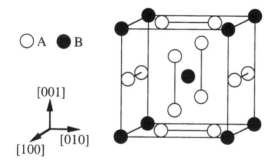

○ A   ● B

[001]

[010]

[100]

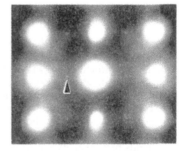

Figure 2 [110] SADP for $Nb_3Al$,
(forbidden 100 spots arrowed)

Figure 1 The unit cell of $Nb_3Al$

Whilst the emphasis in this paper is on the characteristics of the A15 phase it is also interesting to note that additional order has been revealed in SADPs obtained from the A2 solid solution in binary Nb-Al alloys (Kohmoto *et al.* 1993). In melt-spun ribbons a supersaturated A2 phase was observed as reported previously (Schulze *et al.* 1990) but in as-cast ingots or heat-treated ribbons superlattice reflections characteristic of the B2 structure were observed (Fig 3). This is rather unexpected since the B2 structure is normally associated with AB stoichiometry but it could form as a metastable intermediate in the rather sluggish first order A2 to A15 transformation. Much of the recent work on Nb-Al based alloys has been devoted to the addition of transition metals as ternary alloying elements to stabilise this B2 phase as it has been shown to have a good combination of strength and ductility (e.g. Shyue *et al.* 1993).

### 2.2 Defects and deformation mechanisms

The types of defects which occur in the A15 phase $Nb_3Al$ are the same for melt-spun ribbons, as-cast ingots and compressed specimens. It is only the density and distribution of these features which is different.

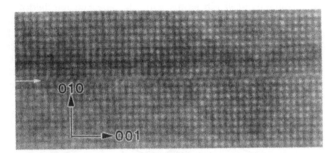

Figure 3 [100] SADP from
Nb-18Al showing B2 order

Figure 4 HREM image from a (001) fault in Nb₃Al

The defects which dominate the microstructure of the A15 phase in cast ingots are large planar stacking faults on {001}(e.g. Aindow *et al.* 1991a). There has been some controversy about the character of these faults, mainly because data from diffraction contrast images were inconclusive. Subsequently, high resolution electron microscopy (HREM) lattice images obtained with these faults edge-on revealed that the displacement across the fault plane is 1/4<021> in each case (e.g. Fig. 4). Since there is a contraction of 1/4 [001] normal to the fault plane these correspond to a missing (004) layer in the structure. We have interpreted this previously as a missing NbAl layer giving an Nb-rich fault but the frequent coupling of these faults in pairs separated by 3/2 [001] has led us to reinterpret these as being a missing Nb layer giving an Al-rich precursor to the formation of the Nb₂Al sigma phase (Smith and Aindow 1997). A further consequence of the contraction across the faults is that the partial dislocations bounding them cannot move conservatively in the fault plane and thus their motion cannot be a main deformation mechanism in the manner suggested by Maruyama *et al.* (1993).

The other defects which are observed in the A15 phase are screw dislocations which have Burgers vectors, **b** = a<100> (e.g. Aindow *et al.*, 1991b). A low density of these dislocations has been observed in as-cast ingots and melt-spun ribbons but they are present in much higher densities in deformed specimens. They must, therefore, correspond to the primary deformation mechanism but there will be insufficient independent slip systems to satisfy von Mises criterion. Moreover, such dislocations are dissociated into partials with **b** = a/2 <100> on {012} and often constrict along their length (Smith *et al.* 1994) as shown in Fig 5. The constrictions correspond to locations where the plane of the dissociation changes from one {012} to another and since the preferred slip plane is {100} this configuration is effectively "locked". It is probably this combination of limited slip systems and locking which is responsible for the brittle behaviour of Nb₃Al.

Figure 5 WBDF image and schematic of a dissociated dislocation in Nb₃Al

## 3. Cr₂Nb

The intermetallic compound Cr₂Nb is a Laves phase which exhibits the cubic C15 (MgCu₂-type) structure below ≈1600°C and the hexagonal C14 (MgZn₂-type) structure between this temperature and ≈1770° C, where it melts congruently. The C15 structure has a face-centred cubic Bravais lattice and a complex basis as shown in Fig. 6; there is a diamond

538

cubic arrangement of Nb atoms with close-packed tetrahedra of Cr atoms located with their centres at the "unfilled" [0.25, 0.25, 0.25] sites in the cell. The high temperature polymorph has the closely related C14 structure in which the same structural units occur but in a hcp arrangement rather than fcc. In contrast to $Nb_3Al$ the phases observed in $Cr_2Nb$ usually correspond to those expected on the basis of the equilibrium phase diagram. The only exception is for samples which have been cooled very rapidly through the C14/C15 transformation temperature when occasional regions of other metastable polymorphs are formed (Kazantzis et al. 1995).

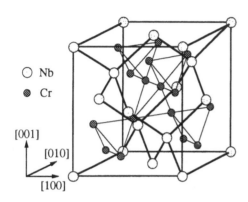

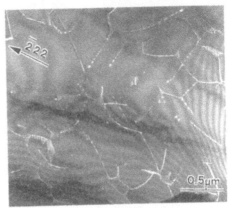

Figure 6  The unit cell of C15 $Cr_2Nb$

Figure 7  A dislocated region in $Cr_2Nb$

The ductile to brittle transition temperature (DBTT) for $Cr_2Nb$ is around 1200° C and samples deformed above this temperature (but below the transformation temperature) have distinctive deformation microstructures whereby each grain is dominated by either slip (Yoshida et al. 1995) or twinning (Kazantzis et al. 1997a). A typical area for a grain which deforms by dislocation glide is shown in Fig. 7. The distribution of the dislocations is consistent with <110>{111} glide as expected for an fcc lattice and most of them are dissociated into Shockley partials in the glide plane. The separations of the partial dislocation pairs and the radii of curvature for the threefold nodes indicate that the stacking fault energy is only ≈$25mJm^{-2}$ (Kazantzis et al. 1996) which is comparable to that for many elemental fcc metals! In those grains which deform by twinning the twinning system is <112>{111} as expected for an fcc lattice but the microstructures are rather unusual with the twins in any one grain being found only on one of the four possible composition planes (e.g. Fig 8). Such behaviour could be due to inhibited nucleation giving twins only from a limited number of regenerative sources, but is more likely to result from an auto-catalytic effect as observed in martensitic transformations.

Figure 8 SADP and WBDF image from a twinned region in $Cr_2Nb$.

In contrast to Nb$_3$Al it is not immediately clear from the character and distribution of the defects why the compound Cr$_2$Nb should exhibit such brittle behaviour. For either of the two microstructures observed, the deformation will proceed by the motion of Shockley partials with **b** = 1/6 <112> on a {111} which contains this vector. It has, however, been shown by Chu and Pope (1995) that the passage of a Shockley partial on {111} in the C15 structure will not produce a simple fault or an additional layer of twin as it would in the fcc structure. This can be shown most clearly by considering the [110] projection of the C15 structure as shown in Figure 9. Each {111} interplanar spacing consists of four layers; a dense Cr layer labelled A,B or C; a less dense Cr layer labelled a, b or c and two low density Nb layers which lie on either side of this labelled α, β or γ. The passage of a Shockley partial through every fourth layer (i.e. on each {111}) would twin the lattice but would disturb the AαcββB-type stacking in the structure. To retain the simple fault or twin character would requires the co-ordinated non-collinear shears of two adjacent atomic sheets within an individual plane {111}. This process is known as synchroshear and can be envisaged as the dissociation of a Shockley partial *out* of the glide plane into two synchro-Shockleys passing between adjacent atomic layers thus:

$$\tfrac{1}{6}\begin{bmatrix}1 & \bar{2} & 1\end{bmatrix} \Rightarrow \tfrac{1}{6}\begin{bmatrix}\bar{1} & \bar{1} & 2\end{bmatrix} + \tfrac{1}{6}\begin{bmatrix}2 & \bar{1} & \bar{1}\end{bmatrix}$$

For C15 the bonding to the ABC - type layers is more rigid so the synchroShockley partials pass between the α & c, and c & β layers, respectively in the "αcβ-type" shearable blocks as shown below

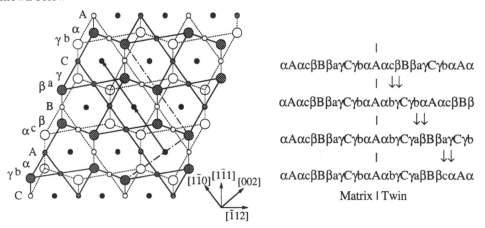

αAαcβBβaγCγbαAαcβBβaγCγbαAα

$\qquad$ ↓↓

αAαcβBβaγCγbαAαbγCγbαAαcβBβ

$\qquad\quad$ ↓↓

αAαcβBβaγCγbαAαbγCγaβBβaγCγb

$\qquad\qquad$ ↓↓

αAαcβBβaγCγbαAαbγCγaβBβcαAα

Matrix I Twin

Figure 9 (a) atomic layers parallel to {111} in C15; (b) synchroshears required for twinning

Attempts have been made to resolve the splitting between the synchro-Shockleys by HREM (e.g. Fig 9) but it has not been possible to distinguish the effects of core splitting from projection effects since the partials in the twin boundaries do not adopt well-defined crystallographic orientations (Kazantzis *et al.* 1997b). Nevertheless, it seems unlikely that these components will be separated widely and it may thus be more appropriate to regard these as zonal cores spread out of the glide plane rather than a true dissociation. Since such features will inevitably be less mobile than simple Shockley partials, the high DBTT may be

Figure 10 HREM image of a twinning dislocation in Cr$_2$Nb

540

related to the thermal activation which is required to overcome the Peierls barriers to the motion of these defects.

## 4. CONCLUSIONS

Transmission electron microscopy is an important tool in the study of novel structural materials, and is particularly useful for exotic intermetallics where it is the peculiarities of the crystallography which can determine the properties. In our work on Nb-Al alloys, diffraction patterns have been used to revealed additional order in the A15 and A2 phases. Defect analysis on $Nb_3Al$ has shown that the planar faults cannot move conservatively and that the dislocations are locked. It is probably these factors, in conjunction with the limited number of independent slip systems, which lead to the brittle character of the A15 phase. In $Cr_2Nb$, however, it was more difficult to establish a link between the microstructures and properties. The grains deformed by dislocation glide or twinning on the usual systems for an fcc lattice but it was necessary to invoke a zonal synchroshear core for Shockley partials to account for the high DBTT. This core structure was, however not resolved directly.

## 5. ACKNOWLEDGEMENTS

The authors would like to thank the many colleagues who have contributed to the exotic intermetallics programme. In particular we are grateful to Professors H.L. Fraser, I.P. Jones and M.H. Loretto for their input. Financial support for most of the work reviewed in this article was provided by EPSRC, DERA and The University of Birmingham.

## REFERENCES

Aindow M, Cheng TT, Beanland R, Shyue J and Fraser HL, 1991a Inst. Phys. Conf. Ser. **90** 249.
Aindow M, Shyue J, Gaspar TA and Fraser HL, 1991b Phil. Mag. Letters **64** 59.
Aindow M, Smith LS, Shyue J, Loretto MH and Fraser HL 1994 Phil. Mag. Letts. **69** 23.
Chu F and Pope DP, 1995 MRS Symp. Proc. **364** 1197.
Jorda JL, Flükiger R and Muller J 1980 J. Less Comm. Metals **75**, 227.
Kazantzis AV, Cheng TT, Aindow M and Jones IP, 1995 Inst. Phys. Conf. Ser. **147** 511.
Kazantzis AV, Aindow M and Jones IP, 1996 Phil. Mag. Lett. **74** 129.
Kazantzis AV, Aindow M and Jones IP 1997a Mater. Sci. Eng. A (in press)
Kazantzis AV, Aindow M, Hutchison JL, Doole RC and Jones IP 1997b unpublished work
Kohmoto H, Shyue J, Aindow M and Fraser H L 1993 Scripta Metall. **29** 1271.
Maruyama Y, Hanada S, Obara K and Hiraga K, 1993 Phil. Mag. A **67** 251.
Schulze K, Müller G and Petzow G 1990 J. Less Comm. Metals **158**, 71.
Shyue J, Hou D-H, Aindow M and Fraser H L 1993a J. Mat. Sci. and Eng. **170** 1.
Sudareva SV, Romanov YP, Popova YN, Prekul AF, Rassokhin VA and Yartsev SV 1986 Fiz. Metall. Metalloved. **61** 1121.
Smith LS, Aindow M and Loretto MH, 1994 in Proc. ICEM13 (Ed. de Physique, Paris) **2** 71.
Smith LS and Aindow M 1997 in preparation.
Yoshida M, Takasugi T and Hanada S, 1995, MRS Symp. Proc. **364** 1395.

*Inst. Phys. Conf. Ser. No 153: Section 12*
*Paper presented at Electron Microscopy and Analysis Group Conf. EMAG97, Cambridge, 1997*
© 1997 IOP Publishing Ltd

# Defect model of martensite formation

**T Nixon and R C Pond**

Materials Science and Engineering Department, University of Liverpool, Liverpool L69 3BX

**ABSTRACT:** A model of the process of martensitic phase transformation is described in which plate growth occurs by the movement of interfacial defects along the interface. Through the quantitative analysis of the topological features of these defects and the material fluxes which accompany their motion and interaction, a model is developed which is consistent with the phenomenological theory. The particular cases of pure Ti and a dilute alloy are analysed and the predictions compared with experimental observations.

## 1. INTRODUCTION

The current understanding of martensitic transformation is based on the phenomenological theory of Wechsler, Lieberman and Read (1955). This model expresses the transformation through a matrix expression comprising a Bain Strain, a lattice invariant shear, a rotation and a dilatation, and successfully describes the crystallographic aspects of martensite plates which exhibit an invariant-plane relationship with the matrix. This approach does not describe the microstructural evolution of the plates, which is modelled here in terms of the motion along the interface of defects traditionally known as transformation dislocations (Bilby 1953). The topological properties of interfacial defects and the diffusive fluxes associated with their motion and interaction can be determined quantitatively. It is then possible to show how transformation dislocations can simultaneously accommodate misfit and effect diffusionless transformation. The crystallographic framework for defining disconnections (interfacial defects with dislocation and step character), the expression describing the material flux associated with their motion (Hirth and Pond 1996), and the transformation mechanism are outlined in section 2. Application of the model to the particular cases of pure Ti and a dilute alloy is addressed in section 3. Finally we comment on the correspondence of our present approach with that of the phenomenological theory, and compare our predictions with microscopical observations.

## 2. MECHANISTIC MODEL

When the orientation relationship between two crystals is known, the topological parameters (Burgers vector, $\mathbf{b}$ and step height, h) of all admissible interfacial defects can be determined (Pond 1989). An interfacial defect exhibiting dislocation and/or step character is defined as a disconnection (finite $\mathbf{b}$ and h), a dislocation (finite $\mathbf{b}$ and h = 0), or a pure step ($\mathbf{b}$ = 0 and finite h). Of the many possible defects that can be identified using this framework, only a few are thought to be energetically and mechanistically plausible and hence likely to accommodate misfit and effect transformation.

Hirth and Pond (1996) have shown that for an elemental system, the flux of material associated with the motion of a disconnection ($\mathbf{b}$, h) is given by the

$$I = v\,[h\Delta X + b_z X], \qquad (1)$$

where I is the number of atoms arriving per unit length of the defect per second, v is the velocity of the defect, $b_z$ is the component of $\mathbf{b}$ normal to the interface, $\Delta X$ is the difference in the number of atoms per unit volume in the two crystals, and X is the density of one of the crystals depending on the relative signs of h and $b_z$. The first term in this expression describes the contribution to I of the step portion of a defect and is associated with the transformation process. The second term relates to the dislocation portion of a defect and corresponds to the growth/evaporation of one or both of the crystals, but not to transformation.

For diffusionless motion of a defect along an interface, the flux given by expression (1) must be equal to zero; it has been shown elsewhere (Hirth and Pond 1996) that if two crystals exhibit an invariant-plane relationship then the two terms in expression (1) are equal and opposite for certain defects and the total flux is zero. In this way in a defect can move along an interface without diffusion and effect a transformation even though there may be a component of **b** normal to the interface. In other words, when an invariant-plane exists, the densities of atomic planes parallel to the interface in the adjacent crystals must be related by an integer factor, although their interplanar spacings may not be so simply related. Thus the atoms of one crystal may be relocated to sites of the other by a combination of shear and shuffle without diffusion by the defect mechanism outlined above.

Additional fluxes may arise when interfacial defects interact because the dislocation portion of one may have to climb up or down the step portion of the other, and vice-versa. As an illustration of a simple interaction, fig.1 shows the case of two orthogonal defects, a disconnection, p (finite **b** and h), and a misfit dislocation, q (finite **b**, h = 0). Imagine that disconnection p moves a distance $y^p$ parallel to y and q moves $x^q$ parallel to x. In this case the interaction flux corresponds to the extension of the misfit dislocation's 'extra half plane', shown hatched in Fig.1, as the disconnection moves in the x direction, i.e. the misfit dislocation has climbed up the step of the disconnection.

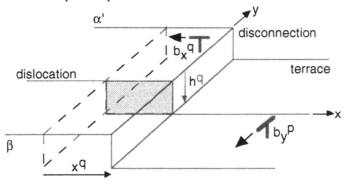

Figure 1. Schematic illustration of the interaction of a dislocation and an orthogonal disconnection.

The martensitic transformation process is modelled by assuming initially the formation of an embryo with low-index faces which is homogeneously strained to fit its surrounding matrix with an invariant-plane relationship. Interfacial defects are then generated which accommodate misfit as the plate grows and effect transformation by their motion. Different arrays of candidate defects can be imagined to accomplish this but only particular combinations would retain the invariant-plane relationship. Accommodation of the misfit between two crystals is seen to be necessary but not sufficient to sustain an invariant-plane relationship as a plate develops. An invariant-plane relation, however, is achieved if the motion and interaction of the defect arrays is diffusionless.

## 3. APPLICATION OF THE MODEL

### 3.1. β(bcc) → α'(hcp) transformation in Pure Ti

The plates are assumed to develop from an embryo orientated at the exact Burgers relationship, $(1\bar{1}00)_{\alpha'}$ // $(2\bar{1}\bar{1})_\beta$ and $[\bar{1}\bar{1}20]_{\alpha'}$ // $[\bar{1}\bar{1}\bar{1}]_\beta$ (Nishiyama 1978), and which is homogeneously strained to exhibit invariant $(1\bar{1}00)_{\alpha'}$ interfaces. An illustration of the formation of a disconnection with line direction parallel to $[0001]_{\alpha'}$ depicted schematically in Fig.2. This defect has relatively small $|\mathbf{b}|$ and h and hence is considered to be energetically and

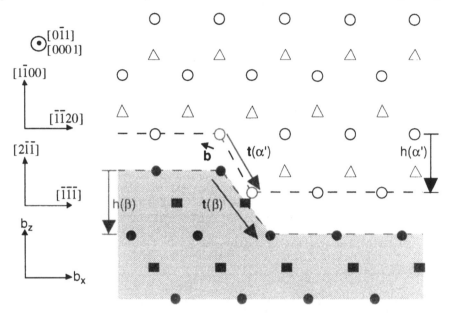

Figure 2. Schematic illustration of the formation of a disconnection with line direction parallel to $[0001]_{\alpha'}$.

mechanistically plausible. The 3.74% misfit parallel to $[\bar{1}\bar{1}20]_{\alpha'}$ could be completely accommodated in a growing plate by an array of such disconnections spaced 1.12nm apart, as illustrated in Fig.3. Since these disconnections exhibit steps, the trace of the interface in Fig.3 will deviate by $\theta_y = 13.55°$ from $[\bar{1}\bar{1}\bar{1}]_{\beta}$, i.e. the habit plane would be $(3.71, \bar{3}, \bar{3})$. Moreover, the normal component of $\mathbf{b}$ would cause a relative rotation of the crystals away from the Burgers orientation by $\phi_y = 0.51°$ about $[0001]_{\alpha'}$.

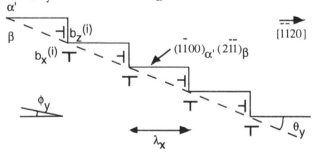

Figure 3. Schematic illustration of misfit accommodation by an array of disconnections of the type illustrated in Fig.2. projected along $[0001]_{\alpha'}$.

Complete accommodation of the 0.86% misfit parallel to $[0001]_{\alpha'}$ could be achieved by an array of crystal dislocations with $\mathbf{b} = \mathbf{c}(\alpha')$ spaced 54.41nm apart. An array of this type would not induce any deviation of the interface plane orientation, $\theta_x$, or misorientation of the crystals, $\phi_x$. No flux is associated with motion of the disconnections, however the interaction between these and the misfit dislocations will involve a flux because the crystal dislocations must climb down the steps of the disconnections. From the conclusions reached above, it appears that the transformation mechanism could not operate martensitically. However, martensitic behaviour is possible if the misfit along $[0001]_{\alpha'}$ is accommodated by $(1\bar{1}01)_{\alpha'}$ and $(1\bar{1}0\bar{1})_{\alpha'}$ twins, as is observed experimentally (Nishiyama 1978).

### 3.2. β(bcc)→α(hcp) Transformation in Ti-6.62 atm %Cr

This alloy has been studied using HREM by Furuhara et al. (1991). Our modelling shows that for complete accommodation of the 4.77% misfit parallel to $[\bar{1}\bar{1}20]_{\alpha'}$, the array of disconnections depicted in Fig.3 would have to be spaced 0.83nm apart. This would induce a rotation of the habit plane of $\theta_y = 17.60°$ and a change of relative orientation of the two crystals of $\phi_y = 0.65°$. These values are in excellent agreement with the experimentally observed ones (Furuhara et al. 1991). Misfit parallel to $[0001]_{\alpha'}$ was not observed to be accommodated by twinning, consistent with the formation of Widmanstätten α (Furuhara 1991). Instead, complete relief of the 2.06% misfit could be achieved by an array of disconnections with $b_x = c(\alpha)/2$ spaced 11.56nm apart, and with step heights and $b_z$ and $b_x$ components which sum to zero (Hirth et al. in Press). Such arrays would not induce any further rotations $\theta_x$ or $\phi_x$. These values are also in good agreement with the experimental observations of Furuhara et al. (1991) which revealed an array of defects with component $b_x = c(\alpha)/2$ spaced $\lambda_y = 10\text{-}13$nm apart and $\theta_x \approx 0$, $\phi_x \approx 0$.

### 4. CONCLUSIONS

The model of martensite formation outlined here provides mechanistic insight into the formation of plate-shaped products through the motion of interfacial defects. The approach is quite different from that of the Phenomenological Theory of Martensite Crystallography (PTMC), but retains its central postulate, the formation of an invariant-plane interface. Misfit accommodation between the two phases is a necessary but not sufficient condition for the formation of such an invariant plane. In our defect model the invariant plane relation can be retained throughout growth from an embryo to the final plate when the interface advances through disconnection motion in a diffusionless manner. Alternative arrays of disconnections can relieve the misfit, but the growth process is not martensitic in general because their motion and interaction is not diffusionless.

### ACKNOWLEDGEMENTS

The authors express their gratitude to Professor J. P. Hirth for many stimulating discussions and to the E.C. under contract CHRX-CT 940407 (R. C. Pond) and EPSRC (T. Nixon) for their financial support.

### REFERENCES

Bilby B A 1953 Phil. Mag. **44**, 782
Furuhara T, Howe J M and Aaronson H I 1991 Acta Metall. Mater. **39**, 2873
Hirth J P and Pond R C 1996 Acta Mater. **44**, 4749
Hirth J P, Nixon T and Pond R C in Press
Lieberman D S, Wechsler M S and Read T A 1955 J. Appl. Phys. **26**, 473
Nishiyama Z 1978 Martensitic Transformations (New York: Academic Press) pp 68
Pond R C 1989 Dislocations in Solids **8**, ed. F R N Nabarro (Amsterdam: North Holland) pp 1

*Inst. Phys. Conf. Ser. No 153: Section 12*
*Paper presented at Electron Microscopy and Analysis Group Conf. EMAG97, Cambridge, 1997*
© *1997 IOP Publishing Ltd*

# Mechanical alloying of FeAl

**E M Knutson-Wedel and M Åsberg**

Department of Engineering Metals, Chalmers University of Technology, SE-412 96
Göteborg, Sweden

**ABSTRACT**: Powders consisting of 60 at% Fe and 40 at% Al were mechanically alloyed in a planetary ball mill. The high energy milling process resulted in small powder particles composed of an extended solid solution, Fe(Al). After up to four hours of milling, laminae of different composition were observed which decreased to a size below detection limit in the SEM after heat treatment or further milling. The heat treatment resulted also in a non-stoichiometric ordered FeAl. A decrease in milling ball size resulted in a decrease of the width of the lamellae.

## 1. INTRODUCTION

Conventional casting of FeAl has proven to be difficult (Vedula 1995). Microstructural segregation results in regions of very high aluminium content which are brittle and cause microcracking during cooling or subsequent processing. An added disadvantage is the large grain size of casted material. Powder metallurgy thus presents the possibility of an improved production route, both in terms of mechanical behaviour and of producing details of near net shape.

Mechanical alloying (MA) is a dry, high energy ball milling process for producing composite powder. Pioneered in the late 1960's by Benjamin (Benjamin 1970) it has proven to have several fields of application; oxide dispersion strengthened alloys, amorphous alloys, solid solutions, intermediate phases, extended solid solutions and nanocrystalline materials amongst others (Benjamin 1990, Froes et al. 1995, Gilman and Benjamin 1983). In principle it is an iterative process where high energy milling balls cold weld, fracture and reweld powder particles by ball-powder-ball collisions. In fact, MA permits a very large departure from equilibrium, opening up the opportunity of large flexibility in microstructural manipulation. In addition, it is a process which is scaleable to commercial production quantities, unlike other "far from equilibrium" processes. Consolidation of the powder produced can be achieved by hot extrusion, sintering or hot isostatic pressing.

The aim of the present study was to investigate the development of a non-stoichiometric FeAl powder during mechanical alloying in a planetary ball mill followed by heat treatment, and to study the effect of some milling parameters. The starting powder mix consisted of pure Fe and Al, thus the process had to take place in an inert atmosphere to avoid oxidation.

## 2. EXPERIMENTAL DETAILS

Milling of Fe powder and Al powder was performed in an argon atmosphere using a planetary ball mill (Fritsch Pulverisette). The Fe powder used was Höganäs, grade ASC100, size < 180 μm , or grade ASC200, size < 150 μm. The Al powder was Goodfellow, size < 60 μm. The starting powder mix, consisting of 60 at% Fe and 40 at% Al was milled at different ball to powder ratios (BPR) and ball size. Small quantities of powder were periodically removed from the container for investigation, see Table 1. Milling was performed at a speed of 360 rpm and 0.5 wt% of stearic acid was used as a process control agent (PCA).

Table 1.    Milling parameters. (* = Fe powder grade ASC200)

| Sample number | 1 | 2 | 2b* | 3* | 4 | 4b* | 5* | 6 |
|---|---|---|---|---|---|---|---|---|
| Time  (h) | 2 | 3 | 3 | 3 | 4 | 4 | 4 | 16 |
| BPR | 4:1 | 4:1 | 4.1 | 4:1 | 4:1 | 4:1 | 4:1 | 8:1 |
| Ball size (mm) | 20 | 20 | 20 | 15 | 20 | 20 | 15 | 20 |

Following milling, a heat treatment for one hour at 570 °C, corresponding to a homologous temperature $T/T_m$ of 0.5, was also carried out for all powder blends (apart from sample no 1) in order to induce a transformation into an ordered FeAl intermetallic.

X-ray diffraction was carried out of all powders, before and after heat treatment, using a Philips PW 1130 diffractometer equipped with a Cu anode. The powder was also characterised by analytical electron microscopy using a Jeol superprobe 733 scanning electron microscope equipped with a Link AN10000 EDX system for quantitative elemental analysis and a Zeiss EM 912 Omega transmission electron microscope, equipped with a Link ISIS EDX system and an energy filter for electron energy loss spectroscopy.

For SEM, powder was mounted in conductive epoxy. The mount was subsequently ground and polished. Thin foils for transmission electron microscopy of powder were prepared by blending the powder into a special epoxy. The blend was hardened inside a 3 mm brass tube, which was subsequently cut into slices. The slice was ground, polished and glued onto a copper support ring. The resultant 30 μm thick powder-epoxy foil was then ion milled at 2-3 kV in a Gatan PIPS ion mill at a low incident angle of 4° which decreased the risk of preferential sputtering of the epoxy.

## 3.    RESULTS AND DISCUSSION

As described by Gilman the mechanical alloying process consists of three stages; early, intermediate and final stage (Gilman and Benjamin 1983). The early process results in cold welded layers of 100 % pure material, but during the intermediate stage fracture and subsequent cold welding of the work hardened particles results in formation of an intermetallic with a structure of lamellae. Finally, the interlamellae spacing decreases and the material becomes more homogeneous. The alloying is considered complete when the interlamellae spacing and diffusion distances are equal (Courtney and Maurice 1996).

In this work, after a short milling time, 2 h, x-ray diffractograms showed a small shift in the Fe peaks and a large decrease of the Al peak heights compared to unmilled powder, implying dissolution of Al in Fe. SEM combined with EDX also showed that the resultant powder particle size distribution was about 10 to 50 μm with an internal structure of lamellae with composition varying from 100 at% Fe to less than 40 at% Fe, see Fig. 1a. The process had thus attained the second of the three stages of processing. After increasing the milling time to 3 hours (sample 2), XRD still implied dissolution of Al into Fe. There were large differences in particle size, from a few μm up to several hundred μm but the lamellae width had decreased, see Fig. 1b. The composition still varied, typically just above 60 at% Fe ± 10 at%.

After  4 hours (sample 4) there was a significant change in the XRD results, now showing lower and broader peaks. The shifts in position had increased to angles corresponding to FeAl. The observed broadening can be a result of several factors; non uniform strain in the milled powder; nanocrystalline particles or possibly partly amorphized FeAl (Cullity 1978). By XRD, it was not possible to separate whether the shift was induced by extended solid solution Fe(Al) or if there existed regions of more or less ordered FeAl. Using SEM it was found that the powder particle size had decreased to about 1-20 μm, the earlier noted large scatter in size also declined as well as the lamellae width, see Fig. 3a. The composition had also become more equal to the starting powder blend with a standard deviation of 2 at%, thus the process could be considered to have entered the final stage.

In sample 6, after a long time of milling,16 h, with a higher BPR, 8:1, SEM studies revealed that the resultant powder particles were about 1-20 μm in size and virtually free from lamellae, see Fig. 2a, but XRD showed no difference from the powder milled for 4 hours. However, using electron diffraction in TEM combined with bright and dark field imaging, it was possible to conclude that the powder consisted of very small nanocrystallites as well as relatively small faulted particles, see Fig. 2b, apart from the larger particles detected in the SEM. It was not possible to completely exclude

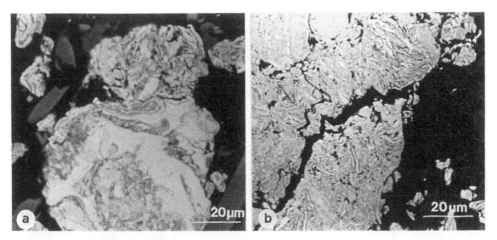

*Figure 1. SEM images (backscattered mode) showing (a) large interlamellae spacings after 2 h of milling (b) lamellae after 3 h of milling (sample 3).*

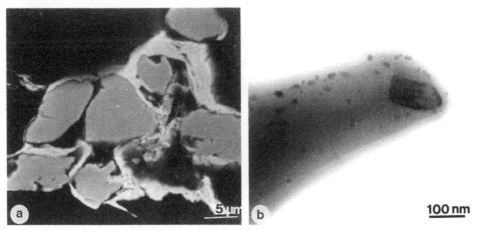

*Figure 2. SEM image (backscattered mode) showing the powder virtually free of lamellae after 16 h of milling. (b) TEM bright field zero loss image showing a small powder particle besides nanocrystalline powder particles, all embedded in epoxy.*

ordering of the structure of the as-milled powder. The corresponding weak superlattice (100) and (111) reflections arising from ordering of the structure were not detected by SAED, but this could be due to their weak intensity. However, residual strains frequently present in milled metal powder (Cullity 1978), could possibly hinder the formation of an ordered FeAl structure. It was not possible to detect amorphous regions, and it has also been pointed out in other work that amorphous material generally demands that the starting powder composition lies in the range of 50-80 at% Al (Dong et al. 1991). EDX studies in both TEM and SEM revealed that the content of Fe was around 70 at% after 16 h of milling even though only 60 at% was added in the starting powder mix. This might be due to uptake of Fe from milling balls made of Cr rich steel which were subjected to wear during milling. It was also possible to detect streaks containing Cr inside the powder particles.

Diffusion during heat treatment resulted in all cases in a non-stoichiometric, ordered FeAl. X-ray diffractograms showed large narrow peaks, including the superlattice (100) reflection and SEM studies showed also particle growth in addition to complete disappearance of the lamellae, see Fig. 3b. Thus, it could not be concluded whether the broadening of peaks from as-milled powder samples was due to nanocrystalline particles or due to a strained non-ordered structure of the as-milled powder.

548

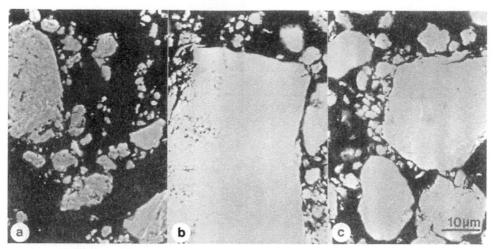

*Figure 3. SEM images (backscattered mode) of (a) sample 4 milled for 4 hours (b) the effect of heat treatment on width of lamellae in sample 4 and (c) the effect of decreasing ball size on width of lamellae in sample 4b\*.*

No large differences in powder particle size distribution or lamellae width could be detected keeping BPR and ball size constant but decreasing Fe powder size from maximum 180 μm to maximum 150 μm. However, keeping BPR and powder size constant but decreasing ball size, resulted in a significant decrease of lamellae width, see Fig. 3c, for both 3 h and 4 h runs. However, no difference in powder particle size distribution could be detected.

## 4. CONCLUDING REMARKS

It is possible to mechanically alloy a non-stoichiometric ordered FeAl by milling in a planetary ball mill followed by heat treatment.

Milling time affects particle size, lamellae width and composition. A long milling time, 16 h, results in an increase in Fe content possibly due to uptake of Fe induced by wear of milling balls.

Heat treatment results in all cases in FeAl free of lamellae and with a homogeneous composition and a smaller spread in particle size distribution.

Keeping BPR constant but decreasing ball size resulted in a significant decrease in interlamellae spacing.

## ACKNOWLEDGEMENTS

P. Hansson and F. Wedberg, undergraduate students, are thanked for their contribution to the results presented here.

## REFERENCES

Benjamin J S 1970 Metallurgical Transactions **1,** 2943-51
Benjamin J S 1990 Metal Powder Report **45,** 122-7
Courtney T H and Maurice D 1996 Scripta Materialia **34,** 5-11
Cullity B D 1978 (USA: Addison-Wesley Publ. Comp. Inc.)
Dong Y D, Wang W H, Liu L, Xiao K Q, Tong S H and He Y Z 1991 Mat. Sci. Eng. **A134,** 867-71
Froes F H, Suryanarayana C, Russell K and Li C-G 1995 Mat. Sci. Eng. 612-23
Gilman P S and Benjamin J S 1983 Ann. Rev. Mater. Sci. **13,** 279-300

*Inst. Phys. Conf. Ser. No 153: Section 12*
*Paper presented at Electron Microscopy and Analysis Group Conf. EMAG97, Cambridge, 1997*
© *1997 IOP Publishing Ltd*

# Martensitic transformation and characterisation of the structure of a NiAl - Ni₃Al alloy

**E Pekarskaya, G A Botton, C N Jones\*, C J Humphreys**

Department of Materials Science and Metallurgy, University of Cambridge
Pembroke street, Cambridge CB2 3QZ, UK
\*ROLLS-ROYCE plc, PO BOX 31, Derby, DE24 8BJ, UK

**ABSTRACT:** A microstructural study of a Ni-32at.%Al-5 at.%Fe intermetallic alloy has been performed by electron microscopy and X-ray diffraction analysis. The martensitic ($L1_0$) and the Ni₃Al ($\gamma'$) phases have been observed in the as-cast alloy. Details of the martensitic transformation are studied by *in situ* heating and cooling experiments. It has been observed that annealing in the ($\beta + \gamma'$) phase region leads to the reverse martensitic transformation of the $L1_0$ phase to the NiAl ($\beta$) phase and to NiAl $\longrightarrow$ Ni₃Al phase transformation. The volume fraction of the Ni₃Al phase increases with increasing annealing temperature.

## 1 INTRODUCTION

Nickel aluminides containing two intermetallic phases, NiAl and Ni₃Al, offer potential for high temperature aerospace applications. These materials have advantages over conventional nickel based superalloys of low density, high thermal conductivity and good oxidation resistance. However, these intermetallic compounds have not found widespread application due to low plasticity at ambient temperatures and low high temperature strength, but in view of the potential applications research into the materials continues. Since the defect structures and microstructure govern ultimately the properties of the material, structural investigations are of vital importance in order to improve the properties of these alloys.

Since ternary alloying additions appear to significantly improve the mechanical properties of NiAl, in particular, Fe, Mo and Ga have been shown to substantially increase the ductility of single crystals (Darolia 1991), our present study concerns the influence of Fe on a NiAl - Ni₃Al alloy microstructure with global composition Ni-32at.%Al-5 at.%Fe.

Ni-Al alloys containing 32 -38 at.% Al, however, have been known to undergo thermoelastic martensitic transformations which are detrimental in structural applications. Control of the martensitic start temperature and thereby of the formation of the martensite phase can be achieved in the ternary multiphase alloys by altering the composition of the $\beta$-phase through an appropriate choice of annealing temperature (Kainuma et al 1992). In the present work two heat treatments have been performed in order to suppress formation of the martensite and also to study the influence of Fe and the effect of annealing on the crystal structure.

## 2 EXPERIMENTAL

The as-cast alloy and the alloys subjected to heat treatments have been studied using X-ray diffraction analysis, conventional transmission electron microscopy (TEM), high resolution electron microscopy (HREM) and dedicated scanning transmission electron microscopy (STEM). After sealing the samples in an argon atmosphere, they were annealed at 800°C for 15 min (sample S2) and 15 hrs (sample S3) in the two phase $(\beta + \gamma')$ region according to the binary phase diagram (Massalski 1986). Annealing was followed by furnace cooling to room temperature. TEM samples were prepared by jet-polishing in a 10 % $H_2SO_4$ - methanol solution at 12 V and temperature $-3\ ^\circ C$. A Philips CM30 TEM operating at 300 kV equipped with energy dispersive X-ray spectrometer (EDX ) LZ4 AN10000 system (thin window detector), JEOL 4000FX (400 kV operating voltage), VGHB501 STEM (equipped with windowless EDX detector) have been used for characterisation.

# 3   RESULTS AND DISCUSSION

## 3.1   The as-cast alloy

TEM and X-ray diffraction analysis have shown that the as-cast alloy consists of two phases, namely the $Ni_3Al$ phase ($L1_2$ crystal structure) and the tetragonal martensitic phase ($L1_0$ structure type). A typical structure of the martensitic phase is presented in Fig. 1. The presence of superlattice reflections in diffraction patterns indicates that both phases are ordered. Apparently, the martensitic phase forms from the parent NiAl phase ($\beta$-phase, B2 crystal structure) during the cooling process (Warlimont and Delaev 1974). The $L1_0$ phase is internally twinned. Two types of twins (Fig. 1) are formed in order to accommodate the large elastic energy of the martensitic transformation. The primary transformation twins are large groups of parallel-sided plates with clearly defined boundaries between them (Fig. 1). They occur on several crystallographic planes. The width of the twins varies from less than one $\mu$m up to $10\mu$m with a mean value about $2\mu$m. The internal striations across the martensitic plates correspond to the secondary nanometer-size twins. These are $\{111\} < 112 >$ twins for which regular straight boundaries are observed in the HREM images (Fig. 2).

EDX analysis performed in TEM and STEM shows that both phases are chemically homogeneous and no iron segregation is detected in the as-cast alloy. The content of the phases are 62 at.%Ni-33 at.%Al-5 at.%Fe and 77 at.%Ni-18 at.%Al-5at.%Fe for the martensite and $Ni_3Al$ phases respectively.

From X-ray diffraction analysis the volume fraction of the $Ni_3Al$ ($\gamma'$) phase is about 0.3. Also, the lattice constants of both phases have been obtained. The lattice parameters are $3.58\mathring{A}$ for the cubic $Ni_3Al$ phase and $3.77\mathring{A}$ $(a)$ and $3.27\mathring{A}$ $(c)$ for the tetragonal martensitic phase. Small deviations from the literature data (Weissman 1978), $3.56\mathring{A}$ for $Ni_3Al$ and $3.78\mathring{A}$ and $3.28\mathring{A}$ for martensite, can be caused by the presence of the Fe atoms.

## 3.2   In situ heating and cooling experiments

To study the features of the martensitic transformation in the alloy, heating and cooling experiments have been performed in the TEM. A foil of the as-cast alloy has been heated up to about 600°C and then cooled down to room temperature. The first changes in the microstructure (the disappearance of the secondary twins) occur at about 200°C and, with continuously increasing temperature, the reverse martensitic transformation occurs first in the thicker areas of the foil and then this process proceeds towards the edge of the foil. There can be two reasons for such behaviour. First, the heating and cooling rates of the sample can be sensitive to the specimen thickness and consequently, the martensitic transformation temperature appears to be dependent on thickness. Also, in thinner regions the free surface will probably affect the transformation mechanisms. Therefore, it is not possible to define exactly the start and the end of the martensitic transformation from TEM observations because of the above effects. However, observations in thicker areas (these would tend to approach the bulk behaviour) allow the start and the end temperatures of the reverse martensitic transformation to be estimated, which are then at about 200 and 450°C respectively. SADPs taken at a temperature of about 600°C from the areas where the martensitic transformation has been

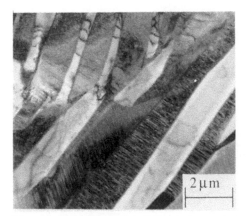

Figure 1: *Bright field image of the typical structure of the martensite in the as-cast alloy.*

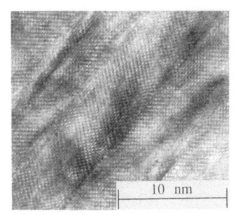

Figure 2: *HREM image of the secondary twin structure in the as-cast alloy. The image was taken with the (111) plane parallel to the electron beam.*

completed show that the tetragonal $L1_0$ phase has been transformed back into the cubic $\beta$-phase.

After cooling an unusual "tweed" contrast is present in the image (Fig. 3). This contrast is likely to be connected with the static lattice distortions which are believed to be premartensitic structures. In fact, similar quasi periodic "tweed" strain contrast has been observed in NiAl - B2 alloys with 56 at.% Ni as described in the work of Schryvers (1995). This "tweed" contrast is believed to appear due to the long range static strain arising from the interacting centers of distortions and has been identified as $\{110\} < 1\bar{1}0 >$ shear (Schryvers 1995). In the present study the internal striations do not have the same orientation and therefore can not be associated with particular directions in the lattice.

### 3.3 Evolution of the microstructure during annealing

During annealing and consecutive cooling, the $L1_0$ phase undergoes a reverse martensitic transformation to the $\beta$-phase, and then the phase transformation $\beta \longrightarrow \gamma'$ phase occurs, although some martensite still remains in both S2 and S3 alloys. A TEM image of the S3 sample structure (Fig. 4) shows the ordered NiAl (B2) phase and the ordered $Ni_3Al$ ($L1_2$) phase as identified from selected area diffraction patterns, X-ray microanalysis in TEM and X-ray diffraction analysis.

An interesting feature, namely the correlation of the Al and Fe concentrations in $\beta$ and $\gamma'$ phases, has been observed in all EDX measurements in the annealed alloys. The iron content appears to increase with increasing Al concentration, consequently, there is more Fe in the B2 phase than in the $\gamma'$-phase.

In both alloys $Ni_3Al$ crystals have elongated shapes (Fig. 4). Annealing for a longer time leads to a change in the width range of the $Ni_3Al$ phases from $0 - 1.8\mu m$ in the S2 alloy to $0 - 2.6\mu m$ in the S3 alloy and also to an increase in the mean value from $0.4\mu m$ (S2) to $0.6\mu m$ (S3). The volume fraction of the $\gamma'$ phase in the alloy increases during annealing from 30% in the as-cast alloy up to 44% in the S2 alloy and 54% in the S3 material.

Dislocations have been observed in both the $\beta$ and the $\gamma'$ phases in the annealed alloys with the dislocation density being higher in the S3 alloy. In the S3 alloy the $Ni_3Al$ phases consist of small fragments with low-angle dislocation boundaries (indicated by arrows in Fig. 4). The dislocations appear due to the fact that the volume change during the reverse martensitic

552

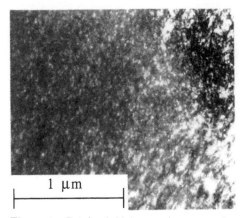

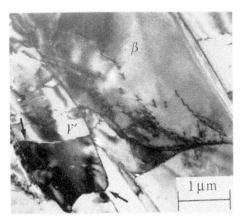

Figure 3: *Bright field image demonstrating the "tweed" contrast in the $L1_0$ phase.*

Figure 4: *Typical TEM image of the S3 alloy. See text for discussion.*

transformation and the $\beta \longrightarrow L1_2$ phase transformation causes stress fields, and their relaxation leads to the formation of the dislocation substructure observed in both alloys.

TEM observations show that the orientation relationships between the two phases, $\beta$ and $\gamma'$ are close to the Nishiyama-Wasserman one (Williams and Carter 1996) being more frequent in the S3 alloy. For the S3 alloy these are

$$[001]_\beta \parallel [\bar{1}01]_{\gamma'}$$
$$(\bar{1}10)_\beta \parallel (111)_{\gamma'} \text{ or } (112_{\gamma'})$$
$$\text{or more rarely } (121)_\beta \parallel (021)_{\gamma'}$$

## 4  SUMMARY

A microstructural characterisation of a Ni-32 at.%Al-5 at.%Fe alloy has been carried out by electron microscopy and X-ray diffraction analysis. An internally twinned martensitic phase formed from the B2-phase is observed in the as-cast alloy. *In situ* heating and cooling experiments performed in the TEM show that the reverse martensitic transformation leads to formation of the $\beta$-phase. After cooling a "tweed" contrast possibly related to the premartensitic static distortions of the lattice appear in the image of the $L1_0$ phase.

It has been observed that in Ni-32 at.%Al-5 at.%Fe annealing at 800°C leads to a decrease of the martensitic transformation temperature below room temperature and to the $L1_0 \longrightarrow B2 \longrightarrow B2+L1_2$ phase transformations. The orientation relationships between $\beta$ and $\gamma'$ phases are close to the Nishiyama-Wassermann orientation relationships.

We would like to thank ROLLS-ROYCE plc for supporting the research.

## REFERENCES

Darolia R 1991 JOM **43**, March, 44

Kainuma R, Ishida K and Nishizawa T 1992 Metall.Trans 23A, 1147

Massalski T B 1986 Binary Alloy Phase Diagrams (American Society for Metals) pp 140-3

Schryvers D 1995 J. De Physique IV **5**, 225

Warlimont H and Delaev L 1974 Progress in Materials Science **18** (Pergamon Press)

Weissman S et al 1978 Metals and Alloys (JCPDS: International Centre for Diffraction Data)

Williams D B and Carter C B 1996 Transmission Electron Microscopy (New-York: Plenum Press) pp 280-5

*Inst. Phys. Conf. Ser. No 153: Section 12*
*Paper presented at Electron Microscopy and Analysis Group Conf. EMAG97, Cambridge, 1997*
© 1997 IOP Publishing Ltd

# Al3Zr Precipitation Behaviour in a Melt-Spun Al-0.2Cu-1.2Mg-0.5Zr Alloy

**C G Jiao[1], M Aindow[1], G F Yu[2] and M G Yan[2]**

1. School of Metallurgy and Materials, The University of Birmingham, Edgbaston, Birmingham, B15 2TT, UK.
2. Beijing Institute of Aeronautical Materials (BIAM), Beijing 10095, P.R.China.

**ABSTRACT**: The $Al_3Zr$ precipitation behaviours of as-quenched and heat-treated samples of a melt-spun Al-0.2Cu-1.2Mg-0.5Zr alloy have been studied by TEM. The results show that metastable f.c.c. $Al_3Zr$ can be formed during solidification, and that initial $Al_3Zr$ precipitation occurred in the subgrain boundaries. Fine spherical $L1_2$ $Al_3Zr$ particles precipitated within the grains after heat treatment at 450°C for 16 hours, and they were stable in this form even after holding at 450°C for 100 hours. The stable tetragonal $DO_{23}$ $Al_3Zr$ phase was found after heat treatment at 500°C for 100 hours.

## 1. INTRODUCTION

The addition of zirconium to commercial aluminium alloys has been used not only for controlling recrystallisation and refining grain structure, but also for improving toughness, stress-corrosion resistance and quench sensitivity. These improvements result from a high density of metastable cubic $Al_3Zr$ precipitates. The phase transformation behaviour of $Al_3Zr$ in aluminium alloys has been studied extensively by Ryum (1969), Nes (1972, 1977), Gayle (1989) and Desch (1991). There are however disagreements on the temperatures at which the various forms of $Al_3Zr$ precipitate and the details of the precipitation sequence itself. In our work we have studied the precipitation behaviour in a Al-0.2Cu-1.2Mg-0.5Zr alloy which we have shown previously to exhibit only $Al_3Zr$ and f.c.c Al solid solution phases (Jiao, 1994). Melt spun ribbons of the alloy have been produced and the precipitation sequence for $Al_3Zr$ as a function of annealing has been examined.

## 2. EXPERIMENTAL PROCEDURE

A pre-alloyed ingot with a nominal composition of Al-0.2Cu-1.2Mg-0.5Zr (in weight percent) was remelted in a sealed silica tube under argon atmosphere in RF melting furnace with a holding time of around 200 to 300s. The molten metal was rapidly solidified into ribbons under an argon atmosphere by using a single roller melt spinning apparatus. The surface velocity of the copper wheel was 20 to 25 $m \cdot s^{-1}$. The average thickness of ribbons was about 80 μm and the width was in range of 6 to 8 mm. The quenched ribbons were heat-treated to temperatures of 300 to 530°C for 2 to 100 hours. Twin-jet electro-polishing with 30% nitric acid in methanol at -30°C and 15V was used to prepare TEM foils from the as-spun and annealed ribbons and these were examined in a JOEL 2000FX TEM.

## 3. RESULTS

Typical microstructures from the pre-alloyed ingot and the cross-section of the as-spun ribbons are shown in figure 1. The plate-like structures in fig.1(a) are precipitates which exhibit the stable $DO_{23}$ $Al_3Zr$ structure in cold water cooled ingots. For the as-spun ribbons, columnar grains were present at the near-roller surface, and cellular structures were observed at the upper free surface of the ribbon. The existence of cellular structures demonstrated that the solidification of the ribbon was not sufficiently rapid to prevent solute diffusion.

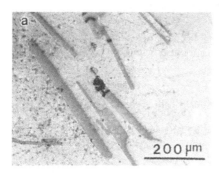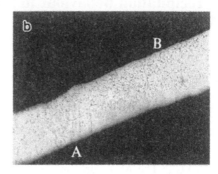

Fig.1 Optical micrographs of (a) the water-cooled ingot (b) the as quenched melt-spun ribbon; the near wheel surface and the top free surface are marked A and B, respectively.

Randomly distributed particles which are 20 to 80 nm in diameter and comprise 3 to 5 vol.% of the alloy are dispersed in the matrix of the as-spun ribbons as shown in figure 2. In qualitative EDX analyses from the particles only zirconium and aluminium peaks were detected. Selected area diffraction patterns (SADPs) from these showed only diffraction maxima corresponding to f.c.c. Al indicating that the particles are coherent. Whilst it is difficult to prove that there is no additional order, it appears that the randomly distributed initial dispersoids in the as-spun ribbons are comprised of a metastable Zr-containing f.c.c. phase.

The amount of the dispersoid in the annealed samples was the same for the as-quenched ones, i.e. 3 to 5% in volume, after annealing for less than two hours at any of the temperatures considered here (300 to 530°C), however their size had increased. SADPs of the dispersoids in these samples showed weak superlattice spots corresponding to the $L1_2$ $Al_3Zr$ phase (figure 3). Diffraction patterns obtained from the matrix showed that no additional precipitation had occurred in the matrix after annealing for 2 hours.

Longer annealing treatments were conducted at 450°C and 500°C to examine the thermal stability of the $Al_3Zr$ phases. TEM micrographs showing the changes in the $Al_3Zr$ dispersoids after annealing for up to 100 hours at 450°C are shown in figure 4. It can be seen clearly that the initial $Al_3Zr$ dispersoids are situated on subgrain boundaries. On annealing the dispersoids first increased in size, and then decreased with increasing annealing time, and finally dissolved gradually in the matrix. The maximum size of these particles did not exceed 500 nm in diameter.

Very fine spherical precipitates with diameters of 4 to 8 nm appeared homogeneously throughout the matrix after annealing at 450°C for 16 hours as shown in figure 5. Strong $L1_2$ f.c.c. $Al_3Zr$ superlattice reflections were observed in SADPs obtained at major zone axes such as those shown inset in figure 5. Not only is a 16 hour anneal needed before these precipitates appear but they are also very stable after they form; after 100 hours at 450°C, the precipitate diameters had only increased to 30 to 50 nm.

TEM images showed that the equilibrium tetragonal $DO_{23}$ $Al_3Zr$ phase was present in samples annealed at 500°C for 100 hours (figure 6). It appeared either in the form of rod-like precipitates or in a fan-shaped arrangement of discontinuous particles that resemble those reported by Nes (1972).

## 4. DISCUSSION

Three types of Zr-containing particles have been observed in the present study. The first is the initial dispersoids, which remained in the as-spun ribbons after various annealing treatments. The second is the fine $L1_2$ precipitates which formed in the matrix after longer annealings at 450°C. The third is the rod-like or discontinuous $DO_{23}$ phase which formed at

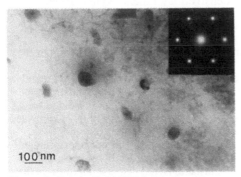

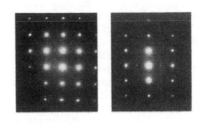

Figure 2 Bright field TEM image and [111]$_{Al}$ diffraction pattern from the as-spun ribbon.

Figure 3 [100] and [112] SAD patterns from particles after annealing at 400°C for 2 hours showing the weak superlattice spots.

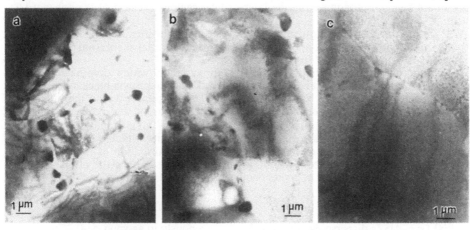

Figure 4 Bright field TEM images showing the distribution of the Al$_3$Zr precipitates in samples annealed at 450°C for (a) 6h, (b) 20h, (c) 100h

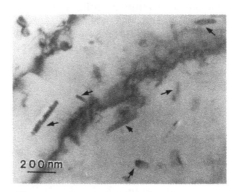

Figure 5 TEM dark field image for the specimen annealed at 450°C for 16 hours with [111] and [100] SADPs inset.

Figure 6 Bright field TEM image of rod-shaped tetragonal DO$_{23}$ Al$_3$Zr in the sample annealed at 500°C for 100 hours with [100] SADP inset.

500°C. The latter two types have been studied extensively by others (Ryum, 1969; Nes, 1972, 1977) and the observations presented here are consistent with this previous work. There has, however, been very little attention given to the dispersoids and therefore in the following discussion we will concentrate on these particles.

There are several possible reasons for the formation of initial $Al_3Zr$ dispersoids in the ribbon including:

(i) The cooling rate for the ribbon is not high enough to obtain a partitionless dispersoid-free solid solution. We estimated that the cooling rate for the present melt-spun experiment is in the range $10^5$ to $10^6$ K·S$^{-1}$. However, a cooling rate of $9 \times 10^{12}$ K·S$^{-1}$ is needed for a planar solidification front in aluminium alloys (Hirth, 1978). So, with such a high Zr content, the metastable $Al_3Zr$ phase may nucleate and grow as solid particles in the melt.

(ii) The Zr content of the present alloy is very high and thus they will be supersaturated solid solutions. Since the growth front will be non-planar, the extent of this supersaturation may be so high in the inter-cellular regions that the metastable $Al_3Zr$ dispersoids form on the sub-grain boundaries in these regions during cooling.

(iii) Segregation may be such that recalescence can occur during the rapid solidification inducing phase separation in the supersaturated solid solution behind the growth front. This effect would be more pronounced in the inter-cellular regions because of the enhanced Zr content.

We are not able to distinguish between these possibilities on the basis of our microstructural observations alone but the location of the initial dispersoids at subgrain boundaries seems to indicate that (ii) is more likely.

It is important to recall that it is the $L1_2$ precipitates whicn dominate the final microstructures of these alloys. It is clear from our work that the growth mechanisms for the dispersoids and the precipitates are different. The $L1_2$ precipitates within the matrix are much more stable than the dispersoids. This may be related to the difference between the diffusivity of Zr in Al which is very low ($D_{zr} = 2.39 \times 10^{-19}$ m$^2$s$^{-1}$) and will control the coarsening of the precipitates, and that in the grain boundaries which will inevitably be higher. The final stage in the precipitation sequence is the formation of the stable $DO_{23}$ phase which occurred after annealing at 500°C in our samples. This stage probably arises because of short circuit diffusion at grain boundaries and boundary migration (Porter 1992).

## 5. CONCLUSIONS

Initial f.c.c. $Al_3Zr$ dispersoids are formed inhomogenously in as-spun Al-0.2Cu-1.2Mg-0.5Zr ribbons. The dispersoids are situated on grain boundaries after annealing at 450°C. A much longer annealing time is needed to decompose the zirconium-containing supersaturated solid solution further to precipitate $Al_3Zr$ in the matrix. Very fine spherical $L1_2$ $Al_3Zr$ precipitates appeared homogeneously throughout the matrix after annealing at 450°C for 16 hours, and they were stable in this form even after holding at 450°C for 100 hours. Rod-like equilibrium $DO_{23}$ $Al_3Zr$ was present in samples after annealing at 500°C for 100 hours.

## REFERENCES

Desch, P. B., Schwarz, R.B., (1991), J. of the Less-Common Metals, **168,** pp. 69-80.
Gayle, F. W., Vandersande, B., (1989), Acta Metall., **37** (4) pp. 1033-1046.
Hirth, J. P. , Metall.Trans., (1978), **9A**, pp. 401.
Jiao, C. G., Yu, G. F., (1994), Acta Aeronautica Sinica, **15** (12), pp. 1450.
Nes,E., (1972), Acta Metallurgica, **20,** pp. 499-506.
Nes, E., Billdal, H., (1977), Acta Metallurgica, **25,** pp. 1031-1037.
Porter, D. A. and Easterling, K. E., (1992), Phase Transformations in Metals and Alloys, 2nd ed., London: Chapman & Hall.
Ryum, N., Acta Metallurgica, (1969), **17,** pp. 269-278.

Inst. Phys. Conf. Ser. No 153: Section 12
Paper presented at Electron Microscopy and Analysis Group Conf. EMAG97, Cambridge, 1997
© 1997 IOP Publishing Ltd

# Intermetallic precipitates in superaustenitic stainless steel welds

**S Heino, E M Knutson-Wedel and B Karlsson**

Department of Engineering Metals, Chalmers University of Technology, SE-412 96
Göteborg, Sweden

ABSTRACT: A superaustenitic stainless steel (Fe-25Cr-22Ni-7.3Mo-3Mn-0.5N) has been studied with special emphasis on welding. The aim of this study was to investigate if intermetallic phases form in the heat-affected zone of welds and to correlate this to a recorded temperature history. After welding discrete elongated grain boundary precipitates of varying length (30-800 nm) were found in the heat-affected zone. The precipitates were identified as σ-phase with a high Mo-content (22 wt%) and the precipitation time was estimated to less than 30 seconds. No additional intermetallic phases or carbides were found.

## 1. INTRODUCTION

Superaustenitic stainless steels possess a very high resistance to localised corrosion in chloride containing environments and to uniform corrosion in halide containing acids. These properties are mainly achieved by careful balancing of the alloying elements chromium, nickel, molybdenum and nitrogen. However, the high content of alloying elements acts as a driving force for precipitation of intermetallic phases, which may impair the corrosion as well as the mechanical properties. The matrix is austenitic (γ) with no ferrite present but intermetallic phases may be present in small amounts in the material. These phases appear as stringers in the sheet centres and will not affect the corrosion properties at the surface (Liljas 1995).

The aim of this work was to investigate if intermetallic phases form in the heat-affected zone during welding, to study their composition and crystal structure and to correlate the precipitation behaviour to a recorded temperature history.

## 2. EXPERIMENTAL DETAILS

The investigated material was the commercial superaustenitic stainless steel grade Avesta 654SMO, with composition given in Table 1. An electro slag refined material was used in the experiments to provide an as segregation free material as possible.

Plates of 12mm thickness with a grain size of 90-100 μm were welded using a nickel based filler material, Avesta P16, of composition Ni-0.02C-23Cr-16Mo-0.5Mn-0.2Si (wt%). The weld geometry was an Y-joint and the weld was produced in three runs. The first run was made with plasma arc welding and the second and third runs with TIG-welding. Argon was used as shielding gas. Heat input during the runs was 2.0 kJ/mm for the first run and 2.1 and 2.6 kJ/mm for the second and third run respectively. These values were higher than recommended, thus providing a possibility to see the precipitation region more clearly. The temperature history in the heat-affected zone was recorded using thermocouples mounted in holes drilled 4 mm from the fusion line at a distance 3 mm from the plate surface. Fig. 1 shows the weld with the heat-affected zone.

558

Table 1. Chemical composition of nominal and as-received material (wt%), determined by Avesta Sheffield AB.

| Avesta 654 SMO | Fe | C | Cr | Ni | Mo | N | Mn | Other |
|---|---|---|---|---|---|---|---|---|
| **Nominal** | Bal. | 0.01 | 24 | 22 | 7.3 | 0.5 | 3-4 | 0.5Cu |
| **As-received** | Bal. | 0.012 | 24.7 | 21.6 | 7.6 | 0.46 | 3.29 | 0.38Cu |

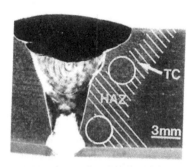

*Figure 1. Optical micrograph of the weld showing the heat-affected zone (HAZ), position of thermocouples (TC) and the regions investigated by TEM (circles).*

Regions with secondary phases were identified using Leica MZ12 and Leitz MM6 optical microscopes and transmission electron microscopy was performed in a Zeiss 912 OMEGA equipped with an OMEGA energy filter and a Link ISIS EDX system for quantitative analysis.

Specimens for optical microscopy were prepared by electrolytic etching in 5% hydrochloric acid in ethanol to reveal intermetallic precipitates and by chemical etching in 5ml HNO$_3$+50ml HCl+50ml H$_2$O+6g FeCl$_3$ to reveal the grain structure. Thin foils for transmission electron microscopy were extracted from the heat-affected zone (Fig. 1) and electropolished in 15% perchloric acid in ethanol at 24V and −30°C.

## 3. RESULTS AND DISCUSSION

TEM studies showed that grain boundaries near the fusion line were decorated with precipitates. The shape was plate-like and the precipitates were uniformly distributed along the grain boundaries as shown in Fig. 2.

From selected area electron diffraction (SAED) patterns the grain boundary precipitates were identified to be σ-phase (Hall and Algie 1966). Fig. 3 and 4 show the different appearance of the precipitates and Fig. 5 shows a dark field electron micrograph from a grain boundary particle containing stacking faults. The amount of distortion was found to decrease with increasing precipitate size.

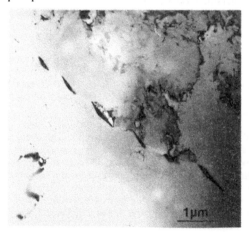

*Figure 2. Grain boundary with σ-phase precipitates. Zero-loss image.*

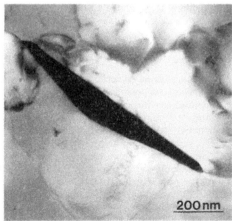

*Figure 3. Large σ-phase precipitate. Zero-loss image.*

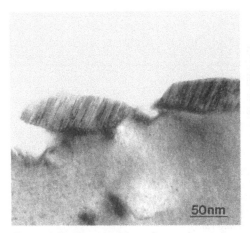

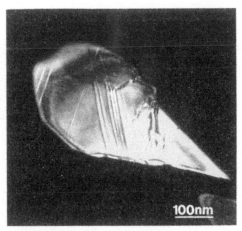

*Figure 4. Small, highly distorted grain boundary precipitate. Zero-loss image.*

*Figure 5. Zero-loss dark field electron micrograph of σ-phase precipitate with stacking faults.*

EDX analysis showed an average molybdenum content of 22 wt% and chromium and nickel contents of 30 wt% and 12 wt% respectively. Only small variations in composition (1-2 wt%) were detected between different precipitates. Compared to previously reported σ-phase compositions in lower alloyed austenitic stainless steels (Slattery et al 1983, Weiss and Stickler 1972) the Mo content was found to be high. High content of Mo in σ-phase was reported by Allten (1954) for a highly alloyed austenitic steel. This σ-phase composition was 34.4Fe-25.3Cr-25.0Mo-15.3Ni with a stability range up to 1300°C. In a 5 wt% Mo superaustenitic stainless steel, Charles et al (1987) reported a σ-phase composition of 34.7Fe-39.9Cr-13.7Mo-11.7Ni, forming at 700-1100°C. Thus, the present work further indicates that the molybdenum content of the σ-phase increases with molybdenum content in the matrix.

All analysed grain boundary precipitates were found to be σ-phase. Grain boundary precipitation in the heat-affected zone has previously been investigated both for AISI 304 (Kokawa and Kuwana 1991a,b) and AISI 316L (Doig and Flewitt 1988) austenitic stainless steels and only formation of carbides was reported in these papers. In the present study, the high content of alloying elements together with the low carbon content explains that no carbides and only σ-phase was found.

The present study concerns nucleation of σ-phase at γ/γ grain boundaries which also has been reported in lower alloyed austenitic stainless steels (Weiss and Stickler 1972). However, in those materials as well as in duplex stainless steels, δ-ferrite is usually present which is richer in alloying elements Mo and Cr that promote intermetallic phase formation. During ageing, intermetallic phases and/or carbides precipitate predominantly at δ/γ interfaces (Farrar 1995, Farrar and Thomas 1983, Lai and Haigh 1979, Slattery et al 1983) or at both δ/γ and δ/δ grain boundaries in the case of duplex stainless steels (Josefsson et al 1991).

The approximate temperature history in the regions investigated by TEM was extrapolated from the one recorded (Fig. 6) by using the Rosenthal equations (Heino et al). This gives a possibility to correlate the temperature history and the precipitation behaviour as shown in Fig. 7. With 800°C as a possible lower limit of intermetallic phase formation, the time for precipitation can be estimated to less than 30 seconds. Thus, the present study concerns shorter precipitation times compared to previously published TTP-diagrams (Charles et al 1987, Weiss and Stickler 1972). The work by Charles et al showed the presence of both σ- and χ-phases. However, that investigation concerned precipitation times longer than 60 seconds.

Further investigations will show whether additional intermetallic phases and/or nitrides form in the heat-affected zone. This is likely since a wide range of maximum temperatures is covered. One phase that might form is the χ-phase which is frequently reported in literature for austenitic (Keown and Thomas 1981, Leitnaker 1982, Ritter et al 1983, Song et al 1996, Weiss and Stickler 1972) as well as superaustenitic stainless steels (Charles et al 1987).

560

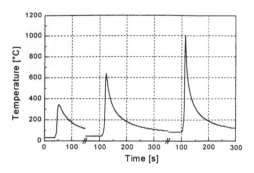

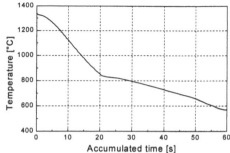

*Figure 6. Temperature history in the heat-affected zone as recorded by thermocouples.*

*Figure 7. Curve showing the accumulated time above a given temperature at a distance 1.5 mm from the fusion line (sum over all three welding runs).*

## 4. CONCLUSIONS

Only the intermetallic σ-phase was detected and no carbides were found in the heat-affected zone after welding. The σ-phase showed a high Mo-content of 22 wt%.

Correlation of precipitation behaviour in the HAZ with the recorded temperature history showed precipitation times less than 30 seconds.

### Acknowledgements

Avesta Sheffield AB is gratefully acknowledged for providing the testing material as well as performing the welding experiments and for technical advise. Personal thanks to Anders Brorson at Avesta Sheffield AB.

This work was financially supported by the Swedish Research Council for Engineering Sciences (TFR).

### REFERENCES

Allten A G 1954 Trans. Am. Inst. Min. Metal. Eng. **200**, 904-5

Charles J, Pugeault P, Soulignac P and Catelin D 1987 Eurocorr '87. European Corrosion Meeting, (Frankfurt am Main: Dechema) pp 601-6

Doig P and Flewitt P E J 1988 EMAG '87. Analytical Electron Microscopy, eds Brown M L (London: The institute of Metals) pp 229-32

Farrar R A 1995 Weld. World **36**, 143-51

Farrar R A and Thomas R G 1983 J. Mater. Sci. **18**, 3461-74

Hall E O and Algie S H 1966 Metall. Rev. **11**, 61-88

Heino S, Karlsson B and Brorson A, To be published

Josefsson B, Nilsson J O and Wilson A 1991 Duplex Stainless Steels '91. Vol. 1, eds J Charles and S Bernhardsson (Les Cedex Ulis) pp 67-78

Keown S R and Thomas R G 1981 Met. Sci. **15**, 386-92

Kokawa H and Kuwana T 1991a Q. J. Jpn. Weld. Soc. **9**, 50-5

Kokawa H and Kuwana T 1991b Q. J. Jpn. Weld. Soc. **9**, 56-61

Lai J K and Haigh J R 1979 Weld. J. **58**, 1s-6s

Leitnaker J M 1982 Weld. J. **61**, 9s-12s

Liljas M 1995 Weld. World **36**, 55-63

Ritter A M, Cieslak M J and Savage W F 1983 Metall. Trans. A **14a**, 37-44

Slattery G F, Keown S R and Lambert M E 1983 Met. Technol. **10**, 373-85

Song Y, Baker T N and McPherson N A 1996 Mater. Sci. Eng. **A212**, 228-34

Weiss B and Stickler R 1972 Metall. Trans. **3**, 851-66

*Inst. Phys. Conf. Ser. No 153: Section 12*
*Paper presented at Electron Microscopy and Analysis Group Conf. EMAG97, Cambridge, 1997*

# Orientation relationships in duplex Nb-Al-V alloys

**D N Horspool, D K Tappin and M Aindow**

School of Metallurgy and Materials and the IRC in Materials for High Performance Applications, The University of Birmingham, Edgbaston, Birmingham, B15 2TT, UK

**ABSTRACT:** A TEM study has been performed of the orientation relationships between lenticular A15 precipitates and the B2 matrix in Nb-10Al-20V and Nb-15Al-20V alloys in the homogenised and cold-rolled & annealed conditions. In most cases these precipitates exhibit the orientation relationship observed previously and the morphology suggests that this may arise because of lattice matching across the habit plane of the precipitates. For the cold-rolled and annealed sample of the Nb-15Al-20V alloy, however, an orientation which has not been reported previously was found.

## 1. INTRODUCTION

The intermetallic compound $Nb_3Al$ exhibits many properties which are desirable for high temperature structural applications but is, however, extremely brittle below 1000°C (Shah and Anton, 1992). Rapid cooling of this alloy from the melt (Kohmoto *et al.*, 1993) produces a metastable phase with the B2 structure. It is possible to stabilise this B2 structure by ternary alloying with additions such as Ti (Shyue *et al.*, 1993a). Mechanical testing of such alloys shows that the B2 phase is inherently ductile, but in the Ti-stabilised alloys embrittlement occurs on long term high temperature exposure due to the formation of omega or orthorhombic phases. In our work we have attempted to overcome this problem by replacing Ti with V, which stabilises the B2, but not the omega or orthorhombic phases. We have examined the phase stability, mechanical properties and deformation mechanisms for a series of five V-modified alloys containing 10-25 at.%Al and 20-40 at.%V (Tappin *et al.*, 1997). These alloys showed extensive ductility, but yield stresses were significantly reduced at elevated temperatures. It may be possible to maintain strength at higher temperatures if the matrix is strengthened by a dispersion of precipitates, such as the A15 phase. Recently we have tried to obtain such duplex microstructures by varying alloy composition and thermal history, and in this paper we present a transmission electron microscope (TEM) study of the orientation relationships (ORs) observed between the B2 and A15 phases in two duplex alloys, Nb-10Al-20V and Nb-15Al-20V.

## 2. EXPERIMENTAL METHOD

Nb-Al-V ingots of 500g weight with nominal compositions Nb-10 at.%Al-20 at.%V and Nb-15 at.%Al-20 at.%V were prepared by plasma melting high purity elemental feedstock under an Ar atmosphere. Portions of each composition were heat-treated at 1500°C for 50hrs in Ar, and for one hour at 1700°C under high vacuum. Cold rolling was performed on homogenised specimens using a laboratory scale mill. Portions of the cold-rolled specimens were heat treated at 1000°C for 50hrs in Ar. Specimens for TEM were prepared by twin-jet electropolishing using a 10% solution of sulphuric acid in methanol at -40°C and 90mA. The thin foils were examined in a Philips CM20 operating at 200kV.

## 3. RESULTS

In the as-cast state, both alloys exhibited a single phase microstructure with columnar and equiaxed zones of large (100-350μm) B2 grains (Horspool *et al.*, 1997). The specimens homogenised at 1700°C also exhibited only the B2 phase in the form of larger equiaxed grains (350-1000μm).

After heat-treating at 1500°C, the Nb-10Al-20V alloy retained a similar B2 microstructure with ≈5% vol. A15 present in the form of occasional large grains. There was no A15 precipitation within the B2 grains, and no well defined orientation relationship between the two phases. Heat-treatment of the Nb-15Al-20V alloy resulted in the formation of ≈ 45% vol. lenticular A15 precipitates (8x60μm) within the B2 grains (100-300μm), and at the grain boundaries (Fig. 1a). Bright field (BF) TEM images from this alloy show that the A15/B2 interfaces tend to be straight and faceted. One example of such an interface is shown in Figure 1b, along with selected area diffraction patterns (SADPs) obtained with the beam direction, B, parallel to [100] and [123] in the B2 matrix (Figs. 1c and d respectively). The extra spots observed in these SADPs correspond to the A15 phase and are consistent with the following OR:

$$\{\bar{1}02\}_{A15} \text{ // } \{011\}_{B2}$$
$$<2\bar{2}1>_{A15} \text{ // } <100>_{B2}$$

(1)

For this OR, $<010>_{A15}$ lies parallel to $<2\bar{1}1>_{B2}$ and therefore (1) is equivalent to that determined by Shyue *et al.* (1993b). Moreover, these latter lattice vectors are parallel to the A15/B2 interface normal for the largest interfacial facets (e.g. Fig. 1b).

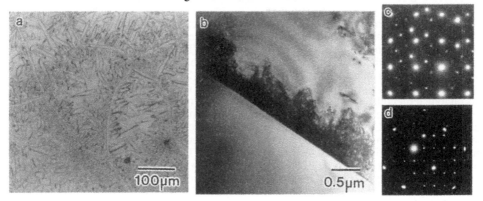

Fig. 1. (a) Optical micrograph from Nb-15Al-20V after 50h at 1500°C; (b) BF TEM image of a B2/A15 interface; (c)&(d) $[100]_{B2}$ and $[123]_{B2}$ SADPs obtained from the region shown in (b).

In an attempt to obtain a finer dispersion of smaller A15 precipitates, homogenised specimens of each alloy were deformed heavily by cold rolling. Reductions in thickness of >80% were achieved with only modest work-hardening. The microstructures of both cold-rolled alloys exhibited fine linear features and TEM analysis (Horspool *et al.*, 1997) revealed that these are regions with an orthorhombic crystal structure and the OR:

$$\{\bar{1}\bar{1}0\}_O \text{ // } \{100\}_{B2}$$
$$<11\bar{1}>_O \text{ // } <011>_{B2}$$

(2)

Similar features have been identified previously in Ti-modified Nb-Al alloys (Yang *et al.*, 1994) and in other B2 compounds (Goo *et al.*, 1985). They are thought to be formed by a stress-induced martensitic transformation referred to as pseudo-twinning. The proposed pseudo-twinning shear is the same as that for true twinning in BCC metals (1/6<111>{112}), but because of the ordered nature of the B2 structure, it results in the formation of the orthorhombic phase. In addition to these pseudo-twins, there was extensive dislocation activity. Diffraction contrast imaging in the TEM showed that these dislocations had Burgers vectors **b** parallel to <111>, tended to adopt screw orientations, and lay in bands on {110}.

Heat-treatment of the cold-rolled specimens at 1000°C produced a fine distribution of A15 precipitates. For the Nb-10Al-20V alloy, small lenticular precipitates (0.3x2μm) formed in bands (Fig. 2a) which were consistent with the A15 phase nucleating preferentially either within the pseudotwins, or at the matrix/pseudotwin interfaces. TEM analysis (Fig. 2b) revealed that these precipitates had clean semi-coherent interfaces with the matrix, and exhibit the same OR as (1).

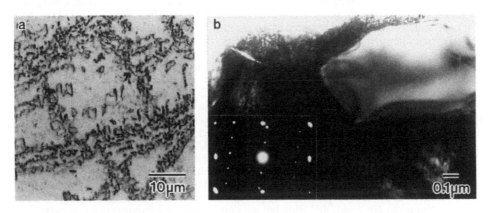

Fig. 2. (a) Optical micrograph from Nb-10Al-20V after cold-rolling and 50h at 1000°C; (b) BF TEM image of an A15 precipitate with inset SADP obtained from the B2/A15 interface.

Heat-treatment of the cold-rolled Nb-15Al-20V alloy at 1000°C also resulted in the formation of small lenticular A15 precipitates (0.5x3μm), but in this case they lay in the matrix, between the previously pseudotwinned regions (Fig. 3a). Here again, TEM analysis (Fig. 3b) revealed clean semi-coherent precipitate/matrix interfaces and a well defined OR of:

$$\{110\}_{A15} \,//\, \{100\}_{B2} \qquad\qquad (3)$$
$$<1\bar{1}2>_{A15} \,//\, <011>_{B2}$$

which is different from that observed previously.

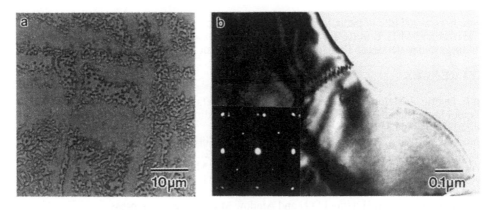

Fig. 3. (a) Optical micrograph from Nb-15Al-20V after cold-rolling and 50h at 1000°C; (b) BF TEM image of an A15 precipitate with inset SADP obtained from the B2/A15 interface.

## 4. DISCUSSION

We consider firstly the specimens annealed at 1500°C for 50 hrs. In the Nb-10Al-20V alloy there was only ≈5% by volume of the A15 phase which had formed as discrete grains between those of the B2 phase, and thus no well-defined OR was observed. For the Nb-15Al-20V alloy, however, there is a much larger proportion of the A15 phase as expected (Tappin *et al.*, 1997) and much of this had precipitated within the B2 matrix as laths, all of which exhibited OR (1) as observed previously (e.g. Shyue *et al.* 1993b). This A15 morphology suggests that lattice matching across the interface may be important in determining the orientation relationship adopted by the precipitates. The interface plane is $\{2\bar{1}1\}_{B2}$ and $\{010\}_{A15}$ and the difference between the spacings of the parallel planes $\{011\}_{B2}$ and $\{\bar{1}02\}_{A15}$ which lie *perpendicular* to the interface is only 1.3%. The misfit in the orthogonal direction is, however much larger (≈20.8%). Indeed in previous studies (Shyue *et al.* 1993b, Hou 1994) subsidiary ORs other than (1) have been observed but in each case these involve planes $\{011\}_{B2}$ and $\{\bar{1}02\}_{A15}$ being parallel.

Cold-rolling and annealing of the alloys at 1000°C produced a much finer dispersion of A15 precipitates, and we have shown elsewhere (Tappin *et al.*, 1997) that this gives greatly improved mechanical properties. Both alloys have cold-rolled microstructures which consist of martensitic pseudotwins and high densities of dislocations. Subsequent heat-treatment appears to produce rather different microstructures. In the Nb-10Al-20V alloy the precipitates exhibit OR(1) and are arranged in bands which correspond to the orientation of the prior pseudotwins, suggesting that they nucleate on or in the pseudotwins during annealing. In the Nb-15Al-20V alloy the precipitates exhibit a different orientation relationship, and indeed one in which $<011>_{B2}$ and $<\bar{1}02>_{A15}$ are not parallel. The precipitates are, however, distributed much more uniformly than in the Nb-10Al-20V alloy but with precipitate-free bands orientated parallel to the prior pseudotwins. We infer that the A15 had nucleated in the matrix between the pseudotwins, perhaps at dislocations, but it is not clear why this results in a different OR.

## 5. CONCLUSIONS

A TEM study of the orientation relationships in two V-modified Nb-Al alloys (Nb-10Al-20V and Nb-15Al-20V) has revealed two distinct orientation relationships between the B2 matrix and the lenticular A15 precipitates. In most specimens the orientation relationship was that observed previously and we propose that this is due to lattice matching in the interface parallel to the habit planes of the laths. The second relationship was observed in the Nb-15Al-20V alloy (but not in the Nb-10Al-20V) after cold-rolling and heat treatment at 1000°C/50hrs, but the reasons for this are not clear.

## 6. ACKNOWLEDGEMENTS

The authors would like to thank Dr P S Bate for assistance with mechanical testing, Professors I R Harris and M H Loretto for the provision of laboratory facilities, and the School of Metallurgy and Materials at The University of Birmingham for financial support.

## REFERENCES

Goo E, Duerig T, Melton K and Sinclair R 1985 Acta Metall. **33** 1725
Horspool D N, Tappin D K and Aindow M 1997 proceedings of ISSI-2, (in press).
Hou D-H 1994 PhD Thesis, The Ohio State University
Kohmoto H, Shyue J, Aindow M and Fraser H L 1993 Scripta Metall. **29** 1271
Shah D M and Anton D L 1992 Mat. Sci. Eng. **A153** 402
Shyue J, Hou D-H, Aindow M and Fraser H L 1993a J. Mat. Sci. and Eng. **170** 1
Shyue J, Hou D-H, Johnson S C, Aindow M and Fraser H L 1993b Mat. Res. Soc. Symp. Proc. **288** 243
Tappin D K, Smith L S, Horspool D N and Aindow M Acta Mater., in press
Yang S S, Hou D-H, Shyue J, Wheeler R and Fraser H L 1994 Mat. Res. Soc. Symp. Proc. **364** 1359

*Inst. Phys. Conf. Ser. No 153: Section 12*
*Paper presented at Electron Microscopy and Analysis Group Conf. EMAG97, Cambridge, 1997*
© 1997 IOP Publishing Ltd

# Intragranular precipitation characterisation within Al-Li-Cu-Mg-Zr(-Ag) alloys.

**WJ Vine.**

Structural Materials Centre, DERA Farnborough, HANTS GU14 0LX, UK

**ABSTRACT:** Silver micro-additions were made to Al-Li-Cu-Mg-Zr in an effort to promote precipitation of the potent strengthening phase $\Omega$ ($Al_2Cu$). Within this alloy system a similar $\{111\}_\alpha$ habit precipitate $T_1$ ($Al_2CuLi$) was however also expected; it remained to be proven whether both would precipitate in Ag bearing systems and if so whether distributions of the two phases could be discriminated from each-other. Plasmon imaging provided a method to distinguish between the two phases and revealed that $T_1$ phase was exclusively formed in all Ag-bearing alloys. Since silver failed to promote $\Omega$ precipitation and had a detrimental influence on alloy ductility its inclusion in Al-Li-Cu-Mg-Zr systems was not recommended.

## 1. INTRODUCTION

Considerable effort is being made into developing new lightweight aluminium-lithium alloys exhibiting significantly higher strength than established systems, such as 8090 (Al-2.4Li-1.2Cu-0.7Mg-(0.06-0.12Zr)). Within the Li-free Al-Cu-Mg system 'micro-addition' of ≤0.4wt% silver stimulates nucleation/ growth of the potent strengthening phase $\Omega$ ($Al_2Cu$), forming as fine slip dispersing platelets on $\{111\}_\alpha$. The mechanism for $\Omega$ precipitation remains unproven, though Ag and Mg atoms are thought to become aligned on $\{111\}_\alpha$ planes such that matrix stacking fault energy is reduced and/or the activation free energy for $\Omega$ nucleation is reduced. Clearly similar addition of minimal quantities of silver to the Li-bearing Al-Li-Cu-Mg-Zr system to produce the strength enhancing $\Omega$ phase was an attractive proposal. However, from limited available phase diagram information, such Ag bearing alloys might be predicted to precipitate two different $\{111\}_\alpha$ habit platelets, namely $\Omega$ or the Li-bearing $T_1$ ($Al_2CuLi$) phase, which does not require silver's presence to form. Precipitation of $\Omega$ phase would be favoured since it would not compete for Li with the most abundant age hardening phase $\delta'$ ($Al_3Li$). To this end, the efficacy of Ag in age hardened Al-Li-Cu-Mg-Zr would be gauged by assessment and discrimination of $\Omega$ from $T_1$ phase in the matrix of a set of Ag-bearing and Ag-free alloys.

## 2. BACKGROUND AND THEORY

This microstructural response of alloys bearing Ag additions was assessed, with specific emphasis placed on discriminating between $T_1$ and $\Omega$ phase distributions. Since the crystallography of $T_1$ phase is still disputed, conventional methods such as HREM or SAD were not favoured; however since $T_1$ and $\Omega$ are compositionally very distinct energy filtered imaging seemed more appropriate. Unfortunately phase discrimination cannot be achieved by mapping using the Li-K edge (54eV), present for $T_1$ but absent for $\Omega$ platelets, since this edge overlaps both with the 4th bulk plasmon (matrix) and Al-L (72eV) edge. Use of the 'low loss' region of the PEELS spectrum was more profitable since Al based intermetallic phases have distinct bulk plasmon energies, and narrow half-widths, readily predicted by the free-electron model. The practical success of plasmon imaging, to identify embedded precipitates in Al-Li-Cu-Mg-Z alloys, required bulk plasmon energies to differ sufficiently from each other ( and from the matrix) such that a narrow energy slit (see section 4.2) could be positioned at a pre-determined energy.

## 3.    EXPERIMENTAL

Al-2.0Li-2.5Cu-0.5Mg-0.12Zr alloy plates bearing 0.0wt%Ag (P0.0), 0.2wt%Ag (P0.2) and 0.5wt%Ag (P0.5) were direct chill cast and thermomechanically processed to 25mm thick plate by conventional procedure. Plates were solution heat-treated and stretched prior to being artificially aged to below the peak-aged temper, to produce a good balance between high strength and fracture toughness. TEM samples were prepared by standard metallographic methods. Conventional TEM utilised a Philips EM-420, operated at 120kV, whilst analytical work was carried out using a 200kV Hitachi HF-2000 TEM (located at the University of Bristol), fitted with cold field-emission source and  a Gatan Imaging Filter (GIF); the energy resolution of this system ≤0.7eV.

## 4.    RESULTS AND DISCUSSION

### 4.1.    Conventional TEM

P0.0, P0.2 and P0.5 predominantly featured large, elongated, unrecrystallised grains which indicated that all plates had retained the {110}<112> hot deformation texture. Grain packets of 2-5μm diameter sub-grains (Figure 1) featured limited amounts of grain boundary precipitation, which was only slightly enhanced for the most Ag enriched plate, P0.5. Intragranular precipitates included a significant volume fraction of δ' phase (15-20nm diameter), small numbers of β' ($Al_3Zr$) dispersoids and large quantities of S' rods ( 120-200nm length) and {111}$_\alpha$ habit platelets (170nm diameter), figure 2. P0.0, P0.2 and P0.5 all exhibited {111}$_\alpha$ habit precipitates (see section 4.2) forming walls at sub-grain boundaries where they *may* offer resistance to planar slip across two adjoining grains and thus improve tensile isotropy. The only discriminating feature between the plates appeared to be a slightly enhanced level of intergranular precipitation within the high silver (0.5wt%) alloy variant. Concerns that silver had largely not been solutionised were discounted by quantitative EDX analysis, which confirmed matrix silver levels to be only slightly below target levels.
The similarities between P0.0, P0.2 and P0.5 in terms of grain size/texture or volume fraction of intragranular reinforcement was consistent with similar measured tensile properties (0.2%Proof Stress~490±5MPa; Ultimate Tensile Strength~550±5MPa). P0.5 plate exhibited slightly lower ductility which might be attributed to segregation of Ag to the grain boundaries. The role of silver, largely remained unclear; certainly its ability to promote widespread {111}$_\alpha$ precipitation was in doubt for such Li bearing alloys.

### 4.2.    Microanalysis.

Plasmon maps of known intermetallic phases were initially produced and their plasmon energies ($E_p$) measured for a range of 'standard' alloy samples (Table 1). It was established that high contrast plasmon images could be readily produced from fine 2-5nm thick platelets in 'edge-on' orientations, so long as a narrow energy slit (1-2eV) was utilised. The signal failed to be dominated by any precipitate-matrix 'interface' plasmon excitation.

Table 1: Theoretical and experimental plasmon energies (and natural half-widths) for some Al intermetallic phases.

| PHASE/ SYSTEM | RELATIONSHIP W. MATRIX | Measured $E_p$ ($\pm 0.1eV$) | $E_p$ (FE) (eV) | $\Delta E_p$ (eV) |
|---|---|---|---|---|
| Al | -- | 15.2 | 15.7 | 0.5 |
| $\delta'$ ($Al_3Li$) | COHERENT $L1_2$ superlattice | 14.2 | 13.8 | 0.6 |
| $T_1$($Al_2CuLi$) | COHERENT $\{0001\}T_1//\{111\}_\alpha$ | 12.5 | 12.6 | 0.6 |
| $\Omega$($Al_2Cu$) | COHERENT $\{001\}/\{0001\}_\alpha // \{111\}$ | 14.8 | 14.0 | 0.6 |

The existence of a 2.3eV plasmon energy difference between $T_1$ and $\Omega$ phases facilitated straightforward discrimination of the phases by energy selected mapping. Energy selected mapping of P0.0, P0.2 and P0.5 plates was then undertaken. Grains within all three alloy variants were aligned near to the [110] zone axis, such that ($1\bar{1}1$) and ($\bar{1}11$) habit intermetallic $T_1/\Omega$ platelets would be very close to edge-on orientation. Figure 3 is a sequence of plasmon maps acquired from P0.0 plate at various energies and features ($1\bar{1}1$) and ($\bar{1}11$) platelets for which the peak plasmon energy was exclusively 12.5eV, i.e. only $T_1$ phase.

Plasmon images were subsequently acquired from the silver bearing P0.2 and P0.5 alloy plates, Figures 4a and b and revealed that all $\{111\}_\alpha$ habit platelets exhibited peak plasmon image intensity at $E_p$=12.5eV, which corresponded *only* to $T_1$ phase. The absence of platelets with a higher associated energy again confirmed the absence of $\Omega$ phase.

The reason for the absence of $\Omega$ phase from all plates has not been confirmed but if Ag and Mg had been prevented, by Li, from becoming aligned on $\{111\}_\alpha$ the $T_1$ phase would be favoured. A significant reduction of Li content would inevitably suppress $T_1$ precipitation, and thus promote $\Omega$ phase, though at detriment to both alloy density and elastic modulus.

## 5.    CONCLUSIONS

Despite the presence of silver 'microadditions' in the Al-Li-Cu-Mg-Zr alloys no changes to the (sub) grain size or texture were observed. Furthermore levels of intragranular precipitation, that is $\delta'$, S' and $\{111\}_\alpha$ habit platelets, remained comparable. These observations showed consistency with the similar tensile properties measured for silver bearing and silver-free variants. Plasmon mapping provided a high spatial resolution means of discriminating between distributions of $\Omega$ ($E_p$=14.8eV) and $T_1$ ($E_p$=12.5eV) precipitates and when applied to experimental plates all were established to be free of $\Omega$ phase. The efficacy of Ag was deemed to be reduced by the presence of Li; Ag appeared to remain either homogeneously distributed through the matrix or to segregate to the grain boundaries.

Figure 1: BF image revealing packets of fine sub-grains

Figure 2: Intragranular precipitation of S' and {111} habit platelets

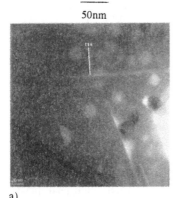

a)

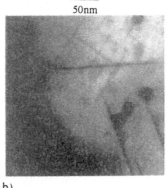

b)

Figure 3: Energy selected images at a)12.5eV and b)14.8eV revealing all $\{111\}_\alpha$ habit platelets to be $T_1$ in P0.0 plate.

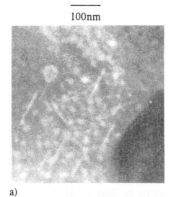

a)

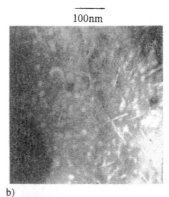

b)

Figure 4: 12.5eV plasmon images revealing $T_1$ platelet distributions in a) P0.2 and b) P0.5 plates.

*Inst. Phys. Conf. Ser. No 153: Section 12*
*Paper presented at Electron Microscopy and Analysis Group Conf. EMAG97, Cambridge, 1997*
© *1997 IOP Publishing Ltd*

# Investigation of the effects of stresses on the formation of ω-phase crystals in β-titanium alloys

**E. Sukedai, H. Nishizawa\*, J. Okuhara\* and H. Hashimoto**

Okayama University of Science, \*: Graduate school, Okayama 700, Japan

**ABSTRACT:**
The effect of a hydrostatic pressure and a tensile stress on aged ω-phase crystal formation in β-titanium alloys was investigated using HREM mthod. It was found that both stressings acted effectively to form homogeneously distributed high density ω-phase crystals. It was considered that a stress activated nucleation-sites of β to ω transformation due to aging.

## 1. INTRODUCTION

Precipitation behaviours of ω-phase crystals in β-titanium alloys such as Ti-Mo, Ti-V and Ti-Cr have been investigated to clarify an influence to their mechanical properties and β to ω transformation behaviours (Duering et al, 1980). The ω-phase crystals are formed by aging and stressing. Especially, it is well known that ω-phase crystals formed by aging make the alloys harder and more brittle (Yukawa et al, 1966). An as-quenched Ti-14Mo (mass%) alloy single crystal shows about 40% strain in tensile test, however, those crystals aged at 623K for 250 s and 100ks became brittle, i.e., the latter one has no plastic strain (Takemoto et al, 1989). Therefore, some efforts to avoid the formation of ω-phase crystals have been made in some industrial processes. Recently, however, it is found that ω-phase crystals become nucleation sites of α-phase crystals in some aged Ti alloys (Takemoto et al, 1993 and Makino et al, 1994), and SP-700 (as shown in Fig. 1). It is considered positively to utilize ω-phase crystals as nucleation-sites of α-phase crystals to prepare (β+α) duplex alloys, which have higher ductility and strength as α-phase crystals are finer (Ouchi et al, 1988). Therefore, it is necessary to develop a new method to precipitate homogeneously distributed fine and high density ω-phase particles in β-titanium alloys.

In the present paper, the effectiveness of applied stresses (hydrostatic pressure and tensile stress) on the formation of such ω-phase particles in Ti-Mo and Ti-Mo-Zr alloys were investigated using high resolution electron microscopy (HREM) and dark field imaging techniques.

## 2. EXPERIMENTAL PROCEDURE

Ti-20Mo (mass%) alloy polycrystals were prepared and a Ti-15Mo-5Zr (mass%) alloy single crystal were prepared by a zone refining method. An isothermal aging was carried out at 623K in air. A Ti-20Mo alloy crystal was aged under a tensile stress of about a half of the yield stress, 413MPa. Ti-15Mo-5Zr alloy crystals were hydrostatically pressurized at 1.5GPa for 10.8ks and some of them were aged at 623K. HREM and dark field images were taken using a JEM 4000 EX operated at 400kV. Detection of the formation of ω-phase particles was carried out using HREM images.

## 3. RESULTS AND DISCUSSION

### 3-1 α-phase initiation from an ω-phase crystal

Fig. 1 shows a HREM image of Ti-4.5Al-3V-2Mo-2Fe (SP-700) alloy crystal aged at 623K for 4.0ks. In the matrix bcc structure, an ω-structure (0.28, 0.40nm spacings) is visible and also the growth of α-phase structure on the top part of ω-particle can be seen. Because at shorter aging period than 4.0ks, no ω-phase structure was observed, it is found that ω-phase particles become formation-sites of α-phase structure in this alloy, too. It seems that α-phase initiation from ω-phase crystals due to aging is a general phenomenon, and it is thought that a method to utilize this phenomenon for preparing (α+β) duplex alloys is valuable.

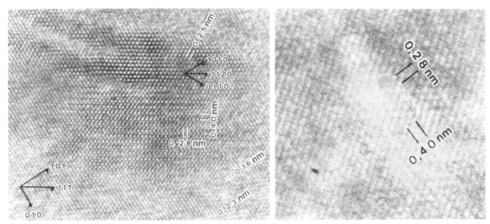

Fig.1 Initiation of α-phase on ω-phase particle. (SP-700 aged at 623K for 4.0ks)

Fig. 2 Formation of ω-phase structure due to a hydrostatic pressurizing at 1.5GPa.

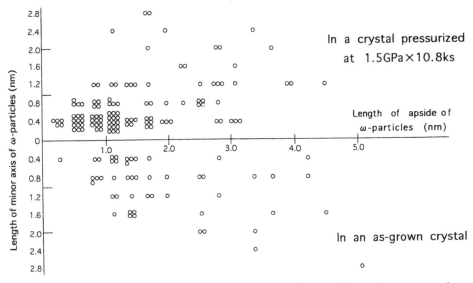

Fig.3 Size distribution of ω-particles in a constant area (40nm x 30nm) of as-grown (lower) and hydrostatically pressurized (upper) Ti-15Mo-5Zr alloys.

## 3-2 Effect of hydrostatic pressure on formation of ω-phase crystals

A Ti-15Mo-5Zr alloy single crystal was hydrostatically pressurized at 1.5GPa for 10.8ks, and HREM observations were carried out. Many small sizes of ω-phase crystals were observed. Fig.2 shows a HREM image. Lattice spacings of 0.28nm and 0.40nm of an ω-phase crystal are visible and the size is about 1.5nm x 2nm. The shape of ω-phase crystals is like an ellipsoid, so the sizes of ω-phase crystals are presented by the lengths of the major and the minor axes. Fig.3 shows the result in an area of 40nm x 30nm together with a result of an as-grown crystal. It is found that the density of small size ω-phase crystals of the pressurized crystal is higher than that of the as-grown crystal. This result indicates that a hydrostatic pressure enhances a nucleation of ω-phase crystals, and the hydrostatic pressure is effective on the formation of fine and high density ω-phase crystals. An age-hardening behaviour of a hydrostatically pressurized Ti-15Mo-5Zr alloy single crystal was also investigated. It was found that the hardness peak of the crystal appeared earlier than an as-grown crystal, however, the peak value was lower than that of as-grown one (Nishizawa et al, 1997). The details of the results will be published soon in elsewhere.

## 3-3 Effect of aging under a tensile stress on formation of ω-phase crystals

Photos.4 (a) and (b) show dark field images of a Ti-20Mo alloy only aged for 12.6ks at 623K and of a Ti-20Mo alloy aged at the same condition under a tensile stress of 413MPa, about a half of the yield stress, respectively. It is found that the density of ω-phase crystals in the specimen aged under the stress is higher than that in the specimen aged only. This result indicates that a tensile stress during aging also enhance to form ω-phase crystals. Since the aged Ti-20Mo crystals was a polycrystal, dark field images from some grains were able to be taken. The number of ω-phase crystals in a constant area of 100nm x 200nm of those grains were counted. The results were summarized as considering relative angles between the stress

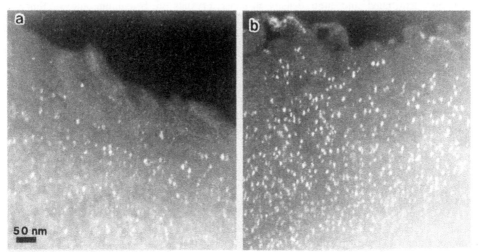

Fig. 4 Dark field images of the formation of ω-phase crystals aged at 623K for 12.6ks: (a) aging only, (b) aging under a tensile stress, 413MPa.

direction and growth directions of ω-phase crystals, <111> directions (Sukedai et al 1991) and the results is shown in Fig. 5. It is found that in the case of only aging, the numbers are 20 to 50, however, in the case of aging under the tensile stress, they are 50 to 220. And also in the latter case, the number for the angles of 0, 60 and 90 degrees are concentrated with 50 to 80,

572

however, for the angles of 33 and 38 degrees, the numbers are 150 to 220, higher than other cases. This results might suggest that a geometrical factor to form ω-phase crystals such as Schmidt factor of a resolved shear stress has an important role. Kawasaki (1964) reported about aging-behaviours of mild steels under tensile stresses, that the stress activated diffusion of carbon atoms, i.e. strain aging was enhanced by stressing. Therefore, in the present study, the effect of a stress on the diffusion of Mo atoms might be also considered.

In case of aging only

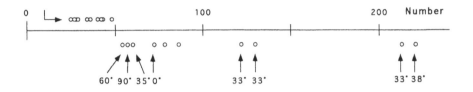

In case of aging under a stress

Fig. 5 Effect of a tensile stress, 413MPa on formation ω-phase crystals and of angles between the tensile direction and growth directions of ω-phase crystals.

## 4. SUMMARY

The results described above suggest that hydrostatic pressure and a tensile stresses contribute the enhancement of the homogeneous nucleation of ω-phase crystals in β-titanium alloys due to aging. Therefore, it is found that the stresses act effectively to prepare a homogeneous distributed and high density ω-phase crystals in β-titanium alloys. Since α-phase crystals initiate from ω-phase crystals, the results of this work suggest a possibility to precipitate high strength and high ductile (β+α) duplex alloys.

## 5. REFERENCES

Duering T W, Terlinde G T and Williams J C 1980 Metall. Trans. A, **11A**, 1987.
Hida M, Sukedai E and Terauchi H 1988 Acta Metall., **36** 1429.
Kawasaki M 1964 Zairyo (inJapanese), **13**, 445.
Makino T, Furuhara T and Maki T 1994 Abstract of Japan Inst. of Metals, **115** 482.
Nishizawa H, Sukedai E, Hashimoto H and Takarabe K 1997 Abstract of Japan Inst. of Metals, **120** 185.
Ouchi C, Suenaga H and Kohsaka Y 1988 Sixth World Conference on Titanium, ed. by P. Lacombe et al., Les Editions de Physique, Paris, 1625.
Sukedai E, Hashimoto H and Tomita M 1991 Philos. Mag. A, **64** 1201.
Takemoto Y, Hida M and Sakakibara A 1993 J. Japan Inst. Metals, **57** 261.
Takemoto Y, Hida M, Sukedai E and Sakakibara A: 1989 J. Japan Inst. Metals, **53** 1004.
Yukawa T, Ohtani S, Nishimura T and Sato T 1966 "Phase transformation of titanium alloys by means of automatic transformation apparatus", The Science, Technology and Application of Titanium, (Pergamon Press), 699.

*Inst. Phys. Conf. Ser. No 153: Section 12*
*Paper presented at Electron Microscopy and Analysis Group Conf. EMAG97, Cambridge, 1997*
© 1997 IOP Publishing Ltd

# Deformation of the Alpha-2 Phase in Fully Lamellar TiAl-based Alloys

**D Hu, A B Godfrey and M H Loretto**

Interdisciplinary Research Centre (IRC) in Materials for High Performance Applications, The University of Birmingham, Edgbaston, Birmingham B15 2TT, UK

**ABSTRACT:** Fully lamellar TiAl-based alloys have been examined with the aim of understanding the mechanism of slip transfer between the soft gamma phase and the hard alpha-2 phase in grains oriented with the basal plane in the alpha-2 either parallel to or perpendicular to stress axis. It has been shown that twinning dislocations in the gamma lamellae give rise to dislocations of $b=1/6<11\bar{2}6>$ and $1/3<11\bar{2}0>$ in alpha-2. Possible reactions which generate dislocations of $b=1/6<11\bar{2}6>$ and $<11\bar{2}0>$ in the alpha-2 from the twinning dislocations in the gamma are proposed.

## 1. INTRODUCTION

The microstructures of TiAl-based alloys can be separated into three categories: fully lamellar, duplex and equiaxed gamma (Kim 1989). Duplex and near gamma microstructures have good tensile strengths but they are less satisfactory in other properties. The fully lamellar microstructure can offer better fracture toughness, creep resistance and a wider range of tensile strengths than the previous two microstructures (Kim 1991), but it is has very limited ductility at room temperature (<0.5%). The large alpha grain size (of the order of mms) resulting from high temperature heat treatments has been suggested as one factor giving rise to the poor ductility of fully lamellar microstructure. Efforts have been made to refine the lamellar colony size in an attempt to improve the room temperature ductility. By decreasing the lamellar colony size down to about 100μm, the room temperature ductility was increased to 1.5% in some TiAl-based alloys (Godfrey et al 1997a). Further improvement in ductility appears to be very difficult without a thorough understanding of the detailed mechanisms occurring during deformation of the fully lamellar microstructure.

Of the two constituent phases in lamellar microstructures the alpha-2 phase is the more difficult to deform. An ordered hcp microstructure together with a high concentration of oxygen make the critical resolve shear stress (CRSS) for the slip systems in the alpha-2 phase much higher than in the gamma phase. In our recent work it has been shown that cracking in the lamellar microstructure occurs preferentially along the alpha-2/gamma or gamma/gamma interfaces for lamellar colonies in hard orientations (Godfrey et al 1997b), i.e. with the stress axis parallel or perpendicular to [0001] in the alpha-2 phase. This cracking indicates a lack of slip transfer between lamellae. Slip on the planes parallel to the interfaces is prohibited when lamellae are stressed perpendicular or parallel to the interface normal because of the zero resolved shear stress. Slip cannot occur in the gamma phase on the {111} planes parallel to the interfaces. If dislocations on the active {111} slip planes in gamma cannot cross the alpha-2/gamma interface, stress concentrations will be built up at the interface and cracks will be initiated. Any deformation in the alpha-2 phase actuated by dislocations in the gamma side will relieve the stress on the interfaces, and therefore retard crack initiation. In this paper observations of deformation in the alpha-2 phase are presented and some possible reactions between dislocations on both sides of alpha-2/gamma interfaces are proposed.

## 2. EXPERIMENTAL

Materials used in this study are Ti45Al4Mn4Nb and Ti48Al2Cr2Nb1B alloys which were either prepared into button form or ingot form by plasma arc melting then isothermally forged at 1150°C and a strain rate of $5 \times 10^{-4}$ s$^{-1}$ to a reduction in height of about 70%. The heat treatments were 1300°C/2h/FC for Ti45Al4Mn4Nb and 1380°C/1h/AC for Ti48Al2Cr2Nb1B, which lead to a fully lamellar microstructure in both of the alloys. After deformation at room temperature either in compression or in tension to a strain of about 1.5%, thin slices were sectioned transversely, giving the foil normal parallel to the stress axis. TEM foils were prepared via ion beam milling or electrochemical polishing. Foils were examined in a JOEL4000FX transmission electron microscope operating at 400kV.

## 3.   RESULTS AND DISCUSSIONS

The orientation relationship between alpha-2 and gamma lamellae is defined as $(0001)//(111)$ and $[1\bar{2}10]//[\bar{1}10]$. The lattice parameters of the alpha-2 phase are $a = 0.578$nm and $c/a=0.802$, and those for gamma phase are $a=0.401$nm and $c/a=1.015$. The tetragonaly of the gamma phase is ignored in vector algebra calculations because it only gives rise to an error less than $0.3°$.

The deformation microstructure of Ti45Al4Mn4Nb lamellae is shown in Fig.1. The stress axis is close to $[10\bar{1}]$ in gamma and $[2\bar{1}\bar{1}0]$ in alpha-2. The imaging condition is **g**=0002 which also excites the (111) reflection in the gamma lamellae. The strong contrast of the dislocations in the alpha-2 lamella indicates that the dislocations have a large $c$-component. These dislocations were identified as $1/6<11\bar{2}6>$ gliding on $\{20\bar{2}1\}$ planes in a previous paper (Godfrey et al 1997b). The twining Shockleys in the gamma phase were identified as $1/6[\bar{1}\bar{1}2]$.

Deformation in the lamellar microstructure starts from the gamma phase because it is softer than the alpha-2 phase. After being generated the twinning Shockleys move across the gamma lamella. Stress concentration may build up at the gamma/alpha-2 interface and will activate dislocations in alpha-2. The dislocations can continue to move within alpha-2 on some planes if the geometry of the two slip systems in gamma and alpha-2 allows them to do so. Of the 12 $<11\bar{2}6>$ directions in alpha-2 $[2\bar{1}\bar{1}\bar{6}]$ and $[11\bar{2}\bar{6}]$ form the smallest angles, 17.5°, with $[\bar{1}\bar{1}2]$. In addition to the small angle between Burgers vectors, the gliding planes $(20\bar{2}1)$ and $(11\bar{1})$ in the two phases are only 8.5° from each other. More importantly the two gliding planes have an intersection axis $[\bar{1}10]$ lying in the interface. This geometry permits slip transfer between the two systems with reasonable ease. Considering the large difference in the magnitudes of the two types of dislocations, 0.544nm for $1/6<11\bar{2}6>$ and 0.163nm for $1/6<112>$, it is presumed that three twinning Shockleys make up one dislocation in alpha-2 as proposed in the following reactions:

$$3 \times 1/6[\bar{1}\bar{1}2]=1/6[2\bar{1}\bar{1}\bar{6}]+1/6[\bar{2}11] \qquad (1)$$
$$3 \times 1/6[\bar{1}\bar{1}2]=1/6[11\bar{2}\bar{6}]+1/6[1\bar{2}1]$$

The $1/6<211>$ dislocations formed during this slip transfer are not twinning Shockleys. They are not permitted in the gamma phase and can be left on and accommodated by the interface.

In gamma lamellae there are four twining systems in total, one on each $\{111\}$ plane. The $1/6[1\bar{1}2](111)$ twining system is not included in this study because it occurs parallel to the interfaces. The other three systems can be separated into groups according to their angles to the interface normal. The angle is 19° for $1/6[\bar{1}\bar{1}2](11\bar{1})$ and about 62° for the other two. As mentioned above the $1/6[\bar{1}\bar{1}2](11\bar{1})$ system leads to $1/6<11\bar{2}6>$ dislocations in alpha-2. For the $1/6[\bar{1}12](1\bar{1}1)$ and $1/6[1\bar{1}2](\bar{1}11)$ twinning systems it is difficult to activate $1/6<11\bar{2}6>$ dislocations in the alpha-2 phase by slip transfer either because the two gliding planes do not have an intersection axis lying on the interface plane, in the case of $[\bar{1}12]$ and $[2\bar{1}16]$ directions or because the angle between the two slip vectors is fairly large, 45.5° in the case of $[\bar{1}12]$ and $[\bar{1}2\bar{1}6]$ directions. Instead, slip transfer from $1/6[\bar{1}12](1\bar{1}1)$ into alpha-2 by $1/3[\bar{2}110](0\bar{1}10)$ dislocations is favoured. The intersection between two gliding planes is $[\bar{1}01]$ and the angle between the slip vectors is 29.6°. This type of reaction have been observed and one example observed in Ti48Al2Cr2Nb1B is given in Fig.2, showing the $1/6[\bar{1}12](1\bar{1}1)$ actuating $1/3[\bar{2}110]$ dislocations in the alpha-2 lath. The stress axis is $[2\bar{1}\bar{1}]$ and $[1\bar{1}00]$. The reactions are proposed as following:

$$3 \times 1/6[\bar{1}12]=1/3[\bar{2}110]+1/2[110] \qquad (2)$$
$$3 \times 1/6[1\bar{1}2]=1/3[\bar{1}\bar{1}20]+1/2[110]$$

For this type of reactions, unit dislocations are generated which can be emitted back into the gamma lamellae.

In gamma lamellae deformation at room temperature is predominantly through twinning and unit dislocations according to research results on the PST crystals (Inui et al 1992). The Schmid factor plays an important role in determining the operating slip systems. It appears to be true in this study as well. In the case shown in Fig.1 the Schmid factors are 0.24 for $1/6[\bar{1}\bar{1}2](11\bar{1})$ twinning in gamma and 0.403 for $1/6[2\bar{1}\bar{1}6](20\bar{2}1)$ in the alpha-2, which are the highest values. The Schmid factors for the systems shown in Fig.2 are 0.39 for $1/6[\bar{1}12](1\bar{1}1)$ and 0.43 for $1/3[\bar{2}110](01\bar{1}0)$, which are also the highest among the systems involved. It is likely that the operating twinning systems in the gamma lamellae determine the operation of dislocations in the alpha-2 lamellae. This is supported by the consistency of the relationships between twinning variants in gamma and the types of dislocations in the alpha-2. Despite the importance of the Schmid factor it is clear that the stress concentration at the intersection of the twins in the alpha-2/gamma interface is also necessary for slip to occur in the alpha-2.

## 4. SUMMARY

Deformation in the alpha-2 lamellae has been observed in fully lamellar TiAl alloys after being strained at room temperature. When the lamellae were stressed perpendicular to the interface normal the dislocations in the alpha-2 actuated by twinning in gamma lamellae are of either $1/6<11\bar{2}6>$ or $1/3<11\bar{2}0>$ types, depending on the gamma twining variants.

Transfer of deformation has been observed when the gamma deforms by twinning and the alpha-2 deforms by the movement of $1/6<11\bar{2}6>$ dislocations on $\{20\bar{2}1\}$ planes and $1/3<11\bar{2}0>$ dislocations on the $\{1\bar{1}00\}$ planes.

## REFERENCES

Godfrey A B, Hu D and Loretto M H 1997a Proc. TMS Annual Meeting Orlando Feb 9-13
Godfrey A B, Hu D and Loretto M H 1997b to be published in Phil. Mag. A
Kim Y W 1989 JOM **41**, n5,
Kim Y W and Dimiduk D M 1991 JOM **43**, n8, 40
Inui H, Nakamura A, Oh M H and Yamaguchi M 1992 Phil. Mag. A **66**, 557

576

Figure 1    Two beam bright field TEM micrograph showing 1/6<11$\bar{2}$6> type dislocations in an alpha-2 lamella actuated by 1/6[$\bar{1}\bar{1}$2] twinning in gamma lamellae in a fully lamellar Ti45Al4Mn4Nb after being compressed perpendicular to the interface normal to 1.5% at room temperature.

Figure 2    Two beam bright field TEM micrograph showing 1/6<11$\bar{2}$0> type dislocations in an alpha-2 lamella actuated by 1/6[$\bar{1}$12] twinning in gamma lamellae in a fully lamellar Ti48Al2Cr2Nb1B after being strained in tension perpendicular to the interface normal to 1.5% at room temperature.

*Inst. Phys. Conf. Ser. No 153: Section 12*
*Paper presented at Electron Microscopy and Analysis Group Conf. EMAG97, Cambridge, 1997*

# An investigation of the 'omega-phase' in Ni-rich Ni-Al

## Y Zheng and C B Boothroyd

Department of Materials Science and Metallurgy, University of Cambridge, CB2 3QZ, UK

**ABSTRACT**    We present an electron diffraction study of a Ni rich Ni-Al alloy. By considering the way the diffuse streaks intersect the reflecting sphere as a function of crystal orientation, we demonstrate that the diffraction patterns conventionally ascribed to the hexagonal 'ω' phase can be explained by the <110> streaking effect. We discuss the implication of the non-existence of the 'ω' phase for the mechanism of the martensitic transformation.

## 1. INTRODUCTION

The $B_2$ Ni-rich Ni-Al alloys undergo martensitic transformation upon quenching when the Ni content is greater than 64at.%. With Ni contents less than this value, the alloy exhibits premartensitic phenomena known as 'tweed' in association with transverse {110} <$\bar{1}$10> shear lattice distortions. The tweed microstructure was found to be responsible for the well known <110> diffuse streaking effect in the alloys' electron diffraction patterns. Also in the electron diffraction patterns, weak diffuse maxima at 2/3 222 positions are found whose origin is conventionally attributed to a metastable 'ω' phase. Although the modelling of the ω-transformation is well established, the structure of the 'ω' phase in the present system has remained an enigma. Unlike systems such as Zr-Nb where these reflections are inevitably found to accompany longitudinal distortions of the matrix in the <111> direction (Kuan and Sass 1977), no trace of such distortions occur in the present material. The fact that the intensity of 'ω' reflections increases in proportion to that of the <110> diffuse streaks as a function of Ni content rules out the possible connection of these reflections with $Ni_2Al$ precipitation. We notice in all existing 'ω' models for the Ni-rich Ni-Al alloys, one considers the origin of the 'ω' reflections as being competitive to that of the <110> streaking and this leads to confusions in both the composition and the displacements present. We suggest here an alternative explanation for the 'ω' reflections by considering the possibility that they originate from the <110> streaks themselves.

## 2. RESULTS AND DISCUSSION

Electron diffraction experiments were carried out on a JEOL 2000 FX electron microscope at an accelerating voltage of 200kV. The sample used was a homogenous $Ni_{62}Al_{38}$ as-quenched thin disc with a near [110] surface normal. A low camera length (250mm) and long exposure times were used in order to enhance the intensity of weak reflections on the TEM plates. The surface area included in the selected area aperture was big enough to represent an average effect from the sample.

Electron diffraction patterns were taken as the sample was tilted off the [110] zone axis along the 002 kikuchi band by 0.1, 4, 5, 6, 8 and 10 degrees so that the incident beam was parallel to the ~[110], [670], [560], [450], [340] and [230] zone axis respectively (Fig. 1). Diffraction patterns are indexed with $B_2$ reciprocal spots and the structures in the vicinity of 330, 5/2 $\bar{5}$/2 0, 2$\bar{2}$0, and 3/2 $\bar{3}$/2 0 positions are framed and named as A, B, C and D respectively. In Fig. 1a patterns in frames A and C correspond to $B_2$ fundamental reflections with diffuse streaks running in <112> directions resulting from the intersection of out-of-plane <110> streaks with the reflecting sphere. Structures in B and D are seen as the conventional 'ω' patterns consisting of diffuse maxima at 2/3 222 positions. The small angles between the successive diffraction patterns enable us to inspect closely the gradual change in the position and intensity of the various reflections. We first inspect the change in the pattern of frame B during tilting. The centre of the pattern which shows as a

578

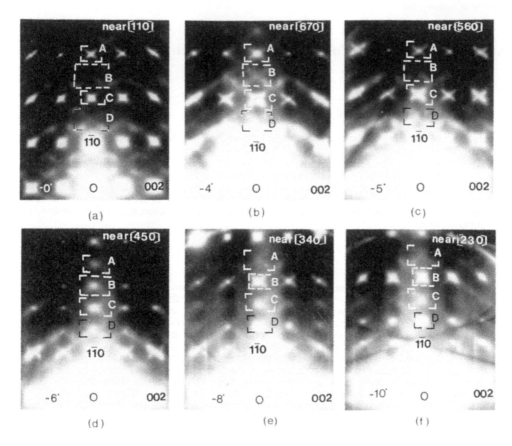

Fig. 1. Electron diffraction patterns taken at 0.1°, 4°, 5°, 6°, 8° and 10° off the exact $[110]_{B2}$ zone axis, showing changes in both $B_2$ and 'ω' reflections as the Ewald sphere moves away from the ZOLZ.

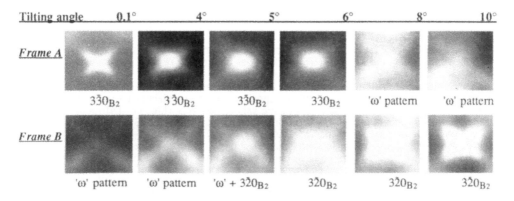

Fig. 2. The Enlargement of the framed structures A and B from Fig. 1 showing the change as a function of tilt.

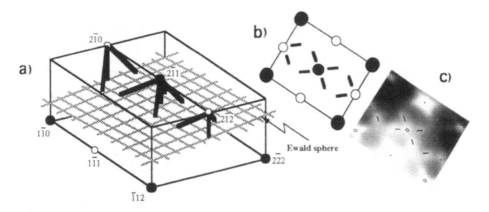

Fig. 3. Formation of 'ω' pattern within a fundamental $B_2$ structure. Open and closed circles represent the $B_2$ superlattice and fundamental reflections respectively. Note the <110> streaks from the ZOLZ are not shown in the map. (a). Imaging condition; (b) predicted pattern; (c) experimental pattern.

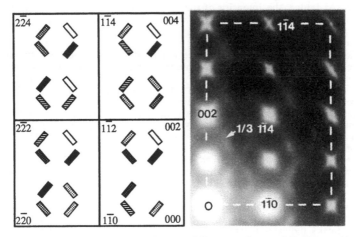

Fig. 4. Calculated and experimental data for the diffuse maxima at 2/3 222 positions in the (110) diffraction pattern. Calculated diffuse intensities increase on a grey scale.

barely visible diffuse maximum at the beginning of the sequence increases in intensity with tilt and eventually turns into a new $B_2$ spot. Correspondingly, the initial 'ω' pattern within this frame shows a reduction in spatial extent during tilting and becomes a <110> streaked pattern surrounding a $B_2$ spot towards the end of the series tilt. Owing to the change in the Ewald sphere position relative to the [110] zone in this process, i.e., as it moves away from the Zero Order Laue Zone and towards the First Order Laue Zone, the new $B_2$ spot can be indexed as $3\bar{2}0$ on the FOLZ. Based on the same reasoning, the framed structure D is found to turn into $\bar{2}10$ $B_2$ upon tilting, although in this case the process is incomplete due to the smaller **g** vector in relation to this position. (At lower **g** values, the Ewald sphere is closer to the ZOLZ thus a complete transformation of the pattern would only be expected at an even higher tilting angle.) The inverse transformation from the $B_2$ spots to the 'ω patterns' can be observed in the structure change within frames A and C as a function of tilt. Again the transformation within frame C is incomplete due to the effect of small **g** vector. The systematic change of the structures A and B from Figs. 1a to 1f are shown enlarged in Fig. 2.

580

The above observations suggest that by taking into account the curvature of the Ewald sphere, anomalous intensity in the (110) rel-plane may be related to the reflections from the FOLZ. It can be seen that the 'ω' reflections have their elongation axis in the same direction as the projections of <110> spikes. This implies that both are due to a similar cause. It is known that each of the $B_2$ spots on the FOLZ, whose projections are located at 1/2 110 positions, has five <110> diffuse spikes pointing towards the (1̄10) plane. The observed 1/2 110 intensities can thus be interpreted as the intersection of the [110] streak, perpendicular to this plane, with the Ewald sphere. Likewise, the 'ω' reflections can be seen as the intersection of the other four <110> spikes, namely the [011̄], [1̄01̄], [01̄1̄] and [1̄01] with the sphere. This concept is represented schematically in Fig. 3. Special notice should be paid to the positions of the 'ω' reflections. Although they are dependent on the precise viewing axis, these reflections possess prominent intensity when located close to the 2/3 222 reflections. This phenomena suggests that there exists intensity maxima on the <110> diffuse streaks at q=1/3<110>. These maxima are consistent with the neutron scattering result of Shapiro et al (1991) who showed that transverse acoustic phonon of Ni rich Ni-Al exhibits a marked softening along the [ζζ0] at ζ=1/6(2π/a). The softening corresponds to the transverse wave with a wavelength of 2.43nm which generates satellite spots at q=n/6<110> in the electron diffraction patterns. .

To test our present interpretation of the origin of the 'ω' reflections, we calculated the intensity distribution of these reflections on the (110) pattern (Fig. 4). The diagram is schematic to the extent that the intensity of all 'ω' reflection are approximated to that of q=1/3<110> and the absolute magnitude of the amplitude of the distortions is unknown. The relative behaviour of each of the streaks from different beams is however potentially prescribed by the type of the distortion as a shear and it is this which has been quantitatively interpreted in the diagrams. The approach used was kinematic and as described by James (1958) for the effect of an elastic wave distortion on X-ray diffraction. As a result of the transverse nature of the {110} <11̄0> static distortions in the Ni-Al lattice, the intensity of the <110> streaks from different beams can be calculated as proportional to $(g.e)^2$ where g is the reciprocal lattice vector of the beam and e the vibration vector of the transverse waves. Relative intensities of diffuse maxima were thus calculated as $\cos^2\alpha$ where $\alpha$ is the angle between g and e. For the reflections in association with $B_2$ superlattice reflections, the intensities were halved in consideration of their low structure factor. Otherwise the effect of structure factors were ignored. Comparing the intensity map and its experimental mirror image, one can see immediately the disparities in the observed and predicted behaviour due to the kinematical approximation adopted in the simulation.. However the experimental 'ω' reflections in general reveal fundamental transverse characteristics which are in good qualitative agreement with the prediction. For instance, the reflections elongated at a larger angle to g exhibit considerably higher intensity than those elongated along g. Another remarkable match is the observed strongest intensities at 1/3 11̄4 positions which have been predicted by the simulation.

It is interesting to recall that when the tweed microstructure in Ni-rich Ni-Al alloys was first reported with its characteristic <110> streaking, it was seen as accompanied by various other effects. Anomalies included the <112> streaking, <100> satellites, {111} diffuse planes and the 'ω' reflections. Effects apart from the <110> streaking were gradually proved artificial. Our present finding shows that the material can be further simplified and the lattice perturbation in the system can be understood as mere {110} <11̄0> shear waves with various wavelengths. The diffuse maxima on <110> streaks indicates a large lattice distortion magnitude corresponding to shear waves with ζ=1/6(2π/a). These waves cause a large fluctuation in the atomic positions in the lattice which may lead to the formation of micro-domain structures. Since the material in this study is within the range of the premartensitic state where the material prepares itself for the oncoming martensitic transformations, the domain structure may serve as martensitic embryos. Evidence of this hypothesis can be found in the neutron scattering experiments which show that the lattice softening at ζ=1/6(2π/a) in [ζζ0] is enhanced as the martensitic transformation approaches (Shapiro et al 1991).

We acknowledge the inspiration of the late Dr. W. M. Stobbs. We thank Professor A. H. Windle for provision of laboratory facilities. Y. Zheng also would like to thank the Cambridge Overseas Trust and the Isaac Newton Foundation for financial support.

**REFERENCES**

Kuan T S and Sass S L 1977 Phil. Mag. **36**, 1473
Shapiro S M, Yang B X, Noda Y, Tanner L E, Schryvers D and Wall M E 1991 Scipta Metall. **24**, 9301
James R W 1958 The Optical Principles of the Diffraction of X-Rays, Bell G and Sons LTD

*Inst. Phys. Conf. Ser. No 153: Section 13*
*Paper presented at Electron Microscopy and Analysis Group Conf. EMAG97, Cambridge, 1997*
© *1997 IOP Publishing Ltd*

# Structural and mechanical characterization of carbon nanocomposite films

**I. Alexandrou**[1,2], **C. J. Kiely**[1], **I. Zergioti**[3], **M. Chhowalla**[2], **H. -J. Sceibe**[4] and **G. A. J. Amaratunga**[2]

[1] Department of Materials Science and Engineering, University of Liverpool, Liverpool L69 3BX, UK.

[2] Department of Electrical Engineering and Electronics, University of Liverpool, Liverpool L69 3BX, UK.

[3] Institute of Electronic Structure and Laser, FORTH, 711 10 Heraklion, Crete, Greece.

[4] Fraunhofer - Institut fur Wekstoffphysic und Schichttechnologie D - 01277 Dresden, Germany.

**ABSTRACT:** A comparative study of carbon based films deposited by carbon arc, laser initiated carbon arc and laser ablation techniques is reported. Films deposited by conventional carbon arc and laser ablation belong to the tetrahedral amorphous carbon (ta-C) category. Those films prepared by a modified carbon arc and laser arc methods consist of a composite ta-C matrix with fullerenelike nanoparticles. These films are hard (45-55 GPa plastic hardness) and elastic (80-85% elastic recovery). Most of the nanoparticles are in the form of deformed or fragmented nanotubes and onions. PEELS and HREM analysis suggest that linking of neighbouring nanoparticles may be important in the observed mechanical properties.

## 1. INTRODUCTION

The increasing interest in applications of carbon based materials has led to their investigation using many deposition and characterization techniques. From the published data, laser ablation and cathodic arc deposition methods are among the most effective for growing hard films. Their excellence is attributed mainly to the high kinetic energy of the depositing species [Voevodin and Donley (1996)]. In this paper we compare the mechanical and structural properties of carbon based films deposited using novel modified laser ablation, carbon arc and a hybrid combination, the laser arc technique. Their mechanical properties vary with the technique and/or the configuration used. The respective change in film microstructure and especially the

582

inclusion of fullerenelike nanoparticles in the films grown by the two arc methods, seems to make a difference in the mechanical properties.

## 2. EXPERIMENTAL

Two configurations of pulsed laser deposition (PLD) are studied here. Firstly, $CN_x$ films were deposited in an $N_2$ atmosphere of various pressures and, secondly, a pulsed $N_2$ nozzle was used to create a gas jet in front of the graphite target when the 248 nm KrF excimer laser ablated the target. In the latter case the rest of the chamber is kept at low pressure.

The cathodic arc method was altered by the use of an $N_2$ jet to create a high gas pressure region in the vicinity of the graphite target [Amaratunga et al (1996)] and/or the arc was pulsed manually at a rate of 1 Hz. Finally, a combination of the above two techniques, the laser arc method was also used. In this method, an excimer laser is used to ablate a graphite target and initiate a cathodic arc through the laser plume. Both the laser and the arc are pulsed at a rate of 200 Hz. Samples suitable for HREM and PEELS analysis were prepared by lifting off the films on to copper grids by chemically dissolving away the Si substrate in an HF / $HNO_3$ solution. The mechanical properties of the films were compared using a Fisher microindenter with ultra low loads of 5 mN.

| Deposition Method | Plastic Hardness (GPa) | Elastic Recovery (%) |
|---|---|---|
| PLD | 10.7 | 55.2 |
| PLD (nozzle) | 16.5 | 56.8 |
| PLD (nozzle 260°C) | 32.1 | 58.4 |
| Arc ($N_2$, 350°C) | 56 | 85 |
| Laser Arc (RT) | 44 | 80 |

**Table I** : Mechanical properties as deduced by microindentation measurements.

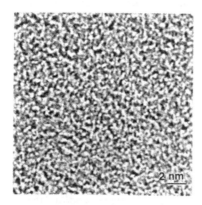

**Figure 1** : HREM micrograph of a carbon nitride film deposited by laser ablation

## 3. RESULTS AND DISCUSSION

Table I shows the plastic hardness and elastic recovery of representative films in this study. The best of the properties for the PLD films, which are the ones shown in table I, were obtained when they were deposited in an $N_2$ atmosphere at the lowest pressure and temperature.

Energetic species deposition favours the growth of hard films [Voevodin et al (1996)]. Increased pressure therefore leads to a reduction in energies of the ablated C ions and reduction in hardness. Also increased substrate temperature has been shown to reduce film hardness because $sp^3$ bonds turn to $sp^2$ upon relaxation. Mechanical properties of the PLD films are enhanced when a pulsed nozzle is used. Since the nitrogen concentration is about the same, the collisions in the $N_2$ jet are sufficient for carbon - nitrogen reactions to occur and the resultant species retain enough energy for hard film growth. Unlike the PLD case, the properties of the arc deposited films are enhanced with temperature. This is very unusual for arc deposited amorphous carbon based coatings and is indicative of a change in the film structure. In addition to the high plastic hardness an additional noteworthy feature is the very high elastic recovery of the films deposited by

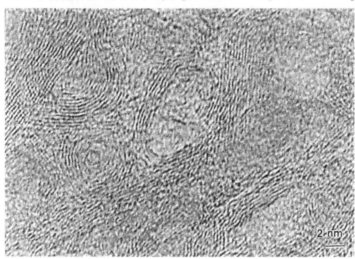

**Figure 2 :** Typical example of a patch of fullerene inclusions seen in the films deposited by the laser arc method.

conventional and laser arc.

Figure 1 shows a representative HREM micrograph for films deposited by pulsed laser ablation. Even though differences in mechanical properties with deposition conditions were found, the films are always amorphous, with a microstructure similar to the amorphous ta-C films deposited by other techniques.

The microstructure of the films deposited by conventional and laser arc methods are considerably different. Both films consist mainly of an amorphous matrix interspersed with patches of strongly diffracting fullerenelike nanoparticles. Figure 2

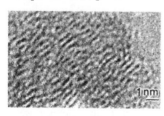

**Figure 3 :** HREM picture of a film deposited by cathodic arc at 350°C.

shows a typical patch of nanoparticles surrounded by the matrix. Many nanoparticles appear deformed or fragmented because of their impingement on the growing film surface. Carbon films deposited by conventional cathodic arc are amorphous and in our study the patches of nanotubes and onions appear only when the gas jet is used and/or the arc is pulsed. In addition, when the arc is pulsed the average nanoparticle size decreases and the patches become wider and appear more frequently within the films. This effect becomes more pronounced in the case of the laser arc deposition where the pulse rate is greater.

Finally, figure 3 shows an HREM image of a film deposited by carbon arc with the gas jet, at a temperature of 350°C. In this film large individual nanoparticles could be traced but at a rather low density. Interestingly, under closer inspection of fig. 3 the

584

film seems to exhibit a definite stripe like texture, suggestive of an intimate mixture of fullerene fragments in the amorphous matrix (as previously observed Sjöstrom *et al* (1995)). Nanoparticle fragmentation is possible via collisions with other plasma species and on impact with the substrate. The lack of large fullerene structures in fig. 3 with respect to fig. 2 can be explained by the more efficient fragmentation of the nanoparticles because of the elevated temperature.

Nanoparticles with their planes almost parallel to the e-beam exhibit strong phase contrast. A TEM micrograph of a cluster of nanoparticles is in fact their projection on the photographic plate along the incident beam direction. In many cases two nanoparticles seem partly one inside the other, but is difficult to ascertain whether they simply overlap in projection, or if they are actually interlinked and interpenetrate. High energy EELS was also used to gain further information from apparent "overlap" areas. In many cases the $sp^3$ to $sp^2$ carbon ratio increases in the region between two nanoparticles, an adequate condition to make covalent bonding between two nanoparticles possible. This possible bonding between two fullerenelike nanoparticles may make these patches the precursors of a uniform material with the mechanical properties of fullerenes. These EELS studies are discussed in more detail in another paper by Yuan *et al* (1997) in this volume.

## 4. CONCLUSIONS

The aforementioned clusters of nanoparticles in fig. 2 can be realised as a three dimensional $sp^2$ network since the bonding along the nanoparticles walls is graphitic. If the nanoparticles can be interlinked as EELS results suggest, because of the high stability of the $sp^2$ bonds at RT and the flexibility of the graphene sheets, the resultant material should exhibit enhanced mechanical properties. This effect is expected to be even more pronounced in the matrix which seems to be a more uniform version of such a structure. Also the improvement in the mechanical properties with substrate temperature is very well explained with this model since the fraction of $sp^3$ bonds needed is small and will be relatively unaffected by an increased growth temperature. At the same time fragmentation of the nanoparticles on impact with the film surface is enhanced.

## REFERENCES

Amaratunga G. A. J. , Chhowalla M., Kiely C. J., Alexandrou I., Aharonov R. and Devenish R. (1996), Nature **338**, 321.
Sjöstrom H., Strafström S., Boman M. and Sundgren J -E. (1995), Phys. Rev. Lett. **75** (7), 1336.
Voevodin A. A. and Donley M. S. (1996), a review paper, Surf. Coat. Technol. **82**, 199.
Yuan J. and Amaratunga G. A. J. (1997), Inst. Phys. Conf. Series, this volume.

*Inst. Phys. Conf. Ser. No 153: Section 13*
*Paper presented at Electron Microscopy and Analysis Group Conf. EMAG97, Cambridge, 1997*

# Surface characterisation of ultrathin Diamond-like Carbon coatings by low voltage SEM/EDX analysis

**P Lemoine and J Mc Laughlin**

NIBEC, School of Electrical and Mechanical Engineering, University of Ulster at Jordanstown, Shore road, Newtonabbey, Co Antrim BT37 OQB, Northern Ireland

ABSTRACT: The intensity of EDX peaks from substrate elements is used to measure the thickness of ultrathin DLC films. It also provide an electron energy loss estimate useful for optimising secondary electron images.

## 1. INTRODUCTION

Nowadays, microscopic techniques are not only compared in term of resolution. Surface sensitivity has also become a crucial asset, especially for the study of overcoats of less than 100nm thickness, hereby called ultrathin films. The Diamond-like Carbon (DLC) coatings studied in this paper are used as wear and corrosion barriers for magneto-recording heads of hard disc drives. In this case, the high read/write density of the device requires very small head/disc distance to minimise magnetic flux losses and DLC layers less than 50nm thick are not uncommon. Corrosion resistance can be measured by AC impedance spectroscopy. The wear characterisations of the DLC layers necessitate microscopic studies on the morphological and mechanical properties of these thin films. The work presented here focuses on the morphological characterisation using SEM/EDX analysis. We examined how EDX analysis can be used to determine the film thickness and study the interaction of the electron beam with the ultrathin DLC film. This information is used to produce SEM images of improved surface sensitivity which are compared to contact mode AFM images. The work was carried out using a Hitachi S3200N SEM linked to an Oxford Instrument Link ISIS 300 EDX microanalysis system. The AFM microscope was a Burleigh personal SPM.

## 2. EDX ANALYSIS

In Electron microscopy and related techniques such as EDX analysis it is common practice to use electron beam of 15kV and higher to obtain good resolution and broad compositional sensitivity. However, at such high energies, the sampling depth is large, especially in the case of DLC which shows low density and light elemental composition. Until recently, low KV imaging (below 5kV) was hindered by the low brightness of thermionic cathode gun. New designs and the use of field emission guns allow for significant improvement in the microscope performances down to 1kV. To avoid working at unnecessary low voltages, the best strategy *prior* imaging is to have a reasonable estimation of the electron range. This sampling depth can be measured using the EDX analysis, observing the emission of characteristic X rays from an element present only in the substrate material, as described by Bentzon (1993). This technique is principally aimed at giving a rapid and simple measurement of the film. However, a proper calibration for known film thickness gives an estimate of the energy loss process in the thin film whereupon the ultrathin DLC films can be imaged using a sensible electron energy. Bentzon analysed relatively thick films (250nm thick). We are interested in testing the method on ultrathin DLC films deposited by PECVD on Si substrates by comparing its predictions with precise glancing angle X ray reflectometry measurements.

The idea is to observe the diminution of characteristic X ray intensity due to primary electron scattering in a DLC film of thickness x (in nm). The X ray intensity $I(E)$ is:

$$I(E_0) = \alpha.i_o \left(E_0 - E_c\right)^n \text{ and } I(E_x) = \alpha.i_x \left(E_x - E_c\right)^n \tag{1}$$

The X ray intensity ratio a is:

$$a = \frac{I_x}{I_0} = \frac{i_x}{i_0}\left(\frac{E_x - E_c}{E_0 - E_c}\right)^n \text{ so that } E_x = bE_0 + (1-b)E_c \text{, with } b = \left(a\frac{i_0}{i_x}\right)^{1/n} \tag{2}$$

where $i_o$ and $i_x$ (in pA)are respectively the specimen current reaching the uncoated and coated substrate , $E_o$, $E_x$ and $E_c$ ( in keV) are respectively the primary incident energy, the energy of the electron after having travelled a distance x in the DLC film and the critical excitation energy for X ray emission and, finally, n is an exponent which depends on the emitting material. The use of specimen current instead of primary beam current is justified since we found that the I/i ratio was independent of beam current.

As the energy and current are being reduced in the thin film, so will $I(E)$. The electron scattering process has been studied by many authors (Cosslett (1964), Landau(1944) and Kanaya(1972)) .The energy loss process seems to follow a square law of the type:

$$-\frac{dE}{dx} \alpha \frac{1}{E} \text{ therefore } kx = E_0^2 - E_x^2 \text{, where k is a constant} \tag{3}$$

therefore, eq. (2) and (3) give: $kx = (1-b)\left[(1+b)E_0 - (2bE_oE_c) - (1-b)E_c^2\right]$ (4)

The specimen current were measured with a picoammeter and the X ray intensity ratio was calculated from the EDX software. The method was calibrated using a DLC film (526nm thickness easily measured by cross sectional SEM imaging) ion beam deposited on a ceramic substrate (AlTiC: 70 Wt% $Al_2O_3$ + 30 Wt%). We used the $TiK_\alpha$ line ($E_c$=4.93 keV, n=1.51). Plotting $a(E_o)$ curves, we found, for instance, a=0.7 for $E_0$=14.25 keV. This yielded a k value of 0.123 $nm^{-1}$ $keV^2$. Taking a density of $2g/cm^3$ from X ray glancing angle measurement, the k value can be transformed into the constant A in equation 5 of Bentzon paper. We found A=3.61 $10^{16}$ at.$cm^{-2}$ $KeV^2$ which compares well with the A=3.41 $10^{16}$ at.$cm^{-2}$ $KeV^2$ found by Bentzon. These calculations were carried out for the ultrathin DLC films on Si substrate using the $SiK_\alpha$ line ($E_c$= 1.84 keV, n=1.35). The agreement with the X ray glancing angle measurements is reasonably good. (see table 1). The slight underestimation of the film thickness could be due to the use of the Thomson-Whiddington law. This law represents the mean energy whereas the X ray emission originates from the most probable energy.

| Deposi. voltage (V) | thickness (nm) ( Xray reflect.) | thickness (nm) (EDX analysis) |
|---|---|---|
| 350 | 87 | 83±3 |
| 200 | 66 | 60±3 |
| 150 | 49.5 | 47±3 |
| 100 | 36 | 31±3 |

**table 1**: Comparison between two different thickness measurements

However, energetic distribution of transmitted electron are asymmetric (Cosslett, 1964), with steeper loss law for the mean energy than for the most probable energy. This would, indeed cause an underrated thickness evaluation.

## 3. LOW VOLTAGE SEM IMAGING

This result was used for imaging DLC thin films typical of magnetic recording head media, i.e. 20nm thick DLC film deposited by PECVD on AlTiC substrate. The films being reasonably hard, it was possible to image them by contact mode Atomic Force Microscopy. In this technique where the tip and sample surface are always in contact (1-2Å) the image is purely topographical with no compositional component. We found that the film and substrate samples have similar topographic features, with roughnesses values of 3nm or less. EDX analysis showed that the film coverage was continuous. Therefore, it seems that the film replicates the substrate own topography, possibly because of preferential growth on TiC. This is also the case for ion beam deposited DLC films (Liu 1997).

The surface signal in an SEM is carried by some of the secondary electrons (SE). Therefore, imaging of these ultrathin films required, first of all, a maximised secondary electron (SE) yield, namely an energy from 1 to 3 kV, as found by Dawson (1966). Moreover, the energy loss law found in the above calculation shows that, on crossing the 20nm thick DLC layer, electrons of 2, 3 and 5keV loose respectively 40%, 15% and 5% of their energy. This means that SE imaging in this voltage range should give some surface information. The SE detector was operated with a positively biased plate to improve the SE collection efficiency. Images were acquired integrating 256 fast TV frames to avoid the build up of charging.

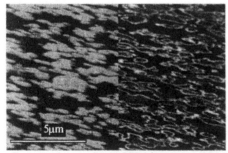 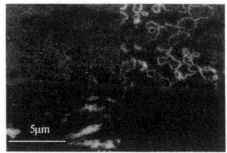

**figure1:** low voltage SEM images: a=DLC at 5kV, b= substrate at 5kV, c=DLC at 2kV and d=substrate at 2kV. a and b are at 75° and c and d at 30°

One can see on fig. 1 that the substrate compositional features (TiC island imbedded in a Al₂O₃ matrix) progressively disappear when the voltage is decreased from 5 to 2 keV. However, very few surface features are seen on the images. This is possibly due to the unavoidable backscatter (BS) component in the signal reaching the SE detector. Using a 30V

588

DC bias on the samples and subtracting the biased image from the unbiased image, it was possible to suppress this BS component. Fig. 2 shows the DLC and substrate images for 3 and 5kV. At 5kV, both the DLC film and the substrate images show compositional contrast. The electrons backscattered by the dense substrate ($Z_{average}$=15.4 ) produce SE electrons with no surface sensitivity. At 3kV, the overall SE yield is higher and the contribution of SE originating from the BE electrons is smaller. Indeed, the substrate compositional features have disappeared from the thin film image revealing the DLC topography already observed by contact AFM microscopy.

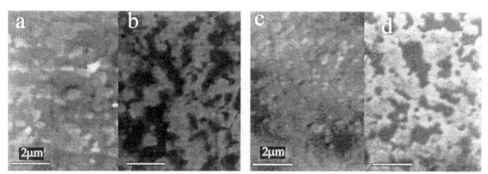

**figure2:** SEM images obtained at O° incidence by subtracting a 30V biased sample from an unbiased sample: a=DLC at 5kV, b=substrate at 5kV, c=DLC at 3kV and d=substrate at 3kV

## 4. CONCLUSION

We carried out morphological characterisation of ultrathin DLC overcoats used as protective layers for magnetic recording hard disc heads. It was shown that the EDX technique developed by Bentzon *et al* can be used to measure the thickness of ultrathin films. It also provide an estimate of the electron energy loss process in the DLC film. This information is useful for obtaining SE images of improved surface sensitivity.

## 5 AKNOWLEDGMENTS

Many thank to Dr. M.J Winter and A. Iberl from Siemens Analytical X ray Syst. for X ray reflectometry measurements, to Dr. J. F. Zhao for preparing the PECVD samples and to William Scanlon for help with the image processing.

## REFERENCES

Bentzon M D , Nielsen P S and Eskildsen S S 1993 Diamond and Related Materials **2** 893
Cosslett V E and Thomas R N 1964 Brit. J. Appl. Phys. **15** 1283
Dawson P H 1966 J. Appl. Phs. **37** 3644
Kanaya K and Okayama S 1972 J. Phs. D Appl Phys. **5** 43
Landau L 1944 J. Phys. Moscow **8** 201
Liu Z H Lemoine P Zhou D M Zhao J F Mailley S Lamberton R W Magill D P McAdams E T Maguire P and McLaughlin J, submitted to Diamond and Related Materials 1997

*Inst. Phys. Conf. Ser. No 153: Section 13*
*Paper presented at Electron Microscopy and Analysis Group Conf. EMAG97, Cambridge, 1997*
© 1997 IOP Publishing Ltd

# Tip-end and bottom structures of BN nanotubes produced by laser heating under high pressure

**T. Tamiya\*, Y. Bando, D. Golberg, M. Eremets, K. Takemura, H. Yusa**

National Institute for Research in Inorganic Materials, Namiki, Tsukuba, Ibaraki 305, Japan
\* also with the Institute of Material Science, University of Tsukuba, Tennodai, Tsukuba, Ibaraki 305, Japan

**ABSTRACT:** HRTEM study has been carried out on BN nanotubes produced by laser heating of $h$BN in a diamond anvil cell under high nitrogen pressure of 8.3 GPa. Special attention has been paid to their tip-end capping features and the epitaxial relationship with the mother phase during their growth. The observed tip-end cone angles ranged from $22°$ to $45°$. The tip-end structural models have been proposed that are relied on the existence of 4 and 8 member BN rings in the nanotube caps instead of 5 and 7 member rings which are typical for carbon nanotubes. The base material in the bottom of the nanotubes mostly revealed cubic BN structure, though, occasionally, hexagonal BN was found as the nanotube mother phase. In the former case, $c$BN in the bottom of the nanotube and the nanotube $h$BN sheets exhibit orientation relationship as follows: $(110)_{cubic}//(0002)_{hex}$ and $[\bar{1}11]_{cubic}//[10\bar{1}0]_{hex}$.

## 1. INTRODUCTION

The discovery of fullerenes and nanotubes has widened our knowledge of possible stable allotropic forms of carbon (Kroto et al. 1985, Iijima 1991). A structural model for carbon fullerenes has been verified by inserting 12 pentagons into hexagonal network. For example, in the case of $C_{60}$ molecule, 3 pentagons are thought to be in touch with every hexagon, while all pentagons are separated from each other. Furthermore, a structural model for carbon nanotubes has been figured out by cutting $C_{60}$ soccer ball onto two halves and by inserting hexagonal rows in between. Since BN is exact structural analogue of C, which exhibits both hexagonal (graphite-type) and cubic (diamond-type) phases, it is reasonable to expect an existence of BN nanotubes and/or fullerenes. Indeed, BN nanotubes were discovered by Chopra et al.(1996) who used the arc-discharge method to synthesize them (similar to that for carbon nanotube production). It is worth noting that these nanotubes always contained internal metal particles located close to the tip-ends. Soon after, Loiseau et al.(1996) obtained pure BN nanotubes without any attachments at the tip-ends, using improved plasma arc-discharge method with HfB$_2$ electrodes. The latter authors proposed a tip-end model to explain flat caps, commonly observed by them, by inserting 3 four member rings into hexagonal BN honeycomb lattice at the cap. Later on, Golberg et al.(1996) succeeded to synthesize pure BN nanotubes by laser heating of $c$BN or $h$BN in a diamond anvil cell (DAC) at high nitrogen pressure. These authors concluded that the nanotubes were grown either from $c$BN or $h$BN matrices and exhibited the stoichiometry of a pure BN compound, B/N=1. As for the tip-ends, they commonly observed tip-end angles varied between 19° to 45°, whereas the tip-end lattice images, as a rule, were not symmetrical with respect to the tube axis.    In the present study the tip-end and the bottom structures of the BN nanotubes are studied more in detail by using HRTEM lattice imaging while applying a nano-beam probe ( $\phi$ 1 nm) for nano-beam diffraction and EELS experiments.

## 2.    EXPERIMENTAL

Polycrystalline hexagonal BN ($h$BN) specimens with a thickness of about 20 $\mu$ m were laser heated in a diamond anvil cell (DAC) under nitrogen pressures of 8.3 GPa. A laser beam spot of about 80 $\mu$ m diameter was focused onto the specimens for a period up to a few minutes. The temperature was estimated to be about 5000 K. The irradiation products were carefully separated from the base material and placed onto holey carbon grids. They were observed by a 300 kV field emission TEM (JEM-3000F). An electron probe of about 0.8 to 1.6 nm diameter was used for nano-diffraction (NBD) and electron energy loss spectroscopy (EELS) observations.

## 3.    RESULTS AND DISCUSSION

In order to investigate the BN nanotube base composition, an electron probe of 1 nm diameter was focused onto the mother phase just below the BN nanotube and corresponding EELS spectra were recorded. Almost all the spectra revealed $c$BN structures, but, rarely, the bases showed $h$BN matrix. These results are shown in Fig. 1a and 1b, respectively. A specific orientation relationship was found to exist between $c$BN base and $h$BN sheets in the nanotube. This relationship was resolved by using NBD experiments. The results are shown in Fig. 2. It is appeared that $(110)_{cubic}//(0002)_{hex}$ and $[\bar{1}11]_{cubic}//[10\bar{1}0]_{hex}$ , though the lattice connection is not perfect due to a marked difference in lattice parameters between $(110)_{cubic}$ and $(0002)_{hex}$ planes.

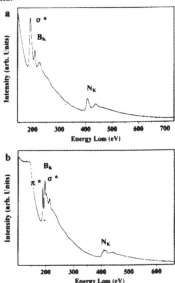

Fig. 1.    EELS spectra obtained from the bases, just below the BN nanotubes, (a) $c$BN base, (b) $h$BN base.

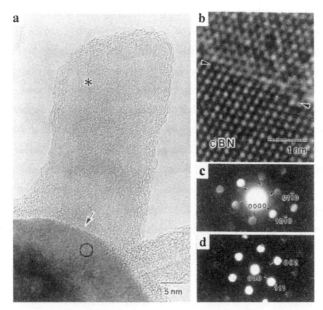

Fig. 2. BN nanotubes grown from cBN base (a), enlarged lattice image of the interface between the nanotube base and the nanotube (region marked with an arrow) (b), corresponding nanodiffraction patterns obtained from the nanotube (region marked with an asterisk) (c), and from the base (region marked with a circle) (d).

Then, HRTEM images of the nanotube tip-ends were considered. Fig. 3 shows some typical 2D lattice images of the nanotube caps. All these images reveal equal intensity of (0002) lattice fringes in various nanotube parts. The tube in Fig. 3a exhibits 8-shells and an outer dimension in cross section of 110 Å. The tubes in Fig. 3b and Fig. 3c display 6-shells and the outer dimensions 45 Å and 55 Å, respectively. The image in Fig. 3a shows flat tip-end, while both in Fig. 3b and 3c exhibit cone-like tip-ends with the cone angles 45° and 22°, respectively.

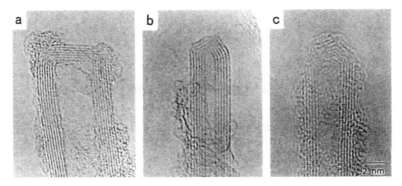

Fig. 3. Typical tip-end capping features of BN nanotubes. Flat tip-end (a), cone-like tip-ends with the cone angles of about 45° (b) and 22° (c).

592

In order to form a nanotube tip-end cap, the polygons other than hexagons should be taken into account. In the case of BN, in order to close the tube, it is preferable to introduce 4 member rings, as Loiseau et al. (1996) did. In such a case, B-B and N-N unfavorable bonds are prevented and perfect atomic ordering of B and N atoms is kept. However, it would be difficult to explain a wide variety of the observed tip-end caps by introducing 4 member rings only. Thus we propose to extend a pre-existing structural model by introducing 8 member rings into hexagonal honeycomb network. In fact, if a single 8 member ring appears, it causes a negative curvature of the hexagonal sheet. On the other hand, by taking into account a combination of the single 8 member ring and multiple 4 member rings, it is possible to construct a positive curvature of the nanotube cap. These results are summarized in Fig. 4. Consequently, we propose here that an introduction of a single 4 member ring may explain the observed tip-end angles varied between $35°$ to $45°$, whereas a combination of single 8 member ring and a couple of 4 member rings may be responsible for the observed tip-end cone angles in the range of $20°$ to $30°$.

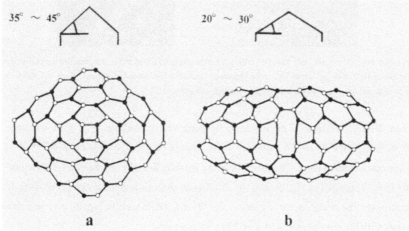

a b

Fig.4 Tip-end structural models of the BN nanotubes. The tip-end consists of single 4 member ring (a) and single 8 member ring and a couple of 4 member rings (b).

**REFERENCES**

Kroto H W, Heath J R, O'Brien S C, Curl R F and Smalley R E 1985 Nature **318**, 162

Iijima S 1991 Nature **354**, 56

Blase X, Rubio A and Cohen M L 1994 Europhys. Lett. **28**, 335

Chopra N G, Luyken R J, Cherrey K, Crespi V H, Cohen M L, Louie S G and Zettl A 1996 Science **296**, 966

Loiseau A, Willaime F, Demoncy N, Hug G and Pascard H 1996 Phys. Rev. Lett. **B76**, 4737

Golberg D, Bando Y, Eremets M, Takemura K, Kurashima K and Yusa H 1996 Appl. Phys. Lett. **69**, 2045

*Inst. Phys. Conf. Ser. No 153: Section 13*
*Paper presented at Electron Microscopy and Analysis Group Conf. EMAG97, Cambridge, 1997*
© *1997 IOP Publishing Ltd*

# TEM study of the crystalline and amorphous phases in $C_{60}$

**V D Blank\*, B A Kulnitskiy\*, Ye V Tatyanin[#] and O M Zhigalina[+]**

\* Research Centre of Superhard Materials, Moscow region, Troitsk, 142092, Russia
[#]Institute for High Pressure Physics, RAS, Moscow region, Troitsk, 142092, Russia
[+]I.P.Bardin Central Research Institute of Iron Steel, Precision Alloys Institute
9/23, 2-nd Baumanskaya St., Moscow, 107005, Russia

**ABSTRACT:** The $C_{60}$-samples after high pressure treatment up to 13 GPa and high temperature treatment up to 1300K have been studied by TEM-methods. The thermobaric treatment of $C_{60}$ has resulted in the appearance of new crystalline states with fcc- and distorted fcc-lattices. Besides, an unusual structure has been found. Diffraction patterns of this structure contain arc-reflections, as well as disordered carbon structure, but with a regular arrangement in reciprocal space. All the structures observed can be explained by moving of molecules closer together and by bonding of molecules.

## 1. INTRODUCTION

A pure $C_{60}$-crystal forms an fcc-lattice at ambient conditions. It was shown by Blank et al. (1994a and 1994b) that among the interesting peculiarities of $C_{60}$ after high pressure and high temperature treatment there were the mechanical properties comparable to natural diamond. The present work was undertaken to study the structure of $C_{60}$-samples after thermobaric treatment in wide ranges of temperatures and pressures by TEM methods using JEM-100C and JEM-200CX Transmission Electron Microscopes.

## 2. RESULTS AND DISCUSSION

The sequence of fcc-phases appears at the first stages of thermobaric treatment. The lattices parameters of the fcc-phases decrease from 1.417 nm to 1.27 nm with growth of applied pressure. Iwasa et al. (1992) and Duclos et al. (1991) presented the data for some of these phases. In addition, some slightly distorted fcc-phases were found in $C_{60}$-objects by Blank et al. (1995). All these distorted crystalline phases exist in small proportion, X-ray diffraction gives no way to identifiy these phases. The inhomogeneity of all the samples was found by TEM-observation. Crystalline and amorphous fragments join smoothly without distinct boundaries.

The different sections of the reciprocal lattice were examined by tilting the sample around the chosen crystallographic directions. The obtained diffraction patterns and their changes during the tilting have been found to correspond to the distorted lattices of the initial phase. The lattice parameters have been defined using the Au-standard. It was found for a number of crystal lattices, studied after treatment in the wide range of temperatures and pressures, that the $a$, $b$, $c$ parameters varied from 1.12 to 1.3 nm, whereas the angles $\alpha$, $\beta$, $\gamma$ differed from the right angle by not more then 6°. The results of TEM phase analysis are summarised in the table.

## 2.1 TABLE

| lattice parameters $a,b,c$(nm); $\alpha,\beta,\gamma$(°) | | distances (nm) in [110]-direction | density (g/cm$^3$) | treatment conditions: (GPa);T(°K) | |
|---|---|---|---|---|---|
| $a$=1.417 | $\alpha=\beta=\gamma$=90 | 1.002 | 1.68 | initial | |
| $a$=1.360 | $\alpha=\beta=\gamma$=90 | 0.962 | 1.90 | 9 | 550 |
| $a$=1.30 | $\alpha=\beta=\gamma$=87.5 | 0.919 | 2.17 | 9 | 750 |
| $a$=1.29 | $\alpha=\beta=\gamma$=90 | 0.930; 0.892 | 2.23 | 9 | 750 |
| $a=b$=1.225 $\alpha$=89; $\beta$=91.5; $c$=1.175 $\gamma$=85 | | 0.856; 0.841; 0.837; 0.859; 0.903; 0.827 | 2.71 | 13 | 900 |
| $a=b$=1.23 $\alpha=\beta=\gamma$=92.5 $c$=1.12 | | 0.813; 0.850; 0.888 | 2.82 | 13 | 1100 |
| $a$=1.17; angles close to 90 | | 0.828 | 2.98 | 13 | 1300 |

The thermobaric treatment forces the molecules to come closer to each other. In this case the number of degrees of freedom , connected with rotation, decreases. The growth of applied pressure can give rise to the connection of molecules. The connection of molecules can be realised only by breaking one of the double bonds between two hexagons in the molecule. The break one of single bonds causes the destruction of the molecule. The free bonds of two molecules form intermolecular connection by linking together.

The initial $C_{60}$ has the fcc-structure with a nearest-neighbour distance of 1.002 nm. As it is evident from the table, the minimal distance between two $C_{60}$-molecules in [110] direction is 0.813 nm. Using the diameter of $C_{60}$-molecule (0.67-0.71 nm) presented in reviews of Dresselhaus et al. (1993) and Eletskiy et al. (1995), we can conclude that the minimal distance in the crystal structure 0.813 nm corresponds to the shortening of intermolecular bonds, indicative of a strong bonding mechanism. The linear chain of molecules with the same orientations corresponds to one of [110]-directions in the triclinic lattice. Different chains vary in bonds between molecules. The distances between molecules in <110> directions are found to be of different lengths, causing distortion of fcc-lattice. The growth of applied pressure and temperature is the reason for the decrease of parameters and distortion of fcc-lattice, as soon as the shortest distances between molecules tend to diminish. As a consequence the calculated density increases from 1.68 in initial $C_{60}$ to 2.98 in the most dense fcc-lattice with parameter 1.17 nm. The appearance of ideal fcc-structure with short distances between molecules (hard intermolecular bonds) is doubtful as the symmetries of a molecule and the fcc-structure are not the same. The angle between hexagons in a free molecule is less than the 70.5°-angle between {111} planes of fcc- lattice. Hence, the different [110] directions vary in bond lengths. Occurrence of linear chains and shortening of intermolecular bonds lead to the improvement of the mechanical properties of $C_{60}$. The discrete values of distances 1.02, 0.919 and some others do not correspond to the connected molecules, but more likely - to the orientational transitions. The van der Vaals intermolecular bondings are available in that case.

Two haloes at 4.5 nm⁻¹ and 8.3 nm⁻¹ are present in the most of Diffraction Patterns (DP). These haloes are the same as the ones in amorphous carbon. Except for these haloes a diffuse ring can be found in some DPs near the most intensive (002)-reflection of the graphite structure. The same diffuse peak was observed in room temperature X-ray powder diffraction for all samples under high pressure treatment up to 9.5 GPa. The position of this peak was found to shift to larger angles θ when the pressure applied increased in the temperature range 750-1050K. The shifted peak position corresponds to a change of the $d$-value from 0.335 nm to 0.314 nm. It is known that $d_{002}$ of different kinds of graphite structures can vary from 0.335 nm to 0.345 nm. The DPs, corresponding to the diffuse peak, differ not only in meanings of $d$ but also in distributions of intensity. The Fig.1 demonstrates the DP containing four arcs resulted from any of specimen fragments under conditions of suitable specimen orientation. In

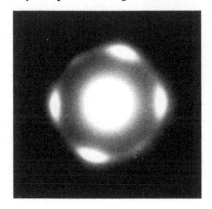

Fig.1 Electron DP of new structure

this case the angle between the nearest arcs appears to be close to 70°. It was possible to observe the joint DP which consists of several elementary DPs. The elementary DP involves four arcs. The dark field image obtained in different arc reflections of elementary DP indicates that these different arcs result from the electron scattering by the same very small area of the specimen. The specimen tilting by ± 60°, when an arc reflection of the elementary DP was situated on the rotational axis of the goniometer, led to the appearance of the same DP, but turned around the electron beam. It has been impossible to observe alternative DPs from the same volume element by specimen tilting. The DP is obtainable in the following manner. The separate arc has been produced by a set of planes with normals close to the <111> direction of the cubic structure. The section of the reciprocal lattice in Fig.1 represents the smeared-out [110] section of the cubic structure, which contains only {111} reflections. The 60°-tilt causes the appearance of new <110> section. The arcs resulting from rotation of specimen appear in the sites of {111} reflections of the cubic lattice after 60°-tilting.

Thus, there is an ordering of planes packing in four directions for any of the volume elements. The angle between any of these directions is close to 70° (Fig.2). The ordering in all alternative directions is unavailable. This is just the case when the reflecting planes form the same tetrahedron as the tetrahedron formed by the {111}-planes of a cubic lattice. The peculiarities of this structure probably can be explained on the basis of the existence of the domain structure with a small disorientation of domains. The obtained DPs show that a molecule stops scattering of electrons as an independently rotatable element, but the connection with the initial cubic structure and probably with the construction of an initial molecule is still traced. Though the microdiffraction pattern is typical for materials with a small size of the coherently scattering areas, some observed characteristics are proper to large crystals: the stability of a DP at large distances, the crystallographic connection between different DPs under tilting, the existence of large volumes with disorientation of the twin type. We can suggest two explanations of DPs obtained. In the first case closely packed {111}-layers of initial fcc-structure have drawn together in one of the <111>-directions in a separately taken volume. The molecules have the chaotic

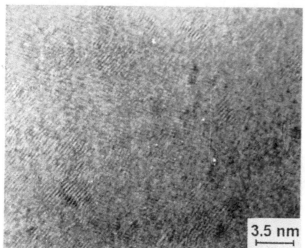

orientations in the layer and the same orientations in the normal direction. This structure must contain four kinds of domains with the 70°-disorientation. In the second case the appearance of new carbon structure can be explained by an ordered packing of curved carbon layers in four <111> directions of the initial cubic lattice, resulting in disordered arrangement of carbon atoms in the other directions. The collapse of molecules is apparently necessary to produce this structure.

3.5 nm

Fig.1b High resolution image of new structure. The angle between lattice fringes in different domains is about 70°. ( made by D.N.Zakharov, Institute of Crystallography, Russian Academy of Sciences)

## 3. CONCLUSION:

The $C_{60}$-samples have been studied after thermobaric treatment by TEM analysis. It was found that crystalline and amorphous fragments join smoothly without distinct boundaries. Distortion of fcc-lattices, decrease of their parameters and growth of density should be explained by the occurrence of the linear chains of $C_{60}$-molecules. The new structure state with DP containing arc-reflections and with a regular arrangement in reciprocal space has been found. The observed structure probably does not have analogies.

## 4. REFERENCES:

Blank V, Buga S, Popov M, Davydov V, Kulnitskiy B , Tatyanin Ye, Agafonov V, Ceolin R, Szwarc H, Rassat A and Fabre C Mol.Mat. 4 1994a 149
Blank V, Popov M, Buga S, Davydov V, Denisov V, Ivlev V, Mavrin B, Agafonov V, Ceolin R, Szwarc H and Rassat A Phys.Let. A 188 1994b 281
Blank V, Kulnitskiy B and Tatyanin Ye Phys.Let. A204 1995 151
Dresselhaus M, Dresselhaus G and Eklund P J.Mater. Res. 8 1993 2054
Duclos S J, Brister K, Haddon R C, Corten A R and Thiel F A Nature 351 1991 380
Eletskiy A and Smirnov B Uspechi Phis. Nauk 165 1995 977
Iwasa Y, Arima T, Fleming R M, Siegrist T, Zhou O, Haddon R C, Rothberg L J, Lyons K B, Carter Jr. H L, Hebard A F, Tycko R, Dabbagh G, Krajevski J J, Thomas G A and Jagi T Science 264 1992 1570

*Inst. Phys. Conf. Ser. No 153: Section 13*
*Paper presented at Electron Microscopy and Analysis Group Conf. EMAG97, Cambridge, 1997*
© 1997 IOP Publishing Ltd

597

# Carbon Deposition on AGR Fuel Pins

**GR Millward, D Cox, M Aindow, PJ Darley\*, HE Evans and CW Mowforth[†]**

School of Metallurgy and Materials, University of Birmingham, Edgbaston, Birmingham, B15 2TT, UK.   \*School of Physics and Space Research, The University of Birmingham. [†]Nuclear Electric Ltd., Barnett Way, Barnwood, Gloucester, GL4 3RS, UK.

**ABSTRACT:**   HREM studies of carbon removed from the surface of an AGR fuel-pin show the presence of at least three forms of carbon. These can be described broadly as: lamellar, microcrystalline and filamentous. The degree of ordering and stacking of the sheets of hexagonally packed carbon atoms observed is unlikely to arise solely from pyrolitic deposition and some catalytic intervention is envisaged. Protection from carbon deposition can be obtained by creating Cr-rich surfaces. Initial experiments demonstrating the nucleation and growth of Cr oxides from grain boundaries are presented.

## 1. INTRODUCTION

When low density carbon deposits are formed on the surfaces of Advanced Gas-Cooled Nuclear Reactor (AGR) austenitic stainless steel fuel-pins, heat transfer from the pins to the coolant gas is impaired. Wood (1980) has discussed how the growth of carbon is thought to be due, in part, to the presence of trace amounts of hydrocarbons formed from $CH_4$ by radiation-induced reactions. The surface of the stainless steel pin is crucial (Horsley and Cairns 1984, Holm and Evans 1987, Barry and Dinsdale 1994) and the tendency for carbon formation can be reduced significantly by the presence of a chromia layer.

The work presented here is part of a laboratory simulated investigation designed to elucidate some features of carbon deposition related to the nature of the fuel-pin surface. Firstly, we report observations on the structure of carbon deposit taken from a fuel-pin of an in-service AGR reactor. Secondly, we describe experiments which give some insight into the mechanisms by which the surface of stainless steel is modified by selective oxidation.

## 2. EXPERIMENTAL METHOD

TEM samples of carbon deposit from an AGR fuel-pin were produced by grinding flakes using an agate mortar and pestle and dispersing the fragments in acetone. The suspension was subsequently dried onto lacy carbon supports on copper EM grids.

Samples of fuel-pin alloy (20Cr/25Ni/0.6Nb austenitic stainless steel) were produced by annealing 3mm discs at 930° C for 1hr in dry argon and then electropolishing to perforation with 10% perchloric acid in ethanol. Selective oxidation was carried out over a range of temperatures and times in an atmosphere of 10% $H_2$ in Ar containing 200-300 ppm $H_2O$.

TEM examinations were carried out using JEOL 200CX and 4000FX instruments. The SEM used was a JEOL 5410.

## 3. STRUCTURAL CHARACTERISTICS OF AGR CARBON DEPOSIT

The carbon in the deposit exhibits a range of morphologies (Fig.1). Lamellar fragments (e.g. Fig. 1a) are typical of turbostratic carbons (Ruland 1968, Ning et al 1990): electron diffraction from region (N), where the beam is normal to the sheets of hexagonally packed carbon atoms, shows random rotation of the basal planes around the c-axis. When the flake is viewed (P) along the lamellar planes, imperfect stacking of sheets is demonstrated by the arcing of the {0002} diffraction spots, and directly in the HREM lattice image (Fig 1b).

598

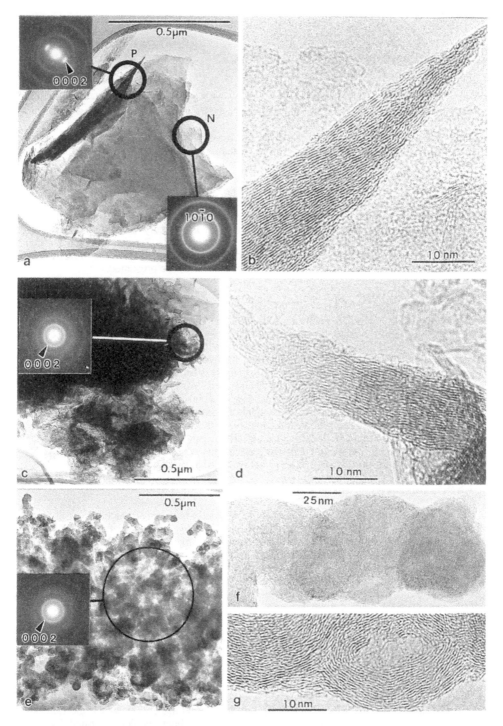

Fig. 1    TEM images and selected area diffraction patterns of various forms of carbon deposited on an AGR nuclear fuel-pin.

The fragment of deposit illustrated in Fig. 1c is different consisting of assemblies of small 'microcrystalline' domains of partially ordered turbostratic carbon. The domains are randomly arranged throughout the fragment (see the inset diffraction pattern), and the stacking of carbon sheets (Fig. 1d) has approximately the same degree of ordering as the first example.

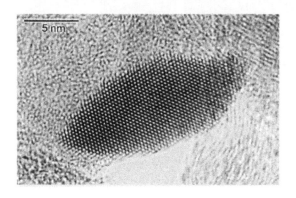

Both types of fragment sometimes have crystalline (iron-rich) particles associated with them (Fig. 2). Common lattice spacings seen are about 0.3nm and 0.26nm.

A third minority component is the filamentous carbon shown in Fig. 1e-g. The formation of such filaments is usually associated with catalytic decomposition of hydrocarbons by small metal particles (Baker and Harris 1978). In this case we could not visualise particulate metal, but the presence of Fe and Ni was detected by EDX analysis.

Fig. 2   Crystalline particle found in reactor deposit

In all of the types of fragment shown above the degree of ordering of carbon sheets is in excess of that expected for carbons formed pyrolytically at the temperatures existing near the surface of the fuel-pins (600-700°C), and it is likely that there has been some form of catalytic influence on the carbon deposition processes.

## 4.   SELECTIVE OXIDATION OF THE SURFACE OF FUEL-PIN STEEL

During exposure in a reactor the coolant gas, of nominal composition $CO_2$/1%CO, has sufficient oxygen potential to oxidise chromium and iron of the major alloy components, but not nickel. Metallic nickel is known (Baker and Harris, 1978) to catalyse filamentary carbon deposits but contact with the coolant gas will arise only if the surface oxide is porous and provides gas-access channels, or if a protective surface oxide cracks or spalls. In the present oxidation experiments, such access to the steel surface is simulated by preventing iron oxidation (apart from formation of $FeCr_2O_4$) through the use of a low oxidation potential -500 kJ mol$^{-1}$.   Thus, at least in the early stages of oxidation, before complete coverage by a chromia layer, metallic nickel and iron will be in contact with the gaseous environment.

At a temperature of about 600°C, the supply of chromium to the surface will occur along alloy grain boundaries (Holm and Evans, 1987), particularly for the current annealed samples where the dislocation density will be low. This is demonstrated in Fig. 3 which indicates that oxide formation does originate at grain boundaries (Fig. 3a and Fig. 3d), and that further oxide growth is dependent upon grain boundary diffusion with the oxide spreading laterally into the grain centres (Fig. 3a-c). EDX analysis of this oxide shows it to be predominantly Cr-rich with some Fe and Mn, but no Ni present. Specimens prepared in this way will be used in future work to investigate the nucleation stage of the carbon deposition process.

An interesting consequence of our method of specimen preparation is that where a hole, caused by electropolishing, has formed in the stainless steel specimen foil, the subsequent oxidising process results in a very thin region of oxide growing out from the edge of the hole. HREM studies of this oxide (Fig. 3e) shows crystals, similar to those observed in the reactor deposit, with lattice spacings indicative of spinel structures. In this case, EDX analysis indicates a high Cr content together with small amounts of Mn and Fe.

## ACKNOWLEDGEMENTS

This work was carried out under contract from Nuclear Electric Ltd, funded by the Industry Management Committee (Nuclear Electric Ltd, Magnox Electric plc, Scottish Nuclear Ltd and British Nuclear Fuels plc).

600

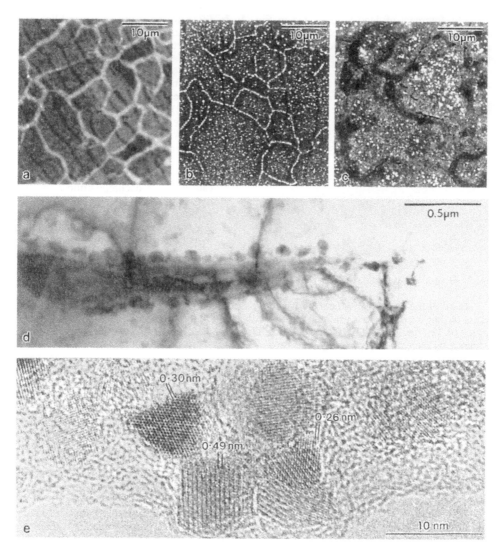

Fig. 3 Oxidation of fuel-pin steel. SEM micrographs of ~200μm thick foils showing grain boundaries after oxidation for 15 min at (a) 600°C, (b) 650°C and (c) 750°C. TEM micrograph of a thinner region (d), showing oxide particles decorating a grain boundary where it emerges at both surfaces (15 min at 650°C). HREM image of an oxide layer growing from the edge of the hole in the foil (e).

## REFERENCES

Baker R T K and Harris P S 1978 Chem. Phys. of Carbon, ed P L Walker and P A Thrower (New York: Marcel Dekker) **14**,83

Barry T I and Dinsdale A T 1994 Mater. Sci. and Techn. **10**, 1090

Holm R A and Evans H E 1987 Werkst. Korros. **38**, 219

Horsley G W and Cairns J A 1984 Applic. Surface Science **18**, 273

Ning X J, Pirouz P, Lagerlof K P D and DiCarlo Jk 1990 J. Mater. Res. **5**, 2865

Ruland W 1968 Chem. Phys. of Carbon, ed P L Walker (New York: Marcel Dekker) **4**, 1

Wood C J 1980 Gas Chemistry in Nuclear Reactors and Large Industrial Plant, ed A Dyer (London: Heyden) pp 6-17

*Inst. Phys. Conf. Ser. No 153: Section 13*
*Paper presented at Electron Microscopy and Analysis Group Conf. EMAG97, Cambridge, 1997*
© *1997 IOP Publishing Ltd*

# Self assembly of nanosized gold clusters into regular arrays

J Fink, C J Kiely, D Bethell* and D J Schiffrin*

Department of Materials Science and Engineering, University of Liverpool, Merseyside, L69 3BX, UK.

* Department of Chemistry, University of Liverpool, Merseyside, L69 3BX, UK.

**ABSTRACT:** Stable solutions of underivatised and thiol derivatised nanosized Au colloids in toluene have been prepared. A drop of the colloidal solution when allowed to evaporate onto a carbon grid results in the formation of self-assembled superstructures which have been subjected to TEM examination. In monolayer form, the Au particles assemble into highly ordered pseudo-hexagonal close packed rafts in which the interparticle separation can be controlled by varying the size of the stabilising species attached to the Au surface. When allowed to form bilayers, the Au particles adopt corrugated chain and ring-like structures in addition to normal close packed stacking sequences. Possible origins of these unusual stacking phenomena are discussed.

## 1. INTRODUCTION

It is now well known that nanosized colloidal metals and semiconductors can have unusual size dependent physical and chemical properties (Schmid 1994). In order to better exploit these properties, it is necessary to devise strategies of assembling such nanoparticles into ordered superstructures. Brust *et al* (1994) have recently had considerable success in the preparation, stabilisation and self-organisation of Au nanoparticles. Their method involves the attachment of quaternary ammonium salts onto the surface of Au nanoparticles in a toluene solution. These *underivatised* colloidal Au solutions are stable for at least a year and can be used to construct superstructures. If a drop of the solution is placed onto a suitable substrate and the toluene allowed to evaporate, weakly ordered two and three dimensional 'molecular crystal' structures can be generated. Brust *et al* (1995) and Whetton *et al* (1996) have shown that the self-assembly process of these particles into ordered superstructures can be greatly improved by *derivatising* the Au surface with various thiol groups prior to deposition of the particles onto a substrate. In this paper we show that transmission electron microscopy studies of such systems can lead to a better understanding of the factors affecting the self assembly processes involved in molecular crystal formation.

## 2. EXPERIMENTAL

Stable solutions of underivatised and derivatised Au nanoparticles in toluene were prepared following the method described by Brust *et al* (1994). A series of tetralkylammonium bromides, ($NR_4^+Br^-$ with R ranging from $C_6$ to $C_{18}$) were used as phase transfer reagents. Solutions of the underivatised Au particles were ruby red in colour, whereas those derivatised with alkanethiol groups were brown in colour. For electron microscopy analysis, a drop of the colloidal gold solution was placed on a carbon coated Cu mesh grid and the toluene allowed to evaporate. Specimens were examined in a JEOL 2000FX transmission electron microscope operating at 200kV.

## 3. RESULTS AND DISCUSSION

### 3.1 Monolayer rafts of particles

A feature very commonly observed in samples prepared from both underivatised and derivatised colloidal Au suspensions were two dimensional rafts in which the nanoparticles have adopted pseudo-hexagonal close packed arrangement (as shown in Fig.1). In general, the rafts found in the thiol derivatised samples were much more ordered than their underivatised counterparts, often covering areas of several μm² on the carbon film. Inspection of Fig.1 shows that there is a narrow particle size distribution between 3 and 5nm. Electron diffraction patterns from such regions reveal only the characteristic 111, 200, 220, 331, 222 and 400 reflections of f.c.c. gold, whereas HREM imaging shows distinct facetting characteristic of truncated octahedral crystallites. It has been proposed by Ohara *et al* (1997) that these rafts are formed as a natural consequence of the solvent evaporation process during which pinholes in the solvent film form and expand; surface tension forces associated with the receding liquid drag the colloidal particles into closer proximity and force the particles to order into a hexagonal array.

A distinctive feature of the pseudo-hexagonal array is that the Au nanoparticles are not in direct contact with each other; they are in fact separated by a region, typically about 1nm in size for the example in Fig.1, which does not exhibit any diffraction contrast. This is because each particles in Fig.1 is coated with a layer of thiol molecules which are anchored in position by the strong Au-S bond as illustrated in Fig.2(a). A similar effect is observed in the rafts of underivatised particles which are covered with a layer of adsorbed tetralkylammonium bromide molecules (Fig 2(b)). In this latter case, the Br⁻ ions are strongly bonded to the Au particle and the tetrahedral $NR_4^+$ ion sits on top of the Br⁻ rather like a 'top-hat'. We have recently been able to demonstrate that varying the length of the carbon chains in the thiol molecule or the tetralkylammonium ion affords a convenient and simple way of controlling the interparticle separation Fink et al (1997)). The interparticle separation in the rafts is approximately equal to the length of one attached molecule suggesting that interpenetration effects between the adsorbed layers on adjacent particles are significant in these systems.

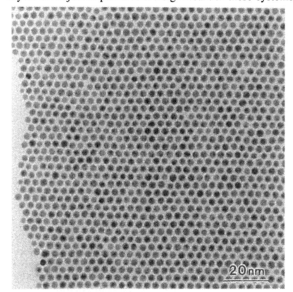

Fig.1. Typical hexagonal raft of alkane thiol derivatised Au particles.

Fig.2 Schematic structures of Au particles stabilised with (a) alkane thiols and (b) tetra-alkylammonium bromide salts.

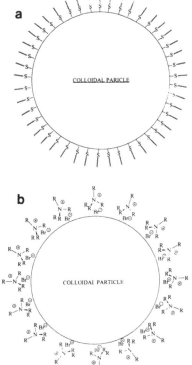

A very recent and exciting development, which occurs with the use of very specific combinations of phase transfer reagents and thiol molecules, is the ability to generate bimodal size distributions of Au particles. An example of a two dimensional raft prepared from a mixture of 9nm and 4.5nm diameter particles is shown in Fig.3. What is remarkable here is the high degree of self assembly where the smaller particles have positioned themselves with great regularity in the interstitial sites formed at the intersection of three larger particles. Such structures can be formed reproducibly and have been observed to cover areas up to 2μm² in size.

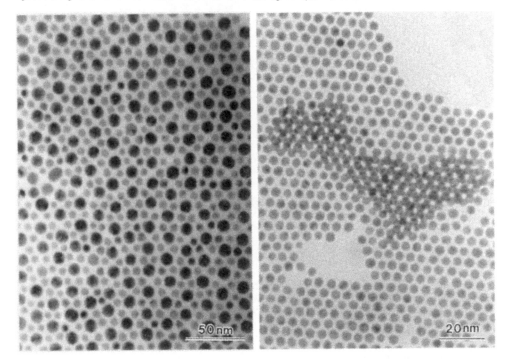

Fig.3. Raft structure formed from colloidal particles of two different sizes.

Fig.4. Typical AB stacking bilayer structure found in thiol derivatised sample.

## 3.2 Stacking sequences in multilayer structures

The stacking sequences formed when successive layers of particles are deposited onto an underlying single hexagonal raft of particles have also been studied. Hexagonal (ABABAB) and cubic (ABCABC) stacking are intuitively the sequences one would expect to observe when building up three dimensional 'crystals' from a single size of sphere. Indeed for the Au particles derivatised with thiol species such simple stacking sequences were observed. A typical area showing such a bilayer region is shown in Fig.4. Here the particles in the second layer reside in the three-fold hollow sites created at the intersection between three particles in the basal layer (ie. AB packing)

However, the underivatised particles *never* exhibited this type of stacking behaviour. Instead, corrugated chain structures of particles, which appear as straight lines in projection, were often visible as shown in Fig.5. Furthermore three discrete domains (labelled 1, 2 and 3) in which lines of clusters are rotated by 60° are clearly discernible. These chain-like arrangements can only be generated if the second layer of particles occupy the two-fold saddle sites formed between pairs of particles in the first layer rather than the usual three-fold depression sites. The three domains of chains that can be observed are then just a consequence of there being three equivalent two-fold saddle positions in the basal layer. The preferred stablization of the two-fold saddle over the three-fold hollow site is surprising and may be related to steric hinderance effects and/or dipole-dipole repulsion effects occurring between adsorbed tetralkylammonium ions on adjacent particles.

604

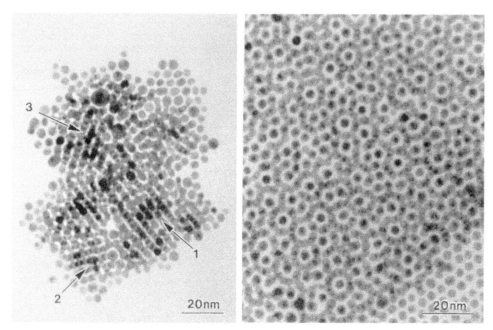

Fig.5. Corrugated linear chain-like features found in bilayers of underivatised Au colloids.

Fig.6. Corrugated closed ring structures (taken from a thiol derivatised sample).

Another very common effect which was observed in bilayers of *both* underivatised and derivatised particles was the striking presence of ring structures as shown in Fig.6. These features are simply closed corrugated ring structures where once more, the second layer of particles preferentially occupy two-fold hollow sites but this time in a circular rather than a linear fashion. At the centre of each ring the particle in the second layer apparently sits vertically above its counterpart in the lower layer which is indeed truly a metastable position! The ring structures observed in the underivatised Au samples are probably again a consequence of the balance between electrostatic repulsion of the surface dipoles present on the Au clusters and dispersion interactions. The co-existence of ring structures and normal ABAB packing in the thiol derivatised samples is much more diffult to explain and is the subject of further investigation.

## 4. CONCLUSIONS

We have demonstrated the propensity of underivatised and thiol-derivatised nanosized Au clusters to self assemble into regular arrays on amorphous carbon substrates. Highly ordered monolayer rafts of Au particles with both single and bimodal size distributions have been fabricated. The stacking sequences in multilayer structures have also been investigated.

Underivatised Au particles stabilised with adsorbed $NR_4^+Br^-$ species only exhibit corrugated chain and ring stuctures formed by filling two-fold saddle sites in the basal layer. Thiol derivatised Au particles exhibit a combination of ring structures as well as more conventional ABABAB and ABCABC stacking sequences.

## REFERENCES

Brust M, Walker M, Bethell D, Schiffrin DJ and Whyman R, 1994, J.Chem.Soc.Chem.Comm, 801.
Brust M, Bethell D, Schiffrin DJ and Kiely CJ, 1995, Adv. Mater. **7**, 795.
Fink J, Kiely CJ, Bethell D and Schiffrin DJ, 1997, Chem. of Mats., in press.
Ohara P, Heath JR and Gelbart WM ,1997, Agnew.Chem.Int.Ed.Engl., **36**, 1077.
Schmid G (Ed), Clusters and Colloids, 1994, VCH, Weineim.
Whetten RL, Khoury JT, Alvarez MM, Murthy S, Vezmar J, Wang ZL, Stephens PW, Cleveland CL, Leudtke WD and Landman V, 1996, Adv. Mater., **8**, 428.

Inst. Phys. Conf. Ser. No 153: Section 13
Paper presented at Electron Microscopy and Analysis Group Conf. EMAG97, Cambridge, 1997
© 1997 IOP Publishing Ltd

# The use of PEELS for the study of precipitate complexes in HSLA steels

## A J Craven, He Kejian[*] and T N Baker[*]

Department of Physics and Astronomy, University of Glasgow, Glasgow G12 8QQ, Scotland.
[*]Metallurgy and Engineering Materials Group, Department of Mechanical Engineering, University of Strathclyde, Glasgow G1 1XN, Scotland.

**ABSTRACT:** Precipitate complexes in three high strength low alloy (HSLA) Ti-Nb steels have been analysed by parallel electron energy loss spectroscopy (PEELS). The complexes consist of a core with overgrowths of a cap phase. The N:Ti atomic ratios in the three steels are 3.1, 1.7 and 0.77 while the Nb:Ti ratios are 1.5, 0.8 and 0.4 respectively. In each steel, the Nb:Ti ratio in the core is ~0.1 while the N/(Ti+Nb) ratio matches the mean value. The cap composition varies from $Nb(C_{0.7}N_{0.3})$ to $NbC$ to $(Nb_{0.7}Ti_{0.3})C_w$ (with $w$ undetermined) as the N content is reduced.

## 1. INTRODUCTION

Precipitation in micro-alloyed high strength low alloy (HSLA) steels frequently results in precipitate complexes. These complexes occur because a phase which comes out of solution at a lower temperature grows on an existing phase which came out of solution at a higher temperature. Such is the case in Ti-Nb micro-alloyed Al killed HSLA steels which have been control-rolled into plate. Here the first phase out of solution is based on TiN while the second is based on NbC. Depending on the mean composition of the steel, the compositions of the phases are modified by substitution on the metal and non-metal sub-lattices. The challenge is to find the composition of the first phase when it is covered by the second.

It has proved possible to analyse, in detail, the parallel electron energy loss spectroscopy (PEELS) data recorded from a few of the precipitates investigated as part of a more general study. The electron energy loss near-edge fine structure (ELNES) is useful for confirming that the C is in the precipitate rather than from the carbon extraction replica or contamination. The use of standards allows the extraction of atomic fractions from spectra which have the closely spaced edges that result from the presence of Nb, C, N and Ti. The high spatial resolution of the VG Microscopes HB5 allows the composition of the outer phase alone to be determined and this then allows a good estimate of the composition of the inner phase to be made in the cases studied.

## 2. MATERIALS AND METHODS

Three laboratory steels A, B and C were vacuum melted at the Swinden Technology Centre of British Steel. The atomic fractions of C, N Ti and Nb are given in Table 1. On going from Steel A to Steel B, the N and Nb content were reduced while maintaining the Ti content. On going from Steel B to Steel C, the N and Nb content were held constant and the Ti content increased. Thus the N:Ti ratio decreased on going from Steel A to Steel B to Steel C, dropping below unity in the last case. Each steel was controlled rolled with a 3:1 reduction below 880°C and finishing at 800°C. The final thickness of the plate was *circa* 16mm. Carbon extraction replicas were made using a four step process involving polishing, etching with 2% Nital, coating with the minimum possible thickness of carbon, stripping in 5% Nital and washing in both distilled water and methanol. These replicas were studied in a VG Microscopes HB5 equipped with post-specimen lenses and a GATAN 666 PEELS spectrometer. The HB5 was operated at 100kV and the probe and collection half angles were 8 and 12.5 mrad respectively. If possible, spectra were recorded from precipitates overhanging the edge of the carbon film at one of the many holes and cracks present in such thin support films.

| Steel | C | N | Ti | Nb |
|-------|------|-------|-------|-------|
| A | 0.33 | 0.031 | 0.010 | 0.015 |
| B | 0.45 | 0.020 | 0.012 | 0.010 |
| C | 0.38 | 0.020 | 0.026 | 0.010 |

Table 1. Atomic % of micro-alloying elements in the three steels

## 3. RESULTS

In this paper we concentrate on precipitate complexes in which there is a core phase, based essentially on TiN, with an overgrowth of a cap phase based essentially on NbC. In some complexes, the overgrowth is in the form of a complete coating of the core while, in others, the cap grows on one or more faces of the cubic core resulting in "spherical" caps i.e. the outer surface of the cap phase has a curved surface. Figure 1 shows an annular dark field (ADF) image of a precipitate complex from Steel B. This has the form of a core coated by a cap phase. ADF images prove very useful in showing the presence of such overlying phases as such images are sensitive to the different scattering powers of the phases. The edge of the carbon support film passes almost along the centre of the complex and can just be seen in the top left hand corner of the image. However, its intensity is very low due to its low elastic scattering power and its thinness.

Figure 2 shows the PEELS spectrum from point $a$ on the cap of the complex. Point $a$ is off the edge of the support film and so there is no contribution to the C K edge from the amorphous C in the support film. A background of the form $AE^{-r}$ has been subtracted from under the Nb $M_{4,5}$ edges. The Nb $M_{4,5}$, the C K and the Nb $M_{2,3}$ edges are clearly seen but there is no evidence of the N or O K edges. The shape of C K edge is very similar to that seen from an NbC standard (Craven and Garvie, 1995). The best way to determine the C:Nb ratio is to compare the spectrum from the cap with that from the NbC standard. Such a comparison is made in Figure 2 where the grey line is the spectrum from a commercial NbC powder. This comparison clearly shows that the cap is close to stoichiometric NbC.

Figure 3 is the spectrum from point $b$ where the core overhangs the edge of the support film. The spectrum contains contributions from the elements in the core and also from those in the overlying cap. A linear combination of spectra from NbC and $TiN_{0.88}$ standards is shown

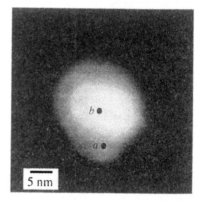

Figure 1. An annular dark field image of a precipitate complex from Steel B consisting of a core surrounded by a cap. PEELS spectra were recorded from point $a$ on the cap and from point $b$ on the core. Both points overhang the edge of the support film.

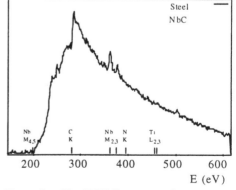

Figure 2. The PEELS spectrum from point $a$ in Figure 1 is shown in black. A background of the form $AE^{-r}$ has been removed from underneath the Nb $M_{4,5}$ edge. The PEELS spectrum from NbC powder is shown in grey.

in grey. The Nb $M_{4,5}$ edges from the NbC have been matched to those from the precipitate immediately prior to the C K edge while the N K edge from the $TiN_{0.88}$ has been matched to that of the precipitate. It is quite clear that the spectrum is not simply the sum of NbC and TiN because there is insufficient C to form NbC. Thus the core must contain some Nb and its change of environment is the likely explanation for the change of shape of the Nb $M_{4,5}$ edges.

The mean elemental ratios in the volume analysed can be found if the intensity from one edge of each element can be determined and an appropriate cross-section is available. To determine the compositions of the individual phases without introducing assumptions, it is necessary to know the relative thickness of the two overlying phases and that each is homogeneous in composition. This information is not available and the best that can be done is to assume homogeneity and also that all the carbon is in the cap phase. Given the solubility

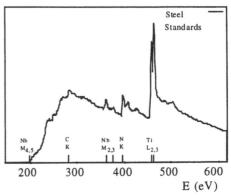

Figure 3. The PEELS spectrum from point $b$ in Figure 1 is shown in black. A background of the form $AE^{-r}$ has been removed from underneath the Nb $M_{4,5}$ edge. A linear combination of the spectra from NbC and $TiN_{0.88}$ is shown in grey. The Nb $M_{4,5}$ edges and the N K edge respectively were scaled to match those of the precipitate.

data (e.g. Houghton, 1993) and the fact that this steel has a N content in excess of that required to form stoichiometric TiN, the latter is likely to be a reasonable assumption. With this combination of edges, the determination of the individual edge intensities is not trivial. The C K edge intensity was separated from the Nb $M_{4,5}$ edge intensity in the spectra from the precipitate and that from the NbC standard using the Nb $M_{4,5}$ edge shape from a NbN standard. The separated intensities from the NbC allow an experimental Nb $M_{4,5}$ cross-section to be determined using the SIGMAK cross-section for C (Egerton, 1986). Comparing the relative intensities in the two cases allows the fraction of Nb in excess of that required to form NbC to be determined. The background shape under the N K and Ti $L_{2,3}$ edges was approximated by the shape of the spectrum from the NbC standard in that region. The Ti $L_{2,3}$ was separated from the N K edge by scaling a SIGMAK N cross-section shape to the N edge. Using the same procedure on the $TiN_{0.88}$ standard allowed the N:Ti ratio to be determined. This value has an estimated error of ~10%. The overall Nb:Ti ratio was determined using the empirical Nb $M_{4,5}$ cross-section and the SIGMAL2 cross-section of Ti. The Nb:Ti ratio in the core was then found by multiplying the overall value by the fraction of Nb which was in excess of that required for the NbC cap. The error in the Nb:Ti ratio in the core is difficult to estimate but is likely to be ~20%. The resulting best estimate of the core composition was $(Ti_{0.85}Nb_{0.15})N_{0.9}$. In Steel A, the cap was found to be Nb(C,N) and so by assuming that the core contained no C due to the very high concentration of N, it was possible to estimate the core composition. In Steel C, the support film was somewhat thicker and no precipitate complex was found overhanging an edge. Thus it was not possible to determine the carbon content of the complexes in this steel. In the complex studied from Steel C, there was no evidence of N in the cap but it contained both Ti and Nb. The spectrum from a point over the core had a strong N edge and also showed that the Nb:Ti ratio was much lower than in the cap. With these data, it is not possible to determine the Ti:Nb ratio in the core alone because there is no information about the relative amounts of cap and core under the beam. Despite this, it is possible to set quite close limits on the core composition because of the large difference in the Ti:Nb ratios. The results for the three steels are presented in Table 2

## 4. DISCUSSION AND CONCLUSIONS

It is clear that the composition of the cap material is markedly affected by the N:Ti ratio. In Steel A, some of the large excess of N appears in the caps. In Steel C, where the N:Ti ratio has

608

| Steel | A | B | C |
|-------|---|---|---|
| Cap | $Nb(C_{0.7}N_{0.3})$ | $NbC$ | $(Nb_{0.7}Ti_{0.3})C_w$ |
| Core | $(Ti_{0.9}Nb_{0.1})N_{1.15}$ | $(Ti_{0.85}Nb_{0.15})N_{0.9}$ | $(Ti_{0.85}Nb_{0.15})(N_{0.45}C_z) \rightarrow Ti(N_{0.55}C_{z'})$ |

Table 2.     Core and cap compositions for a precipitate complex in each of the three steels.  The range of possible core compositions is presented for Steel C where $w$, $z$ and $z'$ are undetermined.

dropped below unity, some of the excess Ti appears in the caps.  Turning to the cores, the Ti:Nb ratio in them varies relatively little compared to its overall variation, the mean values being 1.5, 0.8 and 0.4 in Steels A, B and C respectively.  This result is supported by energy dispersive x-ray spectroscopy on precipitates of similar size which showed only the core phase in their diffraction patterns.  Analysis of ~10 such particles in each steel showed a distribution of values and the results in Table 2 are consistent with these distributions.  The N content of the cores is affected by the N:Ti ratio.  In Steel C, it is so low that that there is likely to be a significant C content as well and it is unfortunate that the available data cannot confirm this.  In Steels A and B, the analysis assumed that no carbon was present in the cores.  This has little effect on the N:Ti ratios found for the cores.  Even assuming that there is no overlying cap at all in the Steel A case (which is unlikely from the morphology of this particular complex), the resulting composition is $(Ti_{0.85}Nb_{0.15})(N_{1.1}C_{0.05})$ giving only a minor change in the N:Ti ratio.  The 10% uncertainty in the N:Ti ratio makes it possible that the metal to non-metal ratio in Steels A and B is unity.  However, the likelihood that the variation is real is strengthened by the close agreement of the N/(Ti+Nb) ratios in the core and in the steel in each case.

Thus, despite the fact that the spectra were not recorded with the intention of such detailed analysis, it is clear that PEELS can provide useful information on the composition of precipitate complexes in HSLA steels even if one phase overlies another.  This is easiest if the cap phase contains an element which is not (or is unlikely to be) present in the core phase.  If this is not the case, then further information is required to determine the relative amounts of each phase under the beam.  In some circumstances, the low loss region of the spectrum may provide the required information but this approach does not look promising in this system.  Alternatively, the ELNES shape may provide the information.  In this system, the shapes of the Nb $M_{4,5}$, the C K and the N K edges change with environment.  If data are acquired with the aim of using these effects, this may prove a viable method of determining the relative amounts of each phase present.

**REFERENCES**

Craven A J and Garvie L A J G 1995 Micrsoc. Microanal. Microstruc. **6**, 89
Egerton R F 1986 Electron Energy-Loss Spectroscopy in the Electron Microscope (New York and London: Plenum) pp358-62
Houghton D C 1993 Acta Metall Mater **41**, 2993

**ACKNOWLEDGEMENTS**

The authors would like to thank EPSRC and British Steel for financial support.

*Inst. Phys. Conf. Ser. No 153: Section 13*
*Paper presented at Electron Microscopy and Analysis Group Conf. EMAG97, Cambridge, 1997*
© 1997 IOP Publishing Ltd

609

# Factors affecting the ELNES of the boron K-edge in titanium diboride in a multilayer coating

**H. J. Davock, G. J. Tatlock, R. Brydson\*, K. J. Lawson# and J. R. Nicholls#**

Department of Materials Science and Engineering and IRC in Surface Science, University of Liverpool, Liverpool, L69 3BX, UK
\*Department of Materials, School of Process, Environmental and Materials Engineering, University of Leeds, Leeds, LS2 9JT, UK
#School of Industrial and Manufacturing Science, Cranfield University, Bedford MK43 OAL, UK

**ABSTRACT:** A PVD multilayer coating of $TiB_2$ / Al has been studied with particular emphasis on the $TiB_2$ layers. PEELS studies highlighted differences in the ELNES of the B K-edge in each $TiB_2$ layer and reasons for these differences are addressed. The effect of ion beam thinning, impurity incorporation and orientation effects are all considered. It is suggested that orientation variation is the prime factor in creating the differences in the ELNES due to the anisotropic nature of the bonding in $TiB_2$.

## 1. INTRODUCTION

Multilayer coatings consist of alternating layers of two (or more) different materials and are attracting increasing interest due to their improved properties over their single layer counterparts. Improvements in wear resistance, hardness and other physical properties have been reported. Crack propagation is often more difficult through a multilayer coating and this is one of the ways wear resistance is improved, although the overall performance is also influenced by layer composition, microstructure and thickness.

The effectiveness of PEELS in analysing the structure of compounds, especially those containing light elements such as boron, has been well reported. Consideration of Electron Energy-Loss Near Edge Structure (ELNES) has proved fruitful in studying boron containing compounds (Sauer et. al.1993) due to the information that it can reveal about bonding and nearest neighbours.

## 2. EXPERIMENTAL

The multilayer system in the present investigation consisted of layers of a hard $TiB_2$ coating sandwiched between soft, ductile layers of aluminium. The coatings were deposited using an r.f. magnetron sputtering technique operated at up to 300 W in a chamber with a base pressure of 7.5 x $10^{-3}$ Torr. Cross-sectional TEM samples were prepared using standard grinding techniques and then thinned to electron transparency with a Gatan dual ion mill using 4kV argon ions. A classical textured ( {0001} $TiB_2$ // to substrate) columnar microstructure is exhibited in the layers of ~200nm in thickness, consistent with predictions from models of PVD growth (Thornton 1974) The structure and

610

composition of the titanium diboride layer in the films were examined using PEELS analysis on a VG 601UX FEG STEM operated at 100 kV. The 1nm spot size was ideal for probing these fine grained structures. In each case the spectra were processed by removing the background and then using a Fourier Ratio deconvolution routine ( Egerton 1989) when necessary.

## 3. RESULTS AND DISCUSSION

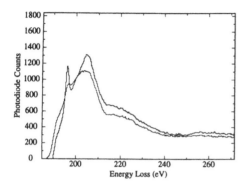 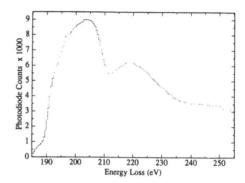

Fig. 1: Differing B-K edge spectra from the multilayer coating

Fig. 2: Boron K-edge from a polycrystalline powder sample

PEELS analysis from different regions within the $TiB_2$ layers gave a wide range of ELNES structure on the boron K-edge (Fig. 1). Some of the spectra collected were similar to the standard spectra collected from $TiB_2$ powders shown in Fig. 2, while others showed features more reminiscent of the fine structure from borate glasses (Schmid et. al. 1995). Reasons for such differences are addressed below.

*Effect of Ion-Beam Thinning*

The main difference between the multilayer and the $TiB_2$ standard was that the multilayer had been subjected to ion beam thinning. In order to check that this was not modifying the structure, a thin PVD film of $TiB_2$ was prepared on a glass slide, from which it could be floated off onto a copper support grid. A typical spectrum from an electron transparent region of this unmilled film is shown in Fig. 3. Once again a range of fine structure with features that are more pronounced than in the powder standard, are observed, suggesting that the ion beam thinning is not the principal cause of the modifications in the spectra.

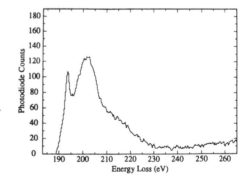

Fig. 3: Boron K-edge from an unmilled sample of sputtered $TiB_2$

*Effect of Impurity Incorporation*

Because ELNES is so dependent on nearest neighbours and bonding, there is a possibility of impurity atoms causing the different spectra. Fig. 4 shows a series of x-ray maps of the multilayer coating which indicates that there is oxygen associated with the $TiB_2$ layer. However this result was treated with caution due to the overlap between the oxygen edge and the titanium L-edge. If oxygen was present it could just be a surface layer. Also XRD work confirmed that within the limits of detection, there was only titanium diboride and aluminium present in the multilayer. So although impurities could not be completely dismissed as a possible explanation, other reasons have to be considered.

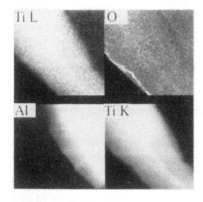

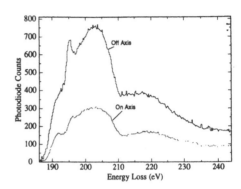

Fig. 4: X-ray maps of several elements across a $TiB_2$ / Al interface

Fig. 5: Boron K-edge spectra from a single crystal of $TiB_2$ (a) along $<1\bar{2}1\bar{3}>$ (b) off axis

*Effect of Orientation*

Orientation effects in PEELS spectra have been previously studied. These have revealed the sensitivity of $\pi^*$ and $\sigma^*$ features to the orientation of the scattering vector and hence direction of the incident beam. BN is such an example, (Garvie et. al. 1995). The relative $\pi^*$ intensity is a maximum for the electron beam perpendicular to the c-axis while the relative $\sigma^*$ intensity is a maximum for an electron beam direction parallel to the c-axis. Large c/a ratios have been thought to be central to such effects. In $TiB_2$ c/a ~1, hence orientation was not initally considered.

In the multilayer coating the grain size of the $TiB_2$ was only 17nm and after tilting (which would produce contrast changes) taking spectra from the exact same place is difficult. We were also concerned about the effect of beam damage or contamination that could be induced in such repetitive analysis in the same position. Therefore all tilting experiments were carried out using a single crystallite in a powder sample of $TiB_2$.

Fig. 5(a) shows a spectrum from the sample that was orientated with the beam along the $<1\bar{2}1\bar{3}>$ direction. Fig. 5(b) shows the spectrum with the beam tilted several degrees away from the $<1\bar{2}1\bar{3}>$ direction. The change in the ELNES between Fig. 5(a) and 5(b) demonstrates the dependency of the B K-edge on orientation. In the B K-edge from $TiB_2$ we assign the feature at ca. 190 eV to transitions to $\pi^*$ states, while features at ca. 195 eV (sharp) and ca. 202 eV (broad) are associated with $\sigma^*$ states. Multiple scattering (MS) simulations of the B K-edge have confirmed this. Additional work on graphite (McCulloch et. al. 1996) suggests that the sharp $\sigma^*$ feature is sensitive to interlayer correlations. This is seen as the feature at ca. 195 eV in the B K-edge of $TiB_2$. Since the B K-edge of trigonally co-ordinated boron oxide gives rise to a $\pi^*$ feature at ca. 194 eV (Sauer et. al. 1993) it is important to compare the respective oxygen levels, especially after the inconclusive work done previously on impurity incorporation. The oxygen edges are shown in Fig. 6. There is a very small level of oxygen in each case and the oxygen/boron ratios remain constant within experimental error.

612

Hence changes in orientation were considered to be the major factor in producing a variation in the ELNES in TiB₂.

It is thought that there are a number of reasons for this effect. Firstly TiB₂ has a hexagonal structure like BN, but in contrast to BN consists of monoelement layers. i.e. boron layers sandwiched between layers of titanium in the a-b plane. It therefore has anisotropic bonding within the whole structure as the bonding within the layers will be different to any interlayer bonding. These differences are very directional and hence it is perhaps not surprising that orientation effects are seen.

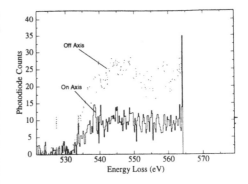

Fig 6: Respective oxygen edges (a) on axis (b) off axis

However, TiB₂ standards in the literature do not show such orientation effects. It is thought that in a polycrystalline fine grained sample of TiB₂ any ELNES differences could be averaged out. But in a highly textured sample such as a PVD coating, using a very small STEM probe, this effect has become apparent. In the TiB₂ {0001} multilayer cross-section the beam is sampling the a-b planes or very close to them. Any buckling of the foil could result in slight changes in orientation and this is thought to be the reason for ELNES changing throughout the layer.

## 4. CONCLUSIONS

It is suggested that ELNES differences seen in the B K-edge are a consequence of the sensitivity of the scattering vector and hence the electron beam to orientation rather than ion beam thinning and impurity incorporation. The anisotropic nature of the bonding in the hexagonal TiB₂ crystal is fundamental in producing the changes. It is suggested that there could also be an additional effect from interlayer correlations which is reflected in the sharp σ* feature at ca. 195eV. The differences seen are thought to be more predominant in the multilayer coating as it is highly textured. In a very fine grained polycrsyalline sample using a larger electron probe size any ELNES changes are possibly averaged out.

## 5. ACKNOWLEDGEMENTS

Thanks are given to EPSRC for funding.

## 6. REFERENCES

Egerton R F 1989 Electron Energy-Loss Spectroscopy (New York: Plenum)
Garvie L A J, Craven A J and Brydson R 1995 American Mineralogist **80**, 1132-1144
Huang R F, Wen L S, Guo L P, Gong J and Yu B H 1992 Surf. Coat. and Technol. **50**, 97-101
Lohmann R, Österschulze E, Thoma K, Gärtner H, Herr W, Matthes B, Broszeit E and Kloos K H
     1991 Material Science and Engineering **A139**, 259-263
McCulloch D G and Brydson R 1996 J. Phys.:Conden. Matter **8**, 3835-3841
Schmid H K 1995 Microsc. Microanal. Microstruct. **6**, 99-111
Sauer H, Brydson R, Rowley P N, Engel W and Thomas J M 1993 Ultramicroscopy **49**, 198-209
Thornton J A 1974 J. Vac. Technol. **11**, 666

*Inst. Phys. Conf. Ser. No 153: Section 13*
*Paper presented at Electron Microscopy and Analysis Group Conf. EMAG97, Cambridge, 1997*
© *1997 IOP Publishing Ltd*

# Microwave resonators in the system BaO•Nd$_2$O$_3$•TiO$_2$

**R Ubic, I M Reaney, and W E Lee**

Department of Engineering Materials, The University of Sheffield, Sir Robert Hadfield Bldg, Mappin Street, Sheffield S1 3JD

**ABSTRACT:** Resonators of the BaO•Nd$_2$O$_3$•TiO$_2$ solid-solution (BNT) phase have been synthesised by a mixed-oxide route. Various dopants, including Ca, Sr, and Pb, were used to tune electrical properties with varying degrees of success. Most samples show a multiphase character, with Nd$_4$Ti$_9$O$_{24}$ and NdTiO$_3$ being the main secondary phases. The space group of BNT may not conform to Pbnm (No. 62) symmetry, and the correct space group is possibly Pb2$_1$m (No. 26). The orientation relationship between NdTiO$_3$ and BNT has also been established.

## 1.    INTRODUCTION

### 1.1    Purpose

The Ba$_{6-3x}$RE$_{8+2x}$Ti$_{18}$O$_{54}$ (RE = rare-earth element) solid-solution is used to make microwave resonators due to its high dielectric constant, low loss, and tunable temperature coefficient of resonant frequency; however, detailed crystallographic data remains incomplete. The aim of this research was to examine the structure of the Nd compound and the mechanisms of its formation during mixed oxide synthesis with a view towards improving its properties.

### 1.2    Crystallography

According to the model proposed by Matveeva *et al.* (1984) and discussed by Ohsato *et al.* (1993) the structure of Ba$_{3.75}$Pr$_{9.5}$Ti$_{18}$O$_{54}$ is orthorhombic with formula A$_4$A'$_{10}$ Ti$_{18}$O$_{54}$. The crystal structure essentially consists of a three-dimensional framework of corner-sharing octahedra joined together in a pattern similar to tetragonal tungsten bronzes. Ten A' positions are randomly occupied by either Pr$^{+3}$ or Ba$^{+2}$, and four A sites are 81.25% occupied by Ba. It is worth noting that during the course of the present work a small error was detected in one of the 4c oxygen positions reported by Matveeva *et al.* (1984). This error has been corrected in this study by substituting 0.104, 0.410,0.487 for the listed coordinates of 0.104,0.485,0.487.

Gens *et al.* (1981) were the first to suggest space groups for the solid solution. Based on systematic absences observed in single-crystal X-ray diffraction of Ba$_{4.5}$La$_9$Ti$_{18}$O$_{54}$, they reported possible space groups Pba2 (No. 32) or Pbam (No. 55). The lattice parameters calculated for the Nd-analogue were $a$ = 22.21 Å, $b$ = 12.30 Å, $c$ = 3.84 Å. Matveeva *et al.* (1984) later confirmed these results by single-crystal X-ray diffraction and X-ray spectroscopic studies of a Ba$_{3.75}$Pr$_{9.5}$Ti$_{18}$O$_{54}$ crystal. They also reported space groups Pba2 or Pbam and lattice constants $a$ = 22.360 ± 0.007, $b$ = 12.181 ± 0.004, $c$ = 3.832 ± 0.004 Å. Using their data of 1,990 measured peaks, atomic positions were calculated based on space group Pba2. Approximately 400 additional peaks corresponding to a doubling of the period along $z$ were also identified, but were left out of the structural refinement. A similar model was proposed by Roth *et al.* (1987) a few years later. Ohsato *et al.* (1992) also observed superlattice peaks by X-ray diffraction in Ba$_{3.75}$Nd$_{9.5}$Ti$_{18}$O$_{54}$ crystals and reported orthorhombic lattice constants $a$ = 12.189, $b$ = 22.319, $c$ = 7.677. The space group of their fundamental lattice agreed with that of Matveeva *et al.* (1984), but that of the doubled cell was reported as either Pbn2₁ (No. 33) or Pbnm (No. 62). Kolar *et al.*'s work (1993) later repeated the findings of Matveeva *et al.* (1984).

Azough *et al.* (1995) published the first electron microscopic study of the solid-solution phase using $Ba_{4.5}Pr_9Ti_{18}O_{54}$. Re-orienting the unit cell, they reported lattice parameters $a$ = 22.2, $b$ = 12.2, $c$ = 7.6 Å. They determined the space group to be Pnam (No. 62). Rawn (1996), also accounting for the $c$-axis doubling and re-orienting the unit cells into the standard setting for each space group, used Rietveld analysis on powder X-ray diffraction data of La and Gd analogues ($x$ = 0.50) to determine the space groups as Pna2₁ (No. 33) in the case of La and Pnma (No. 62) for Gd.

## 2.    EXPERIMENTAL PROCEDURE

Pellets were prepared by a mixed-oxide route with two pre-reaction stages, as described elsewhere (Ubic *et al.* 1996). The pre-reactions involved forming titanates and dehydrating the $Nd(OH)_3$ which forms on wet milling. Calcination was conducted at 1150°C for four hours. Powders were spray dried with organic binders (1 wt% PEG1500 and 1 wt% PVA) and pressed at 360 MPa into rods which underwent a binder burn-out at 700°C prior to sintering at temperatures from 1300° - 1400°C for four hours. The rods were then cut into pellets ≈2 mm thick.

Some pellets were polished, thermally etched at 1200°C for one hour, and carbon-coated for examination in the SEM (JSM 6400, JEOL, Tokyo). Others underwent thinning to electron transparency by conventional ceramographic techniques for observation in the TEM (200CX and JEM 3010, Jeol, Japan). Electron diffraction patterns and high-resolution images were simulated using the EMS software developed by Stadelmann (V3.3, P.A. Stadelmann, 1991).

## 3.    RESULTS AND DISCUSSION

The $x$ = 0.25 and $x$ = 0.50 stoichiometries both yielded single-phase BNT at all sintering temperatures, but the $x$ = 0.75 composition resulted in a two-phase system at temperatures below 1400°C. The second phase was identified by electron diffraction as $Nd_4Ti_9O_{24}$, although the ratio of Nd:Ti was found by EDS to be nearer 1:2, indicating the stoichiometry $Nd_2Ti_4O_{11}$, in agreement with Kolar (1978), who later showed (1981) that this phase is a solid solution ranging in composition between these two stoichiometries. It will be referred to as $Nd_4Ti_9O_{24}$ for the remainder of this work. This phase is gradually consumed by neighbouring BNT grains of a lower $x$ value until an essentially single-phase $x$ = 0.75 BNT material is obtained at ≈1400°C. This transformation is even more apparent in Pb-doped samples. Micrographs of Pb-doped pellets sintered at 1380°C and 1400°C appear in Fig. 1. Both phases are clearly present at 1380° (Fig. 1a), but only BNT is evident at 1400°C (Fig. 1b). Fig. 1c shows the 1380°C pellet after re-sintering at 1400°C, and again only BNT is visible. From these data the conclusion is drawn that the $Nd_4Ti_9O_{24}$ phase either forms during calcination or is formed early in the sintering process and is transformed into BNT at around 1380° - 1400°C. There is no unique orientational relationship between the two phases.

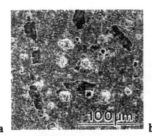 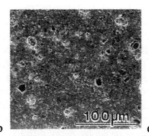 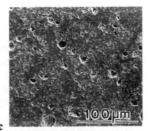

a                                    b                                    c

FIGURE 1 - Secondary-electron images of Pb-doped pellets. The pellets in (a) and (b) were sintered at 1380°C and 1400°C, respectively. The micrograph in (c) was obtained after re-sintering a 1380°C pellet at 1400°C. The large dark grains in (a) are $Nd_4Ti_9O_{24}$.

SADP's of the three principle BNT axes are shown in Fig. 2. As a first approximation, the systematic absences indicate that the space group is Pbnm; however, close examination of the [010] pattern (Fig. 3) reveals faint {100}, {001}, {201} and {102} type spots which are forbidden in Pbnm. If these extra reflections are not due to double diffraction via the FOLZ, then the only selection rules which are observed in the patterns are {0k0}: k = 2n and {0kl}: k = 2n. Assuming the true space group is one of the seven maximal non-isomorphic subgroups of Pbnm, the only two which correspond are $P2_1/b$ (No. 14) and $Pb2_1m$ (No. 26). As no evidence of non-orthogonality was found in the crystals, the monoclinic possibility is excluded from consideration; and the most probable space group for BNT is $Pb2_1m$.

a               b               c

FIGURE 2 - SADP's of BNT at (a) [100], (b) [010], and (c) [001] zone axes, 200 kV.

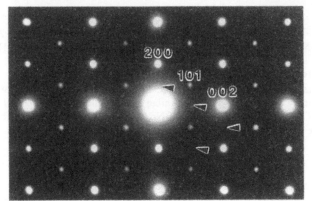

FIGURE 3 - [010] SADP (200 kV) of BNT showing faint {100}, {001}, {102}, and {201} reflections (arrowed) forbidden in Pbnm.

Doping with Ca and Sr resulted in the formation of a third phase, $NdTiO_3$, which existed intragranularly within the BNT. $NdTiO_3$ is orthorhombic but based on a distorted perovskite lattice. For simplicity, the pseudocubic (c) perovskite description of the cell will be used since this allows a more direct comparison with BNT. Examination of the SADP's from $NdTiO_3$ (Fig. 4) shows the

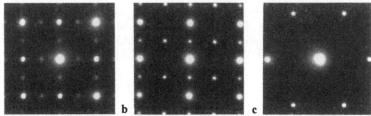

a               b               c

FIGURE 4 - SADP's (200 kV) down the pseudocubic $NdTiO_3$ (a) $[001]_c$, (b) $[110]_c$, and (c) $[111]_c$

616

existence of ½{h00}$_c$, ½{hkl}$_c$, and weak ½{hk0}$_c$ type superlattice reflections, indicating antiparallel Nd$^{+3}$ displacements as well as anti-phase and possibly in-phase rotations of TiO$_6$ octahedra (Reaney *et al.* 1994), doubling the unit cell period in *a* and *b*. Fig. 5 shows a definite orientational relationship between NdTiO$_3$ and BNT, as summarised below.

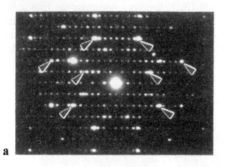

a        b

FIGURE 5 - (a) Coincident [001] zones (200 kV) of BNT and two NdTiO$_3$ variants (arrowed) rotated 36° from each other about their common $\hat{c}$. The $\hat{a}$ and $\hat{b}$ axes of each variant are 18° from the BNT $\hat{a}$ and $\hat{b}$. The NdTiO$_3$ patterns also show weak incommensuration.
(b) High-resolution image of the interface between NdTiO$_3$ and the BNT phase (300 kV)

(100) BNT ∥ (3$\overline{1}$0)$_c$-NdTiO$_3$ ≈∥ (120)$_o$-NdTiO$_3$     [100] BNT ∥ [3$\overline{1}$0]$_c$-NdTiO$_3$ ≈∥ [120]$_o$-NdTiO$_3$
(010) BNT ∥ ($\overline{1}$30)$_c$-NdTiO$_3$ ≈∥ ($\overline{2}$10)$_o$-NdTiO$_3$     [010] BNT ∥ [$\overline{1}$30]$_c$-NdTiO$_3$ ≈∥ [$\overline{2}$10]$_o$-NdTiO$_3$
(001) BNT ∥ (001)$_c$-NdTiO$_3$ ∥ (001)$_o$-NdTiO$_3$     [001] BNT ∥ [001]$_c$-NdTiO$_3$ ∥ [001]$_o$-NdTiO$_3$

## 4. CONCLUSIONS

Most samples show a multiphase character, composed primarily of BNT with Nd$_4$Ti$_9$O$_{24}$ being the main secondary phase and NdTiO$_3$ appearing intragranularly in Ca- and Sr-doped pellets. The symmetry of BNT may not conform to Pbnm (No. 62), and the correct space group is possibly Pb2$_1$m (No. 26). The orientation relationship between NdTiO$_3$ and BNT has also been established.

**REFERENCES**

Azough F, Champness P, and Freer R 1995 J. Appl. Cryst. **28**, 577
Gens A, Varfolomeev M, Kostomarov V, and Korovin S 1981 Russ. J. Inorg. Chem. **26**, 482
Kolar D, Gaberscek S, Barbulescu A, and Volavsek B 1978 J. Less Common Metals **60**, 137
Kolar D, Gabrscek S, and Suvorov D 1993 Third Euro-Ceramics **2**, 229
Matveeva R, Varfolomeev M, and Il'yushenko L 1984 Russ. J. Inorg. Chem. **29**, 17
Ohsato H, Nishigaki S, and Okuda T 1992 Jap. J. Appl. Phys. **31**, 3136
Ohsato H, Ohhashi T, Nishigaki S, Okuda T, Sumiya K, and Suzuki S 1993 Jap. J. Appl. Phys. **32**, 4323
Rawn C 1996 Electroceramics V **2**, eds J Baptista, J Labrincha, and P Vilarinho (Aveiro, Portugal: Eurpean Ceramic Society) pp 67-70
Reaney I M, Colla E, and Setter N, 1994 Jap. J. Appl. Phys. Pt. 1 **33**, 3984
Roth R, Beach F, Santoro A, Davis K, Soubeyroux J, and Zucchi M 1987 Acta Cryst. **A43**, C138
Ubic R, Reaney I M, Lee W E, Evangelinos E, and Samuels J, MRS Proc. **453**, eds P Davies, A Jacobson, C Torardi, and T Vanderah (Boston: MRS) pp. 495-500

*Inst. Phys. Conf. Ser. No 153: Section 13*
*Paper presented at Electron Microscopy and Analysis Group Conf. EMAG97, Cambridge, 1997*
© *1997 IOP Publishing Ltd*

# Transmission Electron Microscopy of the Low Temperature Oxidation of Nb₃Al

**Y Kim and K M Knowles**

University of Cambridge, Department of Materials Science and Metallurgy, Pembroke Street, Cambridge, CB2 3QZ, U.K.

**ABSTRACT:** Parallel arrays of very long, narrow, planar features lying on $\{110\}$ Nb₃Al planes found in a Nb–20at.%Al alloy oxidised in still air for 8 h at 650 °C have been characterised using transmission electron microscopy. Diffuse dark field imaging and HRTEM indicated a lack of crystallinity while EDX analysis showed a high oxygen concentration and a deficiency in niobium relative to Nb₃Al. This local oxidation process enables Nb₃Al to have an improved oxidation resistance relative to pure niobium.

## 1. INTRODUCTION

High melting points of $> 2000$ °C, low density, good high-temperature mechanical strength and relatively good ductility of Nb-rich Nb-Al alloys in the region of niobium solid solution (Nb$_{ss}$) and Nb₃Al make such materials potentially attractive for aerospace applications (Anton and Shah 1991, Steinmetz et al 1993). However, the Achilles heel of such materials is their poor resistance to oxidation (Svedberg 1976, Perkins and Meier 1990). Given the interest over the years in niobium alloys it is perhaps surprising that the details of the oxidation behaviour of Nb-Al alloys is less well understood than might be expected, in contrast to the detailed understanding which now exists about the oxidation of both pure niobium and pure aluminium. Primarily this has been because of the emphasis in research on alloy development. There is a clear need to establish in detail the mechanisms of oxidation in such alloys and the links between these mechanisms and mechanical performance in service.

Prompted by the opportunity offered by transmission electron microscopy to examine cross-sections of oxidised material by careful specimen preparation, we have been studying the early stages of oxidation in Nb-Al alloys in the range Nb–17at.%Al to Nb–20at.%Al. Here we report observations of planar features in a Nb₃Al-rich alloy which are found after a brief low temperature heat treatment in still air.

## 2. EXPERIMENTAL PROCEDURE

A rod of Nb–23at.%Al obtained from the Metal Oxides and Crystals Company (Cambridge, U.K.) was remelted with small buttons of 99.9% pure Nb in an arc furnace to obtain a Nb–20at.%Al alloy ingot. While Nb₃Al was the majority phase in this alloy, traces of Nb$_{ss}$ and Nb₂Al detectable by X-ray diffraction and optical microscopy were found, indicative of ingot inhomogeneity. Difficulties in homogenising alloys at the required temperature of between 1600 °C to 1900 °C (Kattner 1990) meant that the ingot was used as-cast for the oxidation heat treatment. Two strips of alloy 10 mm × 2 mm × 1 mm were ground, polished and heat treated at 650 °C for 8 h in still air. Cross-sections of the oxidised specimens were prepared by standard ion beam thinning procedures, and analysed using conventional bright and dark field TEM, HRTEM and EDX in a Philips CM30 operating at 300 kV. A VG HB501 STEM operating at 100 kV was used for fine probe EDX analysis.

618

## 3. EXPERIMENTAL RESULTS

After heat treatment, a thin and compact multilayered oxide scale ≈ 80 nm thick had formed on the $Nb_3Al$ substrate at the gas/alloy interface. Electron diffraction and EDX analysis of this scale indicated the presence of NbO, $NbO_2$ and an Al-rich oxide having a composition of $AlNbO_4$. These oxides have been reported in the past by investigators studying the oxidation of Nb (Kofstad 1988) and Nb-Al alloys (Perkins et al 1988, Steinhorst and Grabke 1989) by X-ray diffraction.

Immediately adjacent to this oxide scale near a crack in the alloy where extensive oxidation had occurred, the underlying $Nb_3Al$ substrate contained a dense array of long parallel features ≈ 50 nm apart, (Fig. 1). Less dense arrays could be found in regions of the alloy away from main oxide scale. When observed at the edge of the TEM samples these features appeared to be linear, but further examination in thicker areas of the samples and systematic tilting experiments showed that they were planar features lying on {110} $Nb_3Al$ planes.

These features always appeared bright relative to the surrounding $Nb_3Al$ matrix in bright field images. Examination of the Fresnel fringe behaviour in the vicinity of these features indicated that they had a lower scattering potential than $Nb_3Al$.

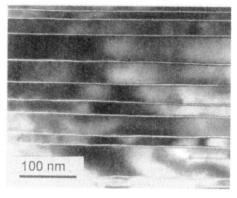

Fig. 1   Low magnification bright field image of the underlying $Nb_3Al$ adjacent to the thin oxide scale.

A careful analysis of electron diffraction patterns from regions containing these features failed to show any discrete spots associated with these defects. Diffuse dark field imaging suggested that the features lacked crystallinity, in that they could be imaged successfully using this technique (Fig. 2).

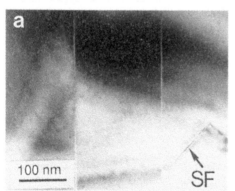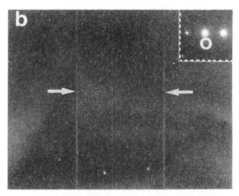

Fig. 2   (a) Bright field image of the planar features in the $Nb_3Al$ shown edge-on and (b) a diffuse dark field image from the same region obtained with the objective aperture set at 'O' in the electron diffraction pattern shown in the inset. The planar features are arrowed. The feature 'SF' in (a) is a {001} stacking fault in the surrounding $Nb_3Al$.

The defects could be imaged using $Nb_3Al$ reflections, but could not be placed out of contrast, suggesting that there was no displacement vector **R** associated with the features. This was a further indication that the features were not intrinsic to $Nb_3Al$, unlike the occasional stacking faults observed on {100} $Nb_3Al$ planes, such as the one indicated in Fig. 2(a). HRTEM imaging with the electron beam parallel to the [001] $Nb_3Al$ zone of one particular defect showed that it appeared very bright relative to the surrounding $Nb_3Al$ and was ≈ 2 nm thick (Fig. 3). On an atomic scale, the interface between the feature and the surrounding Nb₃Al was noticeably rough, even though the macroscopically the feature can be seen to lie on a {110} Nb₃Al plane.

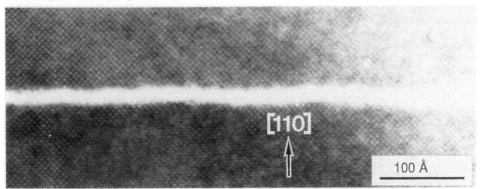

Fig. 3.    HRTEM image of a {110} planar defect in a relatively thick part of a TEM specimen taken with the electron beam parallel to the [001] Nb₃Al zone and using an objective aperture containing the four {100} reflections.

Further analysis of the features in Fig. 2 using EDX in the STEM with a 2 nm probe showed that locally the features were rich in oxygen and deficient in niobium (Fig. 4), while the features were difficult to distinguish from the surrounding Nb₃Al matrix using either X-ray maps or line scans for aluminium. Thus, the aluminium to niobium ratio of the features was noticeably higher than that of the surrounding Nb₃Al.

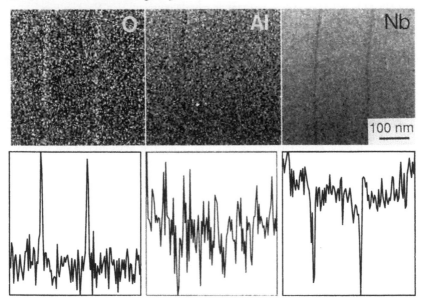

Fig. 4.    X-ray maps of oxygen, aluminium and niobium (top ) together with X-ray line scans of the pair of {110} Nb₃Al features imaged by TEM in Fig. 2.

## 4.   DISCUSSION

The Nb-20 at.% Al alloy exhibited a resistance to oxidation markedly superior to pure niobium (Kofstad 1988) and Nb-Al binary alloys of lower aluminium concentrations (Kim and

Knowles 1997). However, the lack of a continuous protective alumina oxide scale is evident. The presence of the {110} planar features in the Nb$_3$Al substrate is an indication of why Nb$_3$Al is more oxidation-resistant than either pure niobium or Nb$_{ss}$, while at the same being an indication of the mechanism of oxidation attack in Nb$_3$Al.

Our experimental observations of these {110} Nb$_3$Al planar features can be rationalised straightforwardly in terms of oxygen dissolution in the Nb$_3$Al matrix. The absence of such features in as-cast alloys further confirms that the features arise as a direct consequence of oxidation. That they occur on {110} planes of Nb$_3$Al can be rationalised, at least qualitatively. Nb$_3$Al has the A15 crystal structure with $Pm3n$ space group symmetry (Wood et al 1958). This is often described as topologically close packed, with Al atoms occupying b.c.c. positions and Nb atoms forming chains along the three <100> orthogonal directions of the cubic unit cell. Thus, the planes of closest packing of the Al atoms are the {110} planes, on which these features lie, implying that the oxidation process involves aluminium atoms preferentially. The implication from the HRTEM and EDX observations is that the aluminium in Nb$_3$Al can act as a 'getter' for oxygen, forming an amorphous or poorly formed aluminium-rich oxide which prevents a more vigorous attack of the alloy by oxygen without offering sufficient protection to inhibit oxidation. The formation of an amorphous oxide at the temperatures used for heat treatment is consistent with the literature where amorphous oxide production in the early stages of oxidation at low temperatures is well documented (Harvey and Draper 1963-64, Thomas et al 1988).

## 5. CONCLUSIONS

Detailed cross-sectional transmission electron microscopy of samples rich in Nb$_3$Al exposed to low temperature oxidation shows the way in which oxygen attacks Nb$_3$Al by penetrating down {110} planes, forming dense arrays of planar features. These features help to protect Nb$_3$Al from more catastrophic attack. The existence of these features also helps to explain why Nb$_3$Al is not as oxidation-resistant as more aluminium-rich Nb-Al alloys.

## ACKNOWLEDGEMENTS

We would like to thank Prof. A H Windle for the provision of laboratory facilities and Dr C B Boothroyd for his assistance with the STEM experiments.

## REFERENCES

Anton D L and Shah D M 1989 Mat. Res. Soc. Symp. **133**, 361
Harvey J and Draper P H G 1963-64 J. Inst. Met. **92**, 136
Kattner U R 1990 Binary Alloy Phase Diagrams Vol. 2 ed T B Massalski (ASM International), 179
Kim Y and Knowles K M 1997 Microscopy of Oxidation 3 eds J A Little and S B Newcomb (Institute of Materials) in the press
Kofstad P 1988 High Temperature Oxidation of Metals (New York: Wiley) p. 210
Perkins R A, Chiang H J and Meier G H 1988 Scripta Met. **22**, 419
Perkins R A and Meier G H 1990 JOM, **42**(8), 17
Steinhorst M and Grabke H J 1989 Mat. Sci. Eng. A **120**, 55
Steinmetz J, Vilasi M and Roques B 1993 J. de Physique IV **3** Coll. 9, Pt. 2, 487
Svedberg R V Properties of High Temperature Alloys eds Z A Foroulis and F S Pettit, (Pennington, N.J., U.S.A: Electrochemical Society) p. 361
Thomas O, d'Heurle F M and Charai A 1988 Phil. Mag. B **58**, 529
Wood E A, Compton V B, Matthias, B T and Corenzwit, E. 1958 Acta Cryst. **11**, 604

*Inst. Phys. Conf. Ser. No 153: Section 13*
*Paper presented at Electron Microscopy and Analysis Group Conf. EMAG97, Cambridge, 1997*
© *1997 IOP Publishing Ltd*

# Breakdown craters in thin insulating films studied by electron and atomic force microscopy.

**R E Thurstans and P J Harris**

Department of Chemistry and Physics, De Montfort University, Leicester LE1 9BH

**ABSTRACT:** A wide variety of insulating materials are known to exhibit a so called 'formed state' after voltage applications in vacuum. This state is characterised by a region of differential negative resistance and electron emission at low bias. Breakdown craters, which are produced in the forming process, have been studied by electron and atomic force microscopy. The results indicate localised regions of high temperature, causing defects from which the observed properties may arise.

## 1.    INTRODUCTION

When a thin layer of insulator is sandwiched between two metal electrodes, placed in a vacuum, and a voltage applied which is in excess of a certain critical value, the properties of the structure are changed permanently in a profound manner. The device is said to be 'formed' or 'electroformed'. In this new state it has a non-linear current-voltage characteristic with very small temperature dependence and a resistance many orders of magnitude lower than that of its initial state for an applied bias below a initial voltage $V_m$. Beyond $V_m$ there is a region of differential negative resistance until the current falls to a value commensurate with that before forming (see Fig 1).

Other novel properties include electron emission (also shown in Fig 1), electroluminescence, a conductivity sensitive to the gaseous surroundings and a wide range of memory states characterised by different resistance states of the device, which may be read in or out by the applications of a voltage.

It has been known for some time that the anode of a formed device contains defects produced by the forming process and it has been suggested by Biederman (1976) that electron emission is associated with these defects. It was decided that further examination of these defects would be useful.

## 2.    EXPERIMENTAL

A metal-insulator-metal sandwich structure was constructed by successive evaporations of aluminium, silicon oxide and gold onto a glass substrate through a mask. Thicknesses were of the order of 100 nm with appropriate edge thickening layers and the active area of the device was the order of 10 mm$^2$. The devices initially had as resistance greater than 1 MΩ but after location in a vacuum of $10^{-6}$ torr and voltage cycling up to 10V the usual formed characteristic was observed. A typical current-voltage curve is shown in Fig 1 along with the electron emission which was obtained from the device by a collector plate placed a centimetre or so above the active device.

622

The devices were then taken from the vacuum system and studied with an SEM (Leica S430) and AFM (Nanoscope II).

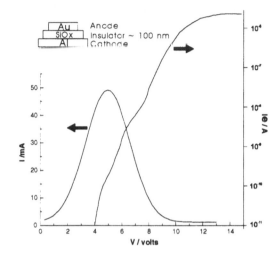

Fig 1. Current I, versus voltage graph of the device (shown inset) and the emitted current, Ie superimposed

## 3.   RESULTS AND DISCUSSION

Both techniques showed that before forming the gold anode had a featureless surface typical of that of any evaporated metal. After forming the anode was covered by a large number of small breakdown craters, typically a few μm across. A number of such craters can be seen in Fig 2 with more detailed views in Fig 3 and 4 and corresponding SEM views in Fig 5 and 6.

It is worth noting that the current-voltage curve of the formed device is non-linear and reproducible and indicative of an electron transport mechanism other than metallic conduction. Thus a breakdown event, which produces a short circuit between the electrodes may be excluded.

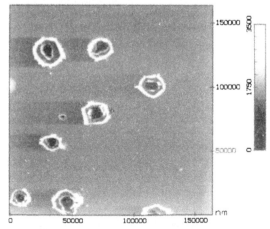

Fig. 2. Breakdown craters on gold anode surface by AFM

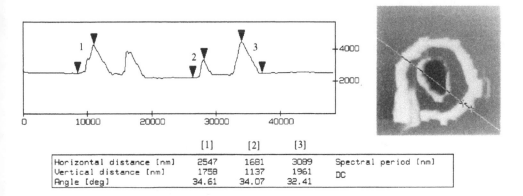

| | [1] | [2] | [3] | |
|---|---|---|---|---|
| Horizontal distance [nm] | 2547 | 1681 | 3089 | Spectral period [nm] |
| Vertical distance [nm] | 1758 | 1137 | 1961 | DC |
| Angle [deg] | 34.61 | 34.07 | 32.41 | |

Fig.3. Line scans across a single breakdown crater

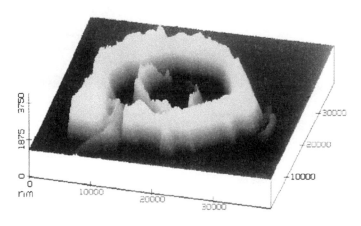

Fig.4. 3-D representation of a single crater

Small globules of solid anode material, which are evident outside the crater, indicate a high temperature localised event. The outer walls of the crater are high, typically 2 μm and are produced by a folding back of the anode surface. Inside the crater is a second, lower wall which contains within it a mixture of molten electrode and insulating material. This was confirmed using the EDX facility of the SEM. If the second crater wall is connected to the cathode of the device either directly or through an intermediate resistive matrix an electric field will be established between this and the anode. This field will be extremely high due to the sharp edge of the wall and its close proximity to the anode. Such a structure can easily give rise to field or enhanced field emission of electrons. The topography of the thin inner wall and its position relative to the anode may be seen in figures 3 to 6 although the line scan in figure 3 and the associated image produced by the atomic force microscope are a convolution

624

of the real topography of the structure and the pyramidal shaped tip. Further evidence in support of this model of electron emission has been provided by Sharpe (1996) et al who measured the emission from such a structure at low voltages and temperatures. They showed that this supported a model of field emission.

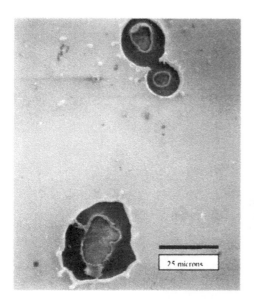 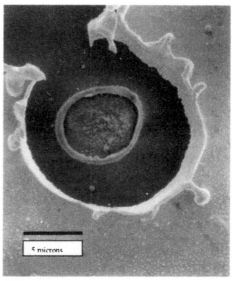

Fig.5. Breakdown craters on gold surface by SEM

Fig.6. Single crater showing inner wall surrounding a matrix of electrodes and dielectric, all surrounded by a high anode wall.

The molten mixture of electrode material and dielectric probably contains particles of the former embedded in the latter and it is this structure which gives rise to the observed current-voltage relationship which has been observed and described by Thurstans (1979).

## REFERENCES

Biederman H 1976 Czech. J. Phys. **B26, 827**
Thurstans R E 1979 Thin Solid Films 57, 153
Sharpe RG and Palmer RT 1996 Thin Solid Films 288, 164

*Inst. Phys. Conf. Ser. No 153: Section 13*
*Paper presented at Electron Microscopy and Analysis Group Conf. EMAG97, Cambridge, 1997*

# A study of carbon particles from engine oil using SEM and atomic force microscopy.

D W Wallis, R E Thurstans, C G Stow and M F Fox

Department of Chemistry and Physics, De Montfort University, The Gateway,
Leicester LE1 9BH

**ABSTRACT:** A significant cause of engine oil degradation is due to particles produced in the combustion process. A study of these particulates is important in the development of lubricating oils, to extend the working life of the oil and to reduce particulate emissions. Using a novel method based upon centrifuging and solvent extraction, particles have been analysed using scanning electron microscopy and the results further confirmed by atomic force microscopy. Particles of less than 1μm have been seen, and agglomerates have been shown to contain surface layers of elements contained within the oil additives.

## 1. INTRODUCTION

The severe conditions in the top piston ring zone of an internal combustion engine are highly degrading to the engine lubricating oil. Degradation, caused by the high temperatures and pressures, and mechanical shearing occurs predominately in this region. The lubricating oil is also subject to the products of the combustion of hydrocarbons (fuel) in the form of particulates, which are abundant in the top piston ring zone.

Current trends in ring pack design increasingly reduce the volume of lubricant present in that area, and combined with demands for increased service life and reductions in sulphur content, degradation due to particulates becomes increasingly more significant. Thus, lubricant and additive technology must progress, along with combustion engine technology to overcome the increasing demands presented by environmental and economic considerations.

An understanding of the process at the top piston ring zone can benefit the development of these technologies. Methods for obtaining samples from the top piston ring zone from an engine running under normal operating conditions have been demonstrated by Fox et al (1988). The particulates must then be separated from the lubricant. A technique for separation has been developed, allowing further analysis by techniques such as Scanning Electron Microscopy (SEM), Energy Dispersive X-ray Microanalysis (EDX) and Atomic Force Microscopy (AFM).

## 2. METHOD

Samples are obtained from the top piston ring zone of the engine, as described by Fox et al (1988). A tube of 0.5 mm internal diameter is attached to a hole drilled in the top ring groove of the piston. This carries the sample outside the engine where it can be collected. The sample is in the form of a high velocity aerosol, and is expelled by the blowby pressure produced during combustion. Thus, samples of degraded lubricant, containing particulates from combustion, wear and corrosion, and which are not diluted by the bulk of oil in the sump are obtained. For a study of the particulates using microscopic techniques, the particles must be separated from the oil for mounting in the instrument. This was achieved by diluting the sample with a solvent (petroleum ether) and separating the particles with a high speed centrifuge. This process is repeated until the oil is diluted enough to leave a negligible amount of oil in the sample, leaving a selection of the particulates in the solvent.

The technique requires that the sample is mixed with a solvent, which removes some of the constituents (most notably, removing the oil, which is the desired effect), but does leave some inorganic components such as metal , salts and carbon particles, which are representative of the particulates absorbed by the lubricant.

Mounting the particles for analytical techniques becomes relatively simple once the oil has been removed. For SEM imaging, the particles were mounted on transparent plastic film by dipping the film in to the solvent containing the particles. This method works well because the plastic film attracts the particles to its surface. After mounting the film on an aluminium stub and coating with gold to aid conductivity, good images were obtained.

EDX results were unobtainable when the sample was gold coated, so results were obtained by mounting the particles on a piece of a silicon wafer, chosen because of its purity. The image quality was reduced in comparison to the gold coated plastic film samples because of charging. The number of particles seen on the surface is also reduced because the silicon does not attract particles as well as the plastic film. Despite this, EDX results were obtained, both as spectra of individual particles and surface scans of particles and agglomerates.

A piece of silicon wafer (chosen for its flatness) was also used to mount the particles for AFM. A light gold coating was necessary to fix the particles to the surface, preventing them from being moved by the tip as it scans the surface.

## 3.    RESULTS

Images of particles were obtained using an SEM, type Lecia S430. Fig.1 shows an image of particles on plastic film. The sample is from a diesel engine, type Petter AA-1, Direct Injection 219cc. The image shows particles up to 30 μm in diameter and numerous particles of approximately 1 μm in diameter. This is typical of the results obtained from this sample, and is in agreement with particle size analysis of similar samples using laser diffraction techniques by Stow C G (1997). Single particles and agglomerates were found with SEM.

SEM also identified the presence of smaller particles, later confirmed with AFM. Fig.2 shows such particles, less than 100 nm in diameter on the plastic film.

EDX analysis indicated the presence of calcium, aluminium, sulphur, potassium, iron and oxygen. Different spectra were seen for different particles, indicating that the surfaces of particles are chemically different. Particles were found to have either a preponderance of calcium or aluminium on the surface. Fig.3 shows the spectrum of a particle with a large peak for calcium, and Fig.4 shows the presence of aluminium. The elements found in the oil reflect the presence of detergents, dispersents, anti-wear additives and metals abraded from the cylinder bore and bearing surfaces e.g. calcium sulphonate. Silicon peaks are due to the presence of the silicon substrate.

Surface scans of agglomerates confirmed these results. Fig.5 shows an agglomerate, with the areas of calcium (blue) and aluminium (green) shown in Fig.6. Concentrations of either aluminium or calcium can clearly be seen to be on separate particles in the agglomerate.

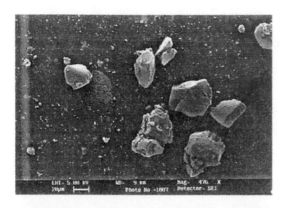

Fig.1 An SEM of particles found in a top ring zone (TRZ) sample of used lubricating engine oil

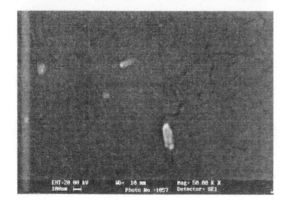

Fig.2 An SEM of particles less than 100 nm in size from a TRZ sample.

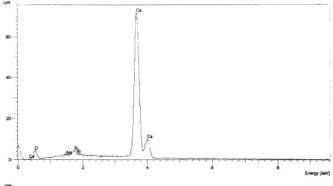

Fig.3 An EDX spectrum of a particle from the TRZ showing the presence of calcium.

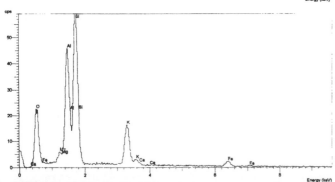

Fig.4 An EDX spectrum of a particle from the TRZ showing a range of elements.

628

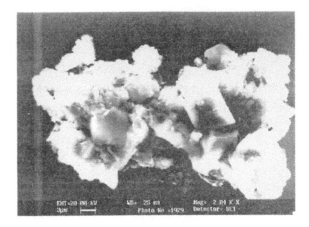

Fig.5  An SEM of an agglomerate from used lubricating engine oil.

Fig.6  The same agglomerate showing calcium (blue) and aluminium (green).

## 4.     CONCLUSIONS

Particles can be separated from used engine lubricating oil using a centrifuge and solvent extraction, leaving a selection of the particles in the solvent.

Images of particles can be obtained using SEM, and particles with a large size distribution were detected.  Particles sizes found were in agreement with results taken with laser diffraction techniques.  In addition, particles of less than 100 nm were found with SEM and their presence confirmed with AFM.

EDX analysis found the presence of calcium, aluminium, potassium, sulphur, iron and oxygen, distributed across different particle surfaces.

### REFERENCES

Fox M, Saville S B, Gouney F D, Cupples S D and Pickers D J.
U.S. Soc. Auto. Eng. Int. Fuels and Lubricants, Portland, Oregon Oct 1988 Paper 881586
Stow C G PhD Thesis, De Montfort University, Leicester, 1997.

*Inst. Phys. Conf. Ser. No 153: Section 13*
*Paper presented at Electron Microscopy and Analysis Group Conf. EMAG97, Cambridge, 1997*
© *1997 IOP Publishing Ltd*

# SEM / EDS - a nearly universal tool for fly ash analysis?

**P Poelt, I Obernberger\* and H Koller**

Research Institute for Electron Microscopy, Technical University of Graz, Steyrerg. 17, A-8010 Graz
\* Institute of Chemical Engineering, Technical University of Graz, Inffeldg. 25/II, A-8010 Graz

**ABSTRACT:**   Computer controlled SEM / EDS enables the unattended analysis of both the geometry and the chemical composition of thousands of individual particles. This method has been applied to a biomass fly ash. The correlations that can be derived from such measurements can help to understand the basic mechanisms of heavy metal deposition on fly ash particles as well as of deposition and corrosion mechanisms in boilers.

## 1.   INTRODUCTION

The knowledge of the chemical composition, the geometrical structure and the mass distribution both of biomass and coal fly ash particles is generally important for various reasons. The deposition of heavy metals in the flue gas on the fly ash, deposit formation and corrosion mechanisms in furnaces and boilers are all influenced by these parameters. All these processes are additionally influenced by the process parameters of the plants.

To be able to fit the process parameters of the plants to special requirements, e.g. to fractionate heavy metals in predetermined ash fractions, first the basic mechanisms of the corresponding processes have to be studied. Both heavy metal deposition and corrosion are dependent on composition, size and surface structure of the fly ash particles.

Computer controlled SEM in connection with image processing and EDX-spectrometry enables the unattended determination of both geometrical parameters and chemical composition of thousands of individual particles. Consequently, correlations between particle size, chemical composition, number of different compounds and their contribution to the overall concentration can be established.

Problems arise in connection with specimen preparation, the inhomogeneous composition of many of the particles and the lacking of a rigorous ZAF correction procedure for particles of arbitrary shape.

## 2.   EXPERIMENTAL

Nucleopore filters with a pore size of around 0.2 µm were coated with a carbon layer of a thickness of some 10 nm. Then a suspension of fly ash in alcohol was filtered through it. Subsequently the filter with the particles was again coated with a carbon layer of 20 nm thickness. The first coating should provide a conductive filter surface also in places shadowed by the particles at the second coating. Measurements were performed in a Philips SEM 505 equipped with a Robinson backscatter detector and an EDAX DX-4 X-ray analysis system (Si(Li) detector, ultrathin window).

## 3.   RESULTS AND DISCUSSION

### 3.1  Specimen Preparation

Specimen preparation is critical, since particle agglomeration has to be avoided. Suspension of the particles in alcohol and subsequent filtering of the solution on a Nucleopore filter gave acceptable results. A more sophisticated procedure has been suggested by Katrinak, Bekke and Hurley (1992). Care has to be taken that the particle size distribution of the measured particles in fact resembles that of the whole specimen. Otherwise great distortions of some of the results have to be expected. For submicron particles, with their high agglomeration tendency, correct sampling is possible by using low pressure cascade impactors (Obernberger 1997)

## 3.2 Measurement

Difficulties in automated particle analysis have been encountered in two respects: particle discrimination from the substrate and the subsequent X-ray analysis. If the substrate is an organic filter, mainly carbonaceous particles, e.g. unburnt biomass fuel, cannot be discriminated by an electron backscatter detector by pure material contrast. In these cases a detector, that gives additionally some topographic contrast, may be preferable.

Biomass fly ash particles are, apart from small submicron particles, very often inhomogeneous, the particle volumes are often smaller than the excitation volumes and thus assumptions for exact ZAF correction procedures are not met. Although special correction procedures for X-ray microanalysis have been developed, they generally demand an analytical expression for the particle shape (Armstrong 1991). Therefore, quantitative results are in all cases only approximative and also dependent on the analysis conditions. Fig. 1 demonstrates the dependence of the calculated concentrations for a single ash particle on the primary energy and the analysis area.

Great particles are often composed of submicrom particles. This is one reason for their inhomogeneity. A typical example can be seen in Fig. 2, where small particles are sintered. Each of these particles could be of a different chemical composition. But also condensation of specific volatile elements on particle surfaces is significant.

oxide conc. / wt%

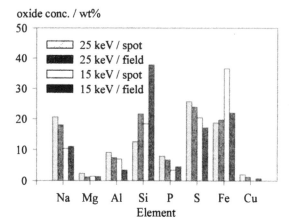

Fig. 2: Biomass fly ash particle (picture width: 2,2 μm).

Fig. 1: Measured oxide concentrations (oxygen by stoichiometry) for a fly ash particle (average diameter 0,9 μm) in dependence upon primary energy and analysis area (spot: point measurement at center of gravity; field: scan across whole particle area).

Particles have also been embedded in resin, cross-sectioned and subsequently polished. X-ray maps at such cross-sections reveal the inhomogeneous build-up of the corresponding particles. Fig. 3 shows, that zinc, potassium and sulphur are often enriched in a thin surface layer.

## 3.3 Automated Particle Analysis

From around 8000 particles of a fly ash sampled from the boiler section (temperature ≈ 350 °C ) of a biomass combustion plant the geometrical parameters and the chemical composition have been measured. Particles with the greatest diameter smaller than 2 μm were neglected. Particles with an average diameter greater than 40 μm had been removed by a preceding sieving. A primary energy of 25 keV was chosen, to be able to quantify zinc with the $K_\alpha$- and lead with the $L_\alpha$- peak. Measurement time per particle was 5 seconds. The converntional ZAF-correction procedure for bulk specimens was used.

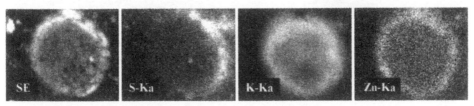

Fig. 3: Secondary electron picture and X-ray maps of the polished cross section of a biomass fly ash particle (picture width: 23 μm).

The overall concentration $c_e$ for a specific element of the fly ash can be calculated by summing up the concentrations of the individual particles with their normalized volume as weight factor:

$$c_e = \sum_{i=1}^{n} c_i * \frac{V_i}{V},$$

with $c_i$ the element concentration of a particle, $V_i$ the corresponding particle volume, calculated from its average diameter, V the sum of all the particle volumes and n the number of the particles. Fig. 4 shows the comparison between a chemical analysis of the ash and the results from the automated particle analysis, the latter for various numbers of particles.

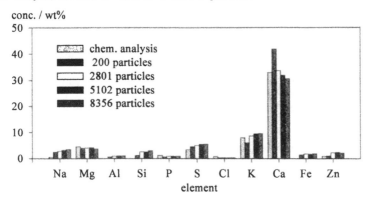

Fig.4: Comparison between the measured overall concentrations of elements in a fly ash by chemical analysis and automated particle analysis in SEM.

Despite the above mentioned shortcomings, with the exception of sodium the results agree surprisingly well. For Al, Si and Fe no values from chemical analysis had been available. Part of the difference between the results from the chemical and X-ray analysis will be due to the exclusion of carbon and also of the submicron particles from the EDX-analysis. The large deviation in the values of sodium may result from errors in peak deconvolution between $Na-K_\alpha$ and $Zn-L_\alpha$ in the very noisy spectra.

In Fig.'s 5a - 5d some of the correlations, that can be derived from the measurements, are shown. According to Fig. 5a a strong correlation between sulphur and potassium does exist. Fig. 5b proves, that calcium is present in the particles over the whole possible concentration range. As a consequence, most of the sophisticated programmes for automated particle classification (e.g. Treiger B et al 1994) cannot be applied. But this is also a consequence of the inhomogeneity of many of the particles.

Around 150 particles had a measurable chlorine content and were sorted out. Fig. 5c does show a

632

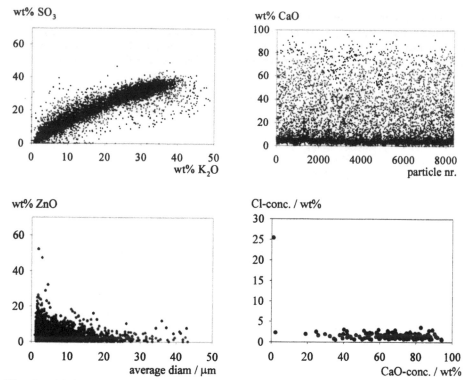

Figs. 5a - 5d (from upper left clockwise) 5a: correlation between particle conc. of $SO_3$ and $K_2O$; 5b: variation of the CaO conc. in the particles; 5c: correlation between the concentrations of Cl and CaO (150 particles with highest Cl conc. only); 5d: dependence of ZnO conc. on particle diameter

strong correlation between chlorine and high Ca-concentrations for these particles. Fig. 5d demonstrates that the Zn-concentration increases with decreasing particle diameter. This may partly be due to the fact, that zinc evaporates during combustion and condenses later on fly ash particles (see Fig. 3.) or forms aerosols.

The concentration of most heavy metals is below the detection limit of EDS. But they can be detected, if they are enriched in a certain particle type with a sufficiently low number of particles. For example, in the biomass fly ash several Cu-rich particles with Cu-concentrations up to 15 wt% have been detected, despite an overall concentration of around 230 ppm (chemical analysis).

As shown in Fig. 5d, for closer investigations into the heavy metal behaviour as well as into aerosol formation processes, submicron particle characterization is essential.

## REFERENCES

Armstrong J T 1991 in: Electron Probe Quantitation, eds KFJ Heinrich and D E Newbury (New York: Plenum Press) pp 261-315

Katrinak K A, Brekke D W and Hurley J P 1992 Proc. 50th Annual Meeting of EMSA, eds G W Bailey, J Bentley and J A Small (San Francisco: San Francisco Press) pp 408-9

Obernberger I 1997 Nutzung fester Biomasse in Verbrennungsanlagen unter besonderer Beruecksichtigung des Verhaltens aschebildender Elemente (Graz: tbV Graz)

Treiger B, Bondarenko I, Van Espen P, Van Grieken R and Adams F 1994 Vol **119** pp 971-4

Acknowledgement: We thank the 'Jubilaeumsfonds der Oesterreichischen Nationalbank' for sponsoring the EDS system.

*Inst. Phys. Conf. Ser. No 153: Section 13*
*Paper presented at Electron Microscopy and Analysis Group Conf. EMAG97, Cambridge, 1997*

# XTEM characterisation of low temperature plasma nitrided AISI 316 austenitic stainless steel

X Y Li, Y Sun, A Bloyce and T Bell

School of Metallurgy and Materials, The University of Birmingham, Birmingham, UK

**ABSTRACT**: Low temperature plasma nitriding of austenitic stainless steels has shown the potential for combined improvement in tribological and corrosion properties. However, several technical problems have been experienced in practice, mainly due to the lack of understanding of the microstructure and phase composition in the nitrided layer. In the present work, a rigorous characterisation of the microstructure and interface of low temperature plasma nitrided austenitic stainless steel has been carried out, employing cross-sectional TEM in conjunction with XRD, EDX and SEM analyses.

## 1. INTRODUCTION

The surface layer formed following low temperature nitriding was named 'S-phase' since it appeared to have a single phase as revealed under the optical microscope. Indeed, both the microstructure and the nature of the phases produced by low temperature plasma nitriding of austenitic stainless steels have not been fully characterised. The nature of the so-called 'S-phase' in low temperature plasma nitrided austenitic stainless steels is still a topic open to debate, representing the complexities and hence difficulties involved in the characterisation of surface modified layers. Ichii et al (1986) first showed X-ray diffraction patterns of the 'S-phase' and suggested that the surface layer is probably composed of a new single phase ('S-phase'), which seemed to have an fcc structure with some peaks shifted to the lower diffraction angle. Dearnley et al (1989) suggested that a stress induced (nitrogen) martensite could be the another possibility for the 'S-phase', Samandi et al (1994) argued that the 'S-phase' is most probably an fcc structured 'nitrogen expanded austenitic phase'. Indeed, all these hypotheses have been based on XRD analysis results. However, it should be pointed out that it is difficult, if not impossible, to fully characterise surface engineered materials only using X-ray diffraction method. Transmission electron microscopy (TEM), especially cross-sectional TEM (XTEM) has emerged as a powerful technique for characterising surface engineered layers because of its high resolution and its ability to provide both compositional and crystallographic information from the same area of the sample. It is therefore the purpose of the present paper to attempt to characterise the microstructure and interface of low temperature plasma nitrided austenitic stainless steels employing cross-sectional TEM technique coupled with XRD, GDS and SEM methods.

## 2. EXPERIMENTAL DETAILS

The material used for plasma nitriding is commercial AISI 316 austenitic stainless steel. The chemical composition of the material is 0.06 wt% C, 19.23 wt% Cr, 11.26 wt% Ni, 2.67 wt% Mo, 1.86 wt% Mn and balance Fe.

Plasma nitriding was carried out using a 60 kW KlocKner DC plasma nitriding unit. The treatment atmosphere was a mixture of 25% $N_2$ + 75% $H_2$ and the total gas pressure was 5 mbar. The specimen was plasma nitrided at 450°C for 7h.

The technique used to prepare cross-sectional TEM discs was adapted from the method reported by Newcomb et al (1985). To overcome the difficult arising from significant difference in thinning rates between the nitrided layer and the substrate, a Gatan Model 691 Precision Ion Polishing System (PIPS) was used.

634

## 3. RESULTS AND DISCUSSIONS

The typical optical microstructure of the specimen shown in Fig. 1. Notwithstanding the fact that optical micrograph showed only a single white layer, detailed XTEM observation on low temperature plasma nitrided layer of AISI 316 austenitic stainless steel revealed that there are three sublayers in the surface modified material. The layer structure is summarised schematically in Fig. 2. Sublayer I is a thin superficial layer; sublayer II is the main layer which was referred to as 'S-phase' in literature, and sublayer III is a interface layer or transition zone.

Fig. 1 Typical optical microstructure.                Fig. 2 Schematic of layer structure.

### 3.1 The Sublayer I

Fig. 3a shows the heterogeneous microstructure of sublayer I with a thickness of about 200 nm. It can also be seen that oriented columnar grains were well developed from the interface (10-30nm in diameter ), whereas very fine, random equiaxed grains dominate near the free surface.

Detailed analyses of the SAD patterns (Fig. 3b and 3c) revealed that there are two phases in sublayer I. The sharp diffraction rings can be assigned to CrN, which is also indicative of fine random equiaxed grains. It is clear, by comparing Fig 3b and 3c, that the intensity of CrN rings decreased with distance from free surface and disappeared at the interface between sublayer I and sublayer II. The $M_4N$ (M=Fe, Cr, Ni, Mo) diffraction patterns exhibited high intensities in 6 positions on the rings (Fig.5(c)) belonging to the {111} and {220} planes, suggesting a [112] texture of the nanocrystals. It should be pointed out that although the superlattice rings in Fig.4 (c) are very weak, they can be clearly identified in the corresponding negatives.

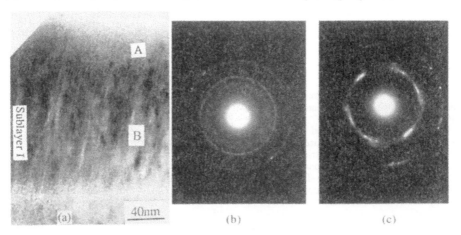

Fig. 3 (a) TEM BF micrograph, (b) SAD pattern in position A in (a),
and (c) SAD pattern in position B in (a).

### 3.2    The Sublayer II

The microstructure of sublayer II is shown in Fig. 4. The nitrided layer exhibited a fine irresolvable random 'dot' contrast under general imaging conditions . No clear regions of second phase can be observed. The boundary between sublayer I and sublayer II was very distinct and easy to determine.

The SAD patterns strongly suggest that sublayer II has an f.c.c structure but tetragonal distortions can be found in the diffraction patterns.  SAD patterns obtained by using longer exposure time ( Fig. 5) showed the presence of forbidden reflections of the f.c.c structure, clearly indicating that this phase is not austenite but primitive cubic $\gamma$-$M_4N$ phase. This is because the nitrogen atoms are ordered in the $\gamma'$-$M_4N$ but are randomly distributed in the expanded austenite. This difference in the distribution of nitrogen atoms leads to a primitive cubic lattice for $\gamma'$-$M_4N$ phase and to an f.c.c structure for the austenite. The lattice parameter is calculated to be a = 3.795 Å by using superlattice reflections, which matches the lattice parameter of the $\gamma'$-$M_4N$.

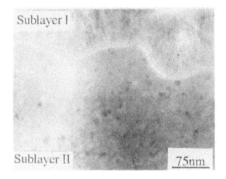

(100)zone

Fig. 4 TEM BF microstructure of sublayer II          Fig. 5  SAD  pattern of sublayer II

However, it has been found that (Li 1997) the X-ray diffraction patterns showed a strong broadening and shift of the peaks, and that the ratio of the d-spacing values, $d_{200} / d_{111}$, significantly deviates from that of $d_{200} / d_{111}$ for an ideal f.c.c structure. Through detailed XTEM observation and diffraction patterns analyses, it has been found that the deviation of the $d_{222} / d_{111}$ ratio of the $\gamma'$-$M_4N$ from the standard value of an ideal fcc structure is likely to be caused by coherent, thin plate precipitates on the (200) planes of $\gamma'$-$M_4N$ or by high residual compressive stresses. However, residual stress itself could not fully accounted for the deviation (Williamson et al 1994).  As shown in Fig. 6a, two sets of spots are superimposed together by diffuse scattering (110) zone. The d-spacing values given by XRD method can be  obtained if it is calculated by using the mean distance of the corresponding diffraction spots (i.e. $R_{200}$ = 1/2 $\{(R_{200})_1 + (R_{200})_2 \}$ of these two diffraction patterns, as indicated by R200 and R111 in Fig.6b. As expected, the ratio $d_{200}/d_{111}$ thus calculated mostly match the deviated value obtained by XRD method. In summary, all these results strongly suggested that so-called 'S-phase' or sublayer II is most probably composed of $\gamma'$-$M_4N$ with  G.P. zone and/or thin precipitates of chromium nitride on the (200) planes.

### 3.3    The Sublayer III

Sublayer III is a transition layer or interface layer between sublayer II and the substrate. XTEM observations revealed high dislocation density (Fig. 7a), and very fine precipitates accompanied with ferrite in this sublayer (Fig. 7b). The fine precipitates are believed to be associated with the dark line between the nitrided case and the substrate in optical micrograph since precipitation of Cr containing particles leads to the depletion of Cr in the austenitic solid solution and thus deteriorated corrosion resistance.

636

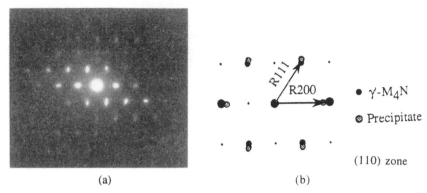

(a)                                          (b)

Fig. 6 SAD pattern of sublayer II showing superimposed two sets of diffraction patterns.

(a)                                          (b)

Fig. 7 XTEM BF micrograph (a) and high dislocation density in sublayer III (b).

## 4.    CONCLUSIONS

(1) The surface 'white layer' observed under optical microscope of low temperature plasma nitrided (350 - 450°C) 316 austenitic stainless steel is not a 'single phase (S-phase)' layer, but consists of three sublayers: sublayer I, sublayer II and sublayer III or interface layer. The thicknesses of these sublayers vary with treatment times and temperatures.

(2) In sublayer I, oriented columnar grains are well developed from the interface, whereas very fine, random equiaxed grains are dominated near the free surface. This sublayer consists of a mixture of CrN and $\gamma(M_4N)$, the amount of $\gamma(M_4N)$ being increased with depth from the surface.

(3) Sublayer II or the so-called "S-phase", is not expanded austenite but most probably a $\gamma$ ($M_4N$) phase with G.P. zone and/or thin precipitates of chromium nitrides, which contributes to the extra hardening effect over conventional $\gamma$ ($Fe_4N$).

## 5. ACKNOWLEDGEMENT

One of the authors (X Y Li) would like to gratefully acknowledge the Committee of Vice-Chancellors and Principals for an Overseas Research Studentship Awards.

## REFERENCES

Dearnly P A, Namvar A, Hibberd G G A and T Bell 1989 Plasma Surface Engineering, eds Broszeit E et al (DGM Informatingesellschaft mbH: Oberursel) pp 219-226
Ichii K, Fujimura K and Takase T 1986 Techn. Rep. Kansai. Univ. **27**, p135
Li X Y 1997 MPhil Thesis The University of Birmingham
Newcomb S B, Boothroyd C B and Stobbs W M 1985 Journal of Microcopy, **140**, pp 195-207
Samandi M et al 1994 J. J.Vac. Sci. Technol. **B12** pp 935-939
Williamson D L, Ozturk O, Wei R and Wilbur P J 1994 Surf. Coat. Technol. **65**, pp15-23

*Inst. Phys. Conf. Ser. No 153: Section 13*
*Paper presented at Electron Microscopy and Analysis Group Conf. EMAG97, Cambridge, 1997*

# Anisotropy in EELS of TiS$_2$

**P Moreau, C Pickard and T Duguet**

Cavendish Laboratory, Madingley Road, Cambridge CB3 OHE, England

**ABSTRACT:** Experiments on the L$_3$, L$_2$ edges of TiS$_2$ have been performed using Electron Energy Loss Spectroscopy. To interpret the small anisotropy observed, band structure and multiplet calculations considering the D$_{3d}$ symmetry of the titanium site have been used. Reasonable agreement is achieved with experiment since the splittings of the empty levels are very well reproduced. It is hoped that this will assist in assessing the importance of the d$_{z^2}$ orbital in the intercalation of species in the van der Waals gap.

## 1.    INTRODUCTION

It is long been known that various species can be intercalated in the van der Waals gap of lamellar compounds. This phenomenon has been used to fabricate lithium batteries since a charge transfer is induced during the intercalation reaction. TiS$_2$ can be intercalated with lithium and the standard picture is that the lithium atom gives an electron to the titanium 3d orbitals to reduce it from Ti$^{4+}$ to Ti$^{3+}$. Furthermore, in the van der Waals gap, the lithium atom is intercalated on the 3-fold axis of the octahedron defined by the six sulphur neighbours, i.e. in the direction along which the d$_{z^2}$ orbital of the titanium atom points. Thus, in addition to an electronic transfer, we expect that a steric factor will modify the 3d orbitals. To test this last assumption, it is necessary to first focus on the L$_3$, L$_2$ edges of titanium in TiS$_2$ before intercalation. In a first approximation, the symmetry of the titanium site is octahedral, but on consideration of the next nearest neighbours the symmetry is reduced to D$_{3d}$. This should give rise to an anisotropy of the 3d orbitals when transitions are made in the c-axis direction (along the 3-fold axis) or perpendicular to the c-axis. We have performed Electron Energy Loss (EEL) measurements and used band structure calculations to study this topic. The 2p$^5$ hole can couple with the d-electron in the excited state which means that a multiplet structure must be considered and for that reason multiplet calculations will also be presented.

## 2.    EXPERIMENTAL

The TiS$_2$ compound was prepared following the method described by Mc Kelvy et al (1987) producing highly stoichiometric single crystals. Part of the powder was placed in a test tube and dispersed in acetone using ultrasound to obtain electron transparent crystals and subsequently deposited on a holey carbon grid. The EEL spectra were recorded using a Scanning Transmission Electron Microscope VG HB501 and a PEELS system. The dispersion was set to 0.17 eV/channel and the resulting FWHM of the zero loss peak through a hole on the grid was 0.85 eV. As the TiS$_2$ crystals are highly bi-dimensional, they all lay flat on the grid and spectra were thus recorded with the c axis parallel to the electron beam. To allow anisotropy measurements we recorded two spectra with different objective apertures to select different ranges of momentum transfer. According to Menon and Yuan (1997) formulae can be obtained to extract the parallel and perpendicular weightings for each configuration of convergence and collection angles. Varying the objective aperture allows us to leave the collection angle fixed and thus the resolution of the microscope is held constant. This is an important point since the changes in the spectra are very small. The weighting coefficients are 41% for the 25 μm objective aperture and 61% for the 50 μm one. Low loss spectra of the same area were also recorded and deconvolved from the high loss spectra. The two spectra were normalised at 490 eV in a region where the anisotropy is negligible. Fig. 1 shows the experimental spectra. The differences between the two spectra are very small (the splittings change from 1.8 to 2.1 eV and from 1.7 to 2.1 eV for the $L_3$ and $L_2$ edges respectively) and are sensitive to data processing. However this difference has been observed on many occasions and is consistent with experiments obtained on the sulphur K edge (Moreau et al 1996). In this latter case, due to the anisotropy of the sulphur 3p orbitals that mix with the titanium 3d orbitals, the splitting changes from 1.6 to 2.2 eV.

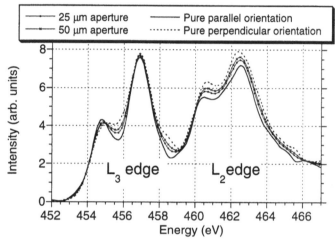

Fig. 1: Experimental ELNES spectra for the titanium $L_3$, $L_2$ edges for two different objective apertures and the retrieval of the pure parallel or perpendicular transitions to the c-axis.

## 3.     BAND STRUCTURE CALCULATION

We have used Density Functional Theory in the Local Density Approximation, optimised non-local pseudopotentials and a plane-wave basis set (Pickard et al 1995 and 1997). The band structure obtained for $TiS_2$ is very similar to the ones calculated previously by other methods (e.g. Umrigar et al 1982). As our purpose is to interpret anisotropy in EEL spectra we will focus on the calculated scattering intensity for an interacting fast electron with the crystal. Fermi's Golden Rule gives the matrix elements with respect to the direction of the fast electron momentum transfer and allows the anisotropy calculation of the titanium L edge. Spin orbit coupling as well as the electrostatic interaction of the core hole with the excited electron are not included. Then, only the crystal field is taken into account and a single "L edge" is calculated. Fig. 2 shows the intensity computed for the titanium L edge for transitions parallel and perpendicular to the c-axis. In a first approximation, the edge is composed of three main peaks at 0.7 eV, 1.6 eV and 3.3 eV above the Fermi level (zero energy). They correspond to the atomic splitting of the d orbitals in a $D_{3d}$ crystal field. Since the six first neighbouring atoms are six sulphurs in an almost perfect octahedral symmetry, the $D_{3d}$ distortion results from the six titanium atoms in the plane perpendicular to the c-axis. The intensities of the transition change quite a lot from one orientation to the other and one would then expect a large anisotropy in the experimental spectra. Nevertheless, the intensities of the transitions with respect to each other are incorrect since no electrostatic interaction is introduced. For example, the integrated intensities of the peaks in the case of the band structure are larger for the group between 0-2 eV compared to the one between 2-4 eV, the opposite way around to the experiment. A multiplet calculation is necessary to take spin interaction into account and to quantify the real amount of anisotropy that should be seen in the spectra.

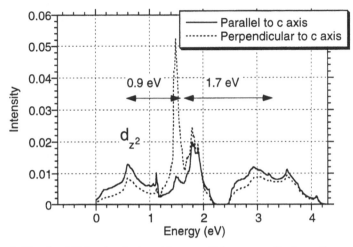

Fig. 2: Simulated Ti L-edge for transitions parallel or perpendicular to the c-axis.

## 4. MULTIPLET CALCULATION

To simulate the $L_3$, $L_2$ edges of titanium we used the programmes developed by Thole (van der Laan and Thole 1991). We used the actual $D_{3d}$ symmetry and performed the calculation for the two orientations. In order to do this, relationships were found in the $D_{3d}$ symmetry that connect the 3 parameters to the splittings of the 3d orbitals (to be published elsewhere).

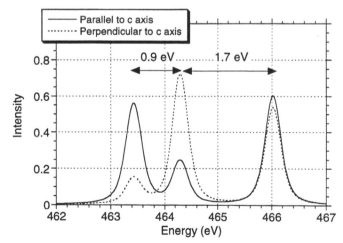

Fig. 3: Intensity of the transitions for the two orientations using the multiplet programme but taking only the crystal field into account.

These parameters can be found to match the splittings extracted from the band structure calculation. For clarity, Fig. 3 shows the result when considering only the trigonal crystal field acting on the 3d electrons. Good agreement is found between the two calculations.

640

Nevertheless, states, after introduction of the electrostatic interaction of the p-hole with the 3d electron mix and this results in an increase of the calculated splittings that no longer agree closely with experiment. We have thus chosen to reduce the splittings to 0.3 eV and 1.7 eV respectively. The agreement with Fig. 1 is then much better apart from the region around 467 eV in Fig. 4. The splittings are increased from 1.9 to 2.1 eV and from 1.7 to 2.1 eV for the $L_3$ and $L_2$ edges respectively. A further step will be to introduce charge transfer from the six sulphur ligands to allow more accurate comparison with experiments (Crocombette and Jollet 1996).

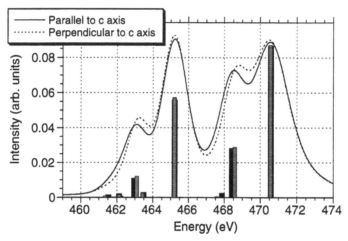

Fig. 4: Multiplet calculation of the titanium $L_3$, $L_2$ edges for transitions parallel or perpendicular to the c-axis. Bars show transitions before the broadening introduced to facilitate comparison with experiment.

## 5.  CONCLUSION

EELS experiments at the $L_3$, $L_2$ edges, band structure calculations and multiplet calculations have been performed on $TiS_2$. The $D_{3d}$ symmetry of the titanium site introduces a small anisotropy that is reasonably well described by the two calculation methods. Better understanding of the pristine material provides valuable information that will be used latter to study intercalated compounds and discriminate charge transfer from steric factors induced by the intercalant.

## AKNOWLEDGEMENTS

We would like to thank the EU grant ERBFMBICT950350 and the EPSRC for funding and N Menon for advice in the multiplet calculations.

## REFERENCES

Crocombette J P and Jollet F 1996 J. Phys.: Cond. Matter **8**, 5253.
Mc Kelvy M J and Glausinger W S 1987 J. Solid State Chem. **66**, 181.
Menon N and Yuan J to be submitted to Ultramicroscopy.
Moreau P, Ouvrard G, Gressier P, Ganal P and Rouxel J 1996 J. Phys. Chem. Solids **57** 1117
Pickard C J, Payne M C, Brown L M and Gibbs M N 1995 Inst. Phys. Conf. Ser. **147**, 211
Pickard C J and Payne M C 1997 "*Ab initio* EELS: beyond the fingerprint" in this volume.
Umrigar C, Ellis d E, Wang D, Krakauer H and Posternak M 1982 Phys. Rev B **26**, 4935
Van der Laan G and Thole B T 1991 Phys. Rev. B **43**, 13401

Inst. Phys. Conf. Ser. No 153: Section 13
Paper presented at Electron Microscopy and Analysis Group Conf. EMAG97, Cambridge, 1997

641

# Characterisation of the scale formed on an Fe,Al,Si alloy, oxidised for short times at high temperature

J Ritherdon, B Ahmad, P Fox

The Department of Materials Science and Engineering, The University of Liverpool, Liverpool L69 3BX

ABSTRACT: STEM and TEM analysis of cross-sections of the metal/oxide interface have been used to examine the transient scales (<200nm) formed on an iron-aluminium-silicon alloy. This alloy behaved differently to most alloys of its type, forming a flat adherent scale rather than a convoluted poorly adhered scale. Analysis showed that the unusual oxidation behaviour was due to the segregation of an impurity element (uranium) to the metal/oxide interface. The analysis also showed how the scale developed during the early stages of oxidation.

## 1.    INTRODUCTION

The importance of using Scanning Transmission Electron Microscopy (STEM) to study high temperature oxidation was first identified during research into 'The Sulphur Effect' (Lees D G (1987)). The Sulphur Effect is a mechanism used to explain why some metals form flat adherent high temperature oxide scales while others form scales that are highly folded and poorly adhered. The Sulphur Effect predicted that poor adhesion was due to the segregation of sulphur or chlorine to the metal/oxide interface. However, conventional TEM/EDX analysis failed to detect this segregation and it was thought that the proposed mechanism was incorrect. Only with the development of high spatial resolution STEM/EDX systems and the use of electron transparent cross-sections of the metal/oxide interface was the layer of segregant detected (Fox P, Lees D G and Lorimer G W (1991)).

The research reported here uses the now well developed STEM and cross-sectioning techniques to study the early stages of scale growth (transient oxidation) of an Fe-Al-Si alloy. Transient oxidation theory (Chattopadhyay B and Wood G C (1970)) developed from studies that used electron-microprobe and X-ray diffraction techniques. The theory predicts that the first oxide scale formed on initial heating will contain all the elements present in the original metal surface, and in the concentrations initially present. However, this theory is based on results obtained using low resolution techniques and, therefore, many of the changes that occur in the scale are not observed. Thus, as in the case of 'The Sulphur Effect', important features may not have been observed.

## 2. EXPERIMENTAL METHOD

Samples of Fe-5wt%Al-1wt%Si, (1cm x1cmx 0.1cm) were oxidised in air at 1000°C for times ranging from 3 to 30 minutes. The samples were rapidly introduced into and removed from the furnace; they were then glued to a supporting sheet and blocks 2mm x 3mm x3mm were cut using a slow speed saw. One side of the specimen was ground to 4000 grit silicon carbide paper, using a tripod thinner, before the sample was turned over and the other side ground to the same finish. The final sample thickness was approximately 20μm. To protect the samples, they were then glued to stiff 3mm slot grids. Thinning to electron transparency was carried out using an Ion Tech thinner with a beam angle of 15°.

## 3. RESULTS

The scale formed at these short oxidation times was flat and retained the original features of the metal surface (fig 1). There were no significant changes to the morphology of the scale during the first 30 minutes of oxidation that could be observed using SEM techniques. The scale formed on this alloy did not convolute or spall which is indicative of either an alloy with a very low sulphur content (<5ppm) or an alloy containing a reactive element (eg yttrium, hafnium). However, low sulphur alloys can only be produced by special processing and SEM/EDX analysis did not detect the presence of a reactive

Fig. 1 SEM Micrograph of the oxide surface.

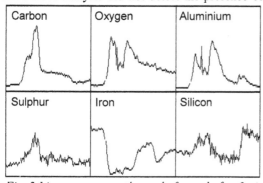

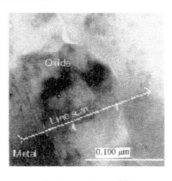

Fig. 2 Linescan across the scale formed after 3minutes oxidation at 1000 °C.

element within the alloy, so its behaviour was something of an enigma.

A STEM micrograph of the scale formed after 3 minutes oxidation is shown in figure 2. Analysis showed that the outer region of the scale was richer in iron, while the inner was almost exclusively aluminium oxide. The grain size was very small and the grain boundaries were poorly defined. Analysis across the metal/oxide interface did not detect sulphur segregation. However, it was noted that sulphur and carbon were present as particles within the scale.

The microstructure of the scale did not change significantly after 10 minutes oxidation and analysis across the metal/oxide interface did not detect the segregation of sulphur or any reactive elements, for which line scans were carried out.

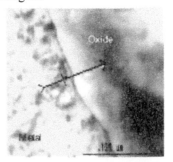

Fig.3 STEM micrograph of the metal/oxide interface.

However, the dark line (fig 3) observed at the interface is indicative of a high atomic weight element segregating to the metal/oxide interface.

After 20 minutes oxidation (fig 4) the structure of the scale changed significantly, with the grains becoming more definite and the grain size increasing in some areas. An area scan along the metal/oxide interface (fig 5) revealed that the element segregating to the interface was uranium, which must have been an impurity within the original alloy.

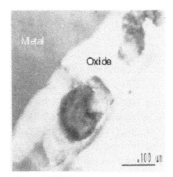

*Fig. 4 STEM micrograph of the oxide scale.*

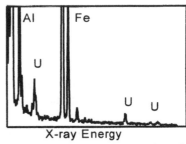

*Fig. 5 EDX analysis of the metal/oxide interface.*

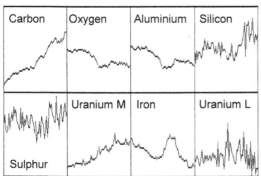

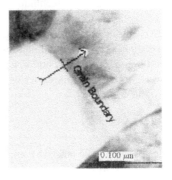

*Fig. 6 STEM line scan and micrograph showing the segregation of iron to the oxide grain boundaries.*

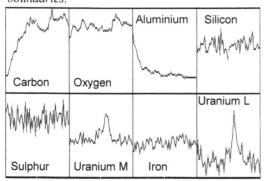

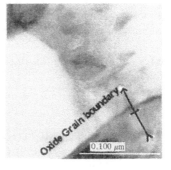

*Fig. 7 STEM linescan and micrograph showing the segregation of uranium to oxide grain boundaries.*

Analysis of the oxide grain boundaries near to the metal/oxide interface showed that uranium was initially only present at the oxide/metal interface but with increasing oxidation time uranium was observed to move out through the scale along the grain boundaries (fig 6).

Analysis of the grain boundaries further out from the metal/oxide interface showed that the changes in microstructure observed after 20 minutes oxidation were accompanied by a rejection of iron from the centres of grains and that the iron built up at the oxide grain boundaries (fig 7).

## 4. DISCUSSION

Although uranium was only present within the alloy at a concentration of a few parts per million and was not detected using conventional SEM/EDX or TEM/EDX techniques, its tendency to segregate strongly to the metal/oxide interface has a significant effect on the behaviour of this alloy during high temperature oxidation. Uranium behaves in a similar way to other reactive element additions, in that it segregates to the metal/oxide interface and to the grain boundaries of the oxide scale. This research has shown that segregation occurs at very short oxidation times (<10 minutes) and that segregation is initially to the metal/oxide interface and only at longer times (20 minutes) does uranium start to move out through the scale via the grain boundaries.

Although using SEM techniques the scale does not appear to change significantly, during these early stages of oxidation, STEM analysis shows that significant changes do occur, with the oxide changing significantly during the first 20 minutes oxidation. Also, it can be seen that although the inner oxide was aluminium-rich, the restructuring of the oxide led to iron being rejected from the oxide grains and building up at the oxide grain boundaries. The increase in iron concentration at these grain boundaries will affect diffusion down the boundaries and thus the growth rate of the scale.

## 5. CONCLUSIONS

1) Uranium can be used as a reactive element to improve the adhesion of alumina scales during high temperature oxidation.
2) Uranium present within the alloy segregates to the metal/oxide interface after very short oxidation times (<10 minutes).
3) Initially some iron dissolves in the aluminium-rich oxide grains but it is rejected to the grain boundaries after about 20 minutes oxidation. The rejection of impurity elements to the boundaries will have a significant effect on grain boundary diffusion and change the growth of the oxide scale.
4) The features observed during this research would not have been detectable using standard TEM techniques.

## REFERENCES

Chattopadhyay B and Wood G C 1970 Oxid. Met. 2(4), 373
Fox P, Lees D G and Lorimer G W 1991 Oxid. Met. 36(5/6), 491
Lees D G 1987 Oxid. Met. 27(1/2), 75

*Inst. Phys. Conf. Ser. No 153: Section 13*
*Paper presented at Electron Microscopy and Analysis Group Conf. EMAG97, Cambridge, 1997*
© 1997 IOP Publishing Ltd

# EELS study of titanium-hydrogen compounds as a function of hydrogen content.

**Y Kihn, A Mazel and G Zanchi**

CEMES / CNRS, BP 4347, 31055 Toulouse cedex 04, France

**ABSTRACT:** Transmission electron microscopy has been used to characterize titanium-hydrogen compounds. Imaging and electron diffraction allow us to identify morphology and lattice parameters of reference samples ($\alpha$ and $\beta$-phases in titanium, TiH, $TiH_2$), while electron energy loss spectroscopy (EELS) has revealed an increase in the bulk plasmon frequency with hydrogen content. By considering hydrogen insertion in the lattice as an increase in the free-electron density, a relationship between hydrogen content and plasmon frequency can be established, from which the hydrogen concentration in $Ti-H_x$ solid solutions can be evaluated up to the solubility limit.

## 1. INTRODUCTION

Interest in titanium alloys has grown recently owing to their potential applications for the aerospace and biomedical industries. Mechanical properties of these alloys are changed by the presence of hydrogen in their structures: hardness and yield strength are increased, but ductility reduced. A strong embrittling effect can appear in the presence of a very small amount of hydrogen (0.3 at.%).

The capacity of metallic titanium to absorb and desorb large amounts of hydrogen was shown as early as 1930. Recent works (Miron 1974, Numakura 1984, Woo 1985) have revealed that the Ti-H system is fully isostructural with the well known Zr-H system. Titanium hydrides are well known, their lattice parameters have been determined by X-ray diffraction on bulk samples.

The main purpose of this work is to characterize these hydrides or Ti-H compounds at the nanometre scale by electron energy loss spectroscopy. In order to establish a relationship between the plasmon frequency increase and the hydrogen content in solid solutions, the EELS studies have been performed with different reference specimens.

## 2. RESULTS

### 2.1 Titanium-hydrogen system

The phase diagram of the Ti-H system is of the eutectoid type, with hydrogen forming either interstitial solid solutions in the $\alpha$ and $\beta$ titanium phases or titanium hydrides depending on the hydrogen content. In hydrides $TiH_x$, different structures are identified depending on titanium phases in which the hydrides precipitate and also on the value of the concentration x. Since 1991 (Lewkowicz 1996), by analogy with the Zr-H system, the main phases are denoted :

$\gamma$, for the metastable fct hydride in $\alpha$-titanium ($x \approx 1$)
$\delta$, for the fcc hydride in $\beta$-titanium ($1.59 < x < 1.94$)
$\epsilon$, for the tetragonal hydride ($x \approx 2$)

In solid solutions, hydrogen adsorption up to the solubility limit has no appreciable effect on the lattice parameters of pure hcp $\alpha$-titanium (Chrétien 1954), while in the $\beta$-titanium phase it expands the bcc lattice (San-Martin 1987).

## 2.2    Experiments

All TEM studies have been performed with a Philips CM20T equipped with a Gatan 666 spectrometer for PEELS and a Tracor device for EDXS. Specimens for electron microscopy observations were prepared either by ion milling at nitrogen temperature or by chemical thinning at low temperature followed by ion milling for a few minutes to eliminate any oxide layer on the surface of thin samples. All thinned samples were kept in methanol. The powder of the $TiH_{1.97}$ reference sample was directly dispersed on a copper grid with a carbon holey film just before TEM observations.

Different specimens have been studied in order to determine reference values for the bulk plasmon frequency : $\alpha$ and $\beta$-phases in pure titanium, $\alpha$ and $\beta$-phases in alloys with aluminium and molybdenum as main alloying elements, the $\varepsilon$ tetragonal $TiH_{1.97}$ and the $\gamma$-TiH phase found in an hydrogenated alloy.

## 2.3    $\alpha$ and $\beta$-phases in titanium and titanium alloys

These phases have been identified by their electron diffraction patterns, according to the lattice parameters given by Wood (1962) for the $\alpha$ and Spreadborough (1959) for the $\beta$-phase in pure titanium. Any oxygen and carbon surface contamination of the thinned samples has been checked by EELS spectra.

For pure titanium, the typical value for bulk plasmon energy in the $\alpha$-phase is 17.4 $\pm$ 0.1eV and in the $\beta$-phase 17.5 $\pm$ 0.1eV. In alloys, the values depend on the nature and on the content of the alloying elements. In the $\alpha$-phase, only aluminium and a small amount of oxygen have been detected by EELS, while in the $\beta$-phase EDXS shows that molybdenum atomic concentration varies from 8 up to 25 at.%. Experimental values of plasmon energy as a function of alloying element content are reported in Table 1, where they are compared with calculated values deduced from the free-electron collective oscillation frequency (Pines 1963).

| $\alpha$-phase | pure Ti | Ti-Al 8% | Ti-Al 12% | Ti-Al 16% |
|---|---|---|---|---|
| measured $\hbar\omega_p$ | 17.4 eV | 17.4 eV | 17.3 eV | 17.3 eV |
| calculated $\hbar\omega_p$ | 17.9 eV | 17.7 eV | 17.63 eV | 17.58 eV |
| $\beta$-phase | pure Ti | Ti-Mo 8% | Ti-Mo 14% | Ti-Mo 25% |
| measured $\hbar\omega_p$ | 17.5 eV | 17.8 eV | 18 eV | 18.6 eV |
| calculated $\hbar\omega_p$ | 17.6 eV | 17.9 eV | 18.1 eV | 18.5 eV |

Table 1: experimental and theoretical values of the bulk plasmon energy of $\alpha$ and $\beta$-phases in titanium and in alloys as a function of alloying element content.

The good agreement between measured and calculated values, particularly in the $\beta$-phases, confirms the validity of the theoretical model of the free-electron collective oscillations in the case of metallic samples and alloys. The small difference (< 3%) between calculated and measured values for $\alpha$-phase in titanium may be attributed to surface oxidization of the samples. The strong influence of Mo as alloying element on the plasmon energy in the $\beta$-phases renders the study of these hydrogenated phases difficult.

## 2.4    Titanium-hydrogen compounds

The lattice parameters of the reference sample $TiH_{1.97}$ given by Yakel (1958) have been confirmed by electron diffraction patterns. The value of the plasmon energy measured on several powder grains is 20.8 $\pm$ 0.1eV as shown in Fig.1.

The $\gamma$-phase of titanium hydride (TiH) has been identified in the $\alpha$-phase of titanium alloy after cathodic hydrogenation. This $\alpha$-monophase alloy, previously studied before hydrogenation, contains aluminium (6 at.%) as the main alloying element (other element content being < 0.8 at.%). The morphologies of TiH shown on the images of Fig. 2, are highly characteristic with long needles (a) or platelets (b) (Bourret 1986). Electron diffraction patterns are interpreted from the lattice parameters given by Numakura (1984) for the tetragonal unit cell. In all cases, the plasmon energy was found to be $19.6 \pm 0.1 eV$. (Fig. 1)

Electron diffraction patterns show that the lattice parameters of the $\alpha$-phase of the titanium alloy are the same as those of the $\alpha$-phase in pure titanium. In the case of the hydrogenated $\alpha$-phases similar to those shown in Fig. 2, the aluminium content is evaluated to be 4 at.% by EDXS. The plasmon energy is found to vary from 17.4 to 17.7 eV without detectable change in the alloying element content. The shift in the plasmon energy may be attributable to the presence of hydrogen in the structures as confirmed by the presence of a broad peak around 12 eV, which is always observed in TiH and $TiH_{1.97}$ as shown by the arrows in the Fig.1.

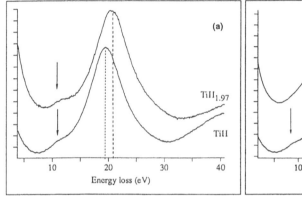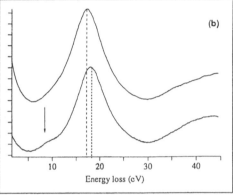

Fig. 1: (a) Plasmon spectra in TiH and $TiH_{1.97}$,
(b) Plasmon spectra in hydrogenated Ti-alloy $\alpha$-phases

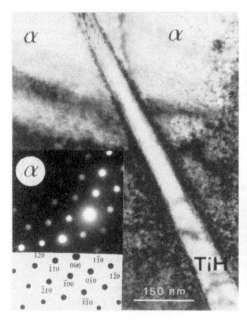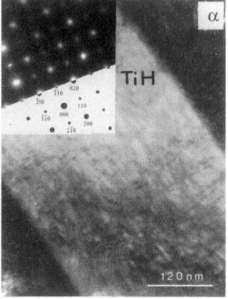

Fig. 2: TEM images of TiH, (a) needle-shaped TiH, (b) lath-shaped TiH, and diffraction patterns of $\alpha$-phase (a) and of TiH (b)

## 2.5 Discussion on EELS results

By considering hydrogen insertion in the Ti-lattice as an increase in the number of free electrons as assumed by Lamartine (1980), the bulk plasmon energy has been calculated for $TiH_{1.97}$ and TiH, taking each unit cell into account. The calculated values are 19.5. and 18.5 eV while the experimental values are 20.8 and 19.6 eV respectively. The important difference (more than 1 eV) between experimental and calculated values can be explained by changes in the density of states at the Fermi level when x is increased, as is shown by the NMR results for $TiH_x$ published by Korn (1983). The simple free-electron model appears is not capable of describing the dielectric function of the the the $TiH_x$ compounds with x>0.5.

In the Ti-alloy $\alpha$-phases, the insertion of hydrogen up to the solubility limit seems to cause no change of the lattice parameters in the limit of the accuracy of the electron diffraction. The solubility limit in the $\alpha$-titanium is 8 at. % at the eutectoid temperature (McQuillan 1950). Assuming that in solid solutions the hydrogen content x is related to the free-electron density, the equation predicting the energy plasmon loss (in eV) allows us to establish the following expression for the bulk plasmon energy in the $\alpha$-Ti as a function of x :

$$\hbar\omega_p = 17.9\sqrt{1+0.25x} \approx 17.9 + 2.23\,x \qquad (1)$$

with 4 free electrons/Ti atom, 2 Ti-atoms in the unit cell with volume $35.10^{-24}cm^3$.
This expression can be adapted to the experimental values found for $\alpha$-Ti-5%Al alloy:

$$\hbar\omega_p = 17.4 + 2.18\,x \qquad (2)$$

This relation is to be compared to that established by Lamartine from experimental results concerning Ti-$H_x$ compounds with 1.5 <x<1.8. ( $\hbar\omega_p = 16.5 + 1.9$ x).

From the expression (2), the maximum plasmon energy loss of 17.7 eV is attributed to a 13.7 at % hydrogen content. This value, which is greater than the solubility limit in pure titanium, is in accord with the fact that this limit can be increased in alloys containing aluminium as observed by Paton (1971).

Given the accuracy of the measurements ($\pm 0.1eV$), the smallest significant shift, about 0.2 eV, would correspond to a 9 at % hydrogen content.

These data show that in solid solution either in titanium or in titanium alloys, it is possible to determine the hydrogen content from the plasmon energy shift. The applicability of this method is limited by a minimal hydrogen content and the effects of alloying elements on the plasmon energy.

## ACKNOWLEDGEMENTS

The authors wish to thank J. Huez (Université de Technologie de Compiègne) for preparing the hydrogenated samples and M. Encrenaz (ENIT Tarbes) for the alloys.

## REFERENCES

Bourret A, Lasalmonie A and Naka S 1986 Scr. Metall. **20**, 861
Chrétien A, Freundlich W and Bichara M 1954 C R Acad. Sci. Paris **238**, 1423
Korn C 1983 Phys. Rev. B **28**, 95
Lamartine B C, Haas T W and Solomon J S 1980 Appl. Surf. Sci. **4**, 537
Lewkowitz I 1996 Solid State Phenom. **49-50**, 239-279
McQuillan A D 1950 Proc. Roy. Soc. London **A204**, 309
Miron N F, Shcherbak V I, Bykov V N and Levdik V A 1974, Kristallografiya **19**, 754
Numakura H and Koiwa M 1984 Acta Metall. **32**, 1799
Paton N E, Hickman B S and Leslie DH 1971 Metallurgical Transactions **2**, 2791
Pines D 1963 Elementary Excitations in Solids, Benjamin, New York.
San-Martin A and Manchester F D 1987 Bull. Alloy Phase Diag. **127**
Spreadborough J and Christian J W 1959 Proc. Phys. Soc.(London) **74**, 609
Woo O T, Weatherly G C, Colean C E and Gilbert R W 1985 Acta Metall. **33**, 1897
Wood R M 1962 Proc. Phys. Soc.(London) **80**, 763
Yakel H J 1958 Acta Crystall. **46**, 11

*Inst. Phys. Conf. Ser. No 153: Section 13*
*Paper presented at Electron Microscopy and Analysis Group Conf. EMAG97, Cambridge, 1997*
© *1997 IOP Publishing Ltd*

# Spatially resolved EELS of anodized Al-Cr alloys.

[1]Y Kihn, [1]G Zanchi , [2]G Thompson and [2]X Zhou

[1] CEMES / CNRS, BP 4347, 31055 Toulouse cedex 04, France
[2] UMIST, PO Box 88, Manchester M60 1QD, United Kingdom

**ABSTRACT:** Energy dispersive systems used in transmission electron microscopy, give a correspondence between a part in the image plane and a part in the spectrum, perpendicular to the dispersion direction. Taking advantage of this property, with the new spectum 2D detection devices (CCD matrix), the spatial resolution of EELS analysis can approach the subnanometre scale. With this technique, ultramicrotomed sections of anodized Al-Cr 4% alloy samples have been studiedto determine the locations of alloying species and consequences of their anodic oxidation.

## 1. INTRODUCTION

In order to inhibit corrosion of aluminium alloys, surface pretreatments with chromium-containing solutions have been shown to be highly effective. Further, protection may also be possible through use of alloys, which requires scrutiny of relatively thin alumina films developed over their macroscopic surfaces (Brown 1992), i. e. Al-4%Cr used here. In order to improve understanding of the relevant physical, chemical and electrochemical phenomena, nanoscale knowledge of the morphology, structure and chemical composition of the surface layers is required as well as their relationship to the alloy substrate.

Transmission electron microscopy (TEM) is used to study ultramicrotomed cross-sections of anodized Al-Cr 4% alloy samples. TEM imaging reveals the morphology of the anodic film attached to the aluminium alloy substrate, showing an external alumina layer with a 2 nm thick interfacial layer. The chemical nature of these layers is characterized by electron energy loss spectroscopy (EELS) by observing spatially resolved spectra in two samples which have been anodized to different voltages.

## 2. EXPERIMENTAL PROCEDURE

### 2.1 Spatially resolved EELS principle

Owing to the geometrical aberrations of the dispersive system, the electron trajectories coming from the object cross-over go through the conjugate spectrum plane inside a figure which appears similar to a very flattened ellipse with the major axis being perpendicular to the gap of the magnetic prism (Zanchi 1982). The correspondence between the points of the image plane and the points of this focal line allows, under some conditions, advantage to be gained from this aberration. If an interface line dividing two parts of the sample with different chemical compositions is set in a parallel direction to the spectrometer gap, the EEL spectra corresponding to the two areas appear well separated in the spectrum display plane (Reimer 1992). This effect can be employed to improve the spatial resolution in chemical analysis and, particularly, in concentration gradient studies.

### 2.2 Experimental device

TEM imaging and EELS experiments have been performed with a Philips CM200 microscope equipped with GIF (Gatan Imaging Filter). The electron image or spectrum detection is performed by a CCD matrix (Gubbens 1993) allowing a 2D parallel spectrum recording on

1024 points in the dispersion direction and in the perpendicular direction on 128 points or more, depending on the spectrometer aperture size.

As an example, the part of the spectrum in the Fig.1 corresponds to a layer of alumina in the upper part and to layer of Al-Cr alloy in the lower part, recorded in the image mode in the following conditions: interface parallel to the spectrometer gap, high voltage 200 kV, microscope magnification of x50000, each line of pixels in the spectrum corresponding to $0.4 \times 40$ nm$^2$ in the object plane. The energy range 510-610 eV presented here allows to see the O-K edge in all the layers and the Cr L$_{2-3}$ edge only in the alloy.

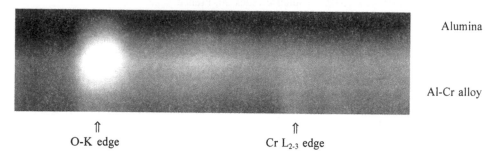

⇑                                           ⇑
O-K edge                          Cr L$_{2-3}$ edge

Fig 1: Spectrum corresponding to a selected area in the sample composed with a layer of Al$_2$O$_3$ (at the top) and a layer of Al-Cr 4% alloy (in the lower part).

## 2.3    Specimen preparation

Electropolished aluminium subsequently anodized to 150V, provides excellent adhesion for magnetron sputtered alloys as shown by Shimizu (1987). Al and Cr were co-sputtered from separate magnetrons in an Ion Tech sputtering unit. The specimens were anodized in 0.01M ammonium pentaborate solution at a constant current density of 5 mA/cm$^2$ to different voltages of 32 and 20V for the named A32 and C20 samples.

## 3.    RESULTS

TEM micrographs of ultramicrotomed cross-sections of the two samples exhibit several layers with different chemical composition as shown in the Fig. 2-a.

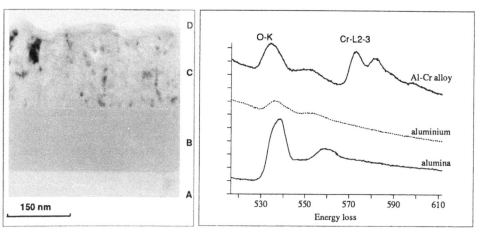

Fig. 2: a-TEM imaging of the different layers in the sample anodized to 32V. : Al substrate (A), alumina layer (B), the Al-Cr 4% alloy (C) and the anodic film (D). A 2 nm thick interfacial layer exists between (C) and (D).

b- EELS profiles of the O-K edge in Al, Al$_2$O$_3$ and Al-Cr 4%

EELS analysis have been performed on different parts in Al, $Al_2O_3$ and Al-Cr 4% with a probe size of 10 nm in diameter. The profiles of O-K edge (and eventually Cr-$L_{2-3}$ edge) in alumina, in the Al-Cr 4% alloy, and in aluminium are drawn in Fig. 2-b. In the last case, oxygen arises from surface oxidation of the thin film prepared for TEM observation.

The main focus of this work is the study of the chemical nature of the external anodic film and of the thin interfacial layer. In order to improve the spatial resolution of the analysis from the external side to the Al-Cr layer, spectra similar to that of the Fig.1 have been exploited.

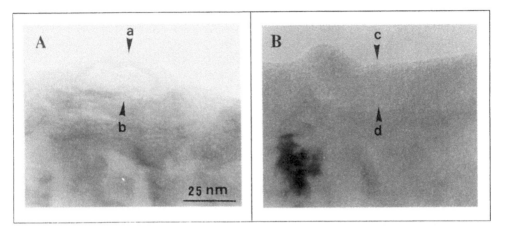

Fig. 3: TEM images of the anodic films in the sample anodized to 32V (A) and to 20V (B).

The micrographs of Fig. 3 show the anodized layer and a part of the Al-Cr alloy, with the thin interfacial layer (in the alloy, immediately below the anodic film). The thickness of the external layer depends on the voltage of the anodizing (near 25 to 35 nm for 20V (A), and 30 to 50 nm for 32V (B)).The morphology of the external layer in the sample (A) shows large voids near the outer side, while in the sample (B) the layer is more uniform with relatively fine voids in some parts. Line-spectra have been recorded in different area in the two samples: Fig. 4 shows an example of the different profiles. The microscope magnification was x 50000 (each line of pixels corresponds to an area of 0.4 x 40 $nm^2$).

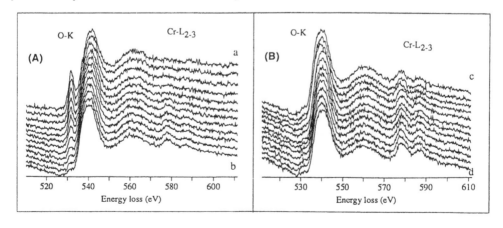

Fig. 4: Profiles of line-spectra. In sample (A) , spectra correspond to the area between the arrows a-b, and in sample (B) to arrows c-d. In the direction perpendicular to the interface, the distance between two spectra is 0.8 nm.

The profiles in sample (A) recorded on a void exhibit a sharp pre-peak at 529 eV behind the O-K edge which corresponds to $O_2$ gas (Ahn 1983).This peak decreases from a to b with a shift of the main O-K peak, while the Cr $L_{2-3}$ edge appears and increases.The profiles in sample (B), from c to d, show that the Cr $L_{2-3}$ edge is always present, with a small increase from c to d, while the O-K edge decreases. Similar profiles have also been found in sample (A) in area without voids, but the Cr content is smaller in the external part. Such profiles drawn from the anodic film to the Al-Cr alloy, show a continuous increase of the Cr peak with a decrease of the O-K edge, the shape of which changes from the O-K peak in alumina to those in the alloy (as shown in Fig. 2), up to the middle of the alloy layer.

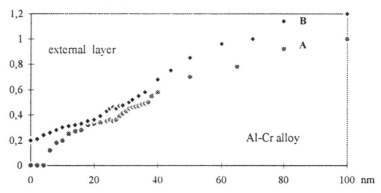

Fig. 5: Profiles of the ratio Cr/O from the external layer up to the middle of the alloy layer

The Cr content from the external part up to the alloy can be quantified by evaluating the ratio of the area under the Cr and the O distributions after background stripping, as a function of the distance from the external side. Fig.5 shows the Cr/O for the two samples. A fine increase in the relative Cr is evident in the interfacial layer (near 25 nm from the side) by a small peak in the curves.

## 4.    CONCLUSION

This spatially resolved EELS method allows elemental content profiles through interfacial layer to be established. The main experimental constraints are the alignment of the interfacial layers to the spectrometer gap in the image mode and the adaptation of the wished microscope magnification to the spatial resolution desired.

Concerning the anodized Al-Cr alloys, these analyses confirm that the external layer is alumina with a small Cr content. They show that voids are filled with oxygen gas, with a small increase of Cr content in the thin interfacial layer.

**ACKNOWLEDGEMENT**

The authors wish to thank Dr. C. Quet of the Elf-Atochem Society (Groupement de recherches de Lacq) who has placed the GIF device at Y. Kihn's disposal.

**REFERENCES**

Ahn C C and Krivanek O L 1983, EELS Atlas (Arizona State University and Gatan)
Brown G M, Shimizu K, Kobayashi K, Thompson G E and Wood G C 1992, Corros. Sci. **33**, pp 1371-85
Gubbens A J and Krivanek O L 1993, Ultramicroscopy **51**, 146
Reimer L, Fromm I, Hirsch P, Plate U and Rennekamp R 1992, Ultramicroscopy **46**, 335-47
Shimitzu K, Kurima Y, Kobayashi K, Thompson G E and Wood G C 1989, Thin Solid Films **173**, 263-268
Zanchi G, Kihn Y and Sevely J 1982, Optik **60, 4,** pp 427-36

*Inst. Phys. Conf. Ser. No 153: Section 13*
*Paper presented at Electron Microscopy and Analysis Group Conf. EMAG97, Cambridge, 1997*

# Measurement of interface roughness independently of interface diffuseness: application to magnetic multilayers

**M J Hÿtch, M G Walls, E Chassaing and J-P Chevalier**

Centre d'Etude de Chimie Métallurgique, Centre National de Recherche Scientifique, 15 rue Georges Urbain, 94407 Vitry-sur-Seine, France. Email: hytch@glvt-cnrs.fr

**ABSTRACT:** We propose a simple formalism which allows the separation of the contributions, due to roughness and chemical interdiffusion, to the total width of an interface by using 2-dimensional information from either images or chemical maps. A definition is also proposed for the roughness of an interface in terms of iso-concentration surfaces. The formalism is based on the relation between projection in real space and the corresponding section in Fourier space, the main hypotheses being that the interface roughness is isotropic and that the composition profile is constant across the interface. The method, although general, will be illustrated with results on Fresnel imaging of Cu-Co magnetic multilayers.

## 1. INTRODUCTION

A persistent problem for a number of interface studies is obtaining separate values for both the physical roughness of the interface and for the composition profiles across the interface. The properties of artificial nanostructures such as semiconducting quantum wells, metallic multilayers, oxide multilayers or capping layers, often depend very differently on these two physical aspects of the interface. Unfortunately, most characterisation techniques measure a width for the interfaces which is a combination of the two. For example, electron microscopy is confronted by the classical projection problem. Only a few studies have been carried out which attempt to obtain reliable estimates for the roughness of interfaces independently of interdiffusion (Goodnick et al 1985, Chen and Gibson 1996).

### 1.1 Definitions

We shall begin by proposing definitions for the roughness and chemical diffuseness of an interface. Fig. 1a shows a schematic view of an interface in cross-section, with the z-axis pointing in the direction of the specimen thickness. The grey levels show how the concentration of a particular element varies across the interface, the contours marking lines of equal concentration. The interface can be seen to be diffuse chemically, and tracing the concentration as a function of position across the interface at a fixed value of y gives a profile such as Fig. 1c. We shall define the chemical diffuseness of the interface as being the width of this profile. At the same time, the position of the interface, which can be defined as the position of one of the iso-concentration contours, varies over the thickness of the foil. The distribution over the foil thickness of the interface position is given in Fig. 1b. The roughness can be defined as the width of this distribution. The concentration profile averaged over the thickness of the foil (Fig. 1d) shows that the projected width is greater than the chemical width at a particular foil thickness (Fig. 1c) due to the varying position of the interface (Fig. 1b). It is the width of this projected profile which is usually measured by characterisation techniques.

The interface can be described mathematically by defining X(z) as the x-co-ordinate of one of the iso-concentration contours, for example at the midpoint of the interface, and C(x) as the concentration or interdiffusion profile (Fig. 1c). For simplicity, we have assumed that the chemical profile is independent of the depth in the foil, so that all the iso-concentration contours are identical. At a given point (x,z) the concentration c(x,z) will be given by:

$$c(x,z) = C(x-X(z)) \qquad (1)$$

The projected profile across the interface, $c_z(x)$, will be given by:

$$c_z(x) = \int C(x-X(z))dz \qquad (2)$$

Given that the 1D concentration profile is independent of 'z' this can be rewritten as:

$$= \int C(x-x')h(x')dx' \qquad (3)$$

where h(x') is the distribution for the function X(z) having a value x' over the thickness of the foil (i.e. the histogram shown in Fig. 1b). The projected chemical profile, $c_z(x)$, can be identified as a convolution of the interdiffusion profile and the roughness:

$$c_z(x) = C(x) * h(x) \qquad (4)$$

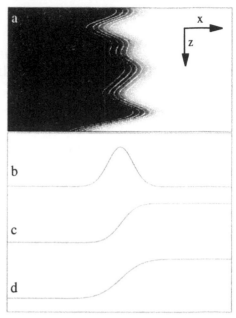

Fig. 1: Definitions of roughness and interdiffusion (a) cross-sectional view of an interface, grey levels correspond to the concentration of a particular element; (b) position of the interface in projection; (c) interdiffusion profile; (d) chemical profile in projection.

The profile seen in Fig. 1d is therefore the convolution of the profiles seen in Fig. 1b and Fig. 1c. The variances of the two contributions are therefore added to produce the total interface width which is measured by electron microscopy.

## 2. SEPARATING ROUGHNESS AND INTERDIFFUSION

From a single measurement of the interface profile, $c_z(x)$, it is impossible to distinguish between the two contributions. However, we have for the moment ignored the third dimension, along the interface in the y-direction.

Fig. 2a shows a plan view of a rough interface, this time in the yz-plane, the height corresponding to the iso-concentration surface X(y,z). In projection, the interface will not be seen as a straight line. Let us assume that the mid-point of the interface will be seen in the image at a position (x,y), shown in Fig. 2b, such that the x-co-ordinate corresponds to the average position of the interface over z, $X_z(y)$. The oscillations will be smaller than for the 2-D surface because of the averaging over the foil thickness. Fig. 2c and Fig. 2d show the power spectrum of the 2-D surface and 1-D line profile respectively. The horizontal line section through the 2-D power spectrum is identical to the power spectrum of the averaged image. The Fourier transform, $X(k_y,k_z)$, of X(y,z) can be defined:

$$X(k_y,k_z) = \int\int X(y,z)\, exp\{-2\pi i\, (k_y y + k_y z)\}dydz \qquad (5)$$

The section corresponding to $k_z=0$ is therefore given by:

$$X(k_y,0) = \int\int X(y,z)\,exp\{-2\pi i\,k_y y\}dydz \tag{6}$$

$$= \int X_z(y)\,exp\{-2\pi i\,k_y y\}dy \tag{7}$$

$$= X_z(k_y) \tag{8}$$

where $X_z(y)$ is the average over zof $X(y,z)$, the line profile shown in Fig. 2b, and $X_z(k_y)$ its Fourier transform (Fig. 2d).

From an image it is possible to plot the position of the interface (Fig. 2b) and therefore to calculate the 1-D power spectrum shown in Fig. 2d. If we assume that the roughness is isotropic we can construct the 2-D power spectrum by rotating the 1-D profile around the

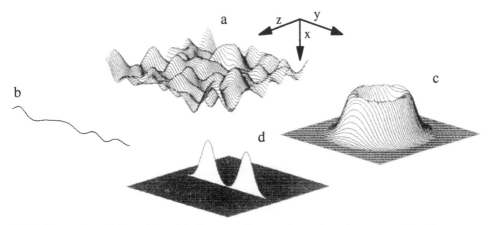

Fig. 2: Illustration of the relationship between the roughness in real space and Fourier space. (a) iso-concentration surface $x=X(y,z)$; (b) average interface position in projection, $X_z(y)$; (c) 2-D power spectrum of (a); (d) 1-D power spectrum of (c)

origin in the 2-D reciprocal space.

There is no way of knowing the phases in the 2-D Fourier transform but fortunately these are not needed for the measurement of the roughness. By Parseval's theorem the variance of the corresponding function in real space is equal to the integral of the power spectrum (excluding the central spot). By integrating the 2-D power spectrum it is therefore possible to estimate the standard deviation of the variations in real space, which is the value for the 2-D roughness.

## 3. APPLICATION OF THE METHOD

We shall illustrate the technique with an example taken from a study of electro-deposited Cu-Co multilayers (Nallet et al 1996). The interfaces between layers were characterised using the Fresnel technique which consists of taking a through focal series of images, in the absence of significant Bragg scattering and using a small objective aperture (Ness et al 1986). Fig. 3 shows an image taken at a defocus of 1440nm on a JEOL 2000FX TEM operating at 200kV (objective aperture radius of 5mrad).

The contrast in a Fresnel image is due to refraction and, to a very good approximation, is determined by the mean inner potential of the material averaged over the foil thickness. The widths of the fringes themselves are, therefore, only indirectly related to the actual interface widths. However, by studying the evolution of the contrast as a function of defocus accurate values for the layer thicknesses and interface width can be determined. In this case the layer period was measured as $6.00\pm0.03$nm, the Cu layers $2.3\pm0.2$nm and the interface width $0.53\pm0.07$nm. The projected profile was modelled in the simulations as an abrupt interface convoluted with a Gaussian distribution having various widths. The interface width is

656

expressed in terms of the standard deviation of the best fitting simulation. The specimen was estimated as 40±20nm thick.

The image shows that the layers are not straight and the position of the layers can be marked as a function of position, an example is shown in Fig. 3. The results for this layer are shown in Fig. 4a. This is equivalent to the 1-D profile shown in Fig. 2b, the assumption being that the fringes follow the mean position of the interfaces viewed in projection. The standard deviation of these oscillations is 0.12±0.01nm. The power spectrum shown in Fig. 4b has been averaged over the results from several such profiles to reduce noise. Given the foil thickness, it is possible to construct the 2-D power spectrum by rotating the 1-D profile around the origin. For simplicity, the 1-D section analysed had a length of 40nm which equals the estimate of the foil thickness. This means that the interface section analysed is square. If a rectangular section had been chosen the rings in the Fourier transform would become ellipses. Integrating the power spectrum gives a result of 0.4±0.2nm for the standard deviation of the 2-D roughness. The error comes directly from the estimate of the foil thickness. Assuming that the interdiffusion is independent of the roughness the variances of the two are added to produce the total interface width. Given that this is 0.53nm, the interdiffusion can be estimated as 0.3±0.2nm.

## CONCLUSIONS

Interface roughness can be defined as the roughness of an iso-concentration surface. For the 2-D roughness to be measured, the main assumptions are that a continuum model is applicable, the interdiffusion profile is independent of roughness, the position of the projected interface corresponds to the projected iso-concentration surface and that the roughness is isotropic. The major source of error is the estimate of the foil thickness but the effect of noise cannot be ignored. The method should be applicable to Fresnel imaging, dark field imaging, chemical maps (STEM) and high angle dark field imaging (HAADF).

## REFERENCES

Chen X D and Gibson J M 1996 Phys. Rev. B 54, 2846
Goodnick S M, Ferry D K, Wilmsen C. W., Liliental Z, Fathy D and Krivanek O L.1985 Phys. Rev. B 32, 8171
Nallet P, Chassaing E, Walls M G and Hÿtch M J 1996 J. Applied Physics 79, 6884
Ness J N, Stobbs W M, and Page T F 1986, Philos. Mag. A 54, 679

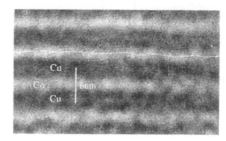

Fig. 3: Fresnel image of Co-Cu multilayers at a defocus of +1440nm.

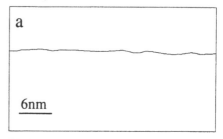

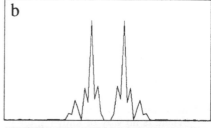

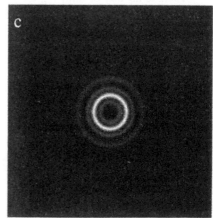

Fig. 4: calculation of the 2-D roughness. (a) position of the layers; (b) 1-D power spectrum; (c) 2-D power spectrum, maximum at a periodicity of 10nm.

# 'Nano-machining' using a focused ion beam

**L C I Campbell, D T Foord, C J Humphreys**

Department of Materials Science and Metallurgy, University of Cambridge, Pembroke Street, Cambridge CB2 3QZ

**ABSTRACT:** A 30 keV $Ga^+$ focused ion beam (FIB) operated with a beam current of 1pA has been used to mill single-pixel width lines in (111) silicon with line doses in the range $10^{10}$-$10^{12}$ ions $cm^{-1}$. The ability of the FIB to produce site specific cross-sectional TEM samples was exploited to observe the depth and profile of the milled structures. Energy filtered images were also obtained to assess the extent of gallium incorporation from the ion source. Groove widths of 50-60 nm and depths up to 120 nm were achieved.

## 1. INTRODUCTION

Focused ion beams have become a useful tool for materials modification on a sub-micron and recently a sub 100-nm scale. They allow maskless implantation and deposition as well as material removal by either purely physical sputtering or gas assisted etching. Applications include the repair and modification of prototype integrated circuits (Komano et al 1991) and masks (Harriott et al 1992) as well as site specific TEM sample preparation (Walker et al 1995). Systems are now capable of producing a beam diameter of less than 10 nm at low currents and lithography has been demonstrated at this scale, notably by Kubena et al (1991) in PMMA. However most studies of milling by physical sputtering alone have been concerned with structures of micron dimensions (for example Pellerin et al 1989 and Lipp et al 1996). In this paper we investigate machining on a sub-100 nm scale by milling grooves with line doses in the range $10^{10}$ - $10^{12}$ ions $cm^{-1}$ and examining the resulting structures using cross sectional TEM and energy filtered imaging.

## 2. EXPERIMENTAL

The FIB instrument used in this work was a computer controlled FEI 200 series workstation with a 30 kV Ga liquid metal ion source. Operating at the lowest current of 1 pA, the instrument is capable of a secondary electron image resolution of better than 10 nm and a full-width half-maximum beam diameter of ~5 nm. The system also has a platinum deposition facility. Deposition occurs by ion-beam induced dissociation of a surface adsorbed, platinum containing gas, which is introduced into the FIB chamber through a fine capillary.

A 3 mm square of silicon was cleaved from a 0.6 mm thick (111) wafer and ground to make a strip approximately 50 μm wide. This was then mounted flat on a TEM grid, modified to allow FIB access to the (111) face, and 10 μm long, single-pixel width lines were milled with a range of ion doses. All milling was carried out with a beam current of 1 pA, a 1μs dwell time per pixel, and a 50% overlap between adjacent pixels.    After the milling had been completed, the sample was removed from the FIB and sputter coated with approximately 50 nm of gold as protection from further gallium implantation. Then the  sample was transferred back to the FIB, a 1 µm thick strip of platinum was deposited across the centre of the lines and a TEM cross section was milled using standard techniques. The spatial distributions of gallium and silicon were mapped using a Gatan imaging filter to obtain energy filtered images for the background stripped Ga-L and Si-K edges of the EELS spectrum. The TEM used for collecting the energy filtered images was a JEOL 4000FX operated at 400 keV.

## 3.    RESULTS AND DISCUSSION

Fig. 1 shows bright field TEM cross sectional images through the centre of the single-pixel width milled lines. The position of the silicon surface is indicated by the arrows.  The largest channel (marked 1 in Fig. 1), milled at a line dose of  $1.6 \times 10^{12}$ ions cm$^{-1}$, is 60 nm wide and 120 nm deep with straight, nearly vertical walls. The shorter milling times resulted in grooves with a Gaussian type profile, remaining fairly constant in width but decreasing in depth. Little or no evidence of sputtering is visible in lines 6 and 7, probably due  partly to the swelling of the silicon surface. In the structures with visible grooves, depth increases with milling time (and ion dose) in a linear fashion as illustrated in  Fig. 2. The linear relation is the same as the behaviour shown for larger milled structures.

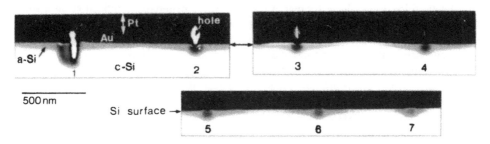

**Figure 1.** Cross sectional TEM micrographs of the milled structures. Line doses for grooves 1-7 respectively are $1.6 \times 10^{12}$,  $6.4 \times 10^{11}$,  $3.2 \times 10^{11}$,  $2.2 \times 10^{11}$,  $1.6 \times 10^{11}$,  $9.4 \times 10^{10}$ and $3.1 \times 10^{10}$ ions cm$^{-1}$.

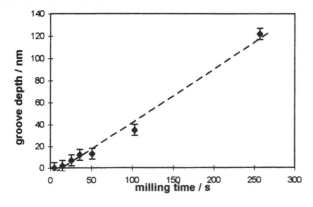

**Figure 2.** Groove depth achieved as a function of milling time.

The width at the top of the grooves remains fairly constant at 50-60 nm. This is much wider than the diameter of the ion beam used, but there is a certain amount of delocalisation in the mechanism of physical sputtering. Atoms are ejected from the surface as a result of an atomic collision cascade, triggered by the incident ion.

The marked asymmetry in the deepest groove is due to sample drift during the long milling time. For grooves 1-4, the holes through the overlying metal layers are believed to have formed during deposition : FIB deposited platinum has a very columnar structure and is heavily influenced by the underlying surface topography (Walker and Broom 1997)

All the milled lines show the formation of a layer of amorphous silicon. The distinctive 'winged' shape of these regions is consistent with the reported beam profile produced by a liquid metal ion source i.e. a central Gaussian distribution with wide exponentially decaying 'tails' (Cummings et al 1986, Chu et al 1991). The depth of the amorphous regions stays approximately constant at ~80 nm measured from the bottom of each channel.

A higher magnification image of line 2 (Fig. 3a) shows that within the amorphised region there are two further levels of damage to the silicon, both showing gallium enrichment (Fig. 3b). The first 'level', with the lesser degree of gallium incorporation, extends to ~50 nm depth and is likely to correspond to ordinary implantation, although it is slightly deeper than the 30-40 nm range quoted by Lipp et al (1996) for 30 keV gallium ions in silicon.. This region is clearly visible in Fig. 1 for lines 1-6 and is fainter, but also present, in line 7. In the same way as the amorphisation, the depth of this region remains fairly constant with milling time.

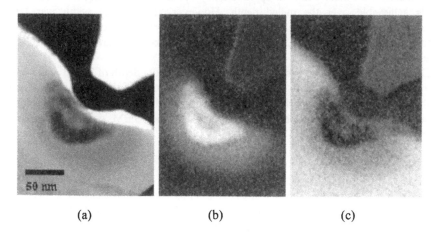

|  (a)  |  (b)  |  (c)  |

**Figure 3.** Energy filtered images of line 2 : (a) Bright field (zero loss) image. (b) Gallium map (Ga-L edge loss) and (c) Silicon map (Si-K edge loss) of area shown in (a).

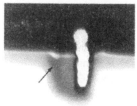

**Figure 4.** Enlargement of line 1 from Fig. 1. The arrow indicates the point where a dark 'core' had started to form under the silicon surface before sample drifting occurred.

The second 'level' is only visible for line doses of $2.2 \times 10^{11}$ cm$^{-1}$ and above (lines 1-4) and appears in the TEM images as a very dark and clearly defined 'core' to the damaged region. In a similar manner to the amorphised and implanted regions, this core increases in width with ion dose but stays fairly constant in depth (20-30 nm), showing that a vertical 'sputter equilibrium' is achieved. In groove 1 the nearly vertical boundaries to each region imply that a similar steady state is achieved in the lateral direction.

The energy filtered images of line 2 shown in Fig. 3b and 3c indicate that this 'core' consists mainly of gallium, interspersed with only a little silicon. In Fig. 4, indicated by the arrow, it is possible to see the dark core had started to form at a depth of ~30 nm underneath the silicon surface before the sample drifted. Gallium has a low solubility in silicon, given as $9 \times 10^{18}$ cm$^{-3}$ at 500°C by Ryssel and Ruge (1986), so it is possible that the core is formed when gallium has reached a high enough concentration to precipitate out.

## 4. CONCLUSIONS

The physical sputtering effect of a 30 keV Ga$^+$ focused ion beam operated at 1 pA beam current has been used to mill line structures with dimensions less than 100 nm in (111) silicon. The largest groove created, with a line dose of $1.6 \times 10^{12}$ ions cm$^{-1}$, was 60 nm wide and 120 nm deep with straight, nearly vertical sides. Grooves milled with line doses from $6.4 \times 10^{11}$ down to $1.6 \times 10^{11}$ ions cm$^{-1}$ showed Gaussian profiles, while no visible channels were formed for line doses of $9.4 \times 10^{10}$ ions cm$^{-1}$ and below. Once formed, the channel depth increased linearly with milling time at approximately 0.5 nm s$^{-1}$, indicating that considerable control is achievable over the milled depth. However, the channel width was largely unaffected by milling time, only increasing from 50 nm to 60 nm over the entire range of milling times used. This is believed to be due to the delocalisation inherent in the mechanism of sputtering.

Damage to the silicon consisted of amorphisation up to a depth of 80 nm and possible gallium implantation up to a depth of ~50 nm below the centre of the milled lines. For line doses of $2.2 \times 10^{11}$ and above, another level of damage, which might be gallium precipitation, was visible as a dark core to the damaged area. Further TEM work is planned for characterisation in greater detail.

## REFERENCES

Chu C H, Hsieh Y F, Harriott L R and Wade H H 1991 J. Vac. Sci. Technol. B **9**, 3451
Cummings K D, Harriott LR, Chi G C and Ostermayer F W, Jr 1986 SPIE **632**, 93
Harriott L R, Garafalo J G and Kostelak R L 1992 SPIE **1671**, 224
Komano H, Nakamura H and Takigawa T 1991 J. Vac. Sci. Technol. B **9**, 2653
Kubena R L, Ward J W, Stratton F P, Joyce R J and Atkinson G M 1991 J. Vac. Sci. Technol. B **9**, 3079
Lipp S, Frey L, Lehrer C, Frank B, Demm E and Ryssel H 1996 J. Vac. Sci. Technol B **14** 3996
Pellerin J G, Shedd G M Griffis D P and Russell P E 1989 J. Vac. Sci. Technol B **7**, 1810
Ryssel H and Ruge I 1986 Ion Implantation (Chichester: John Wiley & Sons)
Walker J F, Reiner J C and Solenthaler C 1995 Inst. Phys. Conf. Ser, **146**, 629
Walker J F and Broom R F 1997 Presented at 10$^{th}$ Int. Conf. on Microscopy of Semiconducting Materials (MSM X), Oxford, March 1997

*Inst. Phys. Conf. Ser. No 153: Section 13*
*Paper presented at Electron Microscopy and Analysis Group Conf. EMAG97, Cambridge, 1997*
© *1997 IOP Publishing Ltd*

# High Angle Grain Boundary Structure in Widmanstätten Alpha Ti Microstructures

S Wang and M Aindow

School of Metallurgy and Materials, The University of Birmingham, Elms Road, Edgbaston, Birmingham, B15 2TT.

ABSTRACT: The structure of a high angle grain boundary between two Widmanstätten α laths in the same prior β grain in pure titanium was studied using TEM. The Burgers vectors of the dislocations were obtained using Pond's topological theory of interfacial defects and the dislocation configurations which would be required to accommodate the deviations from various different reference structures were calculated using the O-lattice algorithm. A good match was obtained between the network which would be expected for a two-dimensional reference structure and that observed experimentally.

## 1. INTRODUCTION

Whilst the coincident site lattice (CSL) model has been used very successfully to model the reference structures exhibited by vicinal high angle grain boundaries (HAGBs) in materials with cubic crystal structures, there is very little experimental evidence to support the constrained CSL (CCSL) model for HAGBs in non-cubic materials (e.g. Chen and King 1989, Shin and King 1991). Moreover, it has been shown recently that there are fundamental weaknesses in the CCSL approach (MacLaren and Aindow, 1997a) and there is some evidence to suggest that reference structures may instead be related to periodicity in, or near to, the boundary plane (Grimmer et al 1990; Lay, Ayed and Nouet 1992). In a recent study of a HAGB in α-Ti (MacLaren and Aindow, 1997b), it was shown that by using the topological theory of interfacial defects (Pond, 1989), the O-lattice algorithm (Bollmann, 1970) can be applied to reference structures other than CSLs or CCSLs and here again a 2-dimensional reference structure was found.

In the present work we have performed detailed TEM investigations of the structure and defect content of HAGBs in pure Ti heat treated in the beta phase field and then cooled to give many Widmanstätten alpha laths in each prior beta grain. The majority of the HAGBs are thus formed between adjacent alpha laths which exhibit different variants of the Burgers orientation relationship with respect to the parent beta grain. An analysis of the reference structure for one example of the most common type of HAGB in this material is presented.

## 2. EXPERIMENTAL PROCEDURE

The as-received titanium was high purity grade (≈1000ppm oxygen) in the form of a sheet, approximately 1.2 mm thick. The specimens were coated with *Deltaglaze* (to prevent oxidation) and solution treated in an air furnace at 1000°C (i.e. within the β phase field) for half an hour, then quenched into water to as quickly as possible to produce the Widmänstatten microstructure. The specimens were then annealed for 20 min. at 700°C to obtain near equilibrium boundary structures.

The TEM foils were produced by grinding to a thickness of approximately 250 μm, punching 3 mm discs and then electropolishing to perforation. To prevent hydride formation a non-acid electropolishing solution of 5.3g LiCl + 11.2g Mg($ClO_4$)$_2$ in 500 ml methanol and 100 ml 2-butoxy-ethanol (Kestel 1986) was used at a temperature of -30°C and a voltage of 30 V. TEM observations were carried out using a Philips CM20 operating at 200 kV.

## 3. RESULTS AND DISCUSSION

The microstructure of the annealed samples consisted of many coarse ($\approx 1\mu m$ across) $\alpha$ laths within each prior beta grain. The relative orientations of these laths can be divided into six types and one example of a boundary separating two laths with the most common orientation is shown in Fig. 1. This boundary was flat over several microns and exhibited a well-defined structure with two distinct sets of periodic linear features. The first set ($U_1$) had a spacing of $\approx 17.3$ nm and was observed in all of the images. The second set ($U_2$) had a spacing of $\approx 7.1$ nm and was only observed for $\mathbf{g} = (\bar{1}011)$ (Fig. 1d). The contrast and orientation of these features mean that they cannot be Moiré or other interference fringes but are, instead probably interfacial dislocations.

Trace analysis was used to determine the line directions of sets $U_1$ and $U_2$ and these were $\mathbf{u}_1 = [0.4686\ -0.5442\ 0.0847\ 0.3007]_\lambda$ and $\mathbf{u}_2 = [-0.6492\ 0.3078\ 0.3414\ 0.1417]_\lambda$, respectively. The boundary plane normal, $\mathbf{n}$, was then calculated from the vector product of $\mathbf{u}_1$ and $\mathbf{u}_2$ giving $\mathbf{n} = [0.0867\ 0.3847\ -0.4713\ 0.4146]_\lambda$. The axis/angle pair $(\mathbf{r},\theta)$ relating the two grains was determined as ($[-0.0074\ 0.5040\ -0.4967\ 0.3143]$ / $179.176°$) from diffraction data. Several different candidate reference structures were considered for the boundary but the most plausible is the $(01\bar{1}1)$ $\Gamma 1$ 2D CSL which corresponds to a rotation of $180°$ about the normal to $(01\bar{1}1)$. The $(01\bar{1}1)$ reference structure plane lies at $14.94°$ to the boundary plane and the deviation of the measured misorientation from that at which the reference structure forms is $(\mathbf{r}_{dev},\theta_{dev}) = ([0.579\ -0.2671\ -0.3119\ 0.3113]$ / $3.04°$).

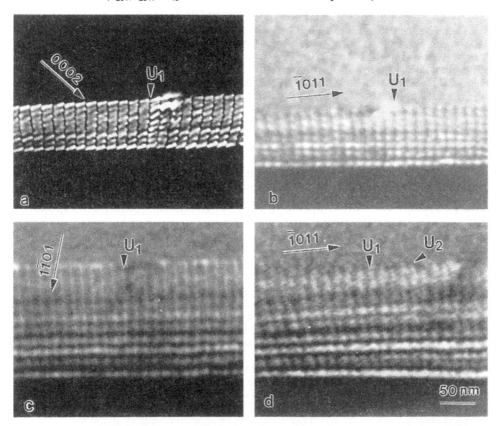

Fig. 1 Weak beam dark field images of a high angle grain boundary in titanium recorded with $\mathbf{g} = (0002)_\lambda$, $(\bar{1}011)_\lambda$, $(1\bar{1}01)_\lambda$ and $(\bar{1}011)_\lambda$, respectively

Following MacLaren and Aindow (1997b) the dislocation configuration in the boundary was modelled using the O-lattice algorithm with Burgers vectors obtained using Pond's topological theory of admissible interfacial defects (Pond, 1989). For the simplest case of Class 1 interfacial dislocations in an interface between two grains related by the transformation **P**, the Burgers vectors, **b**, are given by combinations of translation vectors $t_\lambda$ and $t_\mu$, one from each of the adjacent crystals thus:

$$b = t_\lambda - P\, t_\mu$$

Feasible values for **b** were identified from the infinite number of possible values for **b** by considering only those where **b** is smaller than that of a crystal dislocation (and would thus not be expected to decompose), where $t_\lambda$ and $t_\mu$ are relatively short (and would thus perturb the reference structure over a shorter distance), and where the angle, $\omega$, between **b** and $r_{dev}$ is close to 90° (since these would accommodate the deviation most efficiently). From these feasible values appropriate Burgers vectors for sets $U_1$ and $U_2$ were identified by comparing the observed line directions with those expected on the basis of the O2 lattice algorithm using the method of Shin and King (1989).

Possible values for the Burgers vectors of set $U_1$ are: $b_{1a}$ given by $t_\lambda = [\bar{1}010]$ and $t_\mu = [1\bar{1}0\bar{1}]$, $b_{1b}$ given by $t_\lambda = 1/3[5\bar{1}\bar{4}6]$, $t_\mu = 1/3[\bar{5}2\bar{7}3]$, and $b_{1c}$ given by $t_\lambda = [02\bar{2}\bar{1}]$, $t_\mu = [0002]$. Similarly, possible values for the Burgers vectors of set $U_2$ are: $b_{2a}$ given by $t_\lambda = [02\bar{2}0]$ and $t_\mu = 1/3[\bar{1}2\bar{1}6]$, and $b_{2b}$ given by $t_\lambda = [\bar{2}110]$ and $t_\mu = 1/3[7\bar{2}\bar{5}0]$. A b-net was constructed using all five of these Burgers vectors: a schematic diagram of this net projected onto the plane perpendicular to $r_{dev}$ is shown in Fig. 2(a), and the resultant L-net is shown in Fig. 2(b). The expected dislocation arrangement is shown in Fig. 2(c). The line direction of the dislocations with Burgers vectors $b_{1a}$, $b_{1b}$ and $b_{1c}$ is $[0.3780\ {-}0.5356\ 0.1574\ 0.3553]_\lambda$ which lies at 9.5° to set $U_1$ and the spacing of these features is 6.19 nm. The line direction of the dislocations with Burgers vectors $b_{2a}$ and $b_{2b}$ is $[0.6548\ {-}0.3332\ {-}0.3217\ {-}0.1151]_\lambda$ which lies at 3.3° to set $U_2$ and the spacing of these features is 7.24 nm.

Clearly, there is a good match between the line directions in Figure 2(c) and those observed experimentally, and the spacing of set $U_2$ also agrees well with the model structure. For set $U_1$, however, the spacing in Fig. 1 is roughly three times that in Fig. 2(c). There are two possible reasons for this discrepancy. One is that the dislocations with Burgers vectors $b_{1a}$, $b_{1b}$ and $b_{1c}$ were "grouped" as shown in Fig. 2(d). The spacings in one group might be below the limitation of the conventional TEM resolution, and the spacings between two pairs are still 18.57 nm. Alternatively, the dislocations might combine to give one set with a Burgers vector $b = b_{1a} + b_{1b} + b_{1c}$, and a spacing of 18.57 nm (Fig. 2(e)). In either case the structure could give rise to images such as Fig. 1. The situation depicted in Fig. 2(e) is, however, less likely than that in 2(d) since the magnitude of the combined Burgers vector is about three times $a$ which would be energetically unfavourable, although in the limit, as the spacings in the "groups" decrease, the two situations are equivalent.

## 4. CONCLUSIONS

The structure of one example of a HAGB in Widmanstätten $\alpha$ Ti has been investigated by comparing diffraction contrast images of the dislocation configuration with geometric models of the defect content required to accommodate the deviation from various reference structures. A good match has been obtained for a model structure with a $(01\bar{1}1)$ $\Gamma 1$ 2D CSL as the reference structure.

## ACKNOWLEDGEMENTS

We are grateful to Dr I MacLaren and Prof. R.C. Pond for helpful discussions, Professors I R Harris and J F Knott for the provision of the laboratory facilities, and CVCP and the School of Metallurgy and Materials at The University of Birmingham for financial support under the ORS Scheme.

664

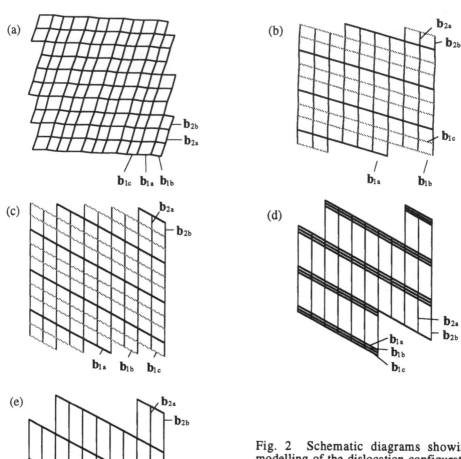

(a)

$b_{2b}$
$b_{2a}$

$b_{1c}$ $b_{1a}$ $b_{1b}$

(b)

$b_{2a}$
$b_{2b}$

$b_{1c}$

$b_{1a}$ $b_{1b}$

(c)

$b_{2a}$
$b_{2b}$

$b_{1a}$ $b_{1b}$ $b_{1c}$

(d)

$b_{2a}$
$b_{2b}$
$b_{1a}$
$b_{1b}$
$b_{1c}$

(e)

$b_{2a}$
$b_{2b}$

$b_{1a} + b_{1b} + b_{1c}$

Fig. 2  Schematic diagrams showing the modelling of the dislocation configuration for the boundary in Fig. 1; (a) projection of b-net perpendicular to r$_{dev}$; (b) corresponding L-net; (c) predicted dislocation network; (d) and (e), as for (c) but having the sets with b$_{1a}$, b$_{1b}$ and b$_{1c}$ grouped and combined, respectively.

## REFERENCES

Bollmann W, 1970, Crystal Defects and Crystalline Interfaces, (Berlin: Springer-Verlag).
Chen F R and King A H, 1988 Metall. Trans. 19A, 2359
Grimmer H, Bonnet R, Lartigue S and Priester L,1990 Phil. Mag. A 61, 493
Kestel B J 1986 Ultramicroscopy 19, 205
Lay, S., Ayed, P., and Nouet, G., 1992, Acta Metall. Mater., 40, 2351.
MacLaren I and Aindow M 1996 Phil. Mag. Lett. 73, 217
MacLaren I and Aindow M 1997a Phil. Mag. Lett. 76, 25.
MacLaren I and Aindow M 1997b Phil. Mag. A 76, 871.
Pond R C 1989 Dislocations In Solids Ed. by F R N Nabarro (North Holland) 8, Ch. 38
Shin K and King A H 1991 Phil. Mag. A 63, 1023

*Inst. Phys. Conf. Ser. No 153: Section 13*
*Paper presented at Electron Microscopy and Analysis Group Conf. EMAG97, Cambridge, 1997*
© 1997 IOP Publishing Ltd

# TEM studies of organic pigments

**P McHendry, AJ Craven, LJ Murphy***

Dept. of Physics & Astronomy, University of Glasgow, Glasgow, G12 8QQ
*Zeneca FCMO, Grangemouth Works, Earls Road, Grangemouth, Stirlingshire, FK3 8XG

**ABSTRACT** This paper reports on the growth and crystallography of organic pigment particles during solvent treatment. Transmission electron microscopy (TEM) and parallel electron energy loss spectroscopy (PEELS) have been used to study crystal growth during the pigmentation process. The action of the solvent has been studied as a function of solvent concentration and time. TEM images indicated that the particles were not of uniform thickness. This was then confirmed using PEELS on a VG HB5 STEM. The information gained gives evidence for growth by both coalescence and ripening

## 1. INTRODUCTION

A good pigment usually takes the form of small particles, typically 50-100nm in size, with a fairly tight particle size distribution. The particles must also be capable of being dispersed uniformly and evenly throughout the chosen application medium. Pigmentation can be via size reduction by physical attrition e.g. bead milling, salt milling or by re-precipitation in a controlled way. This is followed by crystal growth via solvent conditioning. A key requirement for the production of pigments is to achieve a tighter particle size distribution for a suitable range of particle sizes (Murphy, 1995). This project seeks to study the growth and crystallography of these particles during solvent treatment. The size of the pigment particles is of the order of ~100nm and so they are ideally suited to study using electron microscopy. The pigment chosen for study was indanthrone which is a complex organic molecule. It is known to exist in several polymorphic forms with the crystal structure of the α form being monoclinic (a = 3.083nm; b = 0.3833nm; c = 0.7845nm; β=91°55') with space group P2$_1$/a. There are two molecules in the unit cell (Bailey, 1955). The pigment powder is prepared for microscopy by rubbing out with an alcohol/water mixture and a small amount of dispersing agent. This is then sprayed over a hotplate to evaporate the alcohol and water before the pigment is deposited on a carbon support film mounted at the other end of the hotplate. Holey carbon films were used to support the particles in order that PEELS spectra could be taken from particles overhanging the holes. Bright field imaging and diffraction were carried out on a JEOL 1200EX transmission electron microscope. The VG HB5 scanning transmission electron microscope (STEM) was employed for annular dark field (ADF) imaging and parallel electron energy loss spectroscopy (PEELS). Radiation damage led to difficulties obtaining diffraction patterns and recording lattice images. This loss of crystallinity resulted in fading of diffraction spots and deterioration of the pattern within 2 to 6 seconds. However, no obvious mass loss leading to holes in the particles occurred when the probe was focused on the specimen. During lattice imaging, measures were taken to minimise the amount of radiation falling onto the sample although fringes still disappeared within 1 to 4 seconds.

## 2. RESULTS

Figs. 1 to 3 show typical low magnification TEM images of indanthrone at different stages in the pigmentation process. Fig. 1 shows indanthrone in its crude, non-pigmentary state prior to conversion to the amorphous, non-crystalline form shown in fig. 2. The final pigmentary form of indanthrone is shown in fig. 3 and is formed by solvent conditioning of amorphous indanthrone. It is clear that particles can grow at very different rates producing a large range of particle sizes. It can also be seen from this micrograph that growth has been promoted along one main axis.

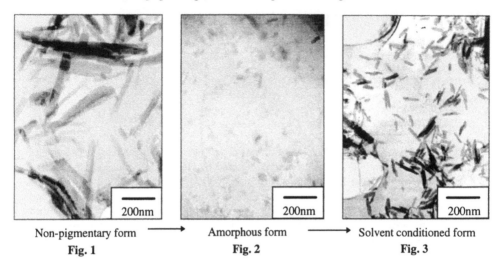

Non-pigmentary form $\longrightarrow$ Amorphous form $\longrightarrow$ Solvent conditioned form

**Fig. 1**          **Fig. 2**          **Fig. 3**

Fig. 4 illustrates schematically how the molecules are arranged within the crystal - the dashed rectangle represents a unit cell.

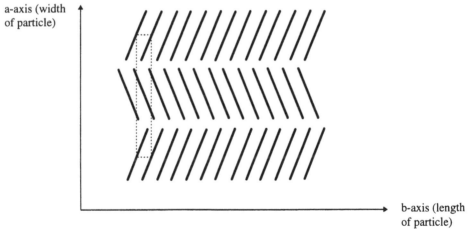

**Fig. 4**: Schematic diagram of indanthrone molecules stacked within the crystal

Convergent beam diffraction from individual pigment particles was used to determine the orientations present. Fig. 5 shows a typical diffraction pattern from a solvent conditioned indanthrone pigment particle. This pattern, showing the [001] orientation, is by far the most common. We are therefore looking down the c-axis in the majority of cases. In fig. 6, the computer simulated diffraction pattern for this orientation is shown. This was calculated by the computer program "Diffract" using the crystal structure information and atom positions as given in Bailey (1955).

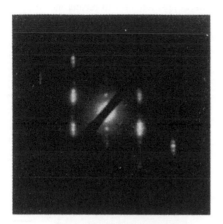

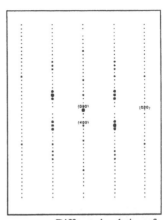

**Fig. 5**: [001] orientation

**Fig. 6**: Diffract simulation of [001] orientation

High magnification lattice images, although difficult to obtain in a radiation sensitive sample like indanthrone, provide good information about the internal structure of the particles. Figs. 7 and 8 show lattice images of solvent conditioned indanthrone. The fringes observed have been measured and correspond to the (100) and (200) planes. The (010) plane spacing is too small to be imaged as the required dose destroys the crystal structure. Those in the (001) planes would not be seen because they are perpendicular to the beam in this crystal orientation. Fig. 7 shows a typical indanthrone particle. At the edge of the particle it can be seen that the width changes in uniform steps, each corresponding to the addition of a further stack of molecules 1.5nm wide. In fig. 8, two apparently separate particles can be seen but their lattice fringes are concurrent indicating that coalescence may have occurred.

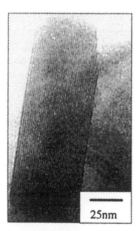

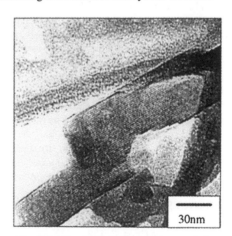

**Fig. 7**: Lattice image of indanthrone

**Fig. 8**: Lattice image of coalescing particles

Even after careful correction of astigmatism, particles imaged in the TEM with a small defocus showed asymmetric Fresnel fringes. It was thought that this may have been the result of thickness variations in the particles and this was confirmed using the scanning transmission electron microscope (STEM). The VG HB5 was used to carry out annular dark field (ADF) imaging which shows thicker areas of the specimen as brighter areas on the display. Parallel electron energy loss spectroscopy (PEELS) gave values for the thickness at various points along particles. In this instance, the low loss region of the spectrum yields a value for the ratio of thickness to inelastic mean free path ($t/\lambda$) after the dark current has been subtracted. By calculating a value for $\lambda$, the thickness can be determined (Egerton, 1986 & 1992). Fig. 9 shows an ADF image of an indanthrone particle supported on a holey carbon grid. This particle is approximately 180nm long by 37nm wide. At position 1, the thickness was found to be 23nm, at 2 it was 24nm whilst at position 3 it had further increased to 32nm. This is reflected in the increased brightness in this area.

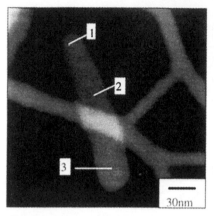

Fig. 9: ADF image of indanthrone particle

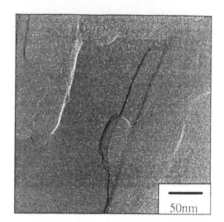

Fig. 10: DPC image of indanthrone particles

Differential phase contrast (DPC) imaging on the HB5 STEM was used to give a topographical image of certain samples and confirmed the presence of ledges and steps on the surface of the particles as well as indications of coalescence. An example of this is shown in fig. 10 with such features visible particularly on the particles to the right of the image.

## 6.    CONCLUSIONS

The low magnification TEM images give useful information on the size, shape and size distribution of pigment particles. They can then be measured and the data is used to form particle size distribution histograms. This allows comparisons to be made between samples at different stages of solvent conditioning. Diffraction studies showed that growth is strongly promoted along the same direction (b-axis) each time whilst lattice imaging confirmed this as well as indicating coalescence. PEELS showed that thickness variations were present in numerous particles due to the way in which ripening and/or coalescence was occurring. The rate of growth was found to be affected by, among other things, the concentration of the solvent and the time it was in contact with the pigment. The most likely method of growth would seem to be ripening followed by coalescence. Initially, larger particles grow at the expense of the smaller particles - the smaller ones go into solution allowing the growth of the larger particles. As they increase in size, their faces become bigger and flatter and so coalescence becomes more likely due to the increased area of contact available. Heat promotes the ripening part of this process as it increases solubility leading to faster ripening.

**REFERENCES**

Bailey M 1955 Acta Cryst. **8**, 182
Egerton RF 1986 Electron Energy Loss Spectroscopy in the Electron Microscope (New York and London: Plenum Press) pp 291-7
Egerton RF 1992 Quantitative Microbeam Analysis: Proceedings of the Fortieth Scottish Universities Summer School in Physics-Dundee pp145-168 (Copublished Scottish Universities Summer School in Physics & IOP Publishing, Bristol and Philadelphia)
Murphy LJ 1995 Private Communication

*Inst. Phys. Conf. Ser. No 153: Section 13*
*Paper presented at Electron Microscopy and Analysis Group Conf. EMAG97, Cambridge, 1997*
© 1997 IOP Publishing Ltd

669

# Phase transformation and oxidation of sputtered MoSi$_2$ thin films

**X. Y. Wang, I.T.H. Chang and M. Aindow**

School of Metallurgy and Materials, University of Birmingham, Birmingham, B15 2TT.

**ABSTRACT:** Amorphous MoSi$_2$ thin films have been produced by magnetron sputtering. The microstructural development during annealing and oxidation has been investigated by XRD, TEM and SEM. Films annealed at 500-1000°C exhibit microstructures consisting of the hexagonal C40 and /or tetragonal C11$_b$ MoSi$_2$ phases depending on the annealing temperature. Oxidation at 500°C produced a mixed oxide which consisted of MoO$_3$ and SiO$_2$ .

## 1. INTRODUCTION

Molybdenum disilicide  has considerable potential for use as electrical contacts in VLSI technology and as a high temperature barrier coating against oxidation, because it exhibits low electrical resistance, high thermal and chemical stability and excellent high temperature oxidation  resistance. ( e.g. Vasudevan and Petrovic, 1992)

It is well known that silicide films produced by sputtering are often amorphous. The microstructure of the films is sensitive to subsequent heat-treatment conditions, because crystallization and polymorphic phase transformations can occur. Moreover, a pest reaction can occur during oxidation of bulk MoSi$_2$ at 400-600°C (Ho et al, 1992). In this paper preliminary observations of the microstructural development in sputtered MoSi$_2$ films after annealing and oxidation are presented.

## 2. EXPERIMENTAL PROCEDURE

MoSi$_2$ thin films have been prepared by D.C planar magnetron sputtering from a hot-pressed (tetragonal, C11$_b$) target. Two types of substrate have been used, (100) single crystal silicon wafers, and lacomit coated glass slides. The silicon substrates were cleaned in a HF-based etch prior to sputtering.

The base pressure prior to sputtering was $2\times10^{-6}$ torr and the argon pressure was $8.5\times10^{-3}$ torr. The target was sputtering using 180W D.C power at room temperature giving a deposition rate of about 60nm/minute. 1 μm thick films were deposited on silicon while 50 nm thick films were deposited on glass. After deposition, the lacomit coating was dissolved from the glass slide and the sputtered films were retrieved on Mo support grids.

Sputtered MoSi$_2$ films on Si substrates and Mo grids were annealed at temperatures in the range 500-1000°C for between 4 and 8 hours in a vacuum furnace backfilled with Ar. Oxidation experiments were carried out on sputtered films in air at a temperature of 500°C for between 1/2 and 48 hours. The microstructures of the as-deposited, annealed and oxidised films on Mo grids were examined in a Philips CM 20 transmission electron microscope (TEM). The structure of the films on Si substrates before and after annealing were studied using X-ray diffraction (XRD). The morphology of the oxidised MoSi$_2$ films on silicon substrates and the oxidation products were examined using a JEOL 6300 scanning electron microscope (SEM) and energy dispersive X-ray spectrometry (EDS).

## 3. PHASE TRANSFORMATIONS

The XRD results (fig. 1) shows the phase transformations which occurred during the annealing. The as-deposited films were amorphous. TEM bright field (BF) images and selected area diffraction (SAD) patterns revealed no crystalline phases in the film (fig. 2a.)

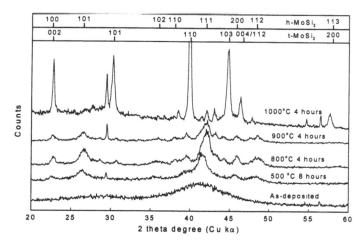

Figure. 1 XRD patterns of $MoSi_2$ films on silicon substrates annealed under vacuum

After annealing for 8 hours at a temperature of 500°C, crystallisation started to occur (figs. 1b&2b). From the XRD peak positions and the SAD patterns, the crystalline phase was identified as the metastable hexagonal polymorph of $MoSi_2$ which exhibits the C40 structure. After annealing at 800-900°C, the structure consists mainly of the C40 phase with some of the equilibrium tetragonal polymorph (C11$_b$) (figs. 1c&2c). This suggests that there is a phase transformation from the metastable C40 phase to the C11$_b$ phase in this temperature range. Both of these structures consist of close packed planes where each Mo atom is surrounded by six Si atoms, these are {110} in the C40 phase and (0001) in the C11$_b$ phase. The C40 to C11$_b$ phase transformation can take place by changing the stacking sequence from ABCABC (hexagonal) to ABAB (tetragonal) without long range transport of Mo and Si atoms.

Finally, after annealing at 1000°C, only the C11$_b$ phase was found in the film (figs. 1d &2d). The XRD data also indicate that the C11$_b$ phase was textured with the {110} planes being parallel to the surface of the film. Since, there was no sign of any texture in the C40 phase, and the TEM data show that large C11$_b$ grains grown from the fine-grained C40, the preferred orientation in the C11$_b$ probably corresponds to a recrystallisation texture.

In contrast to the observations of Chou and Nieh (1992) who found the $Mo_5Si_3$ phase during a C40-C11$_b$ phase transformation in $MoSi_2$ thin films, no such phase was observed in any of our annealed samples.

## 4. PEST OXIDATION OF MOSI$_2$ FILM

The 1μm thick $MoSi_2$ film on silicon began to crack after exposure to air for 1 hour at 500°C. This is probably due to the tensile stresses caused by the mismatch of the thermal expansion coefficients for $MoSi_2$ ($\alpha=8.5\times10^{-6}k^{-1}$) film and the silicon ($\alpha=2.6\times10^{-6}k^{-1}$) substrate during subsequent cooling. After 20 hours exposure, many white-yellowish spots appeared on the surface of the film. After exposure of 48 hours, the film had converted into yellowish powders.

The catastrophic oxidation with the characteristic of disintegration of the sample into powder products is called "pest". Chou and Nieh (1993) suggested that the cause of pest in bulk $MoSi_2$ materials is grain boundary embrittlement produced by the short-circuit diffusion

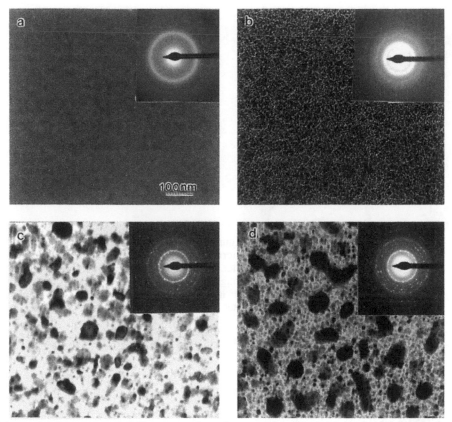

Figure. 2 BF images and SAD patterns from free-standing films. (a) As-deposited film.(b-d) Annealed at 500°C for 8 hours, 900°C for 4 hours and 1000°C for 4 hours respectively.

of $O_2$ and subsequent dissolution into the grain-boundary adjacent areas without the formation of a $SiO_2$ layer to protect the substrate from the further attack by oxygen

Under SEM the pest oxidation products are needle-like whiskers on clusters as shown in fig. 3a. Fig. 3b shows a TEM image of a small fragment of the oxidation products. From the SAD and EDS, the whiskers and clusters were identified as $MoO_3$ and $SiO_2$, respectively. Some residual $MoSi_2$ was also observed during the SEM and TEM investigation. The nature of the oxidation products implies that oxidation at 500°C causes $MoSi_2$ films on silicon to undergo the following reaction (Ho et al, 1992):

$$MoSi_2(g) + 7/2O_2(g) \Rightarrow MoO_3(s+g) + 2SiO_2(s)$$

This suggests a vaporisation of $MoO_3$ and development of a discontinuous $SiO_2+MoO_3$ solid oxide at low temperatures. Without a continuous protective $SiO_2$ layer, the $MoSi_2$ film will continue to oxidise eventually disintegrating into mixed oxide powders.

Many whiskers were observed on the surfaces of free-standing $MoSi_2$ films after 30 minutes exposure at 500°C. Fig. 3c shows a TEM bright field image of the whiskers. These were identified as $MoO_3$ using the corresponding SAD pattern. Unlike the film on Si substrate, the free-standing $MoSi_2$ film did not fracture after cooling from the testing temperature.

672

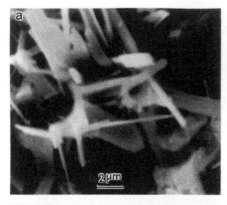

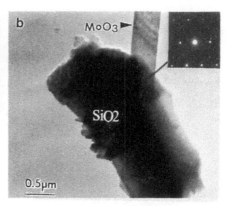

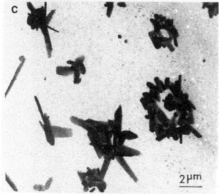

Figure 3: (a) SEM micrograph showing the microstructures of $MoSi_2$ film on silicon after 48 hours oxidation at 500°C. (b) TEM image of the oxidation products in (a). (c) BF image of the free standing $MoSi_2$ film after 30 minutes oxidation at 500°C

## 4. CONCLUSIONS

The microstructural development of sputtered amorphous $MoSi_2$ films during annealing and oxidation were investigated. Annealing of these films at 500-700°C caused the crystallisation of the amorphous film giving a fine-grained polycrystalline microstructure consisting of the hexagonal C40 $MoSi_2$ phase. Annealing at 800°C and 900°C resulted in a duplex microstructure which consisted mainly of C40 $MoSi_2$ phase and some tetragonal $C11_b$ $MoSi_2$ phase. Films annealed above 900°C exhibited the $C11_b$ phase only .

Both the $MoSi_2$ films on silicon substrates and the free-standing films were susceptible to pest oxidation at 500°C. The oxidation products included $MoO_3$ and $SiO_2$.

## ACKNOWLEDGEMENTS

We would like to thank Professor I. R. Harris for provision of laboratory facilities, and CVCP and the School of Metallurgy and Materials at University of Birmingham for financial support under the ORS Scheme.

## REFERENCES

Chou T C and Nieh T G, 1992, Thin Solid Films, **214**, 48
Chou T C and Nieh T G, 1993, Journal of Metal, **119**, 15
Ho C H *et al*, 1992 Thin Solid Films, **207**, 294
Murarka S P 1993, Characterization in Silicon Processing, eds Y Strausser (London
　　:Butterworth- Heinemann) pp. 53-95
Vasudevan and Petrovic J J, 1992, Mater. Sci. & Eng., **A155**, 1

# Author Index

# Subject Index

686

Milton Keynes UK
Ingram Content Group UK Ltd.
UKHW031123141024
449569UK00006B/469